4th International Symposium on Current Progress in Functional Materials (ISCPFM 2019)

IOP Conference Series: Materials Science and Engineering
Volume 902

Badung, Indonesia
6 - 7 November 2019

Editors:

Ivandini T.Anggraningrum
A. Denny Handoko
Young-ill Lee
Djoko Triyono
Yan Mulyana

ISBN: 978-1-7138-4752-6
ISSN: 1757-8981

Printed from e-media with permission by:

Curran Associates, Inc.
57 Morehouse Lane
Red Hook, NY 12571

Some format issues inherent in the e-media version may also appear in this print version.

Institute of Physics
Dirac House, Temple Back
Bristol BS1 6BE UK

Phone: 44 1 17 929 7481
Fax: 44 1 17 920 0979

techtracking@iop.org

Additional copies of this publication are available from:

Curran Associates, Inc.
57 Morehouse Lane
Red Hook, NY 12571 USA
Phone: 845-758-0400
Fax: 845-758-2633
Email: curran@proceedings.com
Web: www.proceedings.com

TABLE OF CONTENTS

IOP Conf. Series: Materials Science and Engineering **902** (2020) 011001 doi:10.1088/1757-899X/902/1/011001

Preface from the Editors

The relevancy of the 4th International Symposium on Current Progress in Functional Materials conference theme -*Functional Materials for Technologies Development*- to an extensive various kinds of disciplines is depicted in this Conference Proceedings volume. We are honored to present this volume to the authors, readers, and all contributors. Hopefully, this volume serves its purpose to offer valuable scientific research findings in an exciting and inspiring manner.

It was in the 6-7 Nov 2019, the conference held in the beautiful Sakala Resort, Bali. The view and scenery were breathtaking, and it is safe to say that the participants have enjoyed the gathering and gained a scientifically rewarding experience in this place. The conference is an annual program organized by the Faculty of Mathematics and Natural Sciences, Universitas Indonesia (FMIPA UI). We continue to maintain the tradition in bringing the academics, students, and researchers together in a suitable platform, where they can share their scientific works and explore new horizons of innovative discoveries to recognize current global challenges and define the best adaptable solutions. The conference standard is to enable the exchange of ideas, methods, and approaches that will preserve and promote the integration as well as a possibility of collaboration in the future.

Prof. Angel Cuesta Ciscar from the University of Aberdeen, UK, introduced the first plenary session, followed by Prof. Hsiang Lin Liu from National Taiwan Normal University (NTNU), Taiwan. On the second day, Prof. Yoshitaka Kitamoto from Tokyo Institute of Technology, Japan opened the session and followed by Prof. Hoeil Chung from Hanyang University, Korea, in the next session. Their lectures have been delivered, and the audiences have gained newfound knowledge and information in a lively and accessible way. Not enough thanks can be expressed for all keynote speakers who have shared their valuable insights on each subject.

It has been an honor to receive more than one hundred abstracts on the conference's website and welcome the participants from Indonesia and foreign countries who have attended the conference. We recorded no less than 90 papers from many respectable institutions. This number demonstrates the popularity of the ISCPFM conference as a scientific gathering. The editorial teams have enjoyed the call of papers, review, and finalize the process of the papers to be included in this volume.

Our highest appreciation is extended for the Universitas Indonesia, which has generously supported the conference financially. We valued the funds, which makes this conference possible. Our gratitude is also shared for the technical organizing committee to secure a substantial input of papers and ensure a successful conference. Finally, we would like to express our thanks to the authors. Without their works, there will be no effectual conference.

We wish all the contributors to the ISCPFM 2019 conference a memorable event in Bali, Indonesia. We are looking forward to the next ISCPFM conference – which hopefully as fruitful and inspiring as 2019's year event.

Depok, June 2020

Editors,
Ivandini T Anggraningrum
Young-ill Lee
Yan Mulyana
A Denny Handoko
Djoko Triyono

IOP Conf. Series: Materials Science and Engineering **902** (2020) 011002 doi:10.1088/1757-899X/902/1/011002

IOP Conf. Series: Materials Science and Engineering **902** (2020) 011003 doi:10.1088/1757-899X/902/1/011003

Organizing Committee

Advisor	:	Prof. Dr. Ir. Muhammad Anis, M.Met
		Prof. Ari Kuncoro S.E., M.A., Ph.D.
		Prof. Dr.rer.nat. Rosari Saleh
		Prof. Dr. rer.nat. Abdul Haris
Steering	:	Dr.Rokhmatuloh, M.Eng
Committee		Dr. tech. Djoko Triyono
Conference Chair	:	Prof. Dr. Ivandini T Anggraningrum
Committee	:	Dr. Ratna Yuniati
		Adibah, S.Si.
		Yustin S Dewi, S.Si
		Aprilia Inggri Astuti, M.Si.
		Fariz Hafizi Roihan, S.Sn.
		Adira Nofeadri Ryofi, S.Si.
		Listya Eka Anggraini, S.Si.
		Millati Hanifah Saprudin, S.Si.

Conference Proceedings Committee

Dr. Djoko Triyono

Prof. Dr. Ivandini T Anggraningrum

Adibah, S.Si.

Yustin S Dewi, S.Si.

Aprilia Inggri Astuti, M.Si.

Annamaria Bupu, M.Si.

Adira Nofeadri Ryofi, S.Si.

Listya Eka Anggraini, S.Si.

Millati Hanifah Saprudin, S.Si.

IOP Conf. Series: Materials Science and Engineering **902** (2020) 011004 doi:10.1088/1757-899X/902/1/011004

ISCPFM 2019

The International Symposium on Current Progress in Functional Materials 2019 (ISCPFM 2019)

The Sponsor

UNIVERSITAS INDONESIA

The conference is held by

FMIPA

Faculty of Mathematics and Natural Sciences
Universitas Indonesia
2019

IOP Conf. Series: Materials Science and Engineering **902** (2020) 011005 doi:10.1088/1757-899X/902/1/011005

Name	Email
A D Rizkita	adendhanarizkita@gmail.com
Abdussalam Balarabe Suleiman	suleiman_abdussalam@yahoo.com
Aden Dhana Rizkita	adendhanarizkita@gmail.com
Adi Riyadhi	adi.riyadhi@sci.ui.ac.id
Adibah	bahfeyn_diba@yahoo.com
Adinda Azkia	adindaazkia@gmail.com
Adira Nofeadri Ryofi	adira.nofeadri51@sci.ui.ac.id
Aflahannisa	anisaaflah136@gmail.com
Agustino Zulys	zulys@ui.ac.id
Albertus William Winata	albertus.willi@gmail.com
Aldes Lesbani	aldeslesbani@pps.unsri.ac.id
Ananda Eprasatya	anandaeprasatya@gmail.com
Andi Ahmad Dzaky Manggabarani	mbdzaky@gmail.com
Andrea Fiorani	andrea.fiorani@keio.jp
Angel Cuesta Ciscar	angel.cuestaciscar@abdn.ac.uk
Annamaria Bupu	annamaria.bupu@sci.ui.ac.id
Anne Zulfia Syahrial	anne@metal.ui.ac.id
Antonius Herry Cahyana	herrykim@ui.ac.id
Aprilia Inggri Astuti	aprilia.inggri@sci.ui.ac.id
Arnia Putri Pratama	arniappm@gmail.com
Asep Saefumillah	asep.saefumillah@sci.ui.ac.id
Atikah Anwar Hasibuan	atikahanwarhasibuan@gmail.com
Azkia Fadhlurrahma	azkiafadhlurrahma@gmail.com
Bayu Alam Herlambang	grouprnd@capcx.com
Brian Yuliarto	brian@tf.itb.ac.id
Budiawan	dr.budiawan@sci.ui.ac.id
Candra Kurniawan	candra.fis43@gmail.com
Choirul Ikhsan	choirulikhsan55@gmail.com
Citra Santikasari	citra.santikasari@ui.ac.id
Dede Djuhana	dede.djuhana@sci.ui.ac.id
Dedi Suyanto	dsuyanto@fisika.ui.ac.id
Dery Putra U Asmarakusuma	deryputra92@gmail.com
Dewangga Oky Bagus Apriandanu	dewanggaoky@ui.ac.id
Dewi Shinta	dewiishintaa@gmail.com
Dita Arifa Nurani	d.arifa@sci.ui.ac.id
Djoko Triyono	djoko.triyono@sci.ui.ac.id
Djonaedi Saleh	djonaedi@gmail.com
Dwi Ayu Setyani	dwiayusetyani25@gmail.com
Dyah Utami Cahyaning Rahayu	dyahutamicr@sci.ui.ac.id
Eka Puteri MEP	mitha141194@gmail.com
Elsa Dara Maylani	elsadaramaylani@yahoo.com
Endang Saepudin	endang.saefudin@sci.ui.ac.id
Fadel Lukman Hakim	fadel.lukmanh@gmail.com
Fadlinatin Naumi	fadlinatinnaumi@gmail.com
Fajriyani	fajriyani16061995@gmail.com
Fariz Hafizi Roihan	farizhafizi@sci.ui.ac.id
Fitria Rahmawati	fitria@mipa.uns.ac.id
Gian Parusa	gian.parusa@gmail.com
Grandprix T. M. Kadja	kadja@chem.itb.ac.id
Hanzhola Gusman Riyanto	hanzhola.gusman71@ui.ac.id
Hendrik Oktendy Lintang	hendrik.lintang@machung.ac.id
Hoeil Chung	hoeil@hanyang.ac.kr
Hsiang Lin Liu	hliu@ntnu.edu.tw
Iman Abdullah	iman.abdullah@sci.ui.ac.id
Iqlima Amelia	iqlima.amelia@ui.ac.id
Irkham	irkham@chem.keio.ac.jp
Is Fatimah	isfatimah@uii.ac.id
Ismi Purnamasari	ismipurnamasari55@gmail.com
Isnaini Rahmawati	isnaini.rahmawati29@gmail.com
Ivandini T Anggraningrum	ivandini.tri@sci.ui.ac.id
Jarnuzi Gunlazuardi	jarnuzi@ui.ac.id
Jessica Elizabeth	jejesjessica40@gmail.com
Khabibil Mustofa	khabibil.mustofa@sci.ui.ac.id
Larasati Denaputri Sudewo	larasatidenaputri@gmail.com
Leny Yuliati	leny.yuliati@machung.ac.id
Listya Eka Anggraini	listya@sci.ui.ac.id
Lizbeth C Christina	lizbethdimonti@gmail.com
Lusitra Munisa	lusitra@sci.ui.ac.id
Maftuhatun Fista Amalia	amaliafistaa@gmail.com
Mansur Sa'id	mansursaid79@gmail.com
Millati Hanifah Saprudin	millati.hanifah@sci.ui.ac.id
Mochammad Arfin Fardiansyah Nasution	marfin.f@sci.ui.ac.id
Mohammed Nasir MNM	nmaharaz@gmail.com

Muhamad Faisal	mfaisal.chairul@yahoo.com
Muhammad Aziz Majidi	aziz.majidi@sci.ui.ac.id
Muhammad Fiqri Andrian	16612012@students.uii.ac.id
Muhammad Hasan	muhammad.hasan62@ui.ac.id
Muhammad Iqbal Syauqi	muhammad.iqbal@sci.ui.ac.id
Muhammad Rafli	muhammad.rafli@sci.ui.ac.id
Muhammad Reinaldo Ongky Billy Anando	muhammad.anando@capcx.com
Muhammad Sujak	muhammad.sujak@sci.ui.ac.id
Munawar Khalil	mkhalil@sci.ui.ac.id
Muqtaf Najich Abdillah	muqtafabdillah@gmail.com
Murni Handayani	handayanimurni2@gmail.com
Mutiara Hapsari	mutiara.hapsari@gmail.com
Nanda Ramadhan Susilo	nandaramadhan7@gmail.com
Nathasya Humaira Adriani	thasyahumaira@gmail.com
Neza Rahayu Palapa	nezarahayu@gmail.com
Nguyen Hong Lien	lien.nguyenhong@hust.edu.vn
Nofentari Putri	nofentari0811@gmail.com
Norhasnita Hassan	noorhae@gmail.com
Nurul Hidayati	hidayati.nrl@gmail.com
Nurul Riezzati	nurulriezzati@gmail.com
Octaviany Magdalena	octavianymagdalena@sci.ui.ac.id
Paulina Taba	ptaba_1511@yahoo.co.id
Phan Thi Thuy Trang	phanthithuytrang@qnu.edu.vn
Prastika Krisma Jiwanti	prastika.krisma@fst.unair.ac.id
Putwi Widya Citradewi	16612112@students.uii.ac.id
R Azsarinka	edsna@sci.ui.ac.id
Ratna Yuniarti	ratnayu67@gmail.com
Ratna Yuniati	ratnayuniati@sci.ui.ac.id
Retno L. Budiman	retno.lestari.budiman@gmail.com
Retno Wulandari	retnoo.wulandari.rw@gmail.com
Ricky Iqbal Syahrudin	srickyiqbal@gmail.com
Rifqi Almusawi Rafsanjani	rifqialmusawi@gmail.com
Rizka Permata Suci	Sucirizkapermata@yahoo.com
Rokhmatuloh	rokhmatuloh.ssi@sci.ui.ac.id
Rosari Saleh	rosari.saleh@gmail.com
Russell F Howe	r.howe@abdn.ac.uk
Sanya Khaerunnisa	sanyaa.khaerunnisa@gmail.com
Shafira Quamila Dewi	shafiraquamila@gmail.com
Shohifah Annur	shohifah.annur@gmail.com
Sri Handayani	yani1964@ui.ac.id
Sukma Alifiana Aziz	sukmaalifianaa@gmail.com
Sumi Hudiyono	sumi.hudiyono@sci.ui.ac.id
Syeila Yusuf	syeilayusuf@gmail.com
Syella Dwi Safitri	syellasafitri29@gmail.com
Syntia Hardianti Oktavia	hardiantioktaviasyntia@gmail.com
Tasiu Zangina	tasiuzangina@gmail.com
Teguh Andika	andikat50@gmail.com
Tiar Asita Baramanda	tiar.asitabaramanda@gmail.com
Tri Wahyuni Widiyas Syahfitri	tri.wahyuni.ws@gmail.com
Tri Yuliani	tri.yuliani31@sci.ui.ac.id
Valentina Sintya	valentina.sintya@gmail.com
Vinny	vinnymulyadi96@gmail.com
Winni Rezki Wulandari	winni.rezki@ui.ac.id
Wulan Tri Wahyuni	wulantriws@gmail.com
Yehezkiel Steven Kurniawan	yehezkiel.steven@machung.ac.id
Yoki Yulizar	yokiy@sci.ui.ac.id
Yoshitaka Kitamoto	kitamoto.y.aa@m.titech.ac.jp
Yulia Mariana Tesa Ayudia Putri	tayudia23@gmail.com
Yuni Krisyuningsih Krisnandi	yuni.krisnandi@sci.ui.ac.id
Yunida	yunida128@gmail.com
Yusraini Dian Inayati Siregar	yusraini.dian@sci.ui.ac.id
Yustin Sinta Dewi	yustin.sd@sci.ui.ac.id
Yustina	nana.biotayuda@gmail.com
Zyulfi Faiz	faiz.zyulfi@gmail.com

IOP Conf. Series: Materials Science and Engineering **902** (2020) 011006 doi:10.1088/1757-899X/902/1/011006

Peer review declaration

All papers published in this volume of IOP Conference Series: Materials Science and Engineering have been peer reviewed through processes administered by the Editors. Reviews were conducted by expert referees to the professional and scientific standards expected of a proceedings journal published by IOP Publishing.

- **Type of peer review:** *Double-blind (the author remains anonymous to the referees throughout the consideration process and vice versa).*

- **Conference submission management system:** *Open Conference Systems by Public Knowledge Project (PKP) installed in the university's server and hosted by Research and Innovation Management Office Universitas Indonesia*

- **Number of submissions received:** 105

- **Number of submissions sent for review:** 87

- **Number of submissions accepted:** 74

- **Acceptance Rate (Number of Submissions Accepted / Number of Submissions Received X 100):** 70.48%

- **Average number of reviews per paper:** 2 or 3

- **Total number of reviewers involved:** 51

- **Any additional info on review process:** each paper undergo a process of checking similarity using iThenticate before sending it to reviewers

- **Contact person for queries:** Ivandini Tri Anggraningrum (ivandini.tri@sci.ui.ac.id)

IOP Conf. Series: Materials Science and Engineering **902** (2020) 012001 doi:10.1088/1757-899X/902/1/012001

Preparation of boron-doped diamond modified by bimetal nickel and zinc

Y M T A Putri[1], P K Jiwanti[2], Irkham[2], J Gunlazuardi[1], Y Einaga[2] and T A Ivandini[1]

[1]Department of Chemistry, Faculty of Mathematics and Natural Sciences (FMIPA) Universitas Indonesia, Depok 16424, Indonesia
[2]Department of Chemistry, Faculty of Science and Technology, Keio University, Yokohama 223–8522, Japan

Corresponding author's email: ivandini.tri@sci.ui.ac.id

Abstract. Nickel has been known as a good catalyst due to its high activity and stability, but the potential formation of NiOOH is quite high. In this work, zinc together with nickel was used to modify the surface of boron-doped diamond (BDD) to lower the formation potential of NiOOH. Electro-deposition (multi pulse amperometry) technique was applied at a potential of -1.6 V for 600 s. The prepared electrode (NiZn-BDD) characterized using SEM-EDX and XPS shows that both nickel and zinc were deposited homogenously on the surface of BDD with the ratio of 1:2 for nickel:zinc.

Keywords: Nickel, zinc, boron-doped diamond, surface modification

1. Introduction

Nickel, a low-cost metal, has been known for its high activity and stability. However, it is reported that nickel has a high over-potential or formation potential of NiOOH, an active catalyst species of nickel [1]. Hence, another metal doping, such as cobalt, zinc, cadmium and manganese, is needed to lower the formation potential of NiOOH [2]. On the other hand, boron-doped diamond (BDD) has been known for its wide potential window, high stability and good durability compared to other conventional electrodes [3].

Nickel-zinc bimetallic has been known for its good characteristics and can be used in many application fields, such as for corrosion protection of steel [4], electrochemical decomposition of urea [5], and for methanol oxidation in alkaline medium [6]. The preparation of nickel-zinc deposited onto BDD is further explained in this work.

The aim of this work is to study the electrodeposition of nickel and zinc deposited on the surface of the BDD using electrodeposition (multi pulse amperometry) technique. The prepared electrode was then characterized using SEM-EDX and XPS to further confirm the nickel and zinc particles' distribution on the surface of BDD.

2. Experimental

2.1. Preparation of NiZn-BDD
BDD 1 % was fabricated by applying a microwave plasma-assisted chemical vapor deposition system (MPCVD, CORNES Technologies/ ASTeX-5400) [7]. To prepare NiZn-BDD, 40 mM nickel solution was prepared by adding 0.58 g of $Ni(No_3)_2\ 6H_2O$ to 50 mL buffer acetate pH 5 and 40 mM zinc solution was prepared by adding 0.136 g of $ZnCl_2\ 6H_2O$ to 25 mL buffer acetate pH 5. The bimetal NiZn solution was prepared by mixing the 40 mM nickel solution and 40 mM zinc solution by the ratio of 4:1. The bimetal NiZn was then deposited onto the surface of BDD using chronoamperometry deposition technique at a potential -1.6 V for 600 s.

2.2. Characterization of NiZn-BDD
The NiZn-BDD electrode was then characterized using scanning electron microscopy (SEM, JCM-6000, JEOL) and X-Ray photoelectron spectroscopy (XPS, JPS-9010TR, JEOL).

3. Results and discussion
Nickel and zinc have been deposited using chrono-amperometry deposition technique. The deposition potential is set at -1.6 V to give enough reduction potential for both nickel and zinc particles to be deposited onto the surface of BDD. Nickel could be deposited on the surface of BDD into two forms through the electrodeposition technique: (1) as nickel metal and (2) nickel hydroxide [8, 9]:

$$Ni^{2+} + 2e^- \rightarrow Ni^0 \tag{1}$$

$$Ni^{2+} + 2OH^- \rightarrow Ni(OH)_2 \tag{2}$$

whereas zinc has been deposited into Zn^0 on the surface of BDD through the electrodeposition technique [8]:

$$Zn^{2+} + 2e^- \rightarrow Zn^0 \tag{3}$$

The prepared electrode was further characterized using SEM-EDX. Figure 1 shows SEM images of bare BDD (a) and NiZn-BDD (b). As shown in figure 1, the nickel and zinc particles were deposited quite homogenously on the surface of the BDD. The EDX spectra in figure 2 shows the distribution of

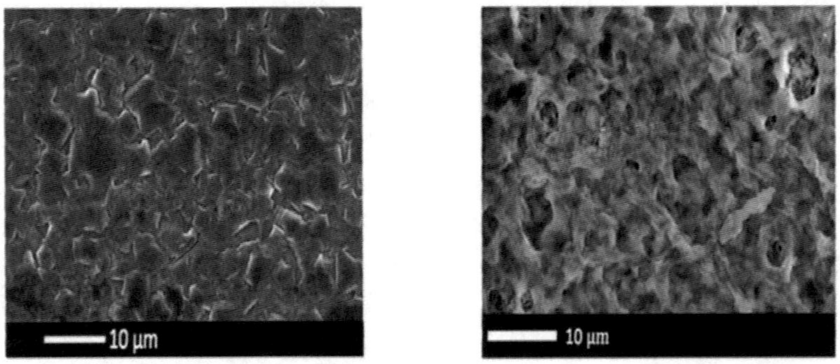

(a) (b)

Figure 1. SEM images of (a) bare BDD and (b) NiZn-BDD.

nickel and zinc particles on the surface of the BDD. The figure implies that both nickel and zinc particles have been deposited on the surface of BDD with the atomic ratio of nickel:zinc is 1:1 according to the EDX result shown in table 1.

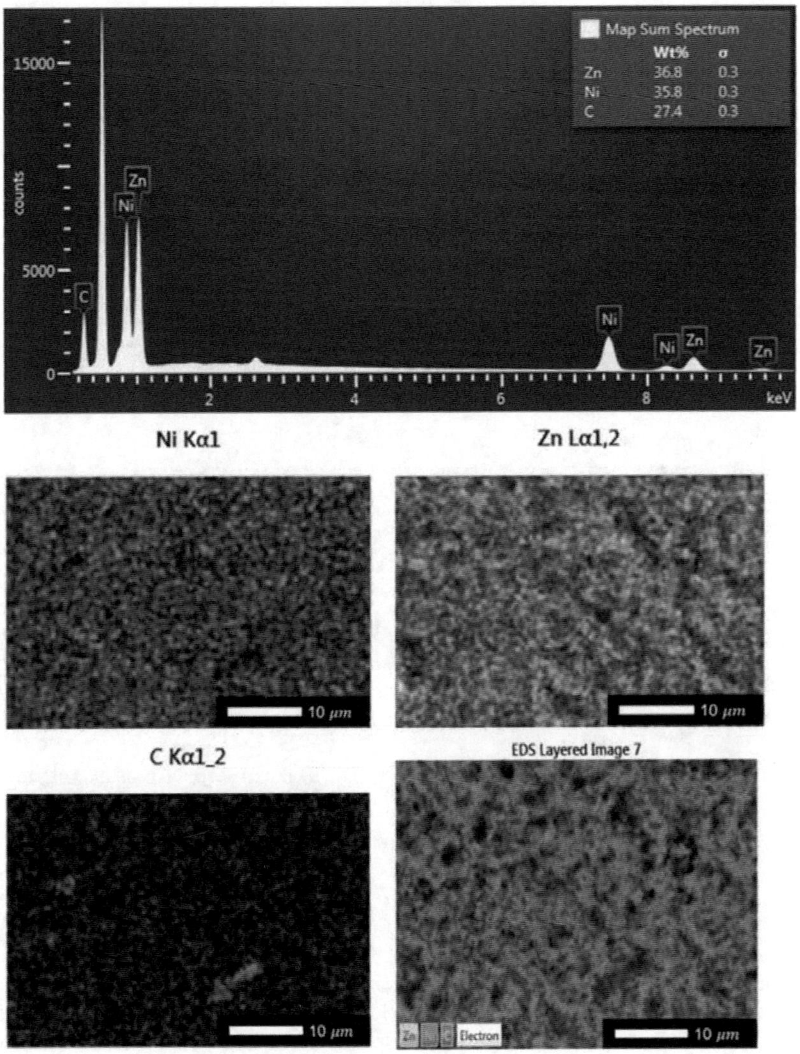

Figure 2. EDX images of NiZn-BDD.

Table 1. EDX characterization of NiZn-BDD.

Nikel		Zinc		Carbon	
% mass	% atomic	% mass	% atomic	% mass	% atomic
35.80	17.65	36.77	16.28	27.43	66.08

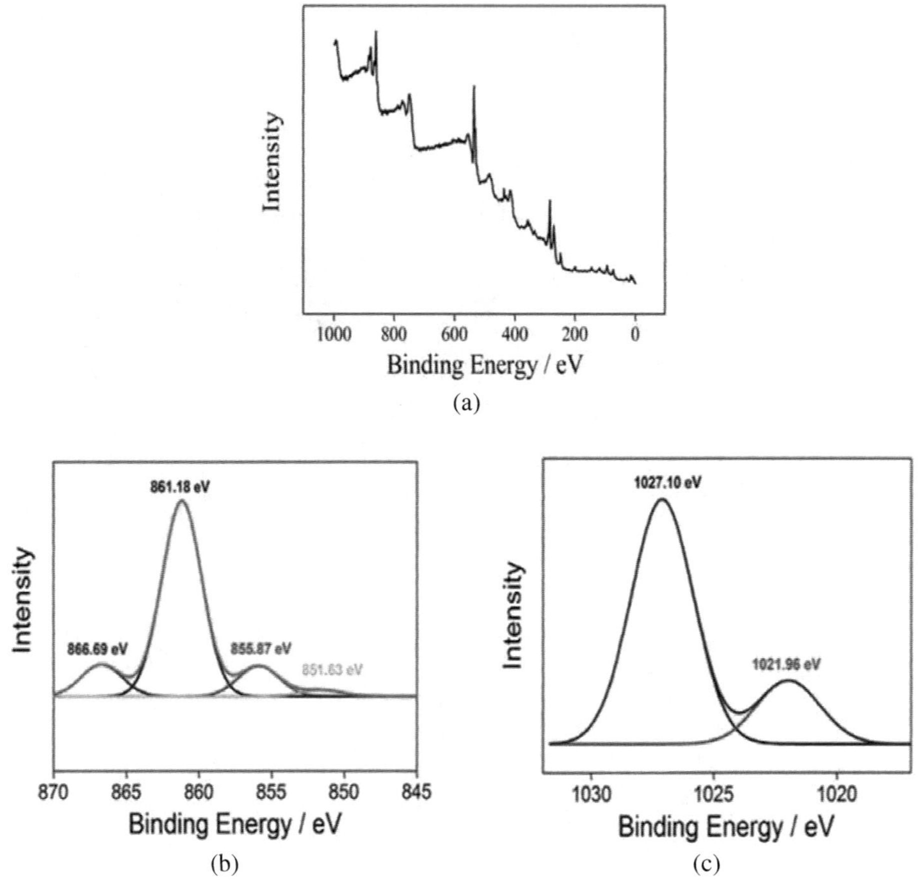

Figure 3. XPS wide spectra of (a) NiZn-BDD, (b) narrow spectra of nickel, and (c) narrow spectra of zinc.

The prepared electrode was then further characterized by XPS as shown in figure 3. The characteristic of metallic nickel of $Ni_{2p3/2}$ was detected at the binding energy (BE) of 851.63 eV and $Ni(OH)_2$ was found at the binding energy of 855.87 eV and 861.18 eV (figure 3b). From figure 3c, it is found the characteristic of metallic zinc of $Zn_{2p3/2}$ at the binding energy (BE) of 1021.96 eV. It is implying that nickel in the form of nickel metal and nickel hydroxide, and zinc in the form of zinc metal have been fully deposited on the surface of BDD at a potential of -1.6 V. The SEM-EDX and XPS characterization conclude that both particles of nickel and zinc have been successfully deposited on the surface of BDD.

4. Conclusion

NiZn particles has been successfully deposited homogenously onto the surface of BDD. SEM-EDX characterization showed the distribution of nickel and zinc particle on the surface of BDD and the ratio obtained from EDX characterization is 1:1 for nickel:zinc. XPS characterization was further confirmed that both nickel and zinc particle was deposited on the BDD surface.

Acknowledgments
This work was partly funded by Hibah PIT 9 Universitas Indonesia Tahun 2019. Contract No. NKB-0028/UN2.R3.1/HKP.05.00/2019.

References

[1] Xu W, Zhang H, Li G and Wu Z 2014 *Sci. Rep.* **4** 5863
[2] Tesfaye R M, Das G, Park B J, Kim J and Yoon H H 2019 *Sci. Rep.* **9** 1-9
[3] Asai K, Ivandini T A and Einaga Y 2016 *Sci. Rep.* **6** 32429
[4] Ebothe J and Hiane M 2001 *Appl. Surf. Sci.* **183** 93-102
[5] Yan W, Wang D and Botte G G 2012 *Appl. Catal. B Environ.* **127** 221-6
[6] Sahin E A and Kardas G 2013 *J. Solid State Electrochem.* **17** 2871-7
[7] Ivandini T A et al. 2019 *J. Phys. Chem. C* **123** 5336-44
[8] Toghill K E, Xiao L, Phillips M A and Compton R G 2010 *Sensors Actuators B Chem.* **147** 642-52
[9] Roventi G, Cecchini R, Fabrizi A and Bellezze T 2015 *Surf. Coat. Technol.* **276** 1-7

Influence of TiO_2's (101) crystal facet exposure on the photoelectroactivity of TiO_2 nanooctahedra/$BiVO_4$ nanocomposite

F Naumi[1], T A Ivandini[1], J Y Mulyana[2], U Pratomo[2] and M Khalil[1]

[1]Department of Chemistry, Faculty of Mathematics and Natural Sciences (FMIPA), Universitas Indonesia, Depok 16424, Indonesia
[2]Department of Applied Sciences, Graduate School of Urban Environmental Sciences, Tokyo Metropolitan University, Tokyo 192-0397, Japan

Corresponding author's email: mkhalil@sci.ui.ac.id

Abstract. This work studies a research regarding the effect of exposing (101) crystal facet of anatase TiO_2 to photoelectroactivity of photoanode composed of TiO_2 nanooctahedra/$BiVO_4$ nanocomposite film. Here, the photoanode was fabricated by depositing the nanocomposite on the surface of FTO via doctor blade technique. In this study, anatase TiO_2 nanooctahedra were synthesized via solvothermal method in the presence of hydrazine hydrate as a directing agent. Furthermore, the as-prepared TiO_2 nanooctahedra was characterized using X-Ray Diffractometer (XRD), Transmission Electron Microscopy (TEM), Brunauer-Emmelt-Tellers (BET) and UV-Vis diffuse reflectance spectra (DRS). Based on the result, the diffraction peaks revealed characteristic for the pure anatase phase with exposure (101) crystal facet. Additionally, photoelectrochemical response of the photoanode was also evaluated using a three-electrode system and exhibited a significantly high current density value of 0.26 mA/cm^2.

Keywords: Anatase TiO_2 nanooctahedra, $BiVO_4$, crystal facet, photoelectrochemical

1. Introduction

Crisis energy issues due to the lack of fossil fuels lead to an idea to substitute petroleum based energy with hydrogen based energy by using water as electron and proton sources. Photoelectrochemical (PEC) devices convert water to oxygen and hydrogen using photon and semiconductor as photoanode. Since the photoanode plays an important role in PEC, visible light responsive, high efficiency, and low cost are crucial to the performance of PEC [1, 2]. One of the promising semiconductors type-n, titanium dioxide (TiO_2), is generally applied, owing to its superior photocatalytic activity, environmentally friendly, and low cost [3]. However, TiO_2 with wide band gap can only absorb ultraviolet energy, which contains only about 4 % of the solar energy [4]. This phenomenon will decrease the efficiency for PEC usage due to the maximum current will increase in the range of photon absorption. Hence, the new materials with photoactive nature in visible light were developed to shift the energy band gap of TiO_2. Bismuth vanadate, ($BiVO_4$) based photoanode is an interesting candidate due to visible-light activity, good stability, and effective charge separation properties. Combining the $BiVO_4$ that has small optical bandgap (2.4 eV) with TiO_2 would improve the light absorption in the visible region[5, 6]. Moreover, the formation of TiO_2/$BiVO_4$ prolongs the carrier lifetime and then promotes the separation of photoelectron and holes that exhibit an excellent performance for photocatalytic water splitting and pollutant degradation [7].

Recently, controlled synthesis of TiO_2 to form a specific morphology with specific exposed crystal facet has gained much attention. Anatase TiO_2 crystal phase is known to have the highest photoactivity among the others phase, rutile and brookite [8]. According to the Wuff construction, the most available facets on the anatase TiO_2 crystals is dominated by (101) facets rather than (001) facets due to its less reactivity and thermodynamically more stable [9]. Therefore, (101) facets found naturally for about 90 % of the total exposed surface on the anatase TiO_2 [10]. In this work, the effect of morphology and exposure (101) crystal facet of anatase TiO_2 nanooctahedra on the photoelectrochemical response of TiO_2 nanooctahedra/$BiVO_4$ nanocomposite film as photoanode were investigated. Here, anatase TiO_2 nanooctahedra were synthesized by a simple solvothermal method with the addition of directing agent into the reaction, followed by loading the as-prepared TiO_2 nanooctahedra onto $BiVO_4$ paste and deposition on the FTO via doctor blade technique to obtain the photoanode film.

2. Materials and method

2.1. Materials

Titanium(IV) fluoride (TiF_4, Sigma-Aldrich), hydrazine hydrate ($N_2H_4.H_2O$, 60 %, Sigma-Aldrich) and ethanol (C_2H_5OH, Merck) were used to synthesize the anatase TiO_2 nanooctahedra. Bismuth(III) nitrate pentahydrate ($Bi(NO_3).5H_2O$, Wako Ltd.), ammonium vanadate (NH_4VO_3, Sigma-Aldrich), polyethylene glycol (PEG, MW 20000, Sigma-Aldrich), and Triton-X (Tokyo Chemical Industry Ltd.) were used to synthesize the photoanode film.

2.2. Synthesis of TiO_2 nanooctahedra

Anatase TiO_2 nanooctahedra were synthesized through a solvothermal method as previously reported [8]. For the typical synthesis, 2 mmol TiF_4 was added to 5 mL distilled water forming a clear solution. Then, 35 mL of hydrazine hydrate was added to the solution with magnetic stirring. The solution mixture was transferred into a Teflon lined autoclave and heated at 200 °C for 24 h. After the reaction, the products were separated by centrifuge and rinsed with distilled water and ethanol for several times. Finally, the products were dried in an oven at 70 °C for 6 h and calcined at 450 °C for 2 h.

2.3. Fabrication of TiO_2 nanooctahedra/$BiVO_4$ nanocomposite

The photoanode was prepared via doctor blade technique according to the previous literature [11]. $Bi(NO_3)_3.5H_2O$ (2 mmol), NH_4VO_3 (2 mmol) were mixed with 2 mL of 13 M HNO_3 and grinded in an agate mortar followed by the addition of as-prepared TiO_2 nanooctahedra (1 mmol). Then, polyethylene glycol (0.2 mmol) was added gradually to the above mixed solution. After the solid was dissolved, one drop of Triton-X was added and grinded to form yellow-brownish paste. The resulting paste was coated on FTO and calcined at 450 °C for 3 h.

2.4. Characterizations

In order to determine the crystallinity of the as-prepared TiO2, the X-Ray diffraction (XRD) analysis was employed by using PANanalytical X'Pert Pro MPD with Cu-Kα radiation. To determine the morphology and the particle size, transmission electron microscopy (TEM) was performed on a TECNAI G2 Spirit Twin High-Resolution. To investigate the porosity and surface area, Brunauer-Emmett-Teller (BET) surface area, Barret-Joyner-Halenda (BJH) pore volume and BET N2 adsorption-desorption isotherm were recorded on a QUADRASOB evo (Quantachrome Instruments). To determine the optical properties, diffuse reflectance spectroscopy (DRS) was carried out using Shimadzu UV-2450 spectrophotometer and the band gap (Eg) was calculated using Kubelka-Munk method.

2.5. Photoelectrochemical measurements

The photoelectrochemical response of the photoanode film was carried out with three-electrode system by using Hokuto Dento Hz-3000 Potentiostat. TiO_2 nanooctahedra/$BiVO_4$ nanocomposite (0.9 cm × 0.9 cm) film was used as working electrode, Ag/AgCl (3 M NaCl) as reference electrode, and coiled Pt as counter electrode. An 500 W Xenon lamp was used as the visible light source with an intensity 160 mW/cm^2 measured by TENMARS TM-208 solar light meter and Na_2SO_4 0.5 M (pH 7) was added into the phosphate buffer 0.1 M as the electrolyte solution [12].

3. Results and discussion

3.1. Fabrication and characterization of anatase TiO_2 nanooctahedra

The crystalline structure of the as-prepared TiO_2 was analyzed by XRD as shown in figure 1a. According to the XRD patterns, the as-prepared TiO_2 was believed to show the anatase crystalline phase, which corresponds to the reference (JCPDS Card No. 21-1272) [13]. Furthermore, the morphology of the as-prepared TiO_2 nanooctahedra was observed by TEM analysis. From the TEM image in Figure 1b, the as-prepared TiO_2 with addition of hydrazine hydrate as directing agent was found to be an octahedral-shape structure with edge width in the range of 200-300 nm and length between the two-pointed ends in the range of 400–500 nm. In addition, to investigate the exposure of the (101) crystal facet, Fast Fourier Transform (FFT) analysis of the TEM images is shown in the inset in figure 1b. As shown, the result confirms that the as-prepared TiO_2 nanooctahedra did expose more (101) crystal facet.

3.2. Surface and optical properties of anatase TiO_2 nanooctahedra

The BET isotherms graphs in figure 2a shows a typical type (III) like with type (IV) hysteresis loops that presence at high pressure [14]. Furthermore, the pore size distribution (PSD) plot was also determined using Barrett-Joyner-Halenda (BJH) method from the desorption of the isotherm (figure 2b). According to the result, the obtained PSD plot exhibited similar features with isotherm plot, having peaks of microporous (< 2 nm) and mesoporous (2–50 nm). Additionally, the result also demonstrated that the average pore volume and BET surface area for the as-prepared TiO_2 nanooctahedra were 0.026 cc/g and 16.133 m^2/g, respectively. Furthermore, the optical properties of the samples were also investigated using UV-Vis spectroscopy. As it can be seen in figure 3, the absorbance of TiO_2 nanooctahedra increases in the UV region and displays an absorption edge at 413 nm, which corresponds to a band gap of 3.0 eV.

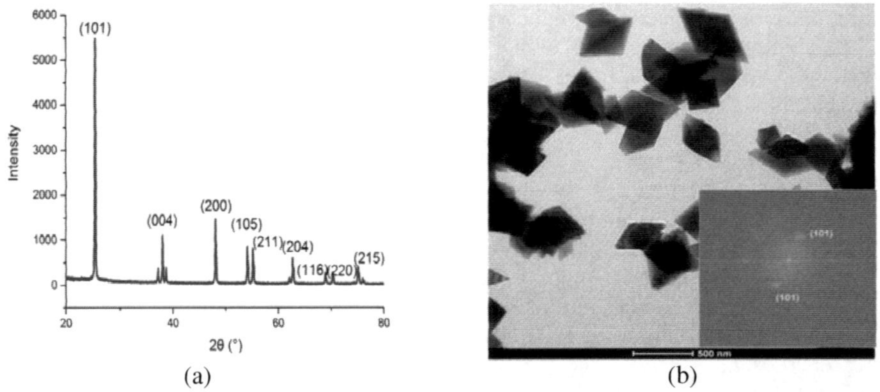

Figure 1. (a) XRD pattern, and (b) TEM images of the as-prepared TiO_2 nanooctahedral (inset: FFT pattern).

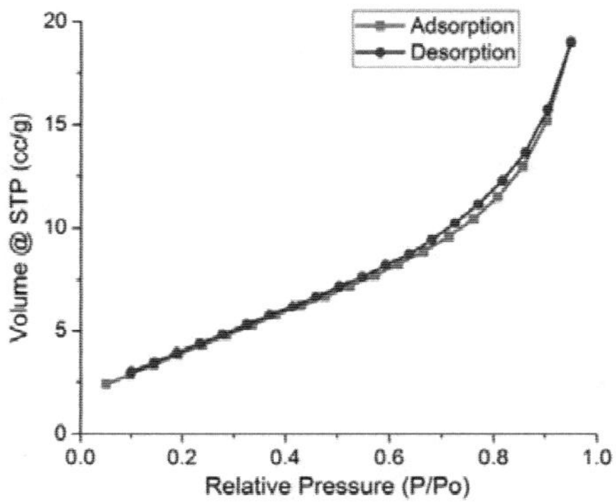

(a)

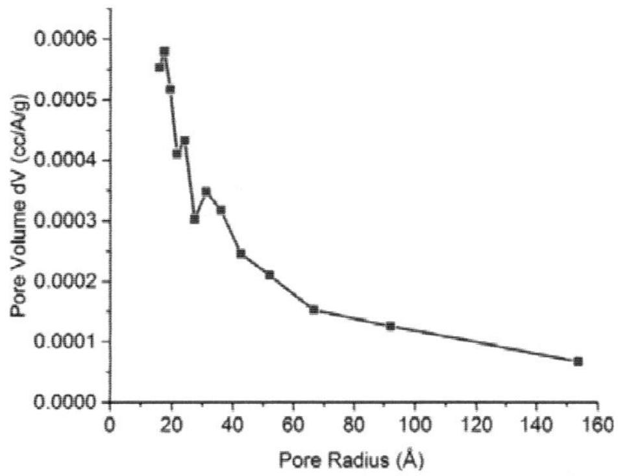

(b)

Figure 2. (a) BET N_2 adsorption-desorption isotherm plot, and (b) BJH pore size distribution (PSD) plot of the as-prepared TiO_2 nanooctahedra.

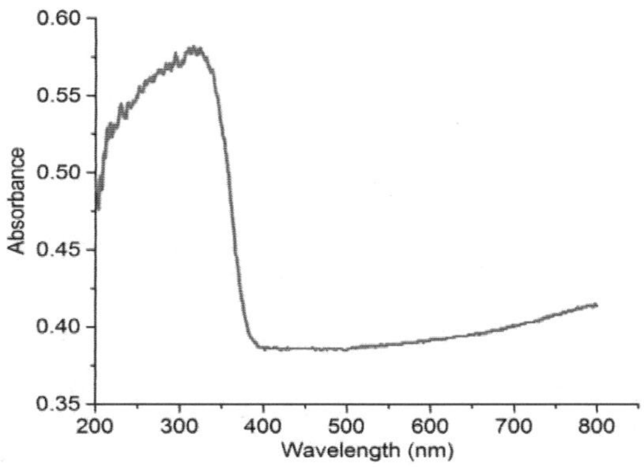

Figure 3. UV-Vis diffuse reflectance spectra of the as-prepared TiO_2 nanooctahedra.

3.3. Photoelectrochemical measurements

In order to gain the information about the film response to the light irradiation, the linear sweep voltammetry (LSV) was carried out both under dark and light exposure conditions. Figure 4a shows the obtained J-V characteristics of the as-prepared TiO_2 nanooctahedra/$BiVO_4$ photoanode. According to the result, it is obvious that there is a significant increase in the photocurrent in light irradiation, indicating the film photoactive response. Additionally, result also showed that the film produced about 0.14 mA/cm^{-2} of photocurrent at the thermodynamic water oxidation potential of 1.23 V. Further investigation was also carried out to evaluate the stability of the photoanode film. Here, photocurrent measurements were performed under the applied potential bias (1.4 V_{RHE}) and light illumination. According to the result in figure 4b, the photocurrent spikes of TiO_2 nanooctahedra/$BiVO_4$ nanocomposite film reached the value of 0.26 mA/cm^2. Nevertheless, it is also worth to note that these photocurrents were quickly decreased during the first 200 sec of light exposure before gradually reached the equilibrium. This photocurrent was found to be significantly higher than the value of bare $BiVO_4$ photoanode film, 0.5 μA/cm^{-2}, under illumination of simulated sunlight (AM 1.5, 100 mW/cm^{-2}) and

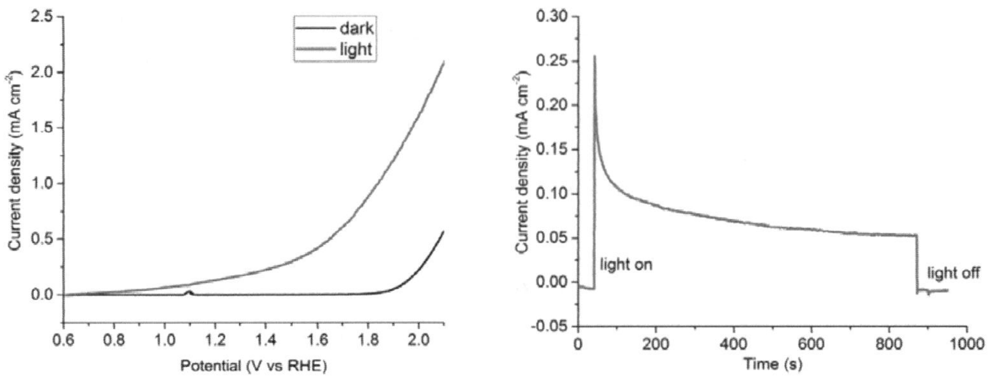

Figure 4. (a) J-V plot under dark and light irradiation, and (b) J-t plot of the as-prepared TiO_2 nanooctahedra/$BiVO_4$ nanocomposite photoanode.

active area of photoanode 2 cm × 2 cm without applied bias [11]. It is believed that the high-energy electrons of $BiVO_4$ that excited by the light would be energetically transferred to TiO_2 and enhanced the lifetimes and separation of visible-excited charge carriers of $BiVO_4$ [7].

4. Conclusion

In summary, anatase TiO_2 nanooctahedra have been successfully synthesized via solvothermal method with the addition of hydrazine hydrate as a directing agent. This conclusion is confirmed by the evidence that the diffraction peaks of TiO_2 powder particles are corresponded to the characteristic peak of pure anatase phase and the morphology of the anatase TiO_2, found to be an octahedral shape, which preferentially exposed (101) facets. The TiO_2 nanooctahedra/$BiVO_4$ nanocomposite film has the potential to become a promising photoanode on PEC for water splitting and pollutant degradation due to its photoactive response and effective charge separation.

Acknowledgments

The authors would like to acknowledge the financial support provided by Directorate of Research and Community Engagement (DRPM), University of Indonesia via Hibah PIT-9 2019 (Contract No. NKB-0028/UN2/R3.1/HKP.05.00/2019).

References

[1] Liu B et al. 2016 *ACS Appl. Mater. Interfaces* **8** 18577-83
[2] Ouyang K, Xie S, Wang P, Zhu J and Zhan P 2019 *Int. J. Hydrogen Energy* **44** 7288-99
[3] Khalil M, Anggraeni E S, Ivandini T A and Budianto E 2019 *Appl. Surf. Sci.* **487** 1376-84
[4] Deng B et al. 2018 *Nano Struct. Nano Objects* **15** 167-72
[5] Tolod K R., Hernández S, Quadrelli E A and Russo N 2019 *Stud. Surf. Sci. Catal.* **178** 65-84
[6] Monfort O et al. 2017 *Electrochim. Acta* **251** 244-9
[7] Xie M et al. 2014 *Adv. Energy Mater.* **4** 1300995
[8] Gai L et al. 2013 *Mater. Res. Bull.* **48** 4469-75
[9] Wang H et al. 2016 *Soft Nanosci. Lett.* **6** 11-30
[10] Yu J, Low J, Xiao W, Zhou P and Jaroniec M 2014 *J. Am. Chem. Soc.* **136** 8839-42

IOP Conf. Series: Materials Science and Engineering **902** (2020) 012002 doi:10.1088/1757-899X/902/1/012002

[11] Nakashima S, Negishi R and Tada H 2016 *Chem. Commun.* **52** 3665-8
[12] Pratomo U, Purnama I and Yan J 2020 *Inorganica Chim. Acta* **500** 119223
[13] Bakre P V and Tilve S G 2018 *J. Phys. Chem. Solids* **116** 234-40
[14] Thommes M et al. 2015 *Pure Appl. Chem.* **87** 1051-69

TiO_2/Co_3O_4 nanocomposite: Synthesis via *Catharanthus roseus* (L.) G. Don leaf extract, characterization and its photocatalytic activity for malachite green degradation

T W W Syahfitri, Y Yulizar, J Gunlazuardi and D O B Apriandanu

Department of Chemistry, Faculty of Mathematics and Natural Sciences (FMIPA), Universitas Indonesia, Depok 16424, Indonesia

Corresponding author's email: yokiy@sci.ui.ac.id

Abstract. TiO_2/Co_3O_4 nanocomposite was successfully synthesized using leaf extract of *Catharanthus roseus* (L.) G. Don. (CRE), precursors of TTIP and $Co(NO_3)_2$. FT-IR characterization shows that TiO_2/Co_3O_4 has been formed due to the presence of secondary metabolites in CRE which have a major role as a weak base source for nanocomposite synthesis. Particle size analysis was employed by PSA and TEM, which were found to be 38.04 nm and 90.80 nm, respectively. Besides, the UV-Vis DRS result shows that TiO_2/Co_3O_4 has bandgap energy at 2.8 eV. According to the morphological analysis, TiO_2/Co_3O_4 nanocomposite has a spherical shape confirmed by SEM. The structural analysis was confirmed by XRD that TiO_2/Co_3O_4 nanocomposite has a diffraction pattern combination of TiO_2 and Co_3O_4 nanoparticle. The photocatalytic activity of TiO_2/Co_3O_4 nanocomposite was examined for malachite green degradation. The result shows that the degradation percentage was 82.61 % under visible light illumination for 120 min.

Keywords: *Catharanthus roseus (*L.) G. Don.; photocatalytic activity; TiO_2/Co_3O_4; nanocomposite, malachite green

1. Introduction

The utilization of plant media in nanoparticle synthesis is one step to reduce the risk in the use of harmful substances to the environment. *Catharanthus roseus* (L.) G. Don. is one of the plants for metal oxide nanoparticle synthesis, containing 11.11 % of alkaloids in the leaves as a source of weak base and capping agents [1]. *Datura metel L., Tinospora crispa,* and *Moringa oleifera* have been reported as tropical plants for metal oxide nanoparticle synthesis [2-4].

On the other hand, TiO_2 is a semiconductor that has a high photocatalytic activity in the form of anatase [5] and has a high melting point, photoactivity, and thermal stability. Also, TiO_2 is an excellent catalyst to be applied in the environment since it is non-toxic, low cost and inert material [6]. TiO_2 has been reported to have a bandgap energy (Eg) around 3.0–3.2 eV [7]. The value of Eg is related to the maximum absorption wavelength in the range of 365–400 nm, which is confirmed to be able to work in the ultraviolet (UV) area. This condition has become one of the significant obstacles in the use of TiO_2-based sunlight because only 4–5 % of the sun rays contain the UV light. There are several thoughts regarding the development of TiO_2 as a catalyst material under sunlight illumination. One of them is to decrease their value of Eg to enhance the absorption of sunlight, especially in the visible area [8].

On the other hand, Co_3O_4 can be used for TiO_2 doping due to its physical and chemical property [9]. Besides, Co_3O_4 has a bandgap value of 2.07 eV, which makes it efficiently absorbs the photons of visible light [10]. TiO_2 was successfully prepared to modify the Au nanoparticles to decrease its bandgap value [11]. This modification has been performed for the degradation of methyl orange [12], methylene blue and crystal violet [13].

In this work, we reported the synthesis of TiO_2/Co_3O_4 nanocomposite using *Catharanthus roseus* (L.) G. Don leaf extract for the degradation of malachite green under the visible light illumination.

2. Materials and method

2.1. Materials

Catharanthus roseus (L.) G. Don. leaves were collected from Indonesia, $Ti(OCH(CH_3)_2)_4$ (TTIP) (98.9 %), $Co(NO_3)_2.6H_2O$ were purchased from Sigma-Aldrich and Merck, respectively. The distilled water, methanol and hexane were in analytical grade.

2.2. Preparation of Catharanthus roseus (L.) G. Don leaf extract

Catharanthus roseus (L.) G. Don leaves were washed by distilled water, then dried at room temperature for 10 days. The dried-*Catharanthus roseus* (L.) G. Don was ground to obtain the powders. Fifty g of leave powders were macerated in methanol for 7 days. The extract was filtrated, and further partitioned by hexane to obtain the methanol fraction. The extract was evaporated to separate its methanol, then dissolved into distilled water to get an aqueous fraction of CRE for nanocomposite synthesis [14-15].

2.3. Synthesis of TiO_2 nanoparticle

TiO_2 nanoparticles were prepared by stirring 2.96 mL $Ti(OCH(CH_3)_2)_4$ (TTIP) and 40 mL isopropanol for 20 min. The mixture was added with 10 mL CRE and stirred for 5 h, then calcined at 500 °C for 150 min to form the nanoparticle powders [16].

2.4. Synthesis of TiO_2/Co_3O_4 nanocomposite

0.1 g of TiO_2 powders were dispersed into 25 mL distilled water, then ultrasonicated for 30 min (Part A). Meanwhile, $Co(NO_3)_2$ 0.01 M was added with 10 mL CRE and stirred for 20 min (Part B). Subsequently, part A and B were mixed and stirred for 5 h, then calcined at 800 °C for 150 min to form the nanocomposite powders [12].

2.5. Characterization of TiO_2/Co_3O_4 nanocomposite

All characterizations of TiO_2/Co_3O_4 nanocomposite comprise UV-Vis diffuse reflectance spectrophotometry (Shimadzu 2450), spectra absorption UV-Vis spectrophotometry (Shimadzu 2600), Fourier transform infrared (FT-IR Prestige 21 Shimadzu), X-ray diffraction (XRD Shimadzu 7000), scanning electron microscope-energy dispersive X-ray spectroscopy (SEM-EDX JEOL JSM 6510 LA), particle size analyzer (PSA Malvern Zetasizer 1600) and transmission electron microscope (TEM JEOL JEM-1400).

2.6. Photocatalytic activity

TiO_2/Co_3O_4 nanocomposite powder was dispersed into malachite green solution and stirred for 1 h under the visible light illumination. The malachite green degradation was observed every 0, 10, 20, 30, 40, 50, 60, 70, 80, 90, 100, 110 and 120 min using a UV-Vis spectrophotometer [17].

IOP Conf. Series: Materials Science and Engineering **902** (2020) 012003 doi:10.1088/1757-899X/902/1/012003

3. Results and discussion

3.1. FT-IR analysis of Catharanthus roseus (L.) G leaf extract

FT-IR analysis was performed to determine the primary functional group of CRE. There were some functional groups such as -OH groups (3373 cm^{-1}), C-H (2945 cm^{-1}), C = C (1589 cm^{-1}), C-H (1459 cm^{-1}), and C-N (1028 cm^{-1}) [18]. The presence of alkaloids was signed by the appearance of C = C and C-N functional groups. The -OH functional group indicates the presence of flavonoids secondary metabolite compound. Figure 1 shows that there are –OH, C=C, and C-N functional group at the wavenumber of 3367, 1641 and 1028 cm^{-1}, respectively. The presence of these groups illustrate that CRE can be utilized in the synthesis of TiO_2/Co_3O_4 nanocomposite.

3.2. Structural analysis of TiO_2/Co_3O_4 nanocomposite

The structural of TiO_2/Co_3O_4 nanocomposite was analyzed by XRD. Figure 2 shows the 2θ value of TiO_2/Co_3O_4, which was a combination of the TiO_2 and Co_3O_4. The 2θ values of TiO_2 were at 27.63; 35.49; 36.13; 36.12 and 49.27°, which were well-matched to the JCPDS card no. 21-1272. Meanwhile, the 2θ values of Co_3O_4 were at 32.90; 32.94; 32.99 and 53.59° which were well-matched to JCPDS card no. 43-1003.

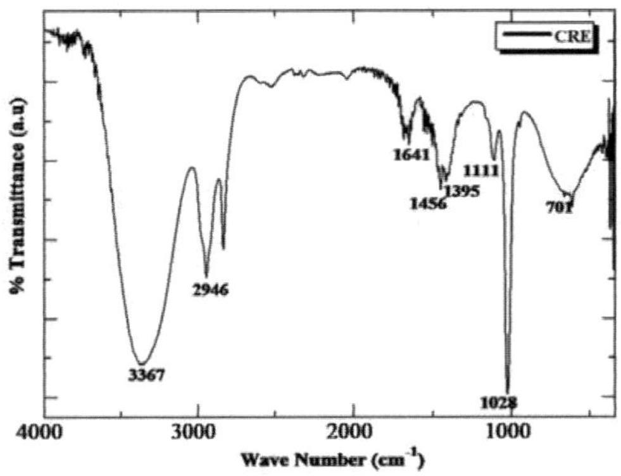

Figure 1. FT-IR spectrum of *Catharanthus roseus* (L.) G Don leaf extract.

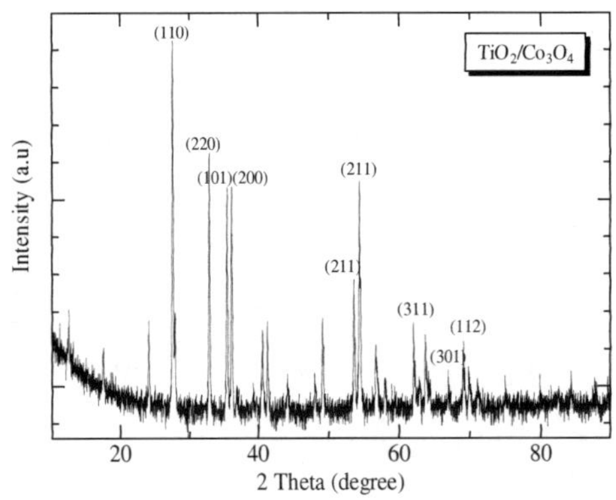

Figure 2. XRD pattern of TiO_2/Co_3O_4.

3.3. Particle size analysis of TiO_2/Co_3O_4 nanocomposite

Particle size of TiO_2/Co_3O_4 nanocomposite was analyzed with PSA and TEM. The average particle size distribution was 38.04 nm, confirmed by PSA as presented in figure 3a. For further characterization, TEM was utilized to confirm the particle size in more details. According to figure 3b, TiO_2/Co_3O_4 nanocomposite exhibits the particle size around 90.80 nm.

3.4. Morphological analysis of TiO_2/Co_3O_4 nanocomposite

TiO_2/Co_3O_4 nanocomposite was characterized using SEM-EDX to determine its morphology and atomic composition. SEM image shows that TiO_2/Co_3O_4 nanocomposite has a spherical shape, as shown in figure 4a. Based on the EDX result, TiO_2/Co_3O_4 nanocomposite contained the atomic composition of Ti, Co and O, which were found to be 19.42; 3.79 and 66.21 %, respectively, as shown in figure 4b.

3.5. UV-Vis DRS analysis of TiO_2/Co_3O_4 nanocomposite

The UV-Vis DRS characterization was conducted to define the bandgap energy of TiO_2/Co_3O_4 nanocomposite by measuring the per cent reflectance, which was then converted to the bandgap value by employing the Kubelka Munk equation. According to the calculation, TiO_2/Co_3O_4 nanocomposite has the bandgap value of 2.80 eV, as presented in figure 5a. Meanwhile, the bandgap energy of TiO_2 nanoparticle was 3.2 eV, as shown in figure 5b.

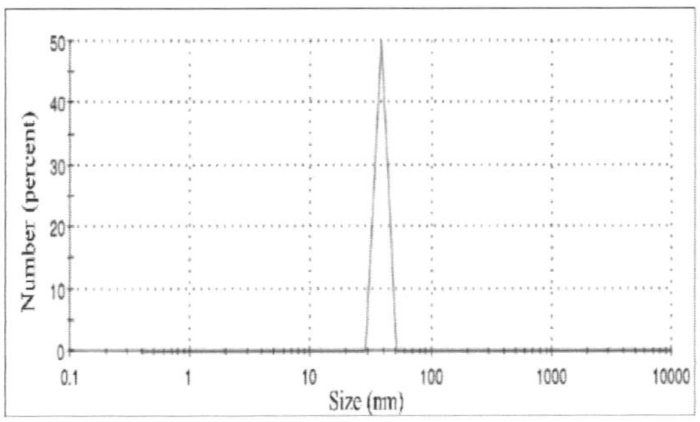

(a)

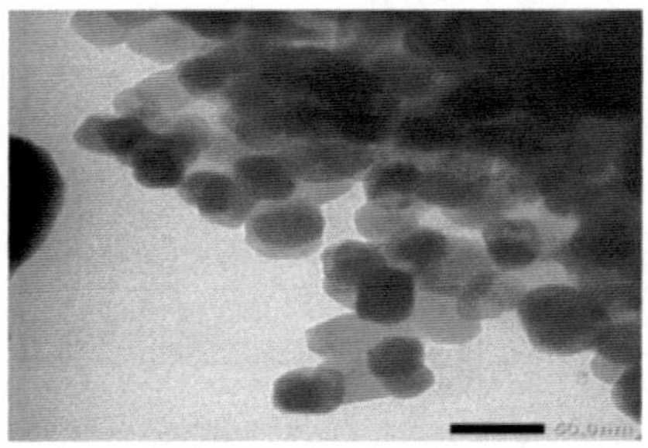

(b)

Figure 3. (a) Particle size distribution and (b) TEM image of TiO_2/Co_3O_4 nanocomposite.

(a)

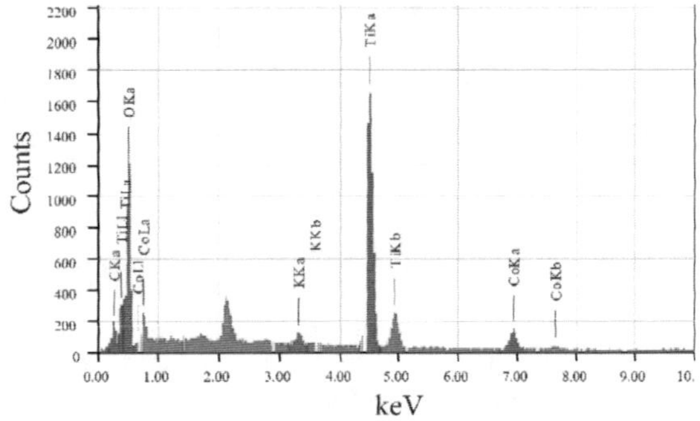

(b)

Figure 4. (a) SEM image and (b) EDX spectrum of TiO_2/Co_3O_4 nanocomposite.

IOP Conf. Series: Materials Science and Engineering **902** (2020) 012003 doi:10.1088/1757-899X/902/1/012003

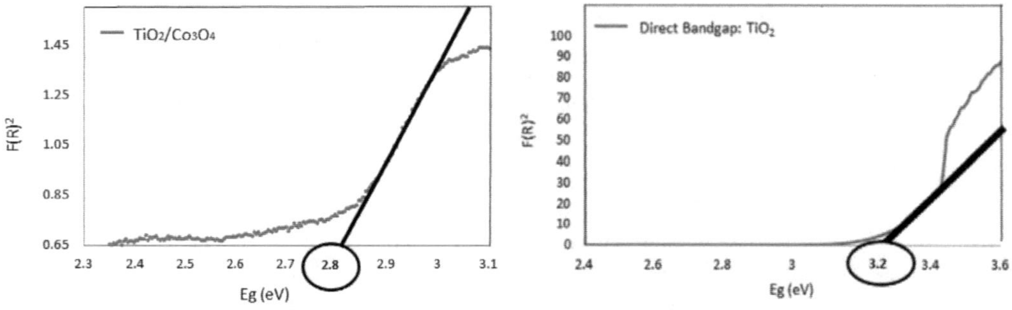

(a) (b)

Figure 5. UV-Vis DRS Spectra of (a) TiO_2/Co_3O_4 nanocomposite and (b) TiO_2 nanoparticle.

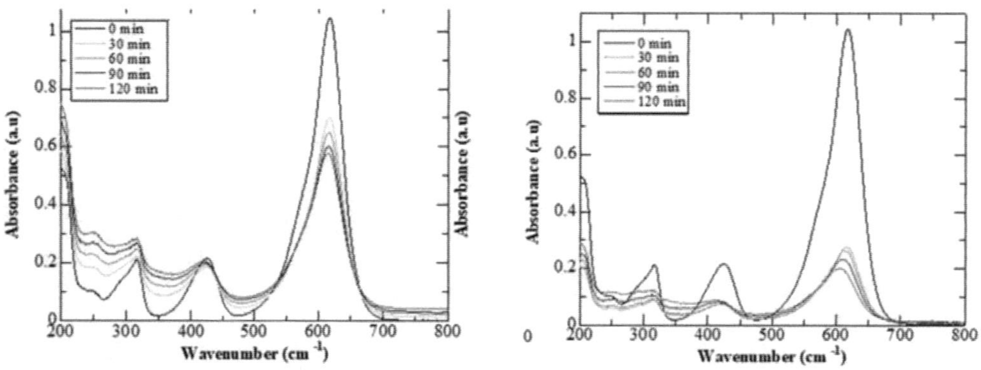

(a) (b)

Figure 6. UV-Vis absorption spectra of malachite green degradation using (a) TiO_2 nanoparticle, and (b) TiO_2/Co_3O_4 nanocomposite.

3.6. Photocatalytic degradation of malachite green using TiO_2/Co_3O_4 nanocomposite

The photocatalytic activity of TiO_2/Co_3O_4 nanocomposite was tested for the malachite green degradation under the visible light illumination. The degradation was observed at the wavelength of 617 nm as a typical peak of malachite green, as shown in figure 6. The results show that the degradation percentages of malachite green using TiO_2/Co_3O_4 nanocomposite and TiO_2 nanoparticle were 82.61 % and 46.03 % for 120 min, respectively.

4. Conclusion

TiO_2/Co_3O_4 nanocomposite has been successfully synthesized using CRE. Alkaloids and flavonoids were the essential secondary metabolites contained in CRE for nanocomposite synthesis. The bandgap value of TiO_2/Co_3O_4 nanocomposite was narrower than the bandgap value of TiO_2 nanoparticle. Also, TiO_2/Co_3O_4 nanocomposite has a good photocatalytic activity rather than TiO_2 nanoparticle, which significantly works as photocatalysts in the visible region. The synthesized nanocomposite has been structurally formed from a combination of TiO_2 and Co_3O_4 confirmed by XRD characterization.

Acknowledgments
This research was granted by Hibah PDUPT 2019 from the Ministry of Research, Technology and Higher Education, Indonesia, through Universitas Indonesia (No. NKB 1617/UN2.R3.1/HKP.05.00/2019).

References
[1] Idrees M, Naeem M, Aftab T, Khan M M A and Moinuddin 2011 *Acta Physiol. Plant.* **33** 987-99
[2] Yulizar Y, Kusrini E, Apriandanu D O B and Nurdini N 2020 *Surf. Interfaces* **19** 100437
[3] Apriandanu D O B and Yulizar Y 2019 *Nano-Struct Nano-Objects* **20** 100401
[4] Yulizar Y, Pratiwi Y, Apriandanu D O B 2020 *Mater. Sci. Forum* **982** 9-13
[5] Michihiro M, Mano T, Nishimoto S and Kameshima Y 2015 *J. Chem. Eng.* **264** 221-9
[6] Smith W, Mao S, Lu G, Catlett A, Chen J and Zhao J 2010 *Chem. Phys. Lett.* **485** 171-5
[7] Kheamrutai T, Pichet L, and Boonlaer N 2008 *Kasetsart J. Nat. Sci.* **42** 357-61
[8] Naik G K, Majhi S M, Jeong K-U, Lee I-H and Yu Y T 2019 *J. Alloys Compd.* **771** 505-12
[9] Suryakant M, Priyanka Y, PR Sagdeo, and Rajesh K 2018 *ACS Appl. Energy Mater.* **1** 790-8
[10] Barreca D, Gasparotto A, Lebedev O I, Maccato C, Pozza A, Tondello E, Turner S, Tendeloo G V 2010 *Cryst. Eng. Comm.* **12** 2185-97
[11] Yulizar Y, Sudirman, Apriandanu D O B, and Wibowo A P 2019 *Compos. Commun.* **16** 50-6
[12] Chowdhury M, Kapinga S, Cummings F, and Fester V 2019 *Water Sci. Technol.* **79** 947-57
[13] Ranjith T, Ranganathan V, Chen S-M, Selvan N S, and Rajam P S 2019 *Ceram. Inter.* **45** 12926-33
[14] Widyaningtyas A L, Yulizar Y, and Apriandanu D O B 2019 *IOP Conf. Ser. Mater. Sci. Eng.* **509** 012022
[15] Mutuma B K, Shao G N, Kim W D and Kim H T 2015 *J. Colloid Interf. Sci.* **442** 1-7
[16] Bojinova A S, Papazova C I, Karadjova I B and Poulios I 2008 *Eurasian J. Anal. Chem.* **1** 34-43
[17] Monireh A, Nasrollahzadeh M and Sajadi S M 2015 *RSC Adv.* **5** 91532-43
[18] Silverstein R M, Webster F X and Kiemle D J 2005 *Spectrometric Identification of Organic Compounds* (Hoboken: John Wiley & Sons)

Fabrication of Gd_2O_3 nanoparticles in hexane-water system using *Myristica fragrans Houtt* leaves extract and their photodegradation activity of malachite green

A Eprasatya, Y Yulizar, R T Yunarti and D O B Apriandanu

Department of Chemistry, Faculty of Mathematics and Natural Sciences (FMIPA), Universitas Indonesia, Depok 16424, Indonesia

Corresponding author's email: yokiy@sci.ui.ac.id

Abstract. This study aims to fabricate the Gd_2O_3 nanoparticles (NPs) using *Myristica fragrans Houtt* leaves extract (MFE) in hexane-water phases system by high-speed stirring method. We performed a phytochemical test to qualitatively confirm that MFE contained the secondary metabolite of alkaloid, steroid and saponin as a weak base source (OH^-), a stabilizing agent, and a capping agent in the Gd_2O_3 NPs fabrication, respectively. According to SEM image, the morphology of Gd_2O_3 NPs was spherical shape in agglomeration. UV-Vis DRS result shows that Gd_2O_3 NPs has a large bandgap which was found to be 5.1 eV. XRD characterization shows that the diffraction pattern of the formed nanoparticle was well-matched with the data source of Gd_2O_3. The particle size of Gd_2O_3 NPs was approximately 40 nm confirmed by PSA. Photocatalytic activity of Gd_2O_3 NPs was tested for dye degradation. In this study, malachite green was utilized as a modelling of dye. The photodegradation percentage of malachite green using Gd_2O_3 NPs catalyst was 75.15 % for 120 min under UV light irradiation.

Keywords: Gd_2O_3 NPs, *Myristica fragrans Houtt*, hexane-water phases, photocatalytic activity malachite green

1. Introduction

Nanoparticle has been developed due to its versatility in many aspects. The synthesis of nanoparticle has been reported with various methods such as co-precipitation, sol-gel method, microemulsion, hydrothermal/solvothermal, or also electrochemical [1]. But mostly, these methods are expensive and not environmentally friendly because it uses a lot of chemical compounds which harms the human body and environment [2]. On the other hand, green synthesis is a nanomaterial synthesis method, designed as a cost-effective alternative in the fields of chemistry and physics [3]. Green synthesis of metal nanoparticles is an environmentally friendly procedure, free from strong, hazardous, and high cost chemicals [4].

The use of plants in the nanoparticle synthesis is related to the content of both secondary metabolites and antioxidant activity. These antioxidants make plants as an alternative material to environmentally friendly reduce the use of hazardous chemicals [4]. Another advantage of using plants for nanoparticle synthesis are readily available, non-toxic, and have a wide variability in metabolites to prevent the agglomeration of nanoparticles and the formation of surfactants, respectively [3, 5, 6].

The hexane-water phase system is a common method to form a complex of metal and ligand since metal ion can dissolve in polar substance, and ligand can dissolve in nonpolar. The advantage of using a hexane-water phase to synthesize nanoparticles are relatively low costs and energy [7].

Gadolinium oxide (Gd_2O_3) is one of the members of lanthanide oxide group which has a high density (7.6 g/cm^3), a crystallographic stability up to 2325 °C, a high mechanical strength, and a large optical band gap (5.4 eV) [8]. Also, Gd_2O_3 is well-known as an n-type semiconductor for photocatalysis. Due to their highly chemical stability, UV absorptivity and efficient photon-to electron conversion, Gd_2O_3 is widely used for the degradation of organic pollutants and water splitting [9-11].

Nanoparticles formation by hexane-water phases system has not been widely reported yet. Therefore, in this research, Gd_2O_3 NPs was synthesized using *Myristica fragrans Houtt* leaves extract (MFE) in hexane-water phase system by high-speed stirring method. Also, their photocatalytic activities were tested for the degradation of malachite green.

2. Materials and method

2.1. Materials

Materials used in this research were *Myristica fragrans Houtt* leaves from *Pusat Studi Biofarmaka Tropika* IPB, Bogor, West Java. The chemicals of gadolinium nitrate hexahydrate ($Gd(NO_3)_3.6H_2O$), malachite green, n-hexane, HCl, Wagner reagent, H_2SO_4, chloroform, lead acetate were all from Merck, and distilled water.

2.2. Method

2.2.1. Myristica fragrans (MF) leaves extraction. MF leaves were washed, cut, dried, then blended to obtain powders [12]. MF leaf powder was macerated into hexane and stirred every day for a week. The mixture was sieved to produce the crude extract, which was further partitioned by distilled water to obtain the hexane fraction as MFE for nanoparticle synthesis [13, 14].

2.2.2. Phytochemcial test. MF leaf extract was tested phytochemically to determine the content of active substances qualitatively. Phytochemical tests were carried out on several classes of compounds, such as alkaloids, steroids, terpenoids, saponins, tannins, polyphenols and flavonoids [15].

2.2.3. Gd_2O_3 nanoparticles synthesis. $Gd(NO_3)_3$ solution was mixed with MFE 0.46 % (w/v) with the high-speed stirring method at 23000 rpm for an hour to form $Gd(OH)_3$ colloids. The result was filtered, rinsed, and dried at 120 °C for 5 h, and calcinated at 600 °C for 5 h. [7].

2.2.4. Gd_2O_3 nanoparticles characterization. Gd_2O_3 NPs were characterized by FT-IR spectroscopy (IR Prestige-21 Shimadzu), UV-Vis spectrophotometer Shimadzu 2600, UV-Vis DRS Shimadzu 2450, scanning electron microscopy-energy dispersive X-ray spectroscopy (SEM-EDX JED-2300), transmission electron microscopy (TEM JEOL JEM 1400), X-ray diffraction (XRD Shimadzu 2700), and particle size analyzer (PSA Zetasizer Nano ZS 90).

3. Results and discussion

3.1. Phytochemical test of Myristica fragrans leaves extract

Phytochemical tests were conducted to qualitatively determine the secondary metabolites in MFE, such as alkaloids, flavonoids, saponins, tannins, polyphenols, steroids, and terpenoids. According to the result, MFE positively contained alkaloids, saponins, and steroids. Positive result of alkaloid compounds in MFE indicated by the appearance of red sediment, saponin was shown from stable bubbles on the solution surface, while the steroid compounds were noticed by the bluish-green color solution.

3.2. Characterization of Gd_2O_3 NPs using FT-IR Spectroscopy

FTIR characterization was conducted to identify the functional groups of the Gd_2O_3 NPs and MFE. FTIR data of MFE have the vibrational spectra at wavenumbers of 3456; 2962; 1645; 1466; 1380 and 720 cm^{-1}, which indicate the presence of O-H stretching, C-H sp^3 stretching, N-H amine bending, C-C stretching, N-O stretching and N-H wagging stretching, respectively, as shown in figure 1. The N-H amine bending indicates the presence of alkaloid. Also, the O-H stretching, C-H sp^3 stretching and C=C stretching show the presence of saponins. These secondary metabolite compounds have an important role in the synthesis of nanoparticles. Alkaloids act as a source of weak bases, and saponins act as a capping agent [16]. FTIR spectrum of Gd_2O_3NPs shows the absorption at wavenumber 556 cm^{-1}, which indicates the bending vibration of Gd-O. This result is in accordance to the previous research that the Gd-O bond has a vibration at the wavenumber range of 563–547 cm^{-1} [8].

3.3. Characterization of Gd2O3 NPs using XRD

XRD characterization was conducted to investigate the diffraction pattern and crystallinity of the synthesized Gd_2O_3 NPs. X-ray diffractograms present the diffraction of 2θ, as shown in figure 2. The 2θ value of synthesized Gd_2O_3 was compared to the data from COD Number 96-101-1289, which were well-matched at 28.71; 33.08; 35.20; 42.48; 47.51; 56.50 and 76.88 deg. This result indicates that Gd_2O_3 nanoparticles have been synthesized using MFE hexane fraction.

3.4. Characterization of Gd_2O_3 NPs using PSA

PSA identification was carried out to study the particle size distribution of Gd_2O_3 NPs. PSA result of Gd_2O_3 NPs is shown in figure 3. The average size distribution of Gd_2O_3 NPs was 40 nm.

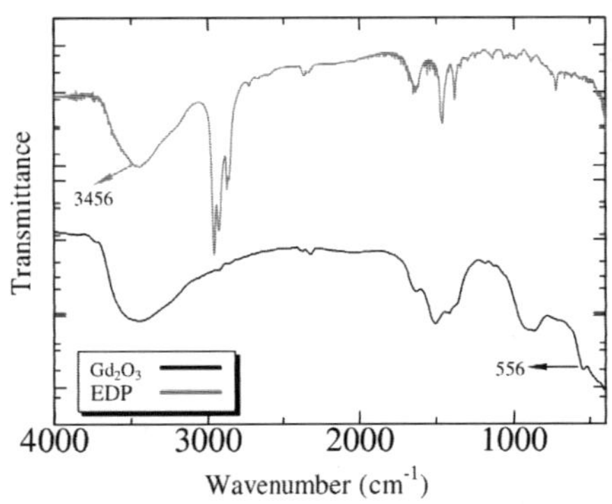

Figure 1. FTIR spectra of MFE and Gd_2O_3 NPs.

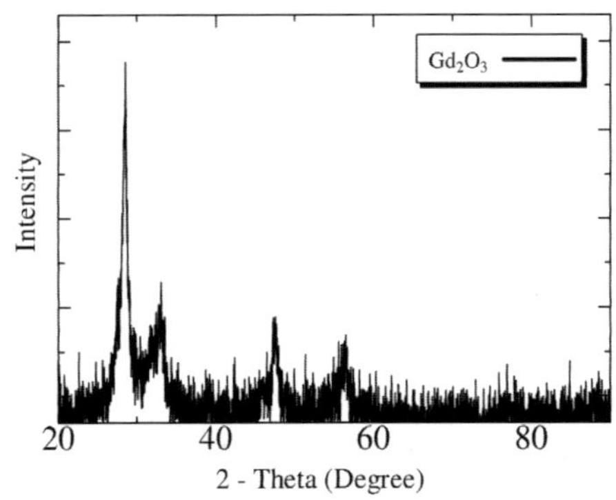

Figure 2. XRD pattern of Gd_2O_3 NPs.

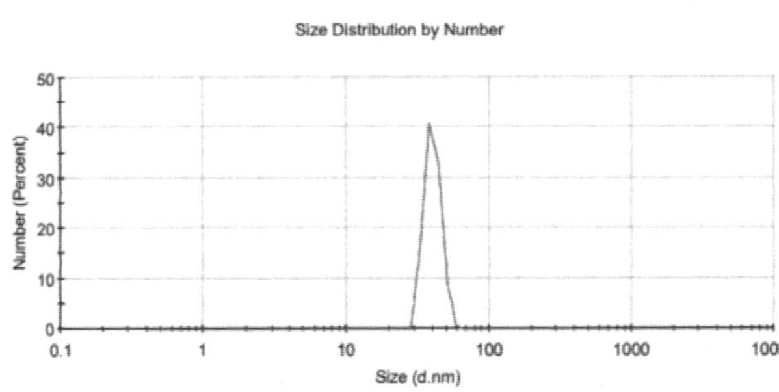

Figure 3. Particle size distribution of Gd_2O_3 NPs.

3.5. Characterization of Gd_2O_3 NPs using SEM-EDX

SEM characterization aims to determine the surface morphology of the Gd_2O_3 NPs, which were appeared as an agglomeration rice-like materials, as shown in figure 4. Meanwhile, EDX analysis aims to determine the atomic composition on the surface of Gd_2O_3 NPs. According to the EDX spectrum, Gd_2O_3 NPs contained Gadolinium atom (Gd) of 46.87 %, and oxygen atom (O) of 32.55 %, as shown in figure 5.

3.6. Characterization of Gd_2O_3 NPs using UV-Vis DRS

UV-Vis DRS identification was carried out to determine the bandgap value of Gd_2O_3, which is defined by converting a % reflectance using Kubelka-Munk function. According to the calculation, the obtained bandgap energy of Gd_2O_3 was 5.1 eV (figure 6), which included in UV light range. This result indicates that Gd_2O_3 has a good absorption in the UV region which has absorption range between 270-325 nm.

3.7. Characterization of Gd_2O_3 NPs using TEM

TEM identification was conducted to define the particle size of Gd_2O_3 NPs. The results were presented at magnification of 43,000x and 97,000x, as shown in figure 7. Based on TEM characterization, Gd_2O_3 NPs were observed with particle size around 150 nm. That size shows that Gd_2O_3 NPs form aggregate of nanoparticles.

(a) (b)

Figure 4. SEM images of Gd_2O_3 NPs with magnification of (a) 30,000x and (b) 10,000x.

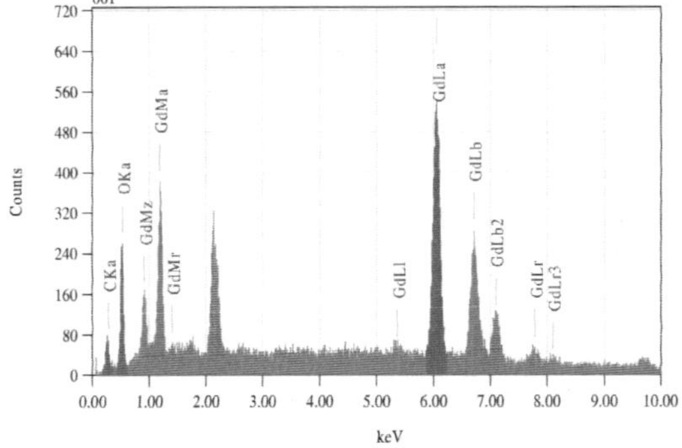

Figure 5. EDX spectrum of Gd_2O_3 NPs.

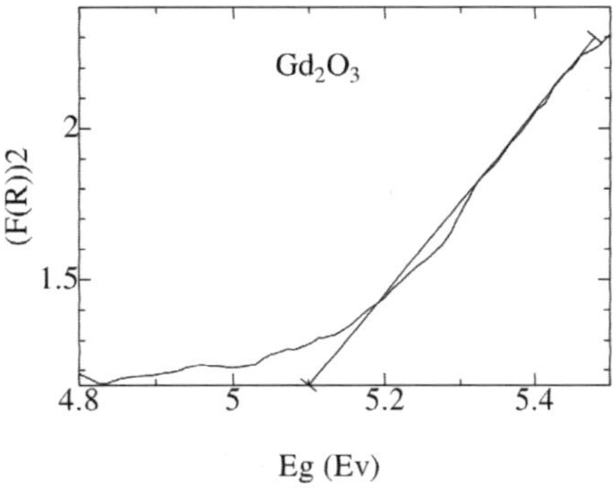

Figure 6. UV-Vis DRS analysis of Gd_2O_3 NPs.

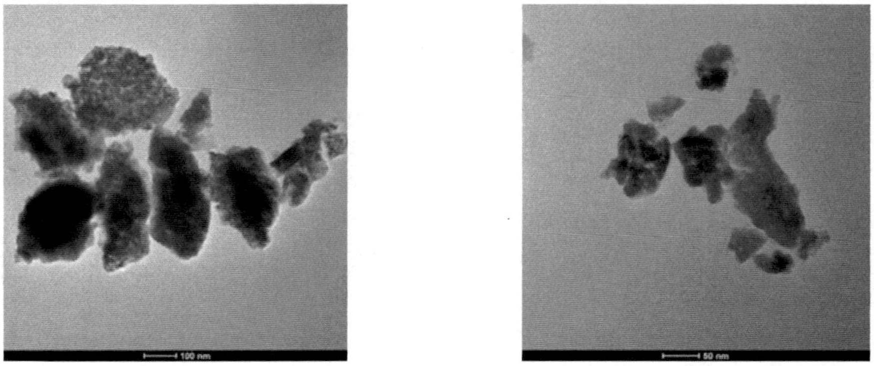

(a) (b)

Figure 7. TEM images of Gd_2O_3 NPs with magnification of (a) 43,000x and (b) 97,000x.

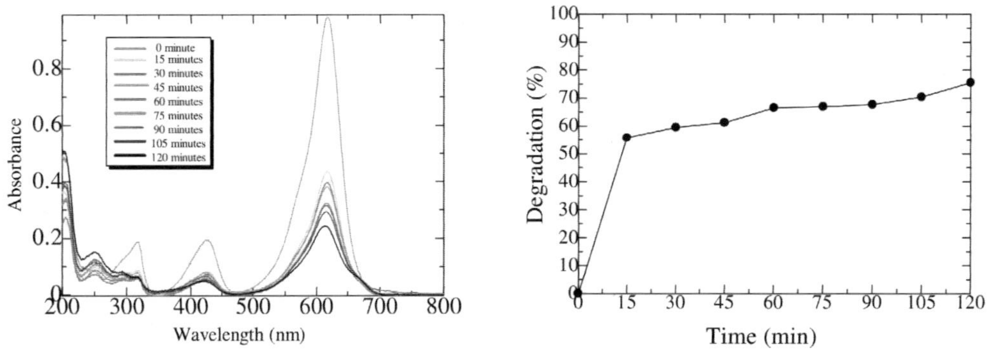

(a) (b)

Figure 8. (a) UV-Vis absorption spectra of malachite green degradation and (b) photodegradation percentage of malachite green against reaction time using Gd_2O_3 NPs.

3.8. Photocatalytic activity analysis of Gd_2O_3 NPs

The photocatalytic activity of Gd_2O_3 NPs was carried out for the degradation of malachite green under the UV light area for 120 min with the observation every 15 min. The results of UV-Vis absorption spectra of malachite green degradation are shown in figure 8. The wavelength shifts were observed at λ_{max} range of 318 to 250 nm. The absorption peak at 250–254 nm indicates the presence of aromatic chromophore and carboxylic acid groups [17]. According to the calculation, the photodegradation percentage of malachite green using Gd_2O_3 NPs for 120 min was 75.15 %.

4. Conclusion

The synthesis of Gd_2O_3 NPs using a hexane fraction of *Myristica fragrans* extract (MFE) has been successfully synthesized. The phytochemical test results show that a hexane fraction of MFE contained alkaloids, steroids, and saponins. PSA characterization shows that the average particle size distribution of Gd_2O_3 was around 40 nm. The results of UV-Vis DRS characterization show that Gd_2O_3 has a bandgap energy of 5.1 eV. The photocatalytic activity of Gd_2O_3 NPs was observed for the degradation of malachite green with the percentage of 75.15 % for 120 min under UV light irradiation.

Acknowledgments
The authors would like to thank Universitas Indonesia for funding this research through PIT-9 Grant Universitas Indonesia with contract number: NKB-0037/UN2.R3.1/HKP.05.00/2019.

References

[1] Jamkhande P G, Ghule N W, Bamer A H, Kalaskar M G 2019 *J. Drug Delivery Sci. Tech.* **53** 101174
[2] Yedurkar S, Maurya C and Mahanwar P 2016 *Open J. Synthesis Theory Applications* **5** 1-14
[3] Parveen K, Banse V and Ledwani L. 2016 *AIP Conf. Proc.* **1724** 020048
[4] Iravani S 2011 *Green Chem.* **13** 2638-50
[5] Abbas M, Sari N, Utari T, Yulizar Y and Apriandanu D O B 2018 *AIP Conf. Proc.* **2023** 020100
[6] Tania T C D, Deastri Y N, Utari T, Yulizar Y and Apriandanu D O B 2018 *AIP Conf. Proc.* **2023** 020103
[7] Yulizar Y, Monjushiro H, and Watarai H 2004 *J. Colloid Interface Sci.* **275** 560-9
[8] Tamrakar R K, Bisen D P, and Brahme N 2014 *Res. Chem. Intermediates* **40** 1771-9
[9] Mamba G, Mbianda X Yand Mishra A K 2016 *Mater Res. Bull.* **75** 59-70
[10] Ayawanna J, Teoh W T, Niratisairak S and Sato K 2015 *Mater. Sci. Semicond. Process.* **40** 136-9
[11] Ledwaba M, Masilela N, Nyokong T and Antunes E 2015 *J. Mol. Catal. A-Chem.* **403** 64-76
[12] Yulizar Y, Kusrini E, Apriandanu D O B and Nurdini N 2020 *Surf. Interfaces* **19** 100437
[13] Yulizar Y, Sudirman, Apriandanu D O B and Wibowo A P 2019 *Compos. Commun.* **16** 50-6
[14] Apriandanu D O B and Yulizar Y 2019 *Nano-Struct. Nano-Objects* **20** 100401
[15] Edrah S M, Elzaedi Y M, Kahel F and Alkhumsi S I 2016 *ICCPGE* **1**, 25-30
[16] Yulizar Y, Latifah I, Bakri R and Apriandanu D O B 2018 *AIP Conf. Proc.* **2023** 020097
[17] Chen C C, Lu C S, Chung Y C and Jan J L 2007 *J. Hazard. Mater.* **141** 520-8

Preparation and characterization of β-Cyclodextrin/Fe_3O_4 nanocomposite

E Saepudin[1], S J Willyam[1] Irkham[2] and T A Ivandini[1]

[1]Department of Chemistry, Faculty of Mathematics and Natural Sciences (FMIPA), Universitas Indonesia, Depok 16424, Indonesia
[2]Department of Chemistry, Faculty of Science and Technology, Keio University, Yokohama 223–8522, Japan

Corresponding author's email: ivandini.tri@sci.ui.ac.id

Abstract. Host-guest mechanism and the formation of inclusion complexes between β-cyclodextrin (BCD) with cholesterol as the electroactive species can indicate a non-enzymatic sensor of cholesterol. For that purpose, in this study, BCD functionalized on Fe_3O_4 magnetic nanoparticles (MNPs) were synthesized by a simple co-precipitation method of Fe^{2+} and Fe^{3+} solutions under alkaline conditions for an easy separation. BCD/MNPs nanocomposite were characterized by XRD, TEM, and FTIR. The average size of MNPs and BCD/MNPs were 16.22 nm and 9.80 nm, respectively showed by both XRD and TEM. FTIR result also indicates that BCD/MNPs was successfully formed.

Keywords: β-Cyclodextrin (BCD), Fe_3O_4, MNP, BCD/MNP, nanocomposites.

1. Introduction

β-cyclodextrin (BCD) is a macrocyclic oligosaccharide formed from seven glucopyranose subunits with glycosidic bonds α-1,4. BCD which has hydrophilic outer cavity and hydrophobic inner cavity can form inclusion complexes with nonpolar molecules [1]. Because of its unique properties and high solubility in aqueous, BCD has been widely used and has a big role in catalyst, supramolecular chemistry, food industry, drug delivery and biosensor [2, 3]. The hydrophobic cavity of BCD can form inclusion complexes with hydrophobic compounds such as cholesterol. The selective inclusion of BCD has been used in the food industry to eliminate cholesterol from animal products. Aside from that, this mechanism can indicate a non-enzymatic sensor of cholesterol. In addition, BCDs in aqueous environments are shaped like truncated cones where the hydroxyl group on the outside leads to a solvent and the hydrophobic group is in the inner cavity. Chemical structure of BCD is shown in figure 1.

On the other hand, Fe_3O_4 magnetic nanoparticles (MNPs) are widely investigated due their special properties and potential use, such as catalyst, as contrast agents in magnetic resonance and magnetic particle imaging, magnetic ink for jet printing, tissue targeting and biomedicine, and biosensing [4, 5]. MNPs are hydrophobic and tend to agglomerate due to the intramolecular interactions of functional groups on the surface and magnetostatics interactions between particles [6]. Surface of MNPs can be easily immobilized by identification agents, so it can be a good point of analytical application by using MNPs.

In this work, β-cyclodextrin (BCD) was modified at MNPs for a biosensor preparation. BCD can also act as a stabilizer and capping agent to form uniform sized particles because BCD has many hydroxyl groups that can be functionalized on MNPs [6]. This causes the BCD/MNPs nanocomposite to be well dispersed in water. The characterization was performed by using FTIR, XRD, and TEM.

2. Materials and method

The materials used were $FeCl_3.6H_2O$, $FeCl_2.4H_2O$, methylene blue ($C1_6H_{18}ClN_3S$), K_2HPO_4, KH_2PO_4, NaOH, HCl, ethanol absolute, NH_4OH 25 %, distilled water, β-cyclodextrin (≥ 97 %).

MNPs and BCD/MNPs were synthesized by co-precipitation of Fe^{2+} and Fe^{3+} solutions under alkaline conditions. To synthesize the BCD/MNPs, 10 mmol $FeCl_2.4H_2O$ and 20 mmol of $FeCl_3.6H_2O$ were dissolved in 100 mL of BCD solution and stirred vigorously at 80 °C. Then, 1 M NH_4OH was added dropwise to the mixture until pH of 11–12 was reached. The obtained solution was stirred continuously for about half an hour at the same temperature. The obtained precipitates were separated with the help of external magnet and washed with water and methanol. The precipitates were dried at 55 °C overnight and grinded. The same procedure was done for MNPs synthesis without an addition of BCD into the solution. The shape and size of samples were observed with transmission electron microscopy (TEM). The crystallinity of material was determined by X-ray diffractometer (XRD). The functional groups were confirmed by Fourier Transform Infrared spectroscopy (FTIR).

3. Results and discussion

The MNPs and BCD/MNPs were synthesized using the co-precipitation method by mixing of Fe^{2+} and Fe^{3+} solutions and adding it with a weak alkaline solution drop by drop following the literature procedure [6]. The obtained MNPs have black-brownish color while BCD/MNPs have a slightly lighter color (figure 2). BCD/MNPs have a better dispersibility than MNPs due to the interactions of hydroxyl groups on the outer surface of BCDs. The hydroxyl groups could facilitate to functionalize onto the MNPs surface [7]. The dispersibility of MNPs and BCD/MNPS were indicated by the length of time the nanocomposite to precipitate. The precipitation of nanocomposites occurs due to the attractive forces between particles caused by magnetic properties and settle due to gravity [6]. MNPs and BCD/MNPs have magnetic properties proven by external magnet as shown in figure 2.

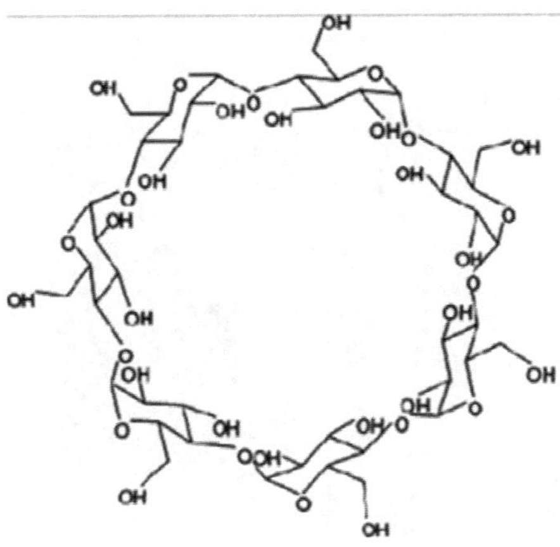

(a)

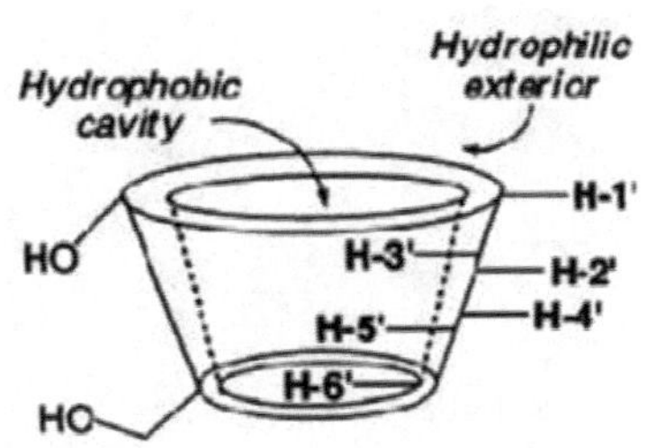

(b)

Figure 1. Chemical structure of β-cyclodextrin (a) top side and (b) cross side.

The shape and size of MNPs and BCD/MNPs were observed using TEM. All materials were of spherical in shape and have homogenous size based on their size distribution graphs (figure 3). The average size of MNPs was 15.16 ± 2.01 nm, while BCD/MNPs was 8.09 ± 1.59 nm. It seems the use of BCD as the modifier in the synthesis process caused a smaller particle size. BCD is a polymer that can act as a capping agent and provide stabilization through a steric effect. MNPs and BCD/MNPs form aggregates that are marked in dark areas due to magneto static interactions.

The XRD spectra of MNPs and BCD/MNPs are shown in figure 4. Based on the theoretical calculation of XRD peaks using Debye-Scherrer equation, the average crystal sizes of MNPs and BCD/MNPs were 16.22 nm and 9.80 nm, respectively. These results are in agreement with the TEM results. The characteristic diffraction peaks at 2θ = 30.43°, 35.75°, 53.55°, 56.76° and 63.05° for (220), (311), (422), (511) and (440) planes match Fe_3O_4 reference (JCPDS No. 19-0629). While in BCD/MNPs the peaks of Fe_3O4 at 2θ = 30.43°, 35.75°, 53.55°, 56.76° and 63.05° for (220), (311), (422), (511), and (440) planes were observed but the intensity of peaks was increased indicating a more crystalline nature of the obtained nanocomposites. The 2θ peaks of MNPs and BCD/MNPs were summarized in table 1.

FTIR spectrum of MNPs and BCD/MNPs (figure 5) shows a strong absorption peak around 586 cm^{-1} indicating the presence of Fe-O stretching vibration. The absorption bands located at 3151 cm^{-1} and 1400 cm^{-1} can be assigned to stretching and bending vibrations of hydroxyl groups from the adsorbed water molecules. The presence of two additional peaks in the spectrum of BCD/MNPs around 1031 cm^{-1} and 1158 cm^{-1} belong to C-C/C-O and C-O-C stretching vibration from the BCD molecules [8]. This observation confirms the formation of BCD/MNPs.

(a) (b)

Figure 2. The physical appearance and separation with external magnet of (a) MNPs and (b) BCD/MNPs.

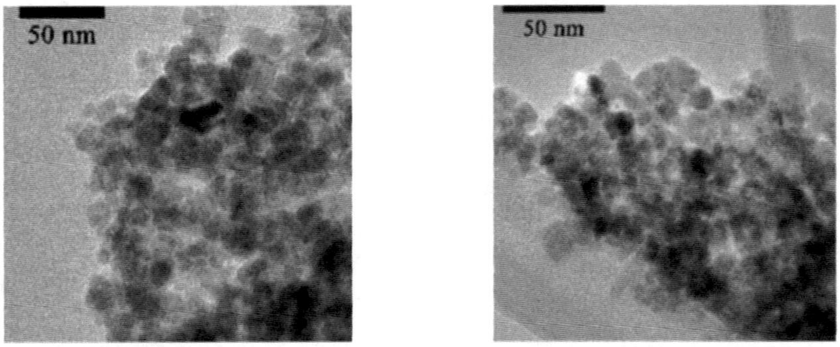

(a) (b)

Figure 3. TEM images and size distributions of (a) MNPs and (b) BCD/MNPs.

IOP Conf. Series: Materials Science and Engineering **902** (2020) 012005 doi:10.1088/1757-899X/902/1/012005

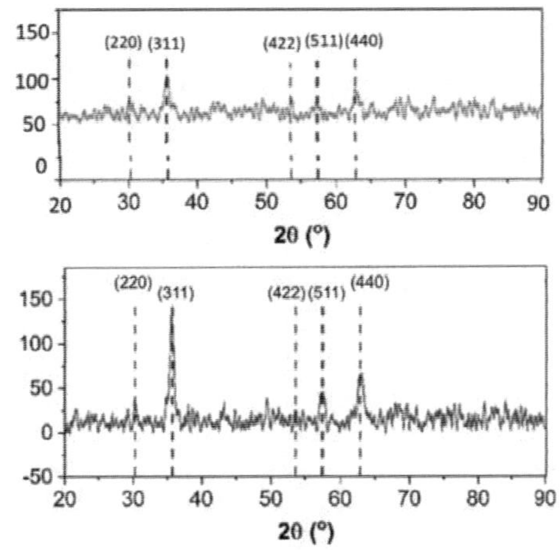

Figure 4. XRD diffractograms of (a) MNPs, and (b) BCD/MNPs.

Table 1. Summarize of the 2θ peaks of MNPs and BCD/MNPs.

JCPDS No. 19-0629	MNPs	BCD/MNPs
30.095	30.343	30.309
35.422	35.757	35.674
53.391	53.554	53.571
56.942	56.763	57.398
62.515	63.047	62.946

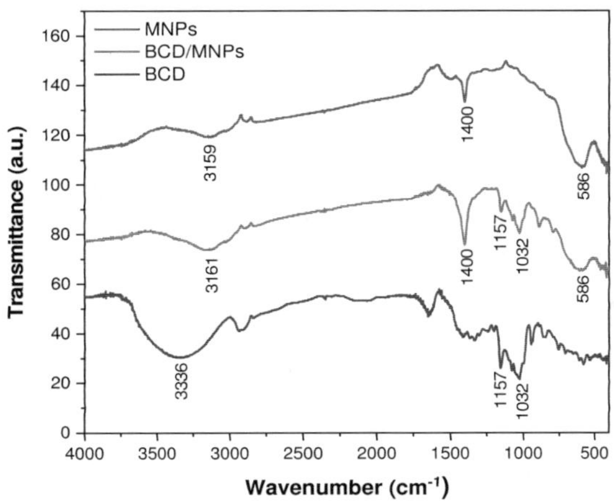

Figure 5. FTIR spectra of MNPs (blue line), BCD/MNPs (red line), and BCD (black line).

4. Conclusion

β-Cyclodextrin/Fe_3O_4 nanocomposite was synthesized using a simple co-precipitation method with an average size of 8.09 ± 1.59 nm. FTIR characterization confirmed the absorption peak at around 1031 cm^{-1} and 1158 cm^{-1} belongs to C-C/C-O and C-O-C stretching vibration from the BCD molecules in addition to the absorption peak around 586 cm^{-1} belongs to Fe-O stretching vibration.

Acknowledgments
This work was funded by PUPT RISTEK DIKTI with contract number NKB-1599/UN2.R3.1/HKP.05.00/2019.

References

[1] Jambhekar S S and Breen P 2016 *Drug Discov. Today* **21** 356-62
[2] Ogoshi T and Harada A 2008 *Sensors* **8** 4961-82
[3] Zhou Y et al. 2016 *Chin. J. Chem.* 34 599-608
[4] Alkire R C, Kolb D M, Lipkowski J and Ross P N 2009 *Chemically Modified Electrodes* (Weinheim: Wiley VCH)
[5] Pino F et al. 2013 *Anal. Sci.* **31** 1061-8
[6] Rostamnia S and Doustkhah E 2015 *J. Magn. Magn. Mater.* **386** 111-6
[7] Maleki A and Kamalzare M 2014 *Catal. Commun.* **53** 67-71
[8] Guo Z, Li Y, Pan S and Xu J 2015 *J. Mol. Liq.* **206** 272-7

IOP Conf. Series: Materials Science and Engineering **902** (2020) 012006 doi:10.1088/1757-899X/902/1/012006

Nickel ferrite nanocrystalline as a green and recyclable catalyst for the synthesis of *Spiro-oxindole-chromene*

S H Oktavia, A H Cahyana and R T Yunarti

Department of Chemistry, Faculty of Mathematics and Natural Sciences (FMIPA), Universitas Indonesia, Depok 16424, Indonesia

Corresponding author's email: herrykim@ui.ac.id

Abstract. A separable $NiFe_2O_4$ nanocrystalline catalyst was synthesized via a simple co-precipitation method. The catalytic activity of this heterogenous catalyst was investigated to synthesize *Spiro-oxindole-chromene* derivates via Knoevenagel condensation, and Michael addition reaction under conventional heating condition. The catalyst showed a prominent catalytic activity in a good yield (86.74 %). This easily separated catalyst could be recycled efficiently for four cycles without significant loss of its activity. Furthermore, the products were investigated for the antioxidant activity using DPPH method. The IC_{50} of the compounds was found to be 182.3 ppm. The use of catalysts has been proven a simple, efficient and environmentally friendly reaction. Based on the results, it is known that the reaction could not be achieved without the presence of $NiFe_2O_4$ catalyst.

Keywords: $NiFe_2O_4$, nanocrystalline, Spiro-oxindole-chromene, antioxidant

1. Introduction

In recent years, the synthesis of magnetic nano-sized has been an interesting objects of catalyst research because of low cost in producing, environmentally friendly, efficient, and recyclable. The nanosized properties are able to maximize contact between reactants and catalyst to improve the catalytic activity. The use of catalysts could be easily separated by external magnet is one of the most promising strategies that make it a useful class of heterogeneous catalysts, because the catalyst could be used for several times [1]. Specifically, nano-sized iron oxide-based magnet such as magnetic (Fe_3O_4) and maghemite (γ-Fe_3O_4) have previously been explored as the catalyst, as well as supporting catalyst in several organic reactions [2]. Another oxide, spinel ferrite (MFe_2O_4; M = Ni, Zn, Mn, and Co) has received much attention because for its application as catalyst. Well-dispersed spinel ferrite is slightly toxic and poisonous accompanied by a high specific surface area. These two compounds not only greatly improve the dispersion and interface structure of bimetal oxide, but also adjust the selectivity, acidity, and durability. Therefore, spinel ferrite catalyst has an amazing synergistic effect to the reaction because of these two kinds of metal on it [3].

Among various spinel ferrite nano-sized that is currently being developed is nickel ferrite $NiFe_2O_4$. This nano-sized posesses an inverse spinel structure, that is, all Ni^{2+} ions are in the octahedral site and Fe^{3+} ions are evenly distributed in the tetrahedral and the octahedral sites. $NiFe_2O_4$ nano-sized shows ferrimagnetic properties from anti-parallel magnetic moment pairs between the magnetic moment of Fe^{3+} ion in tetrahedral site and magnetic moment of Ni^{2+} and Fe^{3+} ion in octahedral site [4]. An inverse spinel structure from $NiFe_2O_4$ is an important soft magnetic material with moderate saturation magnetization, remarkable thermal stability, high magneto crystalline anisotropy, high electrical resistivity, chemical stability, and mechanical hardness [5].

$NiFe_2O_4$ nano-sized has been extensively investigated and showed a better catalytic for organic reactions such as the synthesis of pyridine 2-alkoxy-3-arylimidazo[1,2-α]pyridines [6], 1,4-dihydropyrano[2,3-c]pyrazole [7], spiro[indole-3,2'-pyrrole]-2,5'(1H,1'H)-diones [8], substituted 4H-chromene [9], etc. In this current study, $NiFe_2O_4$ nanocrystalline was used for the first time to synthesize the *Spiro-oxindole-chromene* via multicomponent reaction and the compounds are characterized by TLC, FT-IR and LC-MS/MS, tested for antioxidant activity.

2. Materials and method

All the materials were purchased from Merck and Sigma-Aldrich. Fourier Transform Infrared Spectroscopy (FTIR) was performed by using Shimadzu 8400 spectrometer with KBr background. X-Ray Diffraction patterns were recorded using a Malvern Analytical X-ray diffractometer. Images of Scanning Electron Micrograph (SEM) were obtained with a Sigma Zeiss Microscope. The nanocrystalline elemental analysis was performed using Dispersive Energy X-ray (ZX) Spectroscopy recorded on Amatek. Mass spectrum (MS) was carried out by LC-MS/ MS (Waters). The product was monitored by TLC.

2.1. *General procedure for $NiFe_2O_4$ nanocrystalline synthesis*

$NiFe_2O_4$ was prepared using a modified co-precipitation method from the previous study by Heravi et al. $Fe(NO_3)_2.6H_2O$ (6 mmol) and $Ni(NO_3)_2.6H_2O$ (3 mmol) were dissolved in 25 mL of deionized water and stirred for 30 min at room temperature. The pH was controlled at 10 by adding 0.5 M NaOH dropwise and the brown precipitation formed. Thereafter, the precipitation was washed for several times until pH 7 with deionized water, and dried for 12 h at 70 °C. It was then calcined at 660 °C for 2 h [7].

2.2. *General Procedure for Spiro-oxindole-chromene synthesis*

A mixture of isatin (0.1 mmol), malononitrile (0.1 mmol), 8-hydroxyquinoline (0.1 mmol), and the presence of $NiFe_2O_4$ catalyst in ethanol as a solvent. The mixture was stirred and heated to 70 °C. The reaction was recorded by TLC and an external magnetic field was used to separate the catalyst. The mixture was recrystallized with hot ethanol and cold water.

2.3. *Antioxidant activity test*

The antioxidant assay procedure was determined using 1,1-diphenyl-2-picrylhydrazyl (DPPH•) based on the procedure of Subbareddy and Sumanthi [10]. Briefly, 2 mL of Spiro-oxindole-chromene with various concentrations (0, 50, 100, 150, 200 and 250 ppm) were prepared by dissolving the stock solution (0.0125 g compound in 25 mL ethanol) which was mixed into 2 mL of 0.1 mM DPPH solution and incubated in the dark chamber for 30 min at room temperature. Then, the absorbance of incubated sample was determined at 517 nm by UV-Vis spectroscopy. The free radical scavenging activity was calculated by the equation:

$$\%\ \text{Inhibition} = [\text{AC} - \text{AS}/\text{AC}] \times 100$$

where AC is the absorption of DPPH solution and AS is the absorption of the sample solution. Moreover, the IC_{50} value was determined for the sample.

2.4. *Spectral data of the compound*

2'-amino-2-oxospiro[indoline-3,4'-pyrano[3,2-*h*]quinoline]-3'carbonitrile

Color: red solid. Melting point: 209.4–209.8 °C. Rf: 4 cm.LC-MS/MS [M+H$^+$]= *m/z* 341.10 (Rt 7.97). IR (KBr): υ_{max} = 3319.63, 3203.90, 2199.90, 1724.43, 1614.48, 1473.67, 1380.12, 1115.86, 758.05.

3. Results and discussion

3.1. Characterization of $NiFe_2O_4$ nanocrystalline

The FTIR spectrum of $NiFe_2O_4$ nanocrystalline is shown in figure 1a. FTIR analysis shows two characteristics peak of $NiFe_2O_4$ at 408.92663 cm^{-1} which could be assigned to metal vibration of Ni-O bonds and 614. 35424 cm^{-1} could be assigned to vibration of Fe-O bonds. The bands at 3495.16448 cm^{-1} and 1579.76805 cm^{-1} could be assigned to the stretching and bonding of O-H group [11].

XRD analysis of $NiFe_2O_4$ nanocrystalline is shown in figure 1b. The XRD patterns of $NiFe_2O_4$ nanocrystalline showed peaks at 30.30°; 35.71°; 37.32°; 43.34°; 53.90°; 57.48°; 63.04°. These indicate that $NiFe_2O_4$ nanocrystalline structure is cubic spinel (JCPDS: 01-074-2081). However, there is a small percentage of α- Fe_2O_3 (33.16°) associated with $NiFe_2O_4$ was detected by XRD, it's because of thermal treatment [12]. The crystalline size of $NiFe_2O_4$ was calculated using Debye-Scherrer equation ($D=K.\lambda/\beta.\cos\theta$), based on the peak intensity in the XRD analysis. The crystalline size of $NiFe_2O_4$ nanocrystalline was found 23.56 nm.

Figure 2a and figure 2b show the SEM-EDX analysis of $NiFe_2O_4$ nanocrystalline. SEM-EDX analysis was used to evaluate $NiFe_2O_4$ nanocrystalline composition. According to the result of SEM-EDX, it could be seen that the comparison of Ni, Fe and O in $NiFe_2O_4$ nanocrystalline is suitable with the previous study that is 1:2:4 [13] (table 1).

3.2. Synthesis of Spiro-oxindole-chromene using $NiFe_2O_4$ nanocrystalline

The amount of catalyst becomes the main parameter in obtaining the optimal condition for multicomponent reaction between isatin (0.1 mmol), malononitrile (0.1 mmol), and 8-hydroxyquinoline (0.1 mmol) in figure 3. From the result of the calculation, it was found that by using 6 % (mol) the amount of catalyst gave the optimal result for complete reaction in 4 h and result of 87.84 % in 10 mL ethanol as a solvent and heated at 70 °C. It was discovered that no yield increase noted when the reaction was conducted with the 8 and 10 % (mol) amount of catalyst. $NiFe_2O_4$ nanocrystalline is considered as an efficient catalyst in organic reaction.

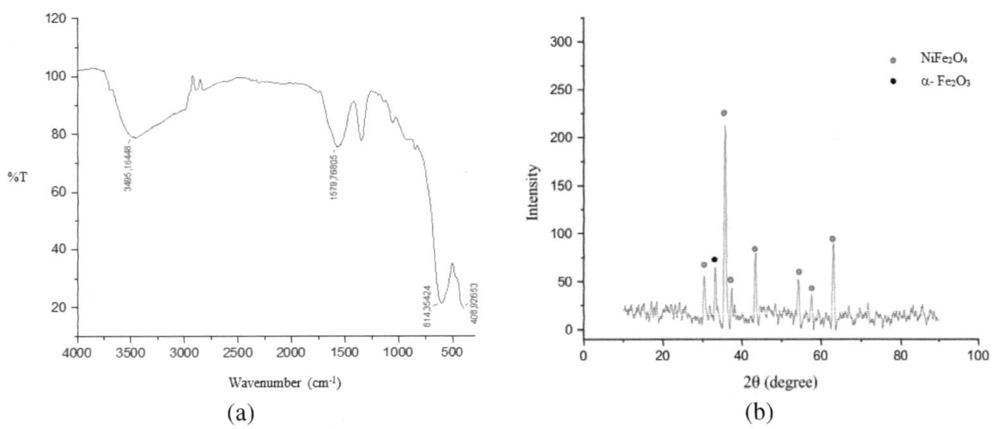

Figure 1. (a) FT-IR spectrum and (b) XRD patterns of $NiFe_2O_4$ nanocrystalline.

IOP Conf. Series: Materials Science and Engineering **902** (2020) 012006 doi:10.1088/1757-899X/902/1/012006

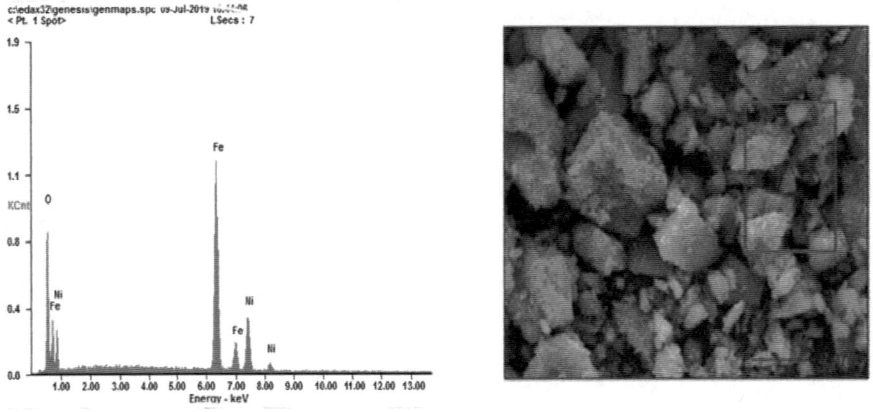

(a) (b)

Figure 2. (a) SEM-EDX patterns and (b) Morphology of $NiFe_2O_4$ nanocrystalline.

Table 1. EDX of $NiFe_2O_4$ nanocrystalline.

Element	wt%	at%
NiK	24.94	15.28
FeK	51.47	32.67
OK	23.60	52.28

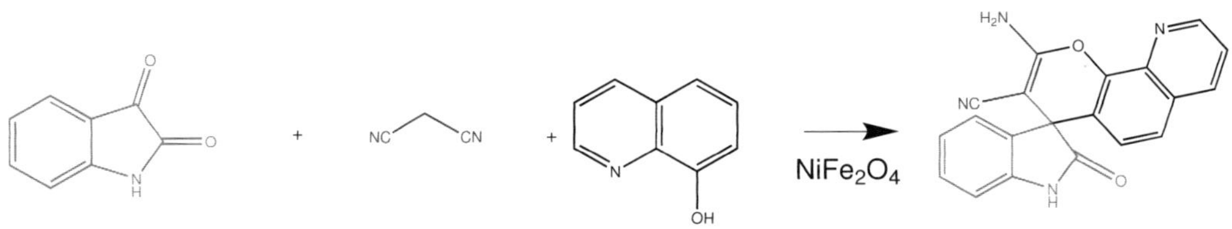

Figure 3. Reaction of *Spiro-oxindole-chromene.*

Table 2 shows that there was no product was obtained in the absence of the catalyst (table 2) indicating that the catalyst plays a necessary role in the reaction. Raising the amount of the catalyst increased the yield of the product even after the optimum amount of the yield decreased. The amount of 6 mol% catalyst gives the highest yield, so the amount of 6 mol% catalyst is used as a reference for further synthesis. Ni is the important active component because of its high Lewis acidity that would lead to the reaction to be more readily achievable [14]. Composition of catalyst is generally affecting the catalytic activity. In the case of ferrite, the activity of catalyst depends on several parameters, namely particle size, redox properties of metal ions, and their distribution between the tetrahedral site and octahedral site of the cubic spinel lattice. The octahedral site takes a significant part in catalyst because the octahedral site is exposed on the surface. In addition, metal ions at octahedral site are located at a greater distance therefore they could interact freely with the reactant molecules [15]. The second parameter is the effect of reaction time (2, 4 and 6 h) on the yield product. The reaction was carried out at 70 °C with the amount of 6 mol % catalyst in 10 mL ethanol as the solvent. The optimal time for the reaction is 4 h, the increase of time decreases the yield of product. The third parameter is the effect of temperature (r.t °C, 50 °C, 70 °C) on the yield product. The reaction was carried out at 70 °C with the amount of 6 mol% catalyst in 10 mL ethanol as the solvent for 4 h.

This catalyst is stable, which could be proven by comparing the fresh and reused $NiFe_2O_4$ nanocrystalline. The proposed mechanism of the reaction using $NiFe_2O_4$ nanocrystalline as a catalyst is likely to follow the Knoevenagel Reaction involving the evident coupling of aldehyde (isatin) with active methylene (malononitrile), and also Michael addition of intermediate gives the target product. The mechanism of reaction is shown in figure 4.

The use of heterogeneously catalyzed reaction has an advantage of the reuse possibility. The recyclable of $NiFe_2O_4$ nanocrystalline catalyst was also investigated using the optimized reaction conditions. The recyclable of the catalyst was tested by running 4 reactions using the same catalyst. The yield indicated that there was no significant change in yield by using the same catalyst for several times as shown in figure 5.

Based on the figure 5, there are no significant change in % yield after the catalyst used for 4 times for synthesis. But, the decline in catalityc ability can be occured due to several factors, such as agglomeration of particles which will reduce the surface area, thereby reducing the catalyst performance. Besides, the activity of the catalys will decrease along with the use of the catalyst. This can be caused by the deactivation of the catalyst. Some of the causes of the catalyst deactivation in this case are (1) poisoning (2) contamination, and solid-state diffusion.

The XRD sepctra in figure 6 show that the typical peaks of $NiFe_2O_4$ still appears although some peaks have decreased, such as the peaks at 37.32° and 53.90° which is the peaks of $NiFe_2O_4$. It can be concluded that the $NiFe_2O_4$ plays the role in catalyzing the organic reaction, even though there is a small percentage of α-Fe_2O_3.

3.3. Antioxidant activity

The antioxidant activity was carried out using the 1,1-diphenyl-2-picrylhydrazyl (DPPH•) assay. The activity of the compounds was compared with 8-hydroxyquinoline as a precursor. The IC_{50} value was obtained using the regression analysis between concentration and % inhibition (figure 7a and figure 7b). The result showed that Spiro-oxindole-chromene has a lower IC_{50} (182.3 ppm) than 8-hydroxyquinoline as a precursor (32.73 ppm).

Table 2. Several optimizing in the reaction condition.

Entry	Catalyst (% mol)	Temperature (°C)	Time (hour)	Yield (%)
1	0	70	4	Trace
2	2.5	70	4	Trace
3	5	70	4	75.57
4	7.5	70	4	86.74
5	10	70	4	81.74
6	7.5	r.t	4	Trace
7	7.5	50	4	Trace
8	7.5	70	2	51.62
9	7.5	70	6	62.93

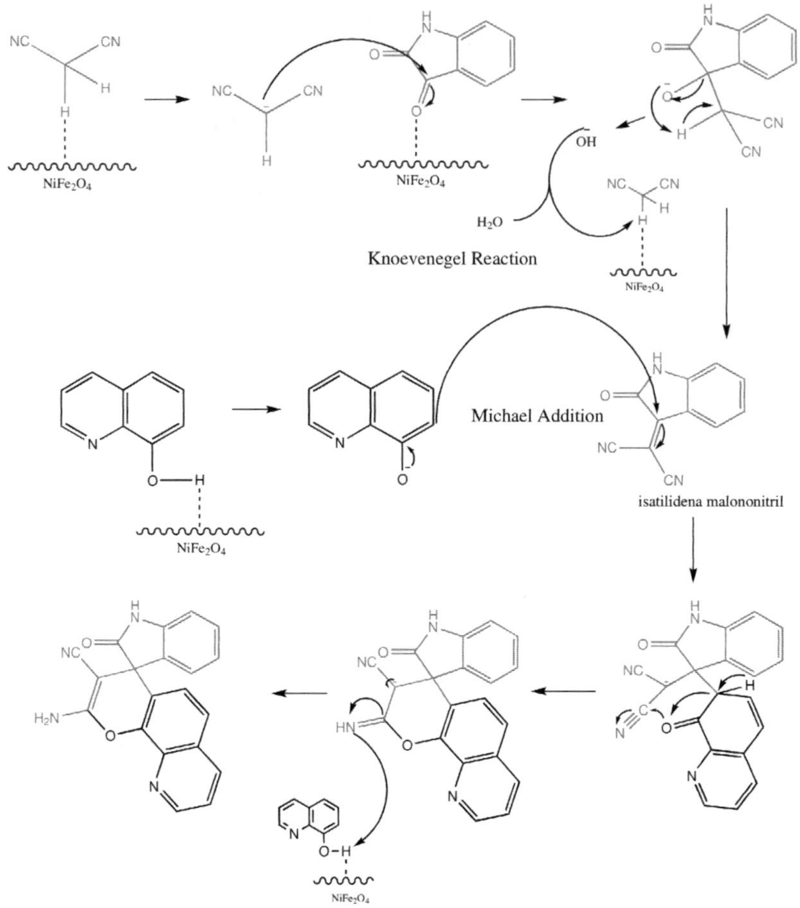

Figure 4. Mechanism reaction of *Spiro-oxindole-chromene* by $NiFe_2O_4$ nanocrystalline.

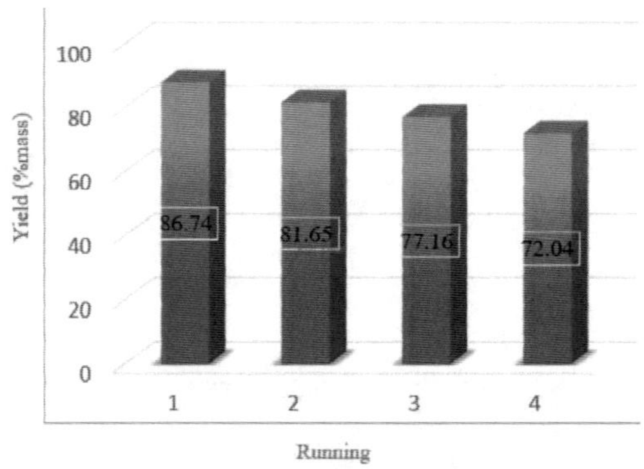

Figure 5. Recyclability catalyst test result in the synthesis of *Spiro-oxindole-chromene.*

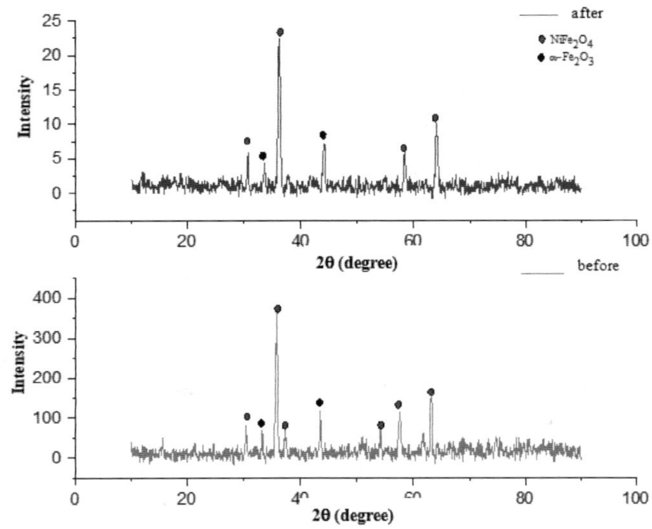

Figure 6. XRD patterns of $NiFe_2O_4$ after reaction.

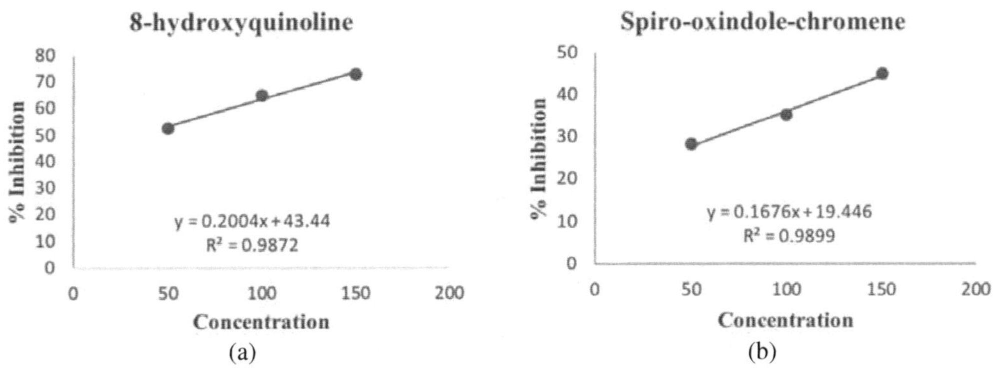

Figure 7. (a) The plot of 8-hydroxyquinoline concentration vs. inhibition percent from DPPH assay (b) The plot of *Spiro-oxindole-chromene* concentration vs. inhibition percent from DPPH assay.

4. Conclusion

A simple, efficient and environmentally friendly $NiFe_2O_4$ nanocrystalline had been synthesized using a simple co-precipitation method as a catalyst in the multicomponent reaction of an isatin, malononitrile, and 8-hydroxyquinoline. The recyclable property of catalyst is the most important property to green chemistry principle. The synthesized compound also has a good antioxidant activity with IC50 261.48 ppm.

Acknowledgments

This work was financially supported by Universitas Indonesia under research grant PITTA A with grant contract number NBK-0432/UN2.R3.1/HKP.05.00/2019.

References

[1] Wang D and Astruc D 2014 *Chem. Rev.* **114** 6949-85
[2] Karimi B and Farhangi E 2011 *Chem. Eur. J.* **17** 6056-60
[3] Luo L et al. 2015 *Catal. Sci. Technol.* **5** 3159-65
[4] Hajalilou A, Hashim M, Ebrahimi-Kahrizsangi R, Kamari H M and Sarami N 2014 *Ceram. Int.* **40** 5881-7
[5] Wang J et al. 2009 *J. Alloys Compd.* **479** 791-6
[6] Payra S, Saha A and Banerjee S 2016 *RSC Adv.* **6** 52495-9
[7] Heravi M R P and Morsalie N 2018 *Iran. Chem. Commun.* **6** 87-96
[8] Meghyasi R, Safaei-Ghomi J and Sharif M A 2016 *J. Chem. Res.* **40** 397-9
[9] Ahankar H, Ramazani A, Slepokura K, Lis T and Joo S W 2018 *Turk. J. Chem.* **42** 719-34
[10] Subbareddy C V, Sundarrajan S, Mohanapriya A, Subashini R and Shanmugam S 2018 *J. Mol. Liq.* **251** 296-307
[11] Moghaddam F M, Tavakoli G and Rezvani H R 2014 *Appl. Organomet. Chem.* **28** 750-5
[12] Gotic M, Czako-Nagy I, Popovic S and Music S 2010 *Philos. Mag. Lett.* **78** 193-201
[13] Naik K M and Sampath S 2018 *Electrochim. Acta* **292** 268-75
[14] Li Q, Wang X, Yu Y, Chen Y and Dai L 2016 *Tetrahedron* **72** 8358-63
[15] Goyal A, Bansal S and Singhal S 2014 *Int. J. Hydrogen Energ.* **39** 4895-908

High stable Hb/Pt/BDD electrode for determining acrylamide in coffee samples

R Wulandari[1, 2], E Saepudin[1] and T A Ivandini[1]

[1]Department of Chemistry, Faculty of Mathematics and Natural Sciences (FMIPA), Universitas Indonesia, Depok 16424, Indonesia
[2]Department of Chemical Engineering, Faculty of Engineering, Universitas Serang Raya, Banten, Indonesia

Corresponding author's email: ivandini.tri@sci.ui.ac.id

Abstract. A novel electrochemical hemoglobin-platinum-modified boron-doped diamond (Hb/Pt/BDD) biosensor was developed for the detection of acrylamide (AA) in coffee samples. The Hb-Pt-BDD electrode was characterized by cyclic voltammetry (CV), scanning electron microscopy with energy dispersive spectroscopy (SEM-EDS), X-ray diffraction (XRD), Raman spectroscopy, and X-ray photoelectrons (XPS). The cyclic voltammetry of Hb-Pt-BDD in 0.2 M sodium acetate buffer (ABS, pH 4.8) containing acrylamide in the concentration range of 0.00213 to 0.00711 ppb, showed linear responses with a detection limit of 0.00155 ppb. The excellent stability of the prepared Pt-modified BDD was proven through the Pt-BDD reusability by removing the Hb adduct without eliminating Pt on the BDD surface. Finally, employing the biosensor was proposed to determine AA in 1 g of coffee showed the AA content of 15.55 ppb. The result was comparable with the reference method using liquid chromatography-tandem mass spectrometry (LC-MS/MS).

Keywords: Acrylamide, biosensors, boron-doped diamond, platinum, hemoglobin, coffee samples

1. Introduction

Previous studies have discussed for the preparation of Hb/Pt/BDD electrodes as stable acrylamide sensors [1]. Acrylamide (AA) is generated from the reaction among reducing sugars namely, glucose and the amino acid, i.e. asparagine. The Maillard reaction mechanism was suggested to promote the AA formation in high starch foods during cooking at high temperatures [2-4]. It is also well-known that acrylamide compounds have been detected in coffee, French fries and some other common foods [5-7].

The high level of coffee consumption in many places in the world, making coffee as a source of daily exposure to acrylamide. Recently, the European Food Control Agency (EFSA) stated that roasted coffee and potato-based foods became one of the main sources of acrylamide exposure in the human body [8]. The acrylamide content in coffee is around 3–68 (μg/kg), while the Tolerable Daily Intake (TDI) or permitted daily AA intake is 40 μg/kg (pbb) per day for neurotoxicity and 2.6 μg/kg (ppb) per day to prevent cancer. Although, according to the Environmental Health Hazard Assessment (OEAHHA), one of the EPA divisions located in California, United States has determined that 0.2 μg/kg (ppb) days of acrylamide are not carcinogenic [8].On the other hand, several studies found that acrylamide has correlation to prostate cancer [9, 10], pancreatic cancer [11, 12], breast cancer [13-15], esophageal cancer [16], ovarian cancer and brain cancer [9, 17]. Only few reports were recorded for the electrochemical analysis on the hemoglobin adducts with modified BDD electrodes [1], therefore, in this work we used Hb/Pt/BDD electrodes for detection of AA in coffee samples.

2. Experimental

2.1. Materials and instrument

The laboratory-made of BDD films was set up in Keio University, Japan by using microwave plasma-assisted chemical vapour deposition (MPCVD, CORNES Technologies/AZTeX-5400)with a 0.1 % boron-to-carbon ratio as the precursor solution [18, 19]. A (100) silicon wafer was used as the support. The BDD film thickness of around 5 μm was observed, $H_2PtCl_6{\cdot}6H_2O$, $NaBH_4$, and NaOH were supplied by Wako Inc. (Japan), while human Hb H7379 and acrylamide were obtained from Sigma-Aldrich. All chemicals were applied as received without further purification.

Electrochemical trials were performed with an EDAQ Potentiostat. A three-electrode system was employed with Ag/AgCl reference and Pt counter electrodes. The working electrode was a BDD modifed with platinum nanoparticles (PtNP) and Hb. SEM images were registered by a EVO MA 10 Carl Zeiss Microscopy Gmbh,Germany. LCMS/MS used were LC (Shimadzu), MS (AB SCIEX), pumps (LC-20AD), Autosampler (SIL-20A HT), column ovens (CTO-20AC), detectors (MS/MS AB SCIEX 3200 Q TRAP).

2.2. Modificationof PtNP and Hb at BDD electrode

The deposition of PtNP on BDD was performed by following the previous reports [20]. Briefly, 10 μL $NaBH_4$ solution was dropped onto the BDD electrode, followed by 40 μL 1 mM H_2PtCl_6 solution. After washed and dried, electrochemical overgrowth of Pt seeds was conducted in 5 mL 1.0 mM H_2PtCl_6 solution gradually for 15 min at a constant potential of -0.2 V (vs. Ag/AgCl). Following the overgrowth, the PtNP coated BDD electrode was handled with thermal annealing at 700 °C for 5 min in N_2 atmosphere. The sample was then electrochemically refreshed by using cyclic voltammetry at the potential between -0.5 V and 1.5 V at a scan rate of 200 mV/s for 100 cycles. A further overgrowth was performed using a deposition voltage of -0.2 V gradually for 15 min to renew the Pt/BDD surface. Then, 10 μL 0.2 M acetate buffer solution (ABS) pH 4.9 containing 0.15 mM Hb was dropped onto Pt/BDD electrodes for 24 h to obtain Hb/Pt/BDD modified electrode. The electrode was rinsed, dried, and kept in air at 4 °C when not used.

2.3. Detection of acrylamide in sample coffee using Hb/Pt/BDD electrode

The electrode was examined for the detection of acrylamide by using 1 g of Luwak Toraja coffee sample dissolved in 10 mL of boiling water. After separation, 1 mL of coffee filtrate was diluted to 50 mL using ABS solution, then, characterized using a Hb/Pt/BDD sensor. Cyclic voltammetry (CV) techniquue in a potential range between -0.5 V and 1.5 V at a scan rate of 100 mV/s was applied. Validation was performed by using LCMS/MS [21, 22]. The mobile phase used was acetonitrile 70 % (B) and 30 % water (A), while the column was PHENOMENEX AQUA 5u C18 125A (50 × 2 mm 5 micron). The flow rate of 0.4 mL min-1 and the constant column temperature at 40 °C were applied.

3. Results and discussion

3.1. BDD electrode modification with PtNP and Hb

The addition of $NaBH_4$ solution and H_2PtCl_6 solution was performed on the BDD surface to implant Pt over the BDD surface. The presence of $NaBH_4$ produces a tensile force between BH_4^- on $NaBH_4$ with H^+ on BDD surface, which will increase the adsorption of $NaBH_4$ on BDD surface. Electrochemical overgrowth of Pt seeds was then performed using electrodeposition to grow and uniform the Pt layer on the BDD surface with a size and density that can be controlled so that Pt was homogeneously distributed on the BDD surface. The thermal annealing at 700 °C was then carried out for 5 min in N_2 atmosphere. N_2 gas was expected to prevent the formation of metal oxides. Then, electrochemically refreshed by using CV at the potential between -0.5 V and 1.5 V at a scan rate of 200 mV/s for 100 cycles in order to eliminate passive surface formation of Pt after high temperature treatment. A further overgrowth was performed using a deposition voltage of -0.2 V gradually for 15 min to renew the Pt/BDD surface. Then, 10 μL 0.2M ABS solution pH 4.9 containing 0.15 mM Hb was dropped onto Pt/BDD electrodes for 24 h to obtain Hb/Pt/BDD modified electrode. This BDD electrode modification with PtNP and Hb has been successfully carried out as evidenced by the results of SEM-EDS in figure 1 and table 1. SEM results in figure 1 shows that the Hb/Pt/BDD electrodes -before being contacted on acrylamide- was covered by Hb and Pt. EDS (table 1) shows the electrode characterization before being used for acrylamide detection. Table 1 confirmed that Pt and Fe could be deposited on the surface of BDD.

3.2. Acrylamide calibration curves with Hb/Pt/BDD electrodes
Hb/Pt/BDD electrodes were then tested for its performance on the standard acrylamide samples with various concentrations using the cyclic voltammetry. Measurements were made from a potential range of -0.5 to 1.5 V using 0.2 M ABS electrolytes containing acrylamide with concentration variation of 0; 0.00213; 0.00284; 0.00355; 0.00426; 0.00497; 0.00568; 0.00639 and 0.00711 ppb.

The results of the Hb/Pt/BDD electrode performance test on the standard acrylamide solutions in figure 2 show that there was a decrease in the peak of the oxidation current at the potential peak of +1.024, which attributed to Fe^{2+} oxidation. Increasing the concentrations of acrylamide decreases this oxidation peak. This is because the interaction between the N-terminal valine amino group of the Hb structure and the acrylamide compound affects the oxidation properties of Hb.

Acrylamide forms adducts with Hb as a result of the reaction between the $-NH_2$ group of valine N-terminal Hb. The formation of this adduct is related to the structure change of Hb. This structural change may be responsible for the decrease in the accessibility of the redox-active center Hb moving on

Figure 1. SEM image of Hb/Pt/BDD electrode.

Table 1. Mass composition of Hb/Pt/BDD in figure 1 Characterized by EDS.

Element	Mass %	Atomic number %
B	ND	ND
C	16.77	72.39
O	1.81	5.88
Fe	0.15	0.14
Pt	81.27	21.60

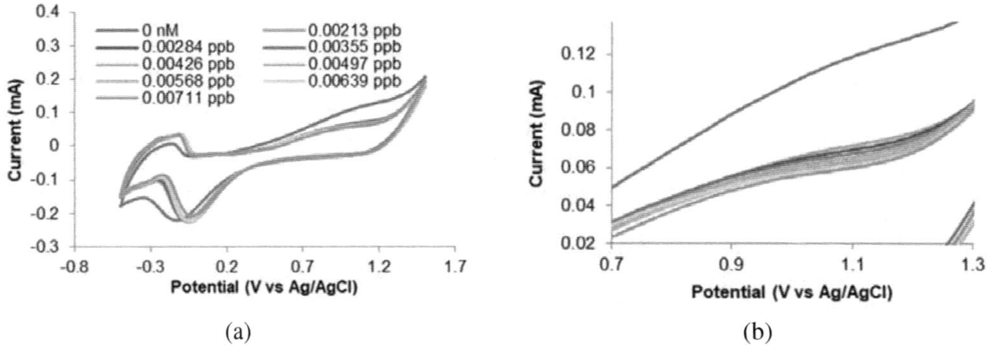

(a) (b)

Figure 2. (a) Hb/Pt/BDD electrode voltammogram with various concentrations of standard acrylamide in 0.2 M acetate buffer solution at a scan rate of 100 mV/s at Hb/Pt/BDD, (b) with the magnification.

the surface of the electrode, which causes a decrease in current. Adsorption of Hb-acrylamide changes the Hb electroactivity and the response that results from biosensors. Investigations of Pt/BDD electrodes modified with Hb displayed the quasi-reversible electrochemical reactions of Hb-Fe^{3+}/Hb-Fe^{2+}. The obtained results indicate that the decrease of the Fe^{2+} oxidation peak can be employed as analytical signal for acrylamide biosensors.

The results of the Hb/Pt/BDD electrode performance test on a standard acrylamide sample are shown in figure 3 in the form of a calibration curve. The results of the measurement of standard acrylamide compounds at various concentrations using Hb/Pt/BDD electrodes obtained R^2 = 0.996. Hb/Pt/BDD electrodes were used to measure standard acrylamide with 6 times repetition. An estimated LOD of 0.00155 ppb was obtained with a standard deviation of 0.00111.

3.3. Hb/Pt/BDD electrode performance test on Toraja Luwak coffee samples
Hb/Pt/BDD electrodes were then tested for their performance on coffee samples using the cyclic voltammetry. Measurements were made from a potential range of -0.5 to +1.5 V and scanrate was 100 mV/s in 0.2 M ABS pH 4.8. The sample was prepared with 1 g of Luwak Toraja coffee powder sample dissolved in 10 mL. Then, 1 mL of the sample was diluted up to 10,000 times in ABS. The solution was then measured using Hb/Pt/BDD electrodes. The voltammogram shown in figure 4 displays a peak oxidation potential of Fe^{2+} around +1.1 V. The measurement results of coffee samples using Hb/Pt/BDD electrodes are used to measure a sample with 3 repetitions. From the results of the Hb/Pt/BDD electrode performance, the calculations using a standard calibration curve acrylamide with CV show acrylamide levels in coffee samples of 15.55 ppb (oxidation current), after multiplied by the dilution factor.

3.4. Validation detection of acrylamide in coffee sample with LC-MS/MS
The measurement results of coffee samples using Hb/Pt/BDD electrodes were then validated by the LC-MS/MS. One g of Luwak Toraja coffee powder sample was dissolved in 10 mL water followed by boiling. The coffee solution was filtered using filter paper, 1 mL was taken, then diluted to 50 mL in a 0.1 % solution of formic acidsolution. The solution was then placed in a sample bottle. Series of standard acrylamide solutions were made with various concentrations (10, 50, 100, 150 and 200) ppb, then characterized using LC-MS/MS. LC-MS/MS was conditioned in the mobile phase of acetonitrile and aquabidest. The analysis time of 2 min was required with a peak flow rate of 0.4 mL/min, the peak was detected at 71.90. In figure 5, it shows the results of detection of acrylamide concentration in the sample of Luwak Toraja coffee amounted to 15.56 ppb, with the analysis conditions as shown in table 2.

Table 3 shows the results of acrylamide level measurements in coffee samples electrochemically using CV and LCMS/MS. In the measurements, the two methods were repeated 3 times. The results of measurements of acrylamide levels in coffee samples electrochemically using CV is not significantly different from the results obtained from the LCMS/MS analysis.

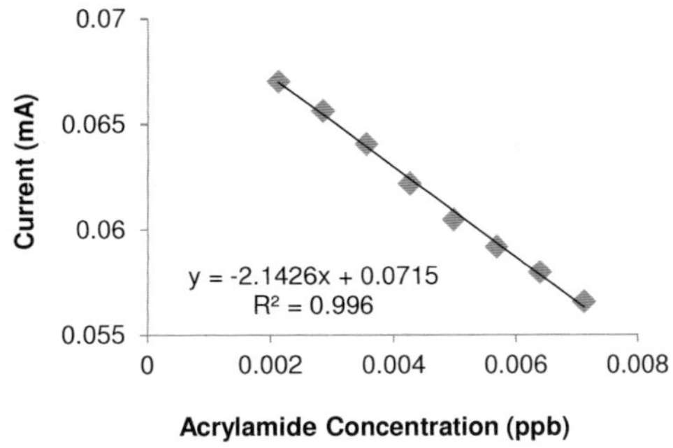

Figure 3. Hb/Pt/BDD electrode calibration curve with standard AA.

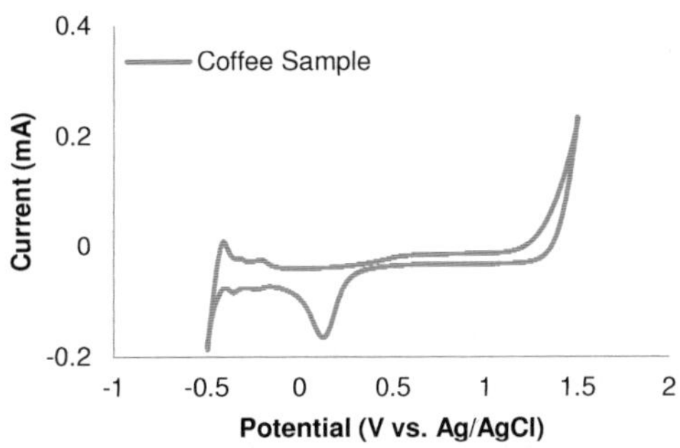

Figure 4. Hb/Pt/BDD electrode voltammogram of acrylamide in Luwak Toraja coffee sample.

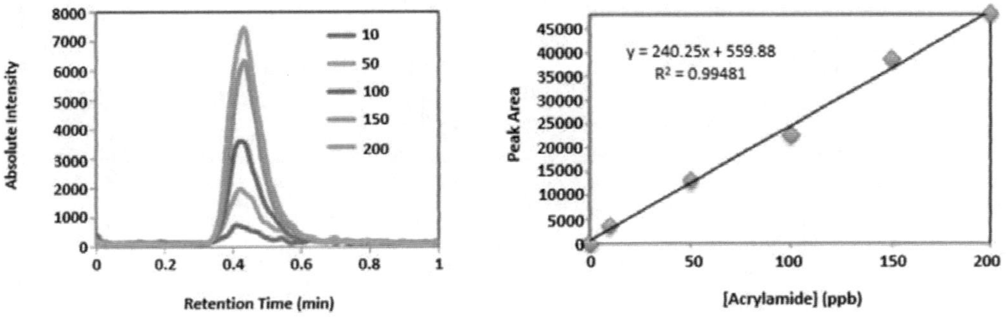

Figure 5. LC-MS/MS in Luwak Toraja coffee samples.

Table 2. Analysis in coffee samples with LC-MS/MS.

Sample	Retention time	Acrylamide 1	Acrylamide 2	Acrylamide 3	Average	Concentration (pbb)
Coffee	0.429	4935	4177	3780	4297.33	15.56

Table 3. Comparison of results obtained by measuring LCMS/MS and CV.

Method	Concentration (pbb)
LCMS/MS	15.56
CV	15.55

4. Conclusion

The Hb/Pt/BDD electrode performance test on the acrylamide standard samples has been proven to be successful with R^2 results of 0.996 and LOD of 0.0015 ppb and standard deviation of 0.00111. The Hb/Pt/BDD electrode was able to detect acrylamide in Luwak Toraja coffee samples 15.55 ppb after multiplied by the dilution factor, in which the results that did not differ significantly when the LC-MS/MS method 15.56 ppb was used. Thsu, it can be concluded that the Hb/Pt/BDD electrode can be used to detect acrylamide in more simple, faster and cheaper manner compared to the LCMS/MS method.

Acknowledgments

This work was financially supported by Universitas Indonesia under research grant PIT 9 with grant contract number NKB-1599/UN2.R3.1/HKP.05.00/2019.

References

[1] Garabagiu S and Mihailescu G 2011 *J. Electroanal Chem.* **659** 196-200
[2] Batra B, Lata S and Pundir C S 2013 *Bioprocess Biosyst. Eng.* **36** 1591-9
[3] Notardonato I, Avino P, Centola A, Cinelli G and Russo M V 2013 *Anal. Bioanal. Chem.* **405** 6137-41
[4] Bortolomeazzi R, Munari M, Anese M and Verardo G 2012 *Food Chem.* **135** 2687-93
[5] Zargar B, Sahraie N R and Khoshnam F 2009 *J. Anal. Lett.* **42** 1407-17
[6] Krajewska A, Radecki J and Radecka H 2008 *Sensors* **8** 5832-44
[7] Li M, Zhao G, Geng R and Hu H 2008 *Bioelectrochemistry* **74** 217-21
[8] Umam K, Saepudin E and Ivandini T A 2016 *IOP Conf. Ser.: Mater. Sci. Eng.* **188** 012006
[9] Wilson K M et al. 2009 *Am. J. Epidemiol.* **169** 954-61
[10] Wilson K M, Giovannucci E, Stampfer M J and Mucci L A 2012 *Int. J. Cancer* **131** 479-87
[11] Pelucchi C, La Vecchia C, Bosetti C, Boyle P and Boffetta P 2011 *Ann. Oncol.* **22** 1487-99
[12] Obón-Santacana et al. 2013 *Ann. Oncol.* **24** 2645-51
[13] Olesen P T et al. 2008 *Int. J. Cancer* **122** 2094-100
[14] Pedersen J R and Olsson J O 2003 *Analyst* **128** 332-4
[15] Burley V J et al. 2010 *Br. J. Cancer* **103** 1749-54
[16] Lin Y, Lagergren J and Lu Y 2011 *Int. J. Cancer* **128** 676-81
[17] Hogervorst J G F et al. 2010 *Crit. Rev. Toxicol* .**40** 485-512
[18] Ivandini T A et al. 2017 *Makara Journal of Science*. **21** 34-42
[19] Ivandini T A and Einaga Y 2015 *Chem. Commun.* **53** 1338-47

[20] Wulandari R, Ivandini T A, Irkham, Saepudin E and Einaga Y 2018 *Sensor. Mater.* **31** 1105-117
[21] Andrzejewski D, Roach J A G, Gay M L and Musser S M 2004 *J. Agric. Food Chem.* **52** 1996-2002
[22] Delatour T, Perisset A, Goldmann T, Riediker S and Stadler R H 2004 *J. Agric. Food Chem.* **52** 4625-31

The performance of sodium ion battery with $NaFePO_4$ cathode prepared from local iron sand

F Rahmawati, Z Faiz, D A N Romadhona, T E Saraswati and W W Lestari

Department of Chemistry, Faculty of Mathematics and Natural Sciences (FMIPA), Sebelas Maret University, Surakarta 57126, Indonesia

Corresponding author's email: fitria@mipa.uns.ac.id

Abstract. The performance of Sodium-Ion Battery, SIB, with $NaFePO_4$ cathode prepared from local iron sand was investigated. The $NaFePO_4$ was prepared through electrochemical sodiation to $FePO_4$ layer on aluminium substrate. $NaFePO_4$ prepared from local iron sand (from NTB, Indonesia) is named as NFP_A. The result is compared to the performance of SIB with $NaFePO_4$ cathode prepared from a commercial FePO4, which is named as NFP_B. The SIB was fabricated in a split cell test type and a pouch type. Battery performance was measured by Cyclic Voltammetry (CV) analysis, Galvanostatic Charge-Discharge (GCD) testing, and Electrochemical Impedance Spectroscopy (EIS) measurements. CV analysis found that sodium intercalation and de-intercalation curve in NFP_A is similar to the CV curve of NFP_B. The reduction peak presents at -2.98 V and -2.99 V for NFP_A and NFP_B, respectively. Those peaks represent Na+ intercalation into $FePO_4$ layer. The oxidation peaks, that represent de-intercalation from $NaFePO_4$ appears at 3.31 and 3.96 V. Split cell SIB with NFP_A as cathode provides 100 cycles under 1 C and 3 C current drawn. Meanwhile, the pouch type only provides 50 cycles before the capacity loss under 1 C current drawn. Each type has 100 % Coulombic efficiency indicates that all of the charged-ion can be completely discharged.

Keywords: Cathodes, iron sand, $NaFePO_4$, sodium ion battery

1. Introduction

A lithium-ion battery is the most used batteries with high operating voltage and efficiency. Nevertheless, low abundance of lithium in the earth's crust significantly restricts the LIBs development [1]. Sodium-ion batteries (SIBs) using sodium ion as the charge carrier, which can be extracted from sea water and earth's crust potentially replaces LIBs as the main electrical energy storage. The lack of cycle ability and structural stability of cathode materials become major problems for SIBs development. NASICON is a phosphate-based electrode which has a stable structure of open-framework, generating a high operating voltage of 3.4 V [2]. Other promising phosphate-based electrodes is $NaFePO_4$. It has a good structural stability and cycle ability because of the PO_4^{3-} existence as an oxygen lattice stabilizer when applied in a high current rate [3]. $NaFePO_4$ has two kinds of phases, olivine and maricite phase. The olivine phase needs a high temperature to synthesize but it has better electrochemical properties than the maricite phase [4]. Former studies used $LiFePO_4$ as the precursor to obtain olivine $NaFePO_4$ through the electrochemical delithiation-sodiation method [5]. The produced materials were used as a cathode for SIBs and delivered a good specific capacity (near the theoretical specific capacity of $NaFePO_4$ = 154 mAh/g) but showed poor cycle ability. This problem was occurred due to the large mismatch ionic

IOP Conf. Series: Materials Science and Engineering **902** (2020) 012008 doi:10.1088/1757-899X/902/1/012008

size during the Li-Na exchange process to obtain $NaFePO_4$ from $LiFePO_4$, causing severe decays in cathode capacity and cycle ability [6]. A previous research in our group conducted an electrochemical-synthesized olivine $NaFePO_4$ from $FePO_4$ layer, in which the $FePO_4$ powder was prepared from iron sand. The produced-$NaFePO_4$ showed a good electrical conductivity. Therefore, this work aims to study the performance of the prepared NaFePO4 while being used as a cathode for SIBs. Split cell test and pouch cell type being used as packages for fabricated SIBs. As a comparison, an electrochemically sodiated commercial $FePO_4$ is also used as a cathode for SIBs fabrication.

2. Experimental

$NaFePO_4$ electrode material was prepared by electrochemical sodiation to a $FePO_4$ layer on Al foil. The $FePO_4$ layer on Al, $FePO_4$/Al was used as working electrode within the electrochemical system by operating cyclic voltammetry mode. Each working electrodes (WE) was prepared using 70 % of the active material ($FePO_4$), 20 % acetylene black (AB) and 10 % polyvinylidene fluoride (PVDF) binder. N-methyl-2-pyrrolidinone (NMP) was being used as solvent to make electrode slurry. This slurry was then evenly coated using a doctor blade onto an aluminum foil as current collector to generate a 0.2 mm coating thickness. After drying in the oven for 2 h, the coated foil was cut into 5 cm x 2 cm each to create the working electrodes. These electrodes were attached onto Ag wire and placed in a three-neck rounded flask using two sodium metals, each attached onto Ag wires as the counter electrode (CE) and reference electrode (RE), and 1 M $NaClO_4$ salt in propylene carbonate solution as the electrolyte. The electrochemical sodiation was performed using cyclic voltammetry mode at a rate of 0.05 mV/s over 2 to 4 V vs Na^+/Na potential window using CorrTest® Electrochemical Workstation CS150. $NaFePO_4$ (NFP) layer dried at 120 °C have been used as cathode for battery fabrication. This method produced NFP_A (from $FePO_4$_synth) and NFP_B (from $FePO_4$_comm).

3. Electrochemical tests

The batteries were fabricated using a split cell test and pouch cell packaging. Split cell test consisting of a coin cell and wrapped by set of a brass shell (figure 1). The electrochemical stack consisting of a $NaFePO_4$ cathode prepared from iron sand extraction, a graphite layer casted on a Cu foil, a polytetrafluoroethylene (PTFE) film separator and 1 M $NaClO_4$ in ethylene carbonate (EC): dimethyl carbonate (DMC) (1:1 by volume) was being used as liquid electrolyte for electrochemical measurements. The pouch cell batteries (figure 2) were fabricated using the same stack as the split cell test batteries, with the addition of 1 cm^2 thin sodium metal attached to Ag wire placed between cathode layer and PTFE film. The stack was wrapped with aluminum plastic foil. For comparison, a split cell test and a pouch cell consisting of a $NaFePO_4$ cathode prepared from commercial $FePO_4$ (NFP_B) have been fabricated using the same stack as the previous fabricated batteries. Thus, the fabricated split cell test batteries were using NFP_A (SIB SC) and NFP_B (SIB KC) as cathode materials. The pouch cell batteries were also using NFP_A (SIB SP) and NFP_B (SIB KP) as cathode materials. Split cell test batteries were then galvanostatically cycled (GCD) at 1 C and 3 C (1 C = 154 mAh/g) and at 1C for the pouch cell batteries for 50 cycles each current rate using CorrTest® Electrochemical Workstation CS150. Cyclic voltammetry (CV) at a rate 0.05 mV/s were performed on pouch cell batteries using CorrTest® Electrochemical Workstation CS150 over a potential window of (-3) to 4 V vs Na^+/Na. The electrochemical impedance spectroscopy (EIS) measurements was carried out following the GCD testing and were conducted over a range of 20 Hz to 300 kHz with 0.2 V AC amplitude using EUCOL® U2826 LCR Meter.

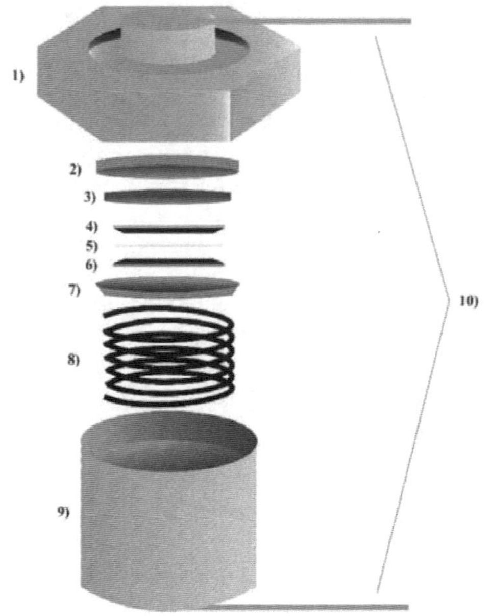

Figure 1. Scheme of split cell test SIBs: 1) Upper brass shell, 2) Cathode plate, 3) coin, 4) $NaFePO_4$/Al foil layer as cathode, 5) PTFE film with 1 M $NaClO_4$ as electrolyte, 6) Graphite/Cu foil layer as anode, 7) anode plate, 8) spring steel, 9) bottom brass shell, 10) Ag wires.

Figure 2. Scheme of pouch cell SIBs; 1) $NaFePO_4$/Al foil layer as cathode, 2) thin sodium metal attached into Ag wire as reference electrode, 3) PTFE film with 1 M $NaClO_4$ as electrolyte, 4) Graphite/Cu foil layer as anode.

4. Results and discussion

Electrochemical studies were performed by cyclic voltammetry (CV) analysis, galvanostatic charge-discharge (GCD) testing and electrochemical impedance spectroscopy (EIS) measurement. CV was conducted to determine the reduction and oxidation potentials of $NaFePO_4$ when cycled in the fabricated batteries. CVs for $NaFePO_4$ batteries are presented in figure 3. The voltammograms need to be analyzed using voltammetry analysis software to determine the reduction and oxidation potentials. The reduction potentials represent Na^+ intercalation into $FePO_4$ layer, as the batteries being discharged (equation 1). The oxidation potentials represent Na^+ de-intercalation from $FePO_4$ layer, as the batteries being charged (equation 2). The prediction of the total electrochemical reaction occurred inside the battery is described by equation 3.

$$NaFePO_4 \rightarrow xNa^+ + xe^- + xFePO_4 + (1-x)NaFePO_4 \quad (1)$$

$$FePO_4 + xe^- + xNa^+ \rightarrow xNaFePO_4 + (1-x)FePO_4 \quad (2)$$

$$NaFePO_4 + C_6 \leftrightarrow FePO_4 + NaC_6 \quad (3)$$

NFP_A reduction-oxidation couple potential was found at (-2.98) and 3.31 V, while NFP_B was at (-2.99) and 3.96 V. These potentials were associated with Fe^{2+}/Fe^{3+} redox couples, which means that sodium ion intercalation/de-intercalation were occurred through 1 step process. As the electrochemical potential is possibly correlated directly to the local atomic environment of the guest ions, particularly the site energy [7]. Another minor peaks could reflect the insertion or de-insertion of Na^+ with lower site energy [8].

Stationary energy storage applications require a long-term electrochemical cycling stability. Figure 4a shows the specific capacity of split cell test batteries which galvanostatically cycled at two charge/discharge current rates (C), 1 C and 3 C. The theoretical capacity of $NaFePO_4$ was calculated to be 154 mAh/g at 1C. SIB KC delivered a stable specific capacity of 76 mAh/g at 1C for 50 cycles, slightly lower than SIB SC which provides 77 mAh/g at 1 C for 50 cycles. These two SIBs delivered a higher specific capacity of 230 mAh/g at 3 C for further 50 cycles. This might be caused by the relation of the charging/discharging rate and electrolyte decomposition rate with the cycle numbers to reach the highest battery capacity [8]. The pouch cell batteries (SIB KP and SIB SP) delivered very small and unstable specific capacity (figure 4b). The main reasons were non-compact electrodes stacking when the batteries were fabricated and trapped gases present between electrodes layer. It also affects the SIBs operating voltage and internal battery resistance (figure 4c and figure 4d). All of the fabricated SIBs provide 100 % Coulombic efficiency that showed good reversibility properties (figure 4e and figure 4f).

The EIS measurement was performed after the GCD testing. There was no semicircle pattern in split cell test SIBs Nyquist plot (figure 5a) which means the absence of solid electrolyte interphase (SEI) layer growth between electrolyte and anode layer [9]. SEI layer provides a "barrier" preventing the electrolyte solvent molecule intercalating the anode structure, increasing the total internal resistance and decreasing the batteries' reversibility process. However, an uncontrollable SEI layer growth would obstruct the sodium intercalation process and then decrease the battery specific capacity. The different pattern is observed at the pouch cell SIBs Nyquist plot (figure 5b). It is believed to be also caused by the electrodes stack bad contact so the ionic migration in the battery was not as good as it is supposed to be. The total internal resistance from pouch cell SIBs determined by Nyquist plot fitting were much higher than split cell test SIBs, as shown in table 1.

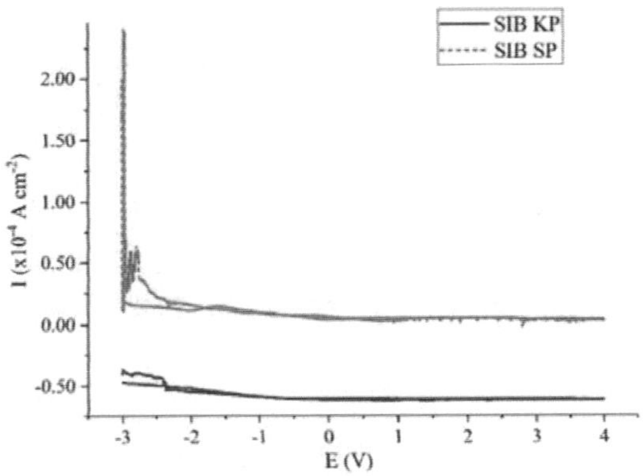

Figure 3. Cyclic voltammogram of SIB KP (blue line) and SIB SP (red line).

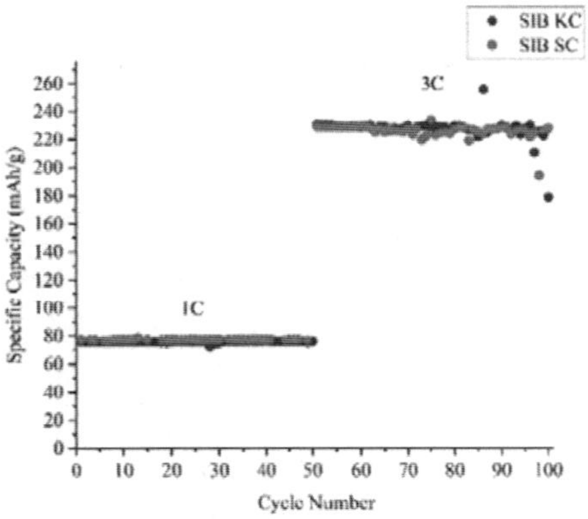

(a)

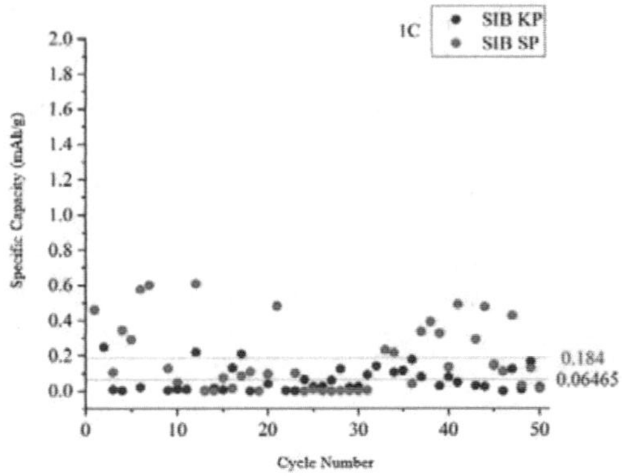

(b)

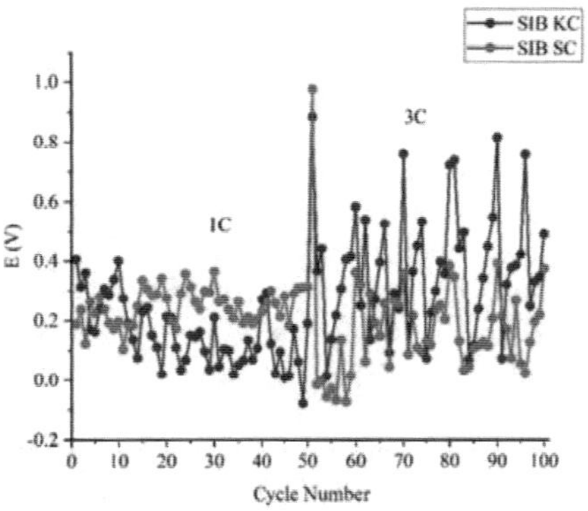

(c)

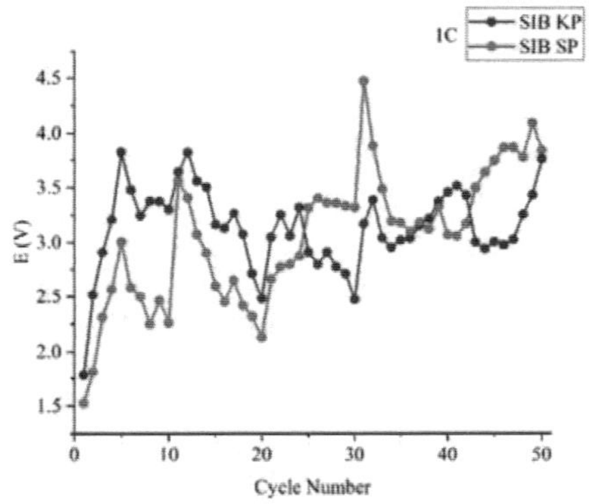

(d)

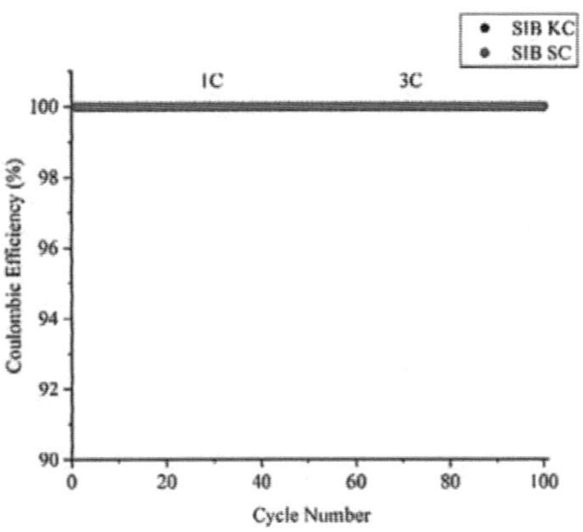

(e)

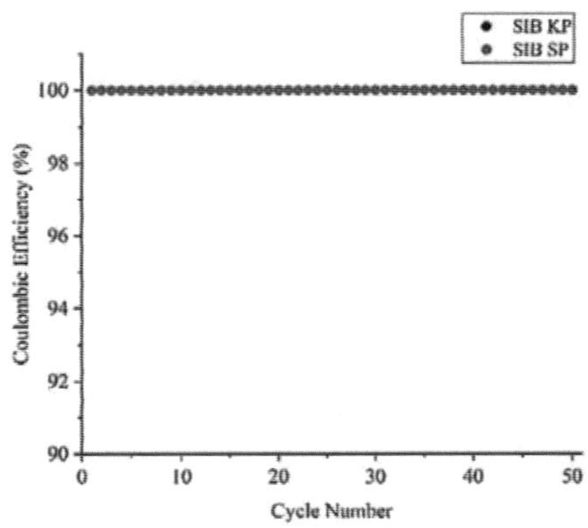

(f)

Figure 4. Specific capacity of (a) split cell test SIBs, and (b) pouch cell SIBs. Operating voltage of (c) split cell test SIBs, and (d) pouch cell SIBs. Coulombic efficiency of (e) split cell test SIBs, and (f) pouch cell SIBs.

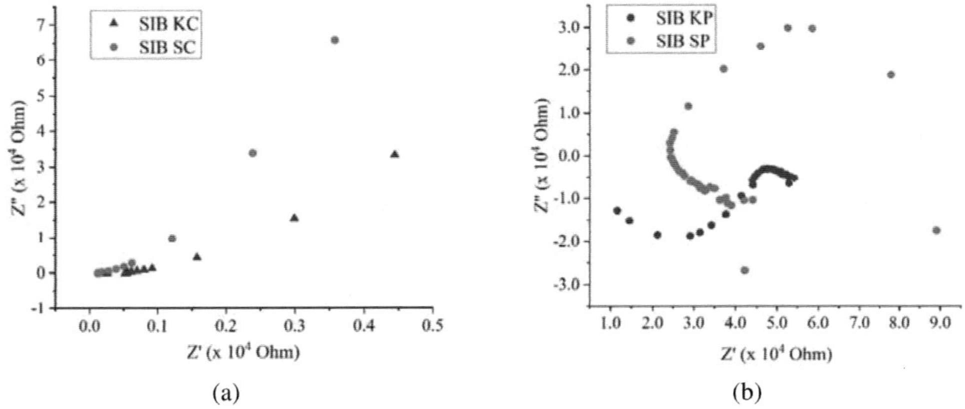

Figure 5. Nyquist plot of (a) split cell test SIBs and (b) pouch cell SIBs.

Table 1. The total internal resistance of all the fabricated SIBs.

Battery	Total Internal Resistance (Ω)
SIB KC	5.6657
SIB SC	6.5297
SIB KP	3.7026×10^4
SIB SP	2.5441×10^{10}

5. Conclusion

Sodium-Ion Battery performance with $NaFePO_4$ cathode from local iron has similar intercalation and de-intercalation trend with the SIB prepared from a commercial $NaFePO_4$. Cyclic voltammetry analysis to the SIB battery shows reduction peaks at (-2.98) and (-2.99) V, while oxidation peaks at 3.31 and 3.96 V. Split cell SIBs provides 50 cycles before dropping, whether under current drawn of 1 C or 3 C. Meanwhile, the pouch cell SIBs only performed under 1C current drawn.

Acknowledgments

The Ministry of Research, Technology, and Higher Education, Indonesia is acknowledged by authors for funding this research under Hibah Penelitian Dasar, contract number 719/UN27.21/PN/2019.

References

[1] Ren W, Zhu Z, An Q and Mai L 2017 *Small* **13** 1-31
[2] Deng J, Luo W-B, Chou S-L, Liu H-K and Dou S-X 2017 *Adv. Energy Mater* **8** 1-17
[3] Palomares V, Serras P, Villaluenga I, Hueso K B, Carretero-Gonzalez J and Rojo T 2012 *Energy Environ. Sci.* **5** 5884-901
[4] Hwang J-Y, Myung S-T and Sun Y-K 2017 *Chem. Soc. Rev.* **46** 3529-614
[5] Oh S-M, Myung S-T, Hassoun J, Scrosati B and Sun Y-K 2012 *Electrochem. Commun.* **22** 149-52
[6] Kundu D, Talaie E, Duffort V, and Nazar L F 2015 *Angew. Chem.* **54** 3431-48
[7] Liu Y, Zhou Y, Zhang J, Zhang S and Ren P 2016 *J. Power Sources* **314** 1-9

[8] Zhu X, Mochiku T, Fujii H, Tang K, Hu Y, Huang Z, Luo B, Ozawa K and Wang L 2018 *Nano Res.* **11** 6197-205
[9] Song J, Lin Y, Xu K and Li X 2018 *Adv. Energy Mater.* **8** 1703082

Preliminary study of electrooxidation isoeugenol into vanillin using platina and boron-doped diamond electrode

Hilyatudini, H G Riyanto, W Wibowo and T A Ivandini

Department of Chemistry, Faculty of Mathematics and Natural Sciences (FMIPA), Universitas Indonesia, Depok 16424, Indonesia

Corresponding author's email: ivandini.tri@sci.ui.ac.id

Abstract. Vanillin was synthesized from isoeugenol through an electrochemical method using Pt and BDD electrodes using chronoamperometry technique in a one-compartment electrochemical cell. The oxidation at C-C bond termination was expected to occur to form an aldehyde group. The oxidation potential of isoeugenol at Pt and BDD electrodes were fixed at +0.61 V and +0.96 V, respectively. The oxidation results were confirmed by GC, GC-MS, and FTIR. Thin layer chromatography result showed that vanillin could be produced at Pt and BDD electrodes with the yield percentage of 2.9 % and 1.33 %, respectively.

Keywords: Isoeugenol, synthesis of vanillin, electrooxidation, BDD and Pt electrode

1. Introduction

Isoeugenol is one of compounds that contained in clove oil (Syzgium Aromaticum) approximately 5-15 %. In Indonesia, the clove production is the largest in the world with a total production of about 98700 tons per year. One of isoeugenol derivate compounds is vanillin (4-hydroxy-3-methoxybenzaldehyde) [1]. The amount of vanillin demand reaches 15,000 tons per year and only 1 % is obtained from the extract of vanilla seeds (Vanilla planifolia) [2]. Vanillin is widely known as scent in perfume and flavoring in food. Moreover, it also acts as an additive to cover the bad taste in food by carrying out oxidative degradation of other compounds in food and has antioxidant activity [3].

Recently, The synthesis of an organic compound by electrochemical method without the addition of catalyst has attracted the attention of researchers [4-8] that usually using catalyst before. Einaga et al. has successfully synthesized (±) -licarin A from isoeugenol in a methanol solvent using BDD electrodes [5]. The yield percentage obtained by using BDD electrodes is relatively greater than using Pt, Ni and GC (Glassy Carbon) electrodes [8-13]. Previously, Parpot et.al. has also carried out Kraft-Lignin electrooxidation into vanillin with Pt, Ni, and Cu electrodes with % yield less than 10 % [6, 13].

In this study, the synthesis of vanillin from isoeugenol was studied by electrochemical method. The double bond in isoeugenol was expected to be broken by direct oxidation to form an aldehyde group. Boron-doped diamond (BDD) and Pt electrodes were used as the working electrodes as it was reported that BDD electrodes have a wide potential window in aqueous solutions as well as resistant to corrosion and have high mechanical strength thus allowing BDD to be used in the synthesis of organic compounds [4, 5, 14]. On the other hand, Pt electrodes are inert and quite stable, but also known for electrocatalysis in many types of reactions. Mubarok et al. reported the oxidation of vanillin at

Pt electrode [8]. In this work, % area in thin layer chromatography (TLC) was used to indicate the success of the conversion. Validation was performed by FTIR and GC-MS.

2. Materials and method

Electrochemical cell was prepared with Ag/AgCl electrode, spiral Pt electrode and Pt or BDD electrode as the reference, counter and working electrodes, respectively. A mixture solution of 2 mL of 0.1 M isoeugenol solution, 2 mL of tetrabutyl ammonium perchlorate (TBAP) salt in 0.1 M methanol, and 0.4 mL distilled water was added into the electrochemical cell in the absence and in the presence of isoeugenol. The cell was measured with cyclic voltammetry method in the potential range from -2.0 V to +2.0 V. The scan rate was varied at 50–250 mV / s, while isoeugenol concentrations were varied from 0.1 M to 0.5 M. The electrolysis was conducted at +0.61 V for Pt electrode and +0.96 V for BDD electrode for 3 h [5, 8]. The product was characterized by FTIR and GC-MS.

3. Results and discussion

3.1. Electrochemical study of isoeugenol with BDD and Pt as working electrode

In electrochemical behavior study of isoeugenol, 0.1 M TBAP solution with and without isoeugenol addition was measured with cyclic voltammetry method at scan rate 100 mV /s and range from -2000–2000 mV (figure 1a).

The oxidation peak of Pt was observed at a potential around -1000 mV and 82 mV that referred to the reaction below:

$$Pt + 2H_2O \rightarrow Pt(OH)_2 + 2H^+ + 2e^-$$

$$Pt(OH)_2 + 2H_2O \rightarrow Pt(OH)_4 + 2H^+ + 2e^-$$

At the potential of -500 mV, electrode was reduced back to the Pt form. On the other hand, the addition of isoeugenol observed an oxidation peak at +0.41 V, suggested an oxidation reaction occurred at the potential. This potential will be used for the next experiment using Pt electrodes. Furthermore, the isoeugenol addition also increased the current of H_2 and O_2 evolution reactions.

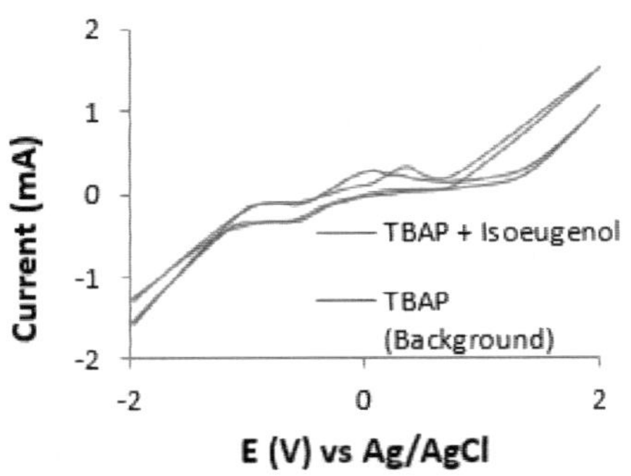

(a)

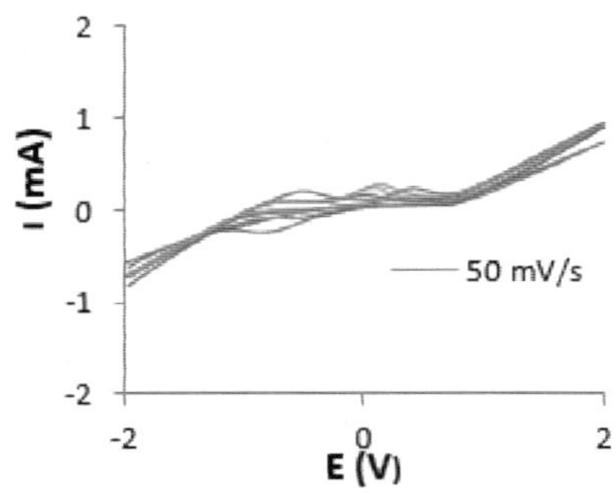

(b)

Figure 1. Cyclic voltammograms of (a) TBAP in 0.1 M methanol in the absence and in the presence of isoeugenol (b) at various scan rates between 50–200 mV/s.

The influence of various scan rate was also measured. The oxidation peak of isoeugenol only appears at the scan rate of 100 mV as shown in (figure 1b) probably because of the low kinetic rates of the reactions.

Increasing the isoeugenol concentrations was expected to increase its current response, however, the cyclic voltammograms (figure 2) show that the increasing of isoeugenol concentrations did not increase the current responses significantly, indicating the limitation of the surface area of the electrode.

In the case of the electrochemical study at BDD electrode (figure 3), an oxidation peak at +0.76 V was observed in the presence of isoeugenol in the system. Therefore, this potential was used for the next experiments. This potential is much higher than that at the Pt electrode, which is normally occurred when comparing metal and carbon electrodes. The lack of empty orbital in the carbon electrode was proposed to be the reason.

Furthermore, the influence of the scan rate variation was studied. Figure 4 shows that the increase of scan rate increases the currents as the diffusion layer that formed on the electrode surface becomes more thinner. This would make mass transfer from and to the electrode surface more easier and increase the current response. The relation between the current response and the square root of scan rate was linear, indicated the mass transfer from the bulk solution to the electrode surface was controlled by the diffusion rate.

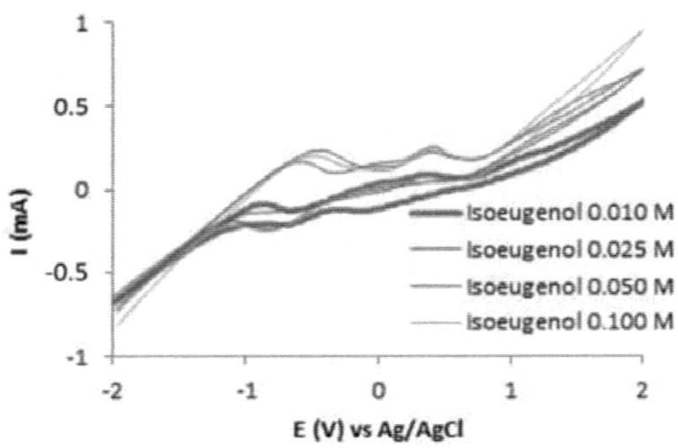

Figure 2. Cyclic voltammogram of isoeugenol in TBAP / methanol at various concentration between 0.01–0.05 M.

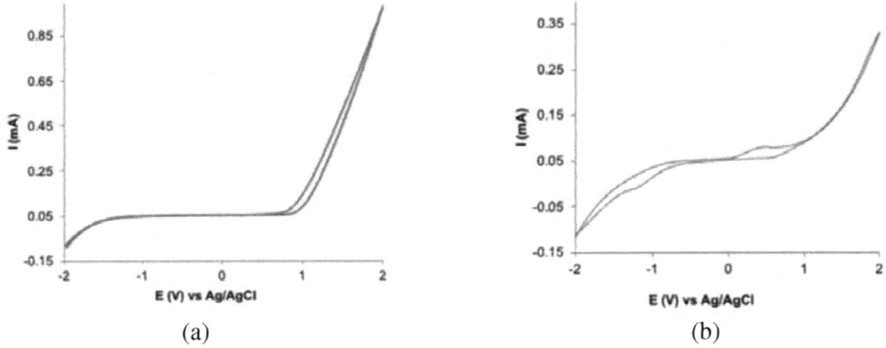

Figure 3. Cyclic voltammogram of (a) TBAP / methanol and (b) isoeugenol in TBAP / methanol at BDD electrode.

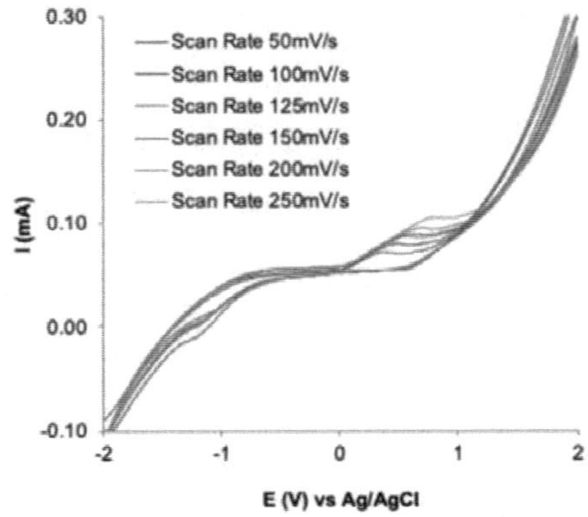

Figure 4. Cyclic voltammogram of Isoeugenol in TBAP/methanol at scan rate 50–250 mV/s in BDD electrode.

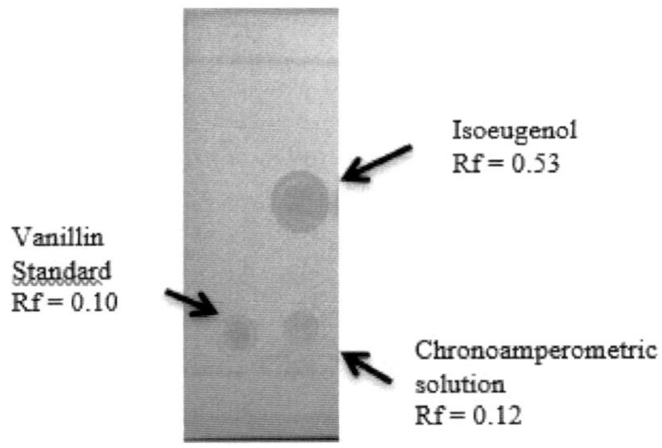

Figure 5. Example of TLC chromatograph of isoeugenol in TBAP /methanol after chronoamperometric treatment.

3.2. Electrolysis of isoeugenol at constant potential in single compartment

The oxidation of isoeugenol was performed in various concentrations of isoeugenol (0.1–0.5 M) in 0.1 M methanol-water solution. The % area of the generated vanillin was calculated by the testing using thin-layer-chromatograpy (TLC) with n-hexane and ethyl acetic (9:1) as the mobile phase. Figure 5 shows 2 spots at R_f = 0.12 and 0.53. The R_f value at 0.12 was almost similar with the R_f of vanillin standard (R_f = 0.10), while R_f at 0.53 was similar to R_f of isoeugenol.

Furthermore, table 1 shows the % area of vanillin generated at TLC, indicating the increase of isoeugenol concentration from 0.1 M to 0.3 M increased the oxidation of isoeugenol to vanillin at Pt electrodes. However, the higher concentration of isoeugenol decreased vanillin production probably due to that formation of isoeugenol layer on the electrode surface that prevents water to transfer onto the electrode surface.

At the BDD electrode, the higher isoeugenol concentrations did not produce vanillin. One of the reason is due to the low reaction rate of oxidation reaction of isoeugenol that was "poisoning" the electrode. The oxidation reaction was also performed in various concentrations of TBAP in 0.1 M methanol. Table 2 shows the produced vanillin increased with the increasing of TBAP concentrations, especially at the BDD electrodes. This probably occurs due to the increasing ion concentrations of electrodes increased the mobility of isoeugenol to the surface of the electrodes.

Pt electrode produces more vanillin than BDD electrode. The better electrocatalytic properties of Pt electrode was caused due to several things:

- The oxidation potential on the Pt electrode was more negative than the BDD electrode which means isoeugenol would be more oxidized on the Pt electrode.
- The orbital electron of Pt was empty that could be easier to be filled by an electron from the oxidation of isoeugenol. While BDD electrode consist C atom that all its orbital had been used to form tetrahedral structure.

3.3. Confirmation of the oxidation results

The chronoamperometric solution of isoeugenol at the optimum condition (0.1 M isoeugenol and 0.3 M TBAP) was characterized using GC-MS. At retention time of 13.953, the peak appeared at mz = 152, indicated vanillin compound. with % area was 2.88 %, while the 6th and 7th peak at m/z 164 was indicated the peak of isoeugenol with % area approximately 73.90 %. The 11th peak at m/z 328 was indicated the peak of isoeugenol dimer with % area was 1.08 %. The dimer formation began with radical oxygen formation of hydroxyl group. The radical was then resonance in all parts of isoeugenol (in the benzene ring and allyl group). The two radicals from the isoeugenol compound were then reacted and formed a stable dimer compound.

The FTIR spectra of chronoamperometric solution at the optimum condition was also performed (figure 6a), showing peaks appeared at 3445 cm^{-1}, 2948 cm^{-1}, 2836 cm^{-1} and 1654 cm^{-1}, which were related to O-H stretching, C-H sp^3 stretching, C-H stretching and C=O stretching, respectively. In addition to that, the peak at 1449 cm^{-1}, 1276 cm^{-1} and 1419 cm^{-1} were also observed related to C=C-C stretching, C-O stretching and C-H vinyl bending, respectively. The effect of conjugation at π bond of aromatic ring and hydrogen bond in methanol solvent would shift the frequency of C=O stretching to the lower frequency [15] The fragmentation of synthesized vanillin with chronoamperometry method is shown in figure 6b.

Gas chromatogram of the synthesized vanillin is shown in figure 6c. The 5th peak was appeared at retention time of 13.953 min with m/z is 152 that indicated vanillin was successfully synthesized with % area is 2.88 %. This synthesized product was too little compared to the volume of solution (55 mL).

Table 1. % area of vanillin with the variation of isoeugenol concentration.

Isoeugenol concentration	Retention time (Rt)		% area	
	Pt	BDD	Pt	BDD
0.1 M	1.633	1.615	1.80 %	0.32 %
0.3 M	1.634	-	2.90 %	-
0.5 M	1.645	-	0.03 %	-

Table 2. % area of vanillin with the variation of TBAP concentration.

TBAP concentration	Retention time (Rt)		% area	
	Pt	BDD	Pt	BDD
0.1 M	1.633	1.615	1.80 %	0.32 %
0.3 M	1.634	1.634	2.90 %	0.65 %
0.5 M	1.645	1.62	0.03 %	1.33 %

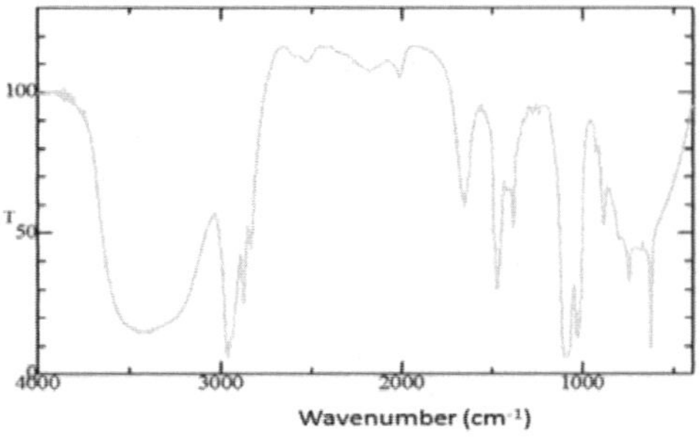

(a)

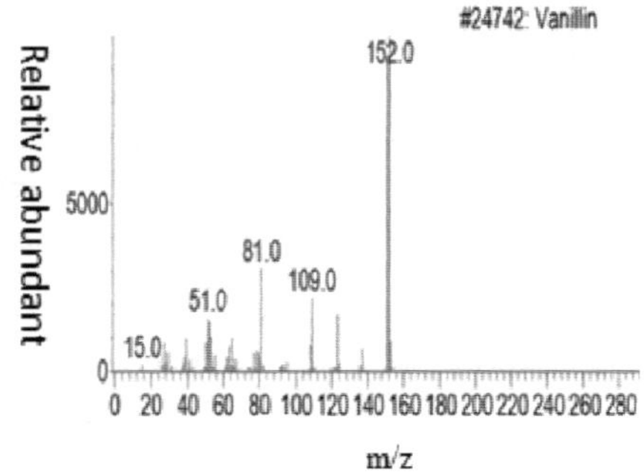

(b)

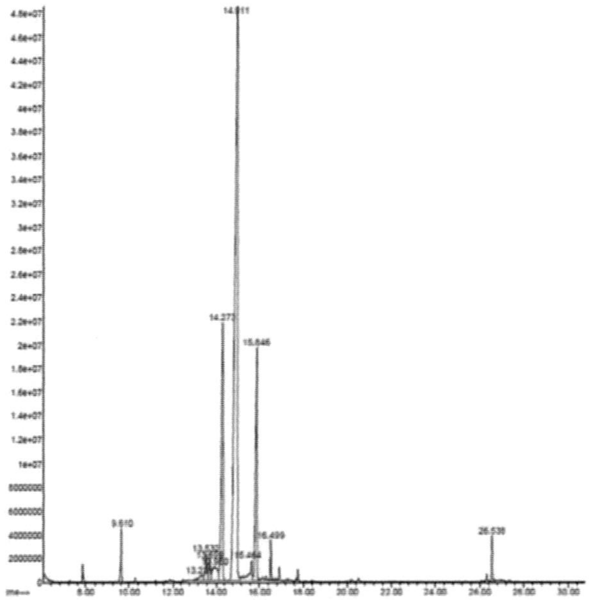

(c)

Figure 6. (a) FTIR and GC-MS, (b) Vanillin fragmentation, and (c) Gas chromatogram after chronoamperometric treatment.

4. Conclusion

The oxidation of isoeugenol performed at +0.41 V and +0.76 V (Vs. Ag / AgCl) in 0.1 M methanol containing of water and of tetrabutyl ammonium perchlorate at Pt and BDD electrode, respectively, produced vanillin about 2.9 % at the Pt electrode as the working electrode, while it was 1.33 % at BDD electrode.

Acknowledgments

This work was financially supported by Universitas Indonesia under research grant Hibah PIT 9 with grant contract number NKB-0028/UN2.R3.1/HKP.05.00/2019.

References

[1] Walton N J, Mayer M J and Narbad A 2003 *Phytochemistry* **63** 505-15
[2] Priefert H, Rabenhorst J and Steinbüchel A 2001 *Applied Microbiology and Biotechnology* **56** 296-314
[3] García J I, Dobado J A, Flores G Cand García H M 2016 *Experimental Organic Chemistry* (London: Elsevier) 71-408
[4] Tian Y, Chen X, Shang C and Chen G 2006 *J. Electrochem. Soc.* **153** J80-J85
[5] Sumi T, Saitoh T, Natsui K, Yamamoto T, Mahito A, Einaga Y and Nishiyama S 2012 *Angew. Chemie-Int. Ed.* **51** 443-6
[6] Parpot P, Bettencourt A P, Carvalho A M and Belgsir E M, 2000 *J. Appl. Electrochem.* **30** 727-31
[7] Shibano S, Ivandini T A, Terashima C, Nakata K, and Einaga Y 2014 *Chem. Lett.* **43** 1292-3
[8] Mubarok H, Hilyatudini, Saepudin E and T A. Ivandini 2016 *IOP Conf. Ser.: Mat. Sci. Eng.* **188** 1-7
[9] Putranto R P, Mubarok H, Umam K, Saepudin E, and Ivandini T A 2019 *IOP Conf. Ser.: Mat. Sci. Eng.* **496** 1-5
[10] Schmitt D, Regenbrecht C, Hartmer M, Stecker F and Waldvogel S R 2015 *Beilstein J. Org. Chem.* **11** 473-80
[11] Möhle S, Zirbes M, Rodrigo E, Gieshoff T, Wiebe A and Waldvogel S R 2018 *Angewandte Chemie - International Edition* **57** 6018-41
[12] Reichert E,Wintringer R, Volmer D A and Hempelmann R 2012 *Phys. Chem. Chem. Phys.* **14** 5214-21
[13] Silva E A B D, Zabkova M, Araujo J D, Cateto C A, Barreiro M F, Belgacem M N and Rodrigues A E 2009 *Chem. Eng. Res. Des.* **87** 1276-92
[14] Muharam S, Jiwanti P, Gunlazuardi J, Einaga Y and Ivandini T A 2019 *Diam. Diamond Relat. Mater.* **99** 107464
[15] Fatoni A, Hariani P L, Hermansyah and Lesbani A 2018 *Indones. J. Chem.* **18** 92-101

Copper oxide modification to improve the photocatalytic activity of titanium dioxide nanoparticles: P25 versus P90

L Yuliati[1, 2], N Hasan[3] and H O Lintang[1, 2]

[1]Ma Chung Research Center for Photosynthetic Pigments, Universitas Ma Chung, Malang 65151, East Java, Indonesia
[2]Department of Chemistry, Faculty of Science and Technology, Universitas Ma Chung, Malang 65151, East Java, Indonesia
[3]Department of Chemistry, Faculty of Science, Universiti Teknologi Malaysia, UTM Johor Bahru, Johor 81310, Malaysia

Corresponding author's email: leny.yuliati@machung.ac.id

Abstract. Titanium dioxide (TiO_2) has been recognized as one of the most active photocatalysts for organic pollutant degradation under ultraviolet (UV) light irradiation. In order to reduce the fast charge recombination in the TiO_2, various methods have been investigated, including the addition of metal oxide co-catalysts. Owing to the characteristic of the nanoparticles, the modification of nanoparticles involving heat treatment is still a challenging task. In this work, two commercial TiO_2 nanoparticles, namely P25 and P90 (Evonik), were modified by copper oxides (CuO) and the photocatalytic activity was evaluated for degradation of 2,4-dichlorophenoxyacetic acid (2,4-D) under UV light. The CuO/P25 and CuO/P90 samples with various loading amounts (0.1, 0.25, 0.5 and 1.0 wt %) were prepared by precipitation of copper(II) nitrate to the P25 or P90 nanoparticles at pH of 9, followed by calcination at 573 K. X-ray diffraction (XRD) patterns indicated that all samples have the characteristics of both anatase and rutile phases. While the addition of CuO did not much affect the structure, crystallite size, and anatase-rutile ratio of the P25 and P90 nanoparticles, the presence of the copper species was confirmed by the scanning electron microscopy (SEM) equipped with the energy-dispersive X-ray (EDX) spectroscopy. Moreover, fluorescence spectra also showed that the CuO quenched the emission intensity of both the P25 and P90 nanoparticles, suggesting the successful decrease of the charge recombination in the TiO_2 nanoparticles. Photocatalytic activity tests showed that the P25 and P90 gave percentage degradation of 90 and 47 %, respectively, after a 1-hour reaction. Even though not much improvement was observed for P25 TiO_2 nanoparticle after the CuO addition (92 %), the activity of P90 nanoparticle was enhanced from 47 % to 86–87 % with the CuO addition of 0.1–0.25 wt %. This study demonstrated that it is feasible to improve the photocatalytic activity of TiO_2 commercial nanoparticles, in this case, the P90, by surface modification using the CuO.

Keywords: Copper oxide, P25, P90, photocatalyst, titanium dioxide

1. Introduction

Titanium dioxide (TiO_2) is one of the most investigated photocatalysts so far since it shows an incredible high photocatalytic activity [1-4]. Although being an active photocatalyst, TiO_2 suffers from recombination of charge carriers under UV illumination, which reduces its photocatalytic activity.

Photocatalytic activity of TiO_2 under UV light irradiation can be further enhanced by depositing co-catalyst such as copper oxide (CuO) on the surface of TiO_2 [5-8]. This modification is essential to achieve high efficiencies in environmental photocatalysis. It has been reported that the presence of CuO co-catalyst deposited on the surface of titania would reduce the electron transfer resistance and increase the photocurrent of the TiO_2 [5], which in turn would enhance the overall photocatalytic activity of TiO_2.

The photocatalytic activity of TiO_2 has been reported to be dependable on the crystal phase of TiO_2. However, there are different photocatalytic activities between the two phases of TiO_2, which are anatase and rutile [9-11]. Previous study reported that anatase exhibited a higher photocatalytic activity compared to rutile because of several good properties such as the good crystallite size, large specific surface area and porosity, as well as low amount of defect formation. All these parameters would affect the charge recombination process. The anatase also possesses 0.1 eV higher Fermi level than the rutile, leading to the higher level of hydroxyl groups on the surface as well as the lower oxygen affinity. It was also stated in literature that the mixture of anatase to rutile, which was usually in the ratio of 70:30 (the Evonik P25 and P90) was found to have very high photocatalytic activity [9, 10].

In this study, two types of commercial TiO_2, which were Evonik P25 and P90, were modified with CuO. The Evonik P25 and P90 consists of both anatase and rutile phases, but with different ratio from each other [9, 10]. The photocatalytic activity of unmodified TiO_2 and CuO-modified TiO_2 were then examined for the degradation of 2,4-dichlorophenoxyacetic acid (2,4-D). The 2,4-D is a low biodegradable herbicide and hazardous to human health and environment [12, 13]. This work demonstrated that the activity of TiO_2 P90 could be more improved than the TiO_2 P25 by the addition of CuO.

2. Experimental

P25 TiO_2 or P90 TiO_2 (1 g) and double distilled water (20 mL) were added into a glass beaker. After the beaker was sealed with parafilm and put onto the stirring plate, the beaker was stirred for 2–3 min, followed by ultrasonication for 5 min. In another glass beaker, a certain amount of copper(II) nitrate, $Cu(NO_3)_2$, was taken and diluted in the double distilled water (5 mL) to give 0.1, 0.25, 0.5, and 1.0 wt % as compared to the amount of the TiO_2. The solution was then added into the beaker consisting of TiO_2 and double distilled water. While stirring the mixture, ammonium hydroxide, NH_4OH, was dropwisely added until the mixture reached pH 9. The pH value was measured in all stages and the stirring was kept for one hour. In order to eliminate the solvent, the mixture was slowly heated at 55 °C. The obtained powder was further dried in an oven before it was calcined at 300 °C for 4 h. The prepared photocatalysts were labelled as CuO(x)/P25 and CuO(x)/P90, where x represents the amount of loaded $Cu(NO_3)_2$ (in wt %). The resulting solids were in white and light blue color, depending on the amount of added CuO. For the treated TiO_2 samples, the P25 and P90 were prepared under the similar conditions, but without the addition of the $Cu(NO_3)_2$.

The properties of the prepared photocatalysts were examined by several instruments. X-ray diffraction (XRD, Bruker, AXS Diffrac plus release 2000) was used to record the diffraction patterns of TiO_2 and the CuO/TiO_2 photocatalysts. A spectrofluorometer (JASCO, FP-8500) was employed to study the excitation and emission sites of the prepared catalysts. Field emission scanning electron microscopy energy dispersive X-ray (FESEM-EDX, JSM-6710) spectrometer was applied to determine the surface morphology, particle size, and elements present in the photocatalyst.

Prior to the photocatalytic degradation of 2,4-D, adsorption test was performed at room temperature. Typically, 2,4-D (28 mL, 5×10^{-4} M) was transferred into a beaker glass containing photocatalyst (0.028 g). The mixture was then sonicated for 5 min and followed by stirring process under dark condition for 1 h. About 3 mL of the mixture was taken by a syringe, filtered, and the obtained solution was measured for its absorbance by using UV-Vis spectroscopy (Thermo Scientific, Genesys 10S) at the monitoring wavelength of 283 nm. The suspension (25 mL) was then irradiated for 1 h under UV light irradiation in a closed box on a stirring plate. After the reaction finished, another 3 mL of the mixture was taken and filtered, prior to the 2,4-D absorbance measurement using the UV-Vis spectrometer. The reaction steps were repeated by using the different types of photocatalyst under the similar conditions. The amount of the degraded 2,4-D was compared to the initial 2,4-D to give the photocatalytic activity as shown in the equation 1.

$$\frac{A_0 - A_1}{A_0} \times 100\ \% \quad (1)$$

where A_0 showed the concentration of the 2,4-D at the initial state and A_t referred to the concentration of the 2,4-D after the photocatalytic reaction.

3. Results and discussion

3.1. Characterizations and activity of P25 and CuO(x)/P25 photocatalysts

The P25 and prepared CuO(x)/P25 series were characterized using XRD in order to study the structure and crystallinity. As shown in figure 1, all samples exhibited clear diffraction peaks, which could be observed at 2θ of 25.35° (101), 27.45° (110), 37.80° (101), 48°(200), 55.10°(211), 62.65°(204) and 68.85°(112) due to the presence of both anatase and rutile phases of TiO_2 (PDF-00.021-1272, PDF-00-021-1276). After either calcination at 300 °C or addition of CuO, the intensities of the diffraction peaks were mostly unchanged, suggesting that the crystallinity and/or crystallite size for all samples were similar to each other. The crystallite size could be calculated using the anatase peak at (101) plane by the Scherrer equation [6] and the calculated results are shown in table 1. There was no diffraction peak of CuO that was detected by XRD. This result suggested that the low amount of added CuO or the well dispersion of CuO on the P25 could be the reason why the CuO could not be detected by the XRD [5-8].

Table 1 also shows the percentage composition of anatase to rutile phases on the P25 and the CuO/P25 samples, calculated using the equation 2, where A is the percentage (%), I_A is the intensity of anatase phase and I_R is the intensity of rutile phase.

$$A\,(\%)\ \frac{1}{1 + 1.265\left(\frac{I_R}{I_A}\right)} \quad (2)$$

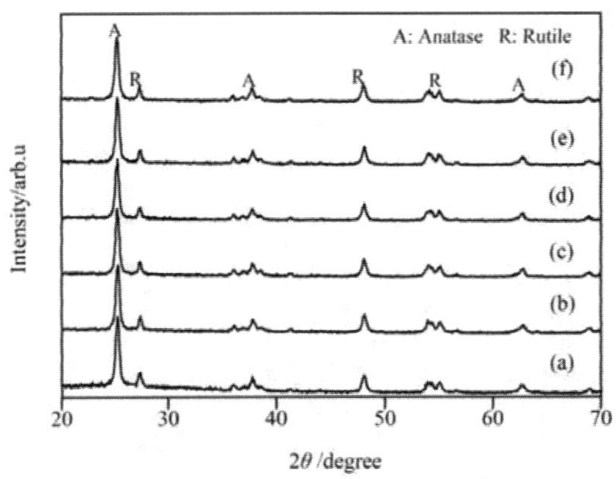

Figure 1. XRD patterns of (a) unmodified P25, (b) treated P25, (c) CuO (0.1)/P25, (d) CuO (0.25)/P25, (e) CuO (0.5)/P25, and (f) CuO (1.0)/P25 samples.

After calcined at 300 °C, the percentage of anatase phase in P25 was slightly increased from 71.7 to 74.4 %. On the other hand, addition of CuO did not much change the percentage of anatase. All CuO/P25 samples showed similar percentage of anatase to each other, which was in the range of 75–76 %. The ratio of anatase to rutile on CuO/P25 samples (3:1) was remained unchanged, might be due to the high stability of crystallite structure of P25 itself.

In order to study the morphology of the CuO/P25 series, FESEM image of the CuO(0.1)/P25 was recorded and shown in figure 2. The CuO(0.1)/P25 has spherical shape and particle size of 30 to 60 nm. The particle size of the CuO(0.1)/P25 was very close to the reported particle size of the P25 [14, 15]. As revealed by SEM, the P25 was reported to have particle size of 19–24 nm [14], while another study by transmission electron microscopy (TEM) gave particle size distribution in the range of 9–53 nm with the average diameter of 23 nm [15]. EDX analysis confirmed the presence of elements, which were Cu, O, and Ti on the surface of the CuO(0.1)/P25. This result suggested the clear evidence for the successful loading of CuO on the P25.

The fluorescence spectra for the P25 and CuO/P25 series are shown in figure 3. The P25 and CuO/P25 samples have excitation at 218 nm and emission at 276 nm. From the spectra, the unmodified P25 showed the highest intensity followed by the treated P25. It can be suggested that after calcination process, the electron-hole recombination was suppressed. For the modified sample, CuO(0.5)/P25 showed the highest intensity followed by CuO(1.0)/P25, CuO(0.1)/P25 and CuO(0.25)/P25. It can be proposed that addition of CuO on P25 might decrease the electron-hole recombination.

Table 1. Crystallite size, percentages of anatase and rutile phases of P25 and CuO/P25 samples.

Sample	Crystallite size (nm)	Percentage of anatase (%)	Percentage of rutile (%)	Degradation of 2,4-D (%)
Unmodified P25	13.2	71.7	28.3	89
Treated P25	14.4	74.4	25.6	90
CuO(0.1)/P25	14.3	76.2	23.8	92
CuO(0.25)/P25	15.8	76.6	23.4	86
CuO(0.5)/P25	15.8	76.5	23.5	84
CuO(1.0)/P25	14.3	75.1	24.9	90

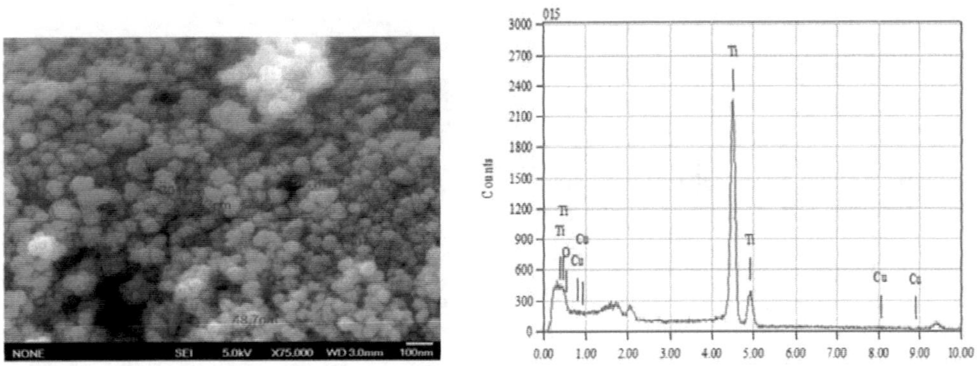

(a) (b)

Figure 2. (a) FESEM image and (b) EDX analysis of CuO(0.1)/P25 sample.

The P25 and CuO/P25 samples were tested for photocatalytic degradation of 2,4-D in order to investigate the photocatalytic performance. table 1 shows the percentage degradation of 2,4-D over the P25 and CuO/P25 photocatalysts after UV light irradiation for 1 h. The unmodified P25 showed high activity toward 2,4-D degradation with percentage of 89 %. The treated P25 that was calcined at 300 °C showed almost similar photocatalytic activity to that of unmodified P25, while the addition of 0.1 wt % of CuO to P25 slightly increased the photocatalytic activity to 92 % degradation of 2,4-D. Addition of CuO to the P25 from 0.25 to 1.0 wt % did not increase the activity, even slightly reduced the activity to 84-90 %. It can be proposed that since the P25 is already an active photocatalyst, further modification did not give much improvement in the photocatalytic activity. The highest photocatalytic activity was obtained with the optimum loading amount of CuO, which was 0.1 wt %.

3.2. Characterizations and activity of P90 and CuO(x)/P90 photocatalysts

The P90 and the prepared CuO/P90 series were characterized using XRD and their XRD patterns are shown in figure 4. All the P90 and modified photocatalysts showed the diffraction peaks at angle of 25.32° (101), 27.34° (110), 37.90° (004), 48.04° (200), 53.82° (211), 62.62° (002) and 68.86° (116), according to the PDF-00-021-1272 and PDF-00-021-1276. The presence of these diffraction peaks showed that all samples have both anatase and rutile phases. The calcination process did not much influence the diffraction peaks intensity, but the intensity was increased with the addition of high loading of CuO (1 wt %). It can be proposed that the addition of CuO with high amount might possibly contribute to the crystal growth. Similar to the CuO/P25 samples, there was no diffraction peak of CuO observed. The absence of CuO peaks might be due to the low amount of metal precursor added to be detected by XRD and well dispersion of CuO on the surface of P90 [5-8].

Table 2 shows the crystallite sizes for the P90 and CuO/P90 samples, calculated by using the Scherrer equation [6]. It can be observed that the calcination process did not have any impact to the crystallite size of the P90, which was calculated to be 9.6 nm when using the anatase peak at (101) plane. When considering all anatase and rutile peaks in the calculation, the average crystallite size of the P90 was calculated to be around 26 nm, which was in good agreement with the SEM measurement that gave 26–33 nm [14]. Addition of CuO up to 0.5 wt % also did not give much influence to the crystallite size. Only when the amount of CuO was 1.0 wt %, the crystallite size of P90 increased from 9.6 to 15.8 nm. This result showed that the addition of high loading of CuO might promote the growth of P90, which was not observed on the P25 TiO_2 series.

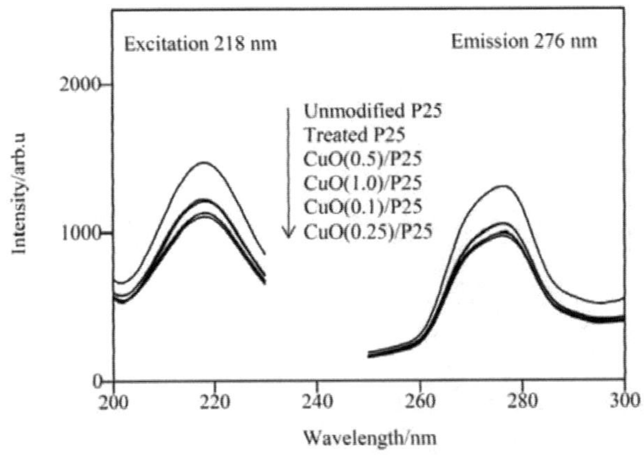

Figure 3. Excitation and emission spectra of unmodified P25, treated P25, and CuO/P25 samples.

Table 2 also displays the percentage of anatase and rutile in P90 and CuO/P90 samples. The percentage ratio of anatase to rutile was calculated using equation 2 as stated previously. From the table, it can be observed that the percentage of rutile was increased from 19.6 to 22.2 % when the P90 was calcined at 300 °C. There was no clear trend on the change of the anatase-rutile composition with the addition of CuO. However, it is clear that when the CuO amount was 1 wt %, the rutile phase increased from 19.6 to 24.9 %. As compared to P25, P90 was shown to be less stable since the anatase-rutile composition could be affected even with low calcination temperature and addition of CuO.

Figure 5 shows the fluorescence spectra for the P90 and CuO/P90 series. From the figure, it was observed that all samples showed similar excitation and emission spectra to each other, which excitation and emission wavelengths were at 218 and 276 nm, respectively. Different from the treated P25, the treated P90 was observed to have a higher intensity than the unmodified P90. This might be due to the different crystal phase composition induced by the calcination process. For the modified samples, CuO(0.25)/P90 showed the highest intensity followed by CuO(1.0)/P90, CuO(0.5)/P90 and CuO(0.1)/P90. It is noted that all the CuO/P90 series showed less intensity compared to the treated P90, suggesting the less electron-hole recombination on these CuO/P90 series.

The photocatalytic degradation of 2,4-D was evaluated using the P90 and CuO/P90 series under UV light irradiation. Table 2 shows the percentage degradation of 2,4-D over the P90 and CuO/P90 series. From the table, it can be observed that the unmodified P90 showed high photocatalytic activity with 68% degradation of 2,4-D but the P90 calcined at 300 °C showed a lower photocatalytic activity than

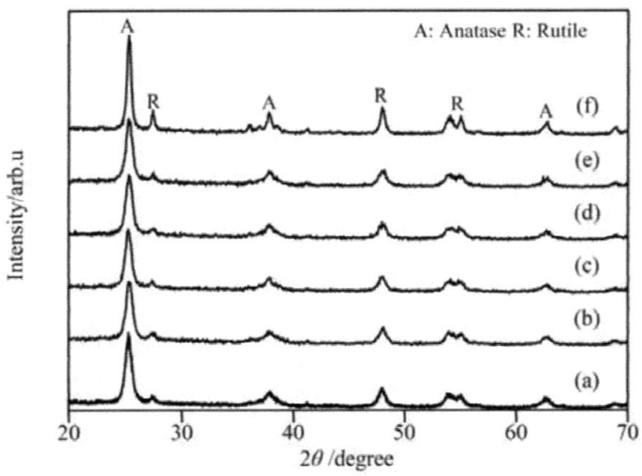

Figure 4. XRD patterns of (a) (a) unmodified P90, (b) treated P90, (c) CuO (0.1)/P90, (d) CuO (0.25)/P90, (e) CuO (0.5)/P90, and (f) CuO (1.0)/P90 samples.

Table 2. Crystallite size, percentages of anatase and rutile phases of P90 and CuO/P90 samples.

Sample	Crystallite size (nm)	Percentage of Anatase (%)	Percentage of Rutile (%)	Degradation of 2,4-D (%)
Unmodified P90	9.6	80.4	19.6	68
Treated P90	8.3	77.8	22.2	47
CuO(0.1)/P90	8.7	78.4	21.6	86
CuO(0.25)/P90	9.9	79.8	20.2	87
CuO(0.5)/P90	8.7	76.9	23.1	70
CuO(1.0)/P90	15.8	75.1	24.9	62

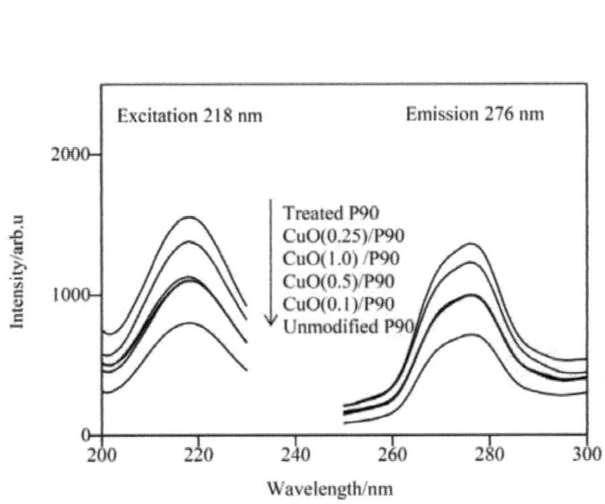

Figure 5. Excitation and emission spectra of unmodified P90, treated P90 and CuO/P90 samples.

unmodified P90, which was 47 % degradation of 2,4-D. The addition of CuO was found to give positive effect to the photocatalytic activity as Cu(0.1)/P90 and Cu(0.25)/P90 gave 86 and 87 % degradation of 2,4-D, respectively. Increasing the amount of CuO from 0.5 to 1.0 wt% decreased the photocatalytic activity of P90 from 70 and 62 %. This result proposed that the highest photocatalytic activity could be achieved when the amount of CuO was optimum, which was 0.25 wt %. This study demonstrated that when the P90 was modified with such a low amount of CuO, it would act as good co-catalyst to trap the photogenerated electron that would catalyze the reduction process, thus promoted the photocatalytic activity of P90. However, the high amount of CuO loading would block the active site of P90, thus decreased the degradation of 2,4-D.

4. Conclusion

The CuO/TiO_2 photocatalysts in various amounts of CuO loading from 0.1 to 1.0 wt % were prepared by deposition-precipitation method using different types of TiO_2, which were Evonik P25 and P90. Based on the XRD result, all the modified photocatalysts clearly showed diffraction peaks of anatase and rutile phases. There was no CuO peak to be observed after the addition of CuO into the TiO_2. The diffraction peaks intensity increased after calcination at 300 °C for CuO/P90 series. However, the peak intensity for CuO/P25 samples was remained unchanged. Accordingly, the crystallite size for CuO/P90 series increased with the calcination process at 300 °C, while no much difference was observed for the CuO/P25. After calcined at 300 °C, the percentage of anatase phase in P25 increased from 71.7 to 74.4 %, while the addition of CuO did not much change the percentage of anatase. On the other hand, the percentage of rutile phase in P90 increased from 19.62 to 24.87 % after addition of CuO. For the fluorescence study, all the series showed similar excitation and emission spectrum to each other. As compared to the treated ones, the addition of optimum amount of CuO on the P25 and the P90 decreased the spectra intensity. It was suggested that the presence CuO might suppress the electron-hole recombination process.

The P25 showed higher activity than the P90. CuO modification gave the highest activity for P25 when the amount was 0.1 wt %, while 0.25 wt % gave the highest activity for P90 series. Even though the CuO(0.1)/P25 gave the highest percentage degradation for 2,4-D (92 %), the CuO(0.25)/P90 showed the highest enhancement (87 %) as compared to the unmodified P90 (68 %).

Acknowledgments

The authors thank for the financial support from Directorate General of Strengthening Research and Development, Ministry of Research, Technology and Higher Education of the Republic of Indonesia via the World Class Research scheme (WCR 2019, No. 041/SP2H/LT/MULTI/L7/2019 and No. 014/MACHUNG/LPPM/SP2H-LIT-MULTI/III/2019).

References

[1] Schneider J et al. 2014 *Chem. Rev.* **114** 9919-86
[2] Reza K M, Kurny ASW and Gulshan F 2017 *Appl. Water Sci.* **7** 1569-78
[3] Shayegan Z, Lee CS and Haghighat F 2018 *Chem. Eng. J.* **334** 2408-39
[4] Moma J and Baloyi J 2018 Modified titanium dioxide for photocatalytic applications *Photocatalysts-Applications and Attributes* pp 37-56
[5] Siah W R, Lintang H O, Shamsuddin M, Yoshida H and Yuliati L 2016 *Catal. Sci. Technol.* **6** 5079-87
[6] Lee S C, Hasan N, Lintang H O, Shamsuddin M and Yuliati L 2016 *IOP Conf. Ser.: Mater. Sci. Eng.* **107** 012012
[7] Yuliati L, Siah W R, Roslan N A, Shamsuddin M and Lintang H O 2016 *J. Analytical Sci.* **20** 171-8
[8] Roslan N A, Lintang H O and Yuliati L 2015 *Adv. Mater. Res.* **1112** 180-3
[9] Siah W R, Lintang H O, Shamsuddin M and Yuliati L 2016 *IOP Conf. Ser.: Mater. Sci. Eng.* **107** 012005
[10] Siah W R, Lintang H O, Shamsuddin M and Yuliati L 2015 *Malay. J. Fund. Appl. Sci.* **11** 106-10
[11] Sun Q and Xu Y 2010 *J. Phys. Chem.* **114** 18911-8
[12] Majzik E S, Tóth F, Benke L and Kiss Z 2006 *Chroma.* **63** S105-9
[13] Barbash J E, Thelin G P, Kolpin D W and Gilliom R J 2001 *J. Environ. Qual.* **30** 831-45
[14] Moro P, Stampachiacchiere S, Donzello M P, Fierro G and Moretti G 2015 *Appl. Surf. Sci.* **359** 293-305
[15] Motzkus C, Macé T, Vaslin-Reimann S, Ausset P and Maillé M 2013 *J. Phys.: Conf. Ser.* **429** 012012

Electroreduction of CO_2 using modified boron-doped diamond electrode as the working electrode

I Rahmawati, N Y Yetri, J Gunlazuardi and T A Ivandini

Department of Chemistry, Faculty of Mathematics and Natural Sciences (FMIPA), Universitas Indonesia, Depok 16424, Indonesia

Corresponding author's email: ivandini.tri@sci.ui.ac.id

Abstract. Electroreduction of CO_2 becomes a popular study to resolve the problems due to the increase of CO_2 in atmosphere. In this work, electroreduction of CO_2 was studied using various working electrodes in 0.1 M NaCl solutions. A reduction peak potential around -1.5, -1.1 and -1.2 V, respectively, was observed at BDD, Cu, Cu-BDD, while Cu_2O-BDD with different deposition times showed typical peaks at around -1.3 V. Liquid chromatography analysis on the solutions after CO_2 electroreduction at the applied potential of -1.3 V (vs Ag/AgCl) showed that by using BDD and Cu electrodes only formic acid and formaldehyde could be produced. However, the use of various Cu_2O concentrations deposited on BDD electrodes can produce not only formic acid and formaldehyde, but also acetic acid and methanol with high selectivity of the products, indicating the electrode is promising to have more various products in the electrochemical reduction of CO_2.

Keywords: Electroreduction, CO_2, electrodeposition, BDD and Cu_2O

1. Introduction

The earth's atmosphere contains enormous amount of carbon dioxide (CO_2) and it is known that the emission of CO_2 is dramatically increase every year [1, 2]. This phenomenon has attracted great attention to reduce the emission of CO_2 and convert it into a more useful compound [1]. Electroreduction of CO_2 becomes a popular study to resolve this problem. One of the focuses is on the use of certain types of electrodes to convert CO_2 into more useful products. The use of electrode in the form of metal or metal oxide is necessary to electrochemically reduce CO_2 [3]. Electroreduction CO_2 is expected not only to reduce emissions of CO_2, but it can produce a variety of beneficial substances such as formic acid, methane, ethylene, methanol, and ethanol [4].

Study of some metals and metal oxides, such as Pb, Cu, Sn has been reported for electrochemical reduction of CO_2 [3]. It was also declared that copper oxide-modified materials increased the formation of hydrocarbons in such reaction [5]. On the other hand, boron-doped diamond (BDD) electrode is established as electrode with many advantages due to its good stability, wide working potential, and very low blank current [1, 6, 7]. Electrochemical reduction of CO_2 at this electrode was reported to be promising in NaCl or sea water [8]. Further, modification BDD electrodes with copper for the electroreduction of CO_2 was reported to produce methanol [9]. In this work, the BDD was modified with copper oxide to study its influence for electrochemical reduction of CO_2. Comparison between BDD, copper, and BDD modified with copper was also observed.

2. Materials and method

2.1. Materials

$Cu(CH_3COO)_2$, CH_3COONa, CH_3COOH, H_2SO_4, NaCl, 2-propanol, $HClO_4$, phosphate buffer, and acetonitrile were supplied from Wako (Japan). The distilled water was prepared by using Millipore®. CO_2 and N_2 gas were supplied from local sources in Indonesia.

2.2. Method

A cell with 3-electrodes system was used with various working electrodes, including BDD, Cu and Cu_2O-BDD. Cu-BDD was also used for comparison. Ag/AgCl was used as the reference electrode, while Pt foil used as the counter electrode. Cu_2O-BDD electrode was prepared using the same procedure that have been reported before [10]. Briefly, cyclic voltammetry in the potential range of -0.2 V to -1.0 V (vs. Ag/AgCl) was performed at BDD electrode in a solution containing 1 mM Cu $(CH_3COO)_2$ and 0.1 M CH_3COONa (ratio was 1: 1) at pH 5.7. Then, amperometry technique at the peak potential of -0.3 V with various deposition time of 60 s, 120 s and 180 s were applied. Meanwhile, Cu-BDD was prepared based on the optimum condition of previous result by using amperometry technique in mM $CuSO_4$ and 5mM H_2SO_4 applied at -0.4 V (vs. Ag/AgCl) for 120 s [9].

The solution for electrochemical reduction of CO_2, i.e. 0.1. M NaCl was pre-treated by bubbling N_2 for 30 min to remove other gas. Further, cyclic voltammetry at the potential range of 0.0 to -2.0 V was conducted after bubbling CO_2 into the solution with various bubbling times of 30, 60 and 90 min. Then, chronoamperometry technique was applied at the potential of -1.3 V for 60 min at all electrodes [10]. Afterward, the solution was analyzed by using HPLC.

3. Results and discussion

Electroreduction of CO_2 was performed at Cu_2O-BDD with various deposition time of 60 s, 120 s and 180 s. Comparison was also performed by using BDD, Cu, and Cu-BDD electrodes. The preparation and characterization of Cu_2O-BDD has been reported in our previous research [9]. Prior to electroreduction of CO_2, the reduction potential of CO_2 was studied using cyclic voltammetry in 0.1 M NaCl solution with without dissolved CO_2 (figure 1).

Cyclic voltammograms in the absence of CO_2 at Cu, Cu-BDD and Cu_2O-BDD electrodes showed typical oxidation-reduction couple peaks, respectively, at -0.2 V and -0.6 V, which related to the potential reduction and oxidation of Cu. In the presence of CO_2, the peak was observed at different potentials for different electrode. The potential reduction measured at BDD, Cu, Cu-BDD, and Cu_2O-BDD with various deposition time were -1.5 V, -1.1 V, -1.2 V and -1.3 V (vs Ag/AgCl), respectively. These reduction potentials were located between Cu-BDD and BDD, indicated the influence of both BDD and Cu (figure 2). Deposition of Cu was prepared only in the optimum condition reported in the previous result [9].

CO_2 electroreduction was performed by using chronoamperometry at -1.3 V for 60 min at various working electrodes. The resulting liquid products were analyzed using HPLC and the Faraday efficiencies were calculated. Figure 3 shows the products observed as the results of CO_2 reduction at various electrodes. It can be seen that formic acid and formaldehyde could be produced at all electrodes, although at the unmodified BDD electrode, very limited products were generated. Acetic acid and methanol were not produced at unmodified BDD, while these products were observed at Cu-based electrode. Furthermore, when the BDD was modified with Cu_2O, these products increased significantly, indicated the influence of Cu_2O in the formation of methanol and acetic acid. The figure also confirmed that Cu_2O-BDD electrodes prepared with longer deposition times increased its selectivity to produce methanol and acetic acid than formaldehyde. The selectivity of the products was calculated by the percentage of a generated product in comparison to all products observed in this reaction [1, 2]. The summary is displayed in table 1.

IOP Conf. Series: Materials Science and Engineering **902** (2020) 012011 doi:10.1088/1757-899X/902/1/012011

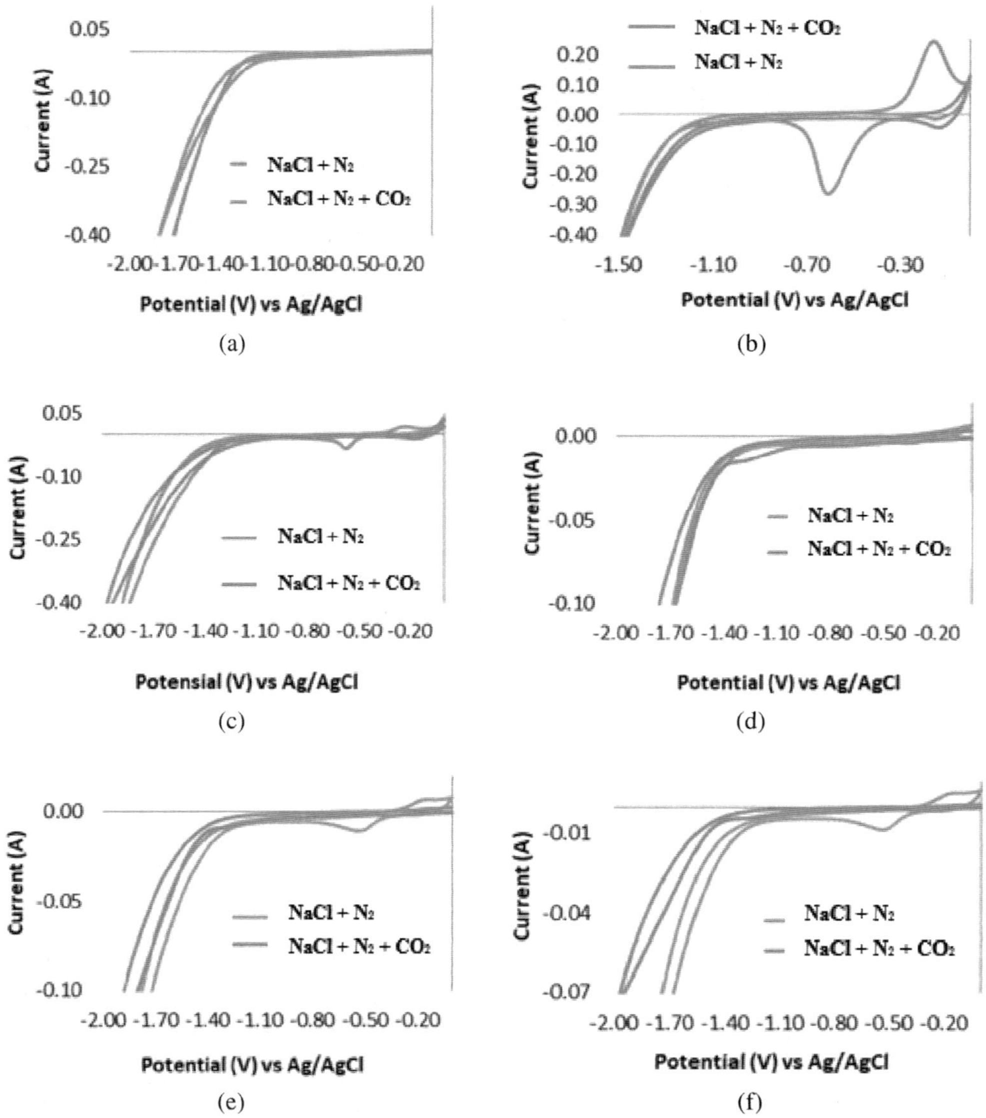

Figure 1. Cyclic voltammograms of 0.1 M NaCl in the absence and in the presence of dissolved CO_2 at (a) BDD, (b) Cu, (c) Cu-BDD as well as Cu_2O-BDD with various deposition times of (d) 60 s, (e) 120 s and (f) 180 s, respectively.

Based on the number of products obtained from CO_2 reduction (table 1), a possible mechanism of the product was determined due to calculations of % Faraday efficiency. The following mechanism is proposed for the CO_2 electroreduction's product [4].

Formic acid, acetic acid, formaldehyde and methanol were produced by CO_2 electroreduction using Cu_2O-BDD electrode with various deposition time. The possible mechanism of methanol production (figure 4) was initiated by the formation of CO_2^- radicals to form $CO_{(ads)}$ intermediate. Further, CO species will be adsorbed and reduced into $HCO_{(ads)}$. HCO that has been adsorbed then will undergo further reduction to become formaldehyde. Formaldehyde continues to be reduced as well to become $H_3CO_{(ads)}$ intermediate. Furthermore, H_3CO is further reduced and reacted with hydrogen atoms so that it will produce methanol.

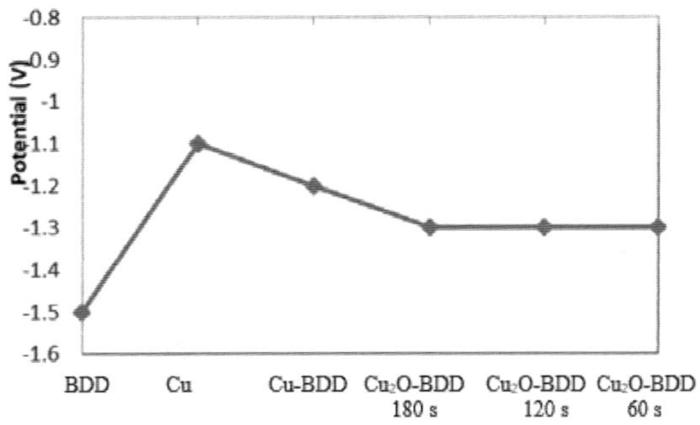

Figure 2. Plots of CO_2 reduction peak potentials at different electrodes.

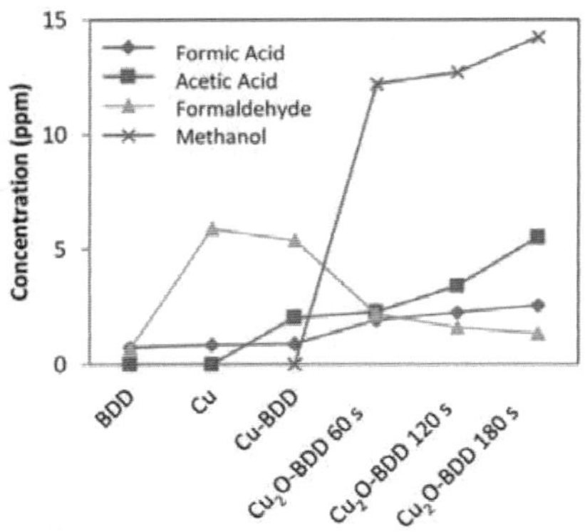

Figure 3. Total products of electroreduction at various electrodes.

Table 1. % Selectivity of the products of CO_2 electroreduction.

Electrode	% Selectivity product			
	Formic acid	Acetic acid	Formaldehyde	Methanol
BDD	52.52	-	47.48	-
Cu	12.46	-	87.54	-
Cu-BDD	8.29	24.49	64.78	-
Cu_2O-BDD 60s	18.54	12.19	11.76	65.7
Cu_2O-BDD 120s	19.92	17.12	7.98	63.65
Cu_2O-BDD 180s	23.6	23.31	5.59	60.25

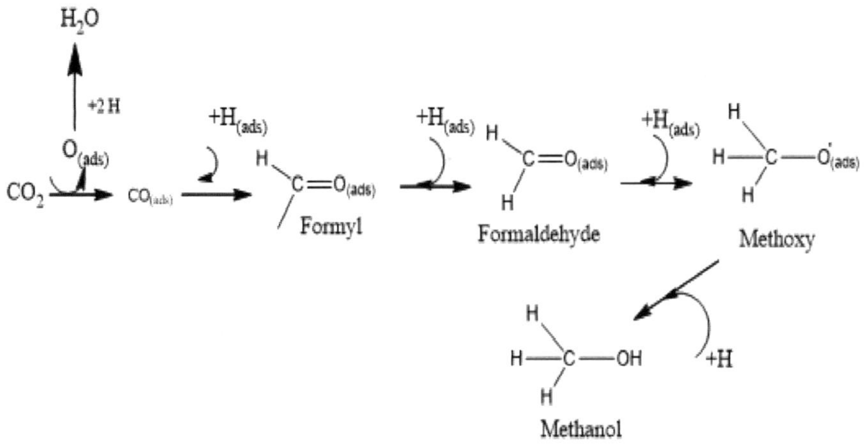

Figure 4. Possible mechanism of methanol formation.

Table 2. Results of % efficiency faraday of electroreduction CO_2.

Electrode	% Efficiency Faraday				
	Formic acid	Acetic acid	Formaldehyde	Methanol	Total
BDD	0.28	-	0.505	-	0.78
Cu	0.04	-	0.555	-	0.59
Cu-BDD	0.06	0.51	0.677	-	1.25
Cu_2O-BDD 60 s	0.24	1.14	0.550	4.61	6.54
Cu_2O-BDD 120 s	0.33	2.03	0.473	5.66	8.50
Cu_2O-BDD 180 s	0.46	3.93	0.472	7.62	12.48

Cu_2O that was deposited on BDD electrode surface supplying Cu (I) species and it takes a significant part in reducing CO_2 for the reaction of the formation of CH_3OH [8]. Cu (I) species can stabilize the intermediates and able to enhance the adsorption of the CO to be selective to methanol as the main product. It is also possible that Cu (I) site more likely to adsorb CO than the Cu sites. By doing determination% Efficiency Faraday the amount of electricity used during the process of CO_2 electroreduction will be known.

Based on the obtained faradaic efficiency (table 2), it appears that the electrical energy supplied to the electrode is used in the formation of methanol. Cu_2O-BDD electrode 180 s slightly has a higher faradaic efficiency of methanol, it is comparable in terms of yield methanol that can be produced at each electrode. Faradaic efficiency of each electrode is still low, this may be due to the formation of other by-products in the form of liquid or gas, so that the parameters of the product is still not yet included into the calculation.

4. Conclusion

Cyclic voltammograms of CO_2 dissolved in 0.1 M NaCl solution at BDD electrode, Cu and Cu-BDD showed reduction peaks at potentials of -1.5, -1.1 and -1.2 V (vs. Ag/AgCl), respectively. However, Cu_2O-BDD electrodes with various deposition times showed a typical reduction peak at the potential around -1.3 V (vs. Ag/AgCl). Applying amperometry technique at this potential using BDD electrode, Cu, Cu-BDD, and Cu_2O-BDD for 60 min for CO_2 reduction showed that modification of BDD with copper oxide produced methanol and acetic acid besides formic acid and formaldehyde. The result indicated that the modification with copper oxide is promising to form hydrocarbon with two carbon numbers.

Acknowledgments
This work was funded by Hibah PIT 9 Universitas Indonesia Tahun Anggaran 2019 No. Contract: NKB-0028/UN2.R3.1/HKP.05.00/2019.

References
[1] Jiwanti P K and Einaga Y 2019 *Phys. Chem. Chem. Phys.* **21** 15297-301
[2] Jiwanti P K, Natsui K, Nakata K and Einaga Y 2016 *RSC Adv.* **6** 102214-7
[3] Albo J, Alvarez-Guerra M, Castano P and Irabien A 2015 *Green Chem.* **17** 2304-24
[4] Kas R, Kortlever R, Yilmaz H, Koper M T M and Mul G 2015 *Chem. Electro. Chem.* **2** 354-8
[5] Ohya S, Kaneco S, Katsumata H, Suzuki T and Ohta K 2009 *Catal. Today* **148** 329-34
[6] Ivandini T A et al. 2019 *J. Phys. Chem. C* **123** 5336-44
[7] Asai K, Ivandini T A and Einaga Y 2016 *Sci. Rep.* **6** 32429
[8] Le M et al. 2011 *J. Electrochem. Soc.* **158** 45-49
[9] Panglipur H S, Ivandini T A, Wibowo R and Einaga Y 2016 *AIP Conf. Proc.* **1729** 020047
[10] Yetri N Y, Ivandini T A and Gunlazuardi J 2017 *IOP Conf. Ser.: Mater. Sci. Eng.* **188** 012011

Sm_2O_3 nanoparticles preparation using caesalpinia pulcherrima leaf extract, characterization and photocatalytic activity

N Putri, Y Yulizar, A Umar and D O B Apriandanu

Department of Chemistry, Faculty of Mathematics and Natural Sciences (FMIPA), Universitas Indonesia, Depok 16424, Indonesia

Corresponding author's email: yokiy@sci.ui.ac.id

Abstract. In this research, Sm_2O_3 nanoparticles (NPs) have been prepared via *Caesalpinia pulcherrima* leaf extract (CPE), whichh contains secondary metabolites contained as a weak base source and a capping agent. The nanoparticles were confirmed by UV-Vis DRS, FT-IR spectroscopy, XRD, PSA and TEM. FT-IR spectrum showed that CPE contains the secondary metabolites which play an essential role in the preparation of Sm_2O_3 NPs. UV-Vis DRS result showed that the bandgap value of Sm_2O_3 NPs was approximately 4.76 eV which works in the UV region. Based on the XRD analysis, Sm_2O_3 NPs have a structure of cubic. PSA characterization showed that the average particle size distribution of Sm_2O_3 was around 84.47 nm. TEM image confirmed that Sm_2O_3 was in nanoscale with the particle size of 73.27 nm. Finally, Sm_2O_3 NPs have been tested for the photodegradation of malachite green (MG) under UV light illumination. The degradation percentage of MG using Sm_2O_3 NPs photocatalyst was 80.14 % for 2 h.

Keywords: Sm_2O_3 nanoparticles, *Caesalpinia pulcherrima* (L.) Sw, characterization, photocatalytic activity

1. Introduction

Nanoparticles have an average size of < 100 nm with some beneficial characteristics such as large surface area and equally distributed composition of cell units. Also, nanoparticles are mostly applied in the fields of health, agriculture, and electronics due to their unique physical, chemical, thermal and electrical properties, which makes the nanomaterial more preferable than the bulk material [1]. Semiconductor-nanoscale materials are important and promising to be developed because of their good optical, photocatalytic, and magnetic properties, especially metal-semiconductors [2]. Also, metal oxides and their semiconductor-based composites are excellent materials for catalytic and antimicrobial activity [3].

Rare-earth metal nanoparticles have unique magnetic, luminescent and electrochemical properties [4]. Samarium oxide (Sm_2O_3) is a wide bandgap rare-earth metal oxide [5]. Recently, metal oxide nanoparticles have been synthesized using fungi, bacteria, or plant extracts as the desired biogenic agent [6] since the presence of toxic chemicals could be adsorbed on the material surface and has an adverse effect on health [7]. Plant extracts are widely utilized to synthesis metal oxide nanoparticles and surfactants such as *Tinospora crispa, Datura metel L., Pandanus amaryllifolius, Ageratum conyzoides L., Orthosiphonaristatus (Blume) Miq, Morinda citrifolia L.* [8-12]. *Caesalpinia pullcherima* has a rich phytochemical content. Some isolated compound from the extract of this plant are alkaloid, saponin, flavonoid [13]. In this work, we reported the synthesis of Sm_2O_3 NPs using *Caesalpinia pullcherima,* the characterization and their photocatalytic activity for MG degradation under UV light illumination.

2. Materials and method

The CPE solution was prepared by an extraction method, as reported in our previous work [9]. Fifty g of dried *Caesalpinia pullcherima* powders were macerated into methanol for 7 days. The mixture was then stirred for 20 min every day. The result was partitioned using n-hexane to obtain the n-hexane and methanol fractions. The methanol fraction was evaporated, then dissolved into distilled water to obtain the CPE solution. All fractions were further tested by phytochemical screening.

Sm_2O_3 NPs were synthesized by adding 0.02 M $Sm(NO_3)_3$ into 10 mL of CPE 1.85 % (w/v). The mixture was stirred for 4 h at 80 °C. The formed colloid was heated at 120 °C to form a gel, then calcined at 600 °C to form a white powder [14]. The obtained Sm_2O_3 NPs was confirmed by UV-Vis spectrophotometer, UV-Vis DRS, FT-IR, XRD¸ PSA and TEM. Photocatalytic activity tests were carried out by adding Sm_2O_3 NPs into 25 mL of malachite green 6×10^{-6} M and stirred up to 120 min under visible light radiation. The degradation result was observed by a UV-Vis spectrophotometer every 10 min [15].

3. Results and discussion

FTIR characterization was performed to identify the functional groups of Sm_2O_3 NPs and a water fraction of CPE. The water fraction of CPE has vibrational spectrum at wavenumbers of 3341, 2941, 1700, 1353, 1231 and 843 cm^{-1}, which were attributed to O-H stretching, C-H stretching, N-H amine bending, N-O stretching, C-N stretching, N-H wagging-stretching, respectively, as shown in figure 1. Amine bending, N-H functional groups, C-N stretching indicate the presence of alkaloid compound [16]. Meanwhile, O-H stretching, C-H sp^3 stretching, and C=C stretching show the presence of saponins [17]. The secondary metabolite compounds in the CPE water fraction have an important role, alkaloids act as a source of weak bases, while saponins act as a capping agent in the nanoparticle synthesis. Also, the absorption at 534 cm^{-1} shows the bending vibration of Sm-O bond. This result has conformity with the previous research, which indicates that the bending vibration of Sm-O bond is in the wavenumber range of 593–475 cm^{-1} [14].

XRD characterization of Sm_2O_3 NPs was performed to determine their diffraction patterns, as shown in figure 2. The diffraction patterns of Sm_2O_3 NPs were presented at 2θ of 25.96°; 28.49°; 30.83°;

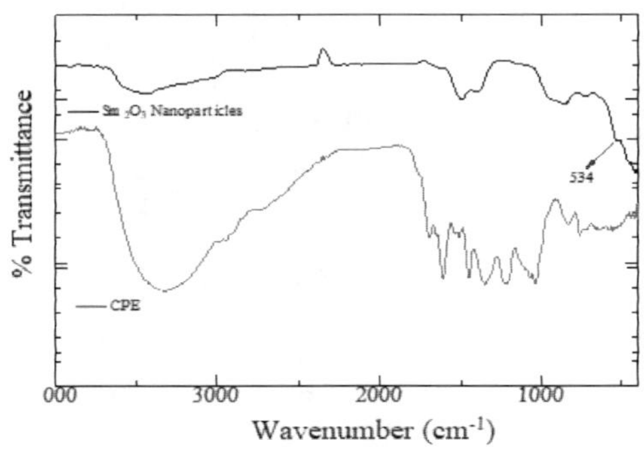

Figure 1. FTIR spectra of Sm_2O_3 NPs (black line) and CPE (green line).

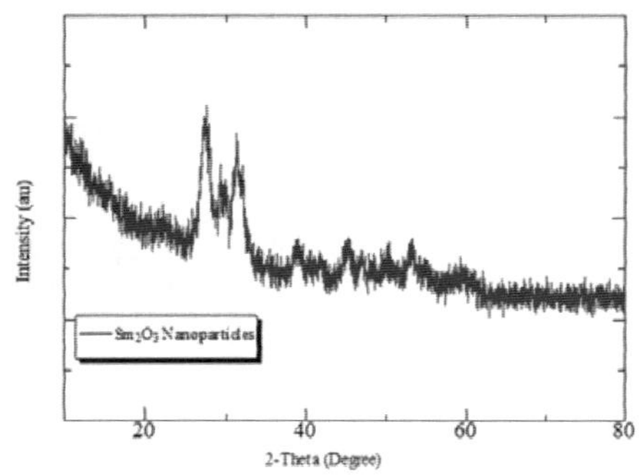

Figure 2. XRD pattern of Sm_2O_3 NPs.

38.93°; 40.74°; 42.48° and 48.90° (COD Entry Number 96-101-0341) with the miller index of 310, 222, 321, 323, 422, 501 and 503, respectively. According to this result, it can be concluded that Sm_2O_3 was well-matched with a cubic crystal structure.

PSA characterization was conducted to determine the average particle size distribution of Sm_2O_3 NPs. The characterization result is shown in figure 3. The average particle size of Sm_2O_3 NPs was 84.47 nm. These results proved that Sm_2O_3 NPs were synthesized using CPE.

The bandgap energy is determined by the UV-Vis DRS characterization. The determination was conducted by converting a % reflectance into a bandgap value using Kubelka-Munk function. The $F(R)^2$ and Eg were plotted by finding and making a linear regression equation with $F(R)^2$ against Eg (eV) [18], as shown in figure 4. From the linear regression equation, the bandgap energy of Sm_2O_3 NPs was 4.76 eV. This result showed that Sm_2O_3 NPs have good absorption in the UV region.

TEM characterization was used to analyze the morphology and particle size of Sm_2O_3 NPs. TEM images were presented in a magnification of 20,000 times, as shown in figure 5. According to the analysis result, Sm_2O_3 NPs had an average of particle size approximately 73.27 nm. This result demonstrates that Sm_2O_3 NPs has the particle size in nano-scale.

The photocatalytic activity of Sm_2O_3 NPs was tested in the degradation of malachite green under UV light illumination for 120 min. The results of UV-Vis absorption spectra for the photodegradation of malachite green using Sm_2O_3 NPs are shown in figure 6.

The wavelength shift occurs at λ_{max} of 318 to 250 nm, as shown in figure 6. A peak shift around 250–254 nm indicates the presence of carboxylic acid and aromatic chromophore groups [19].

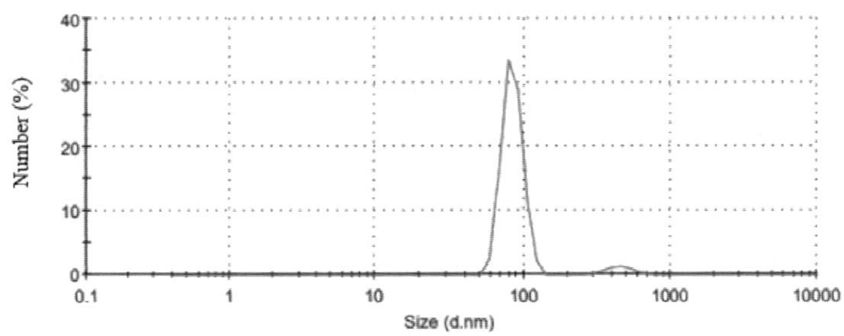

Figure 3. Particle size distribution of Sm_2O_3 NPs.

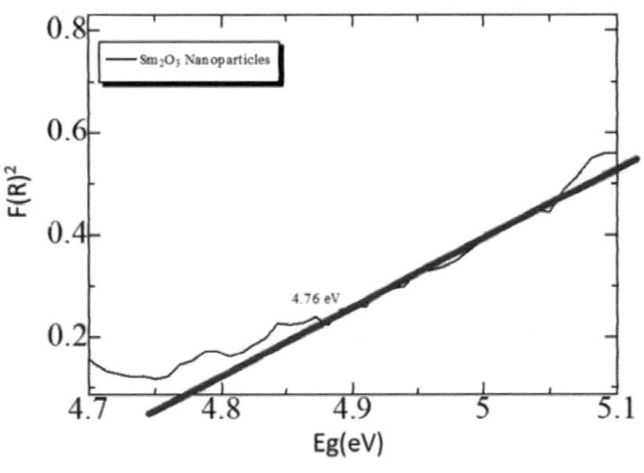

Figure 4. UV-Vis DRS of Sm_2O_3 NPs.

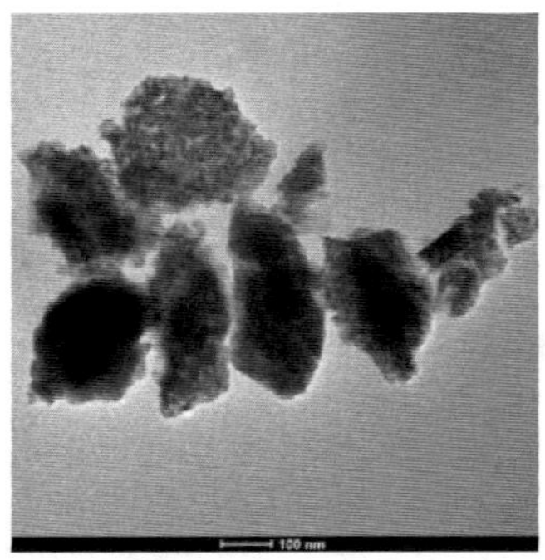

Figure 5. TEM mage of Sm_2O_3 NPs.

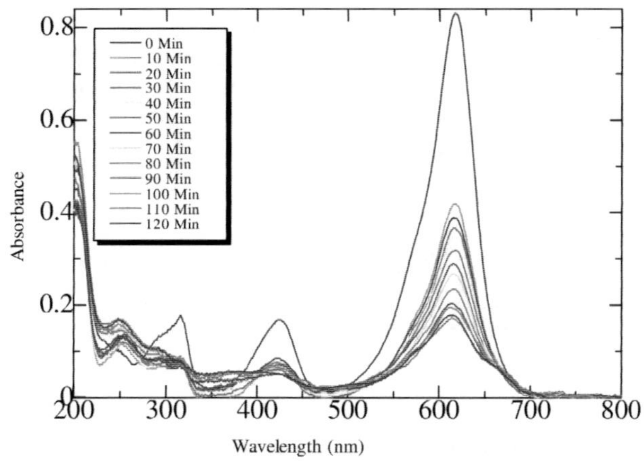

Figure 6. UV-Vis absorption spectra of malachite green degradation.

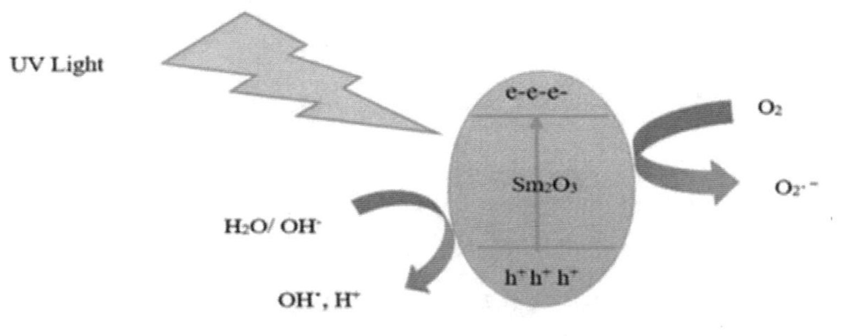

Figure 7. Mechanism of photocatalytic activity.

The absorbance value obtained every 10 min irradiation was used to obtain the concentration of malachite green. Then, the degradation percentage was calculated using equation 1 [20].

$$Degradation\ (\%) = \frac{[C]_0 - [C]_t}{[C]_0} \times 100\ \% \quad (1)$$

$[C]_0$ is the initial concentration of malachite green and $[C]_t$ is the concentration of malachite green at each time interval under UV light illumination. According to the calculation, the degradation percentage of malachite green using Sm_2O_3 NPs was 80.14 % for 120 min under UV light illumination. Furthermore, the photocatalytic activity of Sm_2O_3 NPs is presented in figure 7 [21].

When Sm_2O_3 NPs were irradiated by UV light, it will produce e- (electrons) in the conduction band (CB) and h^+ (hole) in the valence band (VB) due to an electron excitation from the valence band to the conduction band, as presented in figure 7. The hole in the valence band reacts to H_2O to form the hydroxyl radicals ($OH^\cdot$) and H^+ ions, while the electrons in the conduction band react to O_2 dissolved in water to form superoxide radicals ($O_2\bullet$). Both hydroxyl and superoxide radicals can degrade the molecule of malachite green.

4. Conclusion

The green synthesis of Sm_2O_3 NPs was carried out using a water fraction of *Caesalpinia pulcherrima* leaves extract (CPE). Flavanoid and saponin in extract act as capping agent and alkaloid as a weak base source to synthesize the metal oxide nanoparticles. According to TEM characterization, Sm_2O_3 has the particle size in nanoscale. Also, the photocatalytic activity of Sm_2O_3 NPs works at Eg of 4.76 eV, which operates in the UV area. In the next work, we expect that Sm_2O_3 NPs could be utilized as materials, which could be active in the visible area.

Acknowledgments

The authors would like to thank Universitas Indonesia for funding this research through PIT-9 Grant Universitas Indonesia with contract number: NKB-0037/UN2.R3.1/HKP.05.00/2019.

References

[1] Bhagyaraj S M, Oluwafemi O S, Kalarikkal N and Thomas S 2018 *Synthesis of Inorganic Nanomaterilas* 1st edition (London: Woodhead Publishing) pp 237-67
[2] Lee D H, Kim J E, Han T H, Hwang J W, Jeon S, Choi S Y, Hong S H, Lee W J, Ruoff R S and Kim S O 2010 *Adv. Mater.* **22** 1247-52
[3] Magdalane C M, Kaviyarasu K, Vijaya J J , Jayakumar C, Maaza M and Jeyaraj B 2017 *J. Photochem. Photobiol. B Biol.* **169** 110-23
[4] Zhang J, Rahman A Z M S, Li Y, Yang J, Zhao B, Lu E, Zhang P, Cao X, Yu R and Wang B 2013 *Optical Mater.* **36** 471-5
[5] Michel C R, Martínez-Preciado A H, Parra R, Aldao C M, and Ponce M A 2014 *Sens. Actuators B. Chem.* **202** 1220-8
[6] Nasrollahzadeh M, Issaabadi Z and Sajadi S M 2018 *RSC Adv.* **8** 3723-35
[7] Jain P K, Lee K S and El-Sayed I H 2006 *J. Phys. Chem. B* **110** 7238-48.
[8] Yulizar Y, Kusrini E, Apriandanu D O B and Nurdini N 2020 *Surf. Interfaces.* **19** 100437
[9] Yulizar Y, Sudirman, Apriandanu D O B and Wibowo A P 2019 *Compos. Commun.* **16** 50-6
[10] Wardani M, Yulizar Y, Abdullah I and Apriandanu D O B 2019 *IOP Conf. Ser. Mater. Sci. Eng.* **509** 012077
[11] Abbas M, Sari N, Utari T, Yulizar Y and Apriandanu DOB 2018 *AIP Conf. Proc.* **2023** 020100
[12] Tania T C D, Deastri Y N, Utari T, Yulizar Y and Apriandanu DOB 2018 *AIP Conf. Proc.* **2023** 020103
[13] Osuntokun O T, Ju J and Thonda O A 2017 *J. Bioequivalence Bioavailab.* **1** 1-11
[14] Sone B T, Manikandan E, Gurib-Fakim A and Maaza M 2015 *J. Alloys Compd.* **650** 357-62
[15] Li L, Wang X and Zhang Y 2013 *Mater. Res. Bull.* **50** 18-22
[16] Marion L, Ramsay D A and Jones R N 1951 *J. American Chem. Soc.* **73** 305-8
[17] Bajad P N and Pardeshi A B 2016 *Inter. J Sci. Res.* **5** 351-4
[18] Abdullahi S S, Güner S, Koseoglu Y, Musa I M, Adamu B I and Abdulhamid M 2016 *J. Nigerian Assoc. Math. Phys.* **35** 241-6
[19] Korshin G, Chow C W K, Fabris R and Drikas M 2009 *Water Res.* **43** 1541-8
[20] Apriandanu D O B and Yulizar Y 2019 *Nano-Struct. Nano-Objects* **20** 100401
[21] Elumalai K and Velmurugan S 2015 *Appl. Surf. Sci.* **345** 329-36

Modification of gold nanoparticles through allyl mercaptan bridge at boron-doped diamond electrode for an arsen (V) sensor

O Magdalena, N A Fauzillah, H G Riyanto and T A Ivandini

Department of Chemistry, Faculty of Mathematics and Natural Sciences (FMIPA), Universitas Indonesia, Depok 16424, Indonesia

Corresponding author's email: ivandini.tri@sci.ui.ac.id

Abstract. A simple, cheap and fast analytical procedure for arsen (V) detection was developed by applying the BDD (boron-doped diamond) modified by gold nanoparticles (AuNPs-BDD). The gold nanoparticles (AuNPs) with an average particle size of 10 nm was synthesized through reduction reaction of gold ions by $NaBH_4$ in the presence of allyl mercaptan. These nanoparticles could be successfully deposited on the surface of boron-doped diamond (BDD) by an immersion process under the exposure of UV light (λ = 254 nm) for 6 h. An Au/C ratio of around 1:56 was observed at the prepared AuNPs-BDD. Preliminary study for arsen (V) detection by using anodic stripping voltammetry (ASV) technique using this electrode in a solution of 0.1 M HCl containing of $NaBH_4$ showed a good linearity (R^2 = 0.99) of the current flow in the concentration range of 0–100 μM with a limit detection of 0.105 μM.

Keywords: Boron-doped diamond, gold nanoparticles, allyl mercaptan, surface modification, arsenic detection

1. Introduction

Inorganic arsenic contamination in drinking water has attracted a lot of attentions since several million people in various parts of the world are at a risk of arsenic contamination. This contamination can cause health problems, such as chronic diseases or even death [1, 2]. Although arsenic level in natural water are usually low in some area with different geographic, World Health Organization (WHO) established that limit groundwater content level should be no more than 10 μgL^{-1} to reduce carcinogenic effects in humans [1].

The methods of determining inorganic arsenic have been carried out in various ways, such as AAS (Atomic Absorption Spectrometry), AES (Emission Spectrometry), GFAAS (Graphite Furnace Atomic Absorption Spectrometry), HPLC (High Pressure Liquid Chromatography), (ICP-MS) Inductively Coupled Plasma-Mass Spectrometry [3, 4]. Some of methods mentioned above have some drawbacks including the requirement of high operational costs, expensive equipment and complicated sample preparation processes. Another method, ASV (Anodic Stripping Voltammetry) is one of the popular methods used in electrochemistry due to its low detection limit and easy operation [5, 6]. This method was performed by a pre-concentration step of the analyte on the electrode surface followed by a stripping step of the pre-concentrated analyte [6]. Various electrodes, such as mercury, platinum and gold, were reported to be employed for ASV method [5]. Among these electrodes, gold-based electrodes are the most popular used to measure arsen [7].

Meanwhile, BDD (boron-doped diamond) has been reported as an electrode substance with many advantages due to its high chemical, mechanical, and biological innertness, which affect its wide working potential and low blank current [7-9]. However, since BDD has less electrochemical sensitive to arsen, surface modification with gold is necessary [5, 7]. In the previous work, the BDD surface was modified with allyl mercaptan-capped gold nanoparticles (AuNPs) [10]. Under a UV light irradiation, the double bonds in allyl mercaptan was expected to break and form chemical bonds to the BDD surface, while the sulfur sites bind with gold. This AuNP-modified BDD was stable enough for application as As^{3+} detection [10]. In this work, an application for As^{5+} detection was studied by using the same electrodes.

2. Materials and method

2.1. Materials

Hydrogen tetrachloroaurat (III) tetrahydrate ($HAuCl_4.4H_2O$), sodium arsenite ($NaAsO_2$), sodium arsenate ($Na_2HasO_4.7H_2O$), allyl mercaptan (2-propene thiol), and sodium borohydrate ($NaBH_4$) were obtained from Sigma Aldrich, while hydrochloric acid (HCl), 2-propanol and ethanol were supplied from Merck. BDD (Boron-doped diamond) films were laboratory made in Keio University, Japan.

2.2. Synthesis of AuNPs (gold nanoparticles)

AuNPs was prepared by mixing 5 mL of 0.01 % (v/v) $HAuCl_4.4H_2O$ in 50 % ethanol solution. Then, 2.5 mL of 0.002 % allyl mercaptan in ethanol solution was added [10]. After 5 min of mixing, 0.5 mL of 0.1 M $NaBH_4$ solution was added and mixed for 10 min. The nanoparticles was determined by using UV-VIS Spectroscopy, TEM, and PSA.

2.3. Modification of AuNPs-BDD electrodes

AuNPs-BDD was prepared by immersing BDD in the prepared colloidal AuNPs and irradiated under UV light (λ = 254 nm) for 6 h. The surface was identified by using SEM.

2.4. Electrochemical study of AuNPs-BDD as arsenic sensor

Electrochemical study was performed by using cyclic voltammetry (CV) and anodic stripping voltammogram (ASV) for standard solutions of arsenic (III) and (V) using a potentiostat (Dropsens) in an electrochemical cell with 3 electrodes. The prepared AuNPs-BDD was applied as the working electrode, while Ag/AgCl system and Pt wire were used as the reference and counter electrode, respectively. Planar working electrode (estimated geometric area of ~0.26 cm^2) was placed at the bottom of the cell using the Vitton O-ring. The supporting electrolyte was 0.1 M HCl containing of 0.1 M $NaBH_4$.

3. Results and discussion

3.1. Synthesis of AuNPs (gold nanoparticles)

AuNPs was prepared by reduction reaction using $NaBH_4$ as the reducing agent and allyl mercaptan 0.002 % as the capping agent. The change in color of solution from yellow to red purplish indicating the formation of AuNPs with the change of Au^{3+} to be Au^0 [11]. The presence of allyl mercaptan (C_3H_6S) was expected to be around gold nanoparticle as a high affinity of the SH-group performed hydrogen interactions with gold as shown in figure 1 [11]. The interaction between gold and allyl mercaptan was proposed as the complex formation between the ligand and its central metal with gold as the central metal and allyl mercaptan as the ligand [11]. The -SH group on allyl mercaptan has a good affinity with Au based on the nature of the Hard Soft Acid Base.

The initial characterization was performed using UV-Visible Spectrophotometer (UV-VIS), showing the maximum absorption peak at a wavelength of 545 nm (figure 2). In general, AuNPs provides a strong surface plasmon absorption peak in the range 510–580 nm [10].

The stability of this peak was examined for 3 days. The absorbance spectrum for 3 days showed that the peak was relatively stable for 2 days before decreased at the 3rd day. This result indicates the low stability of AuNPs colloids. This is reasonable as the lack of the charge in C=C groups in allyl mercaptan structure did not provide a repulsion effect between the nanoparticles. However, in fact, the stable period of the nanoparticles for several hours was enough to be modified on the BDD surface.

Furthermore, investigation by using Transmission Electron Microscopy (TEM) (figure 3) was performed. The nanosphere shape was obtained from the synthesis of gold using allyl mercaptan [11]. Further confirmation by using Particle Size-Analyzer characterization (PSA) (figure 4) showed that the average particle size was 10 ± 2 nm.

3.2. Modification of AuNPs-BDD electrodes

The prepared AuNPs was expected to be firmly attached to the BDD electrode under photochemical reaction between thiol (-SH) functional groups of the nanoparticles and the surface of BDD (figure 5). Previously, it was reported that covalent bonds between AuNPs and -SH was probably formed under UV irradiation [10, 11]. Confirmation with SEM (figure 6) showed that the nanoparticles were successfully distributed on the BDD surface. Further observation using EDS showed that Au/C ratio of 1:56 was observed (table 1). The modified BDD electrode was then named as AuNPs-BDD electrodes.

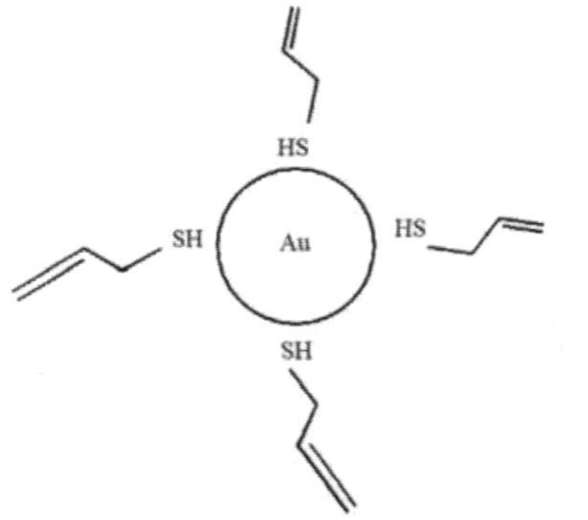

Figure 1. Interaction of Au with capping agent of allyl mercaptan.

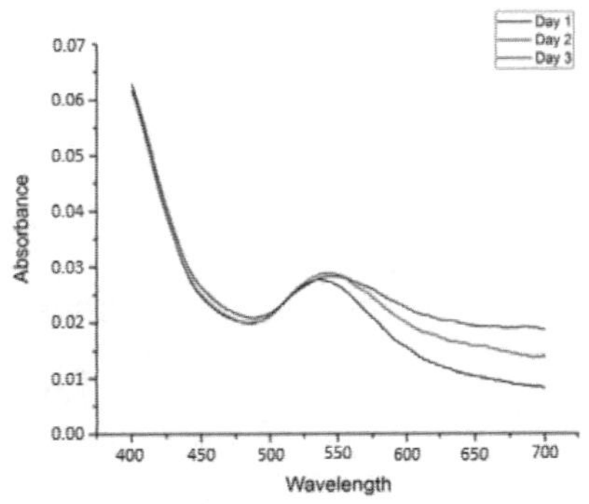

Figure 2. UV-VIS AuNPs spectrum at various times for 3 days after synthesis.

Figure 3. TEM image of AuNP

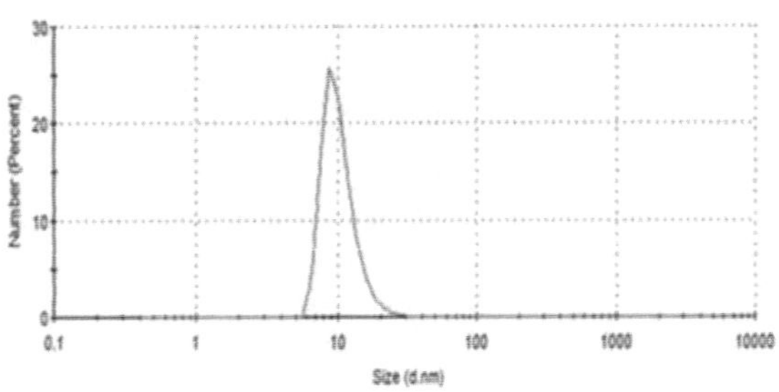

Figure 4. Size distribution of AuNP with PSA.

Electrochemical investigation was conducted in an electrochemical cell with 3 electrodes. CV (Cyclic voltammetry) was performed in 0.1 M HCl to confirm the existence of gold nanoparticles on BDD surface. The voltammograms comparison in the potential range from -0.8 to +2.0 V (vs. Ag/AgCl) at BDD, gold plate and AuNPs-BDD is displayed in figure 7. There was no peak observed on the BDD surface, which confirmed that there was no active species on the BDD surface. On the contrary, the specific gold oxidation peak at around +1.5 V and the reduction peak at around +0.4 V were observed at gold plate electrode. In the case of AuNps-BDD, the oxidation and reduction peaks were found at around +1.33 V and +0.1 V, respectively. These shifts were probably due to the kinetic different between gold plate and AuNPs-BDD surface. Since the surface modification of BDD involved not only gold particles but also allyl mercaptan as the capping agent, the surface different might also influence the kinetics of arsenic ions. However, the different of CVs on the BDD and the AuNPs-BDD surface indicates that the modification has successfully performed.

3.3. Electrochemical study of AuNPs-BDD as arsenic (V) sensor

AuNPs-BDD was then electrochemically examined for arsenic (V) sensors as the working electrode. ASV technique was set up following the previous research at the potential deposition of -500 mV for 120 s [11]. Then, stripping step was performed in the potential range from -0.4 V to +0.8 V (vs. Ag/AgCl). A pretreatment using 0.1 M $NaBH_4$ was applied to reduce As^{5+} to be As^{3+} to perform

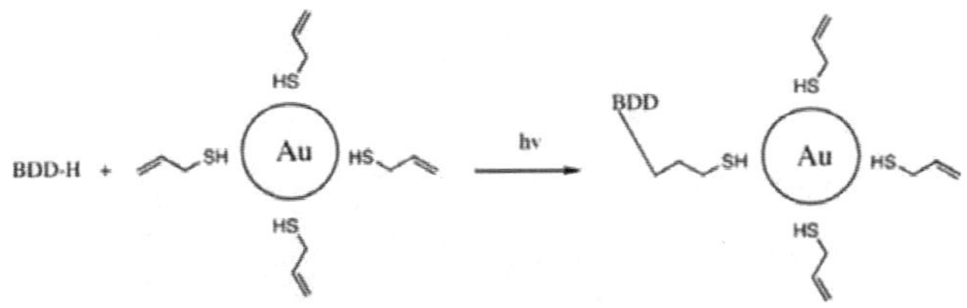

Figure 5. Modification of AuNPs BDD.

Figure 6. Morphology of AuNPs-BDD electrodes at 2500x magnification.

Table 1. Elemental composition of the results SEM-EDS AuNPs-BDD.

Element	wt%
B	7.60
C	90.30
O	0.50
Au	1.60

IOP Conf. Series: Materials Science and Engineering **902** (2020) 012013 doi:10.1088/1757-899X/902/1/012013

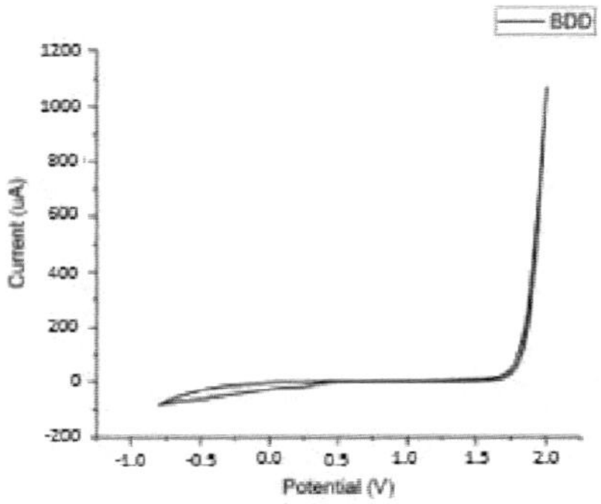

(a)

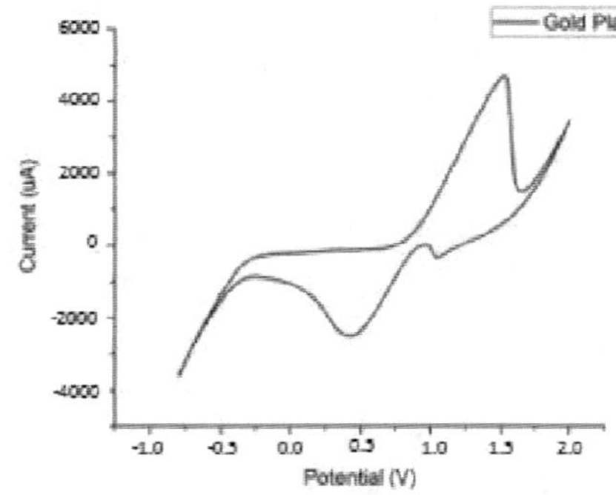

(b)

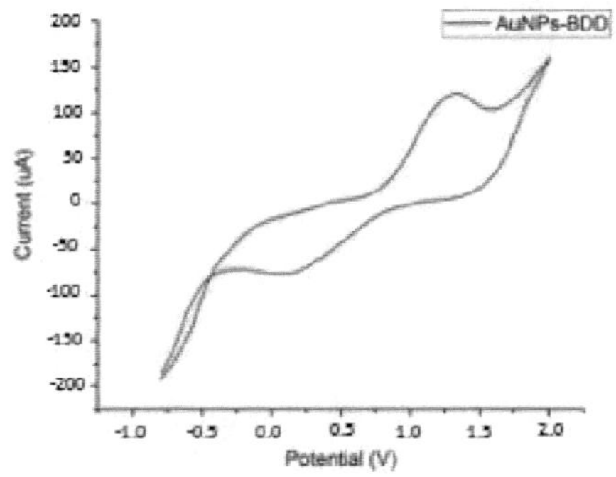

(c)

Figure 7. CV (Cyclic voltammograms) results of 0.1 M HCL solution vs Ag/AgCl on (a) BDD, (b) Gold plate, and (c) AuNPs BDD electrodes.

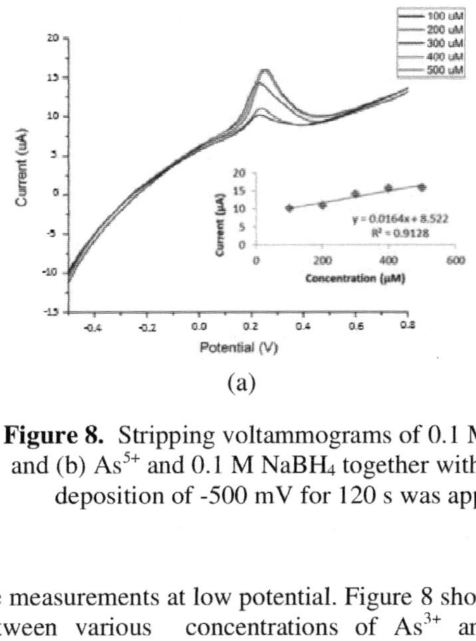

(a)

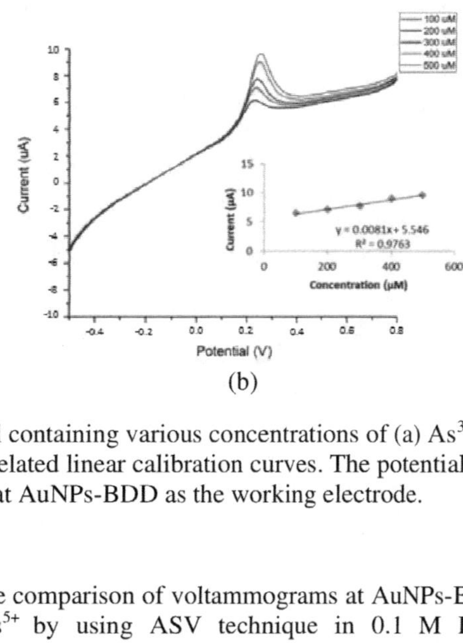

(b)

Figure 8. Stripping voltammograms of 0.1 M HCl containing various concentrations of (a) As^{3+} and (b) As^{5+} and 0.1 M $NaBH_4$ together with the related linear calibration curves. The potential deposition of -500 mV for 120 s was applied at AuNPs-BDD as the working electrode.

the measurements at low potential. Figure 8 shows the comparison of voltammograms at AuNPs-BDD between various concentrations of As^{3+} and As^{5+} by using ASV technique in 0.1 M HCl.

Linear calibration curves for As^{3+} and As^{5+} concentrations of 100-500 μm were obtained at y = -0.0164x + 8.522 (R^2 = 0.91) and y = -0.0081x + 5.546 (R^2 = 0.97) (figure 8). The calibration curve results show that the prepared AuNPs-BDD has the possibility to be applied as arsenic sensors. Similar peak potentials at around +0.2 V were observed at both species. However, lower sensitivity was found in As^{5+} measurements, probably because not all As^{5+} could be reduced to be As^{3+}. Further, more convincing techniques and methods are required to specify the detection for a mixture of As^{3+} and As^{5+} [9].

4. Conclusion

Gold nanoparticles have been successfully synthesized using allyl mercaptan as a capping agent and deposited on the surface of BDD. TEM (Transmission Electron Microscopy) and PSA (Particle Size Analyzer) showed an average size of AuNPs at 10 ± 2 nm. Characterization by using SEM-EDS confirmed that AuNPs-BDD electrodes had an Au:C ratio of 1:56. Application of this material as the working electrode in As^{3+} and As^{5+} sensors using the ASV technique showed a good sensitivity for the detection.

Acknowledgments

This work was financially supported by Universitas Indonesia under research grant PIT 9 with grant contract number NKB-0028/UN2.R3.1/HKP.05.00/2019.

References

[1] Liu P, He K, Li Y, Wu Q, Yang P and Wang D 2012 *Exotox. Environ. Safe.* **79** 90-100
[2] WHO 2018 *Arsenic* available at https://www.who.int/news-room/fact-sheets/detail/arsenic
[3] Yogarajah N and Tsai S S H 2015 *Environ. Sci.: Water Res. Technol.* **1** 426-47
[4] Mays D E and Hussam A 2009 *Anal. Chim. Acta* **646** 6-16
[5] Liu Z-G and Huang X-J 2014 *TrAC-Trend. Anal. Chem.* **60** 25-35
[6] Ivandini T A, Yamada D, Watanabe T, Matsura H, Nakano N, Fujishima A and Einaga Y 2010 *J. Electroanal. Chem.* **645** 58-63
[7] Nagaoka Y, Ivandini T A, Yamada D, Fujita S, Yamanuki M and Einaga Y 2010 *Chem. Lett.* **39** 105-7
[8] Luong J H T, Male K B and Glennon J D 2009 *Analyst* **134** 1965-79
[9] Ivandini T A, Watanabe T, Matsui T, Ootani Y, Iizuka S, Toyoshima R, Kodama H, Kondoh H, Tateyama Y and Einaga Y 2019 *J. Phys. Chem. C* **123** 5336-44
[10] Kurniawan W A, Umar A and Ivandini T A 2018 *2018 3rd International Seminar on Sensors, Instrumentation, Measurement and Metrology* (ISSIMM) available at https://ieeexplore.ieee.org/document/8727741
[11] Tian R H and Zhi J F 2007 *Electrochem. Commun.* **9** 1120-6

IOP Conf. Series: Materials Science and Engineering **902** (2020) 012014 doi:10.1088/1757-899X/902/1/012014

One-pot synthesis of spirooxindole-pyrrolizidine compounds using magnetically separable Fe_3O_4-graphene oxide (Fe_3O_4-GO) catalyst

M Hapsari, A H Cahyana and R T Yunarti

Department of Chemistry, Faculty of Mathematics and Natural Sciences (FMIPA), Universitas Indonesia, Depok 16424, Indonesia

Corresponding author's email: herrykim@ui.ac.id

Abstract. A couple of novel Spirooxindole-pyrrolizidine compounds have been synthesized from chalcone derivatives as α,β-unsaturated carbonyl compounds by the help of heterogeneous magnetically separable Fe_3O_4-GO catalyst. Four steps involved in this research are synthesis of Fe_3O_4-GO catalyst, synthesis of chalcone derivatives from 4-hydroxybenzaldehyde and cinnamaldehyde, synthesis of spirooxindole-pyrrolizidine derivatives by 1,3-dipolar-cycloaddition reaction through the formation of azomethine ylide, and the last is antioxidant activity assay using DPPH method. The result showed that the use of Fe_3O_4-GO catalyst as much as 5 wt% can increase the reaction yield to 86.27 %. From the DPPH assay, it is known that spirooxindole-pyrrolizidine derivatives have an antioxidant activity but spirooxindole-pyrrolizidine derived from 4-hydroxybenzaldehyde is the better one.

Keywords: Spirooxindole-pyrrolizidine, chalcone, Fe_3O_4-GO, 1,3-dipolar-cycloaddition, DPPH

1. Introduction

Spiro is a polycyclic compound which has two or more rings connected by one chiral carbon atom [1]. There are several types of spiro compounds depend on its structure, including spiro-oxindole, spiro-pyrrolidine and spiro-pyrrolizidine, spiro-chromene, spiro-pyrimidine, spiro-xanthene, and spiro-lactone. Among the various types of spiro, spirooxindole attracts the most attention because it promises a good activity [2]. Spirooxindole is the skeleton of many natural alkaloids which possesses biological and pharmacological activities. Besides that, spiropyrrolidine and spiropyrrolizidine also gained much attention because its potential in biological activities such as antimicrobial, antitumor, antiviral, antileukemic, anticonvulsant, and local anesthetic activities [3]. Some further researches also show that the combination of the spirooxindole with pyrrolidine or pyrrolizidine ring has a good activity. Two types of diketopiperazine alkaloids (spirotryprostatin A and B) which were first isolated from Aspergillus fumigatus are examples of spirooxindole-pyrrolidine compounds that have an excellent anticancer activity [4]. They gained much attention for years and finally have successfully synthesized in 1999 [5].

1,3-dipolar cycloaddition reaction is the most widely performed reaction to synthesize spirooxindole-pyrrolidine and pyrrolizidine derivatives. This reaction has been described as the most important method for the formation of five-membered heterocyclic rings in organic chemistry because it has a high level of diastereo-, enantio- and regioselectivity [6].

In recent years, the development of carbon-based heterogeneous catalysts has been carried out because of its environmentally friendly property. Carbon-based catalyst has been known to be used for many organic transformations such as formation of nitriles from primary amines through oxidation reaction, direct C-C bond formation, and synthesis of heterocycles like triazoles and benzimidazoles [7]. The transition metals are slowly replaced by graphene and its functionalized form like graphene oxide (GO) in organic synthesis [8, 9].

From previous research, graphene oxide (GO) played a role in several reactions such as oxidative coupling reaction, Friedel craft alkylation, Diels-Alder reaction, hydration oxidation, and many others [10]. Recently, scientists developed the synthesis of spirooxindole-pyrrolizidine compound using a catalyst GO and aqueous solution medium [10]. This research showed that GO facilitated the 1,3-dipolar cycloaddition reaction through the formation of imine and decarboxylation to become the reactive azomethine ylide which then reacted with dipolarophile compound [10].

The synthesis of spirooxindole-pyrrolizidine using GO catalyst was carried out by reacting isatin, α-amino acids, and alkenes as dipolarophile. By using GO, synthesis of spirooxindole-pyrrolizidine could be done in aqueous medium which was more environmentally friendly. GO was also easy to separate by centrifugation, but it still took a lot of time [10]. Therefore, this study will conduct the synthesis of spirooxindole-pyrrolizidine compounds through a multicomponent reaction between isatin, L-proline, and chalcone derivatives by the help of the Fe_3O_4-GO catalyst. Fe_3O_4 is added just to simplify the separating process of GO from the reaction mixture.

Chalcone derivatives were synthesized first by reacting acetophenone and aromatic aldehyde like 4-hydroxy-benzaldehyde and cinnamaldehyde in alkali medium by Claisen-schmidt condensation reaction. Chalcone is a secondary metabolite compound found in plants. It has three carbons of unsaturated α,β–carbonyl system which connected its two aromatic rings [11, 12]. Chalcone is known to have several bioactivities such as antibacterial, antioxidant, antifungal, anti-inflammatory, and anticancer [13, 14]. Spirooxindole-pyrrolizidine derivatives synthesized by chalcone derivative was characterized and tested for their antioxidant activity analysis.

2. Materials and method

2.1. Materials

The instruments used were Fourier Transform Infrared (FT-IR) IRPrestige-21 Shimadzu, UV-Vis Spectrophotometers UV-2450 Shimadzu, X-Ray Diffractometer (XRD), Energy Dispersive Spectroscopy (EDS), Transmission Electron Microscope (TEM), and LC-MS/MS. Materials used to make Fe_3O_4-Graphene Oxide (GO) were graphite powder, $NaNO_3$, H_2O_2 30 %, $KMnO_4$ powder, concentrated H_2SO_4, $FeCl_3.6H_2O$ and $FeCl_2.4H_2O$. The compound used for the synthesis of chalcone and spiro-oxindole-pyrrolizidine derivatives were acetophenone, 4-hydroxy-benzaldehyde, cinnamaldehyde, NaOH, isatin, L-proline, ethanol, methanol, ethyl acetate, N-hexane, HCl, and distilled water. The material used for the antioxidant test were 1,1-diphenyl-2- picrylhydrazyl (DPPH•).

2.2. Synthesis of graphene oxide (GO)

Graphene Oxide (GO) synthesis was carried out by referring to the Hummers' method that had been carried out before [15]. In the concentrated H_2SO_4 (34.5 mL), an amount of 1.5 g of graphite and 0.75 g of $NaNO_3$ were dissolved with continuous stirring for approximately 20 min, followed by cooling the mixture in an ice bath for 2 h. While maintaining the mixture temperature at below 20 °C, $KMnO_4$ (4.5 g) was added slowly. After that, the reaction mixture was heated and stirred at 35 °C before the distilled water (69 mL) was added to the mixture and stirred for 20 min. Then, the addition of H_2O_2 30 % droplets (1.5 mL) was carried out until the color of the mixture changed from dark brown to yellow. A total 50 mL of distilled water was added again and the mixture was sonicated for 2 h. The solution was centrifuged, separated, and washed several times with distilled water. The product was dried in an oven at 100 °C for one day to get the graphene oxide (GO).

IOP Conf. Series: Materials Science and Engineering **902** (2020) 012014 doi:10.1088/1757-899X/902/1/012014

2.3. Synthesis of Fe_3O_4-GO

Synthesis of Fe_3O_4-GO was carried out by referring to co-precipitation method [16]. An amount of 100 mL of distilled water was poured to 0.1 g of GO followed by sonicating for 1 h. Besides that, an amount of 125 mL of distilled water was poured to 1.1 g $FeCl_3.6H_2O$ and 0.36 g $FeCl_2.4H_2O$ mixture then stirred for 30 min at room temperature. After that, the two mixtures were mixed and stirred at 85 °C for 30 min. This mixture was added by 25 % ammonia solution until pH 10 and the color turned to black as it changed to suspension. After that, the suspension was stirred again for 45 min before it was cooled to room temperature. The solids and liquids were separated by the help of external magnets. Then, this mixture was neutralized using distilled water and it was dried in an oven at 60 °C overnight.

2.4. Synthesis of 4-hydroxychalcone

Synthesis of 4-hydroxychalcone was done by referring to Claisen-Schmidt condensation reaction that previously reported [12, 17]. Into 2.5 g of 4-hydroxy-benzaldehyde (16.43 mmol) in 15 mL ethanol, an amount of 1.98 mL of acetophenone (16.43 mmol) and NaOH 60 % as much as 10 mL were added. The mixture was stirred at 70 °C for 3 hours until the reaction is completed (checked by TLC). After that, ethanol was removed before acidification of the mixture to pH 1 by using 1 M HCl. Anhydrous Na_2SO_4 was added to the organic phase obtained from liquid-liquid extraction (ethyl acetate) to remove the remained water. Furthermore, the solvent was removed and the purification was done by recrystallization. The solids obtained were then analyzed using spectrophotometer UV-Vis, FT-IR and LC-MS/MS.

2.5. Synthesis of 1,5-diphenylpenta-2,4-dienone

Synthesis of 1,5-diphenylpenta-2,4-dienone was done by referring to the procedure that has previously reported [17]. A total of 10 mL ethanol was added to 20 mL NaOH 12.5 % (w/v) with continuous stirring until it cooled down to room temperature. After that, an amount of 15 mmol of acetophenone and 15 mmol of cinnamaldehyde were added to NaOH-ethanol mixture dropwisely with continuous stirring for an hour. The reaction mixture was cooled in an ice bath until it formed a solid. The solution was removed and the solid was washed and recrystallized using ethanol to get the product.

2.6. Synthesis of Spirooxindole-pyrrolizidine derivatives

Synthesis of spirooxindole-pyrrolizidine derivatives was carried out based on the development of the procedure that previously reported [10]. A number of Fe_3O_4-GO catalysts in various mediums were added to a mixture of 1,5-diphenylpenta-2,4-dienone, isatin, and L-proline (1:1:1 mole ratio). The reaction was carried out at 65 °C and monitored using TLC. The catalyst was separated from the external magnetic source, and the solution was evaporated until solids remain. Recrystallization of solids was carried out by washing solids formed using warm ethanol to obtain the derivative of the Spirooxindole-pyrrolizidine-1,5-diphenylpenta-2,4-dienone compounds. Under the same procedure, isatin and L-proline were reacted with 4-hydroxychalcone in 1:1:1 mole ratio to obtain spirooxindole-pyrrolizidine-4-hydroxychalcone.

2.7. Antioxidant activity assay

The antioxidant activity of the organic compounds were calculated by using 1,1-diphenyl-2-picrylhydrazyl (DPPH•) method [18]. Five different concentrations of synthesized products and curcumin as its standard were made by using methanol as the solvent (200, 100, 50, 25 and 12.5 ppm). Then, 2 mL of the test sample was added to 2 mL of DPPH• 0.1 mM solution. The mixture was incubated in the dark place at room temperature for 30 min. All the mixed solution was measured at maximum wavelength 517 nm using a UV-Vis spectrophotometer. The IC50 of all test sample determined by plotting its concentration and % DPPH scavenged in a linear graphic. The equation used to calculate the free radical scavenging activity is:

$$\% \, DPPH \; scavenged = \frac{[A - B]}{A} \times 100 \, \% \tag{1}$$

where A is the absorption of DPPH solution and A is the absorption of the sample solutions.

3. Results and discussion

3.1. Characterization of graphene oxide (GO)

Preparation of Fe_3O_4-GO was done in two step processes. First of all, graphite powder was used to synthesize GO by using Hummer's Method. The synthesized GO was characterized using FT-IR and XRD instruments. There are four peaks appeared at 3578.10 cm^{-1} (O-H), 3101.66 cm^{-1} (C-H sp^2), 1753.36 cm^{-1} (C=O Carboxylate), 1258.60 cm^{-1} (C-O-C Epoxide group) in the FTIR spectrum (figure 1) indicating that the graphite powder was successfully functionalized to GO. These FT-IR data are supported by the XRD spectrum (figure 2) of synthesized GO which exhibits a different sharp peak at Bragg angle 2θ = 10.49° compared to graphite (2θ = 26.56°). The different position of Bragg angle between graphite and GO are indicated the formation of GO. Additionally, the crystallite size of GO was calculated by Debye-Scherrer's equation [19] using XRD data below:

$$D = \frac{k.\lambda}{\beta.Cos\theta} \tag{2}$$

where D = the size of crystallite, λ = 1.5406 nm (wavelength of the X-ray sources), $k = 0.9$ is the Scherrer constant, β = Full Width and Half Maximum/ FWHM (in radian), and θ is the Bragg diffraction angle or peak position (in radian). By this equation, the crystallite size of GO is estimated about 11.21 nm. After GO has been successfully synthesized, the second process is reacting GO with $FeCl_3.6H_2O$ and $FeCl_2.4H_2O$ using co-precipitation method to get the Fe_3O_4-GO composite.

3.2. Characterization of Fe_3O_4-GO

Preparation of Fe_3O_4-GO was carried out by reacting GO with $FeCl_3.6H_2O$ and $FeCl_2.4H_2O$ and the addition of 25 % ammonia solution. The synthesized Fe_3O_4-GO was characterized by FT-IR, XRD, EDS and TEM instruments. In the FTIR spectrum of Fe_3O_4-GO (figure 1), although the peaks are not quite clear, it appears that there are four peaks similar to GO and a peak similar to Fe_3O_4. This spectrum is compared to Fe_3O_4 and GO and it shows there are four peaks appeared at 3610.89 cm^{-1} (O-H), 3138.31 cm^{-1} (C-H sp^2), 1731.18 cm^{-1} (C=O Carboxylate), 1258.60 cm^{-1} (C-O-C Epoxide group) confirming that there is GO in this composite. There is also a peak appeared at 628.82 cm^{-1} (Fe-O bond) confirming that there is Fe_3O_4. The FTIR spectrum of Fe_3O_4-GO shows that there are some intensity reduction such as Fe-O vibration peak of the synthesized Fe_3O_4-GO. The Fe-O vibration peak is not clear and sharp like Fe_3O_4. The peaks indicated GO compound are also not as sharp as like GO. This intensity reduction is assumed because there is an interaction between Fe_3O_4 and GO.

XRD pattern or diffractogram of Fe_3O_4-GO composite is shown in figure 2. This diffraction peaks are shown at Bragg angle 2θ = 30.54°, 35.59°, 43.39°, 57.52° and 63.09° indicated the characteristics of Fe_3O_4 (JCPDS 19–0629) [19]. In this research, Fe_3O_4 was also synthesized for the analysis comparison. The XRD spectrum of the synthesized Fe_3O_4 also showed diffraction peaks at Bragg angle 2θ = 30.17°, 35.51°, 43.09°, 57.14° and 62.57°. This diffraction peaks implied that the Fe_3O_4 core crystal structure was still well-maintained after functionalization. When it examined in more detail, each peak in the XRD spectrum of the synthesized Fe_3O_4-GO has a lower intensity than Fe_3O_4. This further supports the assumption that there is an interaction between Fe_3O_4 and GO.

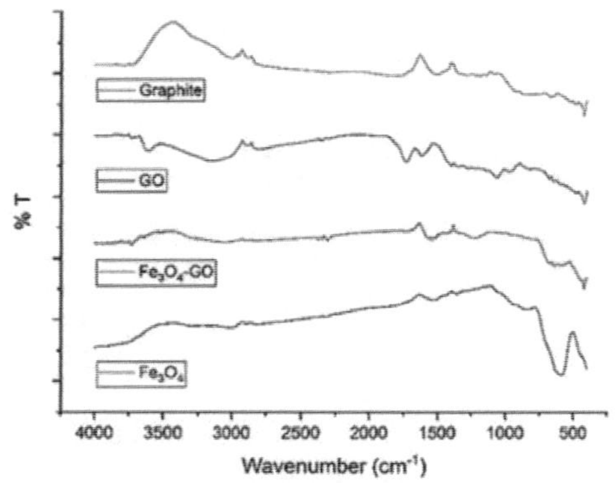

Figure 1. FTIR spectra of graphite, GO, Fe_3O_4 and Fe_3O_4-GO.

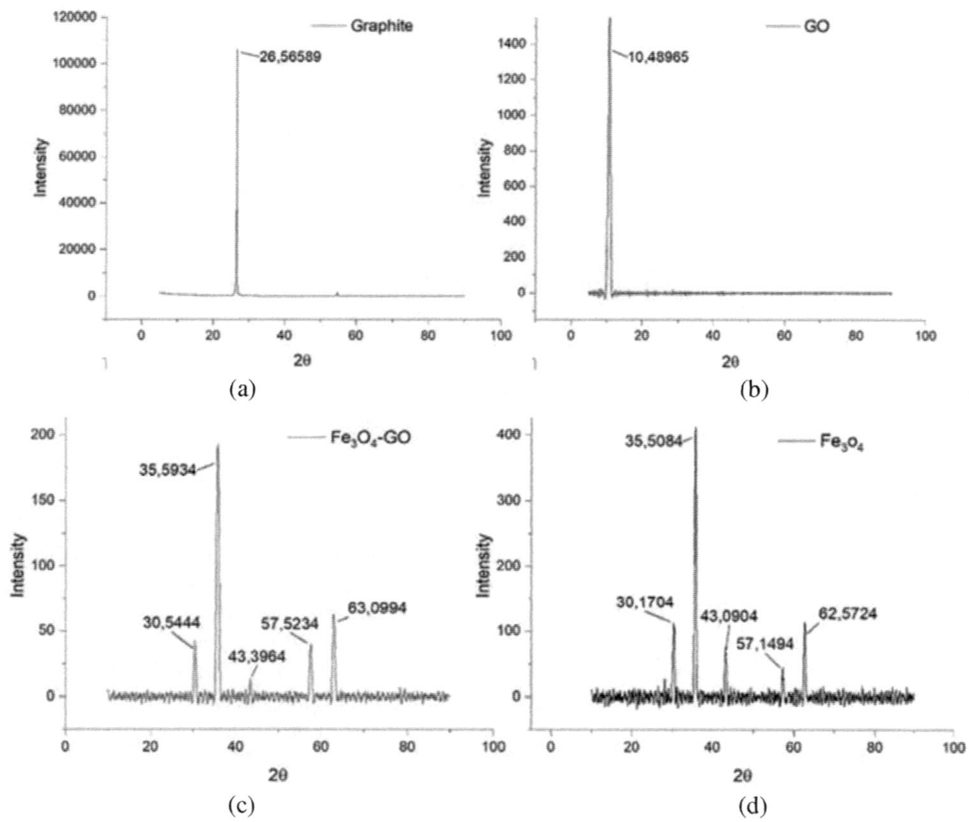

Figure 2. Diffractogram of (a) Graphite, (b) GO, (c) Fe_3O_4 and (d) Fe_3O_4-GO.

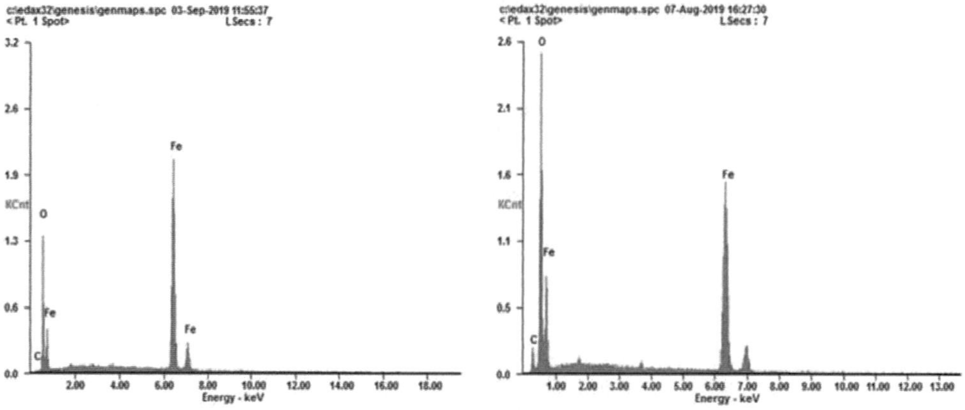

(a) (b)

Figure 3. EDS spectrum of (a) Fe_3O_4 and (b) Fe_3O_4-GO.

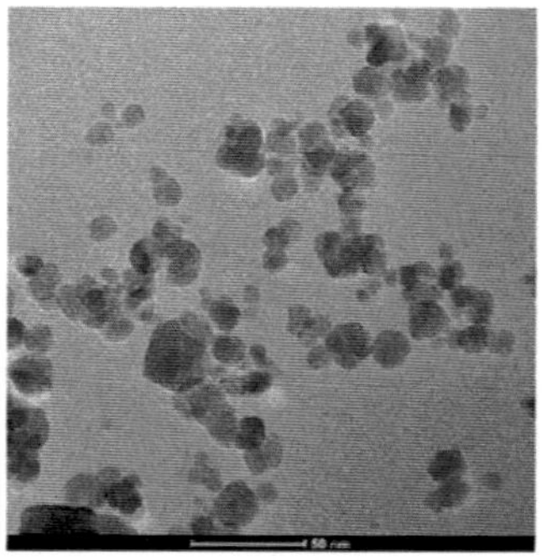

Figure 4. Typically TEM image of Fe_3O_4-GO.

There were no diffraction peaks indicated as GO, due to its low concentration. But the elemental composition of Fe_3O_4-GO confirmed that there is carbon atom in the synthesized Fe_3O_4-GO. The elemental composition is determined by energy dispersive spectrum (EDS) analysis. In the synthesized Fe_3O_4-GO there are 04.49 wt% of C atom, 31.76 wt% of O atom and 63.74 wt% of Fe atom, while in the Fe_3O_4 there are 00.65 wt% of C atom, 18.03 wt% of O atom and 63.74 wt% of Fe atom. This spectrum is shown in figure 3. Typically, TEM image of Fe_3O_4-GO composite in the figure 4 shows that most particles Fe_3O_4-GO have spherical shape and the particle size is 174 nm.

3.3. Application of Fe_3O_4-GO catalyst in the synthesis of spirooxindole-pyrrolizidine derivatives
In this research, synthesis of Spirooxindole-pyrrolizidine derivatives was done by reacting the compounds isatin, L-proline, and two different types of unsaturated α,β-carbonyl compound such as 4-hydroxychalcone and 1,5-diphenylpenta-2,4-dienone. Both of unsaturated α,β-carbonyl compounds were synthesized from the reaction of acetophenone and two different types of aromatic aldehydes like 4-hydroxybenzaldehyde and cinnamaldehyde in an alkaline medium. The reaction of unsaturated α,β-carbonyl compound which represented by 1,5-diphenylpenta-2,4-dienone can be seen in figure 5.

The optimization of reaction condition was done using the model of reaction between isatin, L-proline, and 1,5-diphenylpenta-2,4-dienone. The reaction of spirooxindole-pyrrolizidine derivative

IOP Conf. Series: Materials Science and Engineering **902** (2020) 012014 doi:10.1088/1757-899X/902/1/012014

which is represented by spirooxindole-pyrrolizidine-1,5-diphenylpenta-2,4-dienone can be seen in figure 6. All reaction was done using methanol solvent at 65 °C for approximately 3 h. Quantities of the catalyst used are calculated based on the total weight of all precursors used. The yield of all the reaction are shown in table 1.

From the data in table 1, it can be seen that the use of GO in this reaction has improved the reaction yields compared to reaction with no catalyst. However, the use of Fe_3O_4-GO as much as 5 wt% give the highest result of the desired product. In 3 h of reaction, that condition gives a high yield (86.27 %). The data also showed that increasing amount of catalyst, either Fe_3O_4, GO or Fe_3O_4-GO will decrease the reaction yields. This phenomenon maybe due to the reaction equilibrium between the reactant and the catalyst. Because of this, the excess catalyst can plug the active site of other catalyst so it can be less active. The addition of Fe_3O_4 can also improve the yields, but when compared to GO, there is a confusion that still need to be further investigated. This is because when the GO yields are used as a reference, on the use of 5 wt% catalyst, the addition of Fe_3O_4 to GO can increase the reaction yields, but on the use of 10 wt% catalyst, the addition of Fe_3O_4 to GO actually reduces the reaction yields.

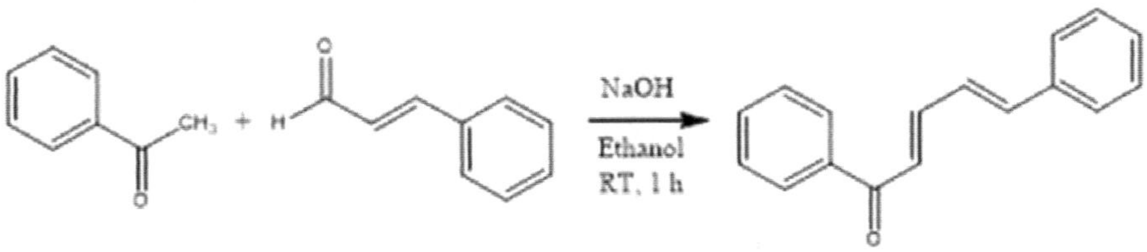

Figure 5. Reaction of 1,5-diphenylpenta-2,4-dienone.

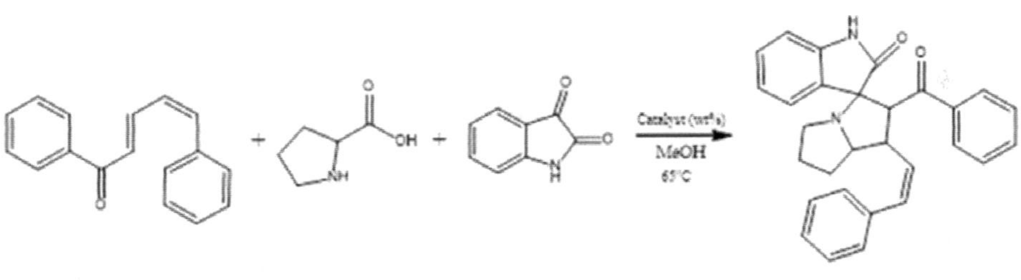

Figure 6. Reaction of Spirooxindole-pyrrolizidine-1,5-diphenylpenta-2,4-dienone.

Table 1. Optimization of the reaction condition of synthesis spirooxindole-pyrrolizidine derivatives.

Catalyst	Quantity (wt%)	Yield (%)*
Fe_3O_4-GO	10	68.51
GO	10	75.87
Fe_3O_4	10	42.85
Fe_3O_4-GO	5	86.27
GO	5	78.00
Fe_3O_4	5	72.33
No Catalyst	-	40.36

**The yield was obtained from the reaction of 1,5-diphenylpenta-2,4-dienone*

GO was also taking part in the optimization because according to the previous report [10], functional group of GO such as -OH and -COOH plays an important role in this reaction. GO facilitated the reaction through hydrogen bonding and π-stacking between the catalyst and the substrates [10]. Furthermore, the addition of Fe_3O_4 was aimed to facilitate its separation from the reaction mixture due to its magnetic properties. It is also known that GO is acidic in nature, so it is facilitated the formation of reactive azomethine ylide. This reactive azomethine ylide leads to the formation of 5 membered heterocyclic compound and the reaction is well known as 1,3-dipolar cycloaddition or Huisgen Cycloaddition. The proposed mechanism of Fe_3O_4-GO catalyst is shown in figure 7.

In this research, characterization of Fe_3O_4-GO and GO after application were done by FTIR. From the FTIR spectrum of Fe_3O_4-GO catalyst after application, it can be seen that the vibration peak of the Fe-O bond has increased in intensity. Meanwhile, from the GO catalyst spectrum after application, it can be seen the overall vibration peaks have decreased in intensity. These results prove that the one that plays a role in the reaction is GO. The peak of Fe-O vibrations has increased due to the decrease in GO. FTIR Spectra of GO and Fe_3O_4-GO after application can be seen in figure 8.

3.4. Antioxidant activity assay.

After all the synthesized product are well characterized (showed in table 2), their antioxidant activity was determined. The free radical scavenging activity which is represented by their IC_{50} value are shown in figure 9. From the result, spirooxindole-pyrrolizidine-4-hydroxychalcone possess the best free radical scavenging activity (IC_{50} 1053.65 ppm) compared to spirooxindole-pyrrolizidine-1,5-diphenylpenta-2,4-dienone and 1,5-diphenylpenta-2,4-dienone as its precursor. Meanwhile, 1,5-diphenylpenta-2,4-dienone possesses the lowest free radical scavenging activity (IC_{50} 2738,6 ppm) compared to the others. This is because 4-hydroxychalcone has a hydroxyl group (-OH) while the others don't have any hydroxyl group. It is known that the antioxidant activity of a compound is related to their hydroxyl groups and conjugation of its double bond or effects of resonance. In this mechanism, a hydrogen atom removed the free radical so the compound becomes a radical [19].

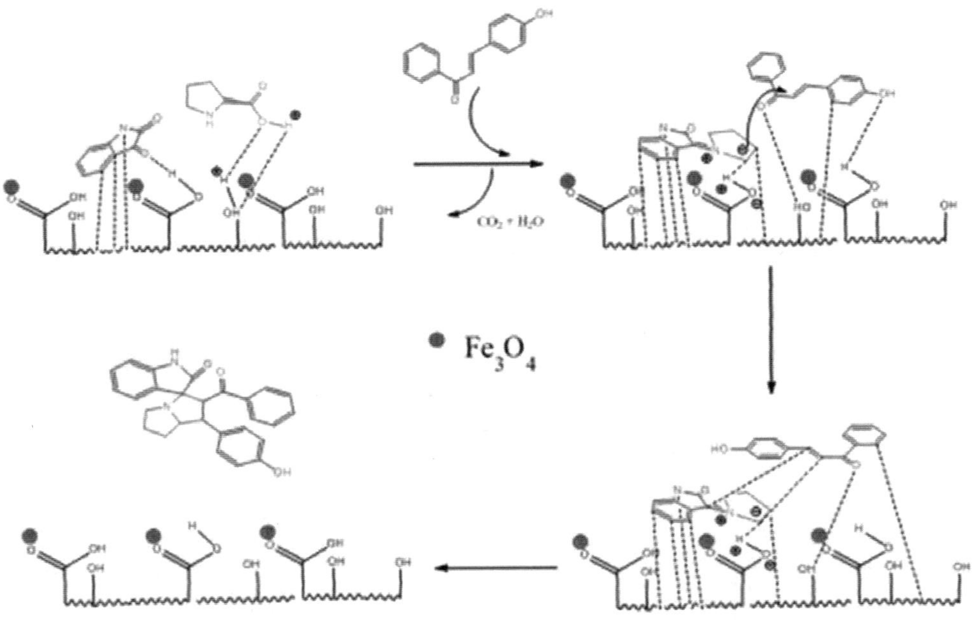

Figure 7. Plausible Mechanism of Fe_3O_4-GO Catalyst.

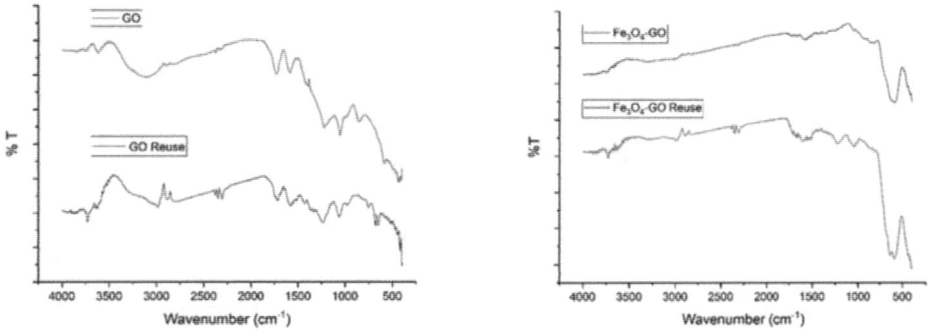

(a) (b)

Figure 8. FTIR Spectra of (a) GO and (b) Fe_3O_4-GO after application.

Table 2. Spectral data of the synthesized organic compound.

No.	Compound	Spectral data	Yield (%)
1.	Spirooxindole-pyrrolizidine-4-hydroxychalcone 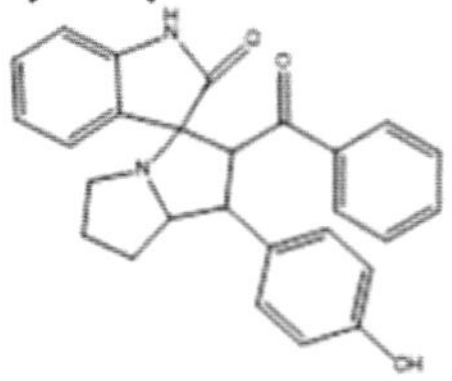	LC-MS/MS $[M+H]^+$ 424.49 FTIR 3517.34 cm^{-1} (N-H amide), 3295.52 cm^{-1} (O-H), 3107.45 cm^{-1} (C-H sp^2), 2980.14 cm^{-1} (C-H sp^3), 1732.15 cm^{-1} (C=O), 1620.27 cm^{-1} (C=C aromatic), 1474.64 cm^{-1} (-CH_2- methylene), 1341.54 cm^{-1} (-CH methine), 762.87 cm^{-1} (ortho substituted benzene).	79.23
2.	Spirooxindole-pyrrolizidine-1,5-diphenylpenta-2,4-dienone 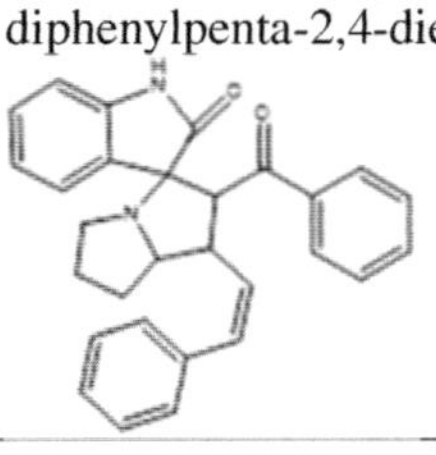	LC-MS/MS $[M+H]^+$ 434. 52 FTIR 3267.55 cm^{-1} (N-H amide), 3089.13 cm^{-1} (C-H sp^2), 2969.53 cm^{-1} (C-H sp^3), 1727.32 cm^{-1} (C=O), 1623.16 cm^{-1} (C=C aromatic), 1468.85 cm^{-1} (-CH_2- methylene), 1341.54 cm^{-1} (-CH methine), 748.41 cm^{-1} (monosubstituted benzene).	86.27

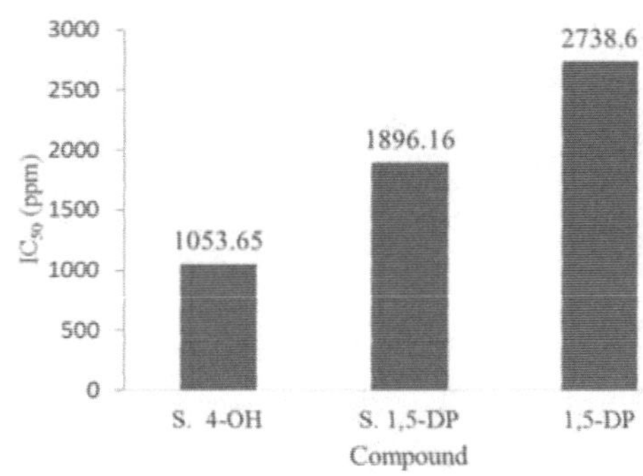

Figure 9. Free radical scavenging activity of spirooxindole-pyrrolizidine derivatives.

4. Conclusion

Synthesis of Spirooxindole-pyrrolizidine derivatives using magnetically separable graphene oxide (Fe_3O_4-GO) catalyst have been reported. The experiment showed that the use of catalyst in 5 wt% can increase the reaction yield (86.27 %). The result of DPPH assay showed that spirooxindole-pyrrolizidine derivatives have an antioxidant activity but spirooxindole-pyrrolizidine derived from 4-hydroxybenzaldehyde is the better one.

References

[1] Borad M A, Bhoi M N, Prajapati N P and Pate H D 2013 *Int. J. Rapid Comm. Synth. Org. Chem.* **44** 897-922
[2] Chen C et al. 2017 *Molecules* **22** 1295
[3] Chande M S et al. 2005 *Eur. J. Med. Chem.* **40** 1143-8
[4] Cui C-B, Kakeya H and Osada H 1996 *Tetrahedron* **52** 12651-66
[5] Edmondson S, Danishefsky S J, Sepp-Lorenzino L and Rosen N 1999 *J. Am. Chem. Soc.* **121** 2147-55
[6] Coldham I and Hufton R 2004 *Chem. Rev.* **105** 2765-2809
[7] Girish Y R, Pandit S, Pandit S and De M 2017 *Chem. Asian J.* **12** 2393-8
[8] Mohammadi O, Golestanzadeh M and Abdouss M 2017 *New J. Chem.* **41** 11471-97
[9] Navalon S, Dhakshinamoorthy A, Alvaro M, Antonietti M and Garcia H 2017 *Chem. Soc. Rev.* **46** 4501-29
[10] Reddy M S, Kumar N S and Chowhan L R 2018 *RSC Adv.* **8** 35587-93
[11] Tran T-D et al. 2012 *Bioorg. Med. Chem. Lett.* **22** 4555-60
[12] Hapsari M, Windarti T, Purbowatiningrum, Ngadiwiyana and Ismiyarto 2018 *IOP Conf. Ser. Mater. Sci. Eng.* **349** 012036
[13] Kumar S and Pandey A K 2013 *Sci. World J.* **2013** 162750
[14] Choudhary A N and Juyal V 2011 *Int. J. Pharm. Pharm. Sci.* **3** 125-8
[15] Marcano D C 2010 *ACS Nano* **4** 4806-14
[16] Yang X, Zhou T, Ren B, Shi Z and Hursthouse A 2017 *J. Anal. Methods Chem.* **2017** 3012364
[17] Anjani O O, Ituen R I and Falomo A 2011 *Pak. J. Sci. Ind. Res. Ser. A. Phys. Sci.* **54** 59-67
[18] Borra S K et al. 2013 *J. Med. Plants Res.* **7** 2680-90
[19] Jaleh B, Khalilipour A, Habibi S, Niyaifar M and Nasrollahzadeh M 2017 *J. Mater. Sci. Mater. Electron.* **28** 4974-83

IOP Conf. Series: Materials Science and Engineering **902** (2020) 012015 doi:10.1088/1757-899X/902/1/012015

Synthesis and characterization of monodisperse core-shell Fe_3O_4@Au nanostructures

T Yuliani[1], M Khalil[1], J W Hong[2] and T A Ivandini[1]

[1]Department of Chemistry, Faculty Mathematics and Natural Sciences (FMIPA), Universitas Indonesia, Depok 16424, Indonesia
[2]Department of Chemistry, University of Ulsan, 12 Tekeunsaneop-ro 55beon-gil, Nam-gu, Ulsan, South Korea

Corresponding author's email: ivandini.tri@sci.ui.ac.id

Abstract. This work reports a synthesis of monodisperse core-shell Fe_3O_4@Au nanostructures. Fe_3O_4 (~4.9 nm) and core-shell Fe_3O_4@Au (5.0–5.6 nm) nanostructures were successfully synthesized via thermal decomposition method of iron compounds in the presence of oxygen-containing ligands. XRD analysis revealed the pure face-centered cubic (FCC) structure of Fe_3O_4 with 2θ peaks at 31°, 38°, 44° and 64°, 77° and 81° attributed to Fe-Au, which can be indexed to lattice planes of gold in a cubic phase. In addition, UV-Vis spectra showed a significant difference absorbance peak between Fe_3O_4NP and Fe_3O_4@AuNP.

Keywords: Fe_3O_4, Fe_3O_4@Au, core-shell nanostructures, thermal decomposition

1. Introduction

It is widely reported that nanoparticles can take a significant part in the improvement of sensor performance, due to the large specific surface area, outstanding conductivity and compatibility. Currently, various types of nanoparticles have been extensively used to fabricate a wide variety of electrochemical sensors [1]. In all sorts of nanoparticles, superparamagnetic iron oxide nanoparticles, especially Fe_3O_4 nanoparticles (Fe_3O_4 NP) have gained much attention because of their particular magnetic properties and their ability to be easily chemically altered to increase biocompatibility and dispersibility [2]. Furthermore, the synthesis of noble metal nanostructures has attracted many attentions because of its unique chemical properties, physical properties, and interesting catalytic properties [3]. In addition, bimetallic nanostructures also exhibit greatly increased optical, catalytic and electrical properties of each monometallic constituent [4]. It was reported that gold-coated Fe_3O_4NP (Au-Fe_3O_4NP) is more active than the single component gold nanoparticles (AuNP) or Fe_3O_4 NP [5]. The increased activity comes from the partial charge transfer between Au and Fe_3O_4 at the nanoscale particle interface. Furthermore, with Au coating, the Fe_3O_4NP can be more efficiently stabilized in corrosive biological conditions and immediately functionalized through the well-developed Au-S chemistry; the coating also renders the Fe_3O_4NP with plasmonic properties [6].

In this study, Fe_3O_4@AuNP with high mono dispersity was synthesized. The synthesis started from Fe_3O_4NP, synthesized by high thermal decomposition methods of iron compounds in the presence of oxygen-containing ligands. The use of acetylacetonates [7] and acetates [8] were normally reported as the ligands. However, oleic acid was used in this method. Together with oleic acid, oleyl amine was also used as surfactants to react with the metallic organic precursor to form an intermediate Fe^{3+} complex, which further decomposed to form Fe_3O_4NP. Then, the coating of Au on the surface of Fe_3O_4NP was conducted by reducing gold ions in chloroform solution containing of oleyl amine at room temperature. The results showed that monodisperse Fe_3O_4@AuNP could be synthesized with an average size of 5.0 nm. This nanoparticle is ready for many applications, such as sensors or drug delivery.

2. Materials and method

2.1. Materials and instruments

Iron-oleate complex, synthesized from Iron (III) hexahydrate ($FeCl_3.6H_2O$, 98 %) and oleic acid ($C_8H_{32}O_2$, OLA, 90). These chemicals as well as Gold (III) chloride trihydrate ($HAuCl_4.3H_2O$, 99) and oleyl amine ($C_{18}H_{37}N$, OAm, 70 %) were supplied by Sigma-Aldrich, while sodium oleate ($C_{13}H_{33}NaO_2$, 97 %) was from Tokyo Chemical Industry, Ltd), All chemicals were reagent grades. The distilled water with a resistivity greater than 18.2 MΩcm was used.

2.2. Synthesis of Fe_3O_4@AuNP

Fe_3O_4NP was synthesized according to thermal decomposition method reported by Kim et al. with some modifications using an Fe-oleate complex as a precursor [9]. Fe-oleate complex (3.0 g) and oleic acid (0.593 mL) were dissolved in 1-octadecene (21.32 mL) in a three-neck bottom round reaction flask under stirring conditions. To initiate the reaction, the temperature was raised to 60 °C for 15 min, then to 320 °C with a speed of 3.3 °C min^{-1} under atmospheric nitrogen. After cooling to the room temperature, the product was washed to produce dark-brown particles (Fe_3O_4NP), which can be drawn by an external magnet. Fe_3O_4@AuNP was synthesized based on the procedure carried out by Xu et al., with some modifications [6]. Fe_3O_4NP acted as the core, while Au was coated as the shell at the surface of Fe_3O_4NP by reducing gold ions in chloroform contained in oleyl amine.

Briefly, chloroform (10 mL) containing of 2 mg Fe_3O_4NP and oleyl amine (2 mmol) were mixed with a dropwise addition of 2.5 mL 10 mM gold solution. The mixture was stirred at 500 rpm to make homogeneous and monodispersed nanoparticles. After 20 h, two layers of brown (bottom) and murky white (top) were formed, then ethanol (10 mL) was added. After centrifuged at 10,000 rpm at 10 °C for 30 min, two separated phases, dark brown (bottom) and clear (top) were formed with precipitate (Fe_3O_4@AuNP). The precipitate was then washed and was dried under vacuum overnight. To make water-soluble Fe_3O_4@AuNP, dried Fe_3O_4@AuNP was dissolved in aqueous solution consists of 0.1 M CTAB and 0.1 mM sodium citrate.

2.3. Characterizations

Transmission Electron Microscope (TEM) images were recorded by using JEOL JEM-2100F. The Inductively Coupled Plasma-Optical Emission Spectrometry (ICP-OES) measurements were performed with a Spectro blue-ICP-OES (Ametek). The X-Ray Diffraction (XRD) measurements were conducted in Rigaku D/MAX2500V/PC scanning 2θ at 30 to 90 degree, while the Scanning Electron Microscope (SEM) was Zeiss EVO M10.

3. Results and discussion

3.1. Synthesis of Fe_3O_4NP

The synthesis of Fe_3O_4 NO was conducted through two steps, including the iron oleate precursor and iron oxide nanoparticles synthesis (figure 1). In order to synthesize the nanoparticles with high-uniform size, complete separation of nucleation, and monodisperse, the iron-oleate complex was produced from the reaction between iron (III) chloride and oleic acid. The use of the iron-oleate complex as the precursor can control the amount of iron in the synthesis to form the monodisperse nanocrystals [10].

TEM image of the synthesized Fe_3O_4NP (figure 2) shows that the average size of Fe_3O_4NP was 4.9 ± 0.05 nm, while XRD pattern (figure 3a) shows a broadening peak, indicates the particle size was very small. The XRD profile of as-synthesized particles show a poor crystallinity in diffractogram; peaks are hardly visible above the noise as it may contain nanoparticles of various oxygen contents (multiple phases) [11]. However, the peaks related a spinel structure was obviously observed with a characteristic peaks for Fe_3O_4 (2θ) at 30°, 35°, 43°, 53°, 57° and 62° [12]. These peaks indicate the pure face-centered cubic structures of Fe_3O_4 [13]. Figure 3b shows the FTIR spectra of Fe_3O_4NP with oleic acid as the capping agent. A strong absorption of Fe-O stretching bond of bulk magnetite at a wavelength of 560 cm^{-1} was observed [12], while the peaks at the wavelengths of 1400 cm^{-1}, 2850 cm^{-1} and 3400 cm^{-1} could be associated with the stretching modes of C=O, C-H and O-H bonds, respectively. The results indicated that the Fe_3O_4 was surrounded by the oleate with oxygen sites interact with iron.

3.2. Synthesis Fe_3O_4@AuNP

ICP-OES data was conducted to confirm the ratio of Fe and Au. Fe:Au mol ratio of 30:1 was observed. TEM image (Figure 4) observed for the two different Au (shell) and Fe_3O_4 (core) ratios showed a narrow particle size distribution (standard deviation below 10 %), indicated that the monodisperse particles were successfully synthesized. TEM images also shows that the average size of Fe_3O_4@Au nanostructures was around 5.0 nm.

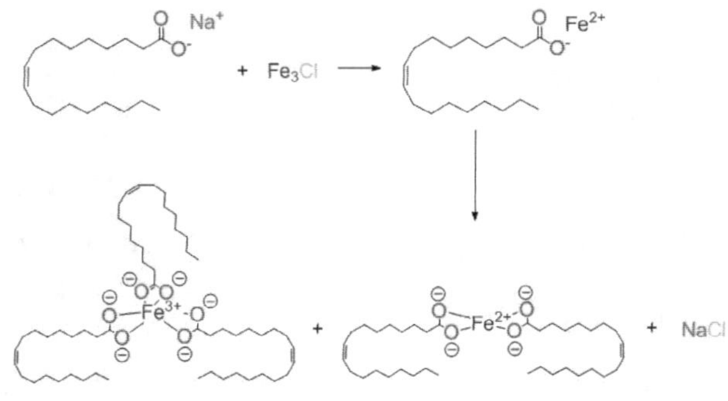

Figure 1. Formation of Fe-oleate complex from sodium oleate [10].

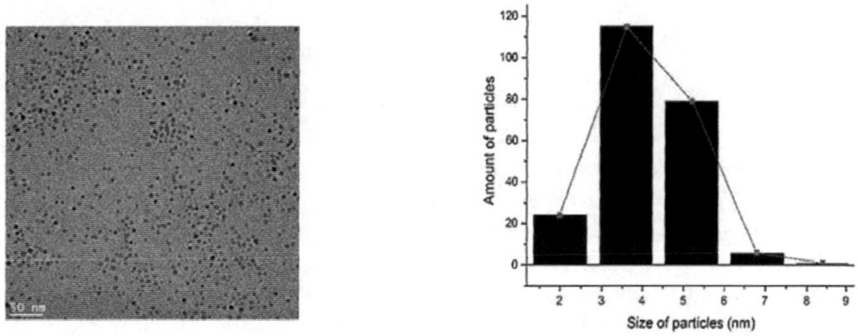

(a) (b)

Figure 2. (a) TEM Image of Fe_3O_4 nanoparticles and (b) its size distribution curve.

The XRD patterns in figure 5a show the Fe_3O_4NP before and after being coated with gold. The diffraction pattern of Fe_3O_4NP was too weak to be detected due to the heavy atom effect of Au. However, the peaks at (2θ) of 31°, 38°, 44° and 64°, 77° and 81° were attributable to Fe-Au, and could be indexed to [111], [200], [311] and [222] lattice planes of gold in a cubic phase, respectively [14].

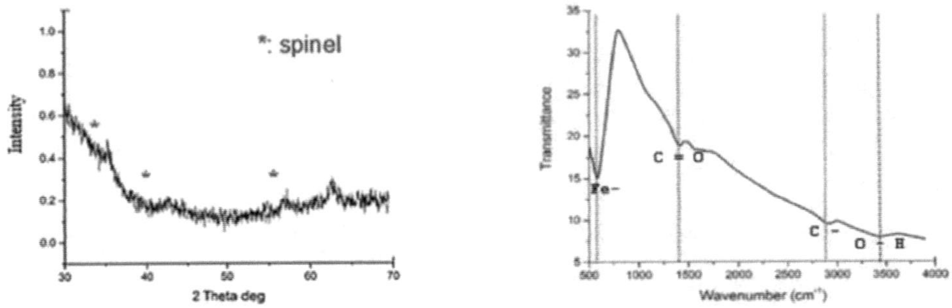

(a) (b)

Figure 3. (a) XRD pattern and (b) FTIR spectra of Fe_3O_4 nanoparticles.

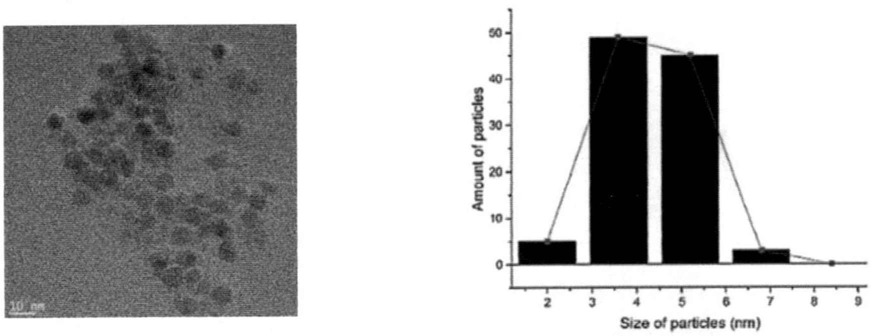

(a) (b)

Figure 4. (a) TEM images of Fe3O4@AuNP and (b) its size distribution curve.

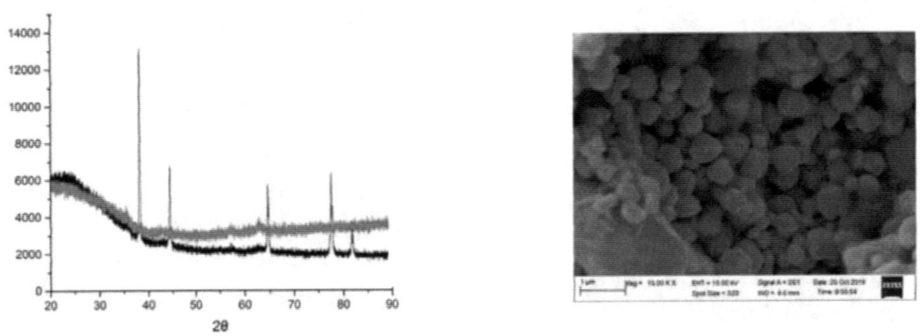

(a) (b)

Figure 5. (a) XRD pattern of Fe_3O_4NP and Fe_3O_4@AuNP, (b) SEM images oleyl amine-capped Fe_3O_4@AuNP and (c) UV-Vis spectra of Fe_3O_4NP and Fe_3O_4@AuNP.

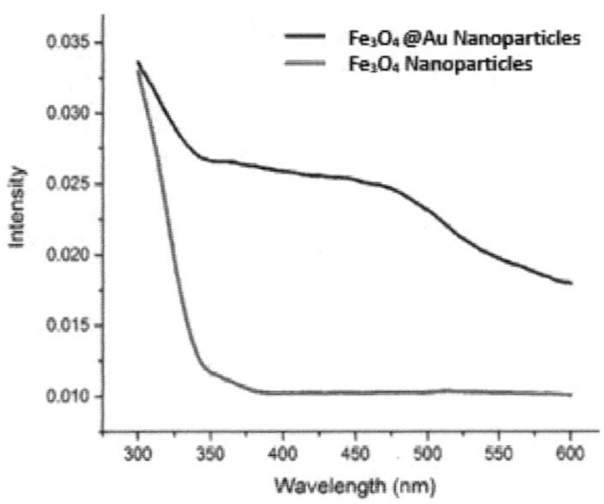

(c)

Figure 5 (continued). (a) XRD pattern of Fe_3O_4NP and Fe_3O_4@AuNP, (b) SEM images oleyl amine-capped Fe_3O_4@AuNP and (c) UV-Vis spectra of Fe_3O_4NP and Fe_3O_4@AuNP.

Figure 5b shows SEM images of Fe_3O_4@Au nanostructures. The morphology of oleyl amine-capped Fe_3O_4@Au nanostructures described spherical shapes as also shown by the TEM characterization in figure 4. Characterization using UV-Visible spectra (figure 5c) shows no absorbance peaks observed in visible region for Fe_3O_4NP, but Fe_3O_4@AuNP displays a broadening peak around 450–500 nm. A blue-shift in comparison to AuNP (520 nm) was observed due to the characteristic of Au. These results indicated that Au shell on the surface of Fe_3O_4NP was successfully formed.

4. Conclusion

Fe_3O_4NP was successfully synthesized by thermal decomposition method with Fe-oleate as the precursor. The nanoparticles have a spherical shape with a particle size of 4.9 nm. Fe_3O_4NP has been successfully coated with Au with Fe:Au ratio of 30:1. The Au coating increases the size of nanoparticles to be around 0.1 nm.

Acknowledgments
This work was financially supported by Universitas Indonesia under research grant PIT 9 with grant contract number NKB-0028/UN2.R3.1/HKP.05.00/2019.

References

[1] Ivandini T A, Naono Y, Nakajima A and Einaga Y 2005 *Chem. Lett.* **34** 1086-7
[2] Li D, Xu Y, Zhang L and Tong H 2014 *Int. J. Electrochem. Sci.* **9** 7217-27
[3] Yu K, Kelly K L, Sakai N and Tatsuma T 2008 *Langmuir* **24** 5849-54
[4] Wilson A R et al. 2013 *J. Phys. Chem. C* **117** 17557-66
[5] Luhur M S P, Ivandini T A and Khalil M 2013 *AIP Conf. Proc.* **2023** 020092
[6] Xu Z, Hou Y and Sun S 2007 *J. Am. Chem. Soc.* **129** 8698-9
[7] Li Z, Chen H, Bao H and Gao M 2004 *Chem. Mater.* **16** 1391-3
[8] Redl F X et al. 2004 *J. Am. Chem. Soc.* **126** 14583-99
[9] Park J et al. 2004 *Nat. Mater.* **3** 891-5
[10] Armijo L M et al. 2012 *Nanomaterials* **2** 134-46
[11] Hufschmid R et al. 2015 *Nanoscale* **7** 11142-54

[12] Bordbar A K et al. 2014 *Biotechnol. Res. Int.* **2014** 705068
[13] Ahmadi M K and Vossoughi M 2013 *J. Nanotech.* **2013** 830151
[14] Salehizadeh H, Hekmatian E, Sadeghi M and Kennedy K 2012 *J. Nanobiotechnol.* **10** 1-7

Phase system of hexane-water for SnO_2 nanoparticles preparation using *Cassia alata* leaf extract and its photocatalytic activity

F L Hakim, Y Yulizar and D O B Apriandanu

Department of Chemistry, Faculty of Mathematics and Natural Sciences (FMIPA), Universitas Indonesia, Depok 16424, Indonesia

Corresponding author's email: yokiy@sci.ui.ac.id

Abstract. SnO_2 nanoparticles (NPs) were prepared using a hexane phase of *Cassia alata* leaf extract (CLE) and a water phase of $SnCl_2$ precursor by high-speed stirring (HSS) method. In this study, HSS was utilized to analyze the nanoparticle formation via phase system of hexane-water. Next, the synthesized SnO_2 NPs were characterized by Fourier transform infrared (FT-IR) spectroscopy, particle size analyzer (PSA), transmission electron microscopy (TEM), scanning electron microscopy energy-dispersive X-ray spectroscopy (SEM-EDX), UV-Vis diffuse reflectance spectroscopy (DRS) and X-ray diffraction (XRD). Phytochemical screening confirms that CLE contains secondary metabolites such as alkaloids as a source of base, saponins and tannin as a capping agent in SnO_2 NPs formation. TEM characterization shows that the particle size of SnO_2 NPs was approximately 25–50 nm. The photocatalytic activity of SnO_2 NPs was tested for methylene blue (MB) degradation under UV light irradiation. As a result, the degradation percentage of MB was 70.29 % for 160 min.

Keywords: SnO_2 NPs, *Cassia alata* leaf extract, high-speed stirring, photocatalytic activity, methylene blue.

1. Introduction

SnO_2 nanoparticles (NPs) are stable semiconductor materials that have been used in various fields of application such as gas sensors [1], catalysts [2], lithium batteries [3]. SnO_2 NPs are n-type tetragonal crystal structure with good photocatalyst materials which have bandgap value around 3.6 eV [4].

Synthesis of nanoparticles has been widely carried out by the method of hydrothermal [5], electrospinning [1] and sol-gel [2]. The sol-gel method is one technique used for nanoparticle synthesis, which has several advantages such as homogeneity, high purity, low processing temperature and the possibility of making new ceramic compositions [6]. Synthesis methods using toxic chemicals are harmful to the environment. Therefore, one of the environmentally friendly methods is green synthesis. Green synthesis is a design to reduce or eliminate the use and formation of hazardous substances that are beneficial to the environment and economy [7]. This method also has advantages such as lower cost, lower energy, temperature, and pressure requirement compared to the physical and chemical methods.

Two-phase system was used to synthesize nanoparticles for controlling the particle size [8]. Two-phase system consisting of water and organic solvents occur a reaction at the interfacial area, and has been reported for metal oxide nanoparticles [9]. In this system, the reaction utilized a mild condition, nucleation and growth occurred together [10]. The two-phase system utilized the high-speed stirring (HSS) method that can detect the interfacial complexation of two solutions [11].

Plant parts such as leaves, roots, seeds, and stems have been used for the synthesis of metal nanoparticles and surfactants [12]. Green synthesis uses the sol-gel method by replacing chemicals into environmentally friendly materials. Green synthesis of nanoparticles using leaf extracts such as *Piper betle* L. [13], *Psidium guajava* [14], *Datura metel L* [15] and *Tinospora crispa* [16] have been reported. The leaf extracts such as *Morinda citrifolia* [17] and *Orthosiphon aristatus (Blume)* [18] plants have been used for the synthesis of surfactants.

Indonesian plants can be used as raw material in SnO_2 NPs synthesis. One of them is *Cassia alata,* a shrub plant known as another *tabankun*, *acon-acon*, and *ketepeng cina*. This plant is widely spread in tropical countries, stretching from tropical America to India, Malaysia and Indonesia [19]. *Cassia alata* leaves have secondary metabolite contents of alkaloids 4.06 %, saponins 4.28 % and tannins 1.54 % [20]. In the present study, SnO_2 NPs were synthesized with *Cassia alata* leaves. Secondary metabolites in *Cassia alata* leaves such as alkaloids, saponins, and tannins can act as a source of bases (OH^-) and capping agent. The SnO_2 NPs were tested for photocatalytic activity of methylene blue degradation under UV light.

2. Materials and method

2.1. Materials

The main raw material was $SnCl_2.2H_2O$ purchased from Merck. The biomaterial of *Cassia alata* plants were obtained from the conservation unit of the tropical biopharmaca center, LPPM IPB.

2.2. Preparation of Cassia alata leaves extract

Two kg of *Cassia alata* leaves were washed with distilled water. They were dried for a week at room temperature by placing them indoor. The dried leaves of *Cassia alata* were milled to obtain a fine powder. Fifty grams of *Cassia alata* leaf powder was macerated in 250 mL of n-hexane for 7–10 days. Then, the filtrated crude extract in hexane was partitioned using water. The hexane fraction was used for nanoparticle synthesis. The phytochemical test was used to determine the secondary metabolite compounds in leaf extract of *Cassia alata* (CLE).

2.3. Synthesis of SnO_2 nanoparticles

SnO_2 NPs were synthesized by sol-gel technique using $SnCl_2$ precursor 0.0025 M and CLE 2.15 % (w/v). High-speed stirring method with rotation rate 23000 rpm was treated for mixing of sol $Sn(OH)_2$ formation. The mixture was heated at 160 °C for 1.5 h to obtain the reaction equilibrium of $Sn(OH)_2$ and $Sn(OH)_4$ gels formation. The gels were dried, and calcined for 2 h at 500 °C to form SnO_2 NPs powder.

2.4. Characterization of SnO_2 NPs

The crystallinity of SnO_2 NPs was identified using X-ray diffraction (XRD) Panalytical X'Pert PRO. The morphology and composition of elements from SnO_2 NPs were carried out by scanning electron microscope-energy dispersive X-ray spectroscopy (SEM-EDS) JEOL JSM 6510 LA instrument. The bandgap value of SnO_2 NPs was determined by UV-Vis diffuse reflectance spectroscopy (Shimadzu 2450). Transmission electron microscopy (JOEL JEM-1400) image and particle size analyzer (PSA Zetasizer Nano ZS 90) were used to know the particle size and its distribution of SnO_2 NPs. Fourier transform infrared (FT-IR) spectroscopy (IR Prestige-21 Shimadzu) was applied to analyze the functional groups of SnO_2 NPs in the range of 400–4000 cm^{-1}.

2.5. Photocatalytic activity

The photocatalytic activity measurement method was used with a slight modification from the previous research [21]. Seven mg of SnO_2 NPs were dispersed in 25 mL of MB $4.0x10^{-6}$ M and stirred for 160 min under UV light radiation. The degradation result of MB was determined by UV-Vis spectrophotometer every 20 min irradiation.

3. Results and discussion

The secondary metabolites of flavonoids, tannins, phenolics, steroids, saponins, terpenoids, alkaloids were tested by a preliminary phytochemical screening. The n-hexane fraction of *Cassia alata* leaves extract showed a positive result of saponins, alkaloids and steroids. Meanwhile, the aqueous fraction showed negative result of all secondary metabolites.

FTIR characterization was performed to identify the functional groups of SnO_2 NPs and CLE. Figure 1 are attributed to the vibrations of (2916 cm^{-1} ν CH sp^3 stretching), (1738 cm^{-1} ν -NH amine group), (1451 cm^{-1} ν C-C stretching), (1377 cm^{-1} ν NO stretching), (1248 cm^{-1} ν C-N stretching) and (837 cm^{-1} ν wagging/bending of the -NH group), respectively. The presence of -NH stretching, C-N stretching, and C-H stretching indicate the presence of alkaloids. Besides, the spectrum of SnO_2 shows that the presence of Sn-O-Sn stretch at 626 cm^{-1} has the conformity with the prior work [22].

Figure 2a shows the XRD pattern, phase and average crystalline size of SnO_2 NPs. SnO_2 NPs had peaks with miller indices on 2θ of 26.7° (110), 34.0° (101), 38.0° (200), 51.9° (211), 54.7° (220), 57.9° (002), 61.9° (310), 64.8° (112), 66.0° (301), 71.3° (202) and 78.7° (321) which have the conformity with the previous research [22]. The SnO_2 NPs had a crystalline phase with the average crystalline size of 19.01 nm calculated by Debye-Scherrer formula.

Figure 2b shows the bandgap energy of SnO_2. The bandgap value was obtained by extrapolating the slope of the curve based on the modified Kubelka-Munk function $(F(r)h\nu)^2$ versus *hv*. The estimated bandgap energy of SnO_2 was 3.21 eV which was lower than the previous report of 3.6 eV [4]. It can be concluded, SnO_2 NPs with the bandgap value of 3.21eV can active to observe in UV light radiation.

Figure 3a shows that the morphology of SnO_2 NPs was investigated by SEM with 500 nm scale bar in aggregate form. Figure 3b shows the elemental compositions of SnO_2 NPs with % atomic of Sn, O and C were 11.08, 32.40 and 43.06 %, respectively. The C element was presented due to the remains of leave extract on the preparation process [12].

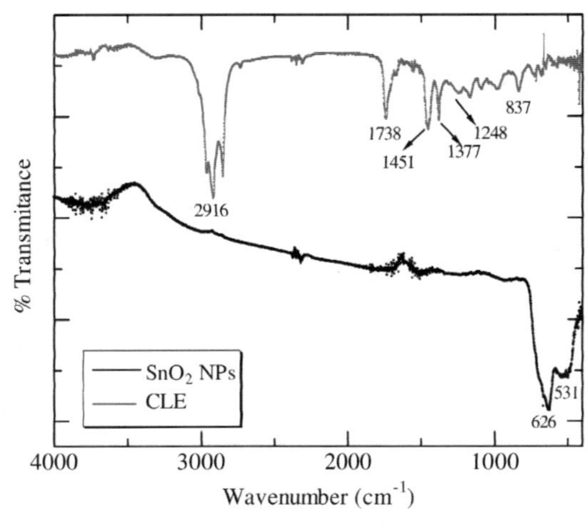

Figure 1. FTIR spectra of CLE and SnO_2 NPs.

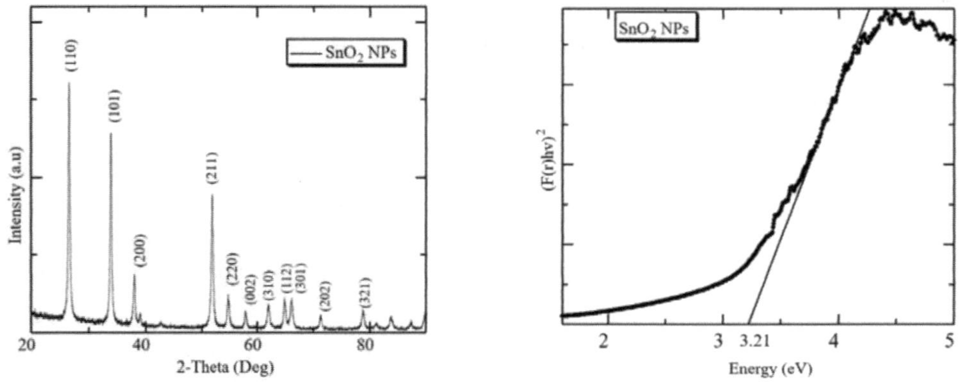

(a) (b)

Figure 2. (a) XRD pattern of SnO_2 NPs, and (b) Band gap value of SnO_2 NPs.

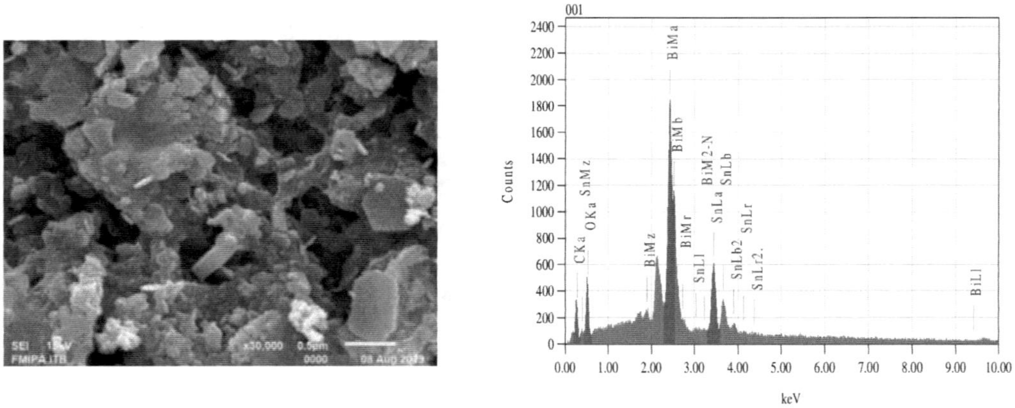

(a) (b)

Figure 3. (a) SEM image of SnO_2 NPs (b) EDS spectra of SnO_2.

TEM characterization was used to assign the shape, morphology, and particle size of SnO_2 NPs. Figure 4a shows the particle size of SnO_2 NPs was within the range of 25–50 nm. The SnO_2 NPs were characterized using PSA to confirm the size distribution and the particle size. According to the result, the average of particle size distribution of SnO_2 NPs was 98.50 nm as shown in figure 4b. The bigger size of PSA result compared the TEM image size due to the SnO_2 NPs was not dispersed perfectly in water.

The photocatalytic activity of SnO_2 NPs was tested for methylene blue (MB) degradation under UV light irradiation. Figure 5a shows UV-Visible spectra for MB photodegradation using SnO_2 NPs. MB displayed a strong absorption band at 664 nm [23]. The MB degradation percentage was 70.29 % for 160 min under UV light as presented in figure 5b.

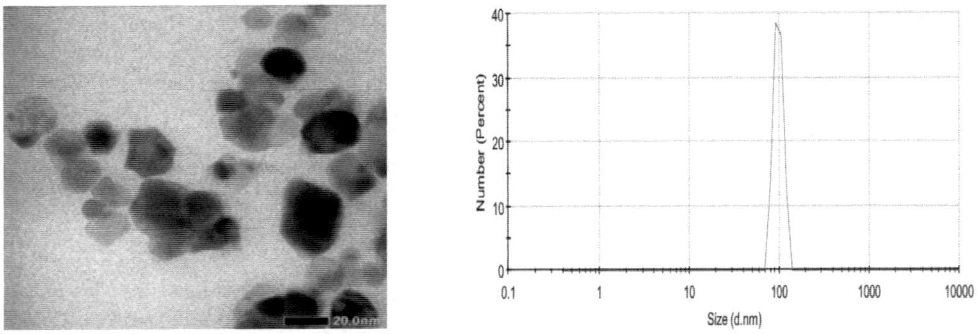

(a) (b)

Figure 4. (a) TEM image of SnO_2 NPs (b) Particle size distribution of SnO_2 NPs.

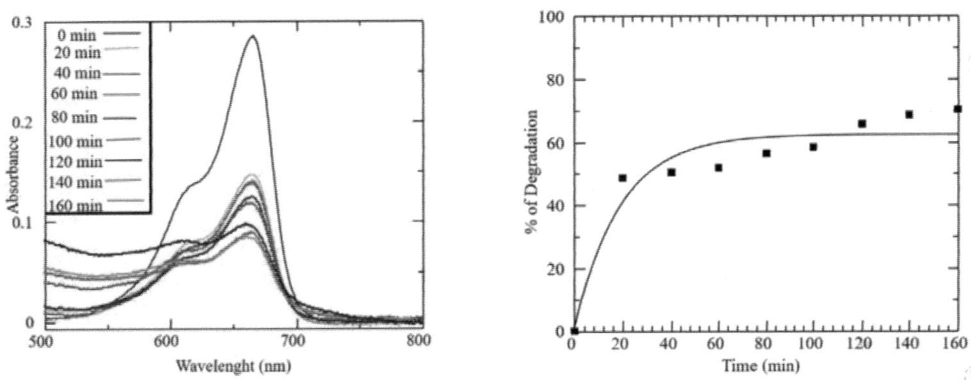

(a) (b)

Figure 5. (a) UV-Vis absorption spectra of MB degradation (b) degradation percentage of MB.

4. Conclusion

SnO_2 NPs were obtained using the hexane fraction of *Cassia alata* leaves extract by the green synthesis method of two-phase system (hexane-water). The CLE in hexane through phytochemical test presented the secondary metabolite compounds such as alkaloids, saponins, and steroids. Based on FTIR characterization, the alkaloid and saponin compounds acted as a weak base source and capping agents in the SnO_2 NP synthesis. The particle size distribution of SnO_2 was confirmed at 98.50 nm in the aggregate form by PSA. TEM images show that the particle size of SnO_2 NPs was in the range of 25–50 nm. The degradation percentage of MB using SnO_2 NPs was 70.29 % for 160 min under UV light irradiation.

Acknowledgments

This research was granted by Hibah PDUPT 2019 from Ministry of Research, Technology and Higher Education, Indonesia, through Universitas Indonesia (No. NKB-1617/UN2.R3.1/HKP.05.00/2019).

References

[1] Li R, Chen S, Lou Z, Li L, Huang T, Song Y, Chen D and Shen G 2017 *Sens Actuators B Chem.* **252** 79-85
[2] Liu A, Zhu M and Dai B 2019 *Appl. Catal., A.* **583** 117134
[3] Zhao Y, Fi X, Dong L, Yan B, Shan H, Li D and Sun X 2015 *Int. J. Hydrogen.* **40** 14338-44
[4] Kim S P, Choi M Y and Choi H C 2016 *Mater. Res. Bull.* **74** 85-9
[5] Hussain S, Jacob J, Riaz N, Mahmood K, Ali A, Amin N, Nabi G, Isa M and Mahmood M H R 2019 *Ceram. Int.* **45** 4063-58
[6] Lutzer T, Charpentier T V J, Barker R, Soltanahmadi S, Taleb W, Wang C, Alejo-Rodriguez A, Perre E, Schneider H and Neville A 2018 *Mater. Chem. Phys.* **216** 272-7
[7] Jorge de Souza T A, Rosa Souza L R and Franchi L P 2019 *Ecotoxicol. Environ. Saf.* **171** 691-700
[8] Dadgostar N, Ferdous S and Henneke D 2010 *Mater. Lett.* **64** 45-8
[9] Yulizar Y, Pratiwi Y and Apriandanu D O B 2020 *Mater. Sci. Forum* **982** 9-13
[10] Li M, Yang Y, Yuan X, Liu Y and Zhang L 2015 *Mater. Lett.* **149** 62-5
[11] Yulizar Y, Ohashi A and Watarai H 2001 *Anal Chim. Acta.* **447** 247-54
[12] Kharissova O V, Dias H V R, Kharisov B I, Pérez B O and Pérez V M J 2013 *Trends in Biotechnol.* **31** 240-8
[13] Singh J, Kaur N, Kaur P, Kaur S, Kaur J, Kukkar P, Kumar V, Kukkar D and Rawat M 2018 *Environ. Nanotechnol. Monit. Manage.* **10** 331-8
[14] Kumar M, Mehta A, Mishra A, Singh J, Rawat M and Basu S 2018 *Mater. Lett.* **215** 121-4
[15] Yulizar Y, Kusrini E, Apriandanu D O B and Nurdini N 2020 *Surf. Interfaces* **19** 100437
[16] Apriandanu D O B and Yulizar Y 2019 *Nano-Struct Nano-Objects* **20** 100401
[17] Tania T C D, Deastri Y N, Utari T, Yulizar Y and Apriandanu D O B 2018 *AIP Conf. Proc.* **2023** 020103
[18] Abbas M, Sari N, Utari T, Yulizar Y and Apriandanu D O B 2018 *AIP Conf. Proc.* **2023** 020100
[19] Ibrahim D and Osman H 1995 *J. Ethnopharm.* **45** 151-6
[20] Azubuike A D 2015 *Sci. J. Chem.* **3** 40-9
[21] Tammina S K, Mandal B K and Kadiyala N K 2018 *Environ. Nanotechnol. Monit. Manage.* **10** 339-50
[22] Bhattacharjee A and Ahmaruzzaman M, 2015 *Mater Lett.* **145** 74-8
[23] Yulizar Y, Sudirman, Apriandanu D O B, and Wibowo A P 2019 *Compos. Commun.* **16** 50-6

IOP Conf. Series: Materials Science and Engineering **902** (2020) 012017 doi:10.1088/1757-899X/902/1/012017

Molecular investigation of modified β-cyclodextrin and cholesterol inclusion complexes through molecular docking simulations

M A F Nasution, H G Riyanto, E Saepudin and T A Ivandini

Department of Chemistry, Faculty of Mathematics and Natural Sciences (FMIPA), Universitas Indonesia, Depok 16424, Indonesia

Corresponding author's email: ivandini.tri@sci.ui.ac.id

Abstract. Hypertension, a medical condition which commonly associated with cardiovascular diseases, the most lethal non-communicable diseases in the world, is indicated by the high blood cholesterol level. Thus, maintaining the blood cholesterol level is essential especially for hypertension-diagnosed patients. In this study, the molecular docking simulations were successfully performed between cholesterol and methylene blue (MB) with the modified β-cyclodextrin (BCD) compounds as the guest and host molecules, respectively, to investigate their molecular interaction when forming the inclusion complexes. The docking results showed that the modification on the -OH hydroxyl group at position 6 of BCD improves the binding affinity of the cholesterol when forming the inclusion complex, where the $-OCH_3$ modification has the highest binding affinity toward cholesterol, with $\Delta G_{binding}$ value of -5.9 kcal/mol, followed by $-OCH_2CHO$, $-OCH_2COOH$ and $-OCOCOH(COOH)_2$, with $\Delta G_{binding}$ value of -5.8 kcal/mol. Moreover, the hydrophobic and van der Waals interactions were observed as the most dominant interactions when both BCD-modified compounds/cholesterol and BCD-modified compounds/MB inclusion complexes were formed. Thus, the electrochemical technique can be employed toward the cholesterol and these BCD-modified compounds to validate the docking results and determine its sensitivity as cholesterol sensor.

Keywords: BCD, cholesterol, non-enzymatic sensor, molecular docking simulation.

1. Introduction

Cardiovascular diseases (CVDs) remain the highest cause of death in the world, with approximately 16 million deaths caused by this disease [1]. In Indonesia itself, this non-communicable diseases rank is at first and second position as the leading cause of death around the archipelago, which according to the latest statistics, about 6.52 million and 5.18 million people died each year because of ischaemic heart disease and cerebrovascular disease, respectively [2]. To date, hypertension is still regarded as the leading cause of CVDs, which is indicated by the irregularly high level of blood pressure, both systolic and diastolic ($\geq$ 140 mmHg and $\geq$ 90 mmHg, respectively), due to the plaque buildup in the blood vessels [3, 4]. In a long term, without proper treatment, the plaque will thicken the blood vessels, disturb the blood circulation, and eventually trigger angina pectoris [5]. Recent studies show that high cholesterol concentration in the blood is directly associated with the higher CVDs risk [6]. Therefore, maintaining the blood cholesterol level is essential, especially for someone who already diagnosed with hypertension to avoid having CVDs risk in the future.

β-cyclodextrin (BCD) is a cyclic oligosaccharide which consisted of seven glucopyranose units that has a hydrophobic region inside its cavity, and hydrophilic interior which is formed by its primary and secondary -OH groups [7]. Due to its unique molecular shape and structure, BCD can entrap a molecule within its hydrophobic cavity, formed an inclusion complex that can be utilized for several applications, such as drug carriers for hydrophobic drugs in the drug delivery system [8]. Recently, BCD-based material has also been widely utilized in the electrochemical sensing field to detect several biomolecules [9, 10]. Moreover, BCD can also be easily modified in its primary -OH groups without removing its entrapment ability by substituting its polar hydrogen atom with citrate or carboxymethyl functional group, enhancing its hydrophilicity for the drug delivery purposes [11, 12]. However, the effect of this modification on BCD for its sensor application has not been fully understood yet. Since both BCD and cholesterol do not produce electrochemical signals that are needed for this complex to be functioned as a sensor using electrochemical method, methylene blue (MB), a dye-based heterocyclic aromatic from thiazine group, can be applied as redox indicator that can undergo redox reactions involving two electrons and one proton in the reduced form of leucomethylene blue (LMB) at physiological pH [13].

In this research, the computational study using molecular docking simulation was employed to predict the binding affinity and conformation of the formed inclusion complex of both cholesterol and MB with BCD-modified compounds. Therefore, the result of this study can be used as the reference before the synthesis, characterization, and application of the BCD-modified compounds as the cholesterol sensor took place.

2. Methodology

The methodology of this research was conducted based on the modification of the previous research [14] using various software that has been widely used and validated through a series of experiments [15, 16]. At first, the BCD structure was obtained from the RCSB-PDB website with PDB ID: 3CK8 [17], followed by the structure optimization by removing the SusD protein and other unnecessary molecules such as ethylene glycol and water, while keeping the BCD structure using BIOVIA Discovery Studio 2019 software. The modification of BCD was took place on the primary -OH group at position 6 using BIOVIA Discovery Studio 2019 software as well (figure 1).

Furthermore, the cholesterol and MB structures were obtained from the PubChem website [18] with Pubchem CID: 5997 and 4139, respectively, and their 3D structure were retrieved in .sdf files. These structures were then prepared and optimized using default minimization protocol in PyRx v0.8.0 software [16], with MMFF94x was selected as the forcefield.

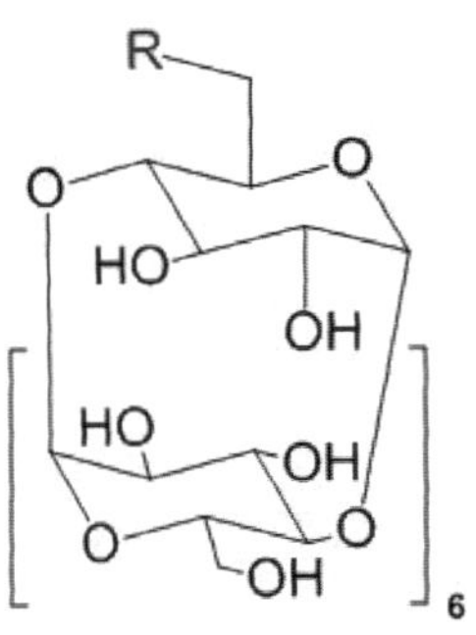

R_0 = -OH

R_1 = $-OCH_3$

R_2 = $-NH_2$

R_3 = $-OCH_2CHO$

R_4 = $-OCH_2COOH$

R_5 = $-OCOCOH(COOH)_2$

Figure 1. The substituted functional group at primary -OH position 6 that used in this study.

The minimized structure was then saved and converted into .pdbqt file format for further use. After that, the standard protocol with adjusted parameters of molecular docking simulation of BCD and BCD-modified compounds against cholesterol and MB was performed using AutoDock Vina module in PyRx v0.8.0 software. Finally, the docking results were then observed and the conformations that have the highest binding affinity of each compound against their respective receptors were visualized using BIOVIA Discovery Studio 2019 software.

3. Results and discussion

According to the previous study, the molecular interaction between BCD and its guest molecule in the hydrophobic cavity are predominately based on hydrophobic interaction, electrostatic interaction, van der Waals interaction, and hydrogen bonds [19]. Thus, the conformation of BCD-cholesterol and BCD-MB complex inclusions can be predicted using molecular docking simulations. To be functionalized as a cholesterol sensor using electrochemical methods, the MB must be included in the sensor system as the electroactive species that can give electrochemical signals, unlike cholesterol and BCD. Moreover, MB was also chosen as a redox indicator because it has a good selectivity for cholesterol even in the presence of other biomolecules that can be potentially acted as interference compounds that can disturb the cholesterol measurements [13]. As depicted in figure 2, the mechanism of BCD-cholesterol inclusion complex formation from BCD-MB complex can be explained as follows: the BCD that has been immersed and dispersed first with MB solution is dissolved with a cholesterol solution. Since cholesterol has more hydrophobicity than MB due to its bulky steroid structure and hydrocarbon tail, it will substitute MB from the hydrophobic cavity of BCD, forming BCD-cholesterol inclusion complex that has a higher binding affinity and more stable complex than its counterpart. The release of MB from BCD will increase the concentration of MB in the solution, which is then assumed to be exact concentration of the added cholesterol in the system. Through electrochemical techniques, such as voltammetry and amperometry, the value of both oxidized and reduced MB can be observed and analyzed to determine the cholesterol level contained in the solution.

In this study, the modification has taken place at the primary -OH hydroxyl group at position 6, this was done because in theory, this -OH group is the most basic, least acidic, and most accessible -OH group among all, which can be easily attacked by electrophilic agent after all the -OH groups are deprotonated using a strong, excess base reagent [20]. For this study, about six BCD-based compounds were drawn and formed, including the unmodified BCD. The other five modified BCDs have their primary -OH hydroxyl group at position 6 substituted with another functional group, namely methoxy

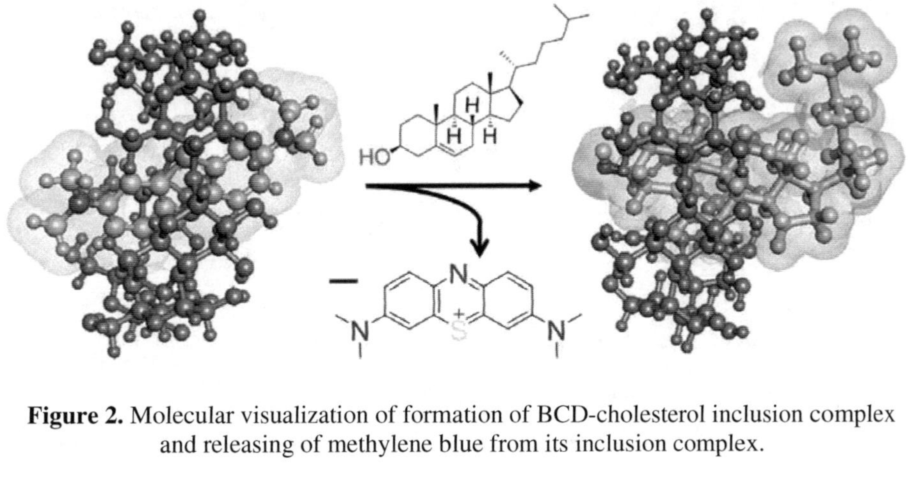

Figure 2. Molecular visualization of formation of BCD-cholesterol inclusion complex and releasing of methylene blue from its inclusion complex.

(-OMe/-OCH_3), amine (-NH_2), formylmethoxy (-OCH_2CHO), carboxymethoxy (-OCH_2COOH), and dihydrogen citrate (-$OCOCOH(COOH)_2$). These functional groups were chosen since they have already been studied and successfully synthesized on BCD in the previous research [12, 20]. These compounds were computationally prepared and optimized using default parameters that have been extensively used and validated for the docking simulations [14]. The docking results in this study can be seen in table 1.

According to the docking results that have been performed, it was determined that the -OCH_3 modification at BCD gave the highest binding affinity toward cholesterol, compared to the other five BCD-based compounds. The formation of BCD-OCH_3 modified compound and cholesterol inclusion complex has $\Delta G_{binding}$ value of -5.9 kcal/mol and pKeq value of 4.3434, this observed ΔGbinding and pKeq value were respectively lower and higher than the BCD-OCH_2CHO, BCD-OCH_2COOH, and BCD-$OCOCOH(COOH)_2$ modified compounds, which were observed at -5.8 kcal/mol and 4.2698, respectively. Finally, BCD-NH_2 modified compound has the highest observed $\Delta G_{binding}$ value toward cholesterol, which determined at -5.7 kcal/mol with pKeq value of 4.1962. Nonetheless, these five BCD-modified compounds have lower $\Delta G_{binding}$ values than the unmodified BCD, which were determined at -5.4 kcal/mol (pKeq value observed at 3.9753). This phenomenon indicated that the modification of primary -OH hydroxy group at position 6 did affect the increase binding affinity of the cholesterol inclusion complexes, mainly due to the increasing number of atoms in the BCD, which contributes to the hydrophobic and van der Waals interactions in the hydrophobic cavity of BCD. Moreover, the docking simulation also revealed that cholesterol has a higher binding affinity toward any BCD-modified compound than MB, where the difference was ranged between -0.1 kcal/mol (at -NH_2) to -0.9 kcal/mol (at -OCH_3). This also indicated that the presence of cholesterol would eventually replace MB in the hydrophobic cavity of BCD, since the formation of BCD-cholesterol inclusion complex is thermodynamically favorable compared to the BCD-MB complex, according to the docking results.

The molecular interactions of cholesterol and MB with BCD and its modified compounds were also observed after the docking simulations were performed. As depicted in figure 3, the interactions between the BCDs with their guest molecules were primarily dominated by the hydrophobic and van der Waals interactions, although π-hydrogen bonds may occasionally occur in the BCD-MB inclusion complex between the aromatic ring of MB and polar hydrogen atoms from BCD. Moreover, according to the docking results, it was also observed that when forming the BCD-cholesterol inclusion complex, the hydrocarbon tail of cholesterol was exposed out on the wider edge of BCD cavity, where the secondary -OHs are located, while the polar hydroxyl and bulky steroid groups of cholesterol were positioned on the narrow edge and hydrophobic cavity of BCD, respectively.

These results were in accordance with the docking results from the previous study [21]. Although the different conformation of BCD-cholesterol inclusion complex was observed as well (polar hydroxyl group and hydrocarbon tail of cholesterol were located on wider and narrow edges, respectively),

Table 1. Molecular docking simulation of BCD-modification with cholesterol and MB results.

-R	Cholesterol		MB	
	$\Delta G_{binding}$	pKeq	$\Delta G_{binding}$	pKeq
-OH	-5.4 kcal/mol	3.9753	-5.3 kcal/mol	3.9017
-OCH_3	-5.9 kcal/mol	4.3434	-5.0 kcal/mol	3.6809
-NH_2	-5.7 kcal/mol	4.1962	-5.6 kcal/mol	4.1226
-OCH_2CHO	-5.8 kcal/mol	4.2698	-5.5 kcal/mol	4.0490
-OCH_2COOH	-5.8 kcal/mol	4.2698	-5.6 kcal/mol	4.1226
-$OCOCOH(COOH)_2$	-5.8 kcal/mol	4.2698	-5.5 kcal/mol	4.0490

IOP Conf. Series: Materials Science and Engineering **902** (2020) 012017 doi:10.1088/1757-899X/902/1/012017

but since it has a higher $\Delta G_{binding}$ and lower pKeq value than the former, the formation of these complex was not likely to be happened due to it was not thermodynamically favorable compared to its counterparts.

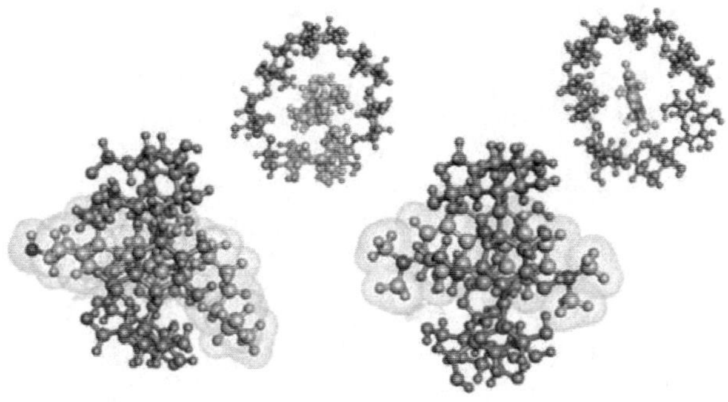

(a)

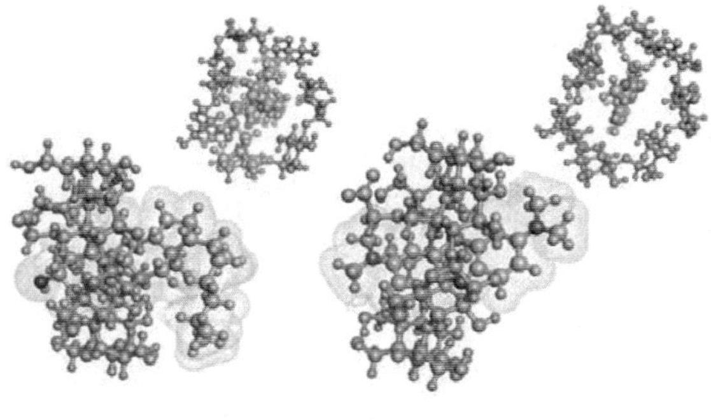

(b)

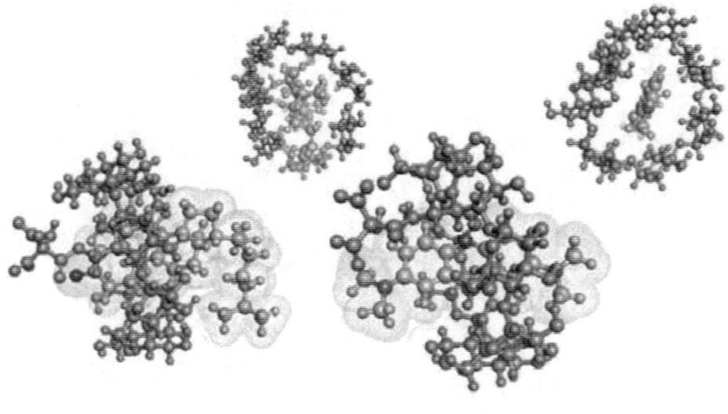

(c)

Figure 3. Molecular interaction of inclusion complex of BCD-modified compound (a) R = -OMe, (b) $-CH_2COOH$ and (c) $-OCOCOH(COOH)_2$ with cholesterol (left) and methylene blue (right).

4. Conclusion

In this study, a series of molecular docking simulations of cholesterol and MB against BCD and its modified compounds were successfully performed. In summary, the modification of BCD in the primary -OH at position 6 did increase the binding affinity of the cholesterol, through some modification also decreasing the binding affinity of MB as well. According to the docking results, the $-OCH_3$ modification has the highest binding affinity toward cholesterol, with $\Delta G_{binding}$ value of -5.9 kcal/mol, followed by $-OCH_2CHO$, $-OCH_2COOH$, and $-OCOCOH(COOH)_2$,with $\Delta G_{binding}$ value of -5.8 kcal/mol. Thus, these compounds can be later synthesized and characterized before the electrochemical application as a cholesterol sensor for these BCD-modified compounds can be performed. Furthermore, the molecular dynamics studies can be carried out as well in the future to know the exact molecular mechanism of the formation of BCD-modified and cholesterol/MB inclusion complexes.

Acknowledgments

This work is financially supported by the Directorate of Research and Community Engagement, Universitas Indonesia, through Hibah Penelitian Unggulan Perguruan Tinggi (PUPT) Tahun Anggaran 2019 No: NKB-1599/UN2.R3.1/HKP.05.00/2019. No conflicts of interest are declared regarding this research and publication.

References

[1] WHO 2017 *Cardiovascular Diseases (CVDs)* available at https://www.who.int/news-room/fact-sheets/detail/cardiovascular-diseases-(cvds)
[2] Mboi N et al. 2018 *Lancet* **392** 581-91
[3] Balakumar P, Maung-U K and Jagadeesh G 2016 *Pharmacol. Res.* **113** 600-9
[4] Goldstein J L and Brown M S 2015 *Cell* **161** 161-72
[5] Van Deventer H E et al. 2011 *Clin. Chem.* **57** 490-501
[6] Deedwania P C, Carbajal E V and Bobba V R 2007 *Clin. Cardiol.* **30** (Suppl. I) I16-24
[7] Jambhekar S S and Breen P 2016 *Drug Discov. Today* **21** 356-62
[8] Loftsson T 2002 *J. Incl. Phenom. Macro. Chem.* **44** 63-7
[9] Huang J et al. 2012 *Anal. Methods* **4** 4264-8
[10] Alarcón-Angeles G et al. 2008 *Carbon* **46** 898-906
[11] Chen P, Yao S, Chen X, Huang Y and Song H 2019 *New J. Chem.* **43** 4282-90
[12] Jayaprabha K N and Joy P A 2015 *RSC Adv.* **5** 22117-25
[13] Arvand M, Sohrabnezhad S, Mousavi M F, Shamsipur M and Zanjanchi M A 2003 *Anal. Chim. Acta* **491** 193-201
[14] Tambunan U S F, Parikesit A A, Wardani F, Nasution M A F and Kerami D 2017 *Rasayan J. Chem.* **10** 910-21
[15] BIOVIA D S 2016 *Discovery Studio Modeling Environment, Release 2017, San Diego Dassault Systèmes* available at https://www.3dsbiovia.com/products/collaborative-science/biovia-discovery-studio
[16] Dallakyan S and Olson A J 2015 *Chem. Biol.* **1263** 243-50
[17] Koropatkin N M, Martens E C, Gordon J I and Smith T J 2008 *Structure* **16** 1105-15
[18] Kim S *et al.* 2016 *Nucleic Acids Res.* **44** D1202-13
[19] Inoue Y, Liu Y, Tong L H, Shen B J and Jin D Sen 1993 *J. Am. Chem. Soc.* **115** 10637-44
[20] Řezanka M 2016 *Eur. J. Org. Chem.* **2016** 5322-34
[21] Choi Y-H, Yang C-H, Kim H-W and Jung S 2001 *J. Incl. Phenom. Macro. Chem.* **39** 71-6

La_2O_3 nanoparticles formation using *Nothopanax scutellarium* leaf extract in two-phase system and photocatalytic activity under UV light irradiation

K Mustofa, Y Yulizar, A Saefumillah and D O B Apriandanu

Department of Chemistry, Faculty of Mathematics and Natural Sciences (FMIPA), Universitas Indonesia, Depok 16424, Indonesia

Corresponding author's email : yokiy@sci.ui.ac.id

Abstract. La_2O_3 nanoparticles (NPs) have been formed using *Nothopanax scutellarium* leaf extract (NSE) and lanthanum precursor in two-phases system (n-hexane/water) with high-speed stirring method. Alkaloids as a secondary metabolite compound in NSE were used as a hydrolyzing agent and the other substances, namely saponin and tannin as a capping agent. All characterizations in this study were undertaken by Fourier transform infrared (FT-IR), UV-Vis DRS, X-Ray diffraction (XRD), scanning electron microscopy energy-dispersive X-Ray spectroscopy (SEM-EDX), transmission electron microscopy (TEM) and UV-Vis spectrophotometer. UV-Vis DRS characterization showed that La_2O_3 NPs has a bandgap value of 5.59 eV. The particle size of La_2O_3 NPs was around 84.3 nm confirmed by TEM. UV-Vis spectrophotometer was utilized to analyze the photocatalytic activity of La_2O_3 NPs for the degradation of malachite green (MG) as a modelling of dye. The degradation percentage of MG using La_2O_3 NPs was 87.32 % for 120 min under UV light irradiation.

Keywords: La_2O_3 nanoparticle, *Nothopanax scutellarium* leaf extract; two-phases system; photocatalytic activity; malachite green

1. Introduction

Recently, research related to the green synthesis of nanoparticles had been well developed. It uses materials such as microorganisms, enzymes, plants, and plant extracts [1]. A biological research reported that plant extract can serve as a reducing and stabilizing agent in metal nanoparticle synthesis [2].

Two-phases system is commonly used for complex formation of metal and ligand to dissolve them in polar and non-polar solvents, respectively. This method has advantages, such as simple and cheaper synthesis process and low energy requirement [3].

Metal nanoparticles green synthesis, namely gold nanoparticles and copper (II) oxide nanoparticles using *Tinospora crispa* plant extracts have been recorded. Besides, metal oxide nanoparticles such as titanium oxide using *Averrhoa bilimbi* then metal oxide transition such as silver oxide and nickel oxide also had been successfully synthesized using *Vernonia amygdalina* and *Ageratum conyzoides,* respectively. Also, Cerium and lanthanum metal oxide nanoparticles had been successfully synthesized using *Datura metel L* and *Physalis angulata* leaf extract, respectively, in one phase system [4-10].

Rare earth elements, as materials with high thermal conductance, melting point, and conductivity, as well as unique physical and chemical properties, have been extensively applied in electronics, biomedical, and agronomic fields [11]. Lanthanide oxide nanoparticles (La_2O_3 NPs) are rare earth metal oxide and semiconductor materials with a bandgap range of 4.3–5.8 eV. La_2O_3 NPs are potentially useful materials for various optical and electronic applications, such as capacitors, optical filters and waveguides [12].

Nothopanax scutellarium is an ornamental plants or hedges grown wild in field and river banks. In the previous study, it is a local medicinal plant for a hair growth contained about 11 % of alkaloids and other secondary metabolites such as flavonoid, saponin, and tannin [13]. Alkaloids served as a hydrolysis agent (weak base source of OH^-). Meanwhile, the other secondary metabolite such as saponins and tannins act as a capping agent for La_2O_3 NPs growth [14].

In this study, the synthesis of La_2O_3 NPs were carried out using *Nothopanax scutellarium* leaf extract (NSE) in a two-phase system (n-hexane / water), which is not reported in the previous study yet. Also, La_2O_3 NPs were tested for the photocatalytic activity for degradation of malachite green under UV light irradiation.

2. Materials and method

Lanthanum nitrate hexahydrate ($La(NO_3)_3.6H_2O$), malachite green, and NSE were applied as the primary raw materials. ($La(NO_3)_3.6H_2O$) and malachite green were purchased from Merck, and *Nothopanax scutellarium* plant obtained from the conservation unit of biopharmaca (UKBB) center study for tropical, LPPM IPB, Bogor, Indonesia.

2.1. Preparation of NSE

The leaves were cleaned and dried at room temperature. Fifty g of leaves powder was macerated in 250 mL of n-hexane for a week by stirring every day. Next, the filtrate was separated and partitioned by aquadest. Then, the formed n-hexane and water phases were identified by the phytochemical test.

2.2. Synthesis of La_2O_3 NPs

La_2O_3 NPs were synthesized by a sol-gel method with slight modification [9]. 2 mmol of $La(NO_3)_3$ precursor was mixed into n-hexane phase of NSE by high-speed stirring method at 23,000 rpm. Then, the water phase changed its color from clear to turbid, and the turbid layer was separated. The result was dried at 120 °C and calcined at 700 °C for 2 h to form white powders of La_2O_3 NPs.

2.3. Photocatalytic activity of La_2O_3 NPs

Three mg of La_2O_3 NPs was dispersed in 25 mL of malachite green 6 x 10^{-6} M and stirred for 120 min under UV light irradiation. The degradation result was monitored by UV-Vis spectrophotometer every 10 min irradiation.

2.4. Characterization of La_2O_3 NPs

Fourier transform infrared (FT-IR) spectroscopy (IR Prestige-21 Shimadzu) was used to identify the functional groups of La_2O_3 NPs in the range of 400–4000 cm^{-1}. UV-Vis diffuse reflectance (UV-Vis DRS, Shimadzu 2600) was utilized to determine the band gap value of La_2O_3 NPs. The degradation of malachite green were observed by UV-Vis spectrophotometer (Shimadzu 2600). The crystallinity of La_2O_3 NPs was determined by X-Ray diffraction (XRD, Shimadzu). The particle size distribution of La_2O_3 NPs was defined by particle size analyzer (PSA, Zetasizer nano ZS 90). The morphology and element composition of La_2O_3 NPs were determined by scanning electron microscopy-energy dispersive X-Ray spectroscopy (SEM-EDX, JED-2300). The particle size of La_2O_3 NPs was investigated by transmission electron microscopy (TEM, JEOL JEM 1400).

3. Results and discussion

3.1. Characterization of La_2O_3 NPs

La_2O_3 NPs and NSE functional groups are presented in FT-IR spectra in figure 1. The spectrum of NSE shows that the vibration of O-H stretch, C-H sp^3 stretch, -NH stretch, C-C stretch, N-O stretch, and N-H Wagg-stretch appeared at the wavenumbers of 3450, 2960, 1646, 1462, 1387 and 810 cm^{-1}, respectively. The N-H Wagg-stretch indicates the presence of alkaloids. Meanwhile, the functional groups of -OH and C-H sp^3 indicate the presence of saponins. Also, the C-C stretch is derived from steroids compound [15-17]. Besides, the spectrum of La_2O_3 NPs shows that there was a La-O stretch at 645 cm^{-1} [18].

The band gap value was determined by using a % reflectance, which is converted into Kubelka-Munk function. A linear regression equation for $F(R)^2 = 0$ is found and made by plotting $F(R)^2$ to Eg [19]. Based on the linear regression equation, the bandgap value of La_2O_3 Nps was 5.59 Ev, as shown in figure 2. XRD pattern of La_2O_3 NPs shows that there were some peaks at the 2θ of 15.72° (1 0 1); 27.36° (2 2 2); 28.12° (3 0 0); 39.62° (4 0 0) and 48.69° (6 2 2), as shown in figure 3.

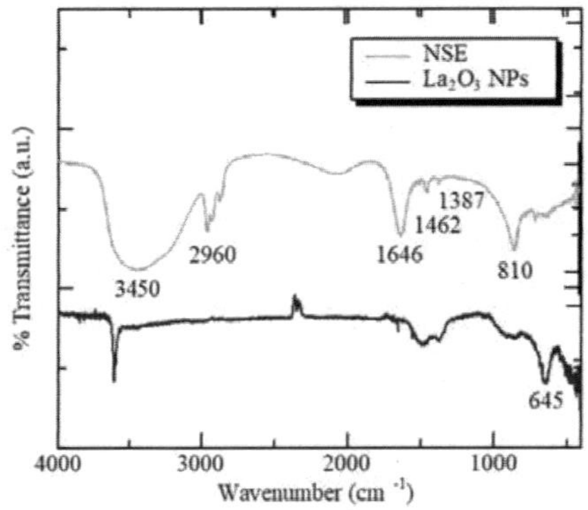

Figure 1. FTIR spectra of NSE (green line) and La_2O_3 NPs (blue line).

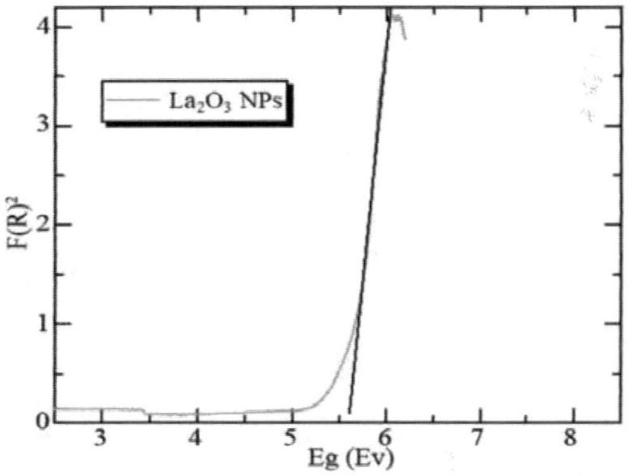

Figure 2. Curve of F(R) to hυ of La_2O_3 NPs.

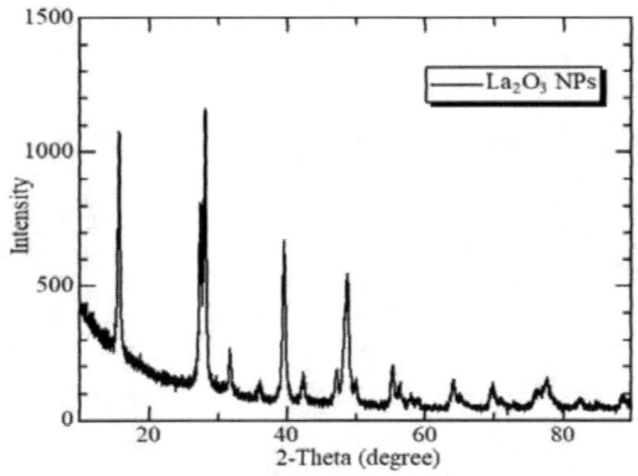

Figure 3. XRD pattern of La_2O_3 NPs.

These values are corresponded to JCPDS data of La_2O_3 no. 04–0856 [20]. According to the data, La_2O_3 had the hexagonal crystal structure. Besides, the crystalline size of La_2O_3 was calculated by Debye-Scherrer equation using the full-width at half maximum β of the diffraction peaks [21], $D = \frac{0.9\,\lambda}{\beta \cos\theta}$, where λ is the X-Ray wavelength and θ is the Bragg angle diffraction. According to the calculation, La_2O_3 had a crystalline size of 29.67 nm. Moreover, La_2O_3 NPs were characterized by PSA to determine the average particle size distribution. The average distribution particle size of La_2O_3 NPs was 55.76 nm, as shown in figure 4. The morphology was investigated by SEM characterization with 500 nm scale bar, as shown in figure 5a. For further characterization, TEM was utilized to determine the particle size of La_2O_3. It can be seen that the particle size of La_2O_3 NPs was about 84.3 nm, as presented in figure 5b.

3.2. Photodegradation activity of malachite green

Figure 6a shows a UV-Vis absorption spectra of malachite green degradation using La_2O_3 NPs. The degradation was observed with UV-Vis spectrophotometer by monitoring the decrease of absorbance at the maximum wavelength of 617 nm. Also, figure 6b shows the plotting of % degradation versus time. The degradation percentage was calculated by equation 1.

$$\%\,D = \frac{[\mathrm{C}]_0 - [\mathrm{C}]_t}{[\mathrm{C}]_0}\ x\ 100\ \% \quad (1)$$

where $[C]_0$ is the initial concentration of MG and $[C]t$ is MG concentration of each time interval by UV light irradiation [22]. According to the calculation, the degradation percentage of MG photodegradation using La_2O_3 NPs was 87.32 % for 120 min under UV light irradiation.

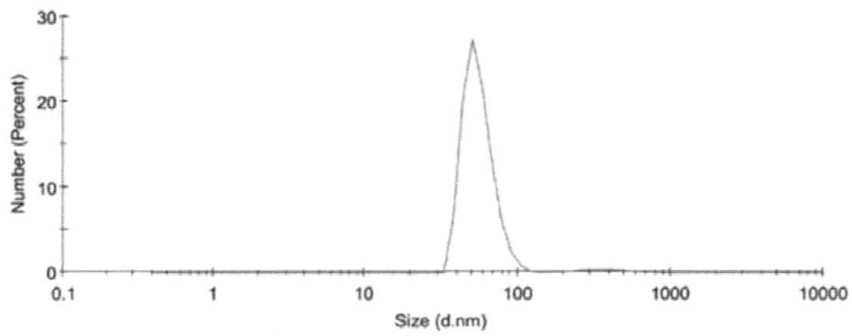

Figure 4. Particle size distribution of La_2O_3 NPs.

(a) (b)

Figure 5. (a) SEM image of La_2O_3 NPs with 30.000x magnification, and (b) TEM image of La_2O_3 NPs.

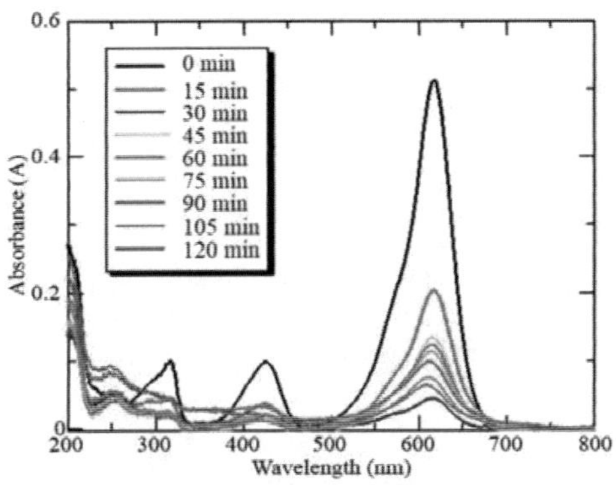

(a)

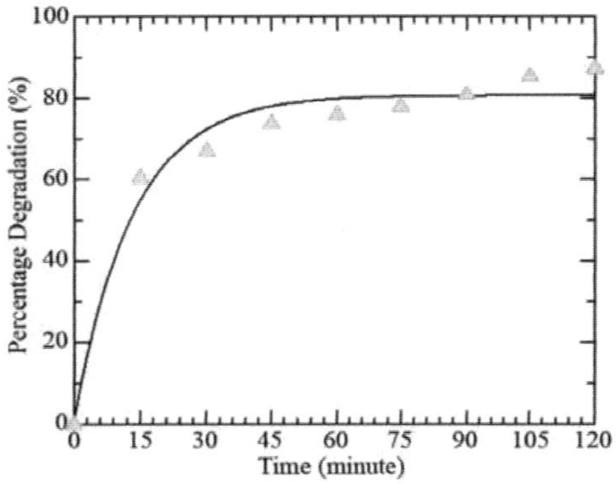

(b)

Figure 6. (a) UV-Vis absorption spectra of malachite green degradation using La_2O_3 NPs, and (b) % degradation of malachite green.

4. Conclusion

La_2O_3 NPs have been synthesized using NSE in the two-phase system (n-hexane/water). The NSE phytochemical test presents the secondary metabolite compounds such as alkaloids, saponin, and steroids, which are used as a weak base source and capping agent for La_2O_3 NPs formation. The XRD analysis confirms that La_2O_3 had the hexagonal crystal structure. Finally, La_2O_3 NPs have a nanometer scale and photodegradation activity on malachite green of 87.32 % under UV light irradiation.

Acknowledgments

This research was granted by Hibah PDUPT 2019 from the Ministry of Research, Technology and Higher Education, Indonesia, through Universitas Indonesia with contract number NKB-1618/UN2.R3.1/HKP.05.00/2019).

References

[1] Mohanpuria P, Rana N K and Yadav S K 2008 *J. Nanopart Res.* **10** 501-17
[2] Anish R, Angel R R and Daizy P 2017 *Open Nano* **2** 1-8
[3] Yulizar Y, Monjushiro H and Watarai H 2004 *J. Colloid Interface Sci.* **275** 560-9
[4] Yulizar Y and Apriandanu D O B 2017 *IOP Conf. Ser. Mater. Sci. Eng.* **188** 012013
[5] Apriandanu D O B and Yulizar Y 2019 *Nano-Struct. Nano-Objects* **20** 100401
[6] Yulizar Y, Sudirman, Apriandanu D O B and Wibowo A P 2019 *Compos. Commun.* **16** 50-6
[7] Widyaningtyas A L, Yulizar Y and Apriandanu D O B 2019 *IOP Conf. Ser. Mater. Sci. Eng.* **509** 012022
[8] Wardani M, Yulizar Y, Abdullah I and Apriandanu D O B 2019 *IOP Conf. Ser. Mater. Sci. Eng.* **509** 012077
[9] Sulaiman N, Yulizar Y and Apriandanu D O B 2018 *AIP Conf. Proc.* **2023** 020105
[10] Yulizar Y, Kusrini E, Apriandanu D O B and Nurdini N 2020 *Surf Interfaces* **19** 100347
[11] Balusamy B, Kandhasamy Y G, Senthamizhan A, Chandrasekaran G, Subramanian M S and Kumaravel T 2012 *J. Rare Earths* **30** 1296-1302
[12] Karthikeyan S, Raj A D, Irudayaraj A A, and Raj D M A 2015 *Materials Today: Proc.* **2 (3)** 1021-1025
[13] Ahdiyah I and Purwani K I 2015 *Jurnal Sains dan Seni ITS* **4** 32-6

[14] Ezhilarasi A, Vijaya J J, Kaviyarasu K, Maaza M, Ayeshamariam A and Kennedy L J 2016 *J. Photochem. Photobiol* **164** 352-60
[15] Moaobe M A G and Nyarango R M 2013 *Global Journal of Pharmacology* **7** 61-8
[16] Almutairi M S and Ali M 2014 *Natural Product Research* **29** 1271-5
[17] Kasal A, Budesinky M and Griffiths W J 2010 *Spectroscopic Methods of Steroid Analysis* (London: Springer) pp 27-161
[18] Moothedan M and Sherly K B 2016 *J. Water Process Eng* **9** 29-37
[19] S. Challagulla, R. Nagarjuna, R. Ganesan and S. Roy 2017 *Nano-Struct. Nano-Objects* **12** 147-56
[20] Fleming P G, Holmes J D, Otway D J and Morris M A 2011 *J. Solid State Chem.***184 (9)** 2595-2600
[21] Patterson A L 1939 *Phys. Rev.* **56** 978-82
[22] Devadi M A H, Krishna M, Murthy H N N and Sathyanarayana B S 2014 *Procedia Mater. Sci.* **5** 612-21

Microwave-assisted synthesis of organic corrosion inhibitor based imidazoline-stearic

D U C Rahayu, S Cahyani, I Abdullah, D A Nurani and Y K Krisnandi

Department of Chemistry, Faculty of Mathematics and Natural Sciences (FMIPA), Universitas Indonesia, Depok 16424, Indonesia

Corresponding author's email: dyahutamicr@sci.ui.ac.id

Abstract. Corrosion often occurs in oil and gas industries' pipelines indicated by a decrease in quality which is harmful to the industries and environment. To slow down the corrosion rate, a corrosion inhibitor is needed. Organic corrosion inhibitors are more widely used than inorganic ones due to its effectiveness and non-toxic. One of the most widely used organic corrosion inhibitors is imidazolines. In this study, imidazoline-stearic was successfully synthesized from triethylenetetramine (TETA) with various purity grade of stearic acid (SA) using 800 W microwave irradiation at 200 °C for 9 min. The obtained products then were separated by a solvent extraction method and identified using thin layer chromatography (TLC). Moreover, the products had also been characterized using FTIR and UV-Vis spectral data. The ability to inhibit corrosion on carbon steel in 1 % NaCl solution was evaluated to obtain inhibition efficiency (%IE). %IE at 500 ppm of imidazoline-stearic from SA pro analysis, technical, and commercial sample were 84.43 %, 75.42 %, and 84.41 %, respectively. Imidazoline-stearic from SA pro analysis revealed a similar corrosion inhibition activity compared with the commercial. Hence, imidazoline-stearic from SA technical grade demonstrated a promising corrosion inhibitor by using economical fatty acid.

Keywords: Microwave-assisted synthesis, organic corrosion inhibitor, and imidazoline-stearic

1. Introduction

Processing equipment for production and petroleum refinery in oil and gas industries requires pipeline operations that susceptible to corrosion [1]. Besides having a potency of gas leaks which triggered the explosions and environmental pollutions, corrosion can also cause industrial losses due to decrease in production and increase in cost production. Until now, carbon steel pipelines are still widely operated to distribute oil and gas since these pipelines are more economical compared with stainless steel modified pipelines [2]. Therefore, compounds having an activity as corrosion inhibitor are needed to slow down the rate of corrosion. This technique is considered as a cost-effective technique which have been developed and widely applied for protecting the metal from corrosion [3].

Imidazoline belongs to heterocyclic organic compound which is widely used as anticorrosive agent in oil and gas industries [4, 5]. This compound is also considered as green inhibitor due to its good biodegradability [3]. Imidazoline represents its ability since this compound has three main parts, i.e. five-ring with two nitrogen atom, long chain of hydrocarbon, and side chain with active functional groups as pendant [6, 7]. Due to a high demand of effective and eco-friendly corrosion inhibitor for industrial purposes, in this present study, we investigate the corrosion inhibition performance of imidazoline-stearic synthesized from various purity grades of stearic acid (SA). Imidazoline-stearic was synthesized under microwave condition and its corrosion inhibition performance was performed under cyclic voltammetry in comparison with a commercial corrosion inhibitor. This study provides a useful information regarding the utilization of technical grade fatty acid which commonly obtained from industrial waste as a reactant for preparing imidazoline-based corrosion inhibitor.

2. Materials and method

2.1. General

Microwave-assisted synthesis was irradiated using Samsung ME-731K/XSE while UV-Vis Shimadzu UV-2450, FTIR Shimadzu IR Prestige 21, and LC/MS Acquity UPLC tandem with Xevo G2-S QT of Waters were used to characterize imidazoline-stearic structures. The measurement of corrosion inhibition activity was carried out using eDAQ 450 potentiostat and VersaSTAT II operated with working electrode (WE) low carbon steel JIS G3123 grade SGD 400D from PT. Citra Tanamas (Fe 99.276 %, C 0.08 %, Mn 0.40 %, Si 0.19 %, P 0.033 %, and S 0.021 %), reference electrode (RE) SCE, and auxiliary electrode (AE) platinum. Chemicals used for synthesis, identification, and corrosion inhibition test in pro analysis grade were purchased from Merck i.e. triethylenetetramine (TETA), SA (C18), CH_2Cl_2, NaCl, methanol, NH_4OH, ant TLC plates, while technical grade of SA was purchased from PT. Wilmar Nabati Indonesia with specification as follows: fatty acid C14 0.4 %, C16 59.2 %, C18 39.8 % and C20 0.6 %.

2.2. Microwave-assisted synthesis of imidazoline-stearic

Imidazoline-stearic from SA pro analysis was synthesized by addition of TETA (5 mmol) and SA (5 mmol) while from technical grade was carried out using TETA/SA molar ratio of 1:1, w/w. Each mixture was irradiated in 800W microwave for 9 min without additional solvent. Temperatures were maintained gradually at 200 °C by using infrared thermometer. After completing the reaction, each mixture was evaporated, identified using TLC with CH_2Cl_2:MeOH (8:2, v/v) and one drop of NH_4OH as mobile phase, and extracted using CH_2Cl_2:saturated NaCl (1:1, v/v). The organic layer then was collected as imidazoline-stearic and further characterized by using FTIR, UV-Vis, and LC/MS.

2.3. Corrosion inhibition activity test

Measurements were completed using cyclic voltammetry method adjusted to the previous study [7]. NaCl 1 % solution was selected as corrosion medium with various imidazoline-stearic concentrations (100, 200, 300, 400 and 500 ppm) while the absence of sample marked as blank solution. The solution was gassed with CO_2 until saturation was reached and three electrodes were arranged. Potentio-dynamic polarization was recorded from -2000 to +500 mV at 50 mV/s scanning rate and Tafel polarization method was chosen to analyze the efficiency inhibition percentage (%IE) using equation:

$$\%IE = \frac{I_{Blank}\left(\frac{mA}{cm^3}\right) - I_{Sample}\left(\frac{mA}{cm^3}\right)}{I_{Blank}\left(\frac{mA}{cm^3}\right)} \times 100\%$$

3. Results and discussion

In this study, imidazoline-stearic from various purity grades of SA were synthesized under microwave irradiation method with solvent-free followed the reaction scheme illustrated in figure 1. Both compounds were obtained as yellowish-brown solid giving R_f value of 0.9 in TLC with CH_2Cl_2:MeOH (8:2, v/v) and one drop of NH_4OH.

FTIR and UV-Vis characterization were carried out to determine if the desired product was formed. Additional LC/MS characterization was also performed to imidazoline-stearic from SA technical grade. FTIR spectra of both imidazoline-stearic compared with SA are shown in figure 2. Imidazoline-stearic from SA pro analysis labeled as TETA-SA pro analysis while from SA technical grade marked as TETA-SA technical. According to figure 2, it can be concluded that both TETA-SA was formed since it showed peaks at the wavenumber of 3300, 2800–3000, 1600, and 1500 cm^{-1} indicating N-H, C-H sp^3, C=N, and C-N-C bonds, respectively. Compared with FTIR spectra of both SA (figure 2), all TETA-SA showed missing peaks at 3300–2700 and 1700 cm^{-1} representing O-H carboxylic and C=O groups, respectively. Furthermore, a new peak at the wavenumber of 1600 cm^{-1} indicating C=N bond compared to FTIR of TETA (data not shown) was an indicator that imidazoline had already produced [8].

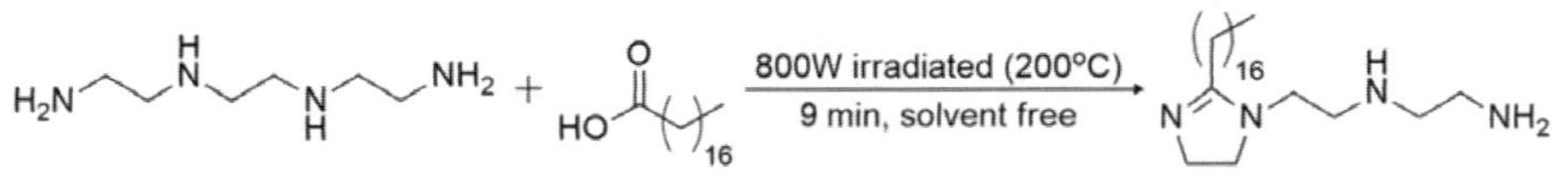

Figure 1. Reaction scheme of imidazoline-stearic synthesis.

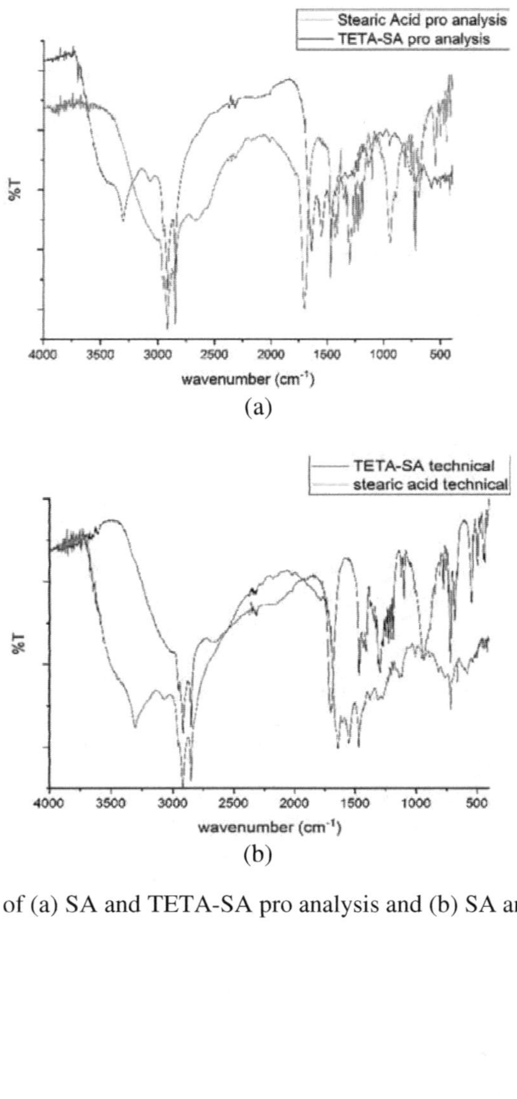

Figure 2. FTIR spectra of (a) SA and TETA-SA pro analysis and (b) SA and TETA-SA technical.

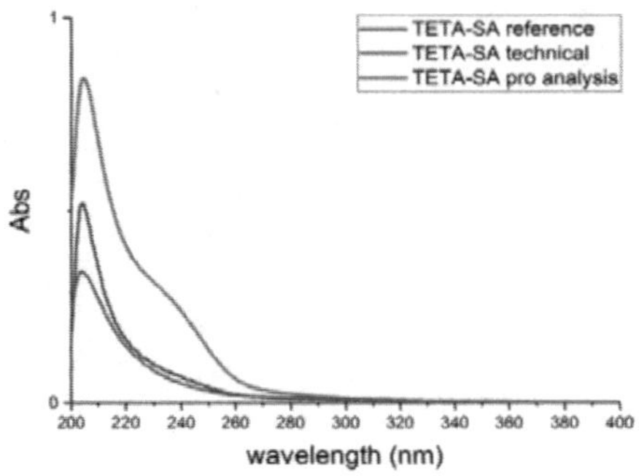

Figure 3. UV-Vis spectra of TETA-SA reference (previous study), technical and pro analysis.

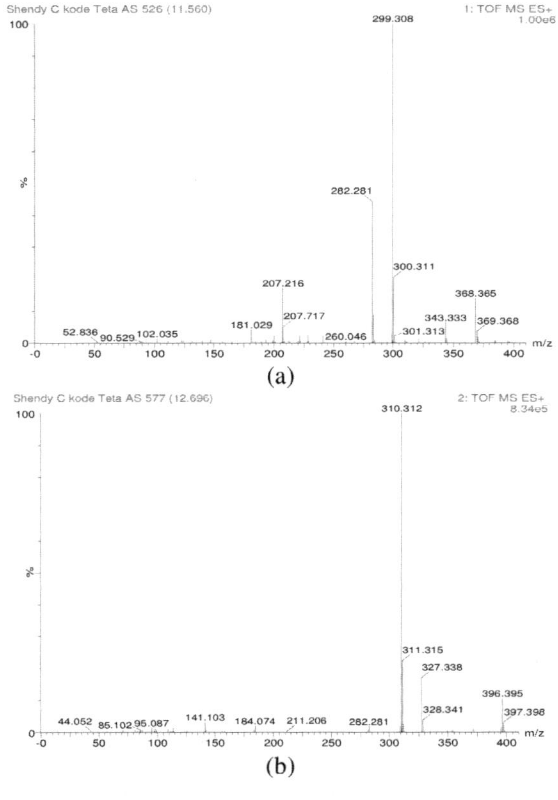

Figure 4. MS spectra of TETA-SA technical which is consisted of (a) TETA-PA and (b) TETA-SA.

Figure 3 summarizes the UV-Vis spectra for both imidazoline-stearic compared with previous study (not yet published). Based on figure 3, all spectra show a similarity in maximum wavelength around 204 nm. Therefore, it indicated that imidazoline-stearic was successfully synthesized.

To analyze the composition of TETA-SA technical, LC/MS spectra was recorded. LC spectra of TETA-SA technical (data not shown) exhibited two dominant peaks at retention time of 11.56 and 12.70 minutes. MS spectra of both dominant peaks are presented in figure 4. According to figure 4, a peak at 11.56 min represents the imidazoline-palmitic (368.365 m/z) derived from TETA with palmitic acid (PA, C16) while 12.70 min belongs to TETA-SA imidazoline (396.395 m/z). TETA-PA exhibited a lower retention time since it was more polar compared to TETA-SA.

The ability to inhibit corrosion of both synthesized imidazoline-stearic had also been investigated. Commercial imidazoline was also tested as comparison. Specifically, corrosion occurs due to the presence of dissolved CO_2 gas in water. When CO_2 dissolves in water, it will produce carbonic acid which causes more corrosive environment to carbon steel [9, 10]. Imidazoline can be adsorbed physically and chemically onto the metal surface due to the presence of N=C-N bond in imidazoline ring having rich π-electron, as well as lone pair electron in both nitrogen atoms. The existence of π-electron and lone pair electron effects the negative charge of imidazoline. In general, metal surface has a positive charge found on its interface. The charge differences between imidazoline and metal surface cause electrostatic interactions between them and form a layer. The existence of long C-H bonds from hydrocarbon chain as hydrophobic site will prevent water molecules from the environment to interact with metal surface. Therefore, the metal will be inhibited from corrosion [11, 12]. The proposed schematic mechanism of imidazoline-stearic adsorption on metal surfaces is illustrated in figure 5.

The results in corrosion inhibition activity from all samples were plotted in figure 6. Based on figure 6, TETA-SA pro analysis exhibited slightly similar activity in inhibiting corrosion compared with the commercial imidazoline. In contrast, TETA-SA technical showed a lower activity compared with others. Technical grade of SA is consisted of more than one fatty acid, having two majors of PA and SA which different in long hydrocarbon chain. As shown in figure 4, the synthesized TETA-SA imidazoline form SA technical grade was consisted of imidazoline-palmitic (TETA-PA) and imidazoline-stearic (TETA-SA). Therefore, it was proven that hydrocarbon chain plays an important role in the corrosion inhibition activity, which prevents water interaction with metal surface [5].

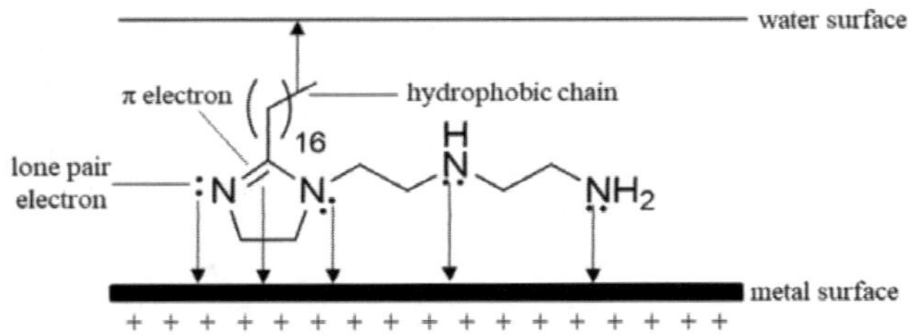

Figure 5. Proposed schematic mechanism of imidazoline-stearic adsorption on metal surfaces.

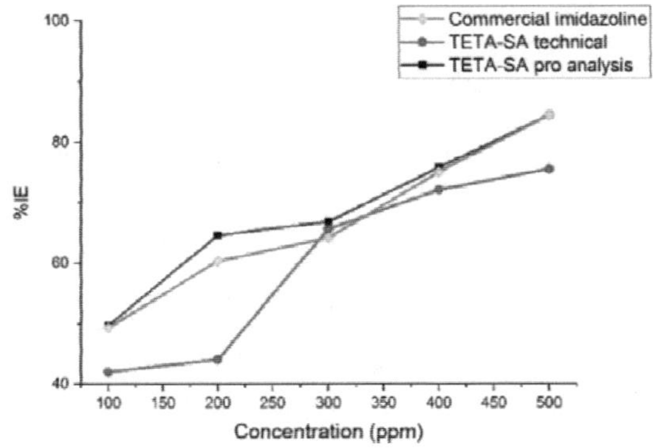

Figure 6. Corrosion inhibition activity of imidazoline-stearic and commercial in NaCl 1 % solution.

4. Conclusion

Imidazoline-stearic from TETA and various purity grades of SA was successfully synthesized using microwave-assisted method. Imidazoline-stearic from SA pro analysis exhibited a similar activity (84.43 %) with the commercial (84.41 %) in inhibiting the corrosion. However, imidazoline-stearic from SA technical grade also revealed a promise as corrosion inhibitor (75.42 %) due to the economical aspect and effectiveness in scale up production. Replication of corrosion inhibition test is needed for further research.

Acknowledgments

This work was financially supported by Universitas Indonesia under research grant PIT 9 with grant contract number NKB-0038/UN2.R3.1/HKP.05.00/2019.

References

[1] Badikova A D *et al.* 2016 *Petrol. Chem.* **56** 651-6
[2] Melati H A, Suratno W, Hidayat R, Syakir N and Fitrilawati 2011 *Bionatura* **13** 1-7
[3] Li Y Z, Xu N, Guo X P and Zhang G A 2017 *Corros. Sci.* **126** 127-41
[4] Jevremovic I, Singer M, Nesic S and Miskovic-Stankovic V 2013 *Corros. Sci.* **77** 265-72
[5] Solomon M M, Umoren S A, Quraishi M A and Salman M 2019 *J. Colloid Interface Sci.* **551** 47-60
[6] Lestari I I, Rahayu D U C, Nurani D A, Krisnandi Y K and Budianto E 2019 *AIP Conf. Proc.* **2168** 020066
[7] Yuniastuti A, Rahayu D U C, Nurani D A, Krisnandi Y K and Budianto E 2019 *AIP Conf. Proc.* **2168** 020067
[8] Solomon M M, Umoren S A, Quraishi M A, Tripathy D B and Abai E J 2020 *J. Petrol. Sci. Eng.* **187** 106801
[9] Zang H-h, Pang X and Gao K 2018 *Appl. Surf. Sci.* **442** 446-60
[10] Mazumder M A J, Al-Muallem H A and Ali S A 2015 *Corros. Sci.* **90** 54-68
[11] Fei F-l, Hu J, Wei J-x, Yu Q-j and Chen Z-s 2014 *Constr. Build. Mater.* **70** 43-53
[12] Qian S and Cheng Y F 2019 *J. Mol. Liq.* **294** 111674

IOP Conf. Series: Materials Science and Engineering **902** (2020) 012020 doi:10.1088/1757-899X/902/1/012020

One-pot four-component synthesis of phenazine derivative using 2-hydroxy-1,4-naphthoquinone

A H Cahyana, N Nazhifah and A R Liandi

Department of Chemistry, Faculty of Mathematics and Natural Sciences (FMIPA), Universitas Indonesia, Depok 16424, Indonesia

Corresponding author's email: herrykim@ui.ac.id

Abstract. Phenazine is one of the most important organic compounds because of its various biological activities. This study aims to synthesize the phenazine derivatives compound using 2-hydroxy-1,4-naphthoquinone (lawsone) with a one-pot four-component method. Phenazine derivative that had been successfully synthesized was carried out by reacting four reactants at once in one flask with the same mole ratio. The reaction occurred between lawsone, o-phenylenediamine, benzaldehyde, and malononitrile. Characterization of the synthesized products was confirmed using FT-IR, UV-Vis Spectrophotometer, and LC-MS. The results of the analysis using several of these instruments obtained one compound, namely 3-amino-2-cyano-1-phenyl-1H-benzo[a]pyrano[2,3-c]phenazine. The reaction conditions were carried out at 50°C for 6 h in ethanol with unassisted catalysts and obtained a product yield of 45%.

Keywords: Phenazine derivative, 2-hydroxy-1,4-naphthoquinone, one-pot four-component.

1. Introduction

1,4-naphthoquinone are generally found in both natural (occur in plants, fungi, and some animals with chromatic pigments) and synthetic products and represent several biological activities [1-3]. Previous studies have revealed that 1,4-naphthoquinone derivated compounds have classified as molluscicide, antiproliferative, antifungal, antimalarial, antibacterial, antileishmanial, and antiviral activities [4-10]. The biological activity of a quinone cannot be separated from the uniqueness of its structure which is capable of transferring electrons and protons [11,12]. Its ability to accept one or two electrons will be useful in the formation of suitable radical anions or radical di-ions [13]. It makes easy to modify the structure by adding the substituents [14]. The existence of these structural aspects, derivatives of naphthoquinone compounds can act as a variety of biological activities. One of the derivatives of naphthoquinone that has an impressive structure and bioactivity is lawsone. Lawsone, 2-hydroxy-1,4-naphthoquinone, is the primary coloring agent found in the henna plant (*Lawsonia inermis* L.) [15]. Lawsone has been reported to have biological activities such as antibacterial, antifungal, and antitumor, and often used as a hair dye and as a sunscreen in ultra-violet (UV) filters [16-20].

In order to increase the wealth of structure and biological activity, the compound 2-hydroxy-1,4-naphthoquinone can be modified into a phenazine derivative compound. Phenazine compound is nitrogen-containing heterocyclic that found mostly as secondary metabolites. It is primarily in Pseudomonas, Streptomyces, and several genera from the soil and marine habitats [21]. Phenazine has a diversity of biological functions like anti-mycobacterial [22], antimalarial [23, 24], and fungicides [25]. Phenazine derivatives that have been produced from a naphthoquinone and aromatic diketone display an activity as antitumor agents [26]. Besides, conjugated heterocyclic pyran has shown an antioxidant activity against DPPH radicals [27]. Moieties of molecules with phenazines and pyrans have gained a significant attention in the discovery of drug.

Various methods have been carried out and reported to synthesize the phenazine-derived compounds. The methods that have been carried out include the one-pot two-step [28] and one-pot four-component methods which have been carried out to synthesize organic compounds with the help of various types of the catalyst [29]. The one-pot two-step method is carried by first reacting two or three precursors. Then, the compound from the synthesis is added to the next precursor to get the targeted compound. The multicomponent domino reaction (MDR) which is carried out by reacting four precursors at once is a useful way for the synthesis of important biological compounds. This method has advantages including environmentally friendly, reduces chemical waste formation, cost-effective, and produces higher yields [30, 31]. Therefore, in this research, the synthesis of phenazine-derived compounds will be carried out with the one-pot four-component method by using lawsone as a precursor.

2. Materials and method

2.1. General

All chemicals were obtained from Merck and Aldrich and applied without further purification. IR spectrum was recorded on a Shimadzu Prestige 21 spectrometer. UV-Vis spectrum was recorded on a Shimadzu 2600 spectrophotometer. The mass spectrum was recorded on an LC-MS (Liquid Chromatography-Mass Spectrometer) Agilent Technologies Type 7890A-5975C. Thin-layer chromatography (TLC) was performed on a GF_{254} plate.

2.2. General procedure for the synthesis of phenazine derivatives

2-hydroxy-1,4-naphthoquinone (1 mmol), *o*-phenylenediamine (1 mmol), benzaldehyde (1 mmol), and malononitrile (1 mmol) were mixed in a round-bottomed flask at 50 °C in ethanol (10 mL). In general, the reaction takes place as illustrated in figure 1. The flask was fitted with a condenser. Then, the mixture was heated to reflux under magnetic stirring and monitored by TLC. Once completed, the reaction was cooled and the product was filtered and cleansed two times with water (2 x 5 mL). Then, the product was dried and immediately recrystallized with hot ethanol. The product was evaluated with instruments FT-IR, UV-Vis Spectrophotometer, and LC-MS to determine the vibration of the functional group, the maximum wavelength, and its molecular weight, respectively.

3. Results and discussion

The synthesis of phenazine-derived compounds using lawsone as a precursor has been performed through the one-pot four-component method. The proposed mechanism of reaction that occurs in the synthesis of this compound is starting from the reaction between 2-hydroxy-1,4-naphthoquinone with o-phenylenediamine. The results of this reaction will form benzo[a]phenazines. On the other hand, an amination reaction also occurs between benzaldehyde and malononitrile simultaneously and will form the aroyl isothiocyanate compound. In the final stage, the reaction occurs between the benzo[a]phenazine compound and aroyl isothiocyanate to form 3-amino-2-cyano-1-phenyl-1H-benzo [a] pyrano [2,3-c] phenazine. Referring to previous studies, the reaction conditions were carried out at 50 °C for 6 h in ethanol with unassisted catalysts and obtained a product yield of 45 %.

Structure evaluation and determination of the compound obtained was conducted by measuring the functional groups, the maximum wavelength, and the determination of the molecular weight with instruments FT-IR, UV-Vis Spectrophotometer, and LC-MS, respectively. Vibration function measurement results with FTIR obtained several specific peaks that identify the functional group of the target compound (figure 2a). Two specific peaks at wave numbers 3415 cm^{-1} and 3321 cm^{-1} are the stretching vibrations of primary amine (NH_2) that will always appear in two absorption peaks.

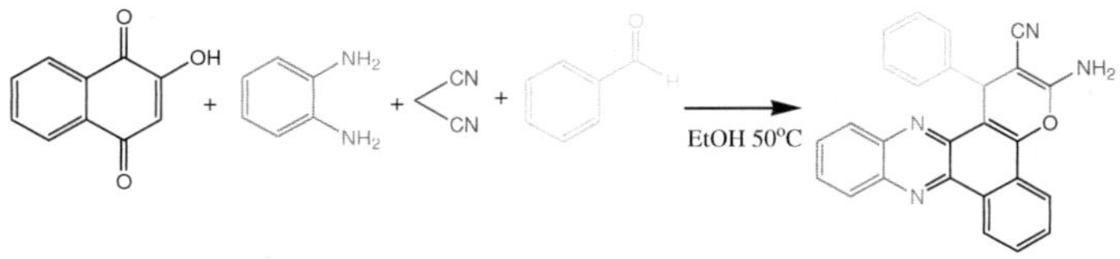

Figure 1. Reaction scheme for the synthesis of 3-amino-2-cyano-1-phenyl-1H-benzo[a]pyrano[2,3-c]phenazine.

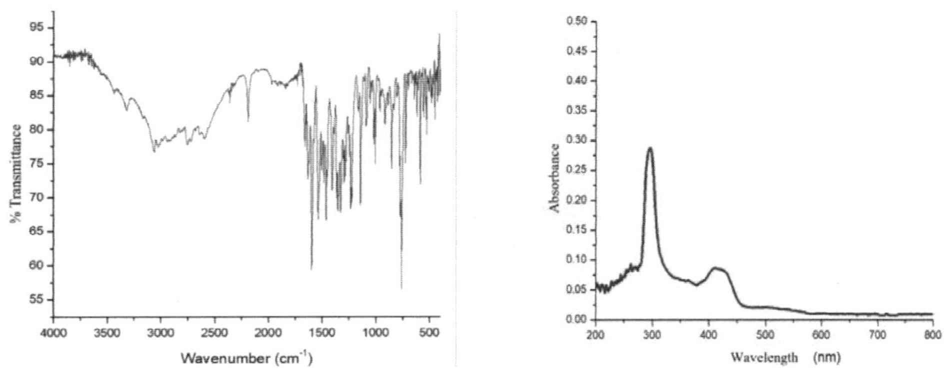

Figure 2. FTIR (a) and Uv-Vis (b) spectrum of phenazine derivative compound.

In addition, there is also an N-H bending vibration at the wavenumber 1403 cm^{-1} with moderate intensity. The stretching vibrations of the C-N, C=N, and C≡N groups are present in the absorption of 1253 cm^{-1}, 1625 cm^{-1}, and 2187 cm^{-1}, respectively. Besides, a peak in 3081 cm^{-1} is a stretching vibration of C-H aromatic (sp^2) contained in the benzene ring, while the peak at 3007 cm^{-1} is a stretching vibration of C-H (sp^3). C = C stretching vibrations from aromatics are found in 1648 cm^{-1} and 1560 cm^{-1}. The peak at 1148 cm^{-1} is the stretching vibration of the C-O-C group. The absorption of functional groups at the peaks, as mentioned earlier, supports that the target compound has all of these groups.

UV-Vis spectrophotometer analysis was performed to detect the chromophore groups from the structure of the obtained compounds. Energy absorbed in the UV region produces a transition of valence electrons in the molecule. This transition occurs from molecular orbitals (usually p and π orbitals) to the higher energy levels (antibonding p* and π*). Measurement of the λ_{max} of the target compound shows the maximum absorption at 295 nm and 412 nm (figure 2b). The peak shifting to the right of the maximum wavelength is called the bathochromic shift or redshift. This shift occurs due to the addition of functional groups that are formed in phenazine-derived compounds. In this case, an electron transition occurs from n orbitals to π* giving an absorption area of λ_{max} more than 300 nm. As a result of this shift occurring in the visible area, the color of the compound can be observed as yellow.

Determination of the molecular weight of the target compound was analyzed using LC-MS by the positive ion method, methanol eluent, and gradient elution system. Figure 3 shows the LC-MS spectrum of the target compound along with the fragments formed from the compound. Based on the analysis results, there is a peak with an m/z value of 401 which is the molecular weight of the synthesized compound. In the LC-MS analysis, the Electrospray Ionization (ESI) setting is used so that the mass that appears is [M+H]+. The m/z value of 401 as [M+H]+ ion corresponds to the compound expected to have

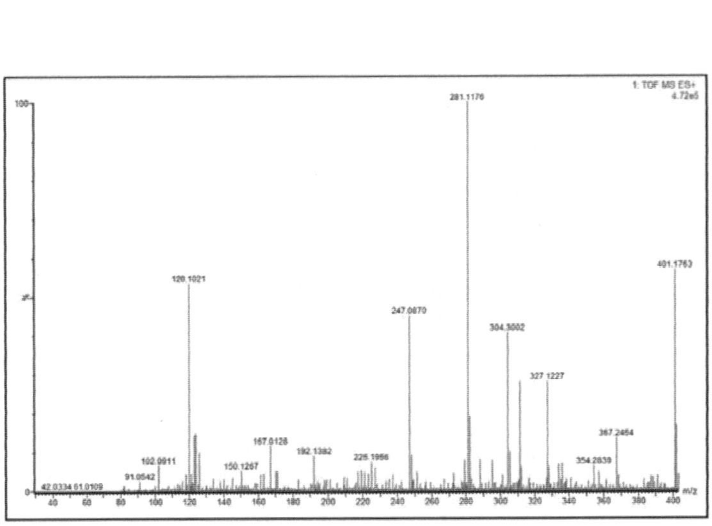

Figure 3. Mass spectra of phenazine derivative compound.

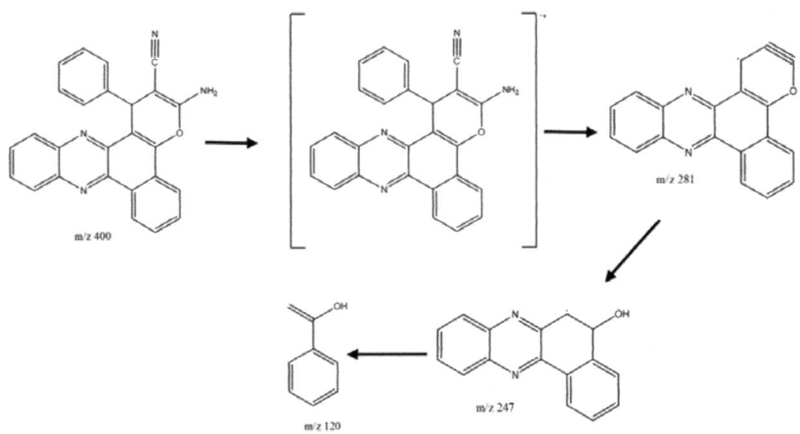

Figure 4. The proposed structure fragment of phenazine derivative compound.

the molecular formula $C_{26}H_{16}N_4O$ with the IUPAC name 3-amino-2-cyano-1-phenyl-1H-benzo [a] pyrano [2,3-c] phenazine. Fragments formed from the results of the analysis using LC-MS can be proposed as seen in figure 4. Several important fragments have a high intensity like m/z = 281, m/z = 247, and m/z = 120.

4. Conclusion

The phenazine derivative was successfully synthesized using lawsone as a precursor through the one-pot four-component method. The synthesized compound was evaluated with some instruments and showed that the analysis of FTIR, Uv-Vis, and LC-MS spectra revealed the characteristics of the target compound. The results of the analysis have identified that the compound formed was a phenazine derivative named 3-amino-2-cyano-1-phenyl-1H-benzo [a] pyrano [2,3-c] phenazine. The target compound was synthesized with a yield of 45 % and the conditions of reaction were conducted at 50 °C for 6 h in ethanol with unassisted catalysts.

Acknowledgments
This work was financially supported by Universitas Indonesia under research grant PITTA with grant contract number NKB-0612/UN2.R3.1/HKP.05.00/2019.

References

[1] Zhou B, Liu Q, Wang H, Jin H and Liu Y 2019 *Tetrahedron* **75** 3815–21
[2] Dar U A, Islam N, Shah S A, Salunke-gawali S, Andrabi S M A and Khujaria Y 2019 *J. Mol. Struct.* **3** 127306
[3] Suhara Y, Watanabe M, Motoyoshi S, Nakagawa K, Wada A, Takeda K, Takahashi K, Tokiwa H and Okano T 2011 *J. Med. Chem.* **54** 4918–22
[4] Ganapaty S, Steve Thomas P, Karagianis G, Waterman P G and Brun R 2016 *Phytochemistry* **67** 1950–6
[5] Silva T M S, Camara C A, Barbosa T P, Soares A Z, da Cunha L C, Pinto A C and Vargas M D 2005 *Bioorg. Med. Chem* **13** 193–6
[6] Biot C, Bauer H, Schirmer R H and Charvet E D 2004 *J. Med. Chem.* **47** 5972-83
[7] Mäntylä A, Garnier T, Rautio J, Nevalainen T, Vepsälainen J, Koskinen A, Croft S L and Järvinen T 2004 *J. Med. Chem.* **47** 188–95
[8] Tandon V K, Maurya H K, Tripathi A, ShivaKeshava G B, Shukla P K, Srivastava P and Panda D 2009 *Eur. J. Med. Chem.* **44** 1086–92
[9] Tandon V K, Maurya H K, Yadav D B, Tripathi A, Kumar M, and Shukla P K 2016 *Bioorg. Med. Chem. Lett.* **16** 5883–7
[10] Suhara Y, Watanabe M, Motoyoshi S, Nakagawa K, Wada A, Takeda K, Takahashi K, Tokiwa H and Okano T 2011 *J. Med. Chem* **54** 4918–22
[11] Liu L and Floreancig PE 2010 *Curr. Opin. Drug Disc. Devel.* **13** 733
[12] Cao L, Long H, Guan H, Bi Y, Bi G, Huang H and Liu L 2019 *Tetrahedron Letters* **60** 1268–71
[13] Kathawate L, Gejji S P, Yeole S D, Verma P L, Puranik V G and Salunke-Gawali S 2015 *J. Mol. Struct.* **1088** 56–63
[14] Delarmelina M, Greco S J and Carneiro J W 2017 *Tetrahedron* **73** 4363–70
[15] Dar U A., Islam N, Shah S A, Salunke-gawali S, Andrabi S M A and Khujaria Y 2019 *J. Mol. Struct.* **3** 127306
[16] Weissenberg M, Meisner J, Klein M, Schaeffler I, Eliyahu M, Schmutterer H and Ascher K R S 1997 *J. Chem. Ecol.* **23** 3–18
[17] Duroux L, Delmotte F M, Lancelin J M, Keravis G and Jay-Allemand C 1998 *Biochem. J.* **333** 275–83
[18] Talcott R E, Smith M T and Giannini D D *Biochem. Biophys.* **241** 88–94
[19] Reese S, Vidyasagar A, Jacobson L, Acun Z, Esnault S, Hullett D, Malter J S and Djamali A 2010 *Fibrog Tissue Rep.* **3** 1
[20] Marchionatti A M, Picotto G, Narvaez C J, Welsh J and Tolosa de Talamoni NG 2009 *J. Steroid Biochem. Mol. Biol.* **113** 227–32
[21] Laursen J B and Nielsen J 2004 *Chem. Rev.* **104** 1663–85
[22] Jardim G A M, Cruz E H G, Valença W O, Resende J M, Rodrigues B L, Ramos D F,Oliviera R N, Silva P E A and da Silva Júnior E N 2015 *J. Braz. Chem. Soc.* **26** 1013–27
[23] de Andrade-Neto V F, Goulart M O F, da Silva Filho J F, da Silva M J, Pinto M D C F R, Pinto A V, Zalis M G, Carvalho L H and Krettli A U 2004 *Bioorg. Med. Chem. Lett.* **14** 1145–9
[24] Makgatho M E, Anderson R, O'Sullivan J F, Egan T J, Freese J A, Cornelius N and van Rensburg C E J 2000 *Drug Dev. Res.* **50** 195–202

[25] Ligon J M, Hill D S, Hammer P E, Torkewitz N R, Hofmann D, Kempf H and Pe K V 2000 *Pest Manag. Sci.* **56** 688–95
[26] Gao J, et al. 2015 *Comb. Chem. High Throughput Screen.* **18** 960–74
[27] Johnson A J, Kumar R A, Rasheed S A, Chandrika S P, Chandrasekhar A, Baby S and Subramoniam A 2010 *J. Ethnopharmacol.* **130** 267–71
[28] Khanna G, Chaudhar A and Khurana J M 2014 *Tetrahedron Lett.* **55** 6652–4
[29] Hasaninejad A, Firoozi and Mandegani F 2013 *Tetrahedron Lett.* **54** 2791–4
[30] Polshettiwar V and Varma R S 2008 *Chem. Soc. Rev.* **37** 1546–57
[31] Jiang B, Tu S J, Parminder K, Walter W and Li G J 2009 *Am. Chem. Soc.* **131** 11660–1

The optical and high-pressure raman scattering study of $La(Fe,Mo)_{0.5}O_3$ ceramics

D Triyono[1], H Laysandra[2] and R A Rafsanjani[1]

[1]Department of Physics, Faculty of Mathematics and Natural Sciences (FMIPA), Universitas Indonesia, Depok 16424, Indonesia
[2]Department of Physics, National Taiwan Normal University, Taipei 11677, Taiwan

Corresponding author's email: djoko.triyono@ui.ac.id

Abstract. This study was focused on the optical-vibrational analysis and high-pressure Raman-scattering spectroscopy of $La(Fe,Mo)_{0.5}O_3$ ceramics at room temperature. This material was fabricated by using the sol-gel method. The element composition was checked by X-ray Fluorescence confirming the atomic ratio of La, Fe and Mo. Fourier Transform Infrared spectroscopy confirmed the (Fe/Mo−O) bonding vibration which is typical perovskite structure. The band gap energy investigated by UV-Vis spectra was found to be 2.03 eV. At room temperature, the Raman scattering spectra under high pressure was characterized by the Raman-scattering spectrum in the range of 0.90 to 16.7 GPa. The high-pressure Raman scattering confirmed the structural transition after the pressure surpasses 7.0 GPa. From this phenomena, it well indicates the lattice structure stability is still preserved under 7.0 GPa.

Keywords: Optical-vibration, raman scattering, $La(Fe,Mo)O_3$, sol-gel method, structural transition

1. Introduction

The research about the effect of Mo substitution on the crystal structure, vibrational, and dielectric studies of $LaFeO_3$ have been reported in our previous work [1]. Mo substitution caused the increase of lattice constant, decrease the average grain size and dielectric parameters, and increase the energy barrier that blocks the mobility of space charges in lattice [1]. Raman scattering analysis at room pressure revealed the local lattice distortions and confirmed the shifting of the T_N to a lower temperature due to the Mo substitution [1, 2].

However, to the best our knowledge, there is no reported research for high-pressure Raman scattering of $La(Fe,Mo)_{0.5}O_3$ ceramics. As comparison, the high-pressure Raman scattering of its parent compound, $LaFeO_3$, have been investigated [2, 3]. The stability of crystal structure of $LaFeO_3$ is still preserved under 5.0 GPa [2]. The high-pressure phase transition in $LaFeO_3$ perovskite material was reported that the second-order structural phase transition from *Pbnm* to *Ibmm* was observed at 21.1 GPa and subsequently a first-order phase transition from high-spin to low-spin Fe^{3+} ions was observed at 38.0 GPa [3]. Based on this, it is interesting to explore more details the structural changes of $La(Fe,Mo)_{0.5}O_3$ ceramics under high pressure. The high pressure Raman scattering measurement is tracked by using Range of Raman equipment usage is starting from 0.90 to 16.7 GPa. The results were fitted using Peakfit software to observe the pressure-dependence on the structural changes of

$La(Fe,Mo)_{0.5}O_3$. In addition, we use the X-ray fluorescence, Fourier transform infrared spectroscopy, and UV-Visible spectroscopy to determine the atomic ratio of the element composition, metal-oxide vibration identification, and direct band gap energy in the $La(Fe,Mo)_{0.5}O_3$ ceramics system, respectively.

2. Experimental

The synthesize and experimental condition for high pressure Raman measurement of $La(Fe,Mo)_{0.5}O_3$ samples have been explained in previous works [1, 2]. The material was synthesized by sol-gel route. The highpressure Raman spectroscopy analysis were measured in Via Renishaw Raman spectrometer system with 512 nm of excitation-laser wavelength. The element composition was checked using X-ray Fluorescence (XRF) type PANalytical Epsilon 1. The molecular and metal-oxide vibration were analyzed by using FTIR (PerkinElmer type SpectrumTwo) in the range of 500–1300 cm cm^{-1}. The optical band gap energy was determined by using UV-Vis (Agilent Cary 100/300).

3. Results and discussion

3.1. X-ray fluorescence analysis

Figure 1 presents the X-ray fluorescence spectra of $La(Fe,Mo)_{0.5}O_3$ ceramics. The spectra revealed the presence of Mo substituting in material. Also, the presence of La and Fe elements were confirmed. The element ratio analysis is tabulated in table 1. The result is insignificantly different from the stoichiometry. However, the differences are still in tolerance limit.

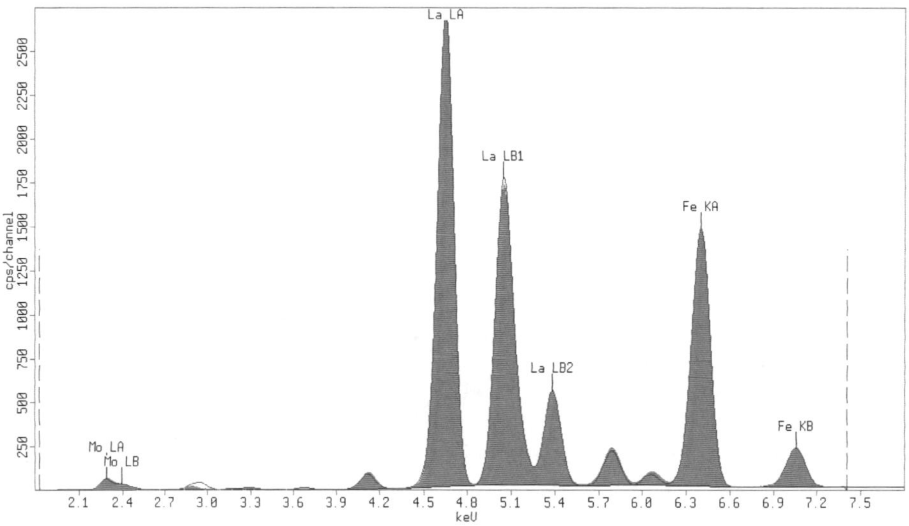

Figure 1. XRF curve of $La(Fe,Mo)_{0.5}O_3$ ceramics.

Table 1. Results of x-ray fluorescence analysis of $La(Fe,Mo)_{0.5}O_3$ ceramics.

Element ratio (design.)			Element ratio (exp.)			Element ratio (calc.)		
La	Fe	Mo	La	Fe	Mo	La	Fe	Mo
64.54	12.64	22.82	1	0.4872	0.5119	1	0.5	0.5

IOP Conf. Series: Materials Science and Engineering **902** (2020) 012021 doi:10.1088/1757-899X/902/1/012021

3.2. FTIR analysis

Figure 2 represents the FTIR Spectra of $La(Fe,Mo)_{0.5}O_3$ ceramics. It shows clear and prominent band at wave numbers 542 cm cm^{-1} confirming the typical perovskite structure formation which is attributed to metal-oxygen (Fe/Mo−O) bond vibration [4]. The band around 800 cm cm^{-1} is possibly attributed to La−O stretching vibration [5]. Additional bands above 1000 cm cm^{-1} might be due to the vibration of water molecules throughout the sample.

3.3. UV-Vis analysis

Figure 3 shows the diffused reflectance spectrum recorded in wavelength range 300–800 nm for $La(Fe,Mo)_{0.5}O_3$ ceramics along with its Tauc plot at room temperature. The diffused reflectance spectrum is used to get the absorption coefficient by using Kubelka Monk (KB) function [4]. We calculate direct band gap energy (E_g) using the Tauc equation [6]. The optical band gap energy estimated by direct allowed transition resulting is about 2.03 eV. It is to be noted that the band gap estimated for $La(Fe,Mo)_{0.5}O_3$ seems lower than $LaFeO_3$ from previously reported [4, 6, 7].

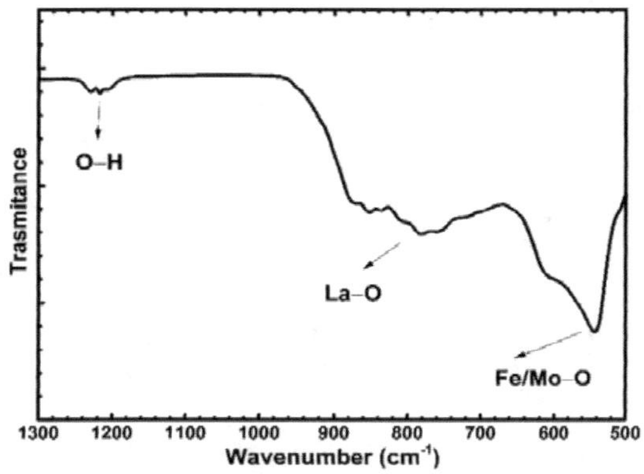

Figure 2. FTIR Spectra of $La(Fe,Mo)_{0.5}O_3$ ceramics.

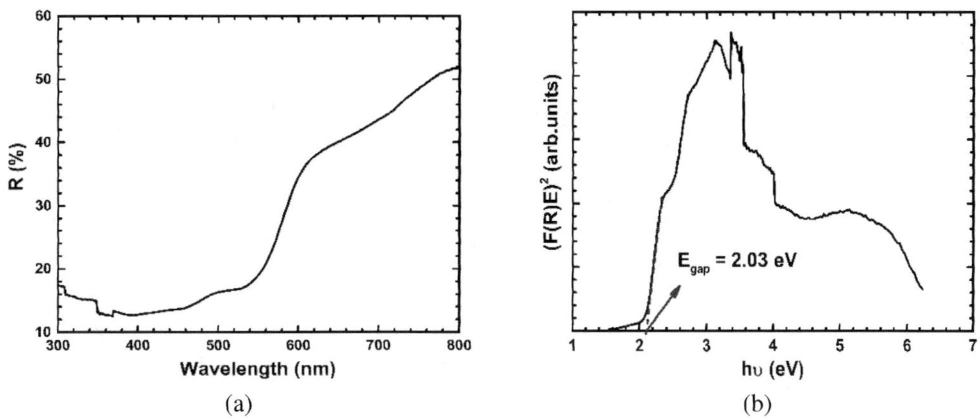

Figure 3. (a) Recorded reflectance spectrum and (b) the plot of $(\alpha h\nu)^2$ vs. $h\nu$ of $La(Fe,Mo)_{0.5}O_3$ ceramics.

Mo substitution leads to the increase of lattice parameters and Fe/Mo−O bond length, so that the atomic bonding becomes weaker than the parent, $LaFeO_3$, and the movement of charge carriers (electron, hole, etc) to another site in lattice become easier. As a consequence, this lowered the energy that needed to move to another site, i.e., gap energy for this material.

3.4. High Pressure Raman Scattering Analysis

Figure 4 shows the Raman scattering spectra $La(Fe,Mo)_{0.5}O_3$ for several pressures at room temperature. The orthorhombic structure with *Pbnm* symmetry of $La(Fe,Mo)_{0.5}O_3$ has been confirmed in our previous work [1]. We are focusing on the Raman modes at low region, below 200 cm^{-1}. As confirmed by previously reported [3, 8-9], the Raman phonon modes under 200 cm^{-1} are related with the vibration of La-atom in lattice. The Raman phonon modes between 100 and 160 cm^{-1} are associated to an A_g mode related to vibration of the La site. The Raman mode around 180 cm^{-1} is related with the A_g mode related to La rotation.

Now, we turn to address the 4 Raman phonon modes below 200 cm cm^{-1}. We identify the modes about 100, 165, 180 and 175 cm cm^{-1} and labelled as M1, M2, M3 and M4, respectively. The Raman phonon modes were fitted with a standard Lorentzian profile [10]. The fitting results are shown in figure 5. The M1 is still preserved until 16.7 GPa. The M2 and M3 mode disappear

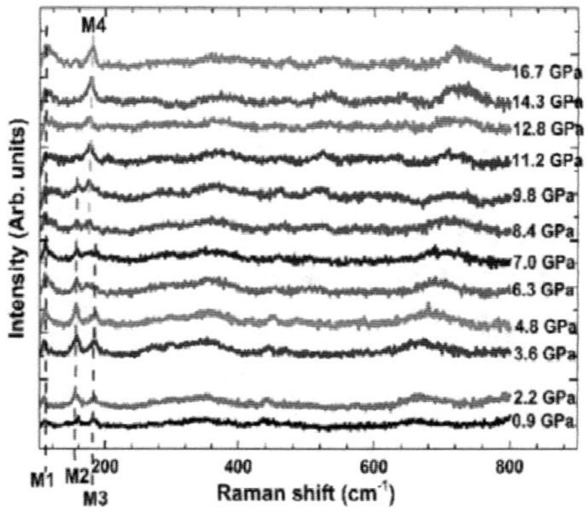

Figure 4. The Raman-scattering spectra of $La(Fe,Mo)_{0.5}O_3$ for several pressures.

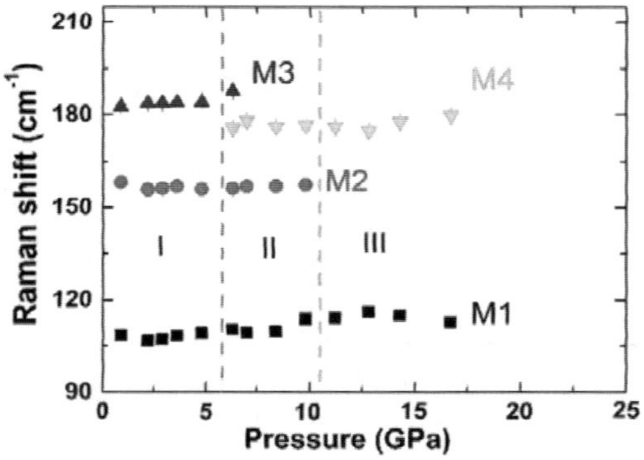

Figure 5. Raman shift as a function of pressure based on fits to Raman modes of $La(Fe,Mo)_{0.5}O_3$ below 200 cm^{-1}.

at 9.8 and 6.3 GPa, respectively. However, the M4 starts to perform under pressure about 6.3 GPa. This confirms the possibility of the local lattice distortion occurring in this sample. The similar trend has been reported for $LaFeO_3$ sample [2].

4. Conclusion

The optical properties of $La(Fe,Mo)_{0.5}O_3$ ceramic have been studied. The XRF analysis confirmed the all elemental compositions (La, Fe, Mo, and O) were in good agreement with their respected molar concentrations. FTIR analysis revealed (Fe/Mo−O) vibration indicating the typical perovskite structure. The direct band gap energy estimated from UV-Vis analysis is about 2.03 eV. The high pressure Raman scattering analysis confirmed the lattice starts to distorted from its initial condition after pressure surpasses 7 GPa.

Acknowledgments

The authors thanks to R A Susilo from Center for High Pressure Science and Technology Advanced Research (HPSTAR) for the assistance of the high-pressure experiments and Hibah Penelitian Dasar Perguruan Tinggi (PDPT) 2020 by Ministry of Research and Technology (KEMRISTEK/BRIN) Republic of Indonesia with contract number NKB-49/UN2.RST/HKP.05.00/2020.

References

[1] Triyono D, Laysandra H and Liu H L 2019 *J. Mater. Sci.: Mater. Electron.* **30** 2512-22
[2] Triyono D and Laysandra H 2019 *Iop Conf. Ser. Mater. Sci. Eng* **496** 012045
[3] Etter M, Muller M, Handfland M and Dinnebier R E 2014 *Acta Crystallogr. B* **70**
[4] Rai A and Thakur A K 2017 *Mat. Sci. Eng. B-Adv* **224** 139-49
[5] Refka A, Karima H-N and Mokhtar F 2012 *Power Technol.* **230** 183-7
[6] Thirumalairajan S, Girija K, Neha Y H, Mangalaraj D, Viswanathan C and Ponpandian N 2013 *RSC Advances* **3** 7549-61
[7] Arima T, Tokura Y and Torrance J B 1993 *Phys. Rev. B.* **48** 17006
[8] Mahapatra A S, Mitra A, Mallick A, Ghosh M and Chakrabarti P K 2016 *Mater. Lett.* **169** 160-3
[9] Coutinho P, Cunha F and Barrozo P 2017 *Solid State Commun.* **252** 59-63
[10] Moreira R L, Lobo R P S M, Ramos S L L M, Sebastian M T, Matinaga F M, Righi A and Dias A 2018 *Phys. Rev. Mater.* **2** 054406

IOP Conf. Series: Materials Science and Engineering **902** (2020) 012022 doi:10.1088/1757-899X/902/1/012022

Sixth-generation of ziegler-natta catalyst for homopolymer polypropylene polymerization: Effect to product properties and regulation

R I Syahrudin and B A Herlambang

Research and Development Department, PT Chandra Asri Petrochemical, Tbk., Cilegon, Banten 42447, Indonesia

Corresponding author's email: grouprnd@capcx.com

Abstract. The development of catalysts for polypropylene (PP) polymerization has begun since Ziegler-Natta (Z-N) discovered the catalyst for polyethylene polymerization. However, the first-generation catalyst which was a combination of $TiCl_3$-$AlEt_2Cl$ (catalyst-co-catalyst) has low productivity and isotacticity. The modifications continue to be made to increase the productivity and isotacticity by combining the Z-N catalyst with $MgCl_2$, internal donors, and external donors. The fourth-generation Z-N catalyst, $TiCl_4$/Phthalate/$MgCl_2$-$AlEt_3$/Silane, has given satisfactory results with high productivity and isotacticity. Another concern arose when phthalate derivatives turned out to be able to initiate cancer formation. This problem has led catalyst development to the sixth-generation Z-N catalyst with a phthalate-free internal donor. This research intends to study the effects of sixth-generation catalysts to polypropylene properties when compared to the fourth-generation catalysts. We produced homopolymer polypropylene (HPP) by fourth- and sixth-generation of Z-N catalyst with the same polymerization technology. Then, the effects on product properties, namely melt flow rate (MFR), xylene soluble, residual catalyst, physical properties, optical properties, and rheological properties were observed. Sixth-generation catalyst exhibited a better stereoregularity and it made a better flexural modulus. The rheological properties showed the sixth-generation catalyst products have lower molecular weight and it generates fewer gels and lower impact strength. Sixth-generation catalyst also exhibited lower catalyst residue which allows lower dosage of an acid scavenger with comparable color performance. With increasingly strict restrictions on the use of phthalates, the sixth-generation catalyst is more convenient to use. In terms of the effect on the polypropylene properties, fourth- and sixth-generation catalysts have their respective advantages.

Keywords: PP Homopolymer, fourth-generation Z-N, sixth-generation Z-N, phthalate

1. Introduction

Polypropylene polymerization has evolved from the past few decades. It started by using the first-generation Z-N catalyst, the catalyst/co-catalyst system, $TiCl_3/Al(C_2H_5)_2$. This first-generation catalyst is only able to achieve the productivity of 0.8–1.2 kgPP/gCat and isotacticity index of 90–94 % (polymerization in hexane slurry, no data for bulk polymerization) [1]. The second-generation was developed by Solvay where the $TiCl_3$ catalyst enhancements involved the reduction of $TiCl_4$ using diethyl aluminum chloride and followed by ether treatment to remove triethyl aluminum and $TiCl_4$ treatment to change the β-$TiCl_3$ to δ-$TiCl_3$ [2]. Solvay claims that his method can produce catalysts with spherical shape and narrow particle size distribution (PSD) resulting in a microporous sponge-like

IOP Conf. Series: Materials Science and Engineering **902** (2020) 012022 doi:10.1088/1757-899X/902/1/012022

structure. The performance of this catalyst can achieve the productivity of 10–15 kgPP/gCat and isotacticity 94–97 % (in bulk polymerization) [1]. Therefore, this second-generation catalyst can be called a porous $TiCl_3$-DEAC system.

In terms of increasing the productivity of the Z-N catalyst for polypropylene polymerization, it was found that $MgCl_2$ supported catalyst gave a higher productivity [3, 4]. However, this catalyst has a low stereospecificity (isotacticity < 50 %) [1]. Answering the problem, Lewis bases were introduced in this catalyst system with the $TiCl_4$, $MgCl_2$, and Lewis base. Lewis base usually has one or more oxygen atoms that can donate a pair of free electrons to the Mg atom in $MgCl_2$. The donors will eventually be bound to the surface of $MgCl_2$ and increase the steric environment so that propylene can only enter on certain sides of the Ti atom [5, 6]. These Lewis bases are differentiated as first Lewis base which is put together with the catalyst, so it is called internal donor (D_i). The second Lewis base is added separately and called external donor (D_e). Ziegler-Natta catalyst systems and its performances are presented in table 1.

Research on the comparison of productivity and stereospecificity of various generations of Ziegler-Natta catalysts has been carried out [1, 7-9]. However, the effect of these catalyst types on polypropylene properties has not been much studied. Therefore, this research focuses on comparing the effect of fourth- and sixth-generation catalysts on polypropylene product.

The consideration of the use of a sixth-generation catalyst began after it was discovered that a phthalate derivative in a fourth-generation catalyst system could initiate breast cancer [10]. The restriction on the use of phthalates has been carried out in the European region with the inclusion of 6 phthalate derivative compounds in the REACH Annex XVII restricted substances list. The compounds in question include Bis (2-Ethylhexyl) phthalate (DEHP), Dibutyl phthalate (DBP), Benzyl butyl phthalate (BBP), Di-isononyl phthalate (DINP), Di-isodecyl phthalate (DIDP), and Di-n-octyl phthalate (DNOP). On 4 June 2015, EU RoHS2 (Directive 2011/65 / EU) was also revised by the publication of new Directive (EU) 2105/863 to add 4 phthalate-derived compounds including DEHP, BBP, DBP, and DIBP. In the EU RoHS2 document, all phthalate derivatives that have been mentioned are limited to less than 0.1% and this prohibition is effective from 22 July 2019 for all electrical and electronic equipment irrespectively from Category 8 and Category 9 [11].

2. Experimental

All experiments have been carried out using homopolymer polypropylene MFR 2, 3, 7, 8 and 10 for biaxially oriented polypropylene (BOPP), thermoforming, yarn, cast film, tubular quenching film, and injection molding application produced by PT Chandra Asri Petrochemical. All the products were produced by two types of catalyst, fourth-generation of Z-N catalyst and sixth-generation of Z-N catalyst with the same reactor and polymerization process (gas phase polymerization).

Table 1. Ziegler-Natta catalyst composition and performance in different generations.

Generation	Composition	Productivity [1] (kgPP/g Cat)	Isotacticity index [1] (%)
1st Gen.	$TiCl_3 + Al(C_2H_5)_2$	N/A	90–94
2nd Gen.	Porous $TiCl_3 + Al(C_2H_5)_2$	10–15	94–97
3rd Gen.	$TiCl_4$/EB/$MgCl_2$ + $Al(C_2H_5)_3$/MPT	15–30	90–95
4th Gen.	$TiCl_4$/Phthalate/$MgCl_2$ + $Al(C_2H_5)_3$/Silane	30–60	95–99
5th Gen.	$TiCl_4$/Dieter/$MgCl_2$ + $Al(C_2H_5)_3$	70–120	95–99
6th Gen.	$TiCl_4$/p-Ethoxy Ester/$MgCl_2$ + $Al(C_2H_5)_3/D_e$	N/A	N/A

2.1. Rheological analysis

The rheological analysis was conducted to see the flow behavior of polymer produced by fourth-generation and sixth-generation of Z-N catalyst. The flow behavior was used as an approach to compare the molecular weight of polymers [12]. It was performed in a twin-bore capillary rheometer (Bagley correction) using die with L/D 16 at temperature 230 °C with total of 9 min of preheating time. The method is in accordance with ASTM D3835. The result was shown as shear viscosity (Pa.s) of polymer melt [13].

2.2. Melt flow rate analysis

Melt flow analysis was performed using an extrusion plastometer according to ASTM D1238. The temperature was set at 230 °C and used a load of 2.16 kg. After the temperature was stable, 4 g of polypropylene pellets were filled into the bore and testing was started after preheating time has been completed (7 ± 0.5 min). The result was shown as the weight of the polymer in grams per 10 min (g/10 min).

2.3. Xylene soluble

Xylene soluble is a method to determine the number of atactic in the polypropylene matrix. The analysis was carried out using NMR equipment because it is more convenience compared to the reflux method in terms of time and steps [14]. Pellet samples were filled into the NMR tube (maximum until 2 cm from the top) and then it was placed into the NMR [14].

2.4. Fines

The analysis was conducted using a sieve shaker apparatus with 125 μ sieve. Polymer resins were weighed and then placed in sieve shaker. The result was shown in % weight.

2.5. Colorimeter

Polypropylene pellet color was rated as Yellowness Index (YI). This parameter was measured by Colorimeter equipment following ASTM D6290. Polypropylene pellet was filled into the transparent sample cup and placed on the scanning area of the instrument [15]. The cup must be covered with a black cap to minimize interference from outside light.

2.6. Gels performance

Gel counting analysis was intended for film grade application following ASTM D3351. It included cast film and inflated film. Before conducting the gel counting, polymer samples were converted to film using a cast film machine (± 40 μm thickness) or tubular quenching machine (± 25 μm thickness). The area of the film being analyzed was 4 × 4 in. with 5 repetitions. The film sample was then placed on the OHP with a 5 times magnification [16]. Gels were categorized as large if it was larger than 4 mm and small if it was between 2–4 mm.

2.7. Catalyst metal residue

The residue of the metal compound of catalyst (Ti) can be measured quantitatively by X-ray fluorescence (XRF) [17]. Polypropylene plaque was prepared by hot press molding using a mold with a thickness of 3 mm and a diameter of 40 mm. The hot press temperature was set at 230 °C.

2.8. Physical properties

Physical properties are included in flexural modulus and Izod impact performance. The test was conducted according to ASTM D790 for flexural modulus and ASTM D256 for Izod impact strength [18]. Polypropylene pellets were first formed into plaques (type I) using injection molding. The plaques were then conditioned according to ASTM D618.

2.9. Film optical properties

Film optical properties are included haze, gloss, and clarity each of which following ASTM D1003, D2457 and D1746. All those ASTM standards are applicable for plastics film and sheet. The method has also been done by Oas for haze, clarity, and gloss analysis of polypropylene bottle samples. Before optical analysis was conducted, fabricated films must first be conditioned at a temperature of 23 ± 2 °C and a humidity of 50 ± 5 %RH for a minimum of 40 h. The films were then made into a specimen 50 mm × 50 mm.

3. Results and discussion

Figure 1a shows the comparison of xylene soluble between polypropylene produced with the fourth-generation of Z-N catalyst and those produced using the sixth-generation Z-N catalyst in several MFR ranges. In all MFR ranges, the sixth-generation shows comparable results towards the fourth-generation even with less catalyst consumption (less internal donor consumption). It means that for the same amount of catalyst consumption, the sixth-generation gives lower xylene soluble values (more isotactic). This also proves that the internal donor of sixth-generation catalysts can match the performance of phthalate [8]. When comparing the external donor consumption, the sixth-generation can save more than 50 % by weight.

Xylene soluble describes the atactic content in polymers. The higher the atactic content, the lower the isotacticity of the polymer. Stereoregularity is very closely related to the physical properties of polypropylene, especially flexural modulus which describes the rigidity of the final product [19]. Figure 1b shows the relationship between xylene soluble and flexural modulus in several grades of PP. As xylene soluble increases, the flexural modulus decreases. This happens to both products produced by fourth-generation and sixth-generation catalysts. The interesting thing is, significantly, the sixth-generation products have a flexural modulus that is higher than the fourth-generation consistently at all grades.

On the contrary to flexural modulus, the impact properties are more influenced by molecular weight [20]. The smaller molecular weight causes the impact strength decreases [18]. The impact energy not only causes deformation/strain but the required energy to cause a fracture. The intermolecular force of the polymer is not enough to hold this energy so that intramolecular interactions are more dominant in this case. The longer the molecular chain (high molecular weight), the higher the intramolecular interactions within it so that the impact strength is greater [21]. On the contrary to low molecular weight polymers with more intermolecular interactions exhibits lower impact strength.

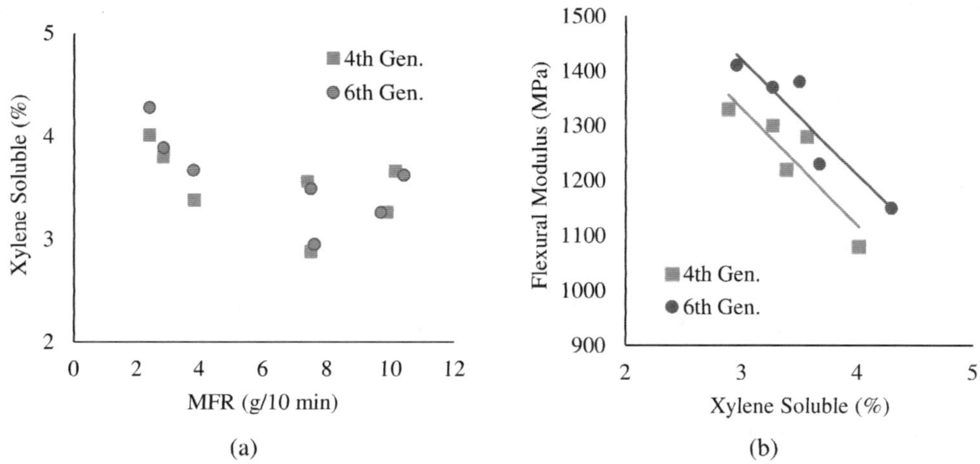

Figure 1. Analysis result for several PP grades, (a) Xylene soluble (% atactic) as a function of MFR, and (b) Flexural modulus as a function of XS.

In figure 2 HPP MFR 10 has lower shear viscosity (less viscous) which means it has lower molecular weight [12]. This proves that with increasing MFR the molecular weight will decrease. This happens during the polymerization process, precisely in the hydrogen termination process. High MFR can be achieved with more hydrogen feeds, meaning that the termination process is more numerous and results in shorter polymer chains (lower molecular weight) [22]. Another fact, in both MFRs, the products produced by the sixth-generation catalyst always have lower viscosity. It means that at the same MFR, the sixth-generation catalyst produces a lower molecular weight.

The relationship between Izod impact and MFR can be seen in figure 3a. In general, the impact strength decreases with the increasing MFR. The data also show that HPP grades produced by sixth-generation catalysts have a lower impact strength than those produced from the fourth-generation catalysts due to lower molecular weight and this is consistent with the results discussed in Kirk-Othmer encyclopedia of Chemical Technology [8]. These characteristics apply to all MFR ranges.

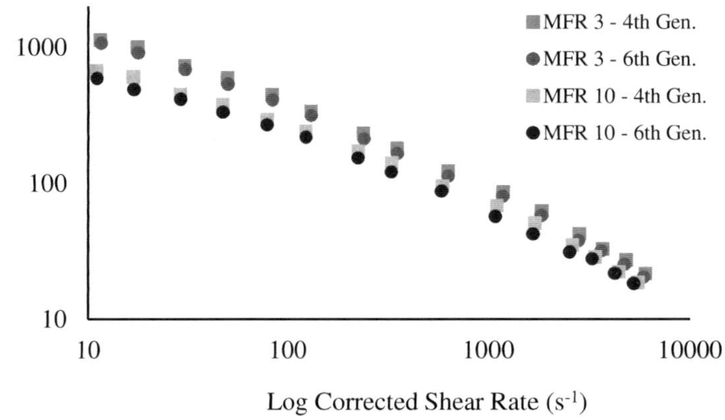

Figure 2. Rheological analysis (shear viscosity) of HPP MFR 3 and MFR 10.

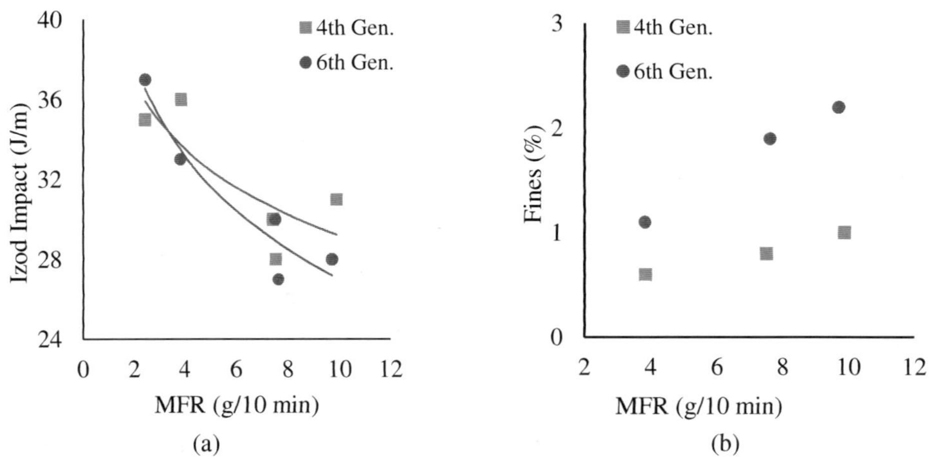

Figure 3. Analysis result for several PP grades. (a) Izod impact strength as a function of MFR, and (b) Fines content of resin as a function of MFR.

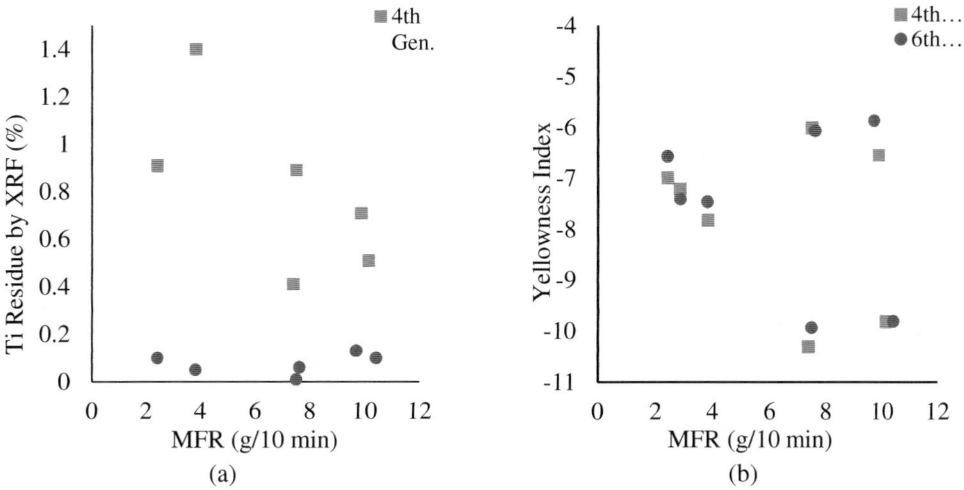

Figure 4. Analysis result for several HPP grades (a) Ti residue, and (b) Yellowness index.

Table 2. Optical film properties of HPP film grades.

Grades	Clarity		Gloss		Gel count			
	4th Gen.	6th Gen.	4th Gen.	6th Gen.	4th Gen.		6th Gen.	
					Small	Large	Small	Large
HPP MFR 7	58	58	109	110	5	2	4	1
HPP MFR 8	51	51	141	142	7	2	5	2

Low molecular weight also affects the physical form of polymer resins as shown in figure 3b Low molecular weight polymer will act as plasticizer and soften the polymer material [23]. The content of fines increases with the increasing MFR due to lower molecular weight. As well as the performance of two different catalysts where the sixth-generation catalysts produce more fines in polymer resins compared to the fourth-generation catalysts.

Polypropylene pellets contain metal residues such as calcium from additives, aluminum from co-catalyst, titanium and magnesium from the catalyst, and other metals from impurities. The metal content can be detected quantitatively by XRF. Metal residue analysis is focused on catalyst residue, Ti, because of its contribution to the polypropylene oxidation process. The results will be displayed in percent by weight of Ti in polypropylene samples. The results of a quantitative XRF analysis of several HPP samples are shown in figure 4a. As the productivity of the sixth-generation catalyst is much higher, this leaves less catalyst residue in the polymer. Catalyst residues (metal compounds) can accelerate the initiation of the oxidation process [25]. One of the parameters that indicates the oxidation in a polymer is discoloration (generally becomes yellowish). In order to neutralize the catalyst residue in the polymer, acid neutralizer/acid scavenger is added which in this case can be from the metal-soap group (calcium stearate, zinc stearate, etc.), metal oxide (CaO, ZnO, etc.), or hydrotalcite compounds [24]. Low catalyst residue can be a reason to reduce the dosage of acid scavenger to reduce costs. In figure 4b, the sixth-generation catalyst product has a typical color similar to a fourth-generation product with a note that the dosage of acid scavenger is lower.

In the case of film properties, especially the clarity and gloss performance, they do not show differences as shown in table 2. However, in gel count properties, the sixth-generation catalyst product shows a better performance than the fourth-generation catalyst. This is related to the molecular weight of the polymer produced. Gels have visual defects and it can be caused by higher-molecular-weight material or contamination [25]. Since sixth-generation catalyst produces lower molecular weight polymer, it can reduce the probability of gel phenomenon.

4. Conclusion

Sixth-generation Z-N catalyst gives a better performance in terms of productivity and stereoregularity. Higher polymer's isotacticity of sixth-generation Z-N catalyst affects physical properties, especially flexural modulus (stiffness). While other physical properties, in this case, the impact strength, are not directly related to stereoregularity. Izod impact is more influenced by the weight of the molecule polymer. From the rheology approach, the sixth-generation catalysts produce polymers with lower molecular weight compared to fourth-generation and it affects to smaller Izod impact of HPP grades. The different catalyst shows an insignificant correlation to clarity and gloss performance but it highly affects the gel performance. Sixth-generation Z-N catalyst shows a better gel performance in terms of lower molecular weight polymer. Higher productivity of sixth-generation Z-N catalyst also affects to the lower metal residue (Ti) so the acid scavenger dosage can be adjusted to lower value with comparable color performance. Thus, with increasingly strict restrictions on the use of phthalates, the sixth-generation catalysts are more convenient to use. In terms of performance, fourth- and sixth-generation catalysts have their respective advantages.

Acknowledgments

The authors are very grateful to Research and Development (RND) team of PT. Chandra Asri Petrochemical, Tbk for their great support to make this study possible.

References

[1] Moore E P 1996 *Polypropylene Handbook: Polymerization, Characterization, Properties, Processing, Applications* (Munich, NY: Hanser Publishers)
[2] Hermans J P and Henrioulle P 1980 U.S. Patent No. 4 210 738
[3] Cui N et al. 2006 *J. Appl. Polym. Sci.* **99** 1399-404
[4] Singh G, Kaur S, Makwana U, Patankar R B and Gupta V K 2009 *Macromol. Chem. Phys.* **210** 69-76
[5] Raj K V and Vanka K 2017 *Resonance* **22** 1025-37
[6] Noto V D et al. 1992 *Makromol. Chem.* **193** 123-31
[7] Jiang X and He A 2013 *Polym. Int.* **63** 179-83
[8] Cecchin G, Morini G and Piemontesi F 2003 *Ziegler-Natta Catalysts* Kirk-Othmer Encyclopedia of Chemical Technology (United States: John Wiley & Sons)
[9] Yaluma A K, Tait P J T and Chadwick J C 2006 *J. Polym. Sci. Part A: Polym. Chem.* **44** 1635-47
[10] López-Carrillo L et al. 2010 *Environ. Health Perspect.* **118** 539-44
[11] The RoHS Directive Electrical and Electronic Waste-Environment-European Commission 2019 available at https://ec.europa.eu/environment/waste/rohs_eee/index_en.htm
[12] Pattamaprom C and Larson R G 2001 *Rheol. Acta* **40** 516-32
[13] Huang J-C and Leong K-S 2002 *J. Appl. Polym. Sci.* **84** 1269-76
[14] Bruker *Xylene Soluble Content in Polypropylene* available at https://public.jenck.com/notijenck/uploads/contenido-de-xileno-soluble-en-polipropileno.pdf
[15] Tocháček J, Jančář J, Kalfus J, Zbořilová P and Buráň Z 2008 *Polym. Degrad. Stabil.* **93** 770-5
[16] Huang Y-B and Wessel T E 1995 *J. Appl. Polym. Sci.* **57** 1227-32
[17] Bichinho K M, Pires G P, Stedile F C, Santos J H Z d and Wolf C R 2005 *Spectrochim. Acta B: Atom. Spectr.* **60** 599-604
[18] Tzoganakis C, Vlachopoulos J, Hamielec A E and Shinozaki D M 1989 *Polym. Eng. Sci.* **29** 390-6

[19] Oas D C and Clairmont D M, Mass., Deacon Machinery, Inc. US4,357,288. 1982 Nov 2.
[20] Phillips R, Herbert G, News J and Wolkowicz M 1994 *Polym. Eng. Sci.* **34** 1731-43
[21] Wei Q 2012 *Functional Nanofibers and their Applications* 1st edition. (United States: Woodhead Publishing)
[22] Ali E M and Ali M A-h 2010 *ISA Trans.* **49** 177-87
[23] Crow Polymer Properties Database Molecular Weight 2019 available at https://polymerdatabase.com/about.html
[24] Allen N S et al. 2009 *J. Vinyl Addit. Techn.* **15** 12-9
[25] Rauwendaal C 2004 *What's Causing Your Gels? (Plastics Technology)* available at https://www.ptonline.com/articles/what's-causing-your-gels

Synthesis and characterization of soft and hard templated mesoporous carbon using phloroglucinol as carbon precursor

I Amelia, Y K Krisnandi and I Abdullah

Department of Chemistry, Faculty of Mathematics and Natural Sciences (FMIPA), Universitas Indonesia, Depok 16424, Indonesia

Corresponding author's email: iman.abdullah@sci.ui.ac.id

Abstract. In this study, both soft and hard templated mesoporous carbon (ST MC and HT MC) were synthesized using phloroglucinol as carbon precursor. The templates used in soft and hard templated MC synthesis were Pluronic 127 (F-127) and SiO_2, respectively. The resulting mesoporous carbon were then impregnated with Ni and characterized using FTIR, SEM-EDS, BET, and XRD to investigate the functional groups, surface morphology, chemical composition, pore character, and crystallinity of the materials. The resulting Ni/MC was then utilized as a catalyst in transformation of CO_2 via carboxylation of phenylacetylene. The catalytic reactions were conducted with a variety of contact time (2, 4, 6 and 8 h) to obtain a reaction condition with the highest yield of phenyl propiolic acid. The resulting final product was tested using HPLC instrument. The phenyl propiolic acid showed best peaks at reaction time of 6 h for ST Ni MC catalyst with the area of 301 mAU and yield of 1.4 %. Meanwhile, HT Ni/MC catalyst showed the best performance in phenyl propiolic acid formation at the reaction time of 8 h with the area of 453 mAU and yield of 2.2 %.

Keywords: Mesoporous carbon, soft template, hard template, phloroglucinol, carboxylation.

1. Introduction

Carbon dioxide is a greenhouse gas that is suspected to be the main cause of global warming. The atmospheric concentration of carbon dioxide (CO_2) has been dramatically raised in the last decades as a result of the industrial development [1]. In 2015, it was recorded that CO_2 emissions from the energy sector in Indonesia reached 261.89 million tons [2]. So, the CO_2 control in the atmosphere needs to be conducted. Currently, there are two main strategies for reducing the CO_2 concentrations, namely: CO_2 capture and storage/sequestration (CCS) and CO_2 capture and its utilization (CCU). CCS is conducted by absorbing CO_2 using various absorbents such as solids, liquids, and membranes, then stored the CO_2 in a closed facility after desorption. While in the CCU strategy, the absorbed CO_2 is used as the raw material of value-added chemical compounds that make the system suitable for the chemical transformation of CO_2 at low pressure (ideally 1 atm), hence no stage of desorption [3]. The CCU's strategy for the study centers on phenylacetylene carboxylation with CO_2 is to produce carboxylic acid and its derivatives. However, it has a big challenge since CO_2 is highly oxidized and thermodinamically stable gas [1]. In fact, most of the reactions concerning CO_2 involve the application of large amount of stoichiometric catalysts and a drastic reaction condition (high reaction temperature and/or high CO_2 pressure). Therefore, the renewable catalytic reactions development is indispensable for the chemical transformation of CO_2 [4]. Due to its stable nature, CO_2 requires low-valence metal assistance e.g. Ni(0) or Pd(0) to react [5]. Among the transition metals, Ni has its own different uniqueness, such as relatively low-cost, its high activity, and its abundant in nature. Therefore, Ni has the potential as an alternative for precious metal catalysts. Nowadays it has been a lot of conversion of carbon dioxide by using the help of a homogeneous catalyst that is in an efficient reaction [3]. However, this homogeneous catalyst is difficult to be separated from the resulting product due to it has the same phase, making it takes a post-treatment. Therefore, the CO_2 transformation method is developed using a heterogeneous catalyst so that the resulting product can be separated more easily and the catalyst can be reused for repeated reactions. On the other hand, mesoporous carbon (MC) has large surface area, pore size, ordered pore structure, and thermal stability [6]. For those reasons, MC has the potential to be used in CCU technology as catalyst support when combined with Ni metal. There are two types of MC synthesis: soft templated and hard templated. Soft templated MC is easier to synthesize while hard templated one has a more ordered structure. In this research, soft templated and hard templated MC were synthesized from phloroglucinol. Phloroglucinol and formaldehyde are able to polymerize via self-assembly around Pluronic F-127 micelle. The resulting polymer is also able to maintain its form during the degradation process, making phloroglucinol suitable as a precursor in the synthesis of mesoporous carbon [7]. The resulting MC was then utilized as a support material for Ni metal by impregnation. The objectives of the present study were to develop aa high surface area templated carbon by pyrolysis of phloroglucinol carbon source in nitrogen environment using both soft and hard templated methods and its application as Ni catalyst support on carboxylation of phenylacetylene with CO_2. The detailed effects of both methods on the development of porous carbon structure were studied. The properties of the precursors and template carbons were studied by Fourier transform infrared (FTIR), X-ray diffraction (XRD), scanning electron microscopy (SEM), energy dispersive x-ray (EDX) and surface area and pore analysis.

2. Materials and method

2.1. Material

Hydrochloric acid (HCl) (37 % v/v) was procured from PT. Smart Lab Indonesia (Jakarta). Silica gel 60, formaldehyde (37 % v/v), N,N-dimethylformamide (DMF), and ethanol (99 % v/v) were purchased from Merck (Germany). Phloroglucinol, Pluronic F127, nickel (II) nitrate, magnesium bromide, phenylacetylene, and zinc powder were provided by Sigma Aldrich, aquadest was purchased from CV Satya Darmawan. Hydrogen, nitrogen, carbon dioxide gases were purchased from CV Retno Gas.

2.2. Pre-treatment of silica gel template

The silica gel 60 received as is from Merck was pretreated by heating at 200 °C for four hours [8]. The silica gel was used as a template in the hard templated mesoporous carbon synthesis. The purpose of the pre-treatment process is to remove water and other impurities from the silica gel.

2.3. Synthesis of soft templated mesoporous carbon

The soft templated mesoporous carbon (ST MC) synthesis adopted the procedure of Liang et al. [9] and Sulistianti et al. [10]. 1.25 g phloroglucinol and 1.25 g Pluronic F127 were dissolved in 9.7 g of 10:9 weight ratio of ethanol and water mixture and stirred at room temperature until the copolymer dissolution was completed. Then, 0.08 mL of 37 % (v/v) HCl was added to the solution and stirred until a light pink color appeared. Then, the addition of 1.25 mL of 37 % (v/v) formaldehyde was conducted to the synthesis solution, and stirred for 2 h until the mixture separated into two layers. The bottom layer was taken and stored while stirring overnight, then cured in an autoclave in an oven at 100 °C for 24 h [11]. Carbonization of the resulting monolith was performed in a tubular furnace with N_2 gas flow using heating profile from 100 to 400 °C at the rate of 1 °C/min, from 400 to 850 °C at the rate of 5 °C/min, and finally samples were kept at 850 °C for 2 h under 10 cc/min nitrogen flow. The samples were later cooled down to room temperature under the same nitrogen environment [12]. The resulting carbons were then kept in desiccator.

2.4. Synthesis of hard templated mesoporous carbon

The synthesis of hard templated mesoporous carbon (HT MC) was conducted by following the recipe of Konwar et al. [8] with some modifications. The original procedure employed furfuryl alcohol as the carbon source and HF to remove the silica template, while this work used phloroglucinol carbon source and NaOH solution for desilication. In the typical procedure, 1.25 g phloroglucinol and 1.25 mL formaldehyde was added to 1.25 g pretreated dried silica gel. Then, stirring was performed on the mixture at room temperature for 72 h. Thereafter the silica gel filled with carbon precursors was filtered and dried at room temperature for 12 h. The pyrolysis of the material was carried out in a tubular furnace under nitrogen environment (100 ml/min) at 650 °C for 3 h. After pyrolysis, the silica gel–carbon composite was stirred in 1 M of NaOH solution at temperature of 90 °C for 10 min to remove the silica template. The desilication step was repeated two times followed with repeated washing with water:ethanol (1:1 w/w) mixture to ensure a complete removal of the template. The carbon was finally dried in an oven at 110 °C overnight.

2.5. Impregnation of Ni

A total of 0.45 g Ni $(NO_3)_2.6H_2O$ was dispersed in 1 mL of water: ethanol solution with a comparative ratio of 1:1 (% w/w). Then, as much as 0.3 g of mesoporous carbon was added to the mixture. The mixture was then stirred for 6 h. The material was then reduced with H_2 gas flow at temperature of 600 °C for 6 h. This procedure allowed the Ni^{2+} reduction to Ni(0) [12]. The procedure was conducted for both soft and hard templated MC. The resulting materials denoted as ST Ni/MC and HT Ni/MC.

2.6. Carboxylation of phenylacetilene

In the catalysis application of each material (ST Ni/MC and HT/Ni MC), the reaction was carried out in the CO_2 atmospheric reactor. A total of 0.0312 g of catalyst, 0.052 g (0.8 mmol) Zn powder and 0.1624 g (0.8 mmol) $MgCl_2.6H_2O$ were inserted into the reactor followed by 0.087 mL (0.8 mmol) phenylacetylene and 4 mL DMF. The flushing was carried out by flowing CO_2 gas for several minutes. Subsequently, a flow of 1 bar CO_2 was inserted into a balloon until it reached the maximum capacity. The reactor was then placed in an oil bath with the temperature of 85 °C and the CO_2 gas was then flowed. The mixture was stirred for various reaction times (2, 4, 6 and 8 h). The resulting products were then added by 1 mL of 10 % HCl solution, filtered and tested using HPLC.

2.7. Result characterization

The HPLC analysis of the resulted carboxylation product was conducted using C18 column (200 nm x 4.6 mm) with 5 µm particle size, UV-vis detector with the wavelength (λ) of 254 nm, eluent solution of methanol:1 % acetic acid with the ratio of 55:45 (v/v). Phenylacetylene standard was measured in the concentration of 50, 100, 250, 500 and 1000 ppm. Meanwhile, phenyl propiolic acid standard was measured in the concentration of 5, 10, 25, 50, 100, 250, 500 and 1000 ppm. Samples with highest result were then co-injected with 250 ppm of phenyl propiolic acid to ensure that the compound measured at the retention time was indeed the desired product.

3. Results and discussion

3.1. Characterized by FTIR

Functional groups and absorption spectra were determined using FTIR as described in figure 1. The spectra of soft templated mesoporous carbon (ST MC) pre, post carbonization, and after impregnation are presented in figure 1a. On the spectrum of MC pre carbonization, there were several peaks attributed to the remaining phloroglucinol precursor and F-127 template at 3300 cm^{-1} from stretching vibration of O-H, 2880 cm^{-1} from stretching vibration of C-H, 1623 cm^{-1} from bending vibration of O-H, 1468 cm^{-1} bending vibration C-H and 1111 cm^{-1} vibration of C-O. Those peaks were then disappeared after pyrolysis, indicating that the carbonization has conducted successfully. Meanwhile, FTIR spectra of hard templated mesoporous carbon (HT MC) are presented in figure 1b. Vibration at 3500 cm^{-1} is attributed to O-H stretching, vibration at 1627 cm^{-1} is from O-H bending, and at 1456 cm^{-1} is attributed to C-H bending. There are also several peaks at 800, 950 and 1071 cm^{-1}, which are respectively contributed to the vibration of Si-O, Si-OH and Si-O-Si formed in the framework of MC [13]. The peaks attributed to silica template were then decreased after desilication indicating that the stirring of HT MC into NaOH solution is effective to remove the remaining template.

3.2. *Characterized by XRD*

The crystallinity of synthesized material was determined using XRD. The XRD pattern of both templated MC before carbonization (figure 2) shows a single signal at 2θ around 21°. After carbonization, the materials show 2θ peaks around 23.67° (002) and 43.56° (100) revealing the presence of graphitic carbon [14]. Widened peak diffraction indicates the material has low crystallinity. This phenomenon can be caused by the collapse of some mesoporous carbon structures during

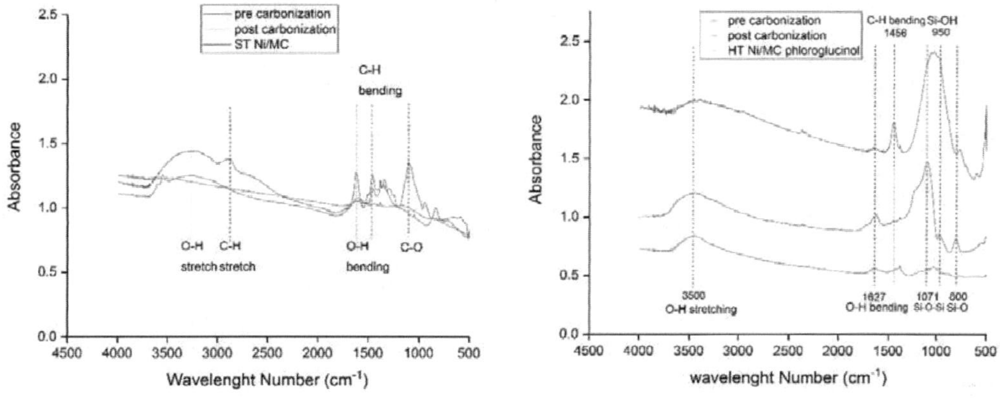

(a) (b)

Figure 1. FTIR spectra of (a) ST MC and (b) HT MC.

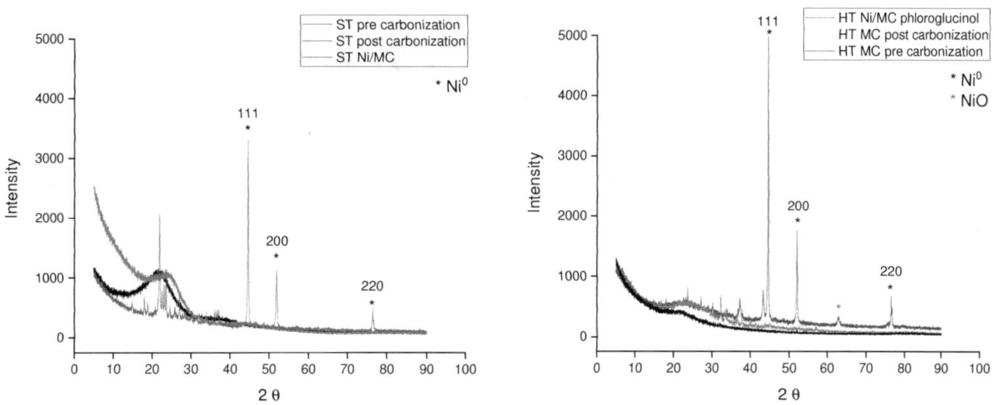

(a) (b)

Figure 2. XRD pattern of (a) ST MC and (b) HT MC before and after carbonization.

the carbonization process [15]. Meanwhile, the diffractograms of both material after impregnation show additional peaks at 44.5° (111); 51.8° (200) and 76.4° (220) which are attributed to Ni^0 [13] and small peak at 62.8° for HT Ni/MC indicating there is a small amount of NiO.

3.3. *Characterized by EDX*
Based on EDX data for soft templated mesoporous carbon, the mean of C elemental measured by the instrument is 86.56 % (table 1). This number then decreases to 64.87 % after Ni impregnation allowing 25.79 % elemental Ni to be added to the percentage. The same phenomenon also happened on the hard templated mesoporous carbon, where the mean of C elemental is 81.86 % and decreased to 13.38 % after Ni impregnation. There is a small amount of the remaining Si template in hard templated mesoporous carbon even after desilication.

3.4. *Characterized by SEM*
The SEM image of Soft Templated MC and Hard Templated MC before and after impregnation with Ni are shown in figure 3. The morphology of soft templated MC is worm hole-like [11] with various pore diameter. The material after impregnation (ST Ni/MC) has a similar morphology with less visible pores. This phenomenon might be caused of Ni metal is being inserted into and/or spread on the surface of MC. Meanwhile, both hard templated MC before and after impregnation have more agglomeration on the surface. This may be attributed to the higher extent of sintering [8].

3.5. *Characterized by BET*
The surface area analysis showed that the ST MC has surface area of 24.079 m^2/g before the pyrolysis and of 431.807 m^2/g after pyrolysis. ST Ni/MC has the surface area of 292.944 m^2/g. The N_2 adsorption/desorption isotherms obtained for ST MC (figure 4a) exhibits the typical type IV isotherm, with signature hysteresis loop revealing the presence of mesoporosity, as also revealed by the BJH pore size distribution (figure 5a). The average pore size of ST MC is 3.59 nm and of ST Ni/MC is 3.15 nm. The decrease of the average pore size might be caused of Ni spread on the surface of the mesoporous carbon. Based on the trend of hysteresis loop, it can be concluded that the pores have a cylindrical shape (type a). Meanwhile, HT MC has the surface area of 72.132 m^2/g and HT Ni/MC of 62.714 m^2/g. The lower surface area of HT MC compared to ST MC is due to the radius of beads silica template used in the synthesis of material. The average pore size of HT MC and HT Ni/MC is 1.93 nm and 1.94 nm. The adsorption isotherm and BJH pore distribution of HT MC are respectively presented in figure 4b and figure 5b.

3.6. *Carboxylation*
Phenyl propiolic acid is the main product of phenylacetylene carboxylation with CO_2. The chromatogram of the carboxylation products is presented in figure 6a for ST Ni MC catalyst and figure 6b for HT Ni/MC catalyst with the retention time of around 4 min for phenyl propiolic acid. Peaks at 12 and 2 min retention time belong to the unreacted phenylacetylene and DMF, respectively. The phenyl propiolic acid showed best peaks at reaction time of 6 h for ST Ni MC catalyst with the area of 301 mAU and yield of 1.4 %. Meanwhile HT Ni/MC catalyst showed the best performance in phenyl propiolic acid formation at the reaction time of 8 h for with the area of 453 mAU and yield of 2.2 %.

Table 1. Elemental composition of soft and hard templated carbon.

No	Element	Percentage (%)			
		ST MC	ST Ni/MC	HT MC	HT Ni/MC
1	C	86.56	64.87	81.86	13.38
2	O	13.44	8.99	17.53	43.74
3	Ni	-	25.79	-	42.87
4	Si	-	-	0.60	-

IOP Conf. Series: Materials Science and Engineering **902** (2020) 012023 doi:10.1088/1757-899X/902/1/012023

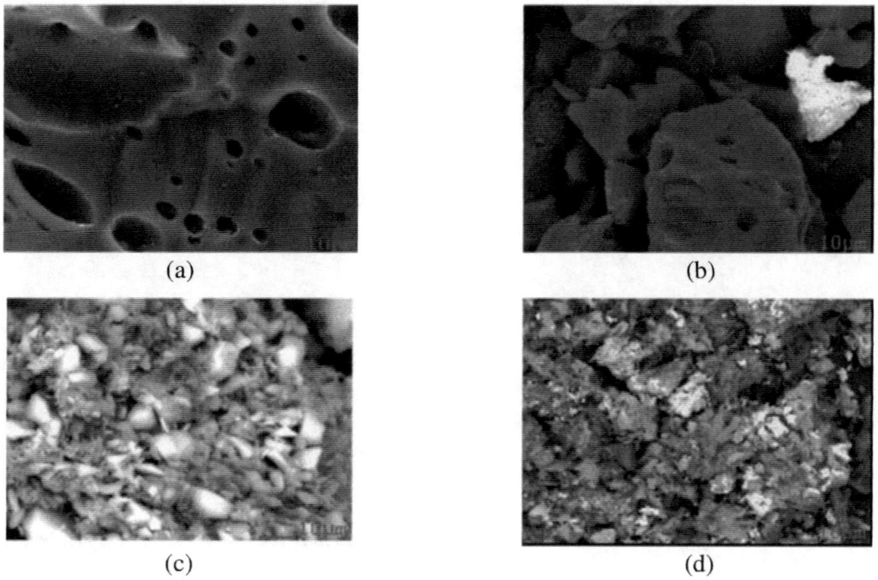

Figure 3. SEM image of (a) Soft Templated Mesoporous Carbon (ST MC), (b) Soft Templated Ni Mesoporous Carbon (ST Ni/MC), (c) Hard Templated Mesoporous Carbon (HT MC), and (d) Hard Templated Ni Mesoporous Carbon (HT Ni/MC).

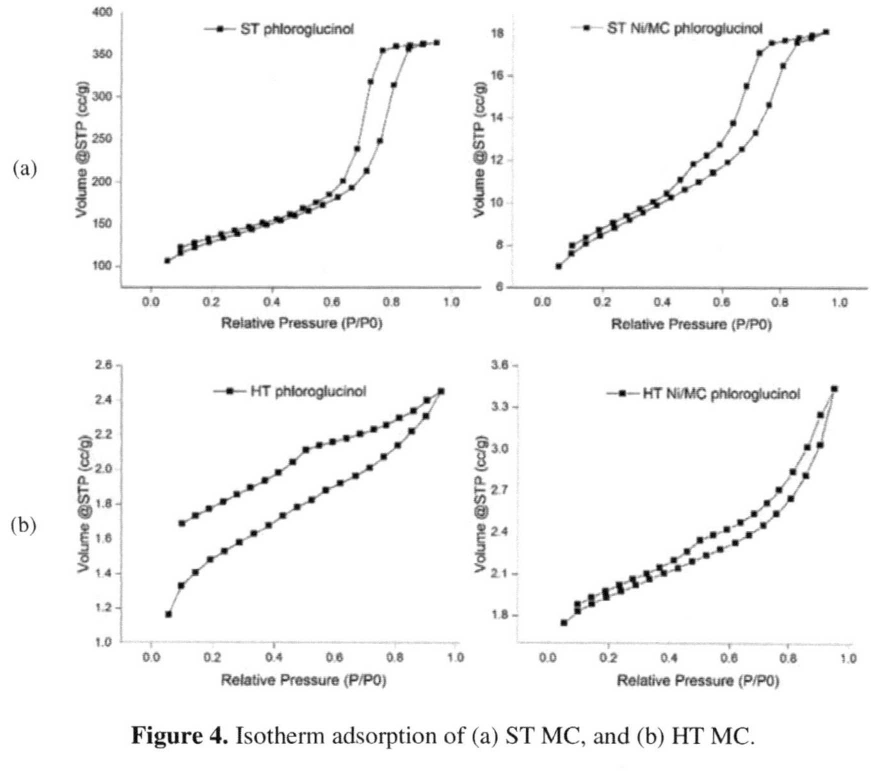

Figure 4. Isotherm adsorption of (a) ST MC, and (b) HT MC.

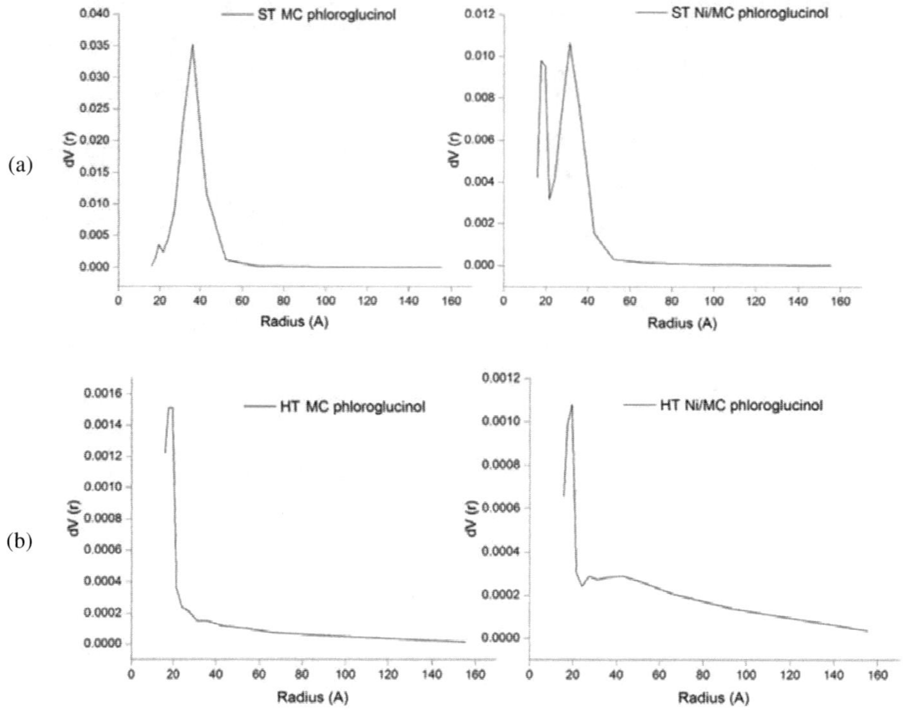

Figure 5. Pore distribution of (a) ST MC and (b) HT MC.

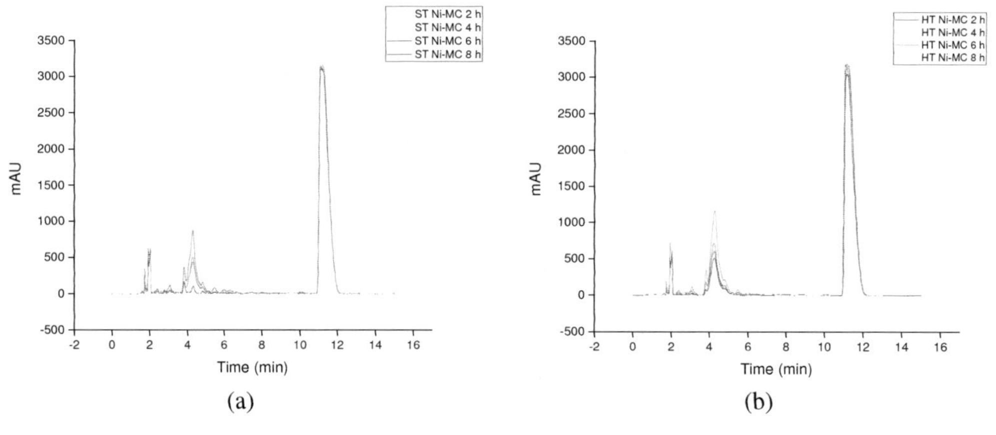

Figure 6. Overlay chromatogram of carboxilation product with (a) ST Ni MC catalyst and (b) HT Ni MC catalyst.

4. Conclusion

Mesoporous carbon from phloroglucinol precursor has been synthesized using soft template and hard template method. The SEM analysis showed the ordered and uniform porous surface area of MC with the worm hole-like pore for ST MC and agglometed surface morphology for HT MC. The BET analysis revealed that the ST MC and HT MC have surface area of 431 and 72 m^2/g, respectively. XRD pattern of both MC showed peaks for 2θ around 23.67° and 43.56° which indicate the graphite crystallinity and additional peaks after Ni impregnation at 2θ of 44.5°, 51.8° and 76.4° which attributed to Ni^0. The Ni/MC catalyst has the potential of being utilized in carboxylation of phenylacetylene with carbon dioxide resulting phenyl propiolic acid with optimum reaction condition: temperature of 85 °C for 8 h contact time with the yield of 2.2 % for HT Ni/MC catalyst, and 6 h contact time with the yield of 1.4 % for ST Ni/MC catalyst.

Acknowledgments

This worked is funded by International Indexed Publication for Student Research grant (PITTA B) NoNKB-0635/UN2.R3.1/HKP.05.00/2019 from Directorate of Research and Community Engagement, Universitas Indonesia (DRPM UI) and assisted by Laboratory Directorate General of Customs and Excise (DGCE) in Jakarta.

References

[1] Juliá-Hernández F, Gaydou M, Serrano E, van Gemmeren M and Martin R 2016 *Top. Curr. Chem.* **374** 45
[2] Supriadi A et al. 2016 Data Inventory Emisi GRK Sektor Energi (Jakarta: Pusat Data dan Teknologi Informasi Kementerian Energi dan Sumber Daya Mineral)
[3] Yu B, Diao Z-B, Guo C-X and He L-N 2013 *Journal of CO_2 Utilization* **1** 60-8
[4] Pamungkas A Z, Abdullah I and Krisnandi Y K 2019 *IOP. Conf. Ser. Mater. Sci. Eng.* **496** 012003
[5] Aziz M A A, Jalil A A, Triwahyono S, Mukti R R, Taufiq-Yape Y H and Sazegar M R 2014 *Appl Catal B Environ.* **147** 359-68
[6] Ma T-Y, Liu L and Yuan Z-Y 2013 *Chem. Soc. Rev.* 2013 **42** 3977
[7] Pal N and Bhaumik A 2013 *Adv. Colloid. Interfaces. Sci.* **189-190** 21-41
[8] Konwar R J and De M 2014 *J. Anal. Appl. Pyrol.* **107** 224-32
[9] Liang C and Dai S 2006 *J. Am. Chem. Soc.***128** 5316-17
[10] Sulistianti I, Krisnandi Y K and Munandar I 2017 *IOP. Conf. Ser. Mater. Sci. Eng.* **188** 012041
[11] Gorka J, Zawislak A, Choma J and Jaroinec M. 2008 *J Carbon* **46** 1159-61
[12] Chen C-S, Budi C S, Wu H-C, Saikia, D and Kao H-M 2017 *ACS. Catal.* **7** 8367-81
[13] Zhang G, Liu J, Xu Y and Sun Y 2019 *Int. J. Hydrogen. Energ.* **44** 4809-20
[14] Chiou J Y Z, Kung H-Y, Wang C-B 2017 *J. Saudi. Chem. Soc.* **21** 205-09
[15] Liu R, Shi Y, Wan Y, Meng Y, Zhang F, Gu D, Chen Z, Tu B and Zhao D 2006 *J. Am. Chem. Soc.* **128** 11652-62

Investigation on the effect of Zirconium doping on the structural and electrical properties of $BiFe_{1-x}Zr_xO_3$ with x = 0.01, 0.03 and 0.05

Aflahannisa and D Triyono

Department of Physics, Faculty of Mathematics and Natural Sciences (FMIPA), Universitas Indonesia, Depok 16424, Indonesia

Corresponding author's email: djoko.triyono@sci.ui.ac.id

Abstract. Multiferroic material was interesting because it has many potential applications in sensors, spintronic devices, actuators, memory devices, microelectronic devices, and convertors. Material multiferroic has two or more primary ferroic properties in the same phase. $BiFeO_3$ is a multiferroic material that has two ferroic properties such as antiferromagnetism and ferroelectricity. $BiFe_{1-x}Zr_xO_3$ with x = 0.01, 0.03 and 0.05 have been prepared by sol-gel method to study effect structural and electrical properties. X-ray diffraction result shows that all of the samples have a hexagonal structure with space group R3c and lattice parameter, volume, density increased when the concentration of zirconium increased. Zr-substitution caused the dielectric constant decrease when the concentration of Zr increases at a lower frequency.

Keywords: $Bi(FeZr)O_3$, sol-gel method, structural, electrical

1. Introduction

Multiferroic material is defined as two or more primary ferroic properties in the same phase, such as ferroelectricity, ferromagnetism, antiferromagnetism and ferroelastic. Two requirements are needed to make multiferroics. First is the coexistence of magnetic and electrical dipoles, and the second is a coupling between different order parameters and need violation and time inversion simultaneously [1]. Multiferroic $BiFeO_3$ (BFO) is one of the most important multiferroic materials, which shows two ferroic properties such as ferroelectric (T_C, Curie temperature 1103 K) and antiferromagnetism (T_N, transition temperature from anti-ferromagnetism to para-magnetism 643 K). BFO having distorted rhombohedral perovskite structure with space group *R3c* and single-phase multiferroic [2]. BFO has several problems include: (i) difficult to synthesis single phase, (ii) the leakage current is too high and (iii) the magnetic properties in BFO are very weak [3].

Several types of research have been reported for synthesized of $BiFeO_3$ doping Zirconium. Mukherjee et al. [4] reported that $BiFe_{1-x}Zr_xO_3$ (x = 0.0 and 0.05), synthesized by the chemical solution decomposition method, exhibited dielectric constant decrease when the concentration of Zirconium increase. Mahajan et al. [5] reported $BaTiO_3$ doping Zirconium at Fe-site assigned dielectric constant increase from ~1675 to ~10586 with diffuse phase transition. Mukherjee et al. [3] reported $BiFe_{1-x}Zr_xO_3$ (x = 0.0, 0.05, 0.10, and 0.15) show two relaxation regimes for the undoped samples while doped samples show only one relaxation. Based on previous several reports, there is no report for synthesized $BiFe_{1-x}Zr_xO_3$ with a concentration of Zr = 0.01, 0.03 and 0.05 with the sol-gel method to investigate electrical properties and structural properties at room temperature.

2. Experimental

$BiFe_{1-x}Zr_xO_3$ (x = 0.01, 0.03 and 0.05) were prepared by the sol-gel method using $Bi(NO_3)_3.5H_2O$, $Fe(NO_3)_3.9H_2O$, $Cl_2H_{12}O_9Zr$ and citric acid monohydrate. All precursors were mixed in water until the gel phase was formed. The gel was dried at 120°C to remove water, citric acid, and other compounds. The sample heated at 700 °C for 3 hours the sample was pressed at 5 kN into bulk and sintered at 800 °C. Phase identification and impurity were characterized by X-ray diffraction with Cu Kα radiation (λ=1.5405 Å) in the range 2θ = 20°– 90° with a step size of 0.02°/s. The chemical composition was checked by X-ray Fluorescence PANalytical with Ag (source energy 50 kV). Dielectric properties of $BiFeZrO_3$ were characterization using RLC-Meter-PM 6303 in a frequency range of 100 Hz–1 MHz at room temperature.

3. Results and discussion

3.1. Structural analysis

Figure 1 shows Le-bail fitting material $BiFeO_3$ doping Zr with concentration x = 0.01, 0.03 and 0.05. The Le-bail fitting plotted from database Crystallography Open Database (COD) such as $BiFeO_3$ with number 96-100-1091, $Bi_2Fe_4O_9$ with number 96-153-0919 and $Bi_{25}FeO_{40}$ with number 96-403. All samples have a hexagonal structure with space group R3c and have some peak impurity, such as $Bi_2Fe_4O_9$ and $Bi_{25}FeO_{40}$. This impurity because of the decomposition of $BiFeO_3$ at a temperature of 790–1040 K during the cooling process from the sintering process [6].

Table 1 shows the crystallographic of $BiFeZrO_3$ as-sintered at 800 °C for 3 hours. Based on table 1, the lattice parameter and volume increased when the concentration of Zr increased. The lattice parameter and volume increased because of ionic radii Zr^{4+} (0.72 Å is greater than ionic radii Fe^{3+} (0.65 Å) [7].

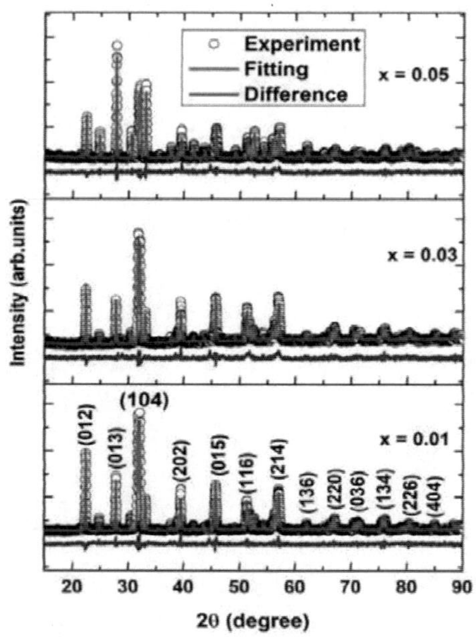

Figure 1. Rietveld refinement of XRD patterns of $BiFe_{1-x}Zr_xO_3$ (x = 0.01, 0.03, and 0.05) using *Highscore Plus* software. The experimental data are indicated by the black dots and the calculated data by the red solid line overlaying them, and blue bottom curves line shows the difference between experimental and calculated data.

The crystallite size decrease when concentration of Zr increased in the range of 76.22–71.04 nm. The reduction in crystallite size indicated the increase in lattice strain inside the lattice due to ionic size mismatch and consequently induced local structural disorder and reduction in nucleation rate [7]. The perovskite structure can be described by Goldschmidt's tolerance factor (t) which is an indicator of stability and structure distortion in the sample following equation:

$$t = (r_{Bi} + r_O)/\sqrt{2}[(1-x)r_{Fe} + xr_{Zr} + r_O]$$

where r_{Bi}, r_{Fe}, r_{Zr} and r_O are the ionic radii of the Bi, Fe, Zr and O [6]. Tolerance factor (*t*) for $BiFeO_3$ doping Zr slight decrease with increasing concentration of Zr. Changes in *t* lead to distortions in BFZO structure and are expected to cause a change in Bi-O and Fe-O bond lengths and subsequently alter Fe-O-Fe angle [8]. According to the above relation, the tolerance factor (*t*) was found to be 0.8399, 0.8393 and 0.8386, indicating the distortion in the lattice parameter.

3.2. X- ray fluorescence

X-ray Fluorescence (XRF) was performed to knowing the elemental composition of $BiFe_{1-x}Zr_xO_3$. Based on table 2, weight experimental slightly difference with weight calculation. This difference is due to the measurement, synthesis, and limitations of XRD detecting elements with small atomic numbers. Table 3 shows a comparison between the atomic ratio experimental with atomic ratio calculation. It can see that slightly different values between experimental and calculation. It may be due to the synthesize process.

Table 1. The crystallographic parameters of $BiFe_{1-x}Zr_xO_3$ (x = 0.01, 0.03 and 0.05).

Structure parameter	x = 0.01	x = 0.03	x = 0.05
Phase	Hexagonal	Hexagonal	Hexagonal
Space Group	R3c (161)	R3c (161)	R3c (161)
Lattice Parameter			
a (Å)	5.581	5.582	5.583
b (Å)	5.581	5.582	5.583
c (Å)	13.879	13.874	13.873
Volume ($Å^3$)	373.879	374.298	375.031
Crystallite Size (nm)	76.22	72.80	71.04
Density (g/cm^3)	8.33	8.35	8.37
R-Factor (%)			
R_e	4.470	4.946	4.837
R_p	4.441	5.092	5.015
R_{wp}	5.699	6.582	6.441
(Goodness of fit) χ^2	1.45	1.77	1.68
Tolerance Factor (t)	0.8399	0.8393	0.8386

Table 2. XRF characterization results for $BiFe_{1-x}Zr_xO_3$ (x = 0.01, 0.03 and 0.05).

$BiFe_{1-x}Zr_xO_3$	(%) weight experimental			(%) weight calculation		
	Bi	Fe	Zr	Bi	Fe	Zr
x = 0.01	75.441	15.725	0.459	78.7272	20.9267	0.3452
x = 0.03	79.106	17.5	1.235	78.5177	20.4491	1.03
x = 0.05	78.955	17.219	2.039	78.3085	19.9742	1.717

3.3. Electrical properties.

Electrical properties for $BiFe_{1-x}Zr_xO_3$ using impedance spectroscopy. The measurement results are impedance (Z) and phase angles (φ). The results of impedance measurements are plotted in the form of Nyquist plots and Bode plots. Figure 2a shows the Nyquist plot of BFZO at room temperature. It shows that semicircle decreased with decreased concentration of zirconium and indicating impedance BFZO decrease when the concentration of Zr decreased. Figure 2b shows the Bode plot of BFZO, indicating impedance value increase with concentration Zr increase and impedance value increase when frequency decrease also impedance close to zero with high frequency. Figure 3a shows impedance real (Z') as a function frequency at room temperature. Impedance real (Z') value increase when the concentration of Zr increased and impedance real (Z') close to zero when frequency > 10^1. Figure 3b shows impedance imaginer (Z") as a function temperature. Impedance imaginer value (Z") increase when the concentration of Zr increased and impedance imaginer (Z") value close to zero when frequency > 10^2. The maxsimum peak for all samples indicated a relaxation mechanism at room temperature.

Figure 4a shows dielectric constant (ε') of BFZO at room temperature as a frequency in the range 100 Hz–1 MHz. The value of dielectric constant decreases with concentration zirconium increase and shows relaxation at a lower frequency. The same result was also obtained by Mukherjee [3], BFZO shows relaxation can be due to the grain boundary condition. Figure 4b shows tan loss at room temperature of $BiFe_{1-x}Zr_xO_3$ with $x = 0.01$, 0.03 and 0.05. Increasing tan loss is observed with an increase of Zr-doping can be due to decreased grain boundary resistance.

Table 3. Comparison between atomic ratio experimental with atomic ratio calculation for $BiFe_{1-x}Zr_xO_3$ ($x = 0.01$, 0.03 and 0.05).

$BiFe_{1-x}Zr_xO_3$	Atomic ration experimental			Atomic ratio calculation		
	Bi	Fe	Zr	Bi	Fe	Zr
x = 0.01	1	0.776	0.01	1	0.99	0.01
x = 0.03	1	0.82	0.03	1	0.97	0.03
x = 0.05	1	0.813	0.049	1	0.95	0.05

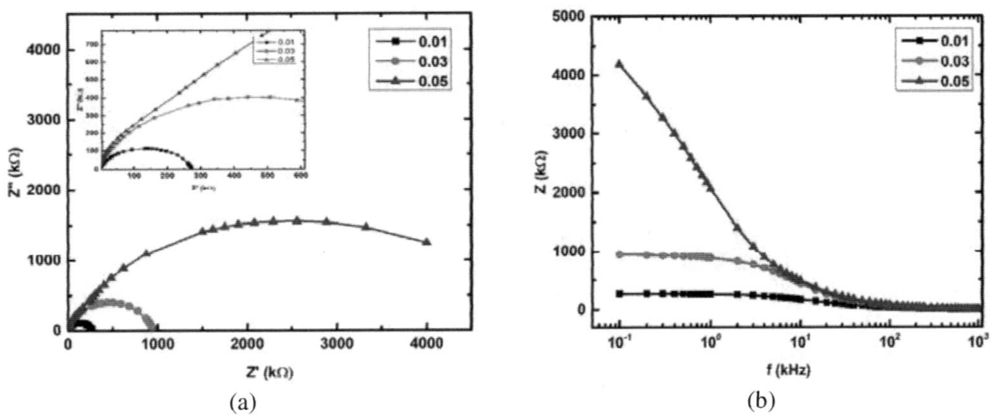

Figure 2. (a) Nyquist plot and (b) Bode plot of $BiFe_{1-x}Zr_xO_3$ ($x = 0.01$, 0.03 and 0.05) at room temperature.

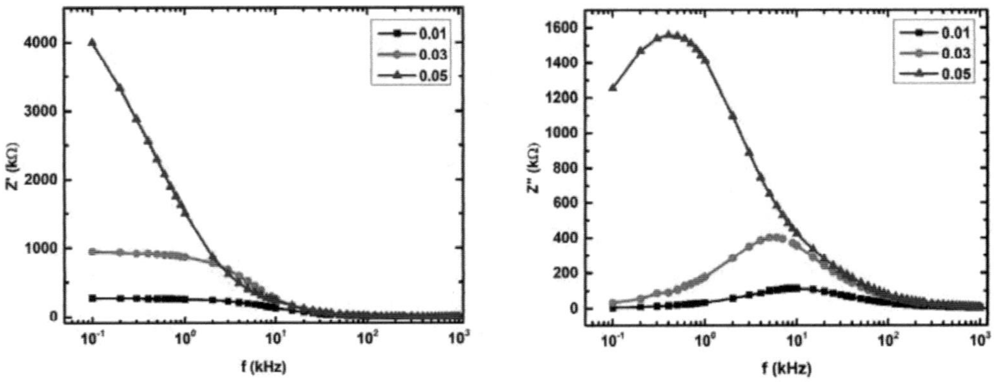

(a) (b)

Figure 3. (a) Real and (b) Imaginary part of impedance as a function of frequency of $BiFe_{1-x}Zr_xO_3$ (x = 0.01, 0.03 and 0.05) at room temperature.

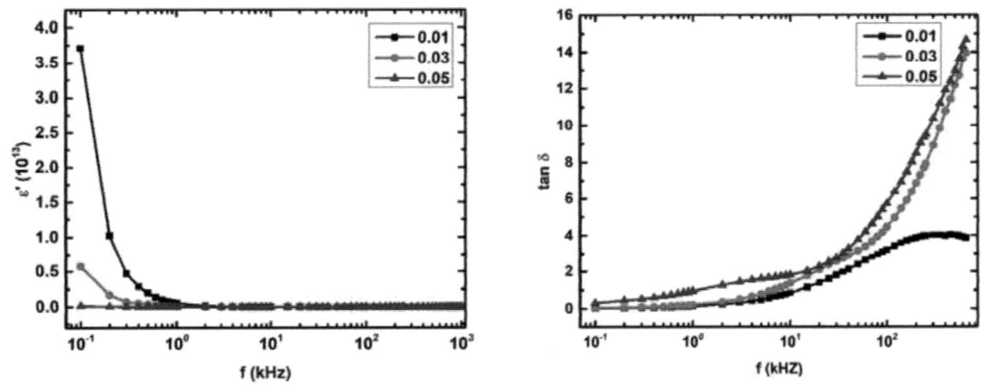

(a) (b)

Figure 4. Frequency dependence on (a) Dielectric constant and (b) Dielectric loss of $BiFe_{1-x}Zr_xO_3$ (x = 0.01, 0.03 and 0.05) at room temperature.

4. Conclusion

$BiFeO_3$ doping Zr at Fe-site ($BiFe_{1-x}Zr_xO_3$, x = 0.01, 0.03 and 0.05) have been successfully synthesized by sol-gel method. The XRD analysis confirms all samples show a hexagonal phase with space group *R*3*c*. Doping of Zr content increased lattice parameter, volume, and density when the concentration of Zr increased. And crystallite size decrease when the concentration of Zr increased. The dielectric constant value increased when the concentration of Zr increased.

Acknowledgments

This work was financially supported by Ministry of Research and Technology (KEMRISTEK/BRIN) Republic of Indonesia under research grant Hibah Penelitian Tesis Magister (PTM) 2020 with contract number NKB492/UN2.RST/HKP.05.00/2020.

References

[1] Tian Y, Xue F, Fu Q, Zhou L, Wang C, Gou H and Zhang M 2018 *Ceram. Int.* **44** 4287-91
[2] Arora M, Chauhan S, Sati P C and Kumar M 2014 *J Supercond Nov Magn* **27** 1867-71
[3] Mukherjee S, Gupta R, Garg A, Bansal V and Bhargava S 2010 *J Appl Phys* **107** 123535
[4] Mukherjee S, Srivastava A, Gupta R and Garg A 2014 *J Appl Phys* **115** 204102
[5] Mahaja S, Thakur O P, Prakash C and Sreenivas K 2011 *B Mater Sci* **34** 1483-89
[6] Xie J, Feng C, Pan X and Liu Y 2014 *Ceram. Int.* **40** 703-6
[7] Sati P C, Kumar M and Chooker S 2015 *J. Arm. Ceram. Soc* **98** 1884-90
[8] Kader S M A, Ruth D E J, Babu M V G, Muneeswaran M, Giridharan N V and Sundarakannan B 2017 *Ceram. Int.* **43**, 15544-50

Manufacturing of Mg-Al5TiB composite and addition of nano-sized Al_2O_3 through stir casting method with T6 heat treatment temperature variation

S Yusuf and A Z Syahrial

Department of Metallurgy and Materials Engineering, Faculty of Engineering, Universitas Indonesia, Depok 16424, Indonesia

Corresponding author's email: anne@metal.ui.ac.id

Abstract. In order to find aluminium replacement for vehicle structure and also application of automotive components, the research on Mg-Al-Ti-B composite was performed in this work. The composite of magnesium is highly compatible to be applied in vehicle body structure because of its low weight and density. Therefore, the lightweight of vehicle body structure, great mechanical properties, and high efficiency can be obtained. Mg-Al-Ti-B serves as the matrix, and 0.20 vf% of nano-sized Al_2O_3 is applied as the reinforcement. The composite of Mg-Al-Ti-B was produced through a stir casting process. Hereafter, T6 heat treatment was employed with variation of aging temperature, such as 170 °C, 200 °C, 230 °C and 260 °C for 6 h, subsequent the previous 1 h, 420 °C for the treatment of solution. The T6 heat treatment impact on microstructure indicates a different morphology of primary $Mg_{17}Al_{12}$ in which spheroidization occurs, the precipitation of Al_3Ti and TiB_2 are also observed from the aging process. In this work, OM, EDS, XRD, OES, density and porosity measurements, as well as destructive tests were performed for the characterization. The T6 heat treated sample's mechanical properties were rectified than the non-heat treated ones. The highest ultimate tensile strength was obtained with 65.31 MPa at 170 °C aging temperature. The optimum value of hardness, impact value and wear rate are observed at 200 °C aging temperature, the values are 92.4 HRH, 0.07 J/mm^2 and 0.00254 mm^3/m, respectively.

Keywords: Magnesium composite, nano-Al_2O_3, stir casting, T6 heat treatment, artificial aging

1. Introduction

Indonesia is one of the developing countries in the Southeast Asia Region, which is always trying to develop technology in various sectors, especially in the industrial sector. In this case, the industrial sector is believed to be one of the keys to national development. Development is one of the efforts to support the national industrial development, one of which is in the automotive sector. Technology in the automotive field is one of the fastest-growing technologies. In the automotive world, the thing that is constantly being developed is the efficiency of fuel in vehicles and reducing the pollution levels. The solution is to save costs and reduce fuel consumption by using a light vehicle mass [1].

One of the technological developments that carried out is the development of new materials that have low density but have good mechanical properties. This is done to improve the fuel efficiency, energy, cost saving and reduce the pollution levels, especially in automotive applications [2]. Material that is in accordance with these characteristics is magnesium, and it becomes one of the materials that has the potential to be applied in the structural components in automotive panel beam instruments, components of steering, transfer cases, gears, bearing gearboxes, radiator support and connecting rods. Magnesium is a metal that has a low density [3].

In this study, the author made a composite with magnesium Al-Ti-B matrix with nano-sized alumina reinforced. In the previous study, it was explained that the volume fraction of 0.20 % is the optimum state of the nano-sized Al_2O_3 in magnesium composites. This process is done through a stir casting method. To increase the strength, especially on the surface of composite materials, the heat treatment process will also be carried out. The magnesium Al-Ti-B composite is expected to have a specific strength, stiffness, good wear resistance, and low density suitable for automotive application materials to improve their fuel and energy efficiency.

2. Materials and method

2.1. Materials

Before the experiment, the author was calculated the mass balance of composite to set up the materials. Pure magnesium 90 wt.% and Al-5Ti-B 10 wt% were applied as the matrix and nano-sized Al_2O_3 particle with 0.20 vf% was used as the reinforcing. 335 g of pure magnesium and 15 g of Al-5Ti-B as matrix, 1.53 g of nano-sized Al_2O_3 as reinforce were prepared.

2.2. Fabrication of magnesium Al-Ti-B with nano-sized Al_2O_3 composite

Argon gas was flowed into the furnace to prevent the reaction between magnesium with oxygen. The air inside the furnace was absorbed by the vacuum furnace. Fabrication of the Mg-Al5TiB composite was carried out by stir casting. The composite of magnesium and Al-5Ti-B was heated at 675 °C. At this temperature, the composite reaches the molten state. Before the molten pouring, Al_2O_3 were also pre-treated by heating at 900 °C for 30 min. After pre-treatment, nano-sized Al_2O_3 were appended to the molten magnesium and Al-5Ti-B, followed by stirring. Before pouring the molten, the preparation of the casting die was firstly conducted by employing a mixture of thinner and zircon. Besides, the furnace was also coated by a mixture of thinner and zircon so that the furnace and the mould will have high thermal resistance to avoid cracking.

After that, as produced cast was passed pass through the T6 heat treatment procedure, start with the solution treatment for 1 h at 420 °C then subsequently performed the water quenching at room temperature [4]. Then, the 6 h artificial aging was conducted at the variation aging temperature of 170 °C, 200 °C, 230 °C, 260 °C, followed by natural aging at room temperature. This is the final step of the composite production [5].

2.3. Characterization

The preparation of all post heat-treated products was performed to undergo the destructive testing and microstructural analysis. Optical microscope was applied to identify the microstructure. The samples were ground and polished by TiO_2. Scanning electron microscope (SEM) was carried out to know the phases in microstructure. X-ray diffraction (XRD) was performed to obtain the different phases in microstructure. The observation of chemical properties was conducted by applying the optical emission spectroscopy. The investigations of mechanical properties were performed with the tensile test based on JIS Z2241; impact test based on Charpy method with ASTM E23 standard, hardness test (Rockwell H) based on ASTM E18 and wear rate test with pin on disc (Ogoshi) method based on ASTMM G99. Each sample's density was also examined with Archimedes Theory. All mechanical tests were carried out at room temperature.

IOP Conf. Series: Materials Science and Engineering **902** (2020) 012025 doi:10.1088/1757-899X/902/1/012025

3. Results and discussion

3.1. Chemical composition

The chemical composition test by Optical Emission Spectroscopy method of the Al-Ti-B magnesium composite sample with the nano-sized Al_2O_3 reinforcement through T6 heat treatment was carried out in order to find out what elements present in the sample so that it can be predicted which phase is formed.

Chemical composition analysis is performed to compare the theoretical chemical composition of magnesium Al-Ti-B composites/nano Al_2O_3 with the actual chemical composition of magnesium Al-Ti-B composite/nano Al_2O_3. During the casting process, according to the mass balance, the composition of magnesium was added by 90 % is 315 g, while the composition of Al-Ti-B added by 10 % is 35 g. Then, the composition of nano Al_2O_3 reinforcement was added as much as 0.2 % which is equal to 1.49 g. However, according to the results from the chemical composition test, the composition of Mg is 95.2 %, which means the composition of magnesium is increased (table 1).

Based on the theoretical composition of aluminum, it should be as much as 9 %, but Al composition was decreased to only 3.98 %. This is caused by the occurrence of dross which is a reaction between aluminum and oxygen. In addition, there was a decrease in Ti levels, in the actual condition of the Ti composition which was only 0.0798 %. This happens because the Ti added is a TiB compound, not a pure Ti element, so there are only a few compositions. Furthermore, as much as 0.0494 % boron can produce TiB_2 phase as a nucleating agent [6].

3.2. SEM and EDS

SEM and EDS tests were carried out on the samples to examine the elements and their composition in Al-Ti-B nano-sized Al_2O_3 magnesium composite samples which were given T6 heat treatment and aging at 200 °C (figure 1 and table 2). In addition, this test is carried out to predict the phases and compounds contained in the composite material.

Table 1. Chemical composition of Mg-as cast and composite material.

	Mg	Al	Si	Ti	B	Zn	Mn	Fe
Mg Al-Ti-B/nano-sized Al_2O_3 composite	~95.2	3.98	0.406	0.0798	0.0494	0.0310	0.0116	0.0087
	Cr	Cu	Ca	Sn	Sr	Na	Ag	Ni
	0.0081	< 0.00010	0.0016	0.0348	0.00017	0.0025	0.00047	< 0.00040

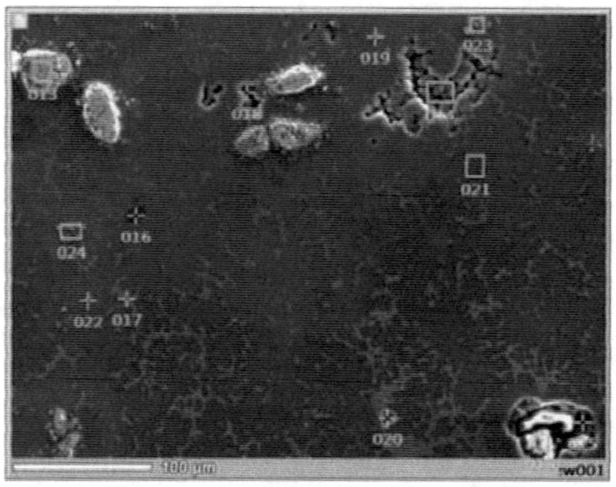

Figure 1. The SEM result of Mg Al-5Ti-B/ nano Al_2O_3 0.20%vf.

Table 2. The EDS result of Mg Al-5Ti-B/ nano Al_2O_3 0.20%vf.

No	Element composition (%wt.)						Phase formed
	Mg	Al	Ti	B	O	Si	
013	4.88	31.34	62.06			1.71	$Mg_{17}Al_{12}$, Al_3Ti
014	81.04	5.74	0.07		11.50	1.65	MgO, Al_2O_3, Al_3Ti
015	33.81	11.74	0.06	0.91	53.49		MgO, $Mg_{17}Al_{12}$, Al_3Ti, Al_2O_3, TiB_2
023	63.94	6.74	20.47		7.89	0.96	$Mg_{17}Al_{12}$, Al_2O_3, Al_3Ti
025	89.68	1.90		1.24	7.16	0.02	$Mg_{17}Al_{12}$, MgO, Al_2O_3

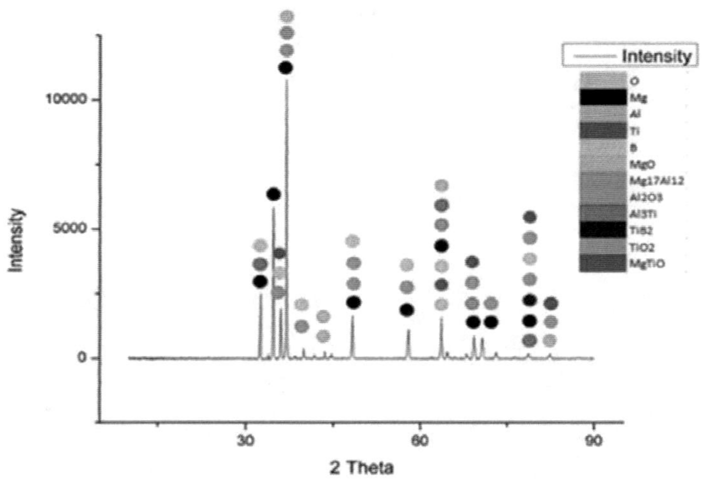

Figure 2. The XRD result of Mg-Al-5Ti-B/ nano Al_2O_3 0.20 %vf.

Based on the SEM-EDS results, point number 013 shows the rounded $Mg_{17}Al_{12}$ phase due to T6 heat treatment. Then, the intermetallic phase of Al_3Ti is formed. Furthermore, at point number 014, there is an oxide compound produced by the reaction between magnesium with nano-sized Al_2O_3 and as a reaction of melting magnesium with oxygen gas to form MgO. At point number 015 indicates the presence of elements Mg, Al, Ti, B and O. The phases formed at this point are the MgO oxide compounds formed, while the precipitate formed is $Mg_{17}Al_{12}$. This is consistent with the solidification reaction of the Mg-Al binary diagram at 436 °C [7]. Then, the formation of TiB_2 phase. TiB_2 particles are also present in every fine equiaxed dendrite based on SEM results [8]. Based on the results of plotting from each point, the distribution of $Mg_{17}Al_{12}$ precipitate phase and Al_3Ti intermetallic phase is evenly distributed.

3.3. XRD

X-ray Diffraction test was performed to find out what compounds are formed in the composite sample. The identification of all phases with optical microscope and scanning electron microscope can be seen in figure 2. Based on the graph, it appears that the most detected compounds are $Mg_{17}Al_{12}$ phase and periclase or MgO. The MgO compound formed is in accordance with equation $3Mg + Al_2O_3 \leftrightarrow 3MgO + 2Al$ where a stable compound is formed when reacting with magnesium [9].

Theoretically, TiB_2 particles act as heterogeneous nucleation of the α-Mg substrate. TiB_2 particles are also present in every fine equiaxed dendrite based on SEM results. Other oxide compounds observed from the XRD test results consist of two types, namely TiO_2 and Mg_2TiO_4.

3.4. Microstructure

Based on figure 3, it can be seen that microstructure (a) composite without heat treatment, and composite samples given heat treatment T6 with variations in temperature aging are (b) 170 °C (HT 170), (c) 200 °C (HT 200), (d) 230 °C (HT 230), (e) 260 °C (HT 260) at a magnification of 500x.

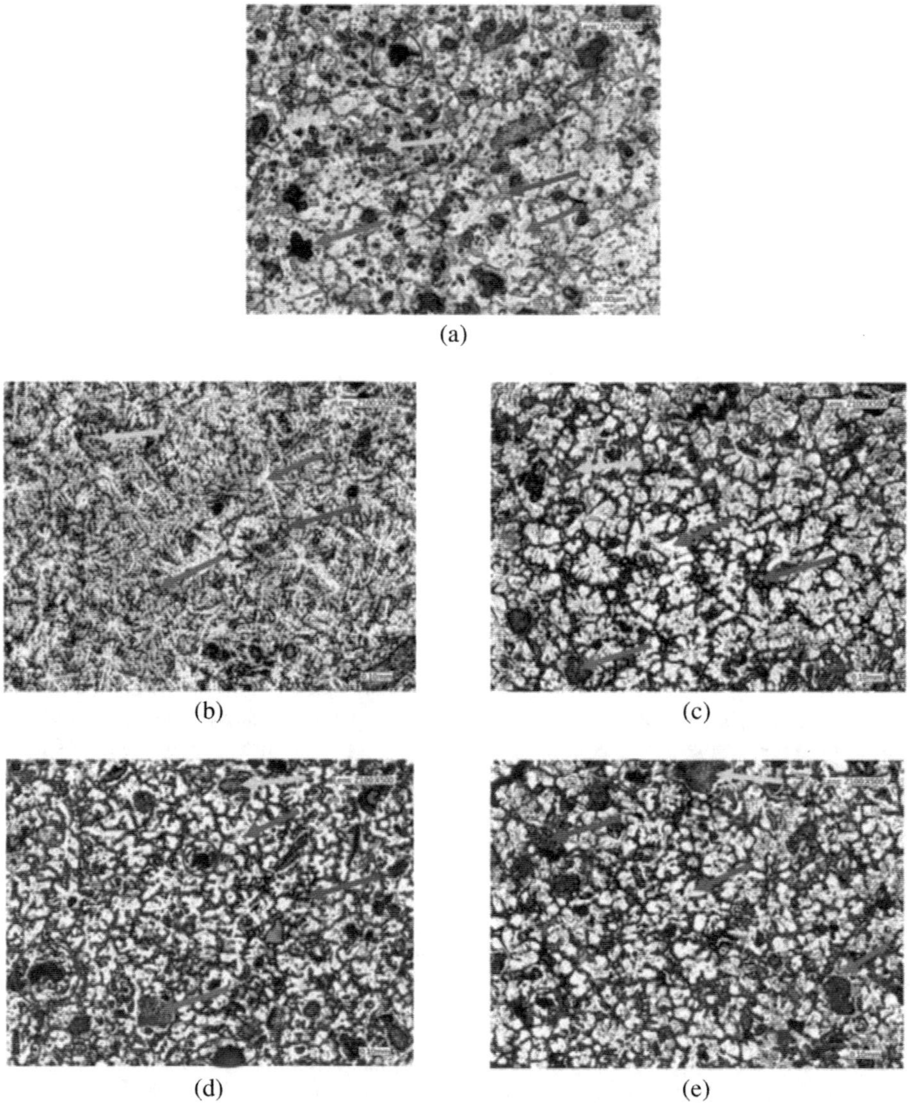

(a) (b) (c) (d) (e)

Figure 3. Microstructure images of (a) non-HT, (b) Mg-Al5TiB/Nano Al_2O_3 heat treated at 170 °C, (c) 200 °C, (d) 230 °C, and (e) 260 °C at a magnification of 500x.

Based on the results of the microstructure, it can be seen based on the colors of the arrow, where the green colored arrow shows the matrix α-Mg. The α-Mg matrix is shown as the basis of the phases mentioned earlier, where α-Mg is brownish white. Blue arrows indicate the phase β-$Mg_{17}Al_{12}$, orange arrows indicate the TiB_2 phase, while yellow arrows indicate the Al_3Ti phase. Meanwhile, the formation of Al_3Ti at 660 °C. Whereas, TiB_2 can be formed at temperatures around 700 °C [9]. Then, there is a black circle which indicates porosity in the microstructure of the composite sample. The transformation of the needle-like structures occurs in this composite. It transforms into spherical. At 200 °C and 230 °C, columnar grain in the samples starts to grow. The higher aging temperature causes the more spherical structure and columnar grain becomes finer. The increasing aging temperature is directly proportional with grain growth. When aging temperature increases, it triggers the columnar grains to become bigger and also the dendrites tend to be more spherical.

3.5. Tensile and elongation

Based on figure 4, the ultimate tensile strength (UTS) value of the composite can be seen on the graph. At 170 °C aging temperature, Mg-Al5TiB composite achieved the highest UTS number. This highest number was occurred at the peak age, which achieved at the 170 °C aging temperature. When the temperature increases until it passed the optimum temperature, the microstructure grows bigger. The $Mg_{17}Al_{12}$, Al_3Ti and TiB_2 phases tend to be incoherent when over aging occurs. The phases reach the semi coherent state at the peak age. Mechanical properties decrease during this phenomenon. The process of T6 heat treatment caused the homogenization of the material structure, which made the value of UTS and elongation percentage became higher. The value of elongation at 230 °C was decreased, this is caused by porosity on that sample.

3.6. Hardness, wear rate and impact

The increasing hardness value of Magnesium Al-5TiB happened when the aging temperature increases, but is decreasing after a particular temperature when over aging occurence takes place. At 200 °C aging temperature, the magnesium composite achieved the highest hardness value. However, sample with 230 °C, 260 °C aging temperature only displays a slight purpose in the reduction of hardness value. When the temperature aging increases to peak age, precipitates will form coarsening which will reduce the strength and hardness of the composite material [10]. Figure 5 impact value data also show that the composite with 200 °C achieved the highest impact value. The trend of graph is quite similar to the hardness test results. Figure 6 shows the wear rate data, based on this graph the high value of hardness in material will have a low wear rate value. This is due to the ability to avoid material loss. If the value of hardness in the material increases, then the rate of wear will decrease. Meanwhile, if the value of hardness of the material decreases, then the rate of wear will increase [11].

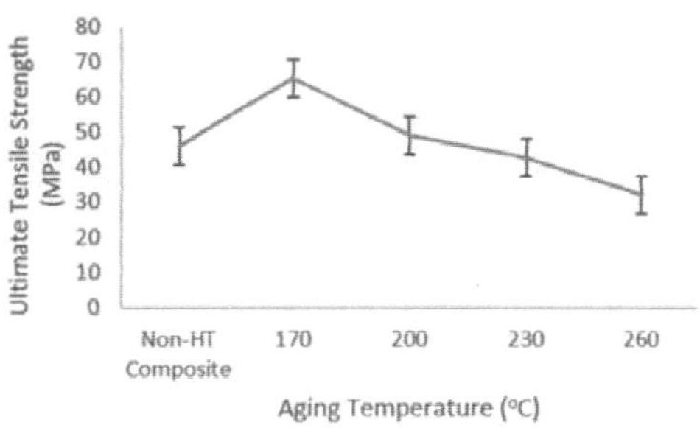

(a)

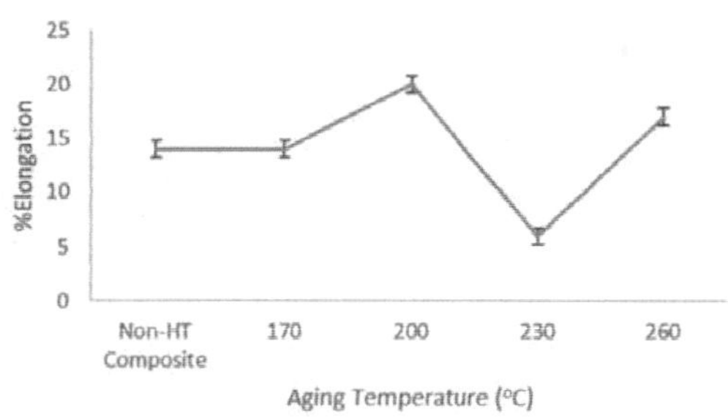

(b)

Figure 4. (a) Tensile and (b) Elongation.

3.7. Density and porosity

According to the density measurement of magnesium composite, the value of density is 1.73 gr/cm^3. This value is higher than the pure magnesium (1.7 gr/cm^3) based on the actual density test. This is due to Al-Ti-B (2.75 gr/cm^3) and Al_2O_3 (3.95 gr/cm^3) that have a higher density. Thus, it affects the density of magnesium composite. The actual density of the composite was decreased if compared to its theoretical density. The decreasing of density value is due to the existence of excess hydrogen in the casting process which enter the molten magnesium during the stirring process. The existence of porosity in the system reduces the tensile strength significantly.

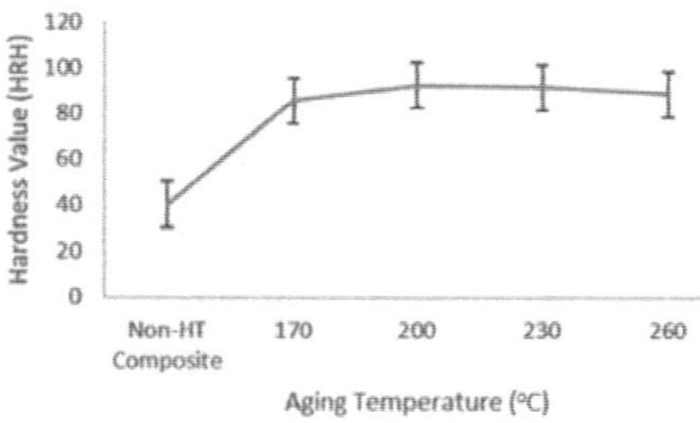

(a)

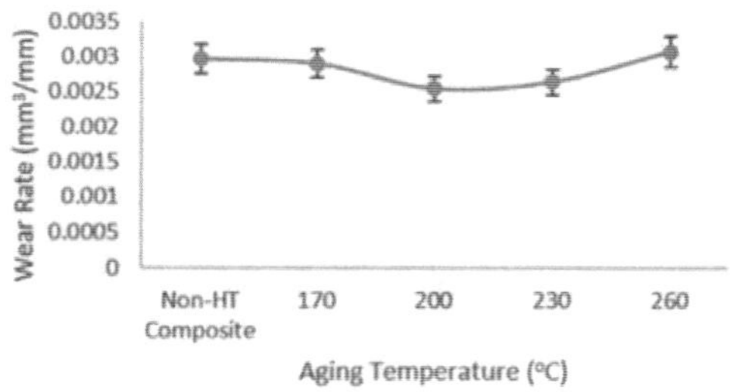

(b)

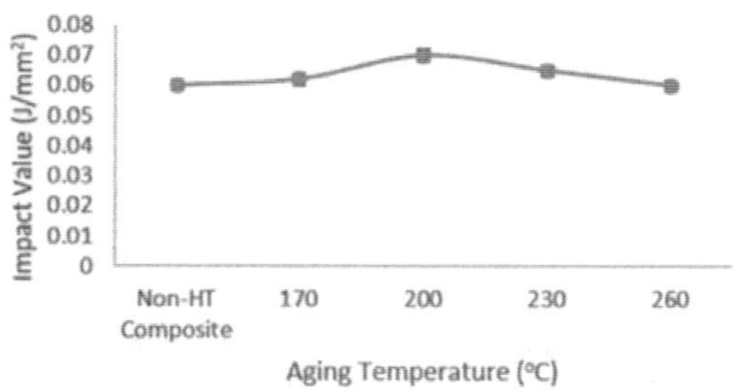

(c)

Figure 5. (a) Hardness, (b) Wear rate, and (c) Impact .

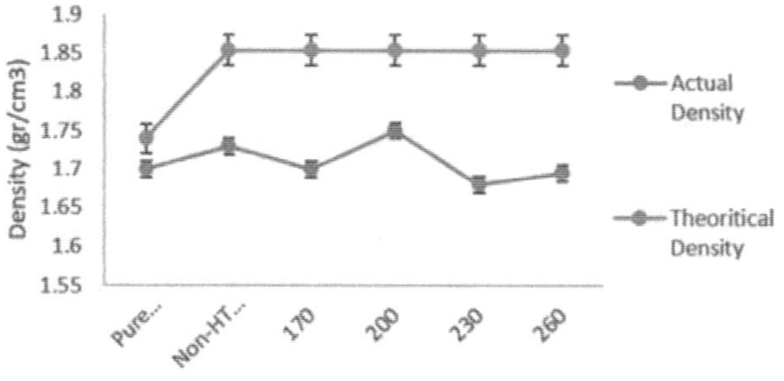

(a)

Figure 6. (a) Density, and (c) Porosity of composite material.

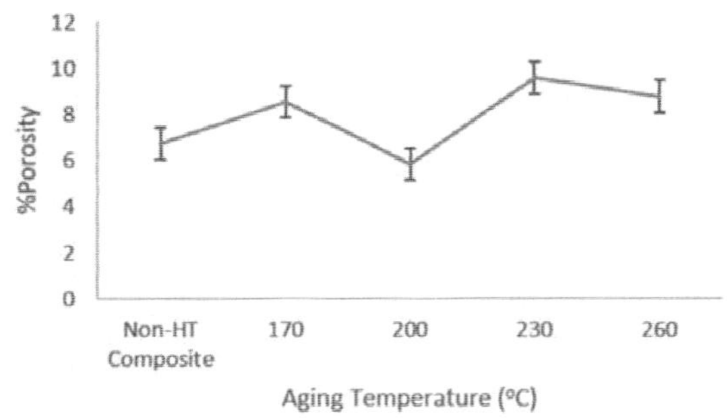

(b)

Figure 6 (continued). (a) Density, and (c) Porosity of composite material.

4. Conclusion

Magnesium Al-Ti-B composite with nano-sized Al_2O_3 with T6 heat treatment achieved the higher values than composite without T6 heat treatment. It can be regarded as a selection for varied uses, especially in the automotive application. This is due to the greater mechanical properties which achieved at the optimum aging temperature of 170 °C in ultimate tensile strength. It reaches 65.31 MPa. Then, at 200 °C is obtained in hardness, toughness and wear rate. It reaches 92.4 HRH in hardness value, 0.07 J/mm2 in impact value, and 0.00254 mm^2/m in wear rate. The T6 heat treatment generates the homogenization structure of material. Also, this treatment triggers the grain and phase transformation which gives the high tensile strength and percentages of elongation. Due to the over aging phenomenon, the material properties decreased, which occurred at a higher aging temperature (230 °C and above). The mechanical properties enhancement is supported by the addition of grain refiner which known as TiB. The result shows the presence of $Mg_{17}Al_{12}$, Al_3Ti and TiB_2 as precipitates which affected the mechanical properties of materials.

Acknowledgments

Authors would like to thank Directorate of Research and Community Services (DRPM), Universitas Indonesia for financial support through International Publications Index (PIT 9), 2019 with contract number: NKB-0042/UN2.R3.1/HKP.05.00/2019.

References

[1] Blawert C, Hort N and Kainer K U 2004 *Trans. Indian Inst. Met.* **57** 397-408
[2] Kainer K U 2006 *Metal Matrix Composites: Custom-made Materials for Automotive and Aerospace Engineering* (Hoboken: Wiley-VCH)
[3] Luo 2013 Applications: Aerospace, automotive and other structural applications of magnesium *Fundamentals of Magnesium Alloy Metallurgy* (United Kingdom: Woodhead Publishing) chapter 8 pp 266-316
[4] Huang X, Han G, and Huang W 2018 *Materials* **11** 1-10
[5] Zeng R 2013 *Precipitation Hardening in AZ91 Magnesium Alloy* PhD Thesis (Birmingham: University of Birmingham)
[6] Chen T J, Wang R Q, Ma Y and Hao Y 2012 *Mater. Design* **34** 637-48
[7] Janz A 2007 *Intermetallics* **15** 506-19

[8] Hosoda H, Inamura T, Shimoyamada T, Noma H and Wakashima K 2010 Phase Reactions of Al-Ti-B Ternary and Al-Ti-B-O Quartenary Systems for Al-based Metal Matrix Composites *Proc. of the 12th Int. Conf. on Aluminium Alloys* (Japan: The Japan Institute of Light Metals) pp 2051-6
[9] Fegley B 2013 *Practical Chemical Thermodynamics for Geoscientists* (Netherlands: Academic Press, Elsevier)
[10] Al-Murshdy J M 2008 *J. Kerbala Univ.* **6** 233-41
[11] Menezes P L, Nosonovsky M, Ingole S P, Kailas S V and Lovell M R 2013 *Tribology for Scientists and Engineers* (New York: Springer)

IOP Conf. Series: Materials Science and Engineering **902** (2020) 012026 doi:10.1088/1757-899X/902/1/012026

Development study of iron-based metal organic frameworks (Fe-based MOFs) as a binding agent in diffusive gradient in thin film method (DGT) for phosphate adsorption

V Sintya, A Saefumillah and A Zulys

Department of Chemistry, Faculty of Mathematics and Natural Sciences (FMIPA), Universitas Indonesia, Depok 16424, Indonesia.

Corresponding author's email: asep.saefumillah@sci.ui.ac.id

Abstract. In this study, Fe-based Metal Organic Frameworks (MOFs) will be developed as a binding agent in Diffusive Gradient in Thin Film (DGT) for phosphate adsorption, where it will be evaluated for the effect of pH, adsorbent concentration and mixing time in phosphate loading capacity of Fe-based MOFs in DGT because these parameters can influence the capacity of phosphate adsorption. Fe-based MOFs were synthesized using ferrichloride as a binding agent and terephthalic acid as a linker in solvothermal conditions. Fe-based MOFs will be characterized by X-Ray Diffraction (XRD), Fourier Transform Infrared Spectroscopy (FTIR), Scanning Electron Microscopy-Energy Dispersive X-Ray (SEM-EDX), Brunauer-Emmelt Tellers (BET). The XRD peaks correspond to MIL-101 for 2θ around 9° and 17°. FTIR analysis indicated the broadening peak at 1595 cm^{-1} for C=O and 1395 cm^{-1} for C-C vibrational. The SEM analysis showed the morphology was rhombus. The BET analysis showed the surface area of Fe-based MOFs was 61.33 m^2g^{-1}. Fe-based MOFs have been successfully synthesized by the solvothermal method and then will be applied as binding agent in DGT. Furthermore, Fe-based MOFs in DGT will be used for phosphate adsorption and measured using Spectrophotometer UV-Vis.

Keywords: Fe-based MOFs, DGT, adsorption, phosphate

1. Introduction

Phosphate is a material, often used in agricultural and industrial fields including for fertilizers, detergents, water softening, beverages, and metallurgy, etc. The use of phosphate in large quantities can cause high waste containing phosphate. The release of phosphate to the surface of water is an environmental concern because it can cause a eutrophication phenomenon that has an impact on water quality degradation [1].

Eutrophication is a process which causes the ecosystem to have a nutrition enhancement from the outside. Phosphorus can be a major element causing the eutrophication which leads to an increase of aquatic plants, algae growth, some of them are toxic and can interfere with the presence of organisms in the water. This can cause a decrease in oxygen due to an increase in biological oxygen demand and acidification [2].

DGT technique is a passive sampling for metal trace. The DGT device consists of a permeable diffusive gel layer for labile contaminant species (as a substitute for aquatic organisms), and a binding gel layer containing binding agent with both layers embedded in the inside of a plastic container. The DGT technique measures the concentration of average time or flux solute during the deployment time. This measurement depends on the solute diffusion control flux from the surface of the DGT device to the binding phase of the device based on Ficks First Diffusion Law [3, 4].

The selection of binding agent is a crucial because it will affect the ability of the analyte ion absorption and affect the selectivity of an analyte species. One binding agent that has a function to absorb phosphate is iron (Fe). In the studies of Li et al. 2016 and Xie et al. showed that iron (Fe) combined with MOFs could play an effective role in phosphate adsorption [5-7]. The interaction between the phosphate and the Fe side of Fe-MOFs is a major factor in the calculation of phosphate adsorption. These Fe-MOFs also show a high selectivity toward phosphates among other anions such as Cl, Br, N, and S. Iron (Fe) is also relatively easy to obtain because of their relatively high availability in the environment [6].

In recent years, the Metal Organic Framework (MOFs) has become a new model in the development of porous materials because of their high porosity, morphology and large surface area [8-10]. The characteristics of MOFs can be easily adjusted with the selection of different metal ions and organic bridging ligands and MOFs have developed a new role as a solid adsorbent [10]. MOFs are stable materials and have a good adsorption to remove various pollutants in water [9].

In this study, Fe-MOFs will be developed as a binding agent in DGT for measuring the dissolve reactive phosphate in water which will be studied and evaluated for the influence of pH, adsorbent concentration, mixing time, diffusion coefficient and phosphate loading capacity in Fe-MOFs-DGT.

2. Materials and method

2.1. Materials

Material used in this study were in analytical grade and without any further purification step. $FeCl_3.6H_2O$, terephthalic acid (H_2BDC), N,N-dimethylformamide (DMF), Acrylamide 40 %, N,N'-methylenebisacrylamide 1 %, N,N,N',N'- tetramethyl ethylenediamine (TEMED) 99 %, cellulose nitrate membrane filter, KH_2PO_4, NaOH, H_2SO_4, ascorbic acid, potassium antimonyl tartrate, ammonium heptamolybdate tetrahydrate, deionized water.

2.2. Method

2.2.1. Synthesis and characterization of Fe-MOF. Fe-MOF was synthesized by solvothermal method. In this study, 0.675 g of $FeCl_3.6H_2O$ and 0.206 g of H_2BDC were dissolved in 30 mL DMF then stirred for 30 min. The solution was then transferred into a Teflon-lined autoclave and heated at 120 °C for 20 h [6]. The product was washed using DMF. The solid was collected by centrifugation and dried at 60 °C. The crystal phases of Fe-MOF were examined by X-Ray diffraction. The morphology, surface area and functional group were characterized using SEM, Brunauer-Emmet-Teller (BET) analysis, and Fourier Transform Infrared.

2.2.2. DGT assembly and deployment. Acrylamide 40 % and N,N'-methylenebisacrylamide 1 % was mixed to form the solution gel. Diffusive gel was then synthesized by mixing 10 mL of solution gel, 20 µL of TEMED and 70 µL of persulphate 10 % then printed in a glass mold and heated in 60 °C for 45 min. The step of binding gel synthesis was the same with diffusive gel by adding 10 mg of Fe-MOF. Binding gel was added into the base of DGT mold, then followed by diffusive gel and cellulose nitrate membrane on the top of the gasket. DGT cap was tightened securely then deployed in DRP solution to determine the DRP concentration.

2.2.3. Determination of elution coefficient. Fe-MOF binding gel with diameter 25 mm was dispersed in 10 mg L^{-1} of P solution for 24 h then eluted with eluent for 24 h. Elution efficiency, described as factor elution was determined using different concentration of H_2SO_4 (0.25–1 $molL^{-1}$) as an eluent then neutralized with NaOH 0.1 $molL^{-1}$.

2.2.4. Analytical method and DRP concentration calculation. Concentration of synthetic DRP adsorbed in binding gel were analyzed by the ammonium molybdate spectrophotometric method, the concentration of DRP solution was measured by UV-Visible spectrophotometer (Shimadzu UV-2450) at the wavelength 713.50 nm. Before analysis, the DRP adsorbed inside Fe-MOF binding gel was eluted by 10 mL H_2SO_4 1 $molL^{-1}$ for 20 h and neutralized by NaOH 1 $molL^{-1}$. The mass (M) DRP inside Fe-MOF binding gel was calculated by equation 1:

$$\frac{M = Ce \times (VFe - MOF + VH_2SO_4 + VNaOH}{Fe} \quad (1)$$

where Ce is the concentration of DRP in eluent solution, VFe-MOF is volume of binding gel FeMOF, VH_2SO4 is volume of eluent used, VNaOH is the volume NaOH added into eluent, and fe is the factor of elution. Factor of elution was obtained by using different concentration of H_2SO_4 from 0.025–1 $molL^{-1}$. The concentration of DRP measured by DGT was calculated by equation 2:

$$C_{DGT} = \frac{M \times \Delta g}{D \times A \times t} \quad (2)$$

where M was calculated by equation 1, Δg is cellulose membrane and diffusive gel thickness, A is exposure area of binding gel which directly contact with DRP solution, t is deployment time in second unit.

3. Results and discussion

3.1. Characterization of Fe-MOF

After the solids are dried, the identification of Fe-MOF powder is carried out using a bioimaging microscope, XRD, and FTIR. The FTIR results shows in figure 1a the IR spectrum of Fe-MOF which is characterized by the presence of a benzene-carboxylic main group which is attributed to C=O bond in the carboxylic group shown at a wavenumber of 1604 cm^{-1} and at a wavenumber of 1396 cm^{-1} indicating the presence of aromatic carbon group C-C vibrational mode. The XRD results of Fe-MOF in figure 1b show that the MOF crystal structure is formed and can be seen from the formation of the same crystal structure with certainty and repetition in the 2 theta position at 5.0–10.9 degree and 15.0–20 degree. The sharp peak with high intensity in XRD result shows that MOF have a high crystallinity. The BET examination results obtained the surface area of 61.33 m^2/g. The BET result is lower because the structure of $FeCl_3.6H_2O$ is amorph. The distribution of binding agent can be seen in figure 1c using the bioimaging microscope and the figure shows the distribution are homogeneous. The SEM analysis shows the morphology is prism-like with rude surface in figure 1d.

3.2. Elusion factor of phosphate adsorption

Before analysis, the DRP adsorbed inside Fe-MOF binding gel was eluted by 10 mL H_2S0_4 1 $molL^{-1}$ for 20 h and neutralized by NaOH 1 $molL^{-1}$. The optimization of eluent was performed by using different concentration of H_2SO_4 from 0.025–1 $molL^{-1}$. After the optimization of eluent is completed, it is obtained the good eluent for analyzing the phosphate in Fe-MOF is H_2SO_4 1 $molL^{-1}$ with elusion value 81.29 %. This elusion value is important because it will be affecting the phosphate adsorption.

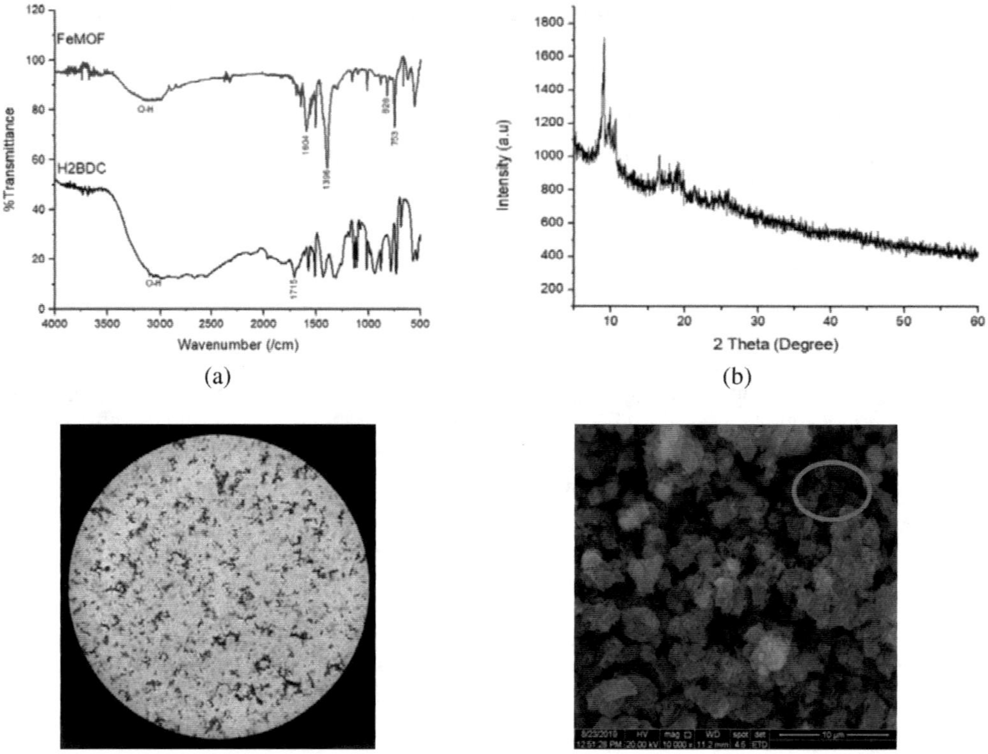

Figure 1. Characterization of F-MOF; (a) FTIR spectra, (b) XRD pattern, (c) Bioimaging microscope image, and (d) SEM image.

Table 1. Results of elution factor from H_2SO_4 1 $molL^{-1}$ as a eluent.

Eluent	Absorbance	Dilution factor	$C_{elusion}$ (ppm)	% elusion	Average
H_2SO_4 1 M (1)	0.016	20	1.037509976	81.25	
H_2SO_4 1 M (2)	0.017	20	1.197126895	68.18	81.29 %
H_2SO_4 1 M (3)	0.018	20	1.356743815	94.44	

3.3. Diffusion coefficient

The diffusion coefficient (D) of DRP was measured by creating graph of time and mass DRP adsorbed in Fe-MOF DGT. Fe-MOF DGT was soaked in 13 mgL^{-1} DRP solution for 4 to 26 h. From figure 2 we can see the diffusion coefficient calculated by the intercept and slope of the linear line that is obtained R^2 value is 0.9921. The D value was 4.74 x 10^{-7} cm^2s^{-1}. The D value is lower than the experimental of Zhang et al. 1998 whereas it obtained 7.39 x 10^{-6} cm^2s^{-1} because of $FeCl_3.6H_2O$ have amorph structure so the porosity is low. From figure 2, it can also be observed that increasing time can cause an increase in the capacity of loading phosphate. The good time to obtain the accumulated phosphate is 26 h.

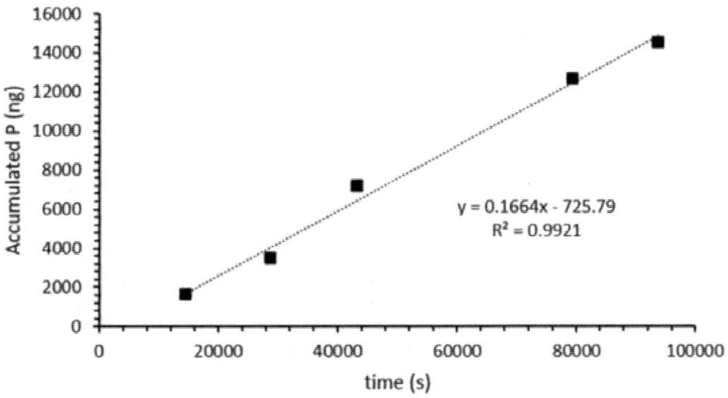

Figure 2. Linearity graph of time and accumulated phosphate.

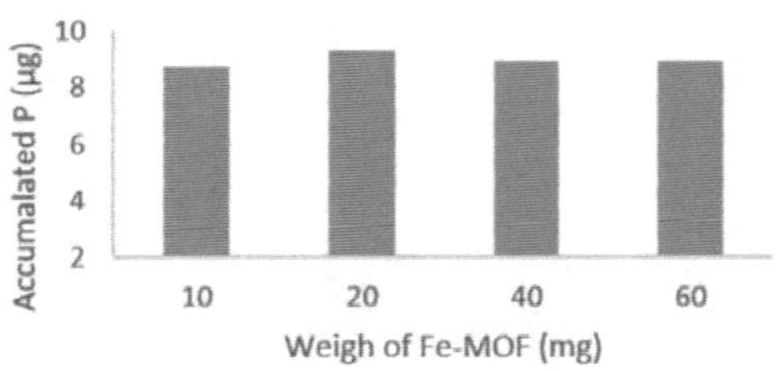

Figure 3. The effect of concentration variation.

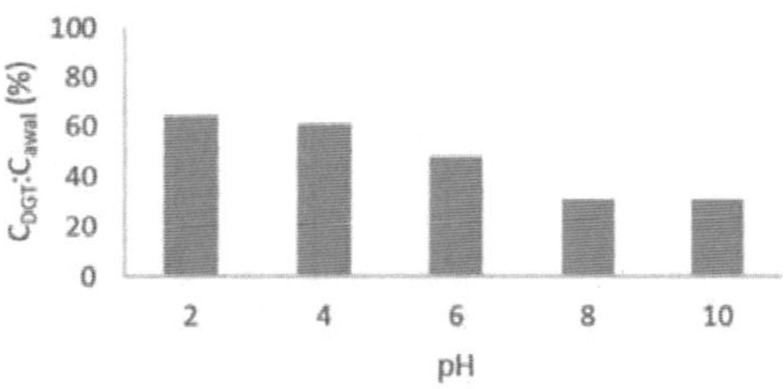

Figure 4. The effect of pH variation.

3.4. Capacity of loading phosphate in Fe-MOF DGT

The uptake of DRP by Fe-MOF DGT was evaluated by deploying the Fe-MOF DGT in 13 mgL^{-1} for 4 to 26 h with variation of Fe-MOF concentration 10 to 80 mg and pH variation from 2 to 10. The capacity of loading phosphate will be affected by the concentration of binding gel and pH variation.

As seen from figure 3, the optimum Fe-MOF concentration of DRP uptake was 20 mg. Higher amount of binding agent will decrease the capacity of loading phosphate because it can cause the binding agent contains in gel to be rigid and the elasticity will decrease, so phosphate will be difficult to interact with material and the capacity of loading phosphate also decrease. The maximum DRP uptake was 9.30 µg P. From figure 4, it is noted that the optimum pH of DRP uptake was pH 2 with capacity 64.35 %. At the lower pH, the charge of Fe-MOF surface is more positive and easier for phosphate bonding, but at the higher pH, the charge will be more negative so it can cause an electrostatic repulsion to negative phosphate (PO_4^{3-}) which is available in alkali condition.

4. Conclusion

This experiment was successfully synthesized the Fe-MOF by solvothermal methods. Material was applied to DGT system for measuring the capacity of loading phosphate. The capacity of loading phosphate in DGT system will be affected by the mixing time, concentration of binding agent, and pH variation. The results revealed that Fe-MOF in DGT system had an activity in the phosphate adsorption.

Acknowledgments
This work was financially supported by Universitas Indonesia under research grant PITTA A .

References

[1] Karageorgiou K, Paschalis M and Anastassakis G N 2007 *J. Hazard. Mater.* **139** 447-52
[2] O'Sullivan P E 1995 *Int. J. Environmental Studies* **47** 173-95
[3] Cusnir R, Steinmann P, Bochud F and Froidevaux P 2014 *Environ. Sci. Technol* **48** 10829-34
[4] Ding S, Wang Y, Zhang L, Xu L, Gong M and Zhang C 2016 *RSC Adv* **6** 88143–56
[5] Li Y, Xie Q, Hu Q, Li Q, Huang Z, Yang X and H Guo 2016 *Sci. Rep.* **6** 30651
[6] Xie Q, Li T, Lv Z, Zhou H, Yang X, Chen J and Guo H 2017 *Sci. Rep.* **7** 3316
[7] Zhao B, Zhang Y, Dou X, Yuan H and Yang M 2015 *Water Sci. Technol.* **72** 2179-86
[8] Furukawa H, Cordova K E, O'Keeffe M and Yaghi O M 2013 *Science* **341** 1230444
[9] Qin W, Gu Y, Wang G, Wu T, Zhang H, Tang X, Zhang Y and Zhao H 2018 *Water Res.* **147** 223-2
[10] Zhou H-C, Long J R and Yaghi O M 2012 *Chem. Rev* **112** 673–4

The influence of calcination temperature on the structural properties of La_2FeMnO_6 double perovskite materials

Yunida and D Triyono

Department of Physics, Faculty of Mathematics and Natural Sciences (FMIPA), Universitas Indonesia, Depok 16424, Indonesia

Corresponding author's email: djoko.triyono@sci.ui.ac.id

Abstract. The double perovskite compound, La_2FeMnO_6 (LFMO), has been synthesized using the sol-gel method. Structural properties were characterized by X-ray diffraction (XRD), and X-ray Fluorescence (XRF), Fourier Transform Infrared Spectroscopy (FTIR) and Raman scattering spectroscopy. Rietveld refinement of calcination powder from X-ray data indicates the compound was a single-phase formation of double perovskite with 750 °C and 900 °C calcination temperatures. XRF showed the weight and atomic ratio of the compound are suitable with double perovskite stoichiometric, respectively. The result of structural analysis confirmed that the La_2FeMnO_6 has a cubic structure with space group *Pm-3m* at both temperatures. Lattice parameters, volume and crystallite size of the samples become larger as higher calcination temperatures. Raman scattering spectroscopy and FTIR have been used to study the peak vibrational modes and various bond formations of the prepared samples.

Keywords: Double perovskite, LFMO, sol-gel method, structural properties

1. Introduction

Single perovskites are the most intensely studied materials in solid-state chemistry and physics in recent years. These materials with ideal cubic compound formula ABX_3 exhibit several fundamentally interesting chemical and physical properties [1]. $LaFeO_3$ is one of kind single perovskite oxide/rare earth orthoferrite which is significant because it is potential to be applied in data storage media, multiple stage memories and sensors [2]. Jung-Hoon *et al* reported that Mn-doped $LaFeO_3$ nanofibers were prepared using the electrospinning technique have a single phase without any trace impurity from XRD analysis [3]. The single perovskite $LaFeO_3$ and $LaMnO_3$ are well known as antiferromagnets (AFM) materials with Néel temperatures (TN) 740 K and 140 K.

Perovskite with substitution of A' for A (or B' for B) cations will take on a disordered arrangement which rises a compound generally labelled as double perovskite, $A_2B'B''X_6$ [4]. La_2FeMnO_6 (LFMO) is a double perovskite that has gained popularity, where A is Alkaline earth or Rare earth element and B′ & B″ is a *3d* transition metal [5]. Lanthanum oxide has an attractive characteristic such as structural, magnetic and electrical properties as a base compound in double perovskite. In La_2FeMnO_6 double perovskite, La^{3+} occupies site A with the ionic radius (1.06 Å) whereas Fe^{3+} (0.64 Å) and Mn^{3+} (0.58 Å) cation occupy site B′ and B″, respectively.

Recent work has been focused on the synthesized method, characterization and magnetic behavior of double perovskite [6]. Hence, this paper has been focused on the influence of calcination

IOP Conf. Series: Materials Science and Engineering **902** (2020) 012027 doi:10.1088/1757-899X/902/1/012027

temperature towards the structure of double perovskite Mn substituted $LaFeO_3$ compounds (La_2FeMnO_6).

2. Experimental

The La_2FeMnO_6 (LFMO) materials were synthesized by the sol-gel method at ambient atmosphere. The composition of lanthanum (III) oxide, iron (II) nitrate nonahydrate, and mangan (II) nitrate tetrahydrate were measured using double perovskite stoichiometric with citric acid and aquadest as a solvent. All precursors were mixed and stirred at 120 °C to remove water and citric acid. The stirring process is also to obtain a gel phase of the sample with a magnetic stirrer. The mixture was dried at temperature 200 °C for 5 hours to get a sample in the form of a sponge, then calcined at 900 °C for 6 h and 750 °C for 5 h. X-ray diffraction type *X'Pert PRO PANalytical* diffractometer with Cu Kα radiation (λ = 1.5405 Å) and range of angle 20°-90° was used for phase and impurity identification whereas the X-Ray Fluorescence (XRF) was used for determining the composition. Raman scattering spectroscopy and FTIR have been employed to study the vibrational modes and various bond formations of LFMO.

3. Results and discussion

Figure 1 shows the XRD pattern of La_2FeMnO_6 powder with calcined 900 °C for 6 h and 750 °C for 5 h (figure 1). The refinement was done by *Highscore Plus* software and the Le-bail fitting was plotted using *Crystallography Open Database* (COD) of $LaFeO_3$ with number COD 96-154-2145. The Le-bail fitting material plotted confirmed that both of samples were the single-phase without any trace of impurities. Sample A has a peak intensity higher than sample B and a smaller full width at half maximum (FWHM). It is due to an enhanced crystallization of perovskite phase as the increasing calcination and the grain size become larger than one on the FWHM [7]. The Rietveld refinement results confirm that both of the prepared LFMO material into a cubic structure with *Pm-3m* (221) space group.

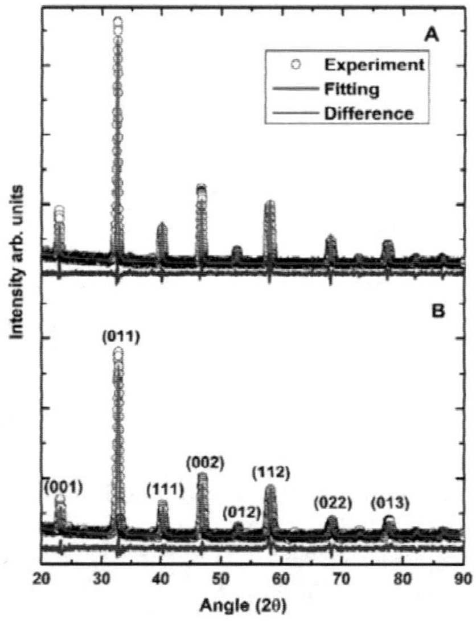

Figure 1. Rietveld refinement of XRD patterns of La_2FeMnO_6 double perovskite with A: calcination of 900 °C and B: calcination of 750 °C.

IOP Conf. Series: Materials Science and Engineering **902** (2020) 012027 doi:10.1088/1757-899X/902/1/012027

3.1. Structural analysis

Table 1 shows the crystallography and geometric parameters of LFMO powders. Sample A and B have Goodness of fit (χ^2) 1.60 and 1.44 respectively. It is indicated that the comparison between Rwp (weighted profile) and Re (expected) from XRD fitting, as a result, Rietveld refinement was suitable to be used [8]. The different crystallite sizes at both samples are due to the diffusivity of the individual grains, sintering temperature and porosity [9].

Table 2 shows the weight and atomic ratio of the La_2FeMnO_6 comparison between experimental and calculation by X-Ray Fluorescence. From the table, it is known that the composition of the sample is a slight different between experimental and calculation results. The weight percentage of Fe and Mn in experimental is less than the one from calculation result. It is due to the deficient mass along the calcination process and correlates to the both components' melting point which is lower than La.

3.2. Raman scattering analysis

Figure 2 shows the Raman spectral analysis of La_2FeMnO_6 with wavenumber 100–2000 cm^{-1}. The peak position was determined by fitting the peak with the Lorentzian function for structure and phase purity analysis. Only a few report mentioning the lattice dynamical calculations and

Table 1. The structural parameters of La_2FeMnO_6 double perovskite with calcination of 900 °C and 750 °C.

Parameter	La_2FeMnO_6	
	Calcination of 900 °C	Calcination of 750 °C
Phase	Cubic	Cubic
Space group Lattice parameters	*Pm-3m* (221)	*Pm-3m* (221)
a (Å)	3.89	3.90
b (Å)	3.89	3.90
c (Å)	3.89	3.90
Volume (Å^3)	59.30	59.33
Crystallite size (nm)	235.20	382
Density g/cm^3	6.79	6.79
Geometrical parameters		
R factor (%)		
Re (expected)	7.07	6.74
Rp (profile)	6.58	6.20
Rwp (weighted profile)	8.97	8.10
Goodness of fit (χ^2)	1.60	1.44

Table 2. XRF characterization of La_2FeMnO_6 double perovskite with (a) calcination 900 °C and (b) calcination 750 °C.

La_2FeMnO_6	Calcination temperature	(%) experimental			(%) calculation		
		La	Fe	Mn	La	Fe	Mn
Weight	900 °C	78.67	11.50	9.83	71.47	14.39	14.13
Atomic ratio		2.76	1	0.86	2	1	1
Weight	750 °C	74.52	13.79	11.69	71.47	14.39	14.13
Atomic ratio		2.17	1	0.86	2	1	1

assignment of LFMO materials, but the Raman Spectra of double perovskite is mostly below the value of 1000 cm^{-1} [10]. Table 3 is the list of the observed Raman active modes. The peak positions are in good agreement with the report in the literature. The peak at 503.72 cm^{-1} and 499 cm^{-1} correspond to A_g vibrational mode and the resonance peak at 653.53 cm^{-1} and 659.89 are attributed to E_1 mode of vibration [11]. The resonance peak at 1325.76 and 1379.29 are the double energy values of E_1. The strong reduction ($2E_2$) is correlated to the existence of magnetic ordering in these materials [12]. It is also shown that the vibrational mode in which appearance at peaks influences the anti-ferromagnetic ordering [13].

3.3. Fourier transform infrared spectra (FTIR) analysis

Figure 3 shows the Fourier Transform Infrared Spectra (FTIR) of LFMO materials with range wave number 500–4000 cm^{-1}. The vibration mode at 502 cm^{-1} is noted and assigned to Fe-O stretching in these materials [14]. The one sharp vibration mode at 569 cm^{-1} is noted due to the internal movement of length change of the bounds Mn-O-Mn. This vibration corresponds to the octahedron MnO_6 which is related to the vibration property of perovskite structure ABO_3 [15]. Sample A has two vibrations evident at 1740 cm^{-1} and 1201 cm^{-1}. The vibration appearance around 1115 cm^{-1} indicates the presence of C-O groups and a strong vibrational band of water (H-O-H) appearance at 1740 cm^{-1} [16].

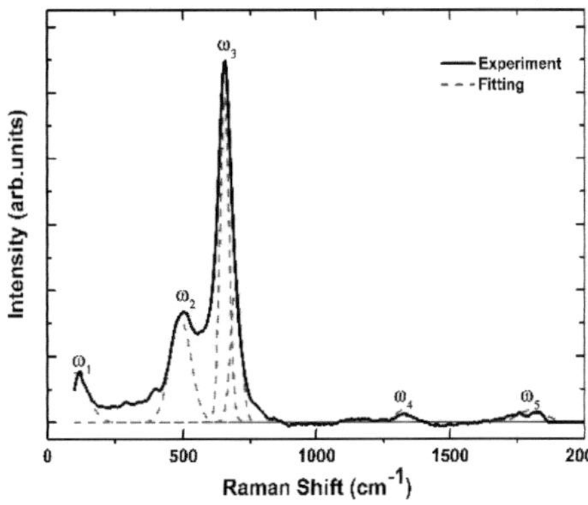

(a)

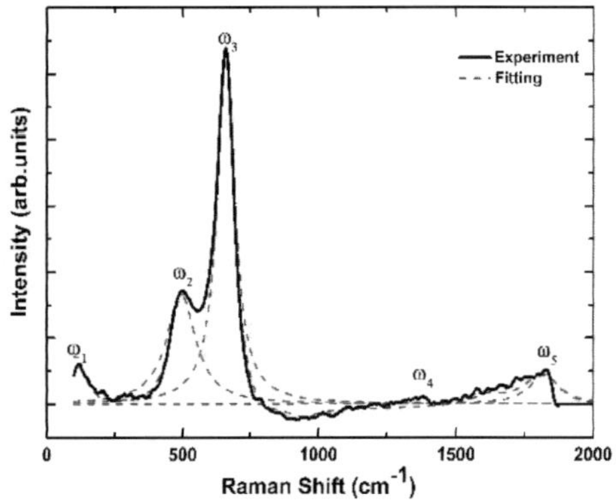

(b)

Figure 2. Raman Scattering of La_2FeMnO_6 double perovskite with calcination of (a) 900 °C and (b) 750 °C.

Table 3. Raman shift (cm^{-1}) of La_2FeMnO_6 double perovskite with calcination of 900 °C and 750 °C.

Raman shift (cm^{-1})		Assignment of modes
Calcination of 900 °C	Calcination of 750 °C	
120.82	119.49	-
503.72	499.27	Ag
653.53	659.89	E_1
1325.76	1379.29	$2E_2$
1758.61	1830	-

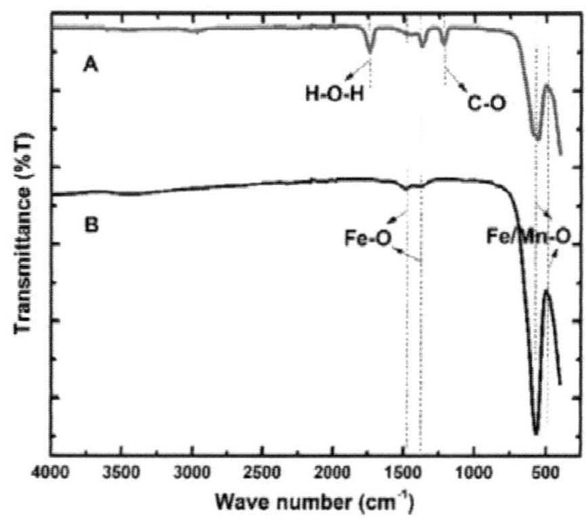

Figure 3. Fourier Transform Infrared Spectra (FTIR) of La_2FeMnO_6 double perovskite with calcination of A: 900 °C and B: 750 °C.

4. Conclusion

Double perovskite La_2FeMnO_6 has been synthesized by sol-gel method at ambient condition. The XRD pattern confirmed that all of the samples have a cubic structure with *Pm-3m* space group. The crystallite size decreases with the increase temperature of calcination. The sample was single phase at both calcination temperatures. The composition of La_2FeMnO_6 was suitable for double perovskite stoichiometric formula. The vibration mode of Raman Spectra and Fourier Transform Infrared Spectra (FTIR) confirms the phase purity of La_2FeMnO_6.

Acknowledgments

This work was financially supported by Ministry of Research and Technology (KEMRISTEK/BRIN) Republic of Indonesia under research grant Hibah Penelitian Tesis Magister (PTM) 2020 with contract number NKB-493/UN2.RST/HKP.05.00/2020.

References

[1] Vasala S and Karppinen M 2015 *Prog. Solid State Chem.* **43** 1–36
[2] Mahmood A, Warsi M F, Ashiq M and Ishaq M 2013 *J. Magn. Magn. Mater.* **327** 64–70
[3] Jeong J –H, Song C –G, Kim K –H, Sigmund W and Yoon J –W 2018 *J. Alloys Comp.* **749** 599-604
[4] Xu X, Zhong Y and Shao Z 2019 *Trends in Chemistry* **1** 410-24
[5] Punitha J S, Dhilip M, Anbarasu V nd Kumar K S 2019 *AIP Conf. Proc.* **2115** 030468
[6] Qian Y et al. 2013 *J. Appl. Phys.* **114** 063713
[7] Sun Y et al. 2014 *Sustain. Energy Techn.* **8** 92-8
[8] Toby B H 2006 *Powder Diffr.* **21** 67–70
[9] Haque M M, Maria K H, Choudhury S, Bhuiyan M and Hakim M 2013 *J. Ceram. Process. Res.* **14** 82-6
[10] Dhilip M, Devi N A, Punitha J S 2019 *Vacuum* **167** 16-20
[11] Iliev M N et al. 1998 *Phys. Rev. B* **57** 2872–7
[12] Hien N T M et al. 2009 *J. Raman Spectrosc.* **41** 983–8
[13] Massey M J, Baier U, Merlin R and W H Weber 1990 *Phys. Rev. B Condens. Matter.* **41** 7822–7

[14] Romero M, Gomez R W, Marquina V, Perez-Mazariego J L and Escamilla R 2014 *Physica B* **443** 90–4
[15] Moghadam L N and Ranjbar Z R 2019 *J. Alloys Compd.* **785** 117–24
[16] Frost R L, Kristof J, Paroz G N and Kloprogge J T 1998 *J. Colloid Interf. Sci.* **208** 216–25.

IOP Conf. Series: Materials Science and Engineering **902** (2020) 012028 doi:10.1088/1757-899X/902/1/012028

Characteristics of composite magnesium with micro SiC particle for automotive applications with stir casting method

N R Susilo and A Z Syahrial

Department of Metallurgy and Materials Engineering, Faculty of Engineering, Universitas Indonesia, Depok 16424, Indonesia

Corresponding author's email: anne@metal.ui.ac.id

Abstract. Composite is a promoted material consists of two or more materials to enhance the properties mechanically. In this study, composites were created by mixing a matrix; magnesium and the reinforcement; SiC microstructure with the addition of a volume fraction of: 2 %vf, 4 %vf, 6 %vf and 8 %vf. This magnesium composite fabrication process is stirred casting. The results of this study show that the addition of SiC micro particles will enhance the properties of the composite–mechanical wise-, namely, wear resistance and hardness. The addition of SiC micro-reinforcing particles that produced the best magnesium composite was by adding 8 %vf volume fraction which produced a hardness value of 83 HB and a wear rate of 0.002015706 mm^3/m. Increased mechanical properties in magnesium composites is caused by the reducing grain size and Orowan strengthening mechanism.

Keywords: Magnesium composite, micro SiC, stir casting

1. Introduction

Technology in the automotive field is one of the many rapidly developing technology fields. In the automotive world, one factor that continues to be developed is the fuel in vehicles. To reduce the fuel requirements on a vehicle, the weight of the vehicle must be balanced [1]. The weight of vehicles takes a significant part in the fuel efficiency. Since, the heavier the vehicle, leads to a higher usage of fuel for the engine to drive. The weight reduction of vehicle by 10 %, can reduce the fuel usage by 6–8 % [2]. The car skeleton is one parameter that could be taken into consideration. One type of metal that is considered to create a skeleton of a car with a low mass is magnesium [3].

Magnesium is a metal with the lowest density. Because the density is only 1.74 g/cm^3 [4], it can reduce the weight and increase the efficiency of the fuel, hence magnesium is the most compatible material to be applied as raw material for high fuel efficiency car skeletons [5]. Magnesium is the sixth most abundant element on earth (2.7 % of the composition of the earth's crust is magnesium) [6], it is possible to use it as a mass production for car skeleton. To make a car skeleton with improved mechanical properties, mixing magnesium with other materials to form magnesium composite material that has better mechanical properties than pure magnesium can be applied.

In composites, two or more materials are involved. Both types the material acts as a matrix and reinforcement. The composite matrix used is magnesium metal. One of the compounds that can be utilized is micro-sized silicon carbide (SiC) ceramic particles [7]. According to the research conducted by Paramshoty et al. [8] ceramic reinforcements have excellent mechanical properties, isotropic properties, chemical hypertension properties, and are stable at high temperatures, as well as the expansion coefficient can be monitored, therefore the particles of ceramic can be regarded as interesting for fabrication of metal matrix composites. In this study, four variable % volume fractions were used from micro-sized SiC reinforcing particles with magnitudes of 2, 4, 6 and 8. With the addition of reinforcing micro-sized ceramic particles, magnesium composites with low density, specific strength and high stiffness were obtained, as well as the high wear resistance [8].

2. Materials and method

2.1. Materials

Before the trial, mass balance calculation of composite was performed to prepare the required materials. Pure magnesium 90 %wt was applied as the matrix and micro sized SiC particle with 2, 4, 6 and 8 % volume fraction was used.

2.2. Fabrication of magnesium with SiC micro-sized particle composite

The magnesium ingot weighed at the same mass balance of magnesium was cleaned with soap and 96 % alcohol in order to remove inclusion particle and oil that might be left on the surface. SiC (reinforce) was heated in a different furnace with a temperature of 900 °C, for 30 min in order to remove inclusions while the furnace and the mold was coated with mixing thinner and zircon so that the furnace and the molt has a high thermal resistance to avoid cracking. After that, the magnesium and micro SiC were inserted into the furnace, before turned on the heater, they were fluxed for 6 min and argon gas was flowed for 20 sec.

The matrix was melted at 650 °C in the furnace and before pouring into the matrix, the reinforce was pre-heated at 900 °C in the muffle furnace to remove the moist. After the pre-heat process, micro SiC particle was inserted into molten magnesium and then stirred for 1 min to get homogenous mixture. Then, the mixture was poured into the mold and the mold was opened after the composite solidified. Finally, this composite sample was being used for further characterization.

2.3. Characterization

To study the features of microstructure, optical microscope (OM) observation was conducted on the composite samples that were grounded and polished by TiO_2. Scanning electron microscope (SEM) micrographs and X-ray diffraction (XRD) patterns were also recorded to obtain the data regarding the presence of different phases in the microstructure. The chemical properties were determined with optical emission spectroscopy (OES). Mechanical properties were studied by conducting the tensile test based on JIS Z2241; impact test based on Charpy method with ASTM E23 standard, hardness test (Rockwell H) based on ASTM E18 and wear-rate test with pin on disc (Ogoshi) method based on ASTMM G99. The density of every sample was also determined with the Archimedes Theory. All mechanical tests were performed at room temperature.

3. Results and discussion

3.1. Chemical composition

OES (Optical Emission Spectrometry) is a test to confirm the presence of elements in magnesium as-cast or non-reinforced magnesium as well as the magnesium composite quantitatively. So that, the effect of the casting process on the metal characterization will be identified. Table 1 below shows the chemical composition of the material. According to the characterization results, magnesium as-cast has 99 % Mg which is in accordance with the mass balance in the first place. Instead, there are other elements that were not supposed to be present in the composite, namely Al, Ti, Be, Cu, Mn, etc. However, these elements would not give a significant effect to the characteristics itself. For the composite, the results show that the increasing of reinforce particles reduces the content of

IOP Conf. Series: Materials Science and Engineering **902** (2020) 012028 doi:10.1088/1757-899X/902/1/012028

magnesium. On the other hand, it decreases in the 2 %vf micro SiC which is caused by the porosity, also due to the speed while pouring the liquid into the mold was not fast and quick enough so that the liquid composite reacted too long with the air.

3.2. SEM and EDS

SEM and EDS tests were performed using magnesium as-cast and magnesium composite with 8 %vf to represent all of the composite sample. Based on the figure 1 above, it shows some elements like Mg, Al, Si and O which probably formed some compounds; MgO and Al_2O_3. The oxygen element at some point indicates of products of rust initiation in the form of MgO at most points in the composition of 8 %vf. Just like other compositions, this is because magnesium is a highly reactive metal and oxidized very easily. This carbon impurity originates from the crucible from the melting furnace which is a graphite-based refractory. There is Al_2O_3 at point 10 which is probably derived from polished Al_2O_3 particles which are still attached onto the surface.

Table 1. Chemical composition of Mg-as cast and composite material.

Element	Mg (as-cast) (JIS)	Mg/2 % micro SiC	Mg/4 % micro SiC	Mg/6 % micro SiC	Mg/8 % micro SiC
Mg	99.8	> 96.0	> 98.0	> 98.0	> 96.0
Al	< 0.005	1.01	0.703	1.190	1.540
Be	0.0001	0.00006	0.00009	0.0001	0.00006
Cu	< 0.001	< 0.0001	< 0.0001	< 0.0001	< 0.0001
Mn	0.0466	0.00690	0.005	0.005	0.0086
Zn	0.0377	0.00710	0.023	0.011	0.0097
Ag	< 0.001	0.00014	< 0.0001	0.0002	0.00019
Ca	0.0023	0.00180	0.002	0.0022	0.0028
Cd	< 0.001	< 0.0001	< 0.00010	< 0.00010	< 0.00010
Sn	0.0421	0.01470	0.019	0.0125	0.018
Sr	0.0007	0.00016	0.0002	0.00014	0.00025
Si	0.134	2.83000	1.210	0.671	2.170

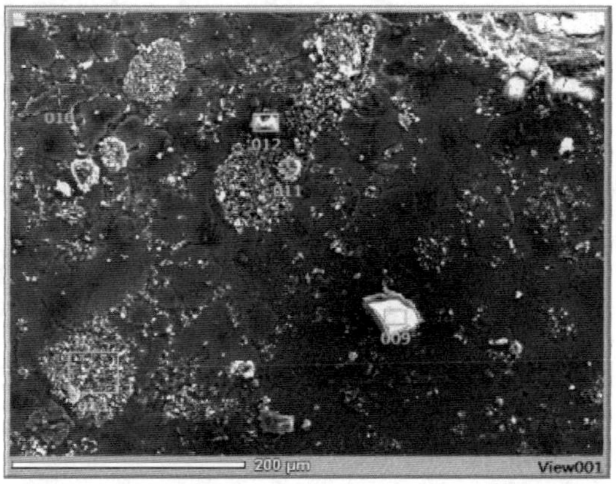

Figure 1. The SEM result of Mg micro SiC.

3.3. XRD
XRD is a test to find out which phases are formed in the composite microstructure. In this research, magnesium composite with micro SiC 8 %vf was observed to represent all the composite samples. From the graph of the XRD test results on the magnesium reinforcing composite SiC 8 %vf, it can be concluded that the compounds formed are Mg, MgO, Si, SiC, SiO. Mg compound appears because the matrix of the micro Mg composite SiC is magnesium. Similarly, Si and SiC appear due to the addition of the amplifier from the SiC (figure 2) Meanwhile, MgO compounds appear because of the reaction of magnesium with oxygen which results in the formation of MgO compounds. SiO compounds are formed due to the reaction of SiC amplifiers with oxygen to form SiO compounds.

3.4. Microstructure
In this study, the addition of SiC micro-booster volume fraction of 2, 4, 6 and 8 %vf resulted in a decrease in grain size in magnesium composites thus increasing some mechanical properties namely hardness and wear resistance. In the microstructure Mg image, there is porosity due to the reaction of the molten magnesium with oxygen so that a porosity phenomenon is formed. This is also seen in the addition of the SiC micro amplifier volume fraction at 2 %vf and 4 %vf because the pouring time tends to be longer than the composite with SiC 6 %vf and 8 %vf micro amplifier volume fraction which has less porosity. In the figure 3, there is no porosity in the SiC 6 %vf and 8 %vf micro amplifier volume fraction because it is not included and also has a little porosity. In Mg composite microstructure images with SiC 2, 4, 6 and 8 %vf micro booster volume fraction occurs SiC reinforcement agglomeration phenomenon.

This is because the gathering of SiC amplifiers is caused by several things, namely the stirring is too fast or too slow when casting, so that the micro amplifier SiC will agglomerate. At 8 %vf micro-SiC amplifier volume Mg_2C_3 precipitates are formed, where precipitates are formed because magnesium as the matrix reacts with SiC as an amplifier where in the 8 % variation SiC micro amplifier, the more SiC amplifier means more carbon will react with magnesium and will form the Mg_2C_3 phase. According to research conducted by Esmaily M et al. [9] the Mg_2C_3 phase has good mechanical properties such as strength and hardness. Therefore, the results of microstructure images in accordance with the literature Mg Micro SiC 8 %vf composite has good strength and wear resistance.

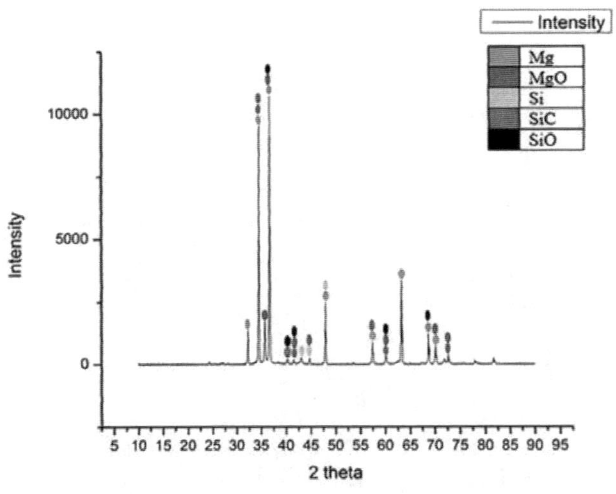

Figure 2. The XRD result of MgSiC 8 %vf.

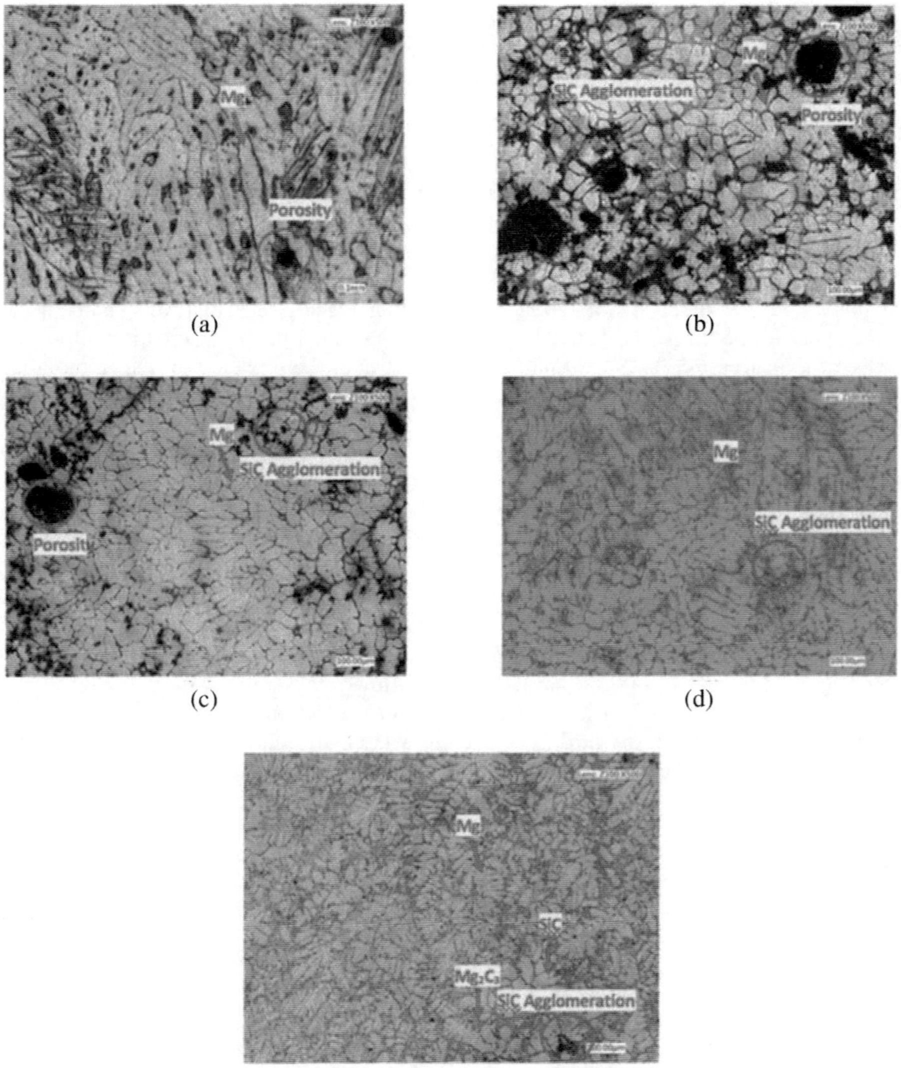

Figure 3. Microstructure images of (a) as-cast 200 x, (b) MgSiC 2 %vf, (c) MgSiC 4 %vf, (d) MgSiC 6 %vf, (e) MgSiC 8 %vf 1000 x.

3.5. Tensile and elongation

Data obtained from the Pull test is Ultimate Tensile Strength (UTS) and % elongation. This test is performed on each variable to determine the effect of changes in SiC content on the tensile strength of the composite. The test results can be seen in figure 4 where in general it can be seen that the more SiC added to the composite, the higher the tensile strength of the composite.

In figure 4a, it can be seen that the tensile strength of magnesium composites added with SiC micro reinforcing particles significantly increases. Besides that, the decrease in strength is caused by the uneven distribution of SiC, so that SiC clumps will appear. SiC accumulation at this point will certainly reduce the number of interfaces formed, so that the SiC strengthening mechanism that occurs is not maximum and the tensile strength will decrease. The uneven distribution of this SiC can be influenced by two factors, namely the speed and duration of the stirring process during the casting process.

In figure 4b we can see that there was a decrease in % elongation in the composite after the addition of SiC compared to the magnesium cast. The decrease elongation value indicates that the addition of SiC into the material will not only make the material stronger but also make the material more brittle. Plastic deformation process will be more difficult to occur in materials that have reinforcing particles in them when compared to pure metals [10].

Research conducted by Song Min [11] also shows that the more SiC levels are added, the lower is the elongation value. The reduced elongation value when increasing the SiC levels can be because the SiC itself has very brittle properties. In addition, SiC can also be a point of crack initiation. When the bond between the matrix and the SiC amplifier is very strong, then when exposed to the load of the dislocation density it will continue to increase until it reaches the maximum limit, and these SiC particles will break and create cracks on the surface of the magnesium matrix where this crack will continue to propagate. Conversely, when the bond between the matrix and the SiC amplifier is weak, the interface will not be broken, but the SiC will detach from the matrix and leave cracks on the magnesium matrix [10]. However, when interface bonds are formed between the matrix and the amplifier, a strength increase will occur without a significant reduction in the elongation value [12].

3.6. Hardness, wear rate and impact

The composite hardness value increases with the increase of the aging temperature. However, it will decrease after certain temperature when over aging phenomenon takes place. Composite with 200 °C aging temperature achieved the highest hardness value. However, sample with 230 °C, 260 °C aging temperature only shows an insignificant hardness value reduction. When the temperature aging increases to peak age, precipitates will form coarsening which will reduce the strength and hardness of the composite material [6]. Figure 5 of impact value data also shows that the composite with 200 °C achieved the highest impact value. This tendency resembles the hardness test results. Figure 5 of wear rate data also displays the tendency of materials having a higher hardness value to show a lower wear rate, since the materials' hardness property is straightly corresponded to the capability in preventing the loss of materials as exposed against abrasion, erosion, or other wear mechanism. The hardness increase in the material leads to a decrease in the rate of wear while the decrease in the hardness in the material creates an increase in the rate of wear [4].

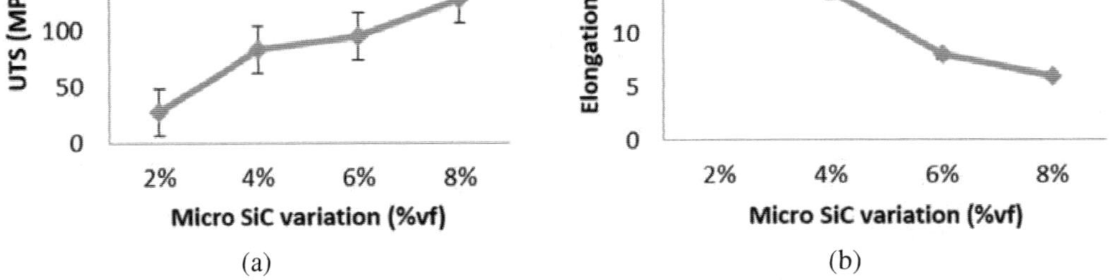

Figure 4. (a) Tensile strength and (b) elongation.

IOP Conf. Series: Materials Science and Engineering **902** (2020) 012028 doi:10.1088/1757-899X/902/1/012028

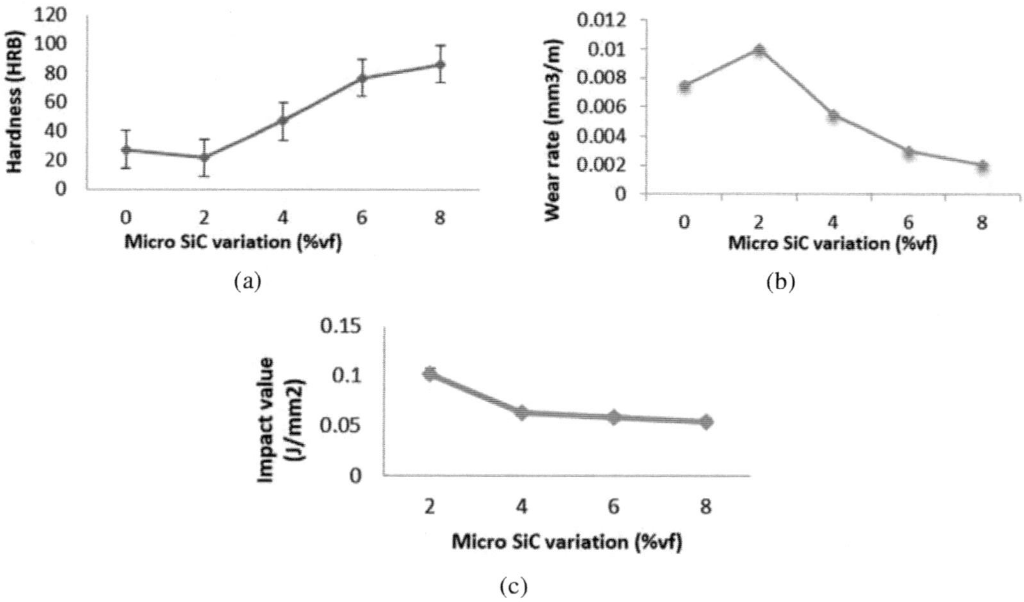

Figure 5. (a) Hardness, (b) wear rate and (c) impact value of composite material.

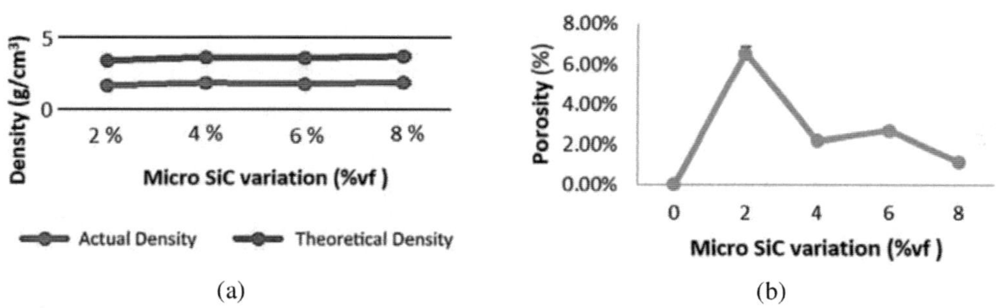

Figure 6. (a) Density and (b) porosity of composite material.

3.7. Density and porosity

Figure 6a shows that the actual density of composite samples of all compositions has a smaller value when compared to the theoretical density. The smaller actual density of the composite sample compared to its theoretical density indicates porosity in the composite sample. The greater the difference between the priority density and the actual density shows the more porosity is formed.

Figure 6b shows magnesium as-cast has a lower porosity value than magnesium composites in all variables adding volume fraction. This is because during the process of making magnesium as-cast, the stirring process is not carried out because there is no addition of SiC micro boosters on liquid magnesium. So that, there is no contamination from the atmosphere or metal stirrers which can cause porosity to occur. The microstructure of SiC reinforcement particle at 2 %Vf has a very large amount of porosity due to external factors such as stirring and agglomeration formed from the SiC micro amplifier [13].

Porosity in composite samples can be formed because there are gases such as hydrogen gas trapped and dissolved in the composite sample during the casting process and during the solidification process. Besides, porosity can also be formed due to several factors, namely the contact between air or the outer atmosphere with molten magnesium metal, materials and equipment that are not clean and not dry, the occurrence of turbulence during the stirring process [14] and because of the high temperature during the stirring process [12]. The degassing process during the casting process can be carried out to reduce the formation of porosity by injecting argon gas into molten metal to remove hydrogen gas content dissolved in liquid magnesium [10]. Previously, the vacuum process was carried out on the furnace which aims to form a vacuum environment in the furnace, and when the melting process of argon Mg gas ingots is continuously fed into the furnace to make the furnace environment inert so that the hydrogen gas is not causing porosity casting defects. Porosity during the casting process occurs because when inserting the amplifier into the furnace for too long so that magnesium and SiC micro reinforcing particles react with air causing porosity.

4. Conclusion

The stir casting method may be used to fabricate Mg composite of the micro SiC particle. The addition of micro SiC can increase the UTS value at 8 %Vf specifically, reduce the impact value, increase the hardness, decrease the wear rate, and increase the density of material. The addition of 8 %wt micro SiC was aimed to refine the material's grain by forming initial nucleation, so that solidification occurs faster and is able to increase the material's strength as well. The result of the microstructure shows the presence of MgO, Al_2O_3 and. MgO.

Acknowledgments

This work was financially supported by Universitas Indonesia under research grant PIT 9 with grant contract number NKB-0042/UN2.R3.1/HKP.05.00/2019.

5. References

[1] Blawert C, Hort N and Kainer K U 2004 *Trans. Indian Inst. Met.* **57** 397-408

[2] U.S. Department of Energy 2011 2010 *Annual Progress Report: Light Weighting Material* available at https://www.energy.gov/sites/prod/files/2014/03/f8/2010_lightweighting_materials.pdf

[3] Kiani M, Gandikota I, Rais-Rohani M and Motoyama K 2014 *J. Magnes. Alloy* **2** 99-108

[4] Gupta M and Sharon N M L 2010 *Magnesium, Magnesium Alloys, and Magnesium Composites* (Singapore: John Wiley & Sons, Inc)

[5] Porro R 1998 *The Innovative Use of Magnesium in Car Design and An Analysis of Cost Versus Weight Savings* (United States: SAE International)

[6] Clark J B and Nayeb-Hashemi A A 1988 *Phase Diagrams of Binary Magnesium Alloys* (Metals Park, Ohio, UnitedSA: ASM International)

[7] Dey A, Pandey K M 2015 *Rev. Adv. Mater. Sci.* **42** 58-67

[8] Paramsothy M, Chan J, Kwok R and Gupta M 2011 *J. Nanopart. Res.* **13** 4855-66

[9] Esmaily M, Svensson J E, Fajardo S, Birbilis N, Frankel G S, Virtanen S, Arrabal R, Thomas S and Johansson 2017 *Prog. Mater. Sci.* **89** 92-193

[10] Ibrahim I A, Mohamed F A and Lavernia E J 1991 *J. Mat. Sci.* **26** 1137-56

[11] Song M 2009 *T. Nonferr. Metal. Soc.***19** 1400-4

[12] Srivastava N and Chaudhari G P 2016 *Mater. Sci. Eng. A* **651** 241-7

[13] Ponappa K, Aravindan S, Rao P V, Ramkumar J and Gupta M 2010 *Int. J. Adv. Manuf. Tech.* **46** 1035-42

[14] Aravindan S, Rao P V and Ponappa K 2015 *J. Magnes. Alloy.* **3** 52-62

Functionalization of titanium dioxide through dye-sensitizing method utilizing red amaranth extract for phenol photodegradation

Y S Kurniawan[1], K Anggraeni[2], R Indrawati[1, 2] and L Yuliati[1, 2]

[1]Ma Chung Research Center for Photosynthetic Pigments, Universitas Ma Chung, Malang 65151, Indonesia
[2]Department of Chemistry, Faculty of Science and Technology, Universitas Ma Chung, Malang 65151, Indonesia

Corresponding author's email: leny.yuliati@machung.ac.id

Abstract. In this work, red amaranth (RA) extract was utilized as the source of the natural pigments for dye-sensitizing of the TiO_2. The natural pigments were isolated from the RA leaves through a maceration technique to obtain the RA extract as a red powder. Since photocatalytic efficiency of titanium dioxide (TiO_2) under visible light is negligible, further modification such as by addition of dye-sensitizer is required. The TiO_2 was sensitized using the natural pigments contained in the RA extract using (3-aminopropyl)trimethoxysilane (APTMS) as the linker agent through an impregnation method to give the dye-sensitized TiO_2 composite (RA-APTMS/TiO_2). After dye-sensitizing, new absorption peaks at 1632–1030 cm^{-1} were detected in the Fourier transform infrared (FTIR) spectra, which were corresponded to the functional groups of chlorophyll originated from the RA extract. Furthermore, the appearance of a new absorption peak at 665 nm in the diffuse reflectance ultraviolet-visible (DR UV-vis) spectra also supported the successful sensitization of the TiO_2 by the RA extract. Photocatalytic tests showed that while the unmodified TiO_2 could not degrade phenol under visible light, the RA-APTMS/TiO_2 composites exhibited photodegradation ability in 9, 13, 22 and 7 % when using 5, 10, 20 and 30 % mass ratio of the RA extract, respectively.

Keywords: Dye sensitizing, red amaranth, phenol degradation, photocatalyst, titanium dioxide

1. Introduction

Titanium dioxide (TiO_2) has been well recognized in the material chemistry field due to its outstanding physicochemical properties [1-4]. Because of that, it has been widely applied as sunscreen, opacifier, paint, and coating materials. TiO_2 can be used as sunscreen materials because it absorbs the ultraviolet light due to its semiconductor properties having band gap energy of 3.2 eV [5]. In photocatalysis field, TiO_2 has been used since 1956 when the photocatalytic activity of TiO_2 for water splitting was discovered by Mashio and coworkers [6]. Many efforts on developing and optimizing the photocatalytic activity of TiO_2 have been reported to improve the photocatalytic activity of TiO_2 under the visible light. Dye-sensitizing of TiO_2 offers a simple solution by combining TiO_2 and dyes to broaden the absorption region of TiO_2 to the visible region [7].

Red amaranth (RA) with the scientific name of *Amaranthus tricolor L.* is a red edible plant that is abundantly found in Indonesia. The RA is rich in natural pigments such as chlorophyll, betalain, anthocyanin and carotenoid. Therefore, it may serve as the dye sensitizer agent to enhance the photocatalytic of TiO_2 under the visible light. It has been reported that the use of solvent affected the type of pigments extracted from the natural sources [8]. In order to eliminate such solvent effect, in this study the RA was processed by a slow juicer and this would make all type of pigments could be extracted and utilized for TiO_2 dye-sensitizing process. After freeze-drying [9], the isolated natural pigments were then impregnated on the commercially available TiO_2 nanoparticles, employing (3-aminopropyl)trimethoxysilane (APTMS) as the linker agent. Furthermore, the mass ratio of the added RA extract to TiO_2 was varied in 5, 10, 20 and 30 % and the composites were evaluated for photocatalytic degradation of phenol under visible light as a representation for the environmental remediation process.

2. Experimental

The RA plant was bought from a supermarket in Malang, East Java, while TiO_2 nanoparticles (TiO_2 P25) and APTMS were purchased from Evonic Industries and Sigma Aldrich, respectively. The extraction method to obtain RA extract was carried out similarly to the previously reported literature [10]. Briefly, the RA leaves were cleaned, dried and then encapsulated with maltodextrin in 5 % wt/wt. The viscous extract was dried using the freeze-drying technique to obtain the desired RA extract.

The extracted natural pigments were immobilized through an impregnation process. Briefly, as much as 0.05, 0.1, 0.2 and 0.3 g of RA extract were separately added into 50 mL of ethanol. The mixture was homogenized and then added by 1 mL of APTMS. The mixture was kept stirred for 1 day at 27 °C. Subsequently, 1 g of TiO_2 nanoparticles was added into the mixture and the mixture was stirred again for 1 day at 27 °C. The composite materials were washed with ethanol and dried to yield RA-APTMS/TiO_2 x (x = 5, 10, 20 and 30 %) materials. The obtained materials were characterized by a Fourier transform infrared (FTIR, JASCO FTIR-6800) and diffuse-reflectance ultraviolet-visible (DR UV-Vis, JASCO V-760) spectrophotometers.

Prior to the photocatalytic reaction, the adsorption process of phenol onto the surface of the composite material was performed. As much as 50 mg of RA-APTMS/TiO_2 x material was added into 50 mL of phenol 50 mg L^{-1} and the mixture was stirred for 1 h in the dark condition. Afterwards, the mixture was irradiated using halogen lamp with a light intensity of 1.1 mmol m^{-2} s^{-1} and a cut-off filter of 420 nm for 24 h to start the photodegradation process of phenol under visible light. The concentration of phenol before ($[PhOH]_0$) and after ($[PhOH]_t$) photodegradation process was measured by using high-performance liquid chromatography (HPLC, Shimadzu LC-20 AT, equipped with a photodiode array (PDA) detector, SPD-M20A). The stationary and mobile phases used were C18 and 100 % acetonitrile (0.8 mL min^{-1}), respectively, and the column oven was fixed at 40 °C. The percentage of phenol photodegradation was calculated using the following equation.

$$\%\ \text{Photodegradation} = \frac{[\text{PhOH}]_0 - [\text{PhOH}]_t}{[\text{PhOH}]_0} \times 100\ \%$$

3. Results and discussion

In the present work, the natural pigments of RA were employed as the sensitizer agents for TiO_2 materials to enhance its efficiency in the phenol photodegradation process under visible light irradiation. After the dye-sensitizing process using APTMS as the linker agent, the RA-APTMS/TiO_2 x (x = 5, 10, 20 and 30 %) materials were obtained and their photographs are shown in figure 1. It was shown that the color of TiO_2 material was changed from white to green indicating that green natural pigments were successfully immobilized on the surface of TiO_2.

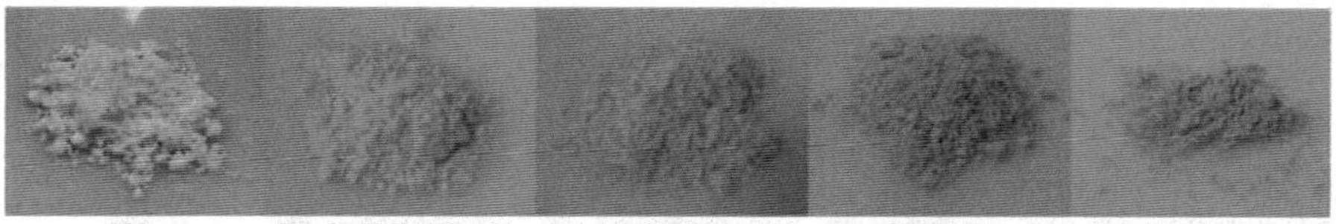

Figure 1. Photograph images of unmodified TiO_2 and its composite materials.

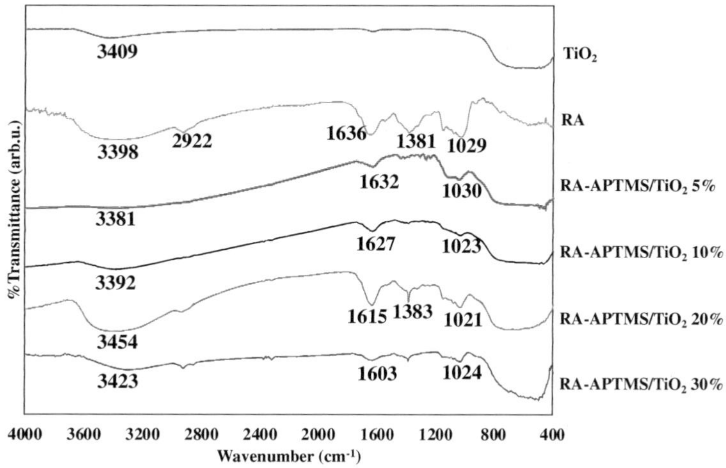

Figure 2. FTIR spectra of unmodified TiO_2, RA and the composite materials.

The obtained composite materials were characterized using FTIR spectrophotometer to identify their functional groups. The FTIR spectra of TiO_2, RA extract and RA-APTMS/TiO_2 x (x = 5, 10, 20 and 30 %) materials are shown in figure 2. The FTIR spectrum of TiO_2 shows two main absorption peaks at 3409 and 800–400 cm^{-1} due to the presence of Ti–OH and Ti–O–Ti moieties, while the FTIR spectrum of RA shows five main absorption peaks at 3398, 2922, 1636, 1381 and 1029 cm^{-1} for O–H, C–H, C=O, C–O and C–N functional groups, respectively. After the dye-sensitizing process, new peaks at 1603–1632, 1381–1383 and 1021–1030 cm^{-1} were observed confirming the presence of green natural pigments on the TiO_2 composite materials.

In order to identify the effect of natural pigments on the absorption properties of the composite materials, the materials were characterized using DR UV-Vis spectrophotometer. The DR UV-Vis spectra of TiO_2, RA extract and RA-APTMS/TiO_2 x (x = 5, 10, 20 and 30 %) materials are shown in figure 3. The absorption peaks of TiO_2 were only observed at 248 and 301 nm (figure 3a), showing that the unmodified TiO_2 could not absorb visible light (400–800 nm), and is in good agreement with the reported literature [11]. In contrast, the RA gave both UV and visible light absorption peaks at 322, 439, 554 and 676 nm (figure 3b). After the dye-sensitizing process (figure 3c), the absorption peak and edge of TiO_2 at 301 nm and *ca.* 400 nm, respectively, were not influenced, indicating that the green natural pigments were located on the surface of TiO_2. Furthermore, a remarkable new peak was observed at 665 nm, which is the characteristic of the chlorophyll pigment [10]. Therefore, it can be suggested that the main immobilized green pigment on the RA-APTMS/TiO_2 x (x = 5, 10, 20 and 30 %) materials was chlorophyll.

The plausible structure of the immobilized chlorophylls on the surface of TiO_2 is shown in figure 4. As reported previously [12], the Ti–OH (from TiO_2) and Si–OCH_3 (from APTMS) would be condensed to form a new covalent bond of Ti–O–Si by losing of CH_3OH (methanol) as by-product. On the other hand, the chlorophylls would be mainly attached to the APTMS through ion-dipole interaction between $-NH_2$ (from APTMS) and Mg^{2+} ion (from chlorophyll). A similar ion-dipole interaction has been suggested to occur between the (3-chloropropyl)trimethoxysilane and betalain [10]. Comparison to the reported work [10], this study obviously showed that the immobilized pigment was strongly depended on the type of the linker. This fact again supported that the immobilized pigment would be attached to the linker, not the TiO_2 surface.

Since the DR UV-Vis spectra of RA-APTMS/TiO_2 x (x = 5, 10, 20 and 30 %) materials show the absorption peak at visible region (665 nm), those materials are promising to be applied as photocatalyst materials under visible light irradiation. In this work, phenol was selected as the photodegradation target because phenol is well-known for its high thermal and chemical stability [13]. Furthermore, phenol pollution seizes the world's attention due to its disadvantages in generating serious diseases to the organism and aquatic environment [14]. The phenol was dissolved in distilled water at 50 mg L^{-1} concentration as the initial concentration and then irradiated for 24 h using a halogen lamp with a cut-off filter ($\lambda > 420$ nm).

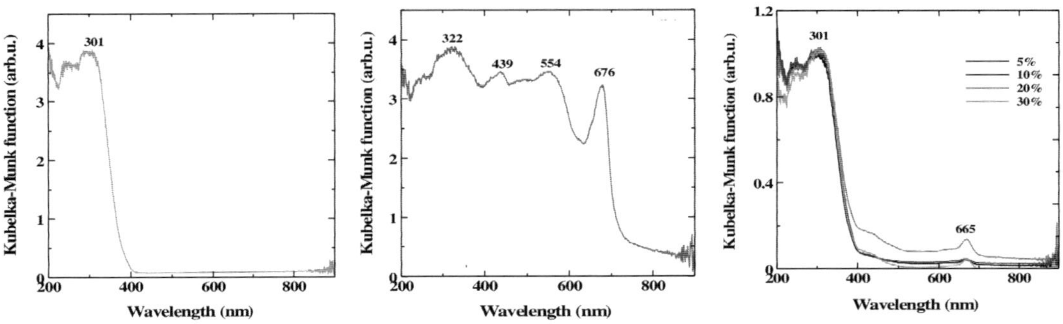

(a) (b) (c)

Figure 3. DR UV-Vis spectra of (a) unmodified TiO_2, (b) RA and (c) the composite materials.

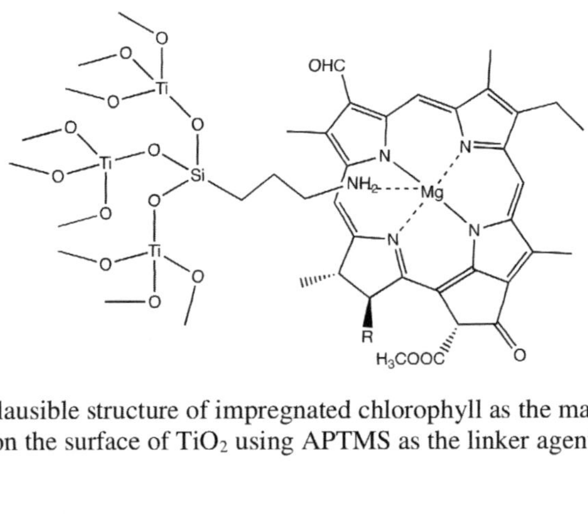

Figure 4. Plausible structure of impregnated chlorophyll as the main sensitizer on the surface of TiO_2 using APTMS as the linker agent.

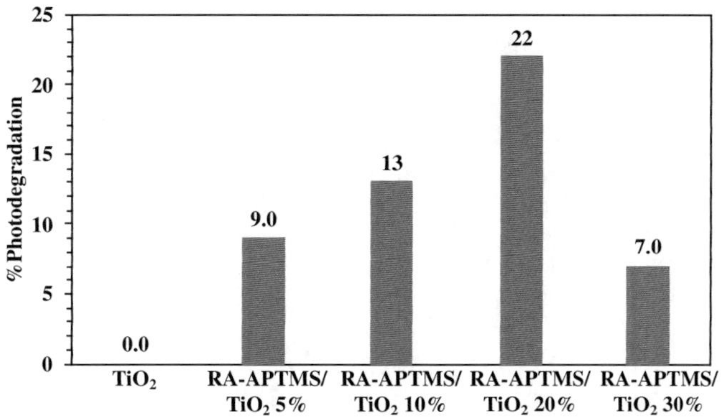

Figure 5. Photocatalytic activities of TiO_2 and the composite materials for phenol degradation process under visible irradiation.

The photocatalytic activities of TiO_2 and the composite materials for phenol degradation process under visible light irradiation are shown in figure 5. It was confirmed that the unmodified TiO_2 hardly degraded phenol under visible light (0 %), and was in good agreement with its DR UV-Vis spectrum which showed no absorption peak at above 420 nm. In contrast, after dye-sensitizing with RA extract, the RA-APTMS/TiO_2 x (x = 5, 10, 20 and 30 %) materials gave 9, 13, 22 and 7 % photodegradation of phenol, which were quite remarkable performances. The photodegradation percentage of phenol was successfully improved by increasing the amount of the impregnated chlorophylls on the RA-APTMS/TiO_2 materials up to 20 %, confirming that the chlorophyll acted as the sensitizer. However, the RA-APTMS/TiO_2 30 % material gave a lower photodegradation percentage (7 %), suggesting that the optimum amount of RA would be *ca*. 20 %. The larger amount of RA might result in the formation of aggregated chlorophyll, which gave less efficiency in electron transfer and thus, inhibited the sensitizing process. However, it was clear that all the composites gave better photocatalytic activity than the unmodified TiO_2, demonstrating that RA could remarkably enhance the photocatalytic activity of TiO_2 under the visible region.

4. Conclusion

In the present work, dye-sensitizing of TiO_2 materials with chlorophyll as the major pigment on the RA extract were shown to successfully enhance the photodegradation percentage of phenol under visible light irradiation. The RA-APTMS/TiO_2 x (x = 5, 10, 20 and 30 %) materials were simply obtained as a greenish powder through an impregnation process employing APTMS as the linker agent in ethanol media. The successful impregnation of chlorophyll onto the TiO_2 surface was reflected by the presence of the FTIR absorption peaks of chlorophyll functional group, *i.e.* C=O, C=N, and C–N, and a new absorption peak at 665 nm in their DR UV-Vis spectra. From the phenol photodegradation experiment, it was revealed that the unmodified TiO_2 gave no photodegradation activity (0 %) while the RA-APTMS/TiO_2 x (x = 5, 10, 20 and 30 %) materials exhibited 9, 13, 22 and 7 % photodegradation activity. These findings offer open research to utilize the tropical plants as the dye sources to sensitize the TiO_2 materials for environmental remediation applications.

Acknowledgments

The authors thank for the financial support from Directorate General of Strengthening Research and Development, Ministry of Research, Technology and Higher Education of the Republic of Indonesia via the Higher Education Excellent Applied Research scheme (PTUPT 2019, No. 041/SP2H/LT/MULTI/L7/2019 and No. 005/MACHUNG/LPPM/SP2H-LIT-MULTI/III/2019).

References

[1] Goutam S P et al. 2018 *Chem. Eng. J.* **336** 386-96
[2] Nishimura S et al. 2003 *J. Am. Chem. Soc.* **125** 6306-10
[3] Nakata K and Fujishima A 2012 *J. Photoch. Photobiol. C* **13** 169-89
[4] Park Y, Lee S H, Kang S O and Choi W 2010 *Chem. Commun.* **46** 2477-9
[5] Khan M M, Adil S F and Al-Mayouf A 2015 *J. Saudi Chem. Soc.* **19** 462-4
[6] Hashimoto K, Irie H and Fujishima A 2005 *Jpn. J. Appl. Phys.* **44** 8269-85
[7] Rehman S, Ullah R, Butt A M and Gohar N D 2009 *J. Hazard. Mater.* **170** 560-9
[8] Kurniawan Y S, Adhiwibawa M A S, Setiyono E, Fahmi M R G and Lintang H O 2019 *Indones. J. Nat. Pigm.* **1** 49-52
[9] Lukitasari D M, Indrawati R, Chandra R D, Heriyanto and Limantara L 2017 *Jurnal Teknologi dan Industri Pangan* **28** 1-9
[10] Kurniawan Y S, Anggraeni K, Indrawati R and Yuliati L 2019 *AIP Conf. Proc.* **2175** 020049
[11] Siah W R, Lintang H O, Shamsuddin M and Yuliati L 2016 *IOP Conf. Ser.: Mater. Sci. Eng.* **107** 012005
[12] Zhao J, Milanova M, Warmoeskerken M and Dutschk V 2012 *Colloid Surf. A Physicochem Eng. Asp.* **413** 273-9
[13] Babich H and Davis D L 1981 *Regul. Toxicol. Pharmacol.* **1** 90-109
[14] Kurnik K, Treder K, Klaput M S, Tretyn A and Tyburski J 2015 *Water Air Soil Pollut.* **226** 1-19

Effect of film fabricating conditions and its implications on mechanical properties of high-density polyethylene film

B Herlambang and M R O B Anando

Research and Development Department, PT Chandra Asri Petrochemical, Tbk., Cilegon, Banten 42447, Indonesia

Corresponding author's email: grouprnd@capcx.com

Abstract. The use of plastics, especially High-Density Polyethylene (HDPE) film has increased rapidly in recent years due to its low cost, lightweight and excellent mechanical properties which is depended on the extrusion condition in the owner's film fabrication machine. The blown film process is the common process used to make a thermoplastic film in large volumes. However, many research performed long ago in the mechanical properties of polyethylene only concerned on the effect of conventional processing parameters such as temperature and pressure in blown film machines and neglected the significant effect of critical extrusion conditions. We did the study to know the effect of critical extrusion conditions such as output rate, blow-up ratio (BUR) and neck-point height (NPH) which was intended to get the optimum film fabrication conditions in a blown film extrusion machine. Each parameter was performed in five different conditions by increasing one parameter and controlling the other. The resin first melted, conveyed, and delivered before finally blown to form the film. The mechanical properties observed were impact resistance, tensile strength, elongation, tear strength, and young modulus. The result indicates that increasing the output rate, BUR and NPH imparted in the film's orientation balance, bubble shape, relaxation, and crystallization effects will lead to the increase of impact resistance, decrease of tear strength, and balance of film direction properties. The excessive and low amount of output rate, BUR and NPH are not desirable as it decreases the mechanical properties from HDPE film whether in mechanical direction or transverse direction. Hence, the optimum operating condition depends on which mechanical properties that needed to be achieved by film fabricators.

Keywords: HDPE film, blown film extrusion, operating condition, mechanical properties

1. Introduction

High-Density Polyethylene (HDPE) is made by low-pressure methods that polymerized in the presence of Ziegler-Natta (ZN) catalyst. The HDPE resins are used for the manufacture of heavy-duty films for fertilizers, planting, and bags/containers [1]. Other applications include pipe, monofilament, and injection. Processing HDPE is somewhat different than processing Low-Density Polyethylene and Linear Low-Density Polyethylene, because of its higher degree of crystallinity, more consistent structure, high strength, and stiffness among polyethylene grades. In addition, HDPE has reasonably good barrier properties owing to its high degree of crystallinity, because its crystalline regions block the travel of gas molecules much better than amorphous regions [2]. The HDPE is preferably to be converted by the film blowing process, instead of cast film due to better property balance, less scrap, and low equipment cost [3]. Converted HDPE films are commonly used in numerous packaging where its mechanical response is excellent, moreover, in today's demanding packaging applications, high strength of films are essentials. It is, therefore, necessary to determine the optimum processing conditions in order to reap the full benefits of HDPE by using blown film machine. However, a lot of research in the mechanical properties of polyethylene concerned on the effect of conventional processing parameters such as temperature and pressure on these properties [4]. Although long ago, Piggot showed that extrusion conditions could have a profound effect on the mechanical properties of polyolefin, studies have neglected the significant effect of critical extrusion conditions [5]. Several authors have attempted to correlate the mechanical properties of blown film with the kinematics and dynamics of the process. Campbell claimed that the plastic strain, as the strain put in the film after the onset of crystallization, correlated with mechanical properties, such correlation based on a limited set of experimental data are easily questionable as the quantitative determination of the plastic strain and strain stress is a very difficult task [6]. Simpson and Horison have investigated the film blowing processing conditions effects on crystalline and amorphous morphologies of HDPE and their relation with film properties. Some changes can lead to the change film's lamellae stacking and orientation both in Machine Direction/MD (where the film direction moves through the machine from the start to finish) and Transverse Direction/TD (where the direction perpendicular to the MD) [7]. The previous study shows that the change of critical extrusion parameters and its effects on film mechanical properties haven't much yet explored. Thus, the purpose of this study is to know the effects of some critical extrusion parameters (output rate, blow-up ratio, and neck-point height) to the properties of HDPE film.

2. Materials and method

2.1. Materials

All of the polymer materials used were HDPE synthesized by ZN Catalyst with the following fundamental properties as shown in table 1.

2.2. Blown film extrusion process

Blown film extrusion is one of the most common used for polymer processing methods. This technique usually produced by using a single screw extruder and temperature of the die is 200 °C and a thickness of 20 μm for each layer. The resin was first melted by subjecting it to heat and pressure inside the barrel of an extruder then conveyed and delivered in a vacuum system before finally forcing the melt through a narrow slit in a die to control its diameter and thickness [8]. The Yoshii Blown Film Machine (BFM) was used and the both materials in the same machine condition were set to prevent the effect of process condition to the final product.

2.2.1. Output rate. In blown film extrusion, screw speed controls the output more than any other factor. Generally, the higher the screw speed, the higher is the output. The output rate of a Yoshii BFM (Q) was calculated by following equation:

$$Q = \pi \times D \times Vf \times Tf \times \rho \qquad (1)$$

where ρ is film density, V_f is the die exit and final film velocities, T_f is the initial and final film thickness.

Table 1. Fundamental properties of HDPE.

Fundamental properties	Test method	Unit	HDPE-1	HDPE-2
Melt Index (190 °C and 21.6 kg load)	ASTM D 1238	g/10 min	15	10.5
Density	ASTM D 1505	g/cm^3	0.950	0.953

2.2.2. Blow-up ratio (BUR). The blow-up ratio (BUR) is described as the ratio of bubble diameter (D_B) per die diameter (D_D). It defines how the bubble has expanded seen in the transverse direction. The BUR is calculated by:

$$\text{BUR} = \frac{D_B}{D_D} = \frac{2 \text{ x Lay flat width}}{\pi \text{ x } D_D} \tag{2}$$

The lay-flat film width (LFW) is the collapsed bubble width before slitting given by LFW = 1.57 D_B.

2.2.3. Neck-point height (NPH). Neck-point height is normally expressed by the distance from the die surface to the point where the necked diameter of the bubble becomes the shortest. The NPH is usually used by a highly skilled person after visually determines the location of the frost line.

2.2.4. Processing variables. To conduct the experiment, several adjustments need to be determined and listed in table 2.

2.3. Testing methods

The testing process of the film will separately differ into two steps for some test items, a machine and transverse direction. The machine direction is where the film set in parallel against the direction of polymer orientation while the transverse direction is where the film set in perpendicular against the machine direction. All the tests were conducted in a room with a temperature of 23 °C and 50 % RH.

2.3.1. Film impact. This method determines the resistance of film to impact-puncture penetration. The energy required for fracture, to burst, and penetrate a with the material or a measurement of the resistance against mechanical shock by pendulum according to ASTM D3420. If film impact strength value is high, the strength is better (stronger).

2.3.2. Dart impact. This method determines the required energy leads to the plastics failing under certain situation of a free-falling dart. The amount energy is expressed by the mass of the falling weight from a specified height which would result in a 50 % failure of specimens tested according to ASTM D1709.

2.3.3. Tensile strength. This method is the load-carrying capability of the film and measured by pulling the end of a rectangular strip at a specified rate in the opposite directions until it breaks and reported as the force at a given point divided by the original cross-sectional area (width multiplied by thickness) of the specimens which mostly reported at the yield point (where the force in the film begins to drop after an initial rise) and the breakpoint. The specimens were tested on a Universal Testing Machine according to ASTM D882.

2.3.4. Elongation. Another property covered by the same ASTM method as that for tensile strength is elongation, described as the ability of the film to stretch prior to breaking or yielding. A material that stretches significantly prior to the breakpoint is called a ductile, while the one that breaks after only

Table 2. Processing variables of blown film extrusion machine.

Parameter	Controlled variable	Manipulated variable
Output rate	BUR 3.8:1; NPH 35 cm	20; 25; 29; 33; 35 kg/h
BUR	Output 29 kg/h; NPH 35 cm	3; 3.5; 3.8; 4; 4.5
NPH	Output 29 kg/h; BUR 3.8:1	20; 25; 30; 35; 40 cm

a small degree of stretching is called brittle. Elongation is reported as a percentage and calculated by dividing the extension achieved in the specimens at the break by the original length and multiplying by 100 according to ASTM D882.

2.3.5. Tear strength. This method defines the force needed to propagate tearing in a film or non-rigid sheet using pendulum swing with initial rips of an Elmendorf-tearing tester according to ASTM D1922.

2.3.6. Young modulus. Another property covered by the same ASTM method as that for tensile strength and elongation is tensile modulus. Young modulus is an index of the stiffness of thin plastic sheeting. The reproducibility of test results is good when the precise control is maintained over test conditions according to ASTM D882.

3. Results and discussion

3.1. Effect of output rate

Increasing the output rate will be affected by increasing the screw speed and maintain the take-up ratio in BFM [2]. In this experiment, we implemented 5 output variations i.e. 20; 25; 29; 33 and 35 kg/h. The fast take-up and screw speed, lead to a short residence time of the molten polymers and gives an effect to higher strength in the film [9]. These variations exhibit a higher film impact and dart impact as the output increased which can be seen in figure 1. Fast take-up speed causes a large neck down in the bubble and gives an effect in short residence time, any processing condition that causes faster cooling will improve the impact strength. Fast cooling yields small spherulites and low crystallinity [10]. It can be seen that HDPE-1 yielded a higher value of film impact and dart impact as the lower density lead to a higher value in impact properties [11].

In figure 2a, it can be seen that as the take-up and screw speed run faster to obtain a higher output rate, lead to a melt drawing that occurs simultaneously in both machine and transverse direction with the former predominating [9]. This will balance the orientation results and increase the tensile strength at yield and at break in both MD and TD properties. But at the freeze line where the residual orientation and stresses are retained as the non-crystalline region, instead of the crystalline, so the elongation in both directions will gradually decrease as the output rate increases during the fabrication process as can be seen in figure 2b [10]. It also can be seen that HDPE-2 yielded a higher value in tensile strength both at yield and at break as the increase of density will result in a higher value, contrary to the elongation which will be lower because the elongation at break tends to respond in the opposite manner to the tensile strength [11].

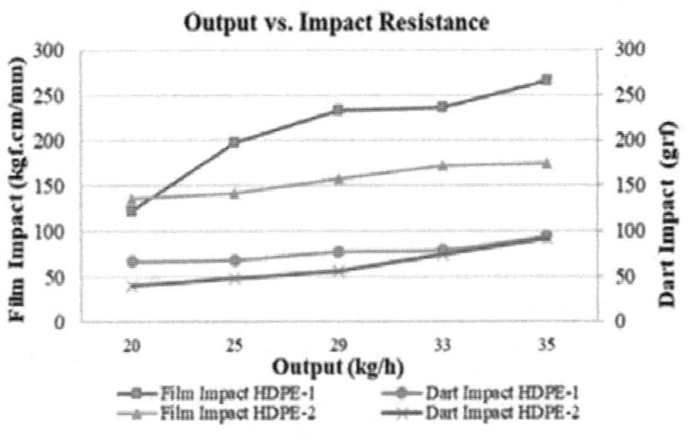

Figure 1. Effect of output rate on impact resistance.

In figure 3, the value of tear strength and young modulus decrease as the output rate increases because the modulus is expected to be governed by the non-crystalline portion [12]. The comparison between HDPE-1 and HDPE-2 exhibits a higher value in HDPE-1, but it is in contrary to the fact that the increase in MI and decrease in density will give a lower value in young modulus property [11].

3.2. Effect of BUR

Increasing the blow-up ratio will be affected by adjusting the bubble diameter and maintaining the bubble stability in BFM. In this experiment, 5 BUR variations i.e. 3; 3.5; 3.8; 4 and 4.5 were implemented. This adjustment leads into the output film width (BUR 3:1 gave 350 mm film width; 410 mm; 450; 470 and 530 mm, respectively [2]. These variations exhibit a higher film impact as the BUR increases from 3:1 to 4.5:1. As seen in figure 4, when the blow-up ratio becomes larger, the transverse direction molecular orientation also becomes higher with the corresponding decrease in the drawdown ratio. This will cause a reduction in cooling time [3]. Any processing condition that causes faster cooling will lead to improve the impact properties (film and dart impact) [10]. The comparison between HDPE-1 and HDPE-2 exhibits a higher value in HDPE-1 as the density increases leading to a lower value in film impact property [11].

In figure 5, tear properties and the young modulus are significantly changing following the change of molecular orientation, when the BUR increase, an insignificant change or constant in MD and lower tear strength in TD [3]. The comparison in tear strength between HDPE-1 and HDPE-2 exhibits a higher value in HDPE-1 as the increase in density will lead to a lower tear strength [11]. As of the young modulus, the comparison gives a comparable value in both grades, but it is in contrary to the fact that as the MI increase and a density decrease, it leads to a lower value in young modulus of HDPE-1 [10].

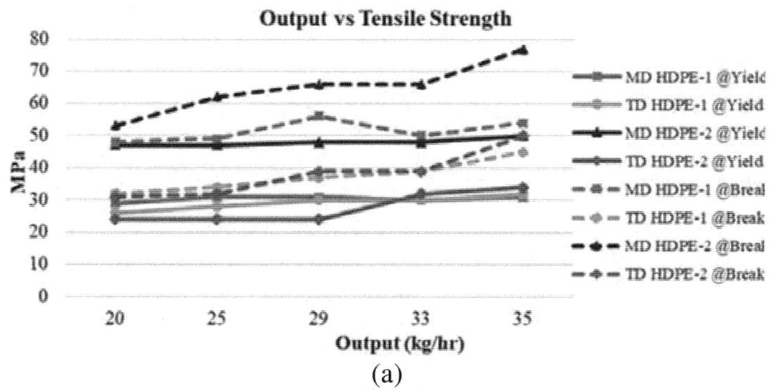

(a)

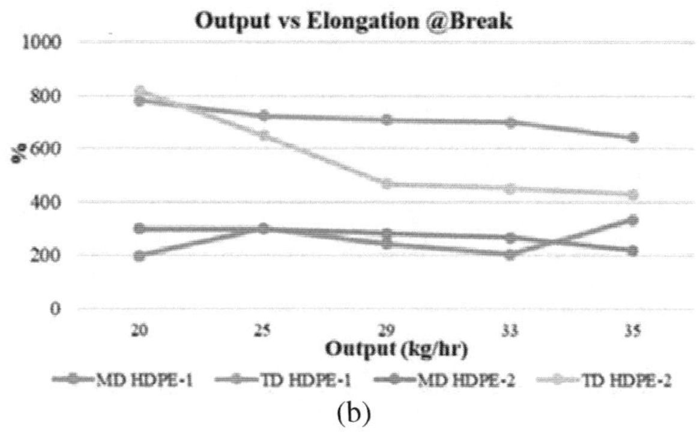

(b)

Figure 2. Effect of output rate on (a) Tensile strength properties, and (b) elongation.

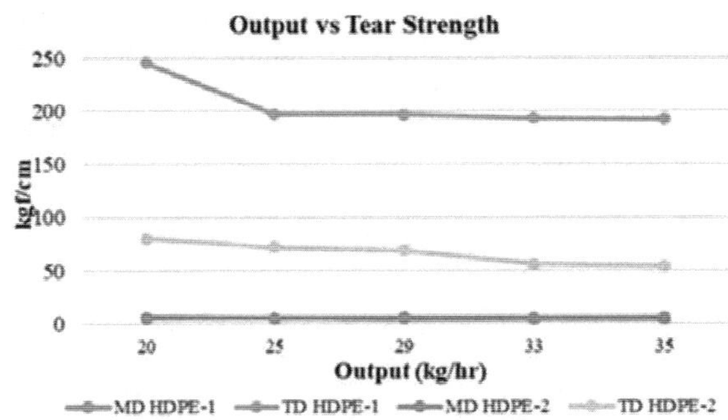

(a)

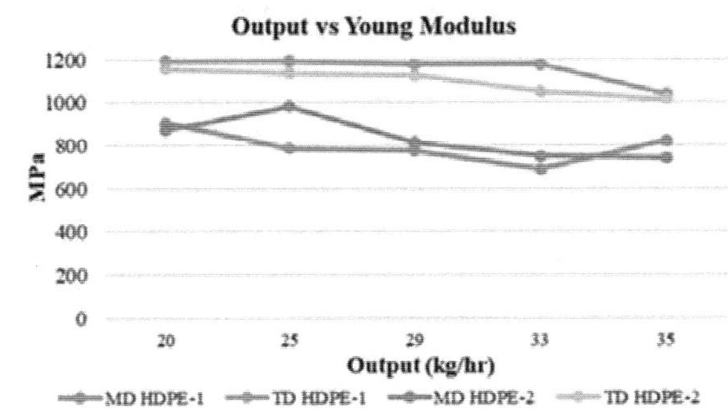

(b)

Figure 3. Effect of output rate on (a) Tear strength, and (b) Young modulus.

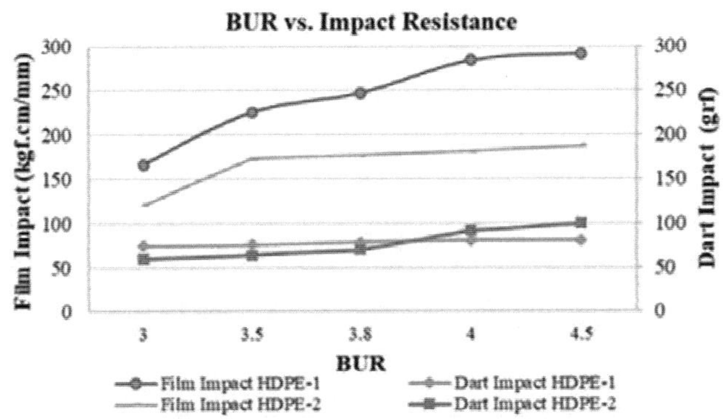

Figure 4. Effect of BUR on impact resistance.

It is acknowledged in figure 6a that the BUR affects several properties like tensile strength at yield and at break. The properties are all corresponded to the molecular orientation and the ability to orient the molecules in more than one direction [11]. As the increase in TD properties is observed,

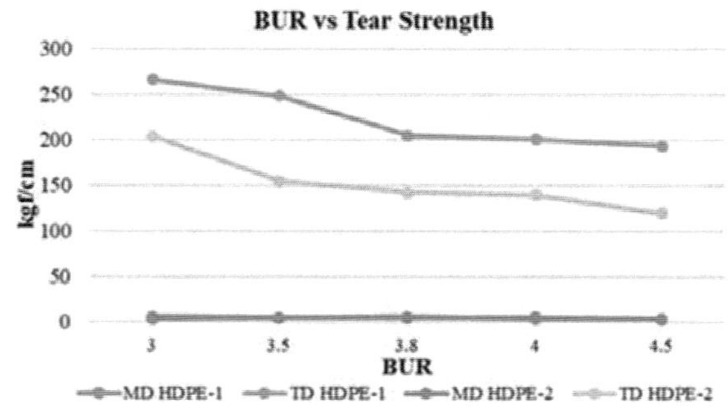

(a)

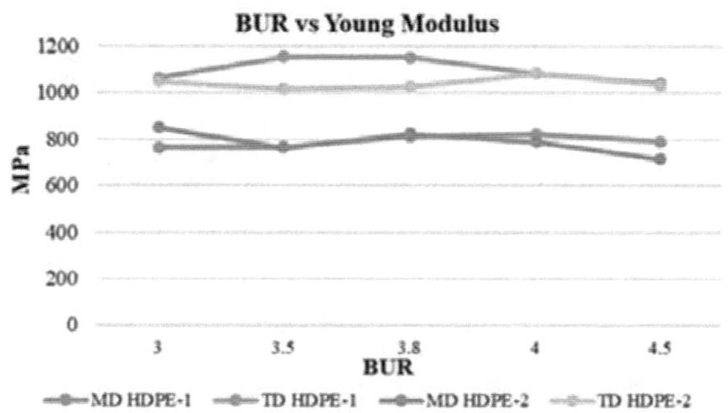

(b)

Figure 5. Effect of BUR on (a) tear strength and (b) Young modulus.

the MD properties will gradually decrease as some molecules that were initially oriented in MD are now oriented in TD [3]. It also can be seen that HDPE-2 yielded a higher value in tensile strength both at yield and at break as the increase of density will result in a higher value, in contrary to the elongation which be lower because the elongation at break tends to respond in the opposite manner to the tensile strength based on figure 6b [10].

A low blow-up ratio will impart molecular orientation in the machine direction and can lead to bubble instability, but the excessive blow-up ratio is not desirable from the viewpoint of maintaining the bubble stability [9]. Also, a larger BUR over a given die gave a thinner gauge film and greater film width at constant take-off [13]. When an ideal BUR cannot be established due to the reasons of available lay-flat film width, adjustments should be made either with the take-up speed and or the frost line height [10].

3.3. Effect of neck-point height (NPH)

Increasing the neck point height will be affected by adjusting the blower valve and maintaining the bubble stability in BFM [2]. In this experiment, 5 NPH variations i.e. 20; 25; 30; 35; and 40 cm were implemented. When the necking point height is raised, a melted film in aired zone is blown suddenly to frost line zone. As a result, it causes a sharp neck down [14]. NPH needs to be adjusted purposefully to change some mechanical properties of film, such as tear strength, tensile strength, and impact resistance [13]. It is considered that the higher the NPH, the higher the film impact and dart impact as seen in figure 7. The comparison between HDPE-1 and HDPE-2 exhibits a higher value in HDPE-1 as the increase in density will lead to a lower value in film impact property [10].

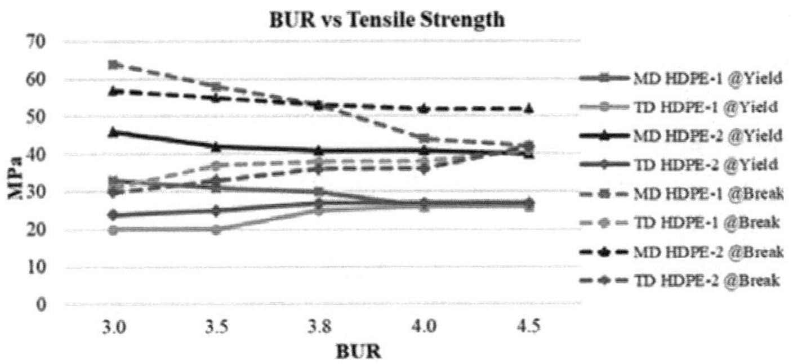

(a)

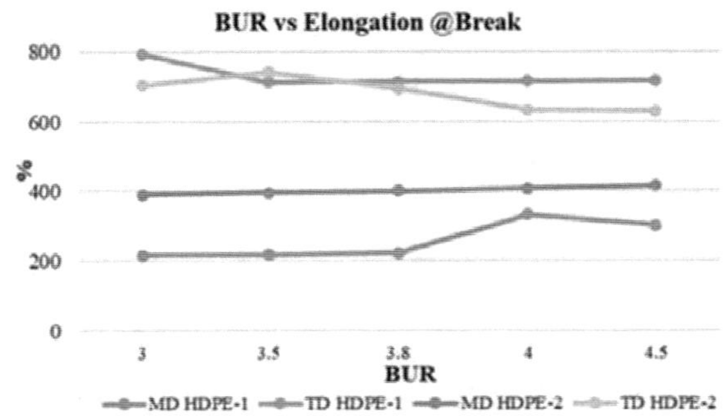

(b)

Figure 6. Effect of BUR on (a) tensile strength and (b) elongation at break.

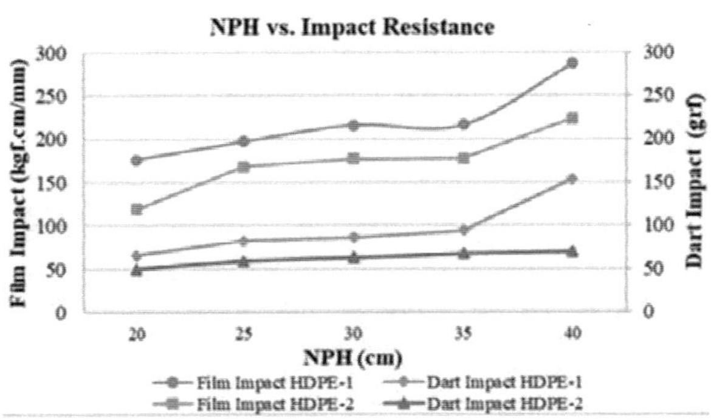

Figure 7. Impact resistance as the effect of NPH.

The effect of NPH in tensile properties as seen in figure 8a gives a better tensile strength at yield and at break in TD and lower tensile strength at yield and at break in MD as the NPH increased, whilst, elongation at break also tends to respond in the opposite manner compared to tensile strength based on figure 8b [11]. A high neck-point height enables the polymer to achieve a balance of film properties by imparting more bi-axial molecular orientation in the film in TD to match the MD orientation [14].

HDPE-2 exhibits higher value in the tensile strength at yield and tensile strength at break as the density increases. While HDPE-1 exhibits a higher value in elongation at break due to its lower density [10].

In figure 9, the tear strength is found to increase in the MD and decrease in TD with the increasing of NPH, the same thing happened to tensile properties. HDPE-1 exhibits a higher value in tear strength compared to HDPE-2 as the density increases [10].

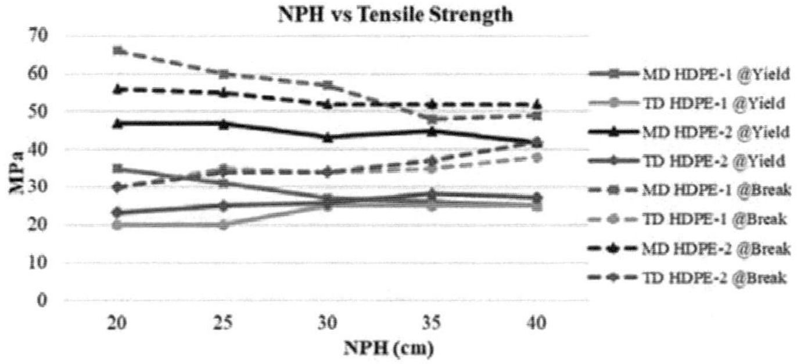

(a)

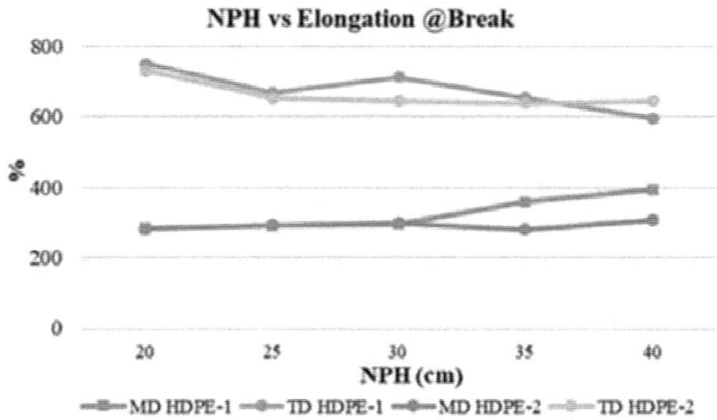

(b)

Figure 8. (a) Tensile strength and (b) elongation at break as the effect of NPH.

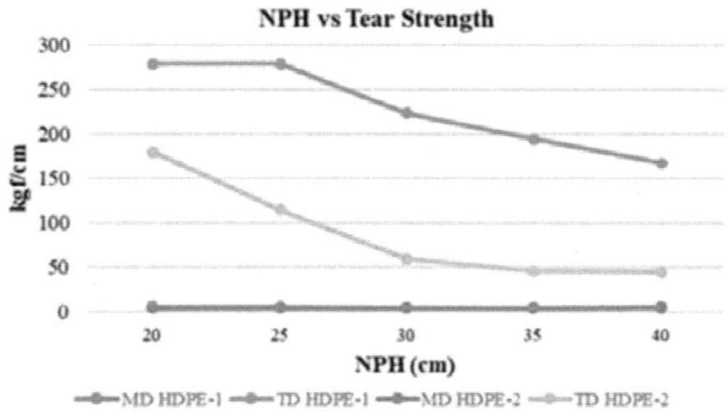

(a)

Figure 9. Effect of NPH on (a) tear strength and (b) Young modulus.

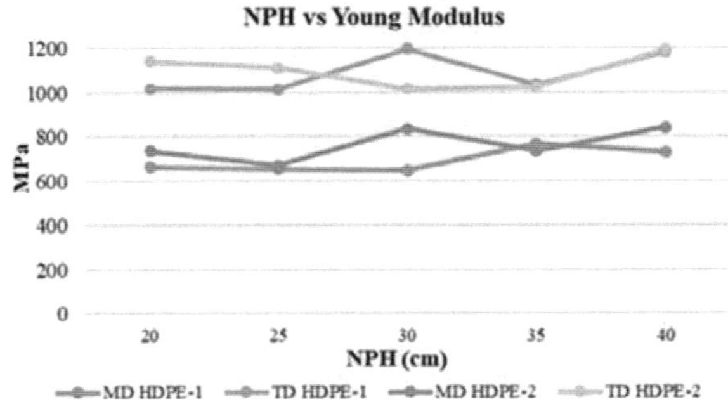

(b)

Figure 9 (continued). Effect of NPH on (a) tear strength and (b) Young modulus.

These 5 variations of NPH exhibit a higher young modulus in both directions as the NPH increased [12]. High NPH allows more time for a film to cool before it reaches the region where the bubble suddenly blown out to its maximum diameter. When it reaches the top stalk will greatly increase the amount of crystallinity and its elasticity [14]. However, since higher NPH tends to reduce other TD properties, sufficient care must be exercised not to raise the NPH too high [10]. In addition, too high NPH may cause the film to stick when it rolled up leading to problems in film gauge control and create a non-uniformity [13].

4. Conclusion

HDPE film was successfully fabricated by using blown film machine with several critical extrusion parameters (output rate, BUR, and NPH) to obtain the optimum film fabricating conditions. These critical extrusion process have affected the film's orientation balance, the bubble shape, relaxation, and crystallization effects. Therefore, yielded a huge difference in mechanical properties. The differences in operating conditions showed that the higher output makes the process difficult to maintain the orientation balance and decreases the tear and tensile in TD direction, higher BUR makes it hard to maintain bubble stability and decreases several aspects in the MD direction, and higher NPH affect the bubble shape and gradually decrease the strength of the film in the TD direction. The optimum operating condition for HDPE film deends on which mechanical properties required to be achieved by film fabricators.

Acknowledgments

The author would like to thank the RND Department of PT. Chandra Asri Petrochemical, Tbk. for their great and continuous support that allows this study possible.

References

[1] Ogah A O and Afiukwa J N 2012 *Int. J. Eng. Manage. Sci.* **3** 85-90
[2] Cantor K 2019 *Blown Film Extrusion* 3rd edition (Cincinnati: Hanser Publications)
[3] Wagner J R Mount E M and Giles H F 2014 *Extrusion: The Definitive Processing Guide and Handbook* (Oxford: William Andrews an Imprint of Elsevier) pp 539-49
[4] Onyeagoro G N 2012 *Acad. Res. Int.* **2** 99-103
[5] Piggot T S 2000 *J. App. Polym. Sci.* **27** 3267-78
[6] Babel A K and Campbell G A 1993 *J. Plast. Film Sheet.* **9** 246-58

[7] Simpson D M and Harrison I R 1994 *J. Plast. Film Sheet.* **11** 216-34
[8] Levy S and Carley J F 2005 *Plastics Extrusion Technology Handbook* 2nd edition (New York: Industrial Press)
[9] Peacock A J 2000 *Handbook of Polyethylene* (Texas: Marcel Dekker, Inc.)
[10] Qenos Technical Guide 2015 *Film Extrusion and Conversion* (Australia: Qenos Pty. Ltd.)
[11] Vasile C and Pascu M 2005 *Practical Guide to Polyethylene* (Shropshire: Rapra Technology Ltd.)
[12] Ghaneh-Fard A 1999 *J. Plast. Film Sheet.* **15** 194-217
[13] LyondellBasell 2017 *A Guide to Polyolefin Film Extrusion* (Cincinnati: LyondellBasell Technology Center)
[14] Formosa Plastics Technical Guide 2014 *Polyethylene Film Processing Guide Quality, Value and Performance* (New Jersey: Formosa Plastics Corporation)

Effect of zirconium substitution on structural and optical properties of lanthanum orthoferrite

I Purnamasari and D Triyono

Department of Physics, Faculty of Mathematics and Natural Sciences (FMIPA), Universitas Indonesia, Depok 16424, Indonesia

Corresponding author's email: djoko.triyono@sci.ui.ac.id

Abstract. Lanthanum Orthoferrite ($LaFeO_3$) is one of the perovskite oxide ABO_3 materials. In this study, $LaFe_{1-x}Zr_xO_3$ (x = 0.01, 0.03 and 0.05) were synthesized by sol-gel method. Synthesized $LaFeO_3$ with substituted Zr on Fe-site was characterized for crystal structures and optical properties. The X-ray diffraction (XRD) analysis confirms that the $LaFe_{1-x}Zr_xO_3$ material has an orthorhombic structure with *Pnma* space group. The local structure of this sample was studied using Raman spectroscopy. Raman analysis shows, there are Zr^{4+} substituted effect on the tilt and stretching phonon modes of $LaFeO_3$. Fourier Transform Infrared (FTIR) characterization test confirmed the effect of substitution on structural properties in terms of the upward shift of tilt, bending, and symmetric stretching of Fe-O-Fe bonds in FeO_6. Optical bandgap has been determined using Kubelka-Munk function with the Tauc connection by the Uv-Vis Spectrophotometer which shows that the energy bands decrease (2.14–2.10 eV) with increased Zr-content.

Keywords: $La(Fe,Zr)O_3$, sol-gel method, structural properties, optical bandgap.

1. Introduction

Lanthanum orthoferrite ($LaFeO_3$) is the most common and very known perovskite material that has been investigated for many years [1]. $LaFeO_3$ has interesting properties because they are the widely used for various applications such as solid oxide fuel cells, sensors, oxygen permeation membranes, environmental catalysts, etc [2]. $LaFeO_3$ is a G-type antiferromagnetic material possessing high Neel temperature (TN) 480 °C and shows a ferroelectric transition at 200 °C [2]. Perovskite properties are changed into ideal materials to get new materials with desired properties, which can be tuned to reach desired applications through substitution or doping [3]. That can happen because the doping and preparation methods can reduce particle size which causes increased optical properties [5]. Synthesis of $LaFeO_3$ samples doped with Titanium to determine the structural, optical and magnetic properties was carried out by other researchers [2]. The results show the average crystallite size (D_c) between 9 nm and 24.6 nm then, the results of Diffuse Reflectance Spectra (DRS) showed an increase in the bandgap (2.05 and 2.61 eV). Therefore, the Zr^{4+} substitution on the Fe-site of $LaFeO_3$ become an interesting topic to study and explore in this part.

2. Experimental

$LaFe_{1-x}Zr_xO_3$ (x = 0.01, 0.03 and 0.05) was prepared by sol-gel method. First, the raw materials were calculated stoichiometrically. Here, we used the lanthanum oxide (La_2O_3) (Merck), Zirconium (IV) chloride octahydrate ($Cl_2H_{12}O_9Zr$) (Merck), iron nitrate nonahydrate ($Fe(NO_3)_3.9H_2O$) (Merck), and citric acid monohydrate as raw materials. Then, the aquadestilation and all of the raw materials were mixed together. Next, using a magnetic stirrer-hotplate, the solvent was stirred until sol-phase changing to gel-phase. After that, the gel was dried using furnace (Vulcan-type 3-1750) at temperature 200 °C for 5 hours, until the sponge-phase was formed. In the next process, the sponge was ground using mortar manually until getting the powder form. Then, the dry-powder calcined at temperature 900 °C along 6 hours till we got the calcined-powder. The calcined-powder then we called as a series of $LaFe_{1-x}Zr_xO_3$ powder sample. The crystallographic structure was checked by X-ray diffraction using X'Pert PRO PANalytical diffractometer. The X-ray diffraction spectra were analyzed using Highscore plus software at room temperature to obtain parameter structure and crystal phase of the perovskite material. The Raman scattering spectra were recorded using Thermo Scientific DXR2 Raman Microscope in the range from 70–1600 cm^{-1} with a laser source of 532 nm and power 2.0 mW. The molecular and metal-oxide vibration were analyzed using FTIR (PerkinElmer type SpectrumTwo) in the range of 400–4000 cm^{-1}. The optical band gap energy was determined by using UV-VIS (Agilent Cary 100/300).

3. Results and discussion

3.1. Structural analysis

Figure 1 shows the XRD pattern of $LaFe_{1-x}Zr_xO_3$ powder sample (x = 0.01, 0.03, 0.05) which calcined at 900 °C. The Rietveld refinement results confirmed that all samples have orthorhombic crystal structure with Pnma space group consistent with structure of $LaFeO_3$ standard from COD (Crystallography Open Database) No. 96-152-6451 and represented by the main peaks in the fields (001), (011), (111), (002), (012), (022) and (013). All samples showed a single phase without additional impurity phase.

Table 1 shows the structural parameter data obtained from the Rietveld analysis. We can see that the lattice parameters slightly increase with increasing Zr content, this is caused by the ionic radii of Zr^{4+} (0.74 Å) ions occupying in Fe-site are larger than Fe^{3+} (0.65 Å) ions [3]. Increasing the value of the lattice parameter causes an increased value of the bond distance. Iron possible vacancies are Fe^{4+}, Fe^{3+} and Fe^{2+}. To ensure the charge neutrality and stable orthorhombic structure, Fe^{4+} ions should necessarily be formed in the lattice structure because of the 4+ valence of zirconium. So, it will remove Fe ions which have a lot of oxygen [4]. When doping with Zr, Fe will be replaced with Zr. The same phenomenon has reported by C. Sasikala et al, when Ti doping on $LaFeO_3$ material, then Ti will replace Fe in the material [5]. The tolerance factor decreases with increasing Zr content indicating Zr-ions occupying in Fe-site causes the crystal becomes more distorted to achieve stability.

3.2. Raman scattering analysis

Figure 2 shows the Raman spectrum for $LaFe_{1-x}Zr_xO_3$ (x = 0.01, 0.03 and 0.05) and some specific modes for Raman phonon modes are tabulated in table 2. Modes A_g, B_{1g}, B_{2g} and B_{3g} are Raman active and modes A_u, B_{1u}, B_{2g}, B_{2u}, B_{3g} and B_{3u} are infrared active [6]. Modes are indicated by the La vibrations which are shown by modes below 200, namely ω_1 and ω_2. Modes between 200 and 400 cm^{-1} are oxygen octahedral tilt modes at La which are ω_3, ω_4 and ω_5. Modes between 400 and 450 are oxygen octahedral bending vibrations which are marked with modes ω6. Modes above 500 cm^{-1} are oxygen stretching vibrations which are marked with modes ω_7, ω_8, ω_9 and ω_{10}.

It can be seen from the spectra that the Zr doping effects of the oxygen tilt, stretching and bending modes of vibration. Then, table 2 shows the changes in the stretching and bending modes of vibration due to the Zirconium doping effect. It appears that the higher the Zirconium doping appears and the higher the intensity of the shutter for modes ω_9 and ω_{10}. This shift in wavenumber and the variation in their intensities may be influenced by lattice strain, defects and crystallite size [7].

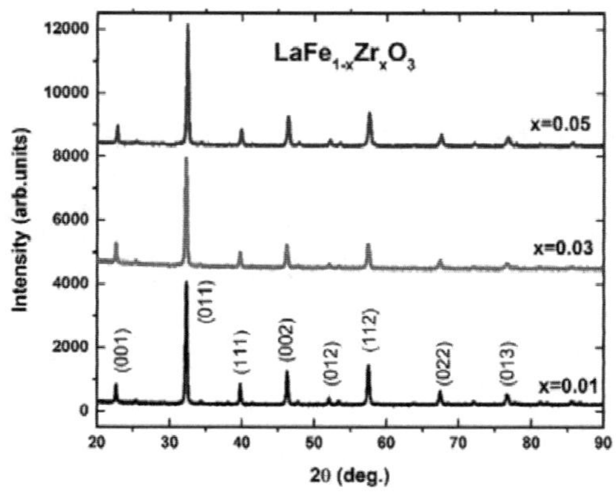

Figure 1. Rietveld refinement of XRD patterns of $LaFe_{1-x}Zr_xO_3$ using HighscorePlus software.

Table 1. The structural parameters of $LaFe_{1-x}Zr_xO_3$ (x = 0.01, 0.03 and 0.05) nanoparticle.

Parameter	$LaFe_{1-x}Zr_xO_3$ x = 0.01	x = 0.03	x = 0.05
Crystallography parameters			
Lattice parameters			
a (Å)	5.569	5.586	5.588
b (Å)	7.856	7.859	7.863
c (Å)	5.561	5.562	5.566
Volume (Å^3)	243.345	244.138	244.559
Crystallite size (nm)	453.2	429.1	394.4
Density g/cm^3	6.635	6.632	6.640
Geometrical parameters			
Bond length (Å)			
Fe/Zr−O1 (s)	1.92616	1.92812	1.92927
Fe/Zr−O1 (l)	2.08910	2.09311	2.09417
Fe/Zr−O2 (m) <Fe/Zr−O>	2.01562	2.01612	2.01738
Bond angle (°)			
Fe/Zr – O1 – Fe/Zr	157.0782	157.0865	157.0860
Fe/Zr – O2 – Fe/Zr	154.043	154.045	154.043
Tolerance factor <t>	0.848583	0.848004	0.847426

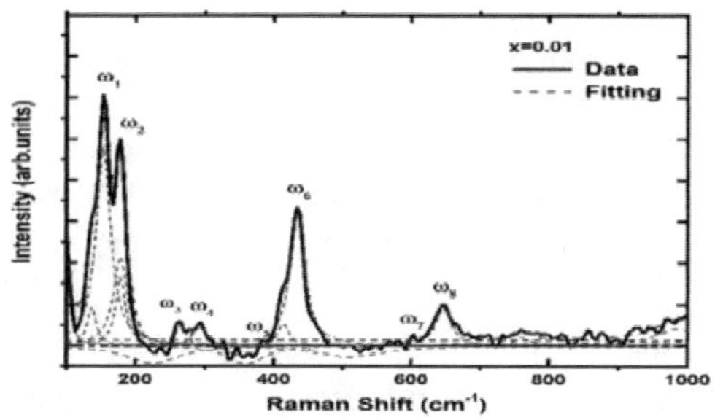

(a)

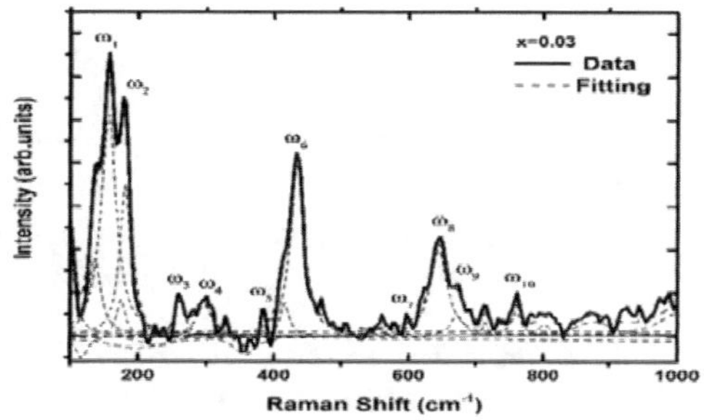

(b)

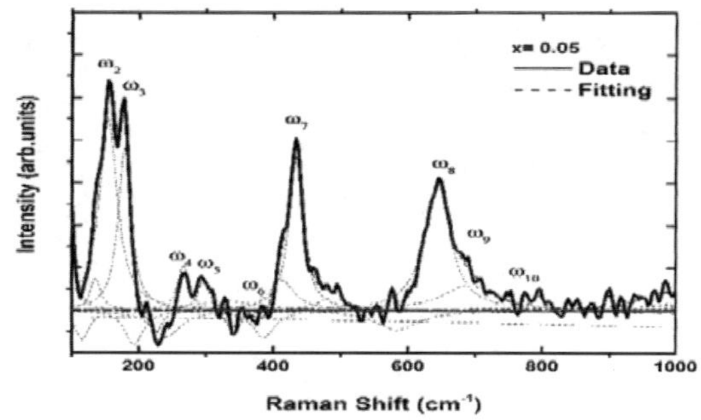

(c)

Figure 2. The Raman Spectra of $LaFe_{1-x}Zr_xO_3$ for (a) $x = 0.01$; (b) $x = 0.03$ and (c) $x = 0.05$; calcined at 900 °C for 6 h.

Table 2. The mode assignment of some specific Raman phonon modes for $LaFe_{1-x}Zr_xO_3$ ($x = 0.01$, 0.03 and 0.05) Raman shift in cm^{-1}.

	$x = 0.01$	$x = 0.03$	$x = 0.05$	Mode
ω_1	156.188	158.387	156.188	B_{1g}
ω_2	177.983	177.983	177.983	B_{1g}
ω_3	262.964	258.765	267.363	A_g
ω_4	295.756	297.956	293.557	A_g
ω_5	382.937	382.937	385.137	A_g
ω_6	433.126	433.124	435.325	B_{3g}
ω_7	603.288	596.689	577.094	B_{1g}
ω_8	646.878	649.078	646.878	B_{1g}
ω_9		675.272	686.269	B_{1g}
ω_{10}		762.653	762.653	

3.3. UV-Vis spectroscopy analysis

The UV-Visible reflectivity spectrum of $LaFe_{1-x}Zr_xO_3$ ($x = 0.01$, 0.03 and 0.05) is shown in figure 3. The UV-Vis spectrum can be used to find bandgaps from the material using the functions of Kubelka-Munk. Bandgap can be described by Tauc plots or extrapolating slop plots $(F)^2$ vs. hυ shown in figure 3. The bandgap value decreases with increasing doping constants, the value is from the range 2.14–2.10 eV. This might be caused by doping Zr^{4+} to Fe^{4+} which has smaller ion radii, bond length increases so that the atomic bond gets weaker so that the energy needed by the electron to detach from the bond, facilitates movement of excited and exciting carriers from the valence band to the conduction band easy so that it will reduce the bandgap energy. The small bandgap of $LaFeO_3$ is interesting for application in photocatalytic [4]. Figure 3a shows the calculation of the bandgap value of $LaFe_{1-x}Zr_xO_3$ (x = 0.01, 0.03 and 0.05) material and summarized in figure 3b.

3.4. Fourier Transform Infrared Spectroscopy (FTIR) analysis

The FTIR trasmittance of $LaFe_{1-x}Zr_xO_3$ (x = 0.01, 0.03 and 0.05) is shown in figure 4. Several types of modes observed are presented in table 3. Changes in metal-oxide stretching and bending vibrations may be related to changes in chemical mass in the material due to Zr substitution [8]. Fe/Zr–O stretching vibration and Fe/Zr–O–Fe/Zr bending showed that Zr was substituted on the Fe site. Also, the change in mode position shows a change in the Fe/Zr–O distance bond and Fe/Zr–O–Fe/Zr bond angle confirmed by XRD data in table 1.

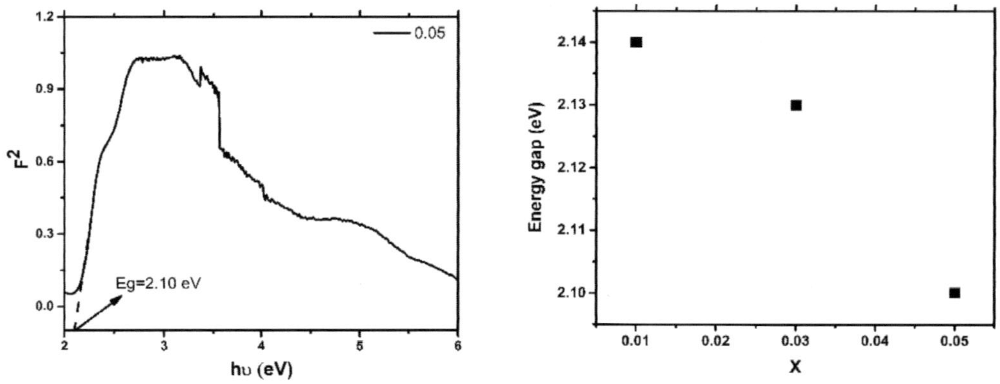

(a) (b)

Figure 3. (a) The plot of $(F)^2$ vs. hν from a series of $LaFe_{1-x}Zr_xO_3$ (x = 0.01, 0.03 and 0.05), (b) Value of bandgap.

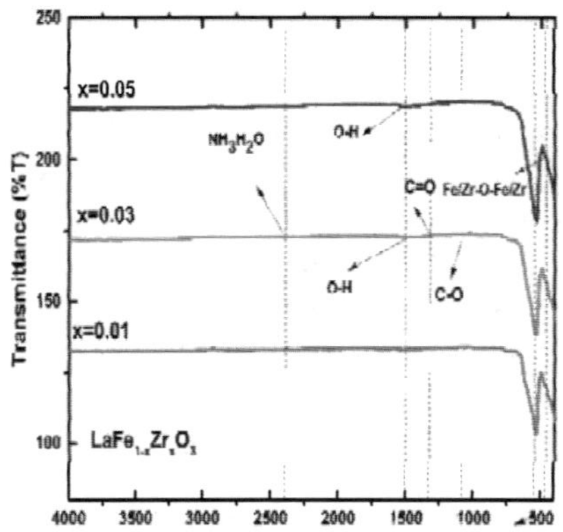

Figure 4. The FTIR patterns of $LaFe_{1-x}Zr_xO_3$ (x = 0.01, 0.03 and 0.05) nanoparticles.

Table 3. FTIR band assignment of some specific absorption modes for $LaFe_{1-x}Zr_xO_3$ (x = 0.01, 0.03 and 0.5).

$LaFe_{1-x}Zr_xO_3$	FT-IR band positions		
	O–Fe/Zr–O Bending	Fe/Zr–O–Fe/Zr Bending	Fe/Zr–O Stretching
x = 0.01	466	508	539
x = 0.03	463	506	540
x = 0.05	460	505	542

4. Conclusion

The series of $LaFe_{1-x}Zr_xO_3$ with (x = 0.01, 0.03 and 0.05) had been successfully synthesized using sol-gel method. The XRD analysis reveals the single phase of the orthorhombic crystal structure with *Pnma* space group. The addition of Zr content leads to the increasing of lattice constant and volume, distortion, and decreasing of crystallite size with increasing Zr content. The Local structure of this sample was studied using Raman spectroscopy. Raman analysis shows, there are Zr^{4+} substituted effect on the tilt and stretching phonon modes of $LaFeO_3$. Fourier Transform Infrared (FTIR) confirmed the effect of substitution on structural properties in terms of the upward shift of tilt, bending, and symmetric stretching of Fe–O–Fe bonds in FeO_6. Optical bandgap has been determined by the UV-Vis Spectrophotometer which shows that the energy bands decrease (2.14–2.10 eV) with increased Zr-content.

Acknowledgments

The author thanks for financial support from Hibah Penelitian Tesis Magister (PTM) 2020 by Ministry of Research and Technology (KEMRISTEK/BRIN) Republic of Indonesia with contract number NKB-489/UN2.RST/HKP.05.00/2020

References

[1] Miyasaka T 2019 *Nat. Energy* **4** 20181221
[2] Liu T and Xu Y 2011 *Mater. Chem. Phys.* **129** 1047–50
[3] Abdalla A, Hossain S, Petra P and Azad A 2018 *Process. Appl. Ceram.* **11** 275–81
[4] Sasikala C, Durairaj N, Baskaran I, Sathyaseelan B and Henini M 2017 *J. Alloys Compd.* **712** 870-77
[5] Sasikala C, Suresh G, Durairaj N, Baskaran I, Sathyaseelan B, Manikandan E, Srinivasan R and Moodley M K 2019 *J Supercond Nov Magn* **32** 1791-97
[6] Jagadeeshwaran C, Selvadurai P B, Pazhanivelu V and Murugaraj R 2014 *Int. J. Innov. Res. Sci. Eng.* **3** 9-14
[7] Mitra A, Mahapatra A S, Mallick A, Shaw A, Ghosh M and Chakrabarti P K 2017 *J. Alloys Compd.* **726** 1195-204
[8] Rai A and Thakur A K 2017 *J. Alloys Compd.* **695** 3579-88

Microstructure and mechanical properties of magnesium Si_3N_4 reinforced composite fabricated by stir casting method

A A D Manggabarani and A Z Syahrial

Department of Metallurgy and Materials Engineering, Faculty of Engineering, Universitas Indonesia, Depok 16424, Indonesia

Corresponding author's email: anne@metal.ui.ac.id

Abstract. Currently, the development of materials to improve the efficiency of fuel usage is increasingly carried out, one of which is a matrix composite. The stages of making composites in this study started from the preparation of magnesium matrix and Si_3N_4 reinforcement to casting with a stirring casting method. The addition of Si_3N_4 reinforce was carried out by varying the volume fraction by 2 vf-%, 4 vf-%, 6 vf-%, and 8 vf-% which was then characterized by metallographic testing, SEM-EDS, tensile testing, hardness, impact, wear, density and porosity, and XRD. The results of this study showed an increase in mechanical properties by adding Si_3N_4 as reinforcing particles. The magnesium composite produced different mechanical properties in each variation of the addition of the reinforce. The best mechanical properties resulted from the addition of a reinforce is with a volume fraction of 8 vf-% which produced a hardness value of 94 HRH and a wear rate of 0.0045 mm^3/m.

Keywords: Efficiency of fuels, matrix composite, Si_3N_4

1. Introduction

The number of vehicle fuel consumption in Indonesia continues to increase every year, therefore improvements need to be made so that the use of fuel becomes more efficient. That way, the costs incurred to buy fuel, and the use of natural resources to produce vehicle fuel can be reduced. This will have a significant impact, especially to the environment, because CO_2 gas emissions are comparable with fuel consumption [1]. Generally, parts of vehicles are made using aluminum because of their good strength. However, to reduce the mass of the vehicle, aluminum replacement materials are needed. Magnesium is considered as feasible to replace the role of aluminum as a base material for vehicle parts with consideration of its relatively light mass, abundance in nature, and easy fabrication process, so as to reduce costs [2]. However, pure magnesium does not have as much strength as aluminum, so engineering process is needed so that good mechanical strength can be achieved. Therefore, it is necessary to do the mixing of magnesium with the reinforcement by using the micro-sized silicon nitride (Si_3N_4) ceramic particles. Silicon nitride reinforcement has a good affinity with magnesium, excellent mechanical properties, good wear resistance, strength, and good toughness so that ceramic particles are considered to be attractive for fabrication of metal matrix composites [3].

IOP Conf. Series: Materials Science and Engineering **902** (2020) 012032 doi:10.1088/1757-899X/902/1/012032

2. Materials and method

2.1. Materials

Materials used in this composite fabrication were pure magnesium ingot, purchased from Xiamen ITG Group Corp. LTD, as the composite matrix, and commercial Si_3N_4 micro-particle in 2, 4, 6 and 8 vf-% variation as the reinforcing particle. Composite mass balance calculation was conducted to prepare the materials needed.

2.2. Fabrication of magnesium with Si_3N_4 micro-particle composite

For the preparation, the surfaces of furnace, mold, and crucible were coated by zircon and graphite compounded with thinner, followed by the cutting of magnesium ingot with bandsaw machine into small blocks, and afterward the blocks were washed and wiped with alcohol to clean inclusion on the ingot surface before the blocks were put into the furnace. Furthermore, the remaining gases in the furnace were emptied out using vacuum pump followed by given of cover gas flux with argon flowed continuously until the casting was finished. On the other hand, the micro particle reinforce Si_3N_4 were heated in the muffle furnace at 950 °C for 30 min as a pre-heating step to remove moisture. Pre-heated reinforce were later put into the furnace and then immediately stirred with metal rod stirrer for 2 min to homogenize the molten metal. The molten metal was then poured into the mold, followed by air cooling at room temperature [4].

2.3. Characterization

Several tests to characterize the chemical composition of elements for each sample of composite were carried out using Optical Emission Spectroscopy (OES) to determine the composition of magnesium ingot, and X-Ray Diffraction (XRD) to find out the compounds formed in the material according to the principle of the difference electrical voltage exists at the anode and cathode. Then, the microstructure was observed using Optical Microscope (OM) to define the composites microstructure, and Scanning Electron Microscope (SEM) testing, then followed by Energy Dispersive Spectroscopy (EDS) to predict the formed phase which can not be observed using OM, and also to determine the distribution of particles in composite. Mechanical properties of each sample of composite were determined using these following step, that is porosity and density testing using a digital machine based on ASTM 378-88 to determine actual density of the composite, tensile testing using GOTECH AL-7000 machine to determine the ultimate tensile strength (UTS), hardness testing through the Rockwell H method using ROCKY machine, impact testing with GOTECH machine to determine each composite toughness, and wear testing using pin-on-disc OGOSHI machine with 100 m sliding distance to determine wear rate on each sample of composite.

3. Results and discussion

3.1. OES testing

Based on the results shown in table 1, the percentage of magnesium in each variable of reinforcing is above 96 % which means that content of magnesium in composites are ideal because the percentage of magnesium are below 98 % [5]. Si_3N_4 is a compound and since the OES testing can only detect chemical elements therefore the Si_3N_4 compound cannot be identified. Ideally, composites only contain magnesium element without the presence of other elements.

However, the OES test results contained several additional elements in the material being tested. This was caused by the casting process of as-cast magnesium and magnesium composites, the condition of the crucible in the furnace and the crucible for pre-heating of Si_3N_4 reinforce was not completely clean, so that other elements were mixed into the matrix as inclusion.

Table 1. Chemical composition of magnesium composite.

Element (%)	Percentage (vf-%)				
	As-cast	2 Si_3N_4	4 Si_3N_4	6 Si_3N_4	8 Si_3N_4
Mg	>96.6	>96.0	>96.0	>96.0	>96.0
Si	0.861	1.23	1.81	>1.40	1.56
Fe	0.0041	0.0046	0.0026	0.0040	0.0045
Cu	0.00070	0.00023	0.00019	0.00025	0.00016
Mn	0.0073	0.0157	0.0096	0.0176	0.0097
Cr	0.0022	0.0020	0.0014	0.0018	0.0018
Ni	<0.00040	<0.00040	<0.00040	<0.00040	<0.00040
Zn	0.0152	0.0189	0.0194	0.0161	0.0153

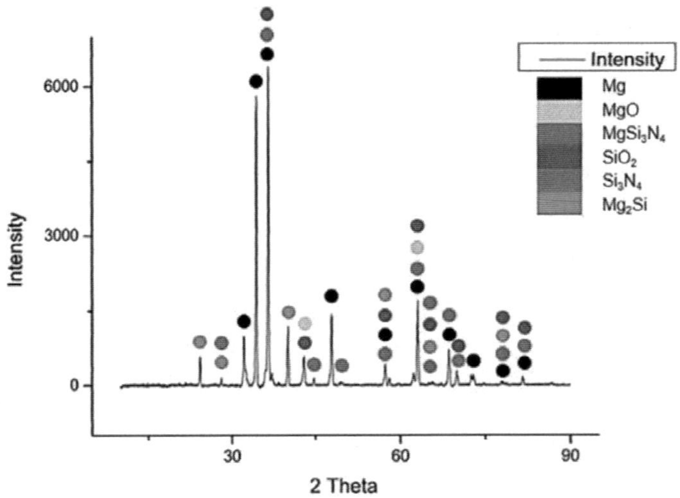

Figure 1. Results of XRD testing of Mg/8 vf-% Si_3N_4 composite.

3.2. XRD testing
Based on figure 1, it appears that at each peak point there is a magnesium phase. This indicates that magnesium as a matrix is the most dominant phase in magnesium composite samples. The Si_3N_4 phase is seen at almost every peak point. This shows the fairly even distribution of Si_3N_4 reinforcing particles in all parts of the pure magnesium matrix in the magnesium composite sample with Si_3N_4 reinforcement. In addition, there is a MgO phase formed, this is due to the characteristic of magnesium which is very reactive and easily reacts with oxygen to form MgO [6]. The presence of Si_3N_4 reinforcing particles in magnesium composites makes magnesium composites is better compared to the magnesium as-cast.

3.3. Microstructure
Based on the results of microstructure observations shown in figure 2, it shows that in the composition of the reinforcement of 2 %, visible brownish-white areas were identified as pure magnesium phase and also grayish-black grains scattered in the magnesium matrix area where the grains are known as the reinforcing phase, while the black grains that clump in one area are known as agglomeration [7].

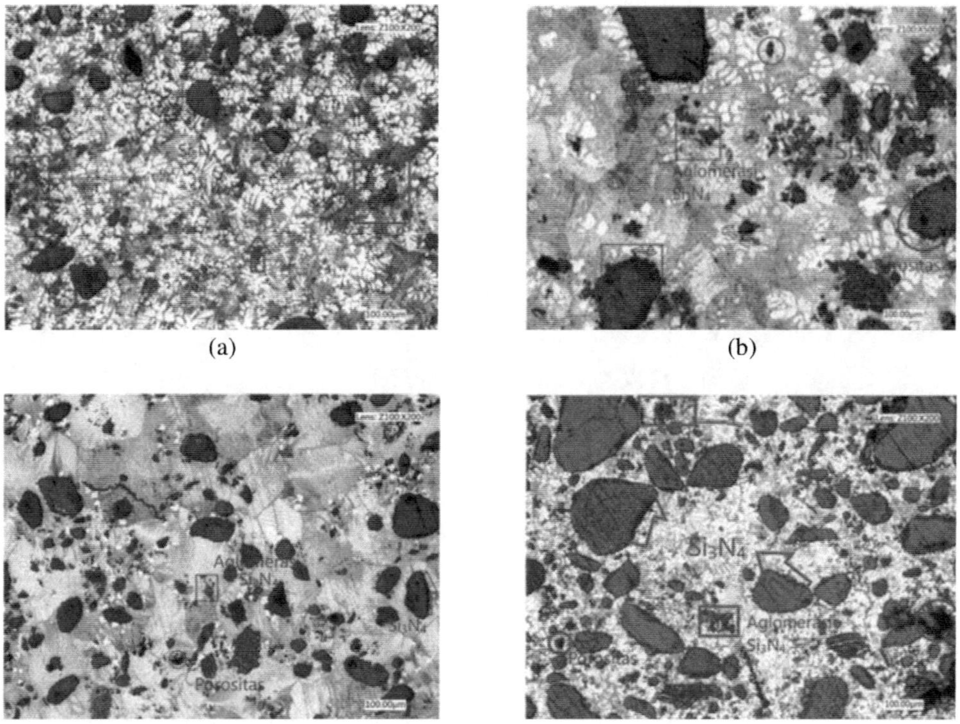

(a) (b)

(c) (d)

Figure 2. Microstructure of magnesium composite with (a) 2 vf-%, (b) 4 vf-%, (c) 6 vf-% and (d) 8 vf-% Si_3N_4 reinforced.

Then, the observation on the composition of the 4 % reinforcement shows visible blackish-gray areas that are lumpy and known as porosity. The observation of the composition of 6 % reinforcement obtained agglomeration of Si_3N_4 particles which is more than other microstructure compositions. Furthermore, the observation of the 8 % reinforcement composition shows that the reinforcing particles are dispersed equally in all area of the magnesium matrix, and only some small regions of porosity and agglomeration in the microstructure are observed.

3.4. SEM-EDS

SEM-EDS testing only performed at 8 % reinforcement composition. In this composition, the resulting composite is the most optimal based on tensile testing, so it is considered as the most representative for SEM testing. The results of this test are shown in images with a magnification of 250 x as shown in figure 3.

Then, from the EDS data shown in table 2, the distribution of N elements in the composite matrix tends to be uniform. However, at several points, namely at points 13, 14, 15 and 22, there is a relatively large content of Si element compared to other points, but the difference is not too significant. In addition, it can be seen that all selected points indicate a high O content, which indicates the presence of MgO compounds which are the rust products or oxidation products. The large amount of MgO compounds is formed due to the very reactive and easily oxidized of magnesium. In addition, the presence of C element is also seen at fairly low levels at all test points [6].

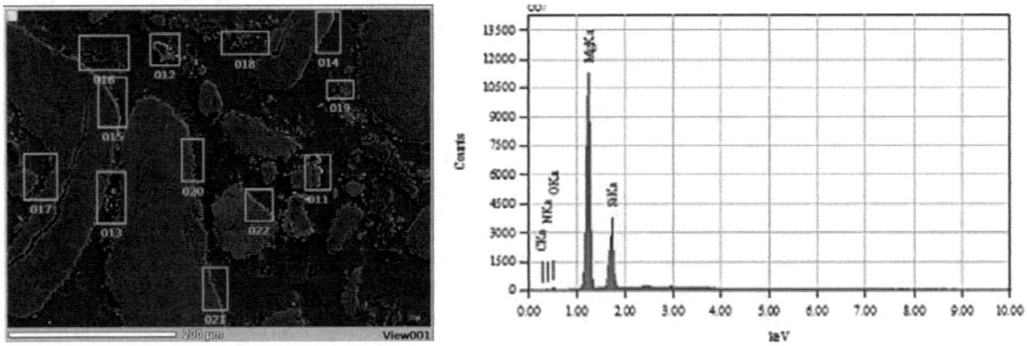

Figure 3. Results of SEM-EDS analysis of Mg/8 vf-% Si_3N_4 composite.

Table 2. Results of EDS testing of Mg/8 vf-% Si_3N_4 composite.

Point	Element (%)					Possible phase formed
	C	N	O	Mg	Si	
11	1.36	0.70	39.62	55.75	2.57	MgO, Mg_2Si
12	1.87	0.00	39.30	57.62	1.20	MgO, SiO_2
14	0.88	1.71	41.98	44.03	11.41	MgO, SiO_2, Mg-Si_3N_4
15	1.45	1.63	41.84	43.47	11.61	MgO, Mg_2Si, Mg-Si_3N_4
16	1.96	0.33	40.27	52.36	5.09	MgO
17	1.62	1.29	40.62	49.30	7.17	MgO, Mg-Si_3N_4
18	2.15	0.01	39.54	55.87	2.43	MgO, SiO_2
19	1.86	0.20	39.17	57.76	1.01	MgO, SiO_2
20	1.46	1.92	41.01	46.45	9.16	MgO, Mg-Si_3N_4
21	1.40	1.98	40.96	46.67	8.99	MgO, Mg-Si_3N_4
22	1.18	1.94	42.19	41.82	12.87	MgO, Mg-Si_3N_4

Table 3. Result of density and porosity testing of Mg/8 vf-% Si_3N_4 composite.

vf-% Mg (%)	vf-% Si_3N_4 (%)	Theoretical density (g/cm^3)	Actual density (g/cm^3)	Porosity (%)
100	0	1.74	1.701	2.251
98	2	1.7686	1.767	0.109
96	4	1.7972	1.772	1.424
94	6	1.8258	1.783	2.337
92	8	1.8544	1.844	0.562

3.5. Density and porosity

The formed porosity in each sample according to the Si_3N_4 reinforce composition variable can be calculated by referring to ASTM 378-88. The results of these calculations can be seen in table 3, where the 6 % reinforcement composition is the variable that has the highest porosity percentage, which is 2.333 %.

Based on the comparison results as in table 3, it can be seen that the actual density value of each Si_3N_4 reinforce composition variable is always smaller than the theoretical density value. This is caused by the formed porosity.

3.6. Tensile and elongation

Based on figure 4a, it is seen that the addition of Si_3N_4 reinforcement can increase the maximum tensile test value compared to magnesium metal without additional reinforcement. This is caused by increased material resistance to the occurrence of deformation which is the impact of an increase in dislocation density on the composite [8]. The strengthening mechanism occurs through the formation of the interface phase between the magnesium matrix and the Si_3N_4 reinforce [9].

On the other hand, the value of maximum tensile strength in the composition of reinforcement 4 % has decreased, even lower than the magnesium metal without reinforcement due to agglomeration that occurs. It can be seen in figure 4b, that in addition to get the maximum tensile test value, tensile testing can also provide information about the percent elongation that occurs in the composite after tensile testing. The percentage of elongation is inversely proportional to the material stiffness, as shown by the results of tensile testing conducted. It can be seen that the highest percentage of elongation is owned by the composition of Si_3N_4 reinforcement as much as 6 % with elongation percent of 17, and at the composition of reinforcement 8 % which has a maximum tensile value which is very high, with a magnitude of percent elongation of 7. Whereas, in the composition of the reinforcement of 2 %, the amount of percent elongation of 14 still tends to be high even though the value of maximum tensile strength is highest. This is caused by the lack of stress concentrations that are formed because the strain is spread evenly due to fine grains that well-occurred.

3.7. Hardness, wear rate and impact

Based on the hardness value obtained in figure 5a, it is seen that the highest hardness value of the composite is obtained from 8 % reinforce composition, which is 94 HRH. It is shown from the result in figure 5a that there is an increase in the hardness value of magnesium as-cast compared to magnesium composites with Si_3N_4 reinforcement.

The more reinforcement particles will increase the hardness value of the composite. This is because the higher the level of Si_3N_4 reinforcement, the grain size in the matrix will be smaller due to obstruction of the growth of the magnesium matrix by reinforcing particles [10], which is followed by an increase in compressive resistance. In addition, the Si_3N_4 particles added to increase the area of the reinforcement and then the particles block the dislocation movement so that the hardness value increases and the material becomes more difficult to deform. Wear rate values in magnesium composites increased at 2 % composition and 6 % composition. This is because in the composition, magnesium composites have decreased the value of toughness so that the wear rate will increase. The increased wear rate is also caused by the agglomeration of the composition. Because the interface strength between the matrix and the agglomerated reinforce is weakened [11], so more volume of material is eroded when tested.

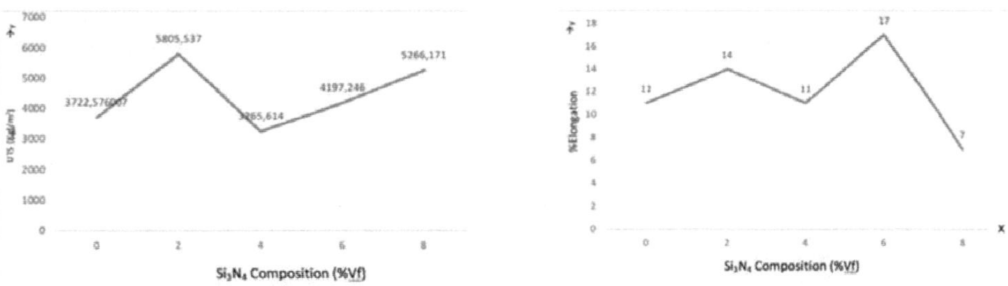

(a) (b)

Figure 4. Graphics of (a) tensile, and (b) elongation of Mg/Si_3N_4 composite.

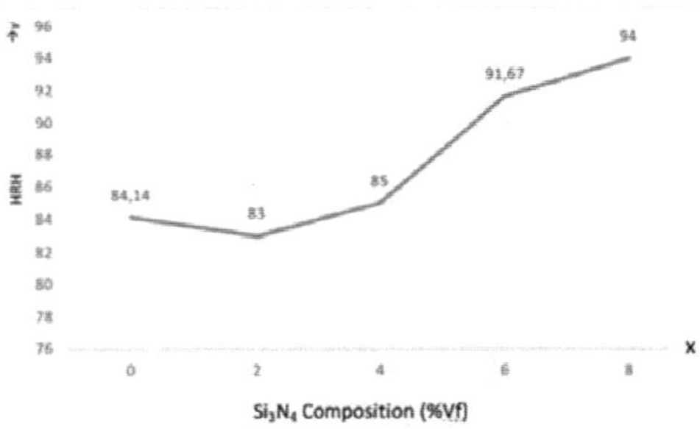

(a)

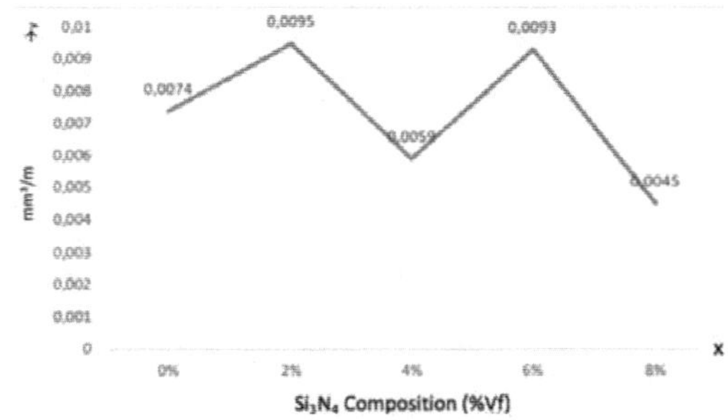

(b)

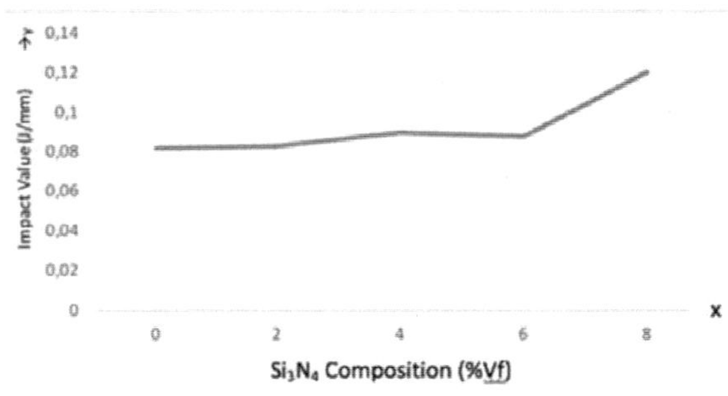

(c)

Figure 5. Graphics of (a) hardness, (b) wear rate, and (c) impact value of Mg/Si_3N_4 composite.

Reduction in wear rate occurs at 4 %, and the highest reduction at 8 % reinforce composition, which is 0.0045 mm^3/m. This is because reinforcing particles contribute to increase the material hardness. The harder the material causes less abrasion [12]. Increasing the volume fraction of the reinforce can also increase more surface protection of the matrix when it comes in contact with other materials due to more interface between matrix and the reinforce so the strength of the interface will prevent the composite from peeling off due to other material friction [13]. Increasing the value of toughness in the Si_3N_4 reinforced magnesium composite sample reinforced evenly dispersed reinforcement in the matrix, and served as a load-bearing element, namely the occurrence of the load transfer process first through a reinforced reinforce, so that a larger load is needed to do deformation [14]. However, the impact price value at the composition of 6 vf-% the value obtained tends to decrease this is because the value of the porosity percentage at the composition is higher when compared to magnesium composites with a reinforcement composition of 4 vf-%, where porosity can reduce the toughness of the resulting composite sample.

4. Conclusion

Pure magnesium composite reinforced Si_3N_4 can be fabricated using the stir casting method. The actual density of magnesium composites has a smaller value than the theoretical density. This is influenced by the percentage of porosity formed in the sample, where the greater porosity percentage will produce a smaller actual density value when compared to the actual density value. The addition of Si_3N_4 particles to the magnesium composite will increase the hardness and impact value of the composite, but it will reduce the wear rate of composite. The maximum hardness value is in the composition of 8 vf-% Si_3N_4, which is 94 HRH, because the more reinforce is added to the composite, the more interfaces between magnesium and the reinforcement are formed, so more strength is needed to deform the material. It will also increase the impact price of the composite. Then, the highest impact price was found in the composition of 8 vf-% Si_3N_4, which was 0.12 Joules/mm^2. The even distribution of the Si_3N_4 particles and the dispersion that occurs in the reinforcing particles plays a role in holding the load, so that the toughness of the composite will increase, whereas minimum wear rate is in the composition of 8 vf-% Si_3N_4, which is equal to 0.0045 mm^3/m, because the harder the material, it will reduce the volume of abrasive material.

Acknowledgments

Authors would like to thank Directorate of Research and Community Services (DRPM), Universitas Indonesia for financial support through International Publications Index (PIT 9), 2019 with contract number: NKB-0042/UN2.R3.1/HKP.05.00/2019.

References

[1] Kulekci M K 2015 *Int. J. Adv. Manuf. Technol.* **39** 851-65
[2] Kumar D S, Sasanka C T, Ravindra K and Suman K N S 2015 *Am. J. Mater. Sci. Technol.* **4** 12-30
[3] Paramsothy M, Chan J and Kwok R 2013 *Metall. Mater. Trans. A* **44** 1123-38
[4] Syahrial A Z and Hannan N 2019 *IOP Conf. Ser.: Mat. Sci. Eng.* **547** 012021
[5] Dey A and Pandey K M 2015 *Rev. Adv. Mater. Sci.* **42** 58-67
[6] Xi Y L, Chai D L, Zhang W X and Zhou J E 2005 *Mater. Lett.* **59** 1831-5
[7] Raturi A, Mer K K S and P K Pant 2017 *Mater. Today Proc.* **4** 2645-58
[8] Akbari M K, Baharvandi H R and Shirvanimoghaddam K 2015 *Mater. Des.* **66** 150-61
[9] Shou-ren W, Hao-ran G and Ying-zi W S 2006 *J. Mater. Sci.* **41** 5751-7
[10] Shen M J, Wang X J, Ying T, Wu K and Song W J 2016 *J. Alloys Compd.* **686** 831-40
[11] Sreekumar V M, Babu N H, Eskin D G and Fan Z 2015 *Mater. Sci. Eng. A* **628** 30-40
[12] Archard J F 1953 *J. Appl. Phys.* **24** 981-8
[13] Sharma S C, Anand B and Krishna M 2000 *Wear* **241** 33-40
[14] H Y Wang, Q C Jiang, Y Wang, B X Ma and F Zhao 2004 *Mater. Lett.* **58** 3509-13

IOP Conf. Series: Materials Science and Engineering **902** (2020) 012033 doi:10.1088/1757-899X/902/1/012033

Study of La(Fe, Zn)O_3 on the structural properties prepared by sol-gel method

F Fajriyani and D Triyono

Department of Physics, Faculty of Mathematics and Natural Sciences (FMIPA), Universitas Indonesia, Depok 16424, Indonesia

Corresponding author's email: djoko.triyono@sci.ui.ac.id

Abstract. Partial substitution of other metallic ions at La and/or Fe sites of $LaFeO_3$ had been applied to modify the structural properties and for enhancing the properties of its material. In this present work, we prepared $LaFeO_3$ by replacing a portion of Fe^{3+} with Zn^{2+} with $x = 0.03$; 0.05; and 0.07 to study the effect on the structural which prepared by sol-gel method. The samples were identified by X-Ray Diffraction (XRD), X-Ray Fluorescence (XRF), Fourier Transform Infrared (FTIR) spectra, Raman spectroscopy, and Ultraviolet-Visible Diffuse Reflectance spectroscopy (UV-Vis DRS). XRD pattern showed all of the samples formed in single crystal and are having an orthorhombic structure with space group Pbnm. The lattice parameters were found to be slightly increased with the increase of Zn concentration. The result of FTIR spectra and Raman spectra also show further information about the formation of $LaFeO_3$ with the distortion due to the replacement of Zn^{2+} into Fe^{3+}. The absorbance and reflectance of samples are measured using UV-Vis DRS and the energy band gap of $LaFe_{1-x}Zn_xO_3$ ($x = 0.03$; 0.05; and 0.07) are determined by Tauc plot using Kubelka-Munk function and was found the energy band gap slightly decreased as the concentration of Zn^{2+} increased.

Keywords: La(Fe,Zn)O_3, perovskite, structural, sol-gel method

1. Introduction

Orthoferrites have attached a considerable attention due to they can be potentially applied in a wide range applications, namely fuel cells, catalysts, gas sensors, semiconductors, and magnetic materials [1]. The unique distorted perovskite structure, especially in orthoferrites, contributes to the properties that make the high applicability of this material. For enlightenment, perovskite ideal structure B-O-B bond angle is 180°, however, for orthoferrites the angle is deviated from 180° to some lower value because of orthorhombicity in the structure [2]. $RFeO_3$ orthoferrites, where R belongs to the rare earth elements, with perovskite structure. As $LaFeO_3$ has an orthorhombic unit cell with the least distortion when compared with other orthoferrites, making this material has the highest Neel temperature ($T_N = 740$ K). The distortion in this material's structure is inspected due to the displace oxygen and rare earth ions from their cubic position, and then the superexchange of Fe^{3+} also contributed to the distortion in the structure [1]. The distorted perovskite $LaFeO_3$ is a canted antiferromagnetic insulator showing a weak ferromagnetism. Consequently, the structure and physical properties of $LaFeO_3$ can be modified by the choice of a suitable doping element at La^{3+} or at Fe^{3+} site in $LaFeO_3$. According to Maria and Gomez [3], doping metals with a 2+ oxidation state and radii identical to that of Fe^{3+} (such as Mg^{2+} or Zn^{2+}) triggers changes in the structural, electrical and magnetic properties of the pure material and with using Mg and Zn as dopants in the $LaFeO_3$ perovskite has been proven as an effective strategy for developing application in gas sensors, oxidation catalysts and photocatalysts. In this present work, we prepared $LaFe_{1-x}Zn_xO_3$ with x = 0.0; 0.03; 0.05 and 0.07 using sol-gel method to analyze the structural of material.

2. Experimental

$LaFe_{1-x}Zn_xO_3$ (x = 0.0; 0.03, 0.05 and 0.07) materials were prepared by sol-gel method. We used Lanthanum (III) oxide [La_2O_3], ferric nitrate nonahydrate [$Fe(NO_3)_3{\cdot}9H_2O$], and zinc nitrate hexahydrate [$Zn(NO_3)_2$. $6H_2O$] as precursors of material. The required starting materials in stoichiometric amounts were dissolved in aquadest. Citric acid served as a chelating agent was added in the solutions and was maintained with stirring continuously at 120 °C until the gel-phase was obtained. The gel-phase of the material then was dried at 200 °C for 5 h. In the end, the powder was calcined at 900 °C for 3 h.

The crystalline structure was identified by X-Ray Diffraction (XRD) type *X'Pert PRO PANalytical* with source radiation Cu Kα (λ = 1.5405 Å). Rietveld Refinement was carried out to find the lattice parameters and bond angles by using Fullprof software. The chemical compounds were measured by X-ray fluorescence to determine the ration of each material which form $LaFe_{1-x}Zn_xO_3$. Raman analysis was performed by *Thermo Scientific DXR2 Raman Microscope* in the range from 70–800 cm^{-1} with the laser source 532 nm and power 2.0 mW. FTIR spectra were obtained on (*PerkinElmer type Spectrum Two*) in the range 250–3500 cm^{-1}. The optical band gap energy was determined by UV–Visible diffuse reflectance spectroscopy (*Agilent Cary 100/300*).

3. Results and discussion

3.1. Structural analysis

The structural properties of the sample have been characterized by X-ray diffraction and were refined by *Fullprof* software. The XRD pattern of $LaFe_{1-x}Zn_xO_3$ (x = 0.0; 0.03; 0.05 and 0.07) is shown in figure 1 with a sharp peak in the pattern indicated a well crystallinity of the prepared samples. The XRD pattern was consistent with the pure $LaFeO_3$ (101), (121), (220), (202) and (240) using the standard from No. COD 1526450. There is no additional peaks for impurities which concluded all of the samples were in a single phase with no impurities and have an orthorhombic structure with the Pbnm space group.

The lattice parameter, goodness of fit, tolerance factor and also geometrical parameters were also analyzed using the *Fullprof* software which is listed in table 1. The lattice parameters of $LaFe_{1-x}Zn_xO_3$ were observed to be increase with the increase of dopant. This may be caused by the ionic radii of Zn^{2+} (r_{Zn} = 0.074 nm) which is larger compared to Fe^{3+} (r_{Fe} = 0.064 nm). According to Bhat et al., because Zn^{2+} is a nonmagnetic cation and does not change its valency, hence, when Fe^{3+} was substituted by Zn it may produce an average radius compatible for a perovskite structure, likewise, the ionic radii of Fe^{3+} is 0.064 nm and for Zn^{2+} is 0.074 nm which is clear that unit cell volume would increase as the Zn concentration increases [4]. The effect of doping Zn also changed the geometrical parameters where bond lengths and bond angles of the material were measured. This is because Zn ion exits in the +2 oxidation state, as it replaces part of the Fe which exits in the +3 oxidation state, to maintain the charge balance, Fe^{3+} changes to Fe^{2+} and created the oxygen vacancy in $LaFe_{1-x}Zn_xO_3$ [2]. As for the oxygen vacancy created, the four octahedra in lanthanum orthoferrite would distort the structure as observed in table 1, where the bond angle and bond length of $LaFe_{1-x}Zn_xO_3$ increase compared with the pure $LaFeO_3$.

3.2. X-ray fluorescence

XRF elemental analysis was conducted to confirm the stoichiometry of the materials or chemical atoms that were synthesized [5]. Table 2 shows the experimental and calculation of weight percent from the atom which formed $LaFe_{1-x}Zn_xO_3$. The table clearly shows that the weight percent of Lanthanum (La)

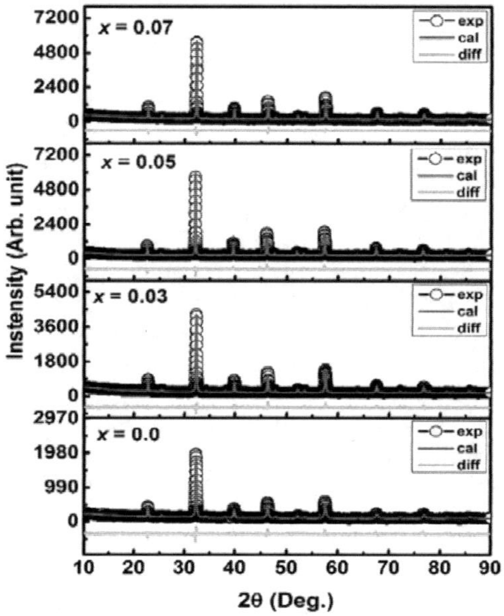

Figure 1. The XRD patterns of $LaFe_{1-x}Zn_xO_3$ (x = 0.0; 0.03; 0.05; 0.07).

and Iron (Fe) elements were near the calculation result but not for Zinc. The difference result we got from the experiment and calculation could be directed to the characteristic of XRF which generally has modest limits of detection and could be failed from particle size effects and XRF only measured the concentration of the sample in a very superficial region of the sample near the sample surface. As for the big difference of weight percent in Zn it may be due to the weight loss caused by the process of calcination where the samples were heated at 900 °C were higher than the melting point of Zinc (Tp ≈ 419.5 °C) for 6 h.

But from the table 2, we could observe the trend of Zn where the weight percentage increases as the value of x increases and no trace amount of any other element was found which confirms the result of XRD which means the purity of $LaFe_{1-x}Zn_xO_3$ (x = 0.03; 0.05 and 0.07).

3.3. Raman spectroscopy

To specify the structure distortion, phase transition, oxygen motion, and the order-disorder effects in the lattice at varied temperatures of perovskite-type materials, the study of the dynamical processes caused by charge carriers, phonons, and spins which affect the properties was conducted [1]. The Raman spectroscopy has been carried out for $LaFe_{1-x}Zn_xO_3$ (x = 0.0; 0.03; 0.05; and 0.07) at room temperature and are shown in figure 2. From figure 2, where we could compare the Raman plot of pure $LaFeO_3$ and the doped of $LaFe_{1-x}Zn_xO_3$. There are some peaks which are formed in range ~130 cm^{-1}, ~255 cm^{-1}, and ~400 cm^{-1}. The additional peaks were formed maybe due to the replacement of Zn^{2+} in Fe^{3+} where the dopant distorted the structure and make the effect on the oxygen vacancy formation.

Table 3 exhibits the modes of Raman in the range 100–800 cm^{-1} where 9 modes were found around ~130 cm^{-1}, ~146 cm^{-1}, ~169 cm^{-1}, ~255 cm^{-1}, ~282 cm^{-1}, ~403 cm^{-1}, ~421 cm^{-1}, ~618 cm^{-1}, and ~771 cm^{-1} in $LaFe_{1-x}Zn_xO_3$ (x = 0.03; 0.05; 0.07). These modes are corresponded with pure $LaFeO_3$ where the peak around ~150 cm^{-1} was attributed to the occupying of La in perovskite unit, the Raman shift around ~283cm^{-1}, ~481cm^{-1} and ~625 cm^{-1} were attributed to the vibration of Fe-O bonds and at ~426 cm^{-1} might be attributed to the existence of the La-O vibration [6]. The modes of Raman for

$LaFe_{1-x}Zn_xO_3$ with $x = 0.0$; 0.03; 0.05; 0.07 show a shifting to the higher value due to the increasing of Zn^{2+}. Although for some modes in Raman only show a small shifting, in fact there are some modes which have the same result in $x = 0.05$ and $x = 0.07$.

The shifting could be directed to the replacement of Zn^{2+} in Fe^{3+}. According to Manzoor and Husain [1], the shift from Raman plot corresponds to the strain in the samples where the shift to the higher value is bound to the compressive stress and tensile stress yields in the shifting to the lower value. The substitution of Zn^{2+} into Fe^{3+} produces a distortion especially in the oxygen bending and stretching vibration. Wanjun et al., also reported the shift in the Raman modes could be due to some factors

Table 1. Refinement $LaFe_{1-x}Zn_xO_3$ with $x = 0.0$; 0.03, 0.05 and 0.07.

Parameter	Sample			
	0.0	0.03	0.05	0.07
Crystallography Parameters				
Phase	Orthorhombic	Orthorhombic	Orthorhombic	Orthorhombic
(Space group)	(P b n m)	(P b n m)	(P b n m)	(P b n m)
A	5.54761	5.55319	5.55440	5.55451
B	5.5372	5.56228	5.56317	5.56516
c	7.82869	7.85461	7.85247	7.85513
χ^2	1.24	1.04	1.44	1.01
t (Tolerance factor)	0.93	0.93	0.93	0.93
Density ($g.cm^{-3}$)	6.705	6.654	6.658	6.659
Volume ($V/ 10^6 pm^3$)	240.485	242.616	242.642	242.816
Geometrical Parameters				
Bond length(Å)				
La-O1	2.751	2.761	2.791	2.762
La-O2	2.565	2.578	2.593	2.579
Fe/Zn-O1	1.919	1.923	1.924	1.924
Fe/Zn-O2	2.008	2.051	2.014	2.015
Bond Angle(°)				
La-O1-La	99.919	99.956	99.868	99.939
La-O2-La	106.849	106.786	105.565	106.785
Fe/Zn-O1-Fe/Zn	157.075	157.070	157.073	157.073
Fe/Zn-O2-Fe/Zn	154.019	154.073	154.061	154.069
R factor (%)				
Re	8.48	6.15	5.84	6.17
Rp	7.34	4.89	5.37	4.88
Rwp	9.44	6.27	7.00	6.21

Table 2. XRF elemental analysis of $LaFe_{1-x}Zn_xO_3$ ($x = 0.00$, 0.03, 0.05 and 0.07).

$LaFe_{1-x}Zn_xO_3$	(%) experimental weight			(%) weight calculation		
	La	Fe	Zn	La	Fe	Zn
$x = 0.0$	63.133	26.66	-	71.325	28.675	-
$x = 0.03$	73.61	26.30	0.09	71.220	27.774	1.005
$x = 0.05$	73.66	26.24	0.10	71.151	27.175	1.674
$x = 0.07$	73. 43	26.46	0.11	71.081	26.577	2.342

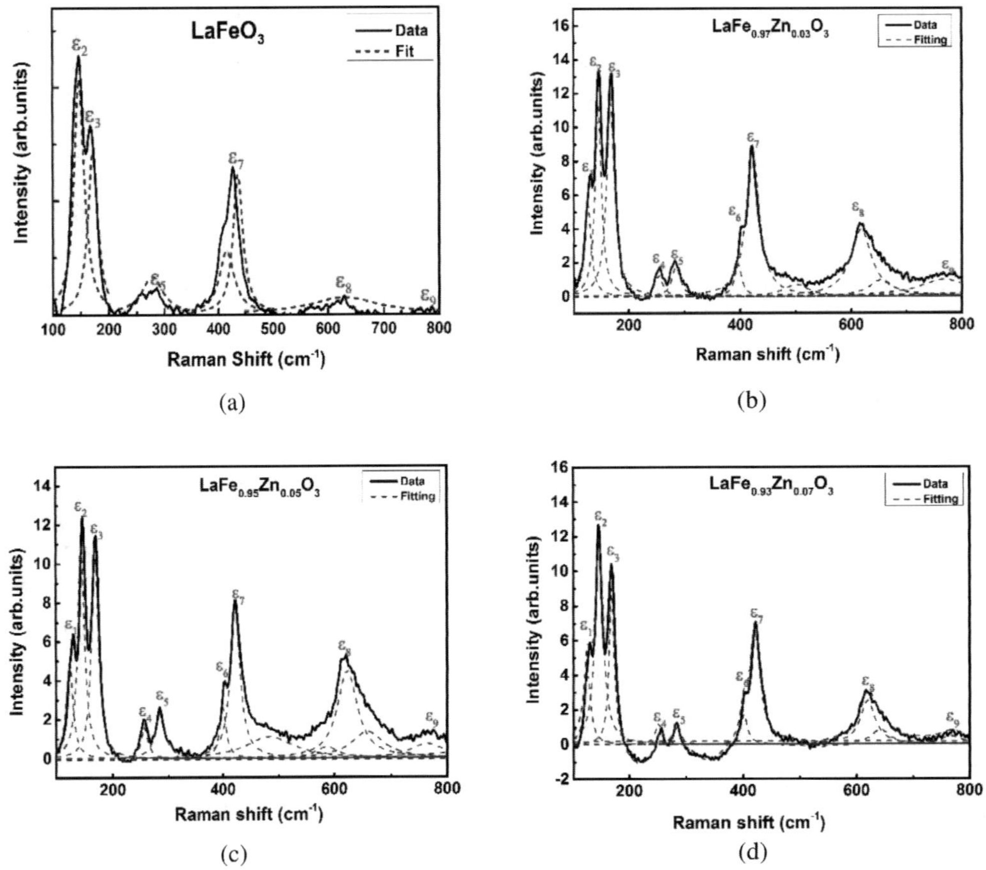

Figure 2. The Raman spectra of $LaFe_{1-x}Zn_xO_3$ with x value (a) 0.0, (b) 0.03, (c) 0.05 and (d) 0.07.

such as the atomic weight, the strength of the corresponding ions and the strength of the chemical bond [6]. The difference of atomic weight between Zn (65.4 g/mol) and Fe (55.845 g/mol) also the difference of ionic radii of Zn (0.074 nm) and Fe (0.064 nm) lead to the strain in the sample which makes the modes shifted to the higher value.

3.4. Fourier transform infrared spectroscopy

FTIR was conducted in all of the samples and are shown in figure 3. Figure 3a shows the result of FTIR in range 250–4000 cm^{-1} where it exhibits peaks around range 490 cm^{-1} and around 540 cm^{-1} then were noted as ε_1 and ε_2. These modes described the establishment of lanthanum orthoferrite where the band around 570 cm^{-1} in the main sample $LaFeO_3$ is attributed to the Fe-O stretching vibration and the band around 410 cm^{-1} corresponds to the O-Fe-O bending vibration [1]. The shifting and the intensity of samples were also observed in the range 450–750 cm^{-1} as shown in figure 3b. The broadening of peaks around 475–525 cm^{-1} become small and sharp as the concentration of Zn^{2+} increase. This is maybe due to the distortion in the structure where the bond angle Fe-O-Fe and bond length change as Zn replaced Fe in $LaFeO_3$. The peaks around 525–550 cm^{-1},

Table 3. Observes and reported raman modes of $LaFe_{1-x}Zn_xO_3$ in range 100–800 cm^{-1}.

Modes	Raman shift $LaFe_{1-x}Zn_xO_3$ (cm^{-1})			
	x = 0.0	x = 0.03	x = 0.05	x = 0.07
ε_1	-	130.212	130.487	130.772
ε_2	146.617	146.140	146.647	147.155
ε_3	167.669	169.026	169.614	169.473
ε_4	-	254.864	255.739	256.456
ε_5	280.902	282.576	284.325	283.082
ε_6	-	402.900	403.219	403.686
ε_7	427.819	421.561	422.491	422.491
ε_8	629.323	617.722	618.783	620.027
ε_9	776.090	775.368	774.978	774.978

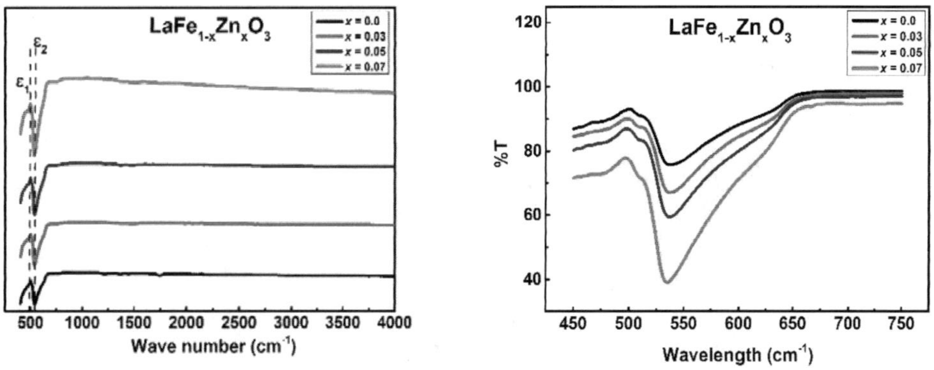

(a) (b)

Figure 3. The FTIR patterns of $LaFe_{1-x}Zn_xO_3$ (x = 0.0; 0.03; 0.05; 0.07), (a) for range 250–3500, and (b) for range 400–1000.

where we could observe the sharp peak with lower intensity. This could be corresponding with the presence of metal-oxygen bond which could be noted as the vibration of Fe-O and O-Fe-O bonding in the octahedral structure of $LaFe_{1-x}Zn_xO_3$. The samples spectra indicate the two bands shifting toward the lower wavenumber for modes ε_1 and ε_2 mode shift to the higher wavenumber which could be assigned to the strain of stretching and bending vibration of the metal-oxygen bond. The modes also exhibit the varied of intensity in all of the samples which is shown in figure 3b. The intensity of samples was decrease and the peaks slightly chanted as the concentration of Zn increases. Table 4 shows us the modes of FTIR in range 450–650 cm^{-1} where the vibrations are formed. The modes slightly move to the lower wavelength as the increase of Zn^{2+}.

3.5. UV-Vis spectroscopy
UV-Vis diffuse reflectance spectroscopy (UV-Vis DRS) was done to determine the band gap of material $LaFe_{1-x}Zn_xO_3$ (x = 0.03; 0.05 and 0.07). The radiation which is impinged on material would penetrate the sample and some portion will be reflected. The part which penetrates through the sample is scattered and transmitted, and the portion of this radiation which is reflected on the surface of the sample is regarded as the diffuse reflectance. Figure 4 shows the absorbance of material $LaFe_{1-x}Zn_xO_3$ (x = 0.0;

0.03; 0.05 and 0.07) in range 350–800 nm. The absorbance for all of samples increases as the concentration of Zn^{2+} increases, may be due to the color of samples which is the darkest by the addition of Zn. The increase of absorbance indicated that all of doped samples could be performed in visible light compared with the pure $LaFeO_3$.

The bandgap energy (Eg) can be determined from the absorbance data using the Tauc relation as:

$$\alpha h\nu = A\,(h\nu - Eg)^{1/n}$$

Table 4. The observes of FTIR spectra modes of $LaFe_{1-x}Zn_xO_3$ in range 450–650 cm^{-1}.

Modes	0.0	0.03	0.05	0.07
ε_1	500.115	498.299	497.627	497.030
ε_2	535.143	536.143	537.337	534.874

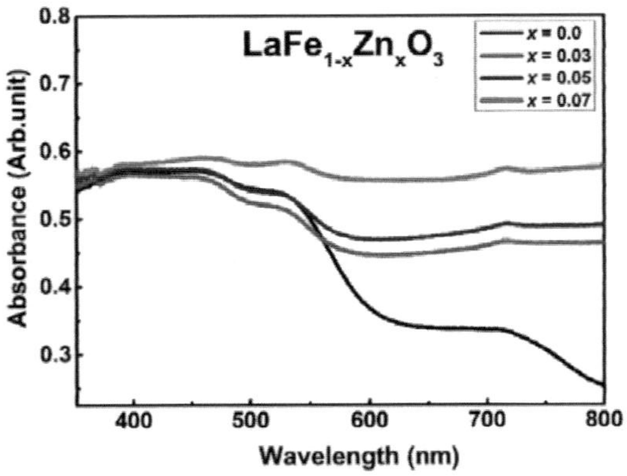

Figure 4. UV-vis (a) absorbance and (b) reflectance spectra of $LaFe_{1-x}Zn_xO_3$ (x = 0.0, 0.03, 0.05 and 0.07).

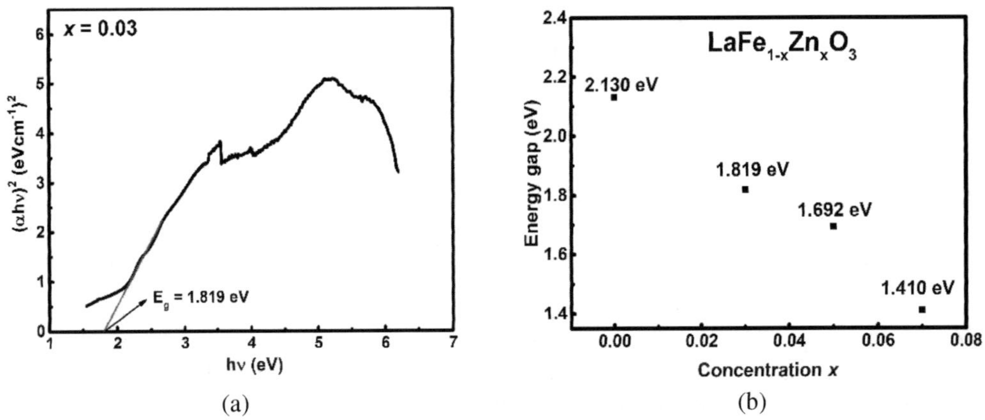

Figure 5. The energy band gap of material $LaFe_{1-x}Zn_xO_3$ (x = 0.0, 0.03, 0.05, 0.07) .

Where A could be noted as a constant called the band tailing parameter, Eg symbolizes the optical energy gap, and n refers to the interband transition which are depended upon the materials nature, in case the material is amorphous or crystalline and the photon transition [7]. By using the absorbance data of all the samples to determine the energy gap and were plotted between $(\alpha h\nu)^n$ and $(h\nu)$ for n = 2 as the best denounced in the literature which $LaFeO_3$ has a direct allowed bandgap [2]. The example of plotted as shown in figure 5a for $x = 0.03$ ($LaFeO_3$ with smallest doped) and figure 5b corresponds to the energy gap for all samples (x = 0.0; 0.03; 0.05; 0.07).

Figure 5b shows the bandgap energyof material $LaFe_{1-x}Zn_xO_3$ (x = 0.0, 0.03, 0.05, 0.07) decreases as the concentration of Zn^{2+} increases, it may be due to the creation of oxygen vacancy which formed because of the replacement of Zn in Fe. A possibility of the oxygen vacancy formation could be the additional energy levels near the conducting band [2]. Further enhancement of the hopping of electron and could be applied as photocatalyst or photocathode.

4. Conclusion

Lanthanum orthoferrites doped by Zinc have been successfully prepared by the sol-gel method. XRD result shows that all of the materials are single phase with an orthorhombic structure and *Pbnm* space group. The lattice parameter, density and also the volume of samples slightly increase due to the concentration of dopant. It might be caused by the ionic radii dopant (Zn^{2+} = 0.074 nm) which is larger than (Fe^{3+} = 0.064 nm). The vibration of the particle in the material is also observed by FTIR and Raman spectroscopy which give further information about the formation of $LaFe_{1-x}Zn_xO_3$ (x = 0.03; 0.05; 0.07). In FTIR results, the vibration modes shift to a lower intensity and slightly chanted as the concentration of Zn increases, corresponds to the presence of metal-oxygen bond which could be noted as the vibration of Fe-O and O-Fe-O bonding in the octahedral structure of $LaFe_{1-x}Zn_xO_3$. Raman spectra detect the Zn^{2+} dopant in Fe^{3+} produces a distortion mainly in the oxygen bending and stretching vibration. The energy band gap of samples is observed to decrease as the concentration of Zn^{2+} increases. It is related to the changing in structural properties, such as lattice parameter increment and shift of vibration modes caused by Zinc dopant.

Acknowledgments

This work was financially supported by Ministry of Research and Technology (KEMRISTEK/BRIN) Republic of Indonesia under research grant Hibah Penelitian Tesis Magister (PTM) 2020 with contract number NKB-489/UN2.RST/HKP.05.00/2020.

References

[1] Manzoor S and Husain S 2019 *J. Appl. Phys.* **124** 6
[2] Manzoor S and Husain S 2018 *Mater. Res. Express* **5** 55009
[3] Díez-García M I and Gómez R 2017 *ChemSusChem* **10** 2457-63
[4] Bhat I, Husain S, Khan W, and Patil S I 2013 *Mater. Res. Bull.* **48** 4506-12
[5] Dhiman M, Tripathi M, and Singhal S 2017 *Mater. Chem. Phys.* **202** 40-9
[6] Li W, Yang F, Xiong P, Jia Y, Liu J, Yan X and Chen X 2019 *J. Mater. Sci.* **54** 7460-68
[7] Manzoor S, Husain S, Somvanshi S, Fatema M and Zarrin N 2019 *Appl. Phys. A* **125** 509

Modification of fly ash using acids and alkali by hydrothermal method and its application as adsorbents material for phosphate adsorption in aquatic system

W R Wulandari, A Saefumillah and R T Yunarti

Department of Chemistry, Faculty of Mathematics and Natural Sciences (FMIPA), Universitas Indonesia, Depok 16424, Indonesia

Corresponding author's email: asep.saefumillah@sci.ui.ac.id

Abstract. In this research, fly ash was modified with acids, such as HCl, H_2SO_4 and mixture of both and alkali with hydrothermal method. XRF analysis showed a decrease of CaO and Fe_2O_3 contents in HMFA, SMFA and HSMFA while AMFAWH has higher CaO content and produced the new zeolite phase (hydroxy-sodalite) in $2\theta = 24.232°$ from XRD test. FTIR analysis showed that the fly ash gave absorption band of O-H group, H-O-H bend group, and Si-O stretching group. SEM images showed that acid-modified fly ash indicated a deposition of gehlenite crystals while AMFAWH showed flat rectangular shaped small particle known as hydroxy-sodalite. The BET method analysis indicated that all adsorbents were classified as mesopores. The experimental data showed that adsorption efficiency and capacity of fly ash was improved significantly after modification. The unmodified fly ash (UFA) has the highest phosphate adsorption efficiency and capacity in 0.16 g was 60.07 % and 0.485 mg P-PO_4/g while modified fly ash (HSMFA) was 85.62 % and 0.705 mg P-PO_4/g. The optimum contact time were obtained at 240 min for UFA, HMFA, SMFA and 480 min for HSMFA and AMFAWH. The optimum of pH were obtained at pH 5 and pH 7 for the others. The fifth adsorbents follow the Freundlich isotherm and pseudo-second-order kinetic model in phosphate adsorption.

Keywords: Phosphate, fly ash, acid activation, hydrothermal, adsorption

1. Introduction

Fly ash is a by-products from burning coal in electric power generating plants [1]. In 2008, the total coal produced in Indonesia reached 215 million tons and produced fly ash with large quantities [2]. To solve this problem, a utilization of fly ash that can prevent the environmental pollution is needed. In recent years, fly ash has become a concern of researchers for its use as adsorbents in phosphate adsorption [3].

The unmodified adsorbent has more pores that closed on the surface of adsorbent [4]. To open these closed pores, a modification on adsorbents such as acid-modified fly ash is required, which can dissolve the impurities minerals in the fly ash so the fly ash pores become wider due to the ability to replace the exchanged cation (alkali or soil alkali) in the fly ash structure with ion H^+ [5]. Fly ash which has high silica and alumina content, can also be used as an ideal precursor for synthesis of zeolites. Conversion of fly ash into zeolite can be done through an alkali reaction with hydrothermal method [6]. The aim of this research was to examine the effectiveness of acid-modified fly ash and alkali-modified fly ash with hydrothermal method as the adsorbent for phosphate adsorption from aquatic system.

2. Materials and method

2.1. Materials
Fly ashes were obtained from PT Smart Tbk in Marunda, Indonesia. Fly ashes were washed using DI- water repeatedly and dried at 120 °C for 6 h. They were grounded and passed through a 160 mesh-sized sieve. The reagents used in this research were HCl (37 %), sulfuric acid (96 %), NaOH, and KH_2PO_4, potassium antimony tartrate, ammonium heptamolybdate tetrahydrate and ascorbic acid. All chemical reagents were of analytical grade and obtained from Merck.

2.2. Preparation of acid-modified fly ash adsorbent
Three types of acid solution used were HCl (2 mol/L), H_2SO_4 (2 mol/L) and H_2SO_4-HCl mixture (1 mol/L H_2SO_4 + 1 mol/L HCl). Twenty g of dried coal fly ash was added to 100 mL of acid solution and mixed for 30 min under ultrasonication. The mixtures were then filtered and washed with DI-water until neutral pH. The solid residue was dried at 120 °C for 6 h. After that, the acid-modified fly ash was grounded and sieved with 160 mesh sieve then stored for further test [7].

2.3. Preparation of alkali-modified fly ash with hydrothermal method
Alkaline solution used was 3.5 M NaOH. Twenty g of dried coal fly ash was mixed with 160 mL of 3.5 M NaOH solution in Teflon autoclave and heated at 100 °C in an oven for 24 h. After that, the solid phase was filtered and washed with DI-water until the filtrate pH reached 10. Then, it was dried in an oven at 55 °C for 24 h and stored for further test [6].

2.4. Characterization
Fly ash before and after modified were characterized using XRF, XRD, FTIR, SEM and SSA instrument.

2.5. Study of phosphate adsorption
One thousand mg/L stock phosphate solution was made by dissolving 2.1993 g of KH_2PO_4 in 500 mL of DI-water. Phosphate adsorption was carried out by mixing the adsorbents with phosphate solution in Erlenmeyer flask and stirred using magnetic stirrer. Then, the solution was filtered with Whatman 42 and the filtrate was analyzed by an UV-vis spectrophotometer with molybdenum blue method.

2.5.1. The effect of initial phosphate concentration. The phosphate concentration used for the effect of phosphate concentration were 4 ppm, 6 ppm, 8 ppm, 10 ppm and 12 ppm. The optimum concentration of phosphate solution that obtained will be used for the next experiments.

2.5.2. The effect of pH. The variation of pH in adsorbate solution that used were 3, 5, 7, 10 and 12. It can be controlled by 0.1 M HCl and 0.1 M NaOH. The optimum pH that obtained will be used for the next experiments.

2.5.3. The effect of contact time. The adsorbate solution mixed with adsorbents was stirred using magnetic stirrer for 0.5, 1, 2, 4 and 8 h. The optimum contact time that obtained will be used for the next experiments.

2.5.4. The effect of adsorbent mass. The variation of the adsorbent mass that used were 0.01 to 0.16 g.

3. Results and discussion

3.1. Characterization of fly ash
The chemical component of fly ash can be seen from XRF characterization in table 1. The content of CaO in acid-modified fly ash (HCl, H_2SO_4 and a mixture of both) is lower than unmodified fly ash. It showed that calcium reacted with acid solutions produce salts that dissolved with water and mostly

removed by filtration process while alkali-modified fly ash with hydrothermal method has a high CaO content due many calcium formed by geopolymerization process [8, 9]. Compared with unmodified fly ash, the content of Fe_2O_3 in acid-modified fly ash is decreased due H^+ replaces the cation from the fly ash particle of structure lead the dissolution of the metals [10] while alkali-modified fly ash with hydrothermal method has a higher Fe_2O_3 content due the most of amorphous aluminosilicate material has dissolved during geopolymerization reaction [11]. The content of SiO_2 in alkali-modified fly ash with hydrothermal method was lower than unmodified fly ash due to the establishment of new products with different microstructures by exchanging some SiO_4 species with AlO_4 lead to a change in the chemical environment of Si-O bond [12] while acid-modified fly ash has a higher SiO_2 content due to the dissolution of metal oxides and silicon include an element that resistant to acid attack [13].

The diffractograms of the fly ash are shown in figure 1. It indicates the presence of quartz (SiO_2), berlinite ($AlPO_4$), gehlenite ($Ca_2Al(AlSiO_7)$) and almandine ($Fe_3Al_2(SiO_4)_3$) in the fly ash. After modified fly ash with acids (HCl, H_2SO_4 and mixture of both), it shows an increase of the minerals in the sample due to the dissolved minor minerals constituents during acid treatment and quartz minerals becomes more intense [14] while AMFAWH has a lower of quartz and higher of gehlenite contents than UFA due to the hydroxide ions attack the silica forming a gel. The product is calcium silicate hydrate if the gel reacts with calcium ions [15]. In addition, AMFAWH produces the new zeolite phase (hydroxy-sodalite) in 2θ=13.913°, 24.232°, 32.778°, 34.605° and 42.739°.

Table 1. The chemical component of the unmodified fly ash (UFA), HCl modified fly ash (HMFA), H_2SO_4 modified fly ash (SMFA), HCl-H_2SO_4 modified fly ash (HSMFA) and alkali modified fly ash with hydrothermal method (AMFAWH).

Composition	UFA (%)	HMFA (%)	SMFA (%)	HSMFA (%)	AMFAWH (%)
Al_2O_3	16.64	16.54	16.20	16.26	12.49
SiO_2	48.46	61.45	65.03	60.43	30.51
CaO	12.10	3.25	2.57	3.52	16.77
Fe_2O_3	14.80	13.10	11.00	13.68	32.52

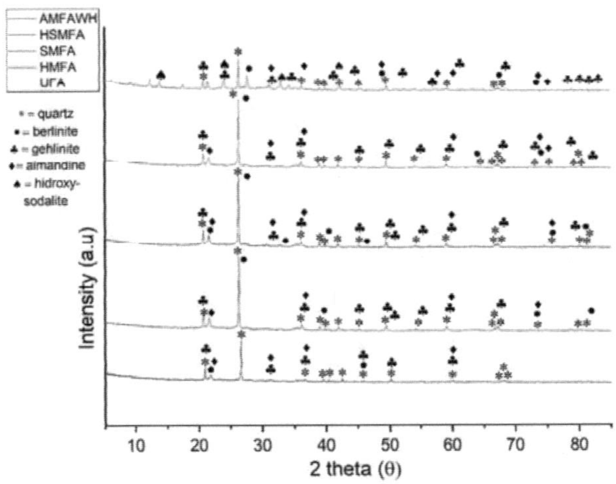

Figure 1. XRD characterization of UFA, HMFA, SMFA, HSMFA and AMFAWH.

FTIR spectra of UFA, HMFA, SMFA, HSMFA and AMFAWH are shown in figure 2. The broad absorption band of 3650–3200 cm^{-1} with the peak around 3282 cm^{-1} in UFA shown as the adsorbed water molecular. After modified with acids (HMFA, SMFA and HSMFA) and alkali with hydrothermal method (AMFAWH) it underwent a shift, respectively at 3294 cm^{-1}, 3309 cm^{-1}, 3382 cm^{-1} and 3370 cm^{-1} produced new -OH groups that will increase the phosphate adsorption [7]. The bands at 1606 cm^{-1} in UFA are demonstrated as H-O-H bending vibration. After modified with acids (HMFA, SMFA and HSMFA) and AMFAWH it underwent a shift, respectively at 1609 cm^{-1}, 1624 cm^{-1}, 1635 cm^{-1} and 1643 cm^{-1} defining the H-O-H deformation [6]. The bands at 1045 cm^{-1} in UFA is defined as Si-O stretching [7]. After modified with acids (HMFA, SMFA and HSMFA) it underwent a shift, respectively at 1058 cm^{-1}, 1097 cm^{-1}, 1101 cm^{-1} indicating that modification with acid can dissolve impurity that stuck in the active sites for adsorption while AMFAWH underwent a shift to lower band and became sharper (1000 cm^{-1}) indicating new products by alkali attack such as hydroxy-sodalite in the result of XRD previously mentioned [6].

The surface morphology of UFA, HMFA, SMFA, HSMFA and AMFAWH can be seen in figure 3. UFA (figure 3a) typically has spherical shapes with different sizes. After modified with HCl (figure 3b), it shows the presence of a more uniform particle size and forms several holes. It was due to the stable glassy layers destroyed by reaction between HCl with fly ash [16]. On the other hand, modification with H_2SO_4 (figure 3c) shows a white layer on the surface which defined that gehlenite crystals were deposited by sulphate anion attack [17]. HSMFA (figure 3d) have morphologies that similar with modification HCl and H_2SO_4 while AMFAWH (figure 3e) shows the presence of rectangular shaped particles known as hydroxy-sodalite [18].

Porous structure of fly ash was analyzed using SSA by BET method and the results can be seen in table 2. It shows that HMFA, SMFA, HSMFA and AMFAWH have a larger BET surface area and total pore volume than UFA. The pore size distribution of UFA, HMFA, SMFA, HSMFA and AMFAWH are classified as mesopores due to their sizes were ranged 2–50 nm. The results indicate that HMFA, SMFA and HSMFA can dissolve the impurity that stuck in the active sites for adsorption while AMFAWH produce new cavities from cracked the Si-O and Al-O bond tetrahedron structure so the surface area can be increased [8].

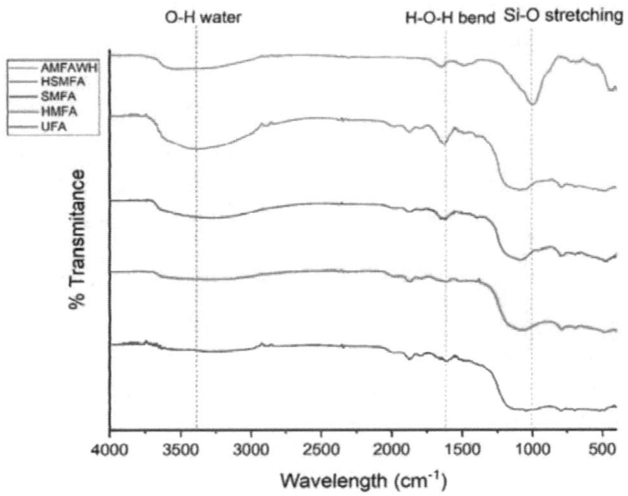

Figure 2. FTIR spectra of UFA, HMFA, SMFA, HSMFA and AMFAWH.

Figure 3. SEM images of (a) UFA, (b) HMFA, (c) SMFA, (d) HSMFA and (e) AMFAWH.

Table 2. Porous structure parameters of UFA, HMFA, SMFA, HSMFA and AMFAWH.

Material	BET surface area ($m^2\ g^{-1}$)	Mean pore diameter (nm)	Total pore volume ($cm^3\ g^{-1}$)
UFA	17.348	3.8602	2.488×10^{-2}
HMFA	21.773	3.8658	2.725×10^{-2}
SMFA	25.661	3.8669	2.997×10^{-2}
HSMFA	45.768	3.8726	4.929×10^{-2}
AMFAWH	35.041	3.8680	5.258×10^{-2}

3.2. Study of phosphate adsorption

Adsorption mechanism that happened in acid-modified fly ash were the ligand exchange between -OH on the surface of fly ash with phosphate (in solution) [19]. Meanwhile, the adsorption mechanism on alkali-modified fly ash with hydrothermal method was the precipitation of calcium phosphate [20].

3.2.1. Effect of initial phosphate concentration. Figure 4 indicates the adsorption efficiency and capacity from the effect of initial phosphate concentration on the adsorbents. UFA has the highest phosphate adsorption efficiency and capacity of 32.01 % and 1.980 mg P/g while HMFA, SMFA, HSMFA and AMFAWH have the highest phosphate adsorption efficiency and capacity, respectively 41.15 %, 48.99 %, 62.70 %, 55.52 % and 2.545 mg $P\text{-}PO_4/g$, 3.030 mg $P\text{-}PO_4/g$, 3.879 mg $P\text{-}PO_4/g$ and 3.434 mg $P\text{-}PO_4/g$. It implies that the amount of adsorbed phosphate increases due to all active sites available on each adsorbents are used for the adsorption at a higher concentration [21].

3.2.2. Effect of pH. The effect of pH on phosphate adsorption is presented in figure 5. It was observed that UFA has the maximum phosphate adsorption efficiency and capacity of 36.71 % and 2.389 mg $P\text{-}PO_4/g$ occurred at pH 5 while HMFA, SMFA, HSMFA and AMFAWH have the maximum phosphate adsorption efficiency and capacity, respectively 51.59 %, 61.15 %, 79.94 %, 78.34 % and 3.336 mg

P-PO_4/g, 3.954 mg P-PO_4/g, 5.066 mg P-PO_4/g and 5.169 mg P-PO_4/g. The maximum phosphate adsorption occurred at pH 5 and 7 due to both of the pH values are lower than the pK_a value of $H_2PO_4^-$ (7.2) lead to more protons donated than hydroxide groups by the solution phase so the surface of adsorbents becomes positively charged and attracting phosphate species which negatively charged [22].

3.2.3. The effect of contact time. The effect of contact time on phosphate adsorption is presented in figure 6. It was observed that UFA, HMFA and SMFA have the maximum phosphate adsorption efficiency and capacity in 240 min, respectively 51.62 %, 63.14 %, 69.25 % and 3.230 mg P-PO_4/g, 4.005 mg P-PO_4/g and 4.393 mg P-PO_4/g while HSMFA and AMFAWH have the maximum phosphate adsorption efficiency and capacity in 480 min, respectively 88.26 %, 82.15 % and 5.599 mg P/g, 5.211 mg P-PO_4/g. The first step of phosphate adsorption was in 0–120 min, the availability of surface area and pore sites of each adsorbent were still abundant so the rate of adsorption increased. In the next step, surface area and pores sites of the fly ash became saturated with phosphates lead to the decrease of rate of adsorption [22].

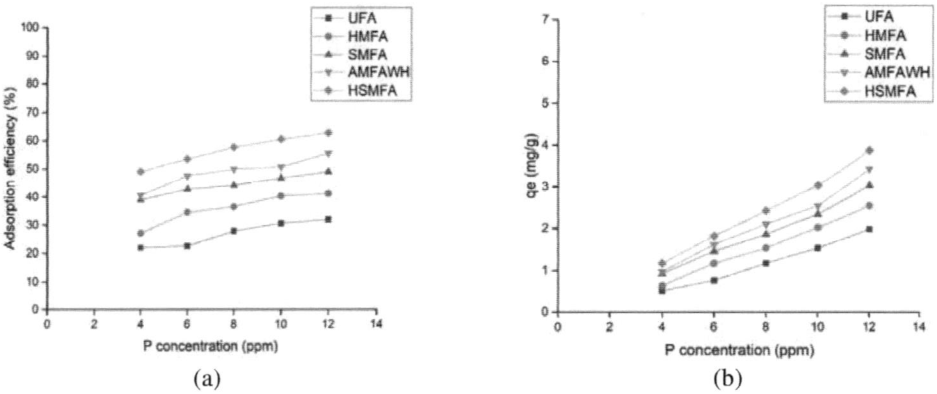

Figure 4. Adsorption (a) efficiency and (b) capacity from the effect of initial phosphate concentration on UFA, HMFA, SMFA, HSMFA and AMFAWH.

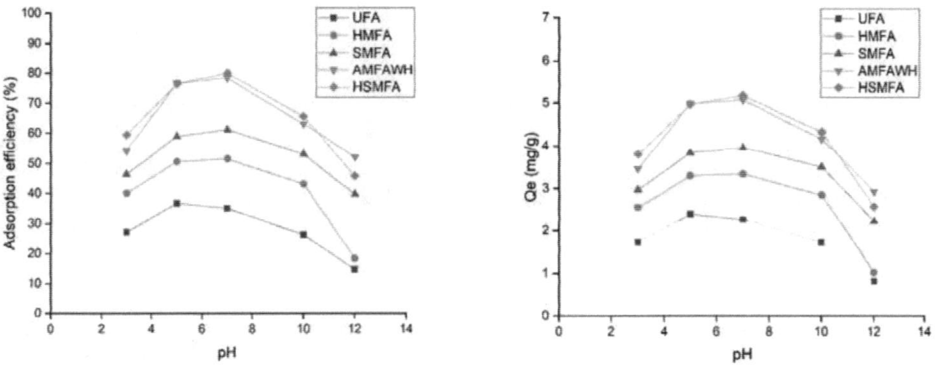

Figure 5. Adsorption (a) efficiency and (b) capacity from the effect of pH on UFA, HMFA, SMFA, HSMFA and AMFAWH.

3.2.4. The effect of adsorbent mass. The effect of adsorbent mass on phosphate adsorption is shown in figure 7. UFA, HMFA, SMFA, AMFAWH and HSMFA have the maximum phosphate adsorption efficiency in 0.16 g, respectively 60.07 %, 73.58 %, 79.60 %, 82.27 % and 85.62 % with adsorption capacity, respectively of 0.485 mg P-PO_4/g, 0.606 mg P-PO_4/g, 0.655 mg P-PO_4/g, 0.677 mg P-PO_4/g and 0.705 mg P-PO_4/g. The availability of adsorption sites for phosphate ions will be more in the higher adsorbent masses. However, the graph also shows that the adsorption capacity reduces with the increase in the mass of adsorbents because the active sites of the adsorbent is more unsaturated during the phosphate adsorption process [23].

3.2.5. Isotherm adsorption. From the varied initial phosphate concentrations, isotherm for this research can be estimated from plotting the curves as comparison among Langmuir and Freundlich. The results are summarized in table 3. The values of R^2 that obtained from Freundlich model was closer to 1 than Langmuir model. In addition, the values of K_F was higher than q_m also. It showed that phosphate adsorption using the fifth adsorbents followed the Freundlich isotherm model which were multilayer adsorption.

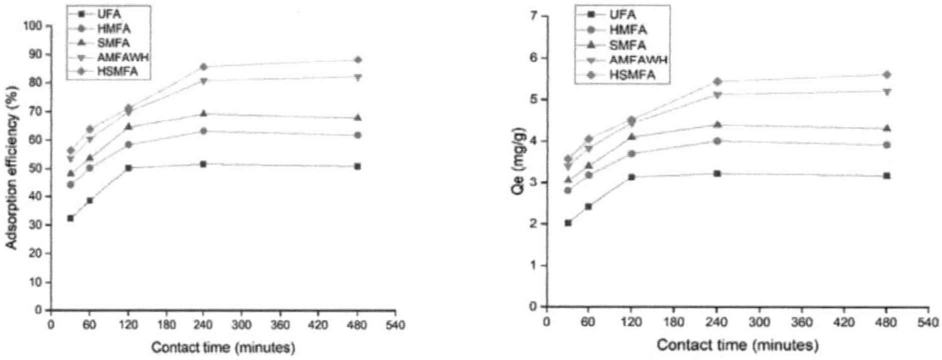

(a) (b)

Figure 6. Adsorption (a) efficiency and (b) capacity from the effect of contact time on UFA, HMFA, SMFA, HSMFA and AMFAWH.

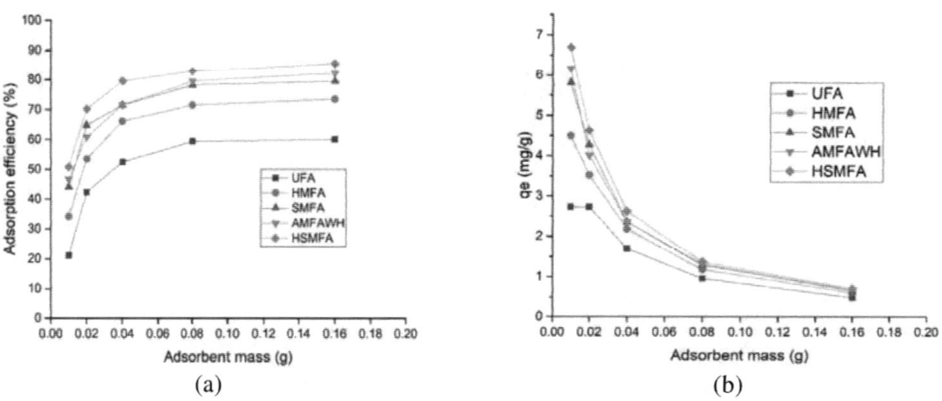

(a) (b)

Figure 7. Adsorption (a) efficiency and (b) capacity from the effect of adsorbent mass on UFA, HMFA, SMFA, HSMFA and AMFAWH.

Table 3. Phosphate adsorption isotherm parameter with fly ash as adsorbent.

Adsorbent	Langmuir Model			Freundlich Model		
	K_L	q_m (mg P/g)	R^2	K_F (mg P/g)	n	R^2
UFA	-0.0749	-1.3037	0.956	0.0524	0.5857	0.972
HMFA	-0.0923	-1.4973	0.934	0.0671	0.5356	0.983
SMFA	-0.0767	-3.3056	0.988	0.1817	0.6567	0.998
HSMFA	-0.1376	-2.3906	0.985	0.2111	0.5233	0.996
AMFAWH	-0.1115	-2.2307	0.966	0.1524	0.5525	0.987

Table 4. The kinetic model parameters for phosphate adsorption on fly ash adsorbent.

Adsorbents	$q_{e,\exp}$ (mg/g)	Pseudo-first-order			Pseudo-second-order		
		k_1 (min^{-1})	$q_{e,cal}$ (mg/g)	R^2	k_2 (g/mg.min)	$q_{e,cal}$ (mg/g)	R^2
UFA	3.230	0.0006909	1.363	0.4889	0.0199365	3.321	0.9982
HMFA	4.005	0.0006909	1.288	0.5602	0.0206685	4.050	0.9991
SMFA	4.393	0.0006909	1.301	0.5641	0.0175417	4.460	0.9989
HSMFA	5.599	0.0009212	1.454	0.7777	0.0064237	5.907	0.9986
AMFAWH	5.211	0.0009212	1.408	0.7240	0.0081536	5.470	0.9993

3.2.6. Adsorption kinetics. The kinetic model parameters obtained by fitting the experimental data are summarized in table 4. The values of $q_{e,cal}$ that obtained from pseudo-second-order kinetic model is closer to the experimental results ($q_{e,exp}$) than pseudo-first-order kinetic model. It showed that adsorption process that occurred was chemisorption involving the precipitation and ligand exchange.

4. Conclusion

Acid-modified fly ash (HMFA, SMFA and HSMFA) and alkali-modified fly ash with hydrothermal method (AMFAWH) can used as adsorbents for the effective phosphate adsorption. The experimental data shows that HMFA, SMFA, HSMFA and AMFAWH have the highest phosphate adsorption efficiency and capacity in 0.16 g, respectively of 73.58 %, 79.60 %, 85.62 %, 82.27 % and 0.606 mg P-PO4/g, 0.655 mg P-PO4/g, 0.705 mg P-PO4/g and 0.677 mg P-PO4/g while the unmodified fly ash (UFA) has the highest phosphate adsorption efficiency and capacity of 60.07 % and 0.485 mg P-PO4/g. The adsorption efficiency was strongly depended on pH. The optimum condition from the effect of pH were obtained at pH 5 for UFA and pH 7 for HMFA, SMFA, HSMFA and AMFAWH. The contact time represents the time needed for the adsorption process to reach the equilibrium. The optimum contact time for phosphate adsorption were 240 min for UFA, HMFA, SMFA and 480 min for HSMFA and AMFAWH. The fifth adsorbents followed the Freundlich isotherm and pseudo-second-order kinetic model in phosphate adsorption.

Acknowledgments
The authors are grateful to Universitas Indonesia for PITTA B 2019 research grant with contract number NKB-0614/UN2.R3.1/HKP.05.00/2019 for financial support.

References

[1] Ugurlu A and Salman B 1998 *Env. Int.* **24** 911-918
[2] Ivalandari 2010 *Penentuan Nilai Wajar Saham melalui Model Discounted Cash Flow dan Relative Valuation: Studi Kasus Saham PT. Bumi Resources, Tbk. Periode 2005-2009* BSc. Projects (Depok: Universitas Indonesia)
[3] Xu K, Deng T, Liu J and Peng W 2010 *Fuel.* **89** 3668-74
[4] Khan I et al. 2015 *Desalin. Water. Treat.* **57** 7232-44
[5] Irawan C, Atikah A and Rumhayati B 2014 *J. Pure App. Chem.* **3** 88-98
[6] de Carvalho T E M, Fungaro D A, Magdalena C P and Cunico P 2011 *J. Radioanal. Nucl. Chem.* **289** 617-26
[7] Li F, Wu W, Li R and Fu X 2016 *Appl. Clay. Sci.* **132-133** 343-52
[8] Qiu M, Wang Y, Niu M, Han Q and Zhang M 2016 *Am. Chem. Sci. J.* **14** 1-6
[9] Dutta D and Ghosh S 2012 *Int. J. Emerg. Technol. Adv. Eng.* **2** 757-63
[10] Sedres G 2016 *Recovery of SiO_2 and Al_2O_3 from Coal Fly Ash* MSc Thesis (Cape Town: University of Western Cape)
[11] Rickard W D A and van Riessen A 2010 *Int. J. Appl. Ceram. Technol.* **7** 81-8
[12] Alehyen S, Achouri M E L and Taibi M 2017 *Journal of Materials and Environmental Sciences* **8** 1783-96
[13] Caroles J D S 2019 *Fullerene Journal of Chemistry* **4** 5-7
[14] Behera S K, Chakraborty S and Meikap B C 2018 *Int. J. Coal. Sci. Technol.* **5** 142-55
[15] Rajabipour F, Giannini E, Dunant C, Ideker J H and Thomas M D A 2015 *Cem. Concr. Res.* **76** 130-46
[16] Bada S O and Potgieter-vermaak 2008 *Leonardo Electron. J. Pract. Technol.* **12** 37-48
[17] Allahverdi A 2005 *Ceram. Silikaty* **49** 225-9
[18] Pahlevi N D, Guo B and Sasaki K 2018 *Ceram. Int.* **44** 8635-42
[19] Xue Y, Hou H and Zhu S 2009 *J. Hazard. Mater.* **162** 973-80
[20] Zhang M et al. 2011 *Sep. Sci. Technol.* **46** 2260-74
[21] Ragheb S M 2013 *HBRC J.* **9** 270-5
[22] Rout P R, Bhunia P and Dash R R 2014 *J. Environ. Chem. Eng.* **2** 1331-42
[23] Gorzin F and Abadi M M B R 2018 *Adsorpt. Sci. Technol.* **36** 146-69

An impedance spectroscopy study of sol-gel prepared $La_{0.9}Sr_{0.1}Fe_{0.7}Mo_{0.3}O_3$ perovskite

D Triyono[1], R A Rafsanjani[1] and H Laysandra[2]

[1]Department of Physics, Faculty of Mathematics and Natural Sciences (FMIPA), Universitas Indonesia Depok 16424, Indonesia
[2]Department of Physics, National Taiwan Normal University, Taipei 11677, Taiwan

Corresponding author's email: djoko.triyono@ui.ac.id

Abstract. In this present work, impedance spectroscopy was applied to investigate the electrical properties of $La_{0.9}Sr_{0.1}Fe_{0.7}Mo_{0.3}O_3$ (LSFMo) perovskite prepared by sol-gel and sintering methods. SEM analysis reveals the non-uniform grain distributed in samples with average grain size 347 nm. The complex impedance has been examined in the frequency range of 100 Hz–1 MHz and temperature range of 75–250 °C. The Nyquist plots were identified to form single semicircular arc described as grain behavior working in conduction mechanism. The dielectric and complex conductivity is shown depending on the temperature and frequency. The evolution of ln σ_{DC} vs $10^3/T$ (K^{-1}) satisfies Arrhenius law and the activation energy is found to be 0.45 eV for this conduction mechanism.

Keywords: LSFMo, sol-gel method, dielectric, conductivity, activation energy

1. Introduction

$LaFeO_3$ material and its derivatives become interesting research topics in recent years due to the potential application in many fields such as solid oxide fuel cells, chemical sensors, etc. [1, 2]. Doping on orthorhombic-LaFeO3 has been investigated in many reports resulting in the increase ionic and electronic conductivity, structure and thermal stability as observed in $La_{1-x}Sr_xCo_{1-y}Fe_yO_{3-\delta}$ [1], $La_{0.8}Ca_{0.2-x}Pb_xFeO_3$ [2], etc. Doping in La- and Fe-site has been reported increasing the oxygen vacancy concentration related to the oxygen ion conductivity and magnetic-electrical conductivity properties, respectively [1].

This research is focused on the double doping of Sr and Mo on La- and Fe-site of $LaFeO_3$, respectively. Our previous works have investigated its structural, room temperature electrical analysis, and optical properties for $La_{0.9}Sr_{0.1}Fe_{1-x}Mo_xO_3$ system [3, 4]. The structural analysis revealed that the orthorhombic system is still preserved under Sr- and Mo-doping variation where the crystallite size and lattice volume increase with addition of Mo-concentration [3]. The room temperature electrical investigation reported the dielectric constant degrades and loss tangent increases with increasing Mo-concentration [3]. Furthermore, the optical analysis revealed the redshift of Fe/MoO_6 octahedral bands with increasing Mo-doping confirming the changes in lattice structure in the systems and the increasing the bandgap energy due to Mo content variation [4]. However, there is less or no systematic study of the electrical mechanism on this such system, so that this paper becomes a continuation of previous studies [3, 4] in which will investigate the temperature and frequency effects on electrical properties of $La_{0.9}Sr_{0.1}Fe_{0.7}Mo_{0.3}O_3$ sample by applying the alternating current (ac) impedance spectroscopy.

2. Experimental

The $La_{0.9}Sr_{0.1}Fe_{0.7}Mo_{0.3}O_3$ (LSFMo) sample was prepared by the sol-gel technique using the highly analytical grade of La_2O_3, $Fe(NO_3)_3.9H_2O$, $Sr(NO_3)_2$, MoO_3, and citric acid monohydrate which have stoichiometric amounts ~99.9 % purity. The starting precursor was dissolved into distilled water and then stirred until the dark brown gel was obtained. The obtained gel was heated with the heating rates to obtain the powder form of the sample. Then, the powder was grinded in an agate mortar, pressed using hydraulic pressure into the pellet, and sintered at 1300 °C. The details about the synthesize process have been explained in our previous work [3, 4]. Finally, the sintered pellet was obtained. A silver paste was applied on both sides of the pellet to form the plate capacitor configuration for electrical measurement. By using RLC-Meter Fluke 6306, the electrical and dielectric properties were examined in the temperature of 75–250 °C and frequency range of 102–106 Hz.

3. Results and discussion

3.1. Surface morphology

The scanning electron microscope (SEM) image of LSFMo sample is shown in figure 1a. The image shows that the grain is very small, non-uniform size and shape, and randomly oriented. The average grain size has been calculated by using ImageJ software. The distribution of grain size is shown in figure 1b. The average grain size is found to be 347 nm. Figure 1c shows a typical EDS spectrum of LSFMo sample. It declares that all of elemental compositions (La, Sr, Fe, Mo, and O) in this sample has present. This EDS result also proves that there is no absence of any integrated element during sintering process.

3.2. Electrical properties analysis

The Nyquist plot of LSFMo at different temperatures, we have plotted in figure 2. It is clear that the plots are formed by a single semicircular arc indicating the existence of only grain contribution as the one primary mechanism for the conduction process [5, 6]. The evolution of Nyquist semicircle diameter for variational different temperature indicates the thermal resistance of the sample. Actually, the decrease in diameter of arc with increasing temperature is occurred with an increase in conductivity.

Figure 3a shows the variation of the real part of impedance for the LSFMo sample as a function of frequency for variational temperatures. The $Z'(f)$ curve reveals that the real impedance value is higher at lower temperature and in the low-frequency range, and then it decreases slowly with increasing frequency. The Z' also become lower with increasing temperature suggesting the increasing in ac conductivity [5]. The merge of the real impedance in the high-frequency range could be caused by space charges release as a consequence of reduction of the barrier properties with increasing temperature [5, 7]. Next, the frequency and temperature dependence on variation of imaginary impedance value is shown in figure 3b. The Z'' value reaches one maximum value and that the peaks broaden with increasing temperature indicating the thermal relaxation mechanism. This mechanism is possible due to the release of charge carriers at lower temperature and ionic vacancies formation at higher temperature [8]. Figure 3c shows the Bode plots of the phase as a function of frequency for different temperatures. It reveals the similar curve for all temperature and only has one peak appear in θ spectrum confirming only one primary contribution to the conduction process [9].

Figure 4 shows the dielectric constant with frequency and temperatures. It reveals the change of dielectric property with frequency and temperature. The dielectric constant value relatively higher in the lower frequency whereas lower and tends to frequency-independent at higher frequency region. This condition might be caused by the influence of higher charge carrier density [5].

(a)

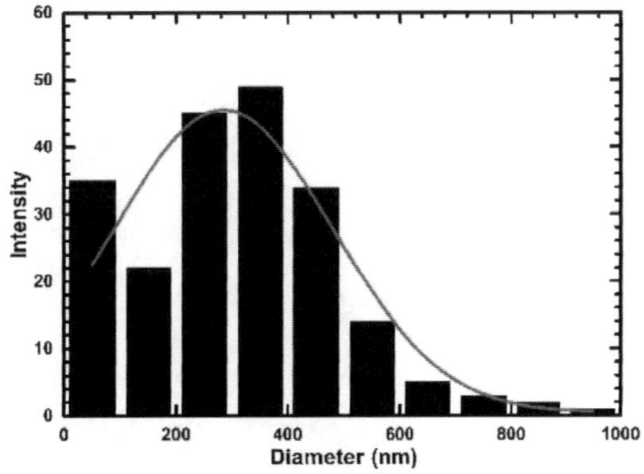

(b)

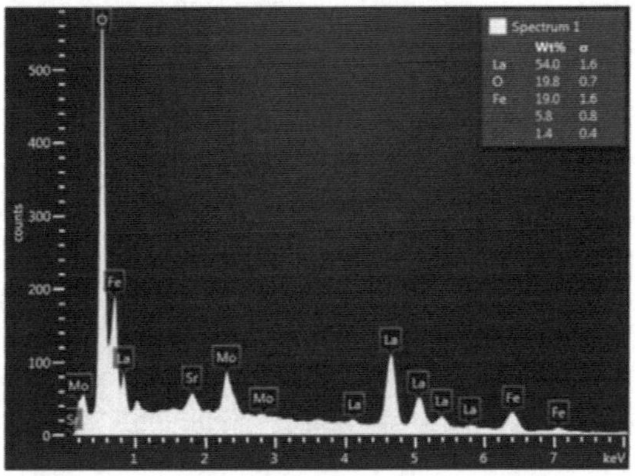

(c)

Figure 1. (a) SEM image, (b) the distribution of average grain size calculation, and (c) EDS spectrum for LSFMo sample.

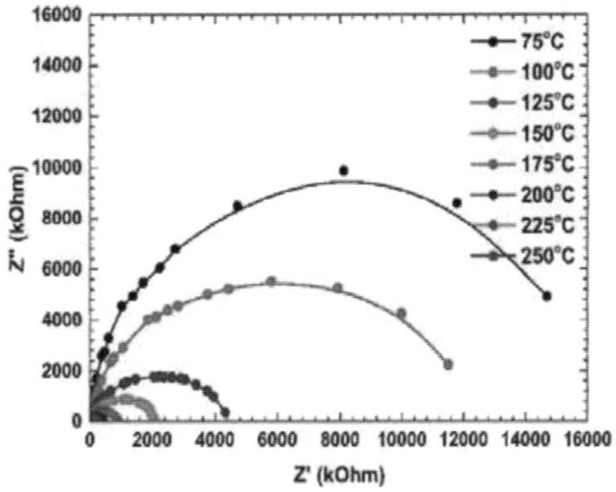

Figure 2. Nyquist plot of LSFMo sample at different temperatures.

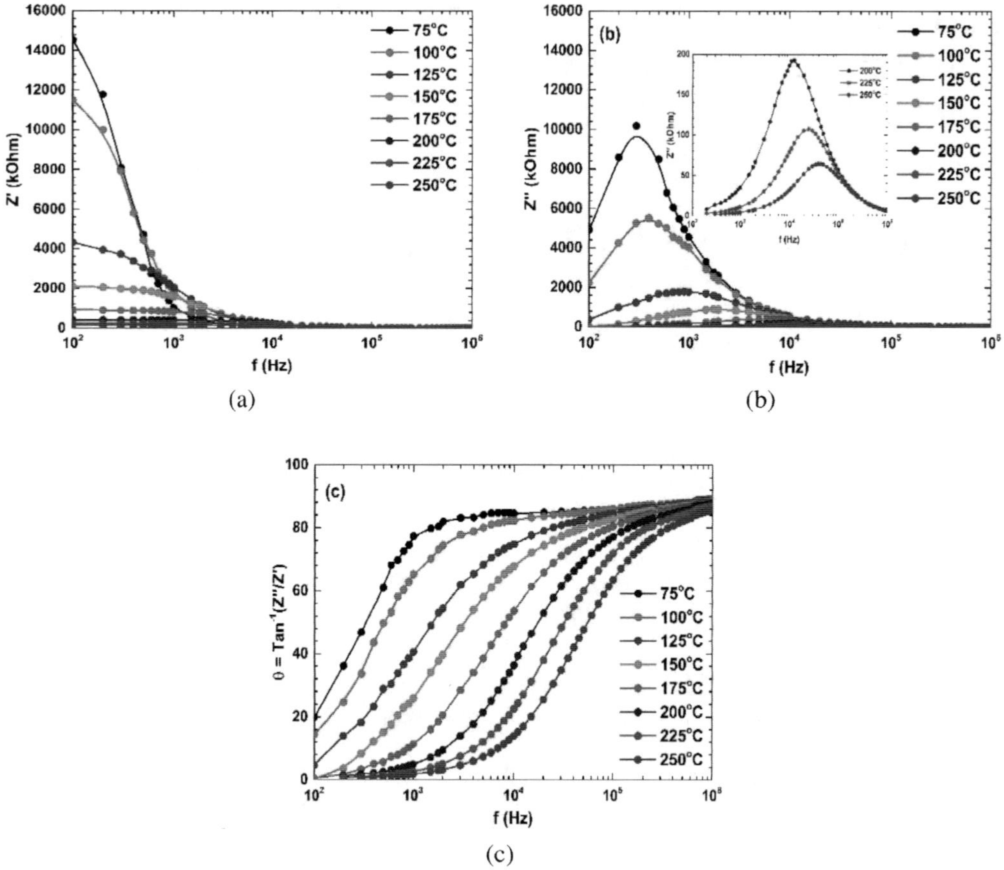

Figure 3. Bode plots of (a) $Z'(f)$, (b) $Z''(f)$, and $\theta(f)$ of LSFMo sample at different temperatures.

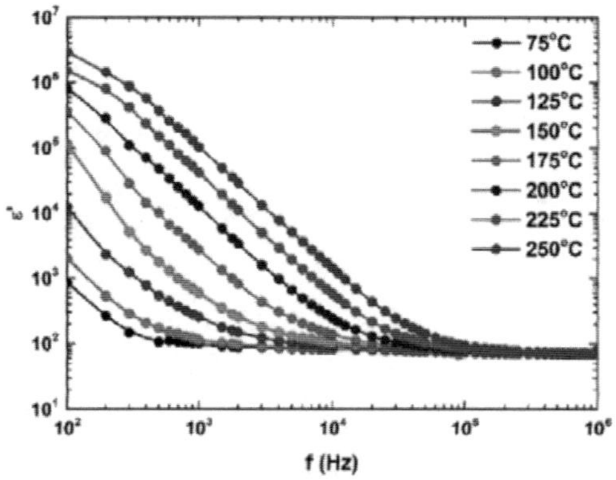

Figure 4. The dielectric constant of LSFMo sample at different temperatures.

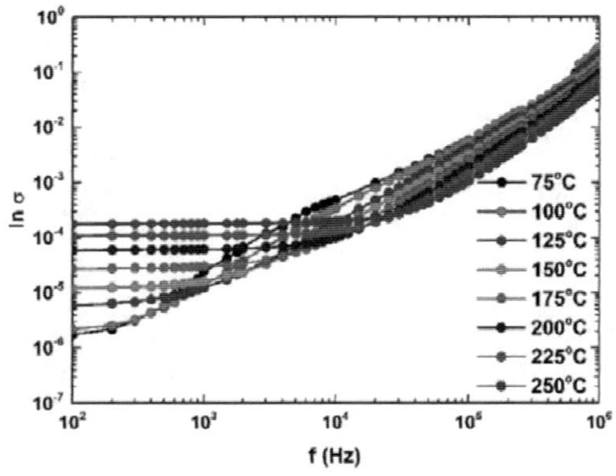

Figure 5. Complex conductivity of LSFMo sample at different temperatures.

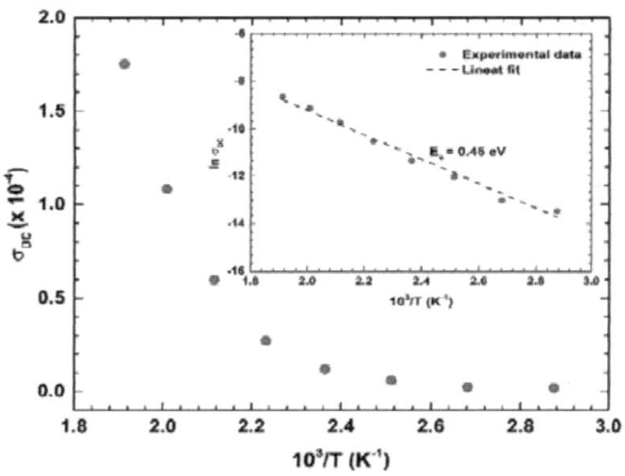

Figure 6. Temperature dependence of the DC conductivity for LSFMo sample. The inset reveals the evolution of ln σ_{DC} vs 10^3/T (K^{-1}) to evaluate the activation energy.

Figure 5 shows the frequency-dependence complex conductivity of LSFMo sample. It is noticed that the complex conductivity is consist of two different trends. The first one is that the independent frequency behaviour conductivity value at low-frequency region which is namely dc conductivity whereas the second one is that the frequency-dependence region occurring in the higher frequency region which increases with increasing temperature called ac conductivity.

Now, we turn to evaluate DC conductivity. Figure 6 shows the evolution of DC conductivity as a function of temperature which is resulted from impedance measurement at low frequencies for LSFMo sample. The conductivity increases with increasing temperature. It is obvious that the conductivity is thermally activated which is correspond to the Arrhenius Law [5, 9]. The inset of figure 6 illustrates the evolution of ln σ_{DC} vs 10^3/T (K^{-1}) and the activation energy is found to be 0.45 eV. As can be seen that only single slope is formed in this temperature range indicating only one mechanism involved in the conduction mechanism which is consistent with the Nyquist plot and bode-phase plot analysis.

4. Conclusion

The $La_{0.9}Sr_{0.1}Fe_{0.7}Mo_{0.3}O_3$ (LSFMo) sample was prepared using sol-gel and sintering methods. The SEM image reveals the non-uniform grains with an average grain size of 347 nm. The impedance analysis identifies only one semicircular arc indicating the grain behaviour as single and dominant contribution in electrical mechanism. The elucidation of variational temperature of the imaginary part of impedance peak and DC conductivity has exhibited that the observed electrical mechanism is thermally activated.

Acknowledgments

The authors would like thanks to Hibah Penelitian Dasar Unggulan Perguruan Tinggi (PDUPT) 2020 by Ministry of Research and Technology (KEMRISTEK/BRIN) Republic of Indonesia with contract number NKB-207/UN2.RST/HKP.05.00/2020.

References

[1] Taylor F H, Buckeridge J and Catlow R A 2017 *Chem. Mater.* **29** 8147-57
[2] Benali A, Azizi, S, Bejar M, Dhahri E and Graça M F P 2014 *Ceram. Int.* **40** 14367-73
[3] Rafsanjani R A, Triyono D and Laysandra H 2018 *AIP Conf. Proc.* **2023** 020024
[4] Utami R W, Rafsanjani R A and Triyono D 2019 *J. Phys. Conf. Ser.* **1153** 012072
[5] Omri A, Bejar M, Dhahri E, Es-Souni M, Valente M A, Graca M P F and Costa L C 2012 *J. Alloys Compd.* **536** 173-8
[6] Sen S, Choudhary R N P and Pramanik P 2007 *Physica B* **387** 56-62
[7] Costa M M, Junior G F M P and Sombra A S B 2010 *Mater. Chem. Phys.* **123** 35-9
[8] Lily L, Kumari K, Prasad K and Choudhary R N P 2008 *J. Alloys Compd.* **453** 325-31
[9] Kharrat A B J, Moutia N, Khirouni K and Boujelben W 2018 *Mater. Res. Bull.* **105** 75-83

Structural and vibrational analysis of $LaFe_{1-x}Mn_xO_3$ (x = 0.05, 0.10, 0.15 and 0.20) perovskite materials

M N Abdillah, D Triyono, A W Anugrah and R A Rafsanjani

Department of Physics, Faculty of Mathematics and Natural Sciences (FMIPA), Universitas Indonesia, Depok 16424, Indonesia

Corresponding author's email: djoko.triyono@sci.ui.ac.id

Abstract. Lanthanum orthoferrite ($LaFeO_3$) and lanthanum manganite ($LaMnO_3$) are the important family of perovskite compound which intensively studied and explored due to its applications as functional materials, such as a fuel cell electrode, photocatalytic and gas sensor. The performance of functional materials is strongly depended on their structural properties. Here, we investigated the structure and vibrational of $LaFe_{1-x}Mn_xO_3$ (x = 0.05, 0.10, 0.15 and 0.20) perovskite materials prepared by sol-gel method. The structural and optical properties of the materials were characterized by X-Ray diffraction (XRD) spectroscopy, Fourier Transform Infra-red (FTIR) spectroscopy, Raman spectroscopy, and Ultraviolet-Visible (UV-Vis) spectroscopy. The analysis of the XRD pattern confirmed the single-phase orthorhombic structure (space group: Pnma) following by the decreasing of lattice parameters, crystallite size, and Fe/Mn – O bond distance with the increasing of Mn-content. The presence of the Fe/Mn–O–Fe/Mn bending and Fe/Mn–O stretching modes has been confirmed by FTIR and Raman investigation. The optical band gap energy was measured by UV-Vis DRS spectroscopy, which obtained in the range of 1.99–2.14 eV. The band gap energy decreases with the increase of Mn-content.

Keywords: Perovskite, sol-gel method; La(Fe, Mn)O_3; structure; band gap energy

1. Introduction

Lanthanum orthoferrite and manganite (chemical formula: $LaFeO_3$ and $LaMnO_3$, respectively) are one of the widely studied and explored materials due to its structural stability and multiferroics behavior [1]. The chemical substitution on La- or Fe-/Mn- sites has been reported to change its physical properties, especially the structural and optical properties which can be useful for improving its performance in applications [2-5]. Kamarkar et al. reported Fe-substitution Mn-sites on $LaMn_{1-x}Fe_xO_3$ ($0 \leq x \leq 1$) revealed the structure transition from rhombohedral to orthorhombic [2]. Y. Janbutrach et al. reported that the band gap energy of $LaFeO_3$ is about 2.05 eV; Al- substitution in $LaFeO_3$ decreases the lattice parameters and crystallite size together with the increase of magnetization and band gap energy [3]. Atma Rai et al. reported that double substitution of Ca and Mn on $La_{1-x}Ca_xFe_{1-x}Mn_xO_3$ increases the metal-oxygen band position at a higher wavenumber side with decreasing lattice parameters and bond distance [4]. The substitution of Li and Mn in $La_{1-x}Li_xFe_{1-x}Mn_xO_3$ has also been reported by Thakur et al. that band gap energy decreases as compared to the $LaFeO_3$ parent in the range value of 2.11–1.61 eV that is useful in various electronics like photo-catalyst [5]. The substitution by Li and Mn improved dielectric, optical, and magnetic properties with maximum enhancement at 20 % concentration.

Following the previous reported work [1-5], a systematic study of structural and optical properties of $LaFe_{1-x}Mn_xO_3$ (x = 0.05, 0.10, 0.15 and 0.20) perovskite materials has been carried out in this work. The $LaFe_{1-x}Mn_xO_3$ perovskite materials were prepared by the sol-gel method. The structural and optical properties investigations were characterized by using X-Ray Diffraction (XRD), Fourier-transform infrared spectroscopy (FTIR), DXR2 Raman microscope and UV-VIS spectroscopy.

2. Materials and method

The $LaFe_{1-x}Mn_xO_3$ (x = 0.05, 0.10, 0.15 and 0.20) perovskite materials were synthesized by the sol-gel method by using the Lanthanum oxide, manganese chloride tetrahydrate, iron nitrate monohydrate, and citric acid monohydrate as a precursor. The precursors were weighed and dissolved with distilled water. Then, the stirring and heating processes were accomplished in the solution until the gel-phase was composed. Then, the gel-obtained was heated with a heating treatment to form the nanoparticle in powder-form. The detailed about the synthesizing process have been explained in our previous work [6].

The lattice structure was investigated by using X-Ray Diffraction (XRD) type PANalytical-X'Pert PRO with source radiation Cu Kα (λ = 1,5418 Å) in the 2θ range of 20–80°. The lattice vibration was examined by Fourier-transform infrared spectroscopy (FTIR) PerkinElmer type Spectrum Two. Raman scattering studied using Thermo SCIENTIFIC DXR2 Raman microscope with an infrared wavelength rate of 538 nm. The band gap energy was measured by UV-VIS (Agilent Cary 100/300) to estimate band gap energy.

3. Results and discussion

3.1. Structural analysis

Figure 1 presents the XRD pattern of $LaFe_{1-x}Mn_xO_3$ nanoparticles plotted from JCPDS Number COD 96-152-6451. The spherical symbol shows the intensity of observations (Iobs) obtained from the results of the XRD characterization test. The red line shows the intensity of fitting results (Ical), and the green line is the difference between the intensity of observation and the intensity of fitting results (Iobs - Ical). The goodness of fit ($\chi 2$) in table 1 from refinement results is closed to 1 which indicates the good agreement between the observed and fitting results [5]. From the Rietveld refinement, the peak shows the hkl as follows: (101), (121), (022), (202), (123), (242) and (161). All peaks have been recognized and suggested the conformation of single-phase $LaFe_{1-x}Mn_xO_3$ (x = 0.05, 0.10, 0.15 and 0.20). There is no secondary phase indicating that Mn^{2+} has successfully occupied Fe^{3+} in B-sites.

The structural data obtained from XRD have been refined by using the *High Score* and VESTA software. The whole pattern refinement results conclude the orthorhombic crystal structure with the *Pnma* space group. The changes in structural parameters caused by the variation of Mn-substitution have been tabulated in table 1. Though the results suggest the changes in the lattice structure, such as a decrease in lattice parameters and/or volume unit cells, bond distance, and crystallite size. This might be associated with the substitution of Mn^{2+} (0.84 Å) for Fe^{3+} (0.64 Å) to make a change in the new valence states of Mn and/or Fe ions because of electric charge compensation, which makes an impact on the lattice volume decrement. This phenomenon can be also caused by the compensation of charge imbalance [4]. The tolerance factor decreases with Mn-substitution indicating the lattice becomes more distorted due to the Mn-substitution. The lattice parameters decreased with the increasing of Mn content. From the table 1, the calculated tolerance factor, *t* value decreased, indicating that the lattice becomes more distorted with the increasing of Mn content. The Fe/Mn–O bond distance is slightly decreased, indicating an escalating distortion of the Fe/MnO_6 octahedra and reduced symmetry to reach stability.

3.2. Fourier transform infrared (FTIR) analysis
Fourier transform infrared (FTIR) spectrums of $LaFe_{1-x}Mn_xO_3$ (x = 0.05, 0.10, 0.15 and 0.20) have been characterized in region 400–4000 cm^{-1}. The FTIR spectrums of the samples reveal the presence of metal oxide bonds which shown in figure 2. The mode position depends on the composition and structure crystal of samples. This spectrum is like the previous report by Yutana et al. The strong modes are clearly observed around 400 cm^{-1} and 500 cm^{-1} [3]. The modes around 400 cm^{-1} and 500 cm^{-1} are assigned to Fe/Mn−O−Fe/Mn bending and Fe/Mn−O stretching vibrations. Each vibration correlates with the internal motion of bond angle and distance in Fe/MnO_6 octahedra, respectively [3, 4]. The metal-oxide stretching mode position increases at a higher wavenumber side with decreasing lattice parameters and Fe / Mn-O bond distance from 2.013 Å to 2.008 Å as shown in table 1. The decreasing metal-oxide bending mode position is maybe due to the increasing Fe/Mn-O–Fe/Mn bond angle from 154.0 Å to 153.9 Å. A similar phenomenon is shown by Rai A [4]. The O−H bending mode was identified in the range of 1485–1635 cm^{-1} suggesting the absorbing and trapping of water from the atmosphere. It is also supported by the modes around 3000 cm^{-1} attributing to the stretching modes of water molecules [3, 7-8].

Some specific modes observed are tabulated in table 2. The changes in metal-oxide stretching and bending vibrations modes might be related to the change in the reduced mass of chemical environment due to the Mn-substitution [5]. Nonetheless, the slight changes in mode position indicate the insignificant changes in bond angle and distance. The presence of Fe/Mn-O-Fe/Mn bending, and Fe/Mn-O stretching confirms the Fe/Mn-O-Fe/Mn bond angle and Fe/Mn-O bond distance in XRD refinement results (table 1).

3.3. Raman spectroscopy analysis
Raman spectroscopy is an effective method to identify and explain the structural phase transition, oxygen motion of perovskite-type samples, and the presence of order-disorder effects in the lattice [9].

Figure 3 describes the Raman spectra of $LaFe_{1-x}Mn_xO_3$ (x = 0.05, 0.10, 0.15 and 0.20). The identical Raman spectrum confirms the unchanging crystal structure for all samples which is consistent with the XRD refinement results (table 1). The Raman spectra of $LaFe_{1-x}Mn_xO_3$ are similar to the previously reported for the orthorhombic structure of perovskite compounds [10, 11]. Some specific Raman modes are collected in table 3. In accordance with previous reports [10, 11], the modes ω1, ω2, and ω3 are assigned to La vibrations, tilting of the oxygen octahedral (Fe/MnO_6), and asymmetric stretching corresponding to Jahn-Teller distortion with Ag symmetry, respectively. The mode ω4 is related to the symmetric stretching of Fe/MnO_6 octahedra with B1g symmetry. Another phonon mode

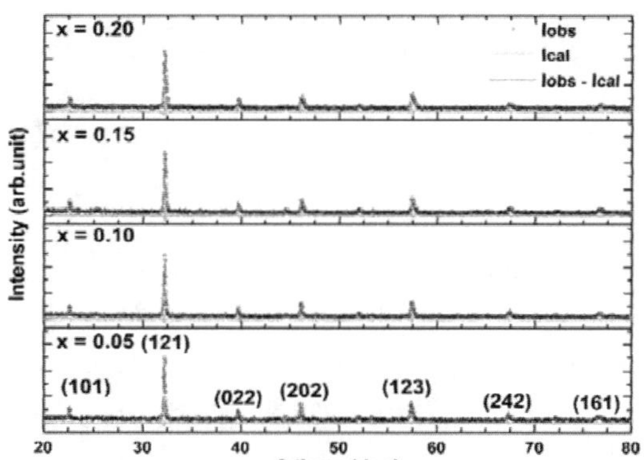

Figure 1. XRD pattern of $LaFe_{1-x}Mn_xO_3$ (x = 0.05, 0.10, 0.15, and 0.20) perovskite materials.

Table 1. Structural parameters from refinement results and calculated the tolerance factor of $LaFe_{1-x}Mn_xO_3$ (x = 0.05, 0.10, 0.15 and 0.20) perovskite materials.

$LaFe_{1-x}Mn_xO_3$	x = 0.05	x = 0.10	x = 0.15	x = 0.20
Lattice parameters				
a (Å)	5.562	5.559	5.554	5.552
b (Å)	7.850	7.845	7.830	7.826
c (Å)	5.554	5.554	5.557	5.554
Volume (Å ^{3})	242.5320	242.2878	241.6958	241.3832
Bond Angles				
Fe/Mn – O1 – Fe/Mn (Å)	157.0748	157.0782	157.0901	157.0910
Fe/Mn – O2 – Fe/Mn (Å)	154.0550	154.0370	153.9800	153.9800
Bond Distance				
Fe/Mn – O1 x 2 (Å)	1.92372	1.92353	1.92334	1.92241
Fe/Mn – O1 x 2(Å)	2.08648	2.08590	2.08470	2.08380
Fe/Mn – O2 x 2(Å)	2.01392	2.01274	2.00915	2.00805
χ^2	1.0689	1.11067	1.25113	1.24851
Crystallite Size (Å)	795.6	765.7	670.1	587.8
Crystallite Strain (%)	4.1 x 10^{-2}	4.3 x 10^{-2}	4.9 x 10^{-2}	2.2 x 10^{-2}
Tolerance factor <t>	0.917654	0.913268	0.908924	0.904621

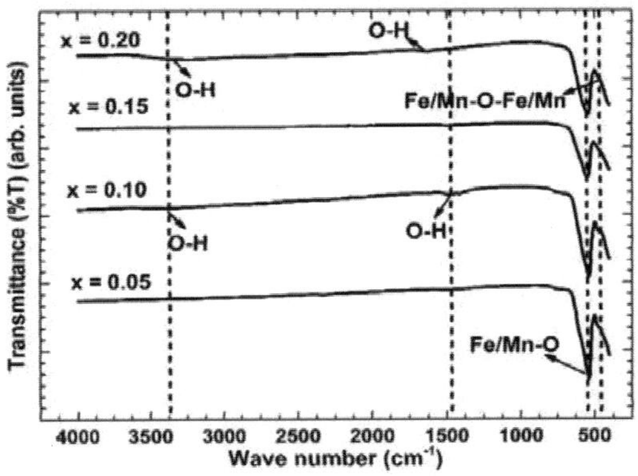

Figure 2. The FTIR patterns of $LaFe_{1-x}Mn_xO_3$ (x = 0.05, 0.10, 0.15 and 0.20) perovskite materials.

Table 2. FTIR modes assignment measured from transmittance spectrum for $LaFe_{1-x}Mn_xO_3$ (x = 0.05, 0.10, 0.15 and 0.20) perovskite materials.

$LaFe_{1-x}Mn_xO_3$	FT-IR band positions	
	Fe/Mn−O−Fe /Mn Bending	Fe/Mn−O Stretching
x = 0.05	477	536
x = 0.10	470	537
x = 0.15	463	547
x = 0.20	470	549

observed for ω5 and ω6 are associated with one-phonon and two-phonon scattering, respectively [12]. The Raman modes shift toward a higher wavenumber with the increasing of Mn-content. This might be affected by the change of crystallite size, lattice strain, and defects in lattice [11].

3.4. Uv-vis spectroscopy analysis

The Kubelka Munk function is used to calculate the absorption coefficient from the reflectance data. The formula is [13]:

$$F(R) = \alpha/s = (1 - R)^2/2R$$

where $F(R)$ is Kubelka-Munk function, α is the absorption coefficient, s is the scattering coefficient, and R is the diffused reflectance. Figure 4a shows the plot of $(F(R)E)^2$ vs. hν for $LaFe_{1-x}Mn_xO_3$ for x = 0.05. The band gap energy is found in the range of 1.99–2.14 eV. This result is compatible with the previously reported for the $LaFeO_3$ compound [3]. Figure 4b describes the band gap energy as a function of Mn-content. The band gap energy decreases with the increasing substitution of Mn-substitution.

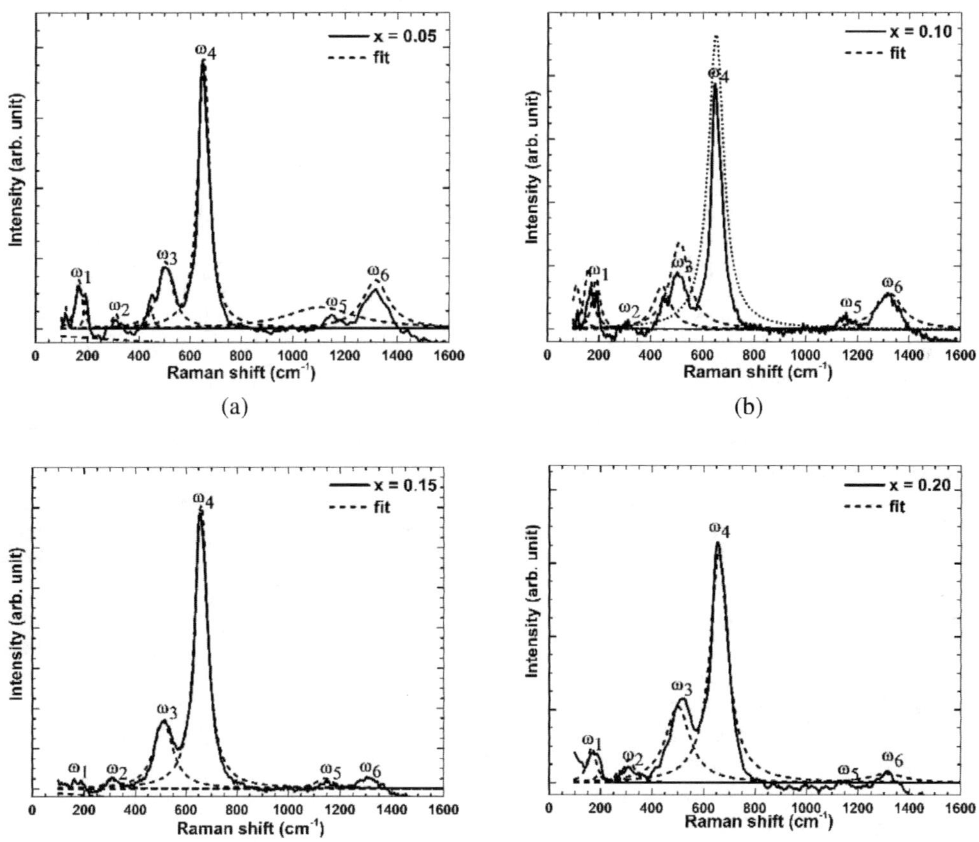

Figure 3. The Raman spectra of $LaFe_{1-x}Mn_xO_3$ (x = 0.05, 0.10, 0.15 and 0.20) perovskite materials. Raman spectra were fitted using the standard Lorentzian profile.

Table 3. Raman shift (cm^{-1}) and mode of $LaFe_{1-x}Mn_xO_3$ (x = 0.05, 0.10, 0.15, 0.20) perovskite materials

	$x = 0.05$	$x = 0.10$	$x = 0.15$	$x = 0.20$	Mode
ω_1	167	170	161	176	A_g
ω_2	313	316	310	313	A_g
ω_3	501	502	515	517	A_g
ω_4	647	648	649	649	B_{1g}
ω_5	1154	1148	1150	1148	Multiphonon process
ω_6	1319	1316	1309	1312	Multiphonon process

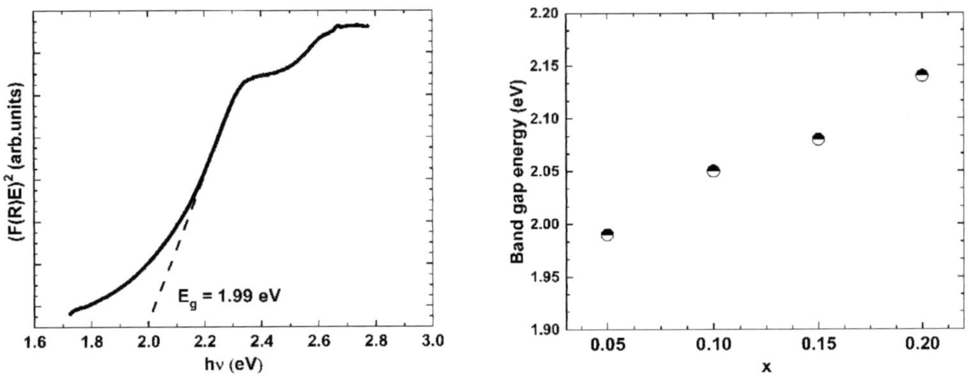

(a) (b)

Figure 4. (a) The plot of $(F(R)E)^2$ vs. hν for $LaFe_{1-x}Mn_xO_3$ (x = 0.05) and (b) band gap energy as a function of Mn-content.

This might be due to the decrease in lattice parameters and/or bond distance as confirmed in table 1 [13]. The decrease of bond distance makes the atomic bonding more bound which causes the increasing band gap in the sample. It can be explained by the Burstein-Moss effect [14]. The Mn-substitution on lattice leads to the exceeding of free carrier concentration in the edge density of the conduction band. Accordingly, the Fermi level moves towards the conduction band, i.e., the increase of band gap energy [14].

The effect of Mn-substitution in Fe-site (has greater ionic radii than Fe-ions) induces the decrease in the lattice parameters and bond distance Fe/Mn−O as presented in table 1. The decrease in lattice parameters and bond distance is confirmed by the shifting in vibrational modes position towards a higher wavenumber as shown in Raman and FTIR. The decrease in the bond distance with Mn-substitution causes the atomic bond to become stronger. It is also confirmed by the band gap analysis that increases with the increasing of Mn substitution.

4. Conclusion

The single-phase $LaFe_{1-x}Mn_xO_3$ (x = 0.05, 0.10, 0.15 and 0.20) nanoparticle studied have been successfully synthesized by sol-gel method. The X-Ray diffraction (XRD) pattern revealed the orthorhombic structure with the *Pnma* space group which is also confirmed by the identical FTIR and Raman scattering spectrums for all samples. The decrease in lattice parameters and bond distance with Mn-content could be interpreted as the consequence of the different ionic radii and valence state occupying and mixing in the lattice. The presence of metal-oxygen bonds, i.e., Fe/Mn−O−Fe/Mn bending and Fe/Mn−O stretching confirmed the formation of the oxide perovskite structure. The band gap energy decreases with increasing of Mn-content.

Acknowledgments
This work was financially supported by Ministry of Research and Technology (KEMRISTEK/BRIN) Republic of Indonesia under research grant Hibah Penelitian Tesis Magister (PTM) 2020 with contract number NKB490/UN2.RST/HKP.05.00/2020.

References

[1] Acharya S, Mondal J, Ghosh S, Roy S K and Chakrabarti P K 2010 *Mater. Lett.* **64** 415-8
[2] Karmakar A, Majumdar S, Singh A K and Giri S. 2009 *J. Phys. D: Appl. Phys.* **42** 092004
[3] Yutana J, Sitchai H, and Ekaphan S 2014 *Nano Scale Res. Lett.* **9** 498
[4] Rai A and Thakur A K 2017 *Mater. Sci. Eng. B.* **224** 139-49
[5] Rai A and Thakur A K 2017 *Ceram. Int.* **43** 13828-38
[6] Anugrah A W, Triyono D and Laysandra H 2019 *IOP Conf Ser Mater Sci Eng* **496** 012033
[7] Haron W, Thaweechai T, Wattanathana W, Laobuthee A, Manaspiya H, veratinisagul C and Koonsaeng N 2013 *Energy procedia.* **34** 791
[8] Sfirloaga P, Poienar M, Malaescu I, Lungu A and Vlazan P 2018 *J. Rare Earth.* **36** 499-504
[9] Mahapatra A S, Mitra A, Mallick A, Shaw A, Greneche J M and Chakrabarti P K 2018 *J. Alloy Comp.* **743** 274-82
[10] Mahapatra A S, Mitra A, Mallick A, Ghosh M and Chakrabarti P K 2016 *Mater. Lett.* **169** 160-63
[11] Mitra A, Mahapatra A S, Mallick A, Shaw A, Ghosh M and Chakrabarti P K 2017 *J. Alloy Comp.* **726** 1195-204
[12] Lee W Y, Yun H J and Yoon J W 2014 *J. Alloy Comp.* **583** 320-24
[13] Ahmad N and Khan S 2017 *J. Alloy Comp.* **720** 502-9
[14] Rajakarthikeyan R K and Muthukumaran S 2017 *Opt. Mater.* **69** 382-91

Synthesis of biodegradable plastic from corn starch and corn husk filler with addition of glycerol as plasticizer and variation of chitosan composition

R Azsarinka[1], D Saleh[1] and E Djonaedi[2]

[1]Department of Physics, Faculty of Mathematics and Natural Sciences (FMIPA), Universitas Indonesia, Depok 16424, Indonesia
[2]Department of Printing Engineering, Faculty of Printing Engineering and Publishing (FTGP), Politeknik Negeri Jakarta, Depok 16424, Indonesia

Corresponding author's email: edsna@sci.ui.ac.id

Abstract. Cellulose, contained in corn husk, is one of natural polymer that can be used as filler for bioplastics. Corn starch flour (maize) as matrix and corn husk powder in different grain size as filler is then synthesized with chitosan and glycerol. The resulting bioplastics is a thin film with muddy-brown color. Alternating the corn husk powder grain size from 150 mesh to 200 mesh modifies the physical characteristics, especially mechanical properties. The most optimal bioplastics were obtained with 200 mesh grain size of corn husk powder and 0.04 % chitosan composition with tensile strength of 286.31 N/cm^2, elongation of 10.19 %, Young modulus of 28.11 N/cm^2 and tear resistance of 705.61 mN. There are changes in molecular structure of some functional groups as compared to the elements. Bioplastics were degraded 30–35 % for 21 days in soil, started moldy after 10 days in open air and endured for one hour in temperature 100 °C.

Keywords: Cellulose, natural polymer, bioplastics, glycerol, chitosan

1. Introduction

Not every industrial and household plastics waste can be recycled. An alternative to alleviate the waste is substitution to biodegradable plastics (bioplastics). According to ASTM sub-committee D20-96, bioplastics is a degradable plastic in which the degradation results from the action of naturally occurring microorganisms such as bacteria, fungi and algae [1]. From National Standard of Indonesia (SNI), materials may be classified as bioplastics if it has tensile strength of 24.7–302 MPa and elongation of 21–220 % [2]. In this research, we investigate the production of such bioplastics using cellulose extracted from corn husk, an abundant byproduct of corn processing, which containing 36.81 % cellulose [3]. High cellulose content in plants acted as filler for bioplastics, which can strengthen the plastics by increasing the tensile strength.

Corn starch flour (maize) as matrix for bioplastics has another function as additive to strengthen the chemical bond in cellulose [4]. Plasticizer is organic solvent that decreases intermolecular force accumulation in long chain molecule [5]. Glycerol, as one kind of plasticizer, is able to attract and hold the moisture from the air [6] so it can improve the mechanical properties of the resulting bioplastics, especially elongation. Chitosan is non-toxic element, biodegradable and antimicrobial, also plays a role in improving the mechanical properties of materials, such as stiffness and strength [7-9]. Acetic acid acted as catalyst during the synthesis process and solvent to chitosan while aquadest (purified water) is used as the solvent due to its neutral characteristic [2, 10].

2. Experimental method

The corn husk is milled and screened to obtain powder with grain size of 150 mesh and 200 mesh. Bioplastics synthesis is performed by varying the grain size and chitosan composition. Corn husk powder and corn starch flour is measured in 1:1 ratio then chitosan is added at variation of 0.02, 0.04, 0.06, 0.08, and 0.1 %wt. According to the previous research [11], the mixture is synthesized with 2.5 mL of 25 % acetic acid, 1.75 mL of glycerol and 70 mL of aquadest. The solution is then heated and stirred until it is changed color and thickened. Then, it is printed on a Teflon paper with a glass plate as supporting base and dried for two hours at 80 °C [11].

Characterization is done to determine the degradation, heat resistance [12], morphological and mechanical properties of bioplastics. Morphological properties can be known by using microscope and FTIR [13]. For mechanical properties, vertical tensometer is used to determine the tensile strength [14] while Elmendorf methods is used to determine the tear resistance [15]. The degradation test is done to determine the degradation rate at certain time. It has two different methods: the first method is performed to know the fungi resistance of bioplastics after being exposed to open air [16] and the second method is performed to know the degradation of bioplastics in the soil [17].

3. Results and discussion

The resulting bioplastics (figure 1) is a thin film with muddy-brown color which texture is smoother by the increased composition of chitosan. The difference of the two types of bioplastics is due to the ratio of corn husk powder and corn starch flour are equal, other than corn husk powder is heated as part of synthesis process so that the powder is dissolved.

Characterization results obtained from microscope (figure 2) showed that the texture of bioplastics is affected by the grain size of corn husk powder. Fifty times magnification showed that bioplastics surface with grain size of 150 mesh have a lot more cracks than grain size of 200 mesh, indicating that corn husk powder and corn starch flour did not blend perfectly [17]. Sample of bioplastics which tested at room temperature exposure for 21 days showed that it has grayish spots on the surfaces at 250x magnification, possibly caused by fungi [5, 17].

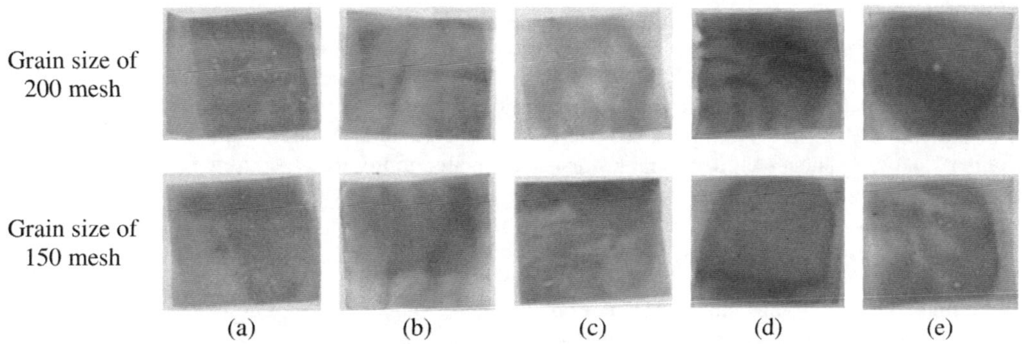

Figure 1. Sample of corn husk filler bioplastics with chitosan composition of (a) 0.02 %wt, (b) 0.04 %wt, (c) 0.06 %wt, (d) 0.08 %wt, and (e) 0.1 %wt.

IOP Conf. Series: Materials Science and Engineering **902** (2020) 012037 doi:10.1088/1757-899X/902/1/012037

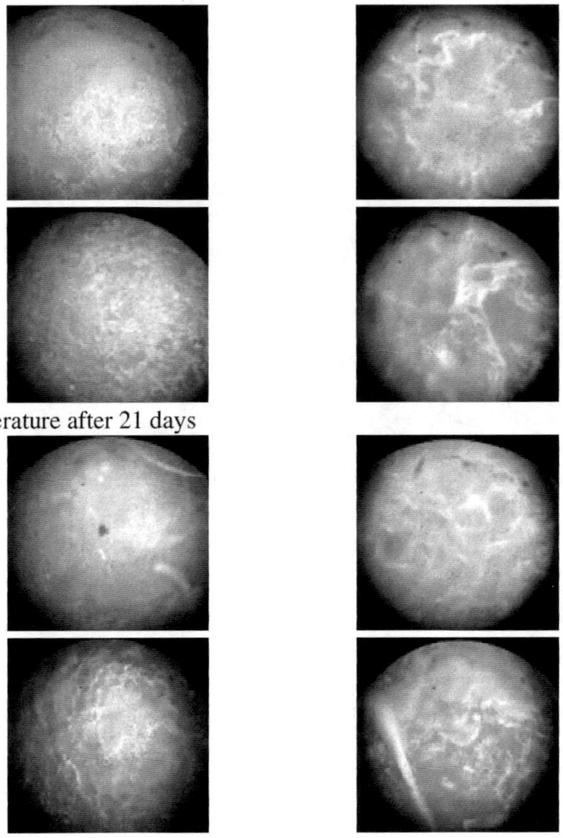

Figure 2. Surface condition of bioplastics with chitosan composition of 0.02 %wt at magnification of (a) 50x and (b) 250x.

FTIR result of corn husk powder (figure 3) and corn starch flour (figure 4) showed that it has a similar spectrum and consists of ether functional groups (C–O), alcohol functional groups (O–H) and alkene functional groups (C=C) as specific functional group.

FTIR results of corn husk bioplastics with 200 mesh grain size (figure 5) and 150 mesh grain size (figure 6) showed changes in functional groups as compared to infrared spectrum of corn husk powder and corn starch flour. Changes in functional groups are caused by gelatinization process that delinked monomers in corn husk and corn starch during synthesis so that the atoms of glycerol and chitosan placed themselves between the starch–corn husk bonds [18-20].

FTIR results below (figure 7) showed shifts of wavenumber from bioplastics elements, loss of alkene (C=C) functional groups and emergence of aromatic (C–N) and amine (C–N) functional groups. Strong hydroxyl, carbonyl and ether bonds indicated that bioplastics have the capability to absorb water so the bioplastics can be degraded by microorganisms. This is because microorganism has an enzyme that can broke down the bonds [18, 19]. The difference in infrared spectrum of corn husk filler bioplastics with grain size 200 mesh and grain size 150 mesh is likely due to the type of chemical bonds between each molecule is changed.

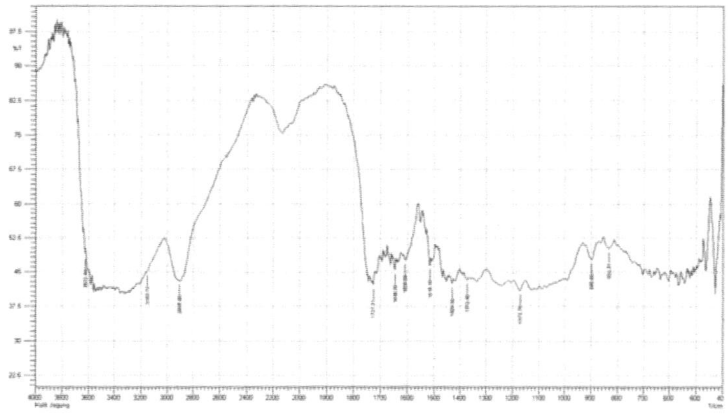

Figure 3. Infrared spectrum of corn husk powder.

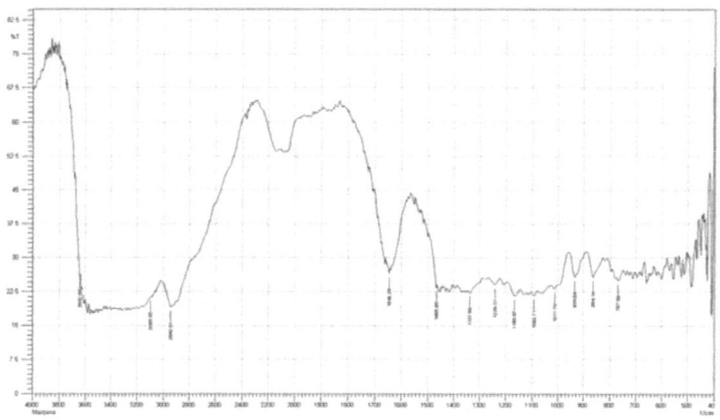

Figure 4. Infrared spectrum of corn starch flour.

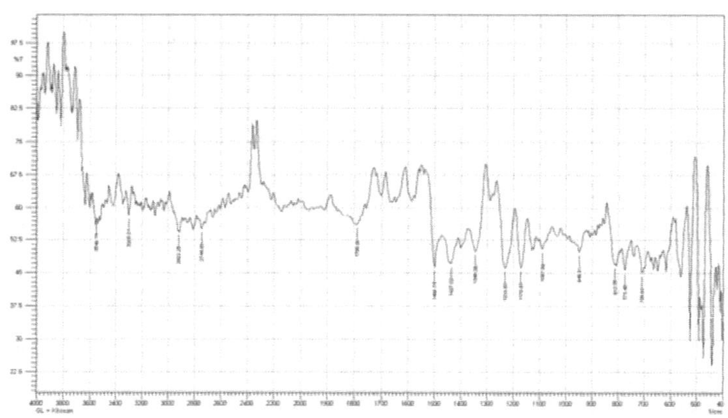

Figure 5. Infrared spectrum of corn husk filler bioplastics with grain size 200 mesh and chitosan composition of 0.1 %wt.

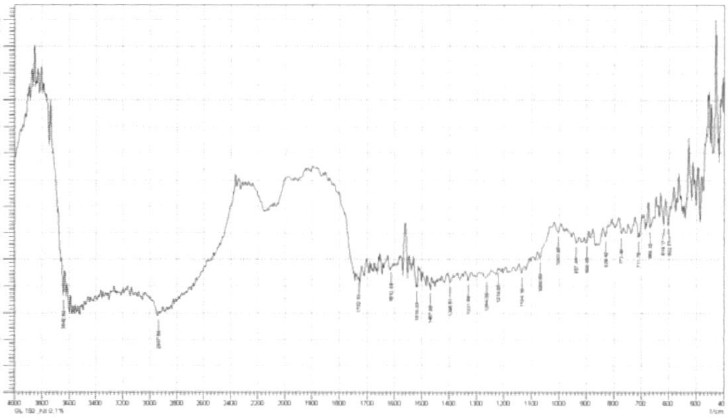

Figure 6. Infrared spectrum of corn husk filler bioplastics with grain size 150 mesh and chitosan composition of 0.1 %wt.

Tensile test results showed that corn husk filler bioplastics with grain size of 200 mesh has higher tensile strength and young modulus than grain size of 150 mesh while the highest elongation received by bioplastics with grain size of 150 mesh (figure 8). Based on the close observation, there are white dots all over the bioplastics, as a result of undissolved chitosan while synthesized. Tensile strength and elongation are hugely affected by grain size and dissolved chitosan. Particles with smaller size are easily inserted into the matrix and bonded with another element so the load distribution is spread evenly [21]. Delignification to the corn husk resulted in starch and glycerol lead to the difficulty to react with cellulose so hydrogen bonds have low quantities. The solubility of chitosan in bioplastics has a major effect to the tensile strength [18]. This is due to the linear chain structure of the chitosan polymer so it is able to arrange the polymer molecules and influence the number of hydrogen interactions, both intermolecular and intramolecular [22].

Referring to SNI for bioplastics standard [2], corn husk filler bioplastics do not meet these criteria. The highest tensile strength only reached 2.86 MPa, drifted with the value of SNI of 24.7–302 MPa. Similarly, according to SNI with range between 21-220 % could not be achieved due to the highest elongation only reached 10.33 % (figure 9).

The tear resistance results with the same thickness are shown in figure 10. Optimization of bioplastics tear resistance occurred at chitosan composition of 0.04 %wt. The graph also showed that the tear resistance of bioplastic with grain size of 200 mesh is bigger than the 150 mesh. It is showed that the torn samples have smoother jagged structure as well. The smoother jagged structure of the 200 mesh is due to the uniform distribution of each component. Another possibility is that the heating temperature corresponds to the synthesis process so that the corn starch gelatinization proceeded well and the chitosan dissolved completely.

The heat resistance test result showed that bioplastics sample with grain size of 200 mesh decreased slightly from its initial mass than grain size of 150 mesh (figure 11). The O–H bonds on the smaller grain size have much more quantity than the bond in larger grain size. Loss of moisture content in bioplastics is demonstrated by a curved and rigid bioplastics form that is easily fractured, as well as darker color along with higher temperatures. This is due to the reaction between hydrogen from hydrocarbon groups with oxygen to produce carbon and water vapor [23].

Table 1 shows that adding chitosan does not have a significant impact to the heat resistance of bioplastic samples. This is determined by the levels of glycerol in bioplastics. The results of bioplastic degradation test for open air degradation is shown by table 2. The bioplastic has fluctuate change

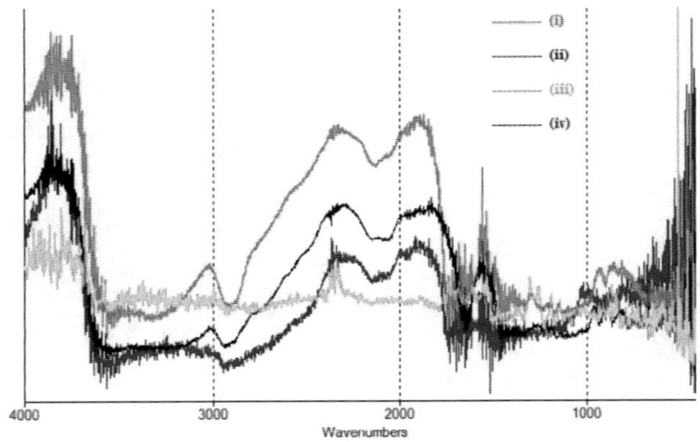

Figure 7. Infrared spectrum of (i) corn husk powder, (ii) corn starch flour, (iii) corn husk filler bioplastics with grain size 200 mesh and chitosan composition of 0.1 %wt, (iv) corn husk filler bioplastics with grain size 150 mesh and chitosan composition of 0.1 %wt.

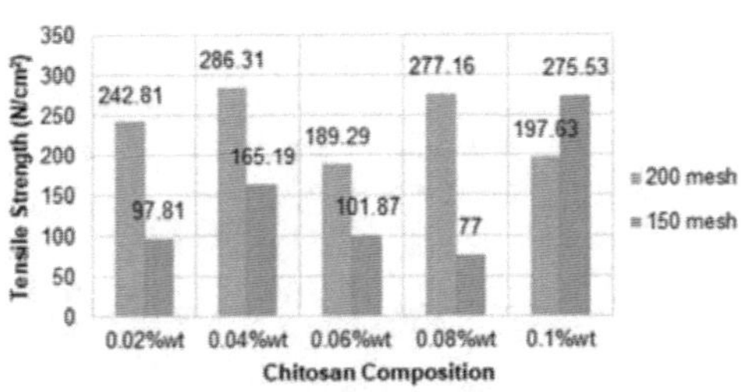

Figure 8. Tensile strength of bioplastics.

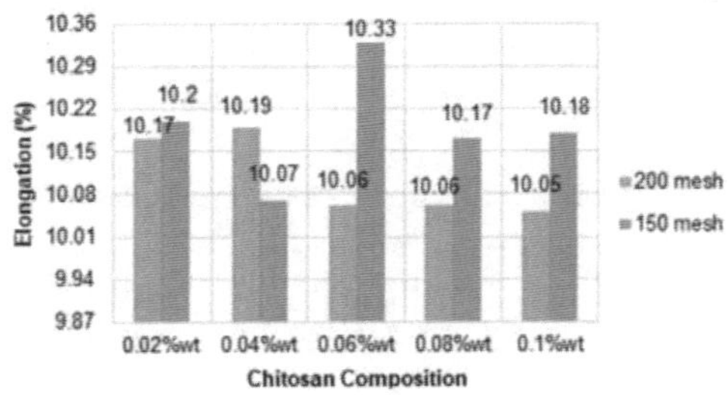

Figure 9. Elongation of bioplastics.

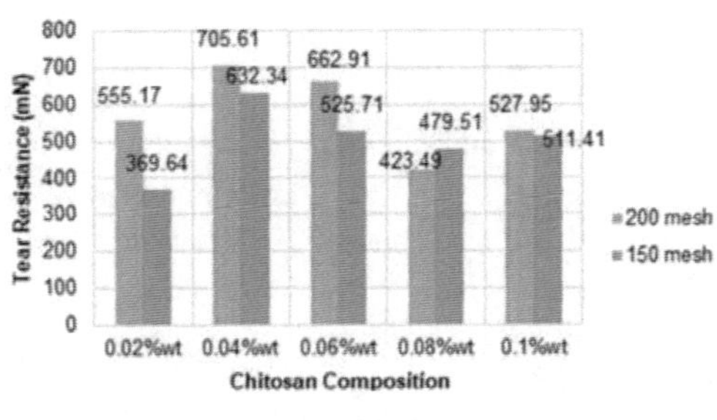

Figure 10. Tear resistance of bioplastics.

of mass in the first week, then subsequent mass reduction in the following two weeks. The increase and decrease in sample mass are due to the ability to bind or release moisture from bioplastics. Grayish spots appeared on the surface of bioplastic samples after day 10 in the 150 mesh has a lot of grayish spots on the surface than the 200 mesh. The grayish spots that appeared on the surface of bioplastics are fungi [16, 24]. Molding in bioplastics is most likely caused by cracks in the bioplastic surface. The addition of chitosan into the mixture increases the bioplastic resistance against the fungi [25].

IOP Conf. Series: Materials Science and Engineering **902** (2020) 012037 doi:10.1088/1757-899X/902/1/012037

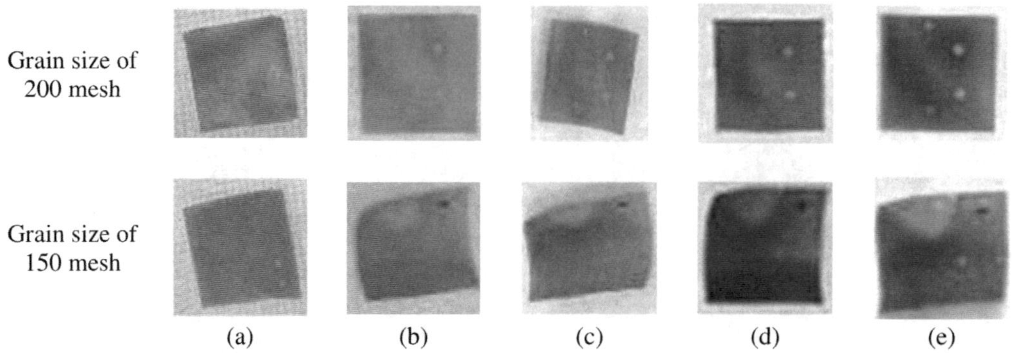

Figure 11. Surface condition of bioplastics with chitosan composition 0.06 %wt at (a) 27 °C, (b) 100 °C, (c) 120 °C, (d) 140 °C, and (e) 160 °C.

Table 1. The heat resistance of bioplastics.

Grain size	Chitosan composition	Initial mass (mg)	Final mass (mg) 100 °C	120 °C	140 °C	160 °C	Decreasing mass (%)
200 mesh	0.02 %wt	118.5	97.9	91	85.3	81.2	31.48
	0.04 %wt	88.5	73.5	67.5	62.3	59.2	33.11
	0.06 %wt	106.4	88.7	83.2	77.4	73.5	30.92
	0.08 %wt	93	76.2	70.7	65.3	61.9	33.44
	0.10 %wt	71.7	57.6	52.4	48.5	46.6	35.01
150 mesh	0.02 %wt	114	88.4	81.5	76.4	72.7	36.23
	0.04 %wt	76.7	62.1	56.3	52.5	50.3	34.42
	0,06 %wt	122.8	90.8	84.6	78.4	74.4	39.41
	0.08 %wt	112.4	91.3	84.7	77.9	73.7	34.43
	0.10 %wt	100.6	81.7	76.3	71.2	67.4	33

Table 2. Degradation of bioplastics in open air (at room temperature).

Grain size	Chitosan composition	Initial mass (mg)	Final mass (mg) 1st week	2nd weeks	3rd weeks	Degradation at 3rd weeks
200 mesh	0.02 %wt	92.7	91.2	83.3	74.4	19.74 %
	0.04 %wt	39.5	36.6	33.6	29.5	25.32 %
	0.06 %wt	61.5	66	56.5	50.4	18.05 %
	0.08 %wt	116.2	117.5	96.8	91.4	21.34 %
	0.10 %wt	71.9	68.5	56.1	43.9	38.92 %
150 mesh	0.02 %wt	92	77.6	67.5	64.7	29.67 %
	0.04 %wt	111.3	103.4	100.2	87	21.83 %
	0,06 %wt	98.2	102.2	77.3	65.6	33.2 %
	0.08 %wt	73.3	68.9	64.7	55.9	23.74 %
	0.10 %wt	100.2	104.9	86.3	72.2	27.94 %

Table 3. Degradation of bioplastics using soil burial test methods.

Grain size	Chitosan composition	Initial mass (mg)	Final mass (mg)			Degradation at 3rd weeks
			1st week	2nd weeks	3rd weeks	
200 mesh	0.02 %wt	95.5	52.8	47.5	31.3	67.23 %
	0.04 %wt	63.1	29.4	24	22	65.13 %
	0.06 %wt	112.7	63	62.2	57.2	49.25 %
	0.08 %wt	148.7	81.7	93.6	109.2	26.56 %
	0.10 %wt	98.2	55.1	62.1	49.3	49.8 %
150 mesh	0.02 %wt	123.1	67.7	70.8	44.3	64.01 %
	0.04 %wt	145.9	85.1	93.6	88.7	39.2 %
	0,06 %wt	133.6	69.8	106.6	59.7	55.31 %
	0.08 %wt	135.8	65.5	107.8	87.9	35.27 %
	0.10 %wt	113.5	64.1	69.6	40.8	64.05 %

The results of bioplastic degradation testing using soil burial test method (table 3) are not significantly influenced by chitosan composition in bioplastic degradation process while grain size generally affect the degradation [26]. In the first week, corn husk filler bioplastics with grain size of 150 mesh lost more mass than the 200 mesh grain size. However, during the second and third weeks, there is a slight cange in mass. This situation is caused by the changes in temperature and moisture conditions from the environment. Uncontrolled soil water levels are also the cause of why changes in the mass of bioplastic samples are fluctuated. If the water content in the bioplastic is excessive, then the bioplastic will be degraded faster.

From table 2 and table 3, it is seen that the degradation process by soil burial test method is faster than bioplastic which is left exposed to open air. Mesh size variations have a greater impact in degradation than chitosan composition.

4. Conclusion

Based on this research, it can be concluded that bioplastics have the form of thin and muddy brown-colored sheets for all variations of grain size and chitosan composition. In average, corn husk filler bioplastics with grain size of 200 mesh has a finer texture, larger tensile strength and tear resistance, and higher endurance of heat than grain size of 150 mesh. Adding more chitosan will make the bioplastics surface texture smoother. Analysis using FTIR indicates that there are shift of wavenumbers but no sign of new functional group compared to the base materials. Bioplastics with the best characteristic is the bioplastics with grain size of 200 mesh and chitosan composition of 0.04 %wt for the reason that it has tensile strength of 286.31 N/cm^2, elongation of 10.19 %, Young modulus of 28.11 N/cm^2, and tear resistance of 705.61 mN. However, these characteristics have not met the SNI standard for bioplastics.

Acknowledgments

The author would like to thank Djonaedi Saleh and Emmidia Djonaedi for their guidance during this research; FTGP Politeknik Negeri Jakarta and FMIPA Universitas Indonesia for supporting this project by giving permission to use the labs.

References

[1] Muller R J 2005 Regulations and Methods for Testing *Biodegradability of Polymers (Biopolymer Online)* pp. 365-74 available at https://application.wiley-vch.de/books/biopoly/pdf_v10/vol10_19.pdf

[2] Anggarini F, Latifah and Miswadi S S 2013 *Indo. J. Chem. Sci.* **2** 173-8
[3] Ningsih E R 2012 *Uji Kinerja Digester Pada Proses Pulping Kulit Jagung dengan Variabel Suhu dan Waktu Pemasakan* Diploma final project (Semarang: D3 Study Program of Chemical Engineering, Universitas Diponegoro)
[4] Ulyarti 1997 *Mempelajari Sifat-sifat Amilografi pada Amilosa, Amilopektin, dan Campurannya* Master thesis (Bogor: Faculty of Agricultural Engineering, IPB University)
[5] Marbun E S 2012 *Sintesis Bioplastik dari Pati Ubi Jalar Menggunakan Penguat Logam ZnO dan Penguat Alami Selulosa* Master thesis (Depok: Department of Chemical Engineering, Faculty of Engineering, Universitas Indonesia)
[6] Tan H W, Aziz A R A and Aroua M K 2013 *Renew. Sustain. Energy Rev.* **27** 118-27
[7] Kumar M N V R 2000 *React. Funct. Polym.* **46** 1-27
[8] Peelman N et al. 2013 *Trends Food Sci. Technol.* **32** 128-41
[9] Murni S W, Pawignyo H, Widyawati D and Sari N 2013 *Pembuatan Edible Film dari Tepung Jagung (Zea Mays L.) dan Kitosan* (Yogyakarta: Prosiding Seminar Nasional Teknik Kimia "Kejuangan")
[10] Hartatik Y D, Nuriyah L and Iswarin 2014 *Pengaruh Komposisi Kitosan terhadap Sifat Mekanik dan Biodegradable Bioplastik* available at https://media.neliti.com/media/publications/159022-ID-pengaruh-komposisi-kitosan-terhadap-sifa.pdf
[11] Djonaedi E and Yuniarti E 2016 Bioplastic From Corn Husk and Corn Starch With Plastisizer Variations *Proceeding of The 5th Annual South East Asian International Seminar (ASAIS) State Polytechnic of Jakarta 2016* pp. 1-53 avaliable at http://asais.pnj.ac.id/attachment/files/Proceeding%20TEC%20ASAIS%202016.pdf
[12] ASTM D5721 2002 *Standard Practice for Air-Oven Aging of Polyolefin Geomembranes (ASTM)* available at https://standards.globalspec.com/std/10388761/ASTM%20D5721
[13] Thermo Electron Corporation 2001 *Introduction to Fourier Transform Infrared Spectrometry* available at http://www.nicoletcz.cz/upload/kc/files/aplikacni-podpora/teorie/IntroductionToFTIR.pdf
[14] ASTM D882 2018 *Standard Test Method for Tensile Properties of Thin Plastic Sheeting (ASTM)* available at https://standards.globalspec.com/std/13050669/ASTM%20D882
[15] Standar Nasional Indonesia 2009 *Kertas – Cara Uji Ketahanan Sobek – Metode Elmendorf (Badan Standar Nasional)* available at https://theadiokecenter.files.wordpress.com/2010/06/9089_sni-0436-2009-kertas-ketahanan-sobek.pdf
[16] ISO 1997 *ISO 846:1997–Plastics: Determination of behaviour under the action of fungi and bacteria, Evaluation by visual examination or measurement of changes in mass or physical properties* avaliable at https://www.iso.org/standard/74599.html (revised) or https://id.scribd.com/document/390876490/ISO-846-pdf
[17] Ardiansyah R 2011 *Pemanfaatan Pati Umbi Garut untuk Pembuatan Plastik Biodegradable* Master thesis (Depok: Department of Chemical Engineering, Faculty of Engineering, Universitas Indonesia)
[18] Agustin Y E and Padmawijaya K S 2016 *J. Tek. Kim.* **10** 40-8
[19] Septiosari A, Latifah and Kusumastuti E 2014 *Indo. J. Chem. Sci.* **3** 158-62
[20] Sinaga R F, Ginting G M, Ginting M H S and Hasibuan R 2014 *J. Tek. Kim. USU* **3** 19-24
[21] Njoku R E, Okon A E and Ikpaki T C 2011 *Niger. J. Technol.* **30** 87-93
[22] Sanjaya I G and Puspita T 2008 *Pengaruh Penambahan Khitosan dan Plasticizer Gliserol pada Karakteristik Plastik Biodegradable dari Pati Limbah Kulit Singkong* (Surabaya: Institut Teknologi Sepuluh Nopember) available at http://digilib.its.ac.id/public/ITS-Undergraduate-17047-2305100060-paperpdf.pdf
[23] Bourtoom T 2008 *Songklanakarin J. Sci. Technol.* **30** 149-65

IOP Conf. Series: Materials Science and Engineering **902** (2020) 012037 doi:10.1088/1757-899X/902/1/012037

[24] Jecu L, Grosu E, Raut I, Ghiurea, Constantin M, Stoica A, Stroescu M and Vasillescu G 2011 *Fungal Degradation of Polymeric Materials: Morphological Aspects* available at http://www.inginerie-electrica.ro/acqu/2011/P_1_Fungal_degradation_of_polymeric_materials_Morfological_aspects.pdf
[25] Martínez-camacho A P, et al. 2010 *Carbohyd. Polym.* **82** 305-15
[26] Emadian S M, Onay T T and Demirel B 2017 *Waste Manag.* **59** 526-36

Inhibitor effect of PCMB and PMSF towards partially purified product from pineapple core (*Ananas comosus* [L.] Merr) using gel filtration column chromatography method

M F Amalia, S Hudiyono and S Setiasih

Department of Chemistry, Faculty of Mathematics and Natural Sciences (FMIPA), Universitas Indonesia, Depok, 16424, Indonesia

Corresponding author's email: sumi.hudiyono@sci.ui.ac.id

Abstract. The research aims to isolate, purify, and examine the inhibitors effect of a partially purified product from pineapple core with Palembang type pineapple (*Ananas comosus* [L.] Merr). The isolation process obtained crude enzymes with specific activity 42.60 Units/mg. Purification of crude enzymes using fractionation method with ammonium sulfate obtained bromelain fraction with the highest specific activity at 0–50 % saturation level of ammonium sulfate, which is 230.22 Units/mg with a purity level of 5.40 times from its crude extract. Further purification of 0–50 % ammonium sulfate fraction by gel filtration chromatography method using DEAE-Sephadex G-50 matrix obtained enzyme with specific activity 275.46 Units/mg with purity level 6.50 times from its crude extract. The optimum substrate concentration was at a concentration of 1.0 % (w/v). The purified bromelain has a Michaelis-Menten (Km) constant value and the maximum speed of the enzymatic reaction (Vmax) i.e. 0.52 % (w/v) and 0.51 Units/min. The proteolytic activity of bromelain is strongly inhibited by PCMB compound and activated by PMSF at 0.1 mM concentration of 67.27 %.

Keywords: Pineapple core, bromelain, proteolytic activity, specific activity, inhibitors

1. Introduction

Pineapple is a tropical fruit from the *Bromeliacieae* family that widely cultivated in the tropics and sub-tropical regions. Based on the data of pineapple production development in Indonesia, in 1980 Indonesia has produced 180.64 thousand tons of pineapple and by 2015 pineapple production has reached 1.73 million tons or an increase of 13.46 % per year [1]. The increase in pineapple production is correlated with high pineapple demand for both direct and industrial consumption. The impact of the number of pineapple requests produces a by product or pineapple waste in the form of fruit peels, stems, and bulbs that are plentiful to reach 35–40 % of the pineapple mass. The pineapple waste can be utilized optimally as a source of bromelain which has a higher economical value.

Pineapple fruit contains water, carbohydrates, fats, fiber, sugars, vitamin A, vitamin C, β carotene, a number of proteins, ash, low fiber content, flavonoids, and protease enzyme complexes of bromelain [2]. The enzyme Bromelain has a sulfhydryl group (cysteine-histidine) on its active side [3]. Bromelain is an abundant cysteine of endopeptidase in the stems, leaves and pineapple skin. The main component of bromelain is the sulfhydryl proteolytic fraction [4]. Bromelain can be inhibited by Hg^{2+}, Ag^{++}, Cu^{2+}, anti-trypsin, estatin A and B and iodoacetic acid [4].

This research used pineapple waste which is pineapple core as bromelain enzyme source. This research includes several steps, i.e. bromelain isolation from pineapple core, purification and characterization. The purification process was conducted using the method of fractionation with ammonium sulphate, subsequent with dialysis process and further purification by gel filtration column chromatography method. Bromelain characterization was observed on several parameters: proteolytic activity of enzyme, protein content, enzyme specific activity, purity level, % yield, determination of Km value, Vmax value, and inhibitor effect on the proteolytic activity of enzyme.

2. Materials and method

2.1. Materials

2.1.1. Pineapple. Pineapple Palembang (Ananas comosus [L] Merr) was purchased from East Jakarta central market. Pneapple core was separated from its skin and meat before used in this research.

2.1.2. Chemicals. The chemicals used were phosphate buffer pH 7.0, ammonium sulphate, Tris HCl buffer pH 8.0, casein, DEAE-Sephadex G-50, HCl, NaOH, TCA (trichloroacetic acid) solution, BSA standard solution (bovine serume albumin) , Lowry reagent, Follin ciocalteu 2 N, PMSF (Phenylmethylsulfonyl fluoride), PCMB (4-Chloromeruribenzoic acid), ice cubes, aluminum foil, cellophane pouch, filter paper, and aquades. Other chemical reagents utilized in this research were obtained from Merck and are in analytical grade.

2.2. Extraction of crude enzyme bromelain from pineapple

To obtain crude enzymes, 500 g of pineapple core were cut into pieces, refined in a cold state and filtered with gauze. The filtrate was centrifuged at (5000 rpm, 45 min, 4 °C) to obtain the crude enzyme. The crude enzyme was stored in the refrigerator until used.

2.3. Enzyme fractionation by ammonium sulphate

The crude enzyme was fractionated gradually using ammonium sulfate (0–20 %, 20–50 % and 50–80 %). The protein fraction was then allowed to settle overnight at 4 °C and then centrifuged using a refrigerated centrifuge (Thermo Scientific HANIL COMBIL 514 R). The addition of ammonium sulfate with saturation level S1–S2 % can be calculated by using the formula:

$$ammonium\ sulphate(g) = \frac{533\ (S2 - S1)}{100 - 0.3\ x\ S2} x \frac{V\ (mL)}{1000\ mL}$$

with S1 is the saturation level of the initial salt (%), S2 is the saturation level of the target salt (%) and 533 is the solubility of ammonium sulfate in 1 L of water at 30 °C.

2.4. Dialysis

The remaining ammonium sulfate was separated by dialysis. Replacement of the buffer was carried out periodically until the enzyme fraction is free of ammonium sulfate. Dialysis was stopped when there is no white sediment of $BaSO_4$. Proteolytic activity and protein content of enzyme fraction were tested.

2.5. Gel filtration column chromatography

The bromelain enzyme fraction was further purified using the DEAE-Sephadex G-50 matrix in a 30 cm column system with diameter of 2.3 cm. Purification begins by introducing a sample of bromelain enzyme into chromatographic column which has been filled by DEAE-Sephadex G-50 matrix then eluted using Tris-HCl 0.05 M buffer pH 8.0 and carried out by 1 mL/min flow rate. The elute was accommodated every 5mL in a vial bottle, then its proteolytic activity was tested by Kunitz method.

2.6. Determination of proteolytic activity of bromelain enzyme
Proteolytic activity of bromelain enzyme was then measured using Kunitz method (1947).

2.7. Determination of protein levels
The protein levels of each enzyme fraction can be measured using the Lowry method (1951).

2.8. Determination of specific activity of bromelain enzyme
Determination of specific activity of enzyme was obtained by comparing the number of enzyme activity (U/mL) with protein levels (mg/mL).

$$Specific\ activity\ \left({}^{U}/_{mg}\right) = \frac{Total\ of\ proteolitic\ activity\ (U)}{Total\ of\ protein\ levels\ (mg)}$$

2.9. Determination of Km and Vmax
The enzyme kinetic parameters, the Michaelis-Menten (Km) constant and the maximum velocity of the enzymatic reaction (Vmax) can be determined by transforming the Michaelis-Menten equation $\left(v = \frac{V_{maks}[S]}{K_M + [S]}\right)$ into Lineweaver- Burk $\left(\frac{1}{V_0} = \frac{K_M}{v_{maks}}\frac{1}{[S]} + \frac{1}{V_{maks}}\right)$ by mapping 1/V to 1/[S]. The enzyme reaction rate of bromelain enzyme was done with variation of casein substrate concentration 0.25; 0.50; 0.75; 1.0; 1.25 and 1.50 (w/v) [4].

2.10. Determination of inhibitor effect towards bromelain enzyme activity
Determination of inhibitor effect was done by addition of 0.1 mL inhibitor with concentration 0.1; 0.2; 0.3; 0.4 and 0.5 mM at optimum casein substrate concentration, then, the proteolitic activity of each fraction was determined and % inhibition and% activation were calculated.

$$\%\ \text{Inhibition} = \frac{\text{Proteolitic activity of enzyme fraction without inhibitor} - \text{Proteolitic activity of enzyme fraction with inhibitor}}{\text{Proteolitic activity of enzyme fraction without inhibitor}}\ \text{x}\ 100$$

$$\%\ \text{Activation} = \frac{\text{Proteolitic activity of enzyme fraction with inhibitor} - \text{Proteolitic activity of enzyme fraction without inhibitor}}{\text{Proteolitic activity of enzyme fraction without inhibitor}}\ \text{x}\ 100$$

3. Results and discussion

3.1. Extraction and purification of bromelain enzyme
The crude enzyme extract of bromelain from the pineapple core was fractionated with ammonium sulphate, dialyzed, and further purified using gel filtration column chromatography method with DEAE-Sephadex G-50 matrix. Proteolytic activity, protein levels, and specific activity of bromelain enzyme fractions can be seen in table 1. The bromelain crude enzyme extracts increased the specific activity after fractionation using ammonium sulfate salt from 42.60 U/mg to 230.22 U/mg with 5.40 times purity level from the crude enzyme extract and increased again in the process of dialysis to 238.21 U/mg with a purity level 5.60 times from the crude enzyme extract.

Purification by gel filtration column chromatography method was used to separate bromelain enzyme with other proteins based on its molecular size. Samples that have relatively small molecular sizes would be diffused from the solution into the pores of the gel, whereas the relatively large molecules could not penetrate in the pores of the gel so it will come out faster than small molecules as it passes through the gel matrix [5]. Bromelain has a molecular weight between 30,000–36,000 Da [6]. Therefore, the use of DEAE-Sephadex G-50 matrix could separate the protein with molecular weight 1,500–30,000 Da.

Table 1. Specific activity of bromelain enzyme from Palembang pineapple core.

Sample	Protein (mg/ml)	Enzyme activity (U)	Specific activity (U/mg)	Purity levels
Crude enzyme	0.28	2067.70	42.60	1.00
Ammonium sulphate fraction (20–80%)	0.05	238.71	230.22	5.40
Dialysis fraction	0.03	150.49	53.78	5.60
DEAE-Sephadex G-50 fraction	003	385.50	60.67	6.50

The elution used in the gel filtration column chromatography was 0.05 M tris-HCl buffer pH 8.0. Elute produced 60 fractions that were accommodated in a vial bottle every 5 mL. Each fraction was measured for absorbance of protein absorption at 280 nm wavelength and tested for its proteolytic activity using Kunitz method. The results of the measurements were plotted in chromatograms as shown in figure 1. The highest specific activity of the DEAE-Sephadex G-50 fraction was found in KR1 which had 275.46 U/mg with a purity level 6.50 times of crude enzyme.

3.2. Determination of Km and Vmax

The determination of Km and Vmax of PA1 fraction. It determined its enzyme activity based on the variation of casein substrate concentration 0.25 %, 0.50 %, 0.75 % 1.0 %, 1.25 % and 1.50 % (w/v) using the Kunitz method. The rate of casein hydrolysis increased to a substrate concentration of 1.0 % and did not increase significantly when added a substrate with higher concentration. This is because at high concentrations all of the enzyme's active side has been filled by the substrate (already saturated) and the catalytic activity of the enzyme has reached the maximum [5]. The graph of the relation of the enzymatic reaction rate to the substrate concentration is displayed in figure 2.

Michaelis-Menten Constants (Km) value of PA1 fraction is 0.52 % (w/v) and the value of the maximum speed of enzymatic reaction (Vmax) is 0.51 U/min.

3.3. Determination of inhibitor effect towards bromelain enzyme activity

This experiment was to determine the effect of addition of compound 4-(hydroxymercuri)benzoic acid (PCMB) and phenylmethylsulfonyl fluoride (PMSF) to the proteolytic activity of bromelain enzyme.

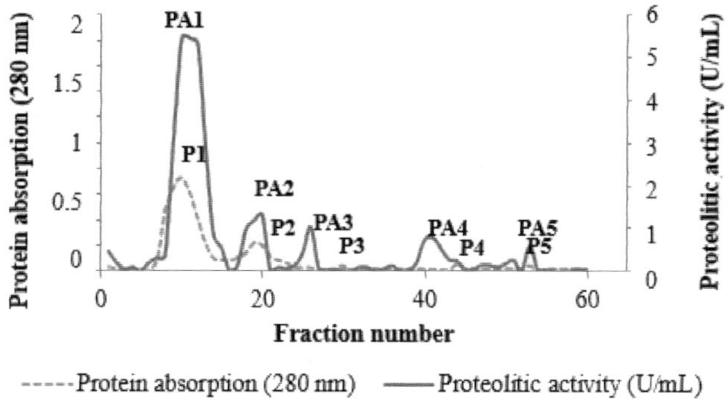

Figure 1. Chromatogram of gel filtration column chromatography.
(PA: Protein absorption, P: proteolytic activity).

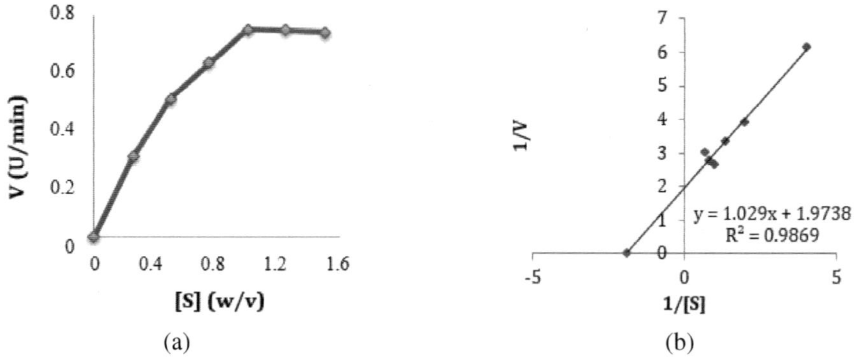

Figure 2. (a) The relationship of the enzymatic reaction rate to the substrate concentration and (b) Lineweaver-Burk plot of hydrolysis of bromelain enzyme casein. (S: substrate concentration, v: enzyme reaction rate).

Table 2. Effect of PCMB dan PMSF towards bromelain enzyme.

Concentration (mM)	% Inhibition of PCMB	% Activation of PMSF
0.0	-	-
0.1	76.45	67.27
0.2	81.74	63.42
0.3	91.35	35.06
0.4	93.27	32.90
0.5	96.40	25.45

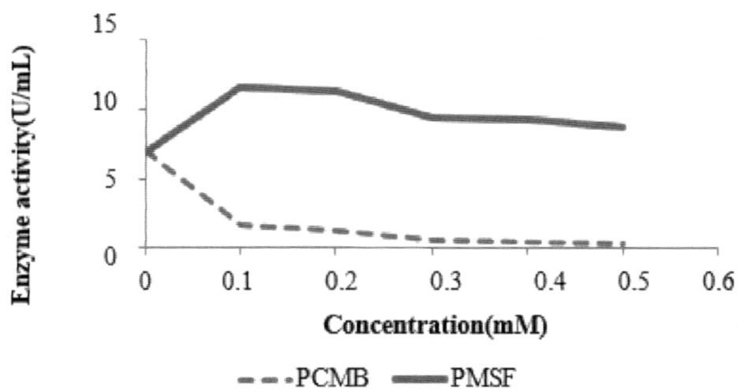

Figure 3. Inhibitor effect of PCMB and PMSF toward bromelain enzyme activity (U/mL).

In table 2 and figure 3, the influence of PCMB on bromelain activity showed a significant decrease in the activity along with the increasing concentrations of PCMB compounds with variations: 0.1; 0.2; 0.3; 0.4 and 0.05 mM. It may decrease bromelain activity due to a conformational change that causes the enzyme's active side to undergo a modification so that the enzyme interaction with the substrate is inhibited. The results of this test indicated that PCMB is a strong inhibitor of bromelain enzyme, in which the PCMB concentration of 0.1 mM has been shown to have a great inhibitory effect on enzyme the proteolytic activity, which is 76.45 %.

Based on the literature, PCMB compounds are cysteine protease inhibitors [7], such as bromelain from pineapple core. In this case, bromelain contains a sulfhydryl group which reacts with heavy metal Hg^{2+} to the PCMB compound and forms a mercaptide salt which results in the inhibition of an enzyme.

Sulfonyl fluoride has been widely studied as a serine protease inhibitor compound since the early discovery by Powers et al. [8]. The effect of PMSF on bromelain activity showed a relatively significant increase of proteolytic enzyme activity. At the concentration of PMSF 0.1 mM, it increased the proteolytic activity of bromelain up to 67.27 %. This is estimated because of the modification of conformation that facilitates the interaction of enzymes with substrate. Thus, an increase in proteolytic activity by PSMF compounds may indicate that bromelain congestive enzymes are not a serine protease group.

4. Conclusion

The crude enzyme bromelain from Palembang pineapple core (Ananas comosus [L.] Merr) has been effectively isolated. The gradually purification has a purity level of 6.50 than the crude enzyme. The optimum condition of casein substrate is at 1.0 % (w/v). The determination of Michaelis-Menten constants (Km) and maximum velocity of enzymatic reaction (Vmax) obtained values 0.52 % (w/v) and 0.51 U/min. Inhibitor effect of PCMB and PMSF compounds showed a change in proteolytic activity of enzymes along with increasing high concentration. In addition to the PCMB compound, the proteolytic activity was decreased along with the increase of concentration. It indicates that the proteolytic activity was strongly inhibited by PCMB compounds. While on the addition of PMSF compound, the proteolytic activity was activated at 0.1 mM concentration with 67.27 % activation.

References

[1] Pusat Data dan Sistem Informasi Pertanian 2016 *Outlook Komoditas Pertanian Sub Sektor Hortikultura* (Jakarta: Pusat Data dan Sistem Informasi Pertanian Kementrian Pertanian)

[2] USDA Food Composition Databases (n.d.) *Full Report (All Nutrients): 09429, Pineapple, raw, traditional varieties* available at https://ndb.nal.usda.gov/ndb/foods/show/2430?format=Full

[3] Gautam S S, Mishra S K, Dash V, Goyal A K and Rath G 2010 *Thai J. Pharm. Sci.* **34** 67-76

[4] Kurnia S D 2017 *Uji Parameter Kinetik Bromelain Hasil Pemurnian Parsial dari Bonggol Nanas (Ananas Comosus [L.] Merr) dengan Kromatografi Penukar Ion dan Uji In Vitro Aktivitas Antiplateletnya* BSc Final Project (Depok: Department of Chemistry, FMIPA, Universitas Indonesia)

[5] Lehninger A L 1982 *Dasar-dasar Biokimia Jilid 1* (Jakarta: Erlangga)

[6] Soares P A G, Vaz A F M, Correia M T S, Pessoa A and Caeneiro-da-Cuncha M 2012 *Sep. Purif. Technol.* **98** 389-95

[7] Overall C M dan Blobel C P 2007 *Nat. Rev. Mol. Cell Biol.* **8** 245-57

[8] Powers J C, Asgian J L, Ekici Ö D and James K E 2002 *Chem. Rev.* **102** 4639-750

Efficiency and potential synthesis of pyrimido [4,5-d] pyrimidine derivatives by iodine catalyst as antioxidant agent

A H Cahyana, A R Liandi and M Z R Zaky

Department of Chemistry, Faculty of Mathematics and Natural Sciences (FMIPA), Universitas Indonesia, Depok 16424, Indonesia

Corresponding author's email: herrykim@ui.ac.id

Abstract. Pyrimido [4,5-d] pyrimidine is an uracil compound that has similarities to the structure of pteridine and purines and has significant bioactivity. The aim of this research is to synthesize pyrimido [4,5-d] pyrimidine derivatives and test their activity as antioxidant. The reaction was carried out with the support of the iodine catalyst to get an effective reaction. Experimental evidence in support of the formation products was confirmed using FTIR, UV-vis Spectrophotometer, and GC-MS. The results of the analysis produced four pyrimido [4,5-d] pyrimidine derivatives, 5-Styryl-5,8-dihydro-1H,6H-pyrimido[4,5-d]pyrimidine-2,4,7-trione (compound 1), 5-Styryl-7-thioxo-5,6,7,8-tetrahydro-1H-pyrimido[4,5-d]pyrimidine-2,4-dione (compound 2), 5-Phenyl-5,8-dihydro-1H,6H-pyrimido[4,5-d]pyrimidine-2,4,7-trione (compound 3), and 5-Phenyl-7-thioxo-5,6,7,8-tetrahydro-1H-pyrimido[4,5-d]pyrimidine-2,4-dione (compound 4). Reaction optimization was conducted by varying reaction time, reaction temperature, amount of catalyst and solvent. The optimum conditions of the reaction were obtained under conditions of 10 % mol of catalyst, temperature 70 °C, for 4 h with a water solvent. Compound 1 has a better activity than 3 other compounds with an IC_{50} value of 10.11 ppm.

Keywords: pyrimido [4,5-d] pyrimidine derivatives, iodine catalyst, cinnamaldehyde, barbituric acid

1. Introduction

Uracil derivatives containing nitrogen groups in their heterocyclic make it to have an attractive bioactivity ability. One of the significant uracil annulated is pyrimido[4,5-d]pyrimidine because it is similar to purine and pteridine system [1, 2]. Several pyrimido [4,5-d] pyrimidine derivative compounds have been reported to have a diverse bioactivity. Among them are hepatoprotective, antitumor, antimicrobial, anti-allergic effects, antioxidants, antiviral, antiallergic, antifungal, antihypertensive, anticancer, vasodilators, and cardiotonics [3-5].

The formation of pyrimido compounds [4,5-d] pyrimidine is obtained by the Biginelli reaction [6]. Previous research has done it by reacting aromatic aldehyde with barbituric acid and urea/thiourea [7, 8]. In this research, cinnamaldehyde is used as an alternative to aromatic aldehydes which will enrich the pyrimido [4,5-d] pyrimidine compounds. Cinnamaldehyde is an active compound found in cinnamon having bioactivity as an antibacterial, antifungal, antidiabetic, antiviral, anticancer, anti-inflammatory, and antioxidant [9, 10]. In this study, the reaction is supported by iodine catalyst. The application of molecular iodine has gained much attention as a cheap, easy-to-obtain catalyst for various organic transformations. This catalyst has also been exploited for various types.

2. Materials and method

2.1. *Chemical reagents and instrumentations*

All materials and reagents were obtained from the company of Merck, Fluka, and Sigma-Aldrich and used without further purification. Merck grade Thin Layer Chromatography (TLC) plate was employed for thin layer chromatography to evaluate the progress of the reaction, visualized by UV light. The UV–Vis spectra was registered by Agilent 8453 spectrophotometer, Fourier transform infrared spectroscopy (FTIR) spectra was registered in the 4000–400 cm^{-1} region, on a Shimadzu infrared spectrophotometer in KBr discs and Mass spectra (MS) was performed by GC-MS (Agilent).

2.2. *General procedure for the synthesis of pyrimido [4,5-d] pyrimidine*

The synthesis of pyrimido [4,5-d] pyrimidine was carried out by mixing barbituric acid (1 mmol), aldehyde (1 mmol), urea/thiourea (1 mmol), and iodine 10 % mol. Then, the mixture was heated in distilled water at variation temperature under reflux conditions. The reaction was monitored and evaluated with TLC. After completion, the product was washed with ethanol, crystallized, and purified by hot ethanol. Then, the the products were characterized for its absorbance, functional group, and molecular weight using UV/Visible Spectrophotometry, FTIR, and GCMS. Simple illustration of the reaction that occurred in this study is described in figure 1.

2.3. *Antioxidant activity test*

The antioxidant activity test used 1,1-diphenyl-2-picryl hydrazyl (DPPH) and the sample solution was prepared in a series of concentrations ranging from 5–1000 ppm in a methanol solvent. Absorbance of the control solution was measured for DPPH solution using methanol as a blank. A total of 2 mL of each solution was reacted with 1 mL DPPH in a tube. The reaction mixture was then incubated at room temperature for 30 min. After that, the absorbance of the solution was measured at a wavelength of 517 nm using a UV-vis spectrophotometer. The calculation of % free radical DPPH follows the following equation:

$$\%\, Inhibition = \frac{(Ac - As)}{Ac}\, x\, 100\, \%$$

where, *Ac* is the absorbance of DPPH control, and *As* is the absorbance of sample. IC_{50} values were determined using a linear regression correlation between concentration and % free radical DPPH. IC_{50} is a concentration value that scavenging 50 % of DPHH free radicals [11, 12].

3. Results and discussion

The synthesis of pyrimido [4,5-d] pyrimidine-derived compounds using iodine as a catalyst has been synthesized through a one-pot three-component method. Analysis of the maximum wavelength formed, compound 1 was measured with a Uv-vis spectrophotometer and obtained a maximum wavelength at 246 nm and 377 nm (figure 2a). The rightward shift is caused by additional substituents and called as a bathochromic shift or redshift. Experimental evidence in support of the formation pyrimido [4,5-d] pyrimidine compound is also obtained from the analysis of the functional groups of the target compound through the FTIR spectrophotometer. FTIR spectrum of compound 1 (figure 2b) shows several peaks that identify the vibrations of the functional group of this compound. A peak at wavenumber of 3170 cm^{-1} is the stretching vibrations of secondary amines (N–H). In addition, there is also a bending vibration N-H at wave number 1571 cm^{-1} with strong intensity. Besides that, the peak at 3083 cm^{-1} is a stretching vibration of C-H aromatic (sp^2) found on the benzene ring. While the peak at 2887 cm^{-1} is stretching vibration C–H (sp^3). The stretching vibration of the C–N is at 1420 cm^{-1}, C=C from aromatics at 1620 cm^{-1} and C=O at 1756 cm^{-1}.

The main support in determining the structure of formation pyrimido [4,5-d] pyrimidine-derived compounds in this study is the use of GC-MS for the determination of the molecular weight of the target compound. Figure 3 shows the GC-MS spectrum of compound 1 with a parent peak m/z 284. Compound 1 which is a product of the reaction between cinnamaldehyde, barbituric acid, and urea would produce a compound with the molecular formula $C_{14}H_{12}N_4O_3$, with a molecular weight of 284 g/mol. The name of compound 1 is 5-Styryl-5,8-dihydro-1H,6H-pyrimido[4,5-d]pyrimidine-2,4,7-trione. Some fragments formed by compound 1 are found at m/z 242, 199, 176.8, 155, and 127.

Evaluation of the ability of iodine as a catalyst in the synthesis reaction of pyrimido [4,5-d] pyrimidine derivatives was conducted by varying the temperature, reaction time, type of solvent, and iodine concentration as in table 1. From some of these experiments, it is known that the optimum conditions of the reaction are achieved at conditions of 10 % mol I_2, for 4 h at 70 °C with a water solvent. The yield obtained at this optimum condition is 65.45 %. Based on the optimum condition information in table 1, the reaction to enrich the derivative of compound pyrimido [4,5-d] pyrimidine was conducted by varying the aromatic aldehyde and replacing urea into thiourea. Data on the characteristics of pyrimido [4,5-d] pyrimidine-derived compounds can be seen in table 2.

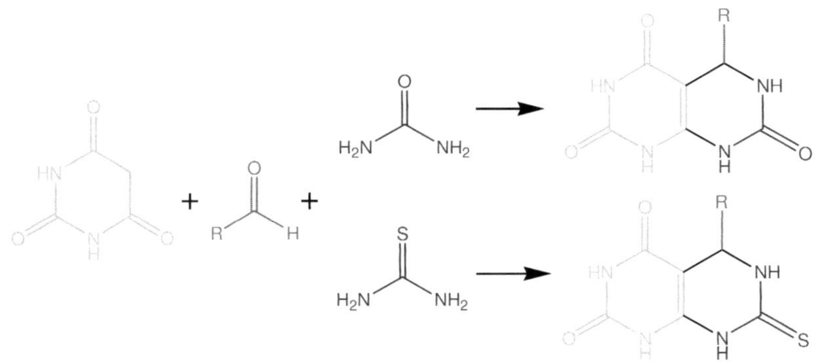

Figure 1. Reaction scheme for the synthesis of pyrimido [4,5-d] pyrimidine derivatives.

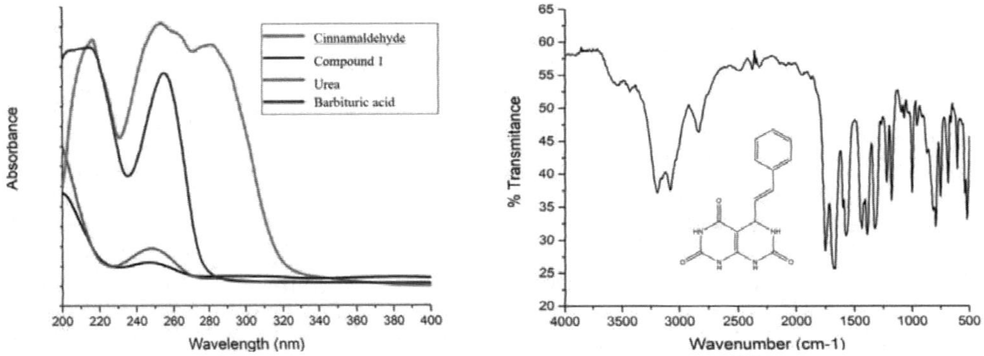

(a) (b)

Figure 2. (a) UV-vis and (b) FTIR spectra of compound 1.

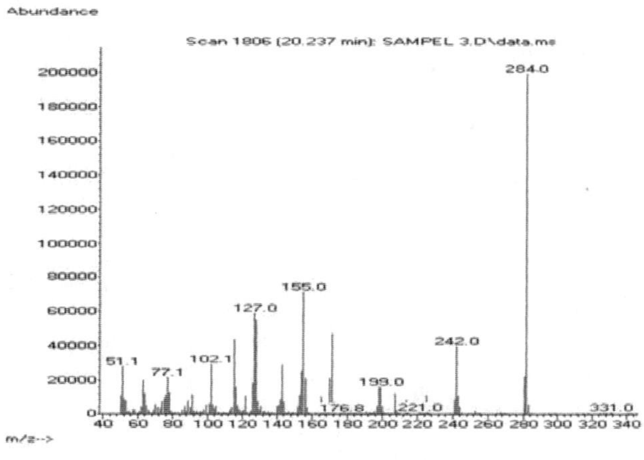

Figure 3. Mass spectrum of compound 1.

Table 1. Optimization of the reaction condition for the synthesis of pyrimido [4,5-d] pyrimidine (compound 3).

Entry	Catalyst	Time (h)	Temp. (°C)	Solvent	% yield
1	I_2 (10 mol%)	2	100	H_2O	48.74 %
2	I_2 (10 mol%)	4	100	H_2O	51.22 %
3	I_2 (10 mol%)	5	100	H_2O	51.80 %
4	I_2 (10 mol%)	6	100	H_2O	51.77 %
5	I_2 (10 mol%)	8	100	H_2O	41.64 %
6	I_2 (10 mol%)	4	90	H_2O	53.54 %
7	I_2 (10 mol%)	4	80	H_2O	56.54 %
8	I_2 (10 mol%)	4	70	H_2O	65.45 %
9	I_2 (5 mol%)	4	70	H_2O	55.51 %
10	I_2 (15 mol%)	4	70	H_2O	63.45 %
11	I_2 (10 mol%)	4	70	Ethanol	49.76 %
12	I_2 (10 mol%)	4	70	No solvent	52.00 %

Based on the structure formed, the name of these compounds are 5-Styryl-5,8-dihydro-1H,6H-pyrimido[4,5-d]pyrimidine-2,4,7-trione (compound 1), 5-Styryl-7-thioxo-5,6,7,8-tetrahydro-1H-pyrimido[4,5-d]pyrimidine-2,4-dione (compound 2), 5-Phenyl-5,8-dihydro-1H,6H-pyrimido[4,5-d]pyrimidine-2,4,7-trione (compound 3), and 5-Phenyl-7-thioxo-5,6,7,8-tetrahydro-1H-pyrimido[4,5-d]pyrimidine-2,4-dione (compound 4).

The antioxidant activity test was carried out using the DPPH method. Each sample solution was prepared with a concentration range of 5 to 1000 ppm in a methanol solvent. Each sample solution was added with 2 mL DPPH, then allowed to stand for 30 min in a dark room and absorbance of sample solution was measured at a wavelength of 517 nm using a UV-Vis spectrophotometer. The antioxidant activity of compounds 1, 2, 3 and 4 was measured along with its precursor. Absorbance values obtained from measurements are used to get % inhibition. Table 3 shows the comparison of IC_{50} of each compound. IC_{50} is a concentration value which can capture 50 % of DPPH free radicals.

IOP Conf. Series: Materials Science and Engineering **902** (2020) 012039 doi:10.1088/1757-899X/902/1/012039

Table 2. Characterization data and yield of pyrimido [4,5-d] pyrimidine derivatives.

No	Product	Characterization data	Yield (%)	Photo
1	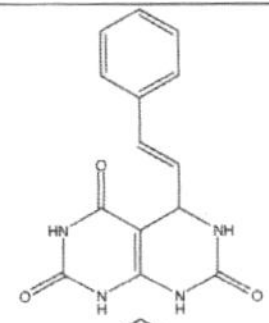	Yellow powder. IR (cm^{-1}): 3170 (N-H stretching), 1571 (N-H bending), 3083 (C-H sp^2), 2887 (C-H sp^3), 1420 (C-N), 1620 (C=C), and 1756 (C=O). UV-Vis (nm): 246 and 377. MS (m/z): 284	79.58	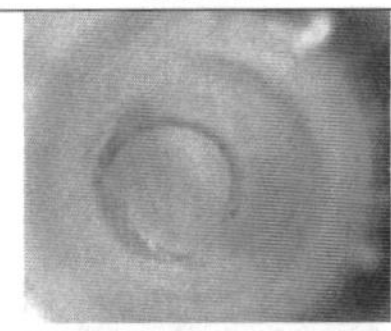
2	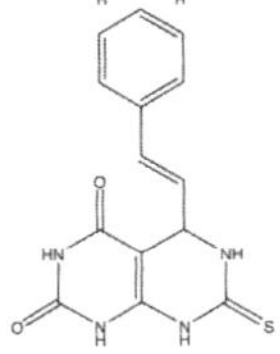	Yellow powder. IR (cm^{-1}): 3201 (N-H stretching), 1539 (N-H bending), 3077 (C-H sp^2), 2844 (C-H sp^3), 1416 (C-N), 1622 (C=C), and 1763 (C=O). UV-Vis (nm): 384. MS (m/z): 300	86.70	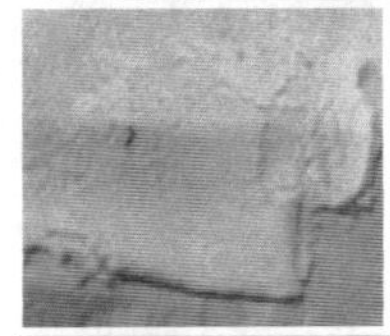
3	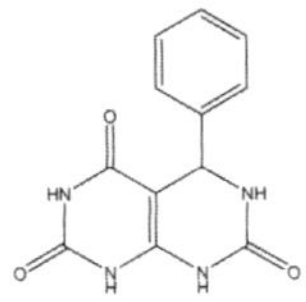	White powder. IR (cm^{-1}): 3236 (N-H stretching), 1521 (N-H bending), 3004 (C-H sp^2), 2890 (C-H sp^3), 1407 (C-N), 1625 (C=C), and 1725 (C=O). UV-Vis (nm): 248. MS (m/z): 259.	69.48	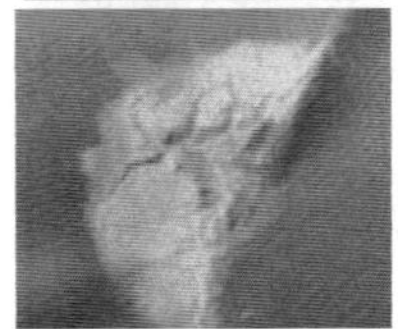
4	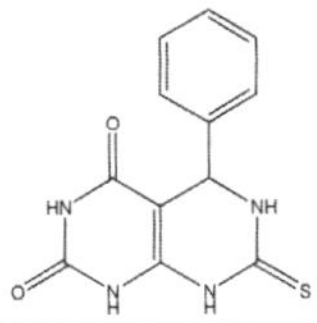	Pale yellow powder. IR (cm^{-1}): 3173 (N-H stretching), 1500 (N-H bending), 3004 (C-H sp^2), 2866 (C-H sp^3), 1411 (C-N), 1618 (C=C), and 1724 (C=O). UV-Vis (nm): 262. MS (m/z): 275	65.45	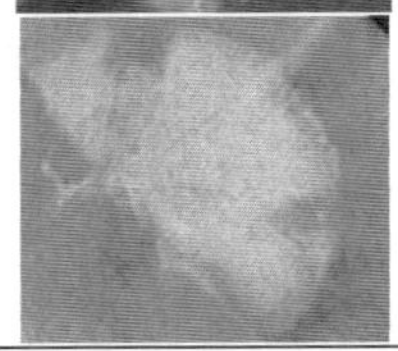

Table 3. Antioxidant activity test of pyrimido [4,5-d] pyrimidine derivatives.

Sample	IC_{50} (ppm)
Compound 1	10.11
Compound 2	32.74
Compound 3	168.09
Compound 4	109.11
Benzaldehyde	797.22
Cinnamaldehyde	579.32

Antioxidant activity (table 3) shows that compound 1 has a better activity than 3 other compounds. Its low IC_{50} value means that the compound can inhibit the free radicals by as much as 50 %. The smaller the concentration needed to inhibit it, the better its ability as an antioxidant. Compound 1 was synthesized using cinnamaldehyde which also has antioxidant activity. However, when compared to its precursors, compound 1 has a significant increase in antioxidant activity.

4. Conclusion

The pyrimido [4,5-d] pyrimidine derivatives were successfully synthesized using cinnamaldehyde and benzaldehyde as a precursor through the one-pot three-component method with iodine catalyst. The characterization of synthesized compounds was evaluated with some instruments like FTIR,

UV-vis, and GC-MS and showed the characteristics of the target compound. The results identified that the compound formed was pyrimido [4,5-d] pyrimidine derivatives namely 5-Styryl-5,8-dihydro-1H,6H-pyrimido[4,5-d]pyrimidine-2,4,7-trione (compound 1), 5-Styryl-7-thioxo-5,6,7,8-tetrahydro-1H-pyrimido[4,5-d]pyrimidine-2,4-dione (compound 2), 5-Phenyl-5,8-dihydro-1H,6H-pyrimido[4,5-d]pyrimidine-2,4,7-trione (compound 3), and 5-Phenyl-7-thioxo-5,6,7,8-tetrahydro-1H-pyrimido[4,5-d]pyrimidine-2,4-dione (compound 4). The optimum condition reaction was carried out at 10 % mol I_2, for 4 h at 70 °C with a water solvent. Compound 1 has better activity than 3 other compounds with an IC_{50} value of 10.11 ppm.

Acknowledgments
This research was financially supported by Ministry of Research and Technology, Republic of Indonesia with grant contract number NKB-1586/UN2.R3.1/HKP.05.00/2019.

References

[1] Prajapati D, Gohain M and Thakur A J 2006 *Bioorg. Med. Chem. Lett.* **16** 3537-40
[2] Aksinenko A Y, Goreva T V, Epishina T A, Trepalin S V and Sokolov V B 2016 *J. Fluor. Chem.* **188** 191-5
[3] Dabiri M, Arvin-nezhad H, Khavasi H R and Bazgir A. 2007 *Tetrahedron* **63** 1770-4
[4] Wang H, Wang C and Bannister T D 2015 *Tetrahedron Lett* **56** 1949-52
[5] Bakhshali-dehkordi R, Ghasemzadeh M A and Ghomi J S 2020 *J. Mol. Struct.* **1206** 127698
[6] Kappe C O 2000 *Acc. Chem. Res.* **33** 879-88
[7] Shinde S V, Jadhav W N, Karade N N 2010 *Orient. J. Chem.* **26** 307-17
[8] Gupta R, Jain A, Joshi R and Jain M 2011 *Bull. Korean Chem. Soc.* **32** 899-904
[9] Liandi A R, Yunarti R T, Fajri M and Cahyanaa A H 2020 *Materials Today: Proc.* **22** 193–8
[10] Zhu R, Liu H, Liu C, Wang L, Ma R, and Chen B 2017 *Pharmacol. Res.* **122** 78–89
[11] Subbareddy C V and Sumathy S 2017 *New J. Chem* **41** 9388-9396
[12] Febriantini D, Cahyana A H, Yunarti R T, Liandi A R and Ardiansah B 2019 *AIP Conf. Proc.* **2168** 020068

IOP Conf. Series: Materials Science and Engineering **902** (2020) 012040 doi:10.1088/1757-899X/902/1/012040

A preliminary study on modified chitosan-curcuminoids as material active food packaging with antioxidant and antibacterial activities

A Azkia, A I Baihaqi, H A Prabowo, Hendri, S Handayani and T Agustiany

Department of Chemistry, Faculty of Mathematics and Natural Sciences (FMIPA), Universitas Indonesia, Depok 16424, Indonesia

Corresponding author's email: yani1964@ui.ac.id

Abstract. The compound of inorganic nanoparticles is commonly mixed with polymeric materials for food packaging that has antibacterial function. However, the inorganic nanoparticles are toxic to human and the polymeric materials used are not easily biodegraded. The curcuminoids compound from turmeric extract has antioxidant and antibacterial activity which can be a candidate as polymer combination. The polymer used is chitosan which isolated from crab shell with the yield of 4.54 %. In this study, the modified chitosan-curcuminoids was synthesized by in-situ loading method and moulded into a film. Interaction between chitosan and curcuminoids is hydrogen interaction which determined using FTIR. Then, antioxidant activity was tested using radical DPPH scavenger. Antioxidant test results showed that modified chitosan with 20, 40, 60 80 and 100 ppm of ascorbic acid have inhibition levels of 15.49, 16.20, 16.90, 23.94 and 27.47 %, respectively. Modified chitosan with 25 ppm of curcuminoids have a comparable antioxidant activity to chitosan modified ascorbic acid 80 ppm with 24.65 % inhibition. Meanwhile, in the antibacterial test, chitosan modified with curcuminoids showed antibacterial activity toward *E. coli* and *S. aureus* bacteria with inhibition zone of 9 and 6 mm, respectively.

Keyword: Food packaging, chitosan-curcuminoids, antioxidant, antibacterial

1. Introduction

In the food packaging field, development of polymer material with antioxidant and antibacterial activity is an important aspect. Materials with antioxidant activity are able to increase the lifespan of the food by preventing the oxidation process in the food [1]. Inorganic nanoparticle materials such as copper [2], zinc oxide [3] and silver [4], are often used as packaging material additives to enhance the antibacterial activities of polymeric materials. However, inorganic nanoparticles materials are heavy metal, which can be toxic to humans' body and the polymer material used are non-biodegradable [5]. Modified natural polymer materials have a potential to replace the function of inorganic nanoparticles used in the synthetic polymer blends. The ability of the natural polymer, which is to be naturally degradable, makes it a highly potential biomaterial to be developed and modified in various sectors [6]. The example of natural polymers is chitin and its derivatives, such as chitosan, which has a potential for packaging materials. Chitin can be converted into chitosan by deacetylation process [7].

Chitosan isolation from crab shells waste has biocompatibility property, antibacterial, antioxidant activity and also good to form film. Thus, chitosan can be used as a coating material, packaging and wound cover in the fields of food, health and chemical industries [8]. Curcuminoids are natural compounds from turmeric (*Curcuma domestica* Val.) extract. Curcuminoids are known for their antibacterial and antioxidant activities. Curcuminoids has potential as an antibacterial agent towards foodborne pathogens namely *S. aureus* and *E. coli* which represent gram-positive and gram-negative bacteria, respectively [9]. Curcuminoids can be a strong candidate to suppress oxidative propagation chains and minimize lipid oxidation in food systems, as well as maintaining the nutritional quality and extending the age of food by controlling the process of decaying microorganisms. The aim of this study is to make the material active food packaging from the modified chitosan-curcuminoids with antioxidant and antibacterial activities.

2. Methods

2.1. Preparation of crab shells
Crab shells were washed with clean water to remove physical impurities. Then, dried at 60 °C and mashed using a mortar

2.2. Demineralization of crab shells
Powder of crab shell was dissolved in HCl 1 M with ratio of 1:15 and stirred for 30 min. The residue was washed with mineral-free water until neutral (pH 6–7), then dried at room temperature [10].

2.3. Deproteination of crab shells
Powder from demineralization process was dissolved in NaOH 3.5 % with ratio of 1:10. The solution was refluxed for 2 h at 65 °C while being stirred. Then, the solution was filtered and washed with mineral-free water until the pH was neutral and dried at 60 °C for 4 h [10].

2.4. Conversion of chitin into chitosan
Chitosan was acquired by deacetylation of chitin by adding 50 % NaOH with ratio of 1:15 and then heated to 70–75 °C for 1 h. The obtained solid was washed with water until neutral pH. Then, chitosan was dried at 80 °C for 24 h. The isolated chitosan was weighed and stored at room temperature. Chitosan was analyzed using Fourier transform infrared (FTIR) [11].

2.5. Isolation of curcuminoids
Turmeric extract was obtained by using maceration extraction method. Turmeric powder was made by drying turmeric and made it into 80 mesh powder. Furthermore, 100 g of turmeric powder was dissolved in 400 mL of acetone for 3×24 h. Then, the solvent was evaporated until the extract of turmeric remained. Turmeric extracts were tested using silica thin layer chromatography with eluent of chloroform and methanol (95:5) [9]. The isolation results were also analyzed by mass spectrometry instrumentation to confirm the structure.

2.6. Preparation of chitosan-curcuminoids film
Chitosan film control was obtained by dissolving 2 g of chitosan in 90 mL glacial acetate 1 % and stirred at room temperature until became homogeneous. Then, propylene glycol of 0.5 mL was added to the chitosan gel. Followed by the addition of 2 mL additives compounds to obtain the modified chitosan film. The additives compound used in this study were curcuminoids 25 ppm from turmeric extract and commercial antioxidant as comparison was ascorbic acid with variation of 20, 40, 60, 80, 100 ppm. The gel solution of films was dried at 60 °C and stored in a desiccator before test. Chitosan and modified chitosan film were analyzed by FTIR.

2.7. In vitro antioxidant assay
The antioxidant activities of film were obtained by putting 0.14 g of film into 5 mL of DPPH radical solution (10 ppm). Then, the mixture was shaken and incubated in dark condition for 1 h. The absorbance

IOP Conf. Series: Materials Science and Engineering **902** (2020) 012040 doi:10.1088/1757-899X/902/1/012040

of the DPPH radical solution was analyzed with UV-visible spectroscopy instrument at 515 nm. The antioxidant capacity was measured on the basis of the following equation [6].

$$\text{DPPH Scavenging ratio} = \frac{ABS_c - ABS_s}{ABS_c} \times 100$$

2.8. Antibacterial activity assay

The antibacterial activity of the chitosan-curcuminoids films was investigated by the inhibition zone method. *E. coli* as gram-negative bacteria and *S. Aureus* as gram-positive bacteria were used in this study. Bacterial suspensions were activated in the nutrient broth medium. The diluted bacterial suspension mentioned above was pipetted as much 200 μL onto a sterilized plate, then 15 mL of nutrient agar was added and coated uniformly. After the NA medium became solid, the chitosan film sample with diameter 6 mm was placed on top of the medium and incubated at 32–37 °C for 24 h. Antibacterial activity was indicated by the inhibition zones of bacterial growth.

3. Results and discussion

3.1. Isolation chitosan from crab shells

Isolation chitin from crab shell was done by deproteination and demineralization. At the demineralization stage, mineral compound such as calcium that contained in the crab shells would react with water-soluble hydrochloric acid. After the demineralization stage, the precipitate obtained was washed with distilled water to neutral pH, universal pH was used as indicator. This washing was intended to prevent the product degradation during the drying process. Then, the deproteination aimed to break the bond between protein and chitin, by adding NaOH. During immersion, the protein would be extracted in the form of sodium-fatty acid (Sodium-proteinate) due to saponification reaction between the fat that contained in the crab shells with NaOH solution, where the Na^+ ions bonded to the end of the negatively charged protein chain and then settled. Chitosan was produced through the deacetylation step by adding NaOH. Deacetylation aimed to break the acetyl group ($COCH_3$) that present in chitin. Based on this research, 4.54 % of chitosan was obtained and then analyzed using FTIR. The FTIR spectrum of chitosan is presented in figure 1.

Based on the spectrum in figure 1, the functional group can be identified by the absorption of the wave number 3250–3500 cm^{-1} which is the absorption bands of the -OH group. The -OH group is used as a standard in chitosan [12], while the spectrum of chitosan compound is seen in the wave number

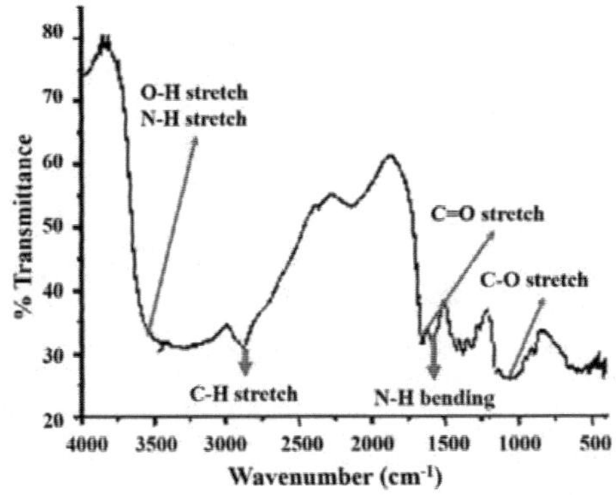

Figure 1. Chitosan Powder FT-IR spectrum.

IOP Conf. Series: Materials Science and Engineering **902** (2020) 012040 doi:10.1088/1757-899X/902/1/012040

1600–1700 cm^{-1} which is the N-H bending swell that indicates the presence of primary amine (NH_2). The wave number 1690 cm^{-1} is the absorption bands of the vibration that remaining C=O acetyl group on the chitosan chain. The wave number 1160 cm^{-1} shows the vibration of the C-O-C bridge which is the typical absorption of the saccharide structure.

3.2. Isolation and characterization of curcuminoids

Isolation of curcuminoids was done by maceration method, because the tool that was used is simple and can be used for substances that resistant and not resistant to heating. Maceration was carried out with acetone, because it was obtained more curcuminoids compounds than hexane, chloroform, ethyl acetate, and methanol [13]. After obtaining that crude extract, the curcuminoids was tested by thin layer chromatography in chloroform and methanol (95:5), and then compared with standard curcumin. The result of TLC for the separation of standard curcumin and isolation result are shown in figure 2. Based on TLC spot, retention factor (Rf) were obtained for each spot contained in table 1.

Order of the spots appeared were determined by its polarity. Compound with more polar characteristic will appear earlier than others. The order of the spots from high to low are: curcumin, demethoxy curcumin, and bis-demethoxy curcumin. Data in the table 1 has a similar trend with the data reported before [13].

Analyses were also performed by using the MS instrument to confirm the structure of the compounds. These compounds have $[M+1]^+$ peak at MS spectra (figure 3). The bisdemethoxy curcumin spectra appear as positive ion mode with m/z 309.11, demethoxy curcumin m/z 339.12, and curcumin m/z 369.13 [14].

3.3. Control chitosan film and modified chitosan-curcuminoids film

The film was produced by mixing chitosan in 1 % acetic acid and added with propylene glycol as plasticizer. In general, the addition of propylene glycol aims to form hydrogen bonds

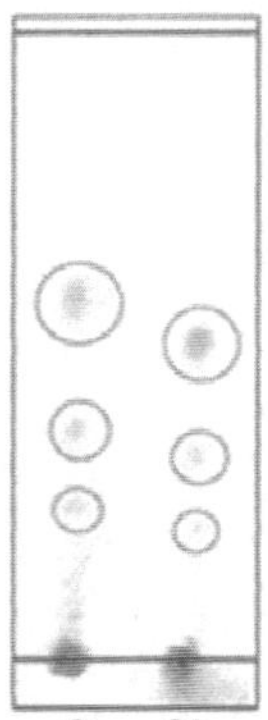
a) b)

Figure 2. The result of thin layer chromatography between a) standard and b) turmeric extract.

Table 1. Comparison of Rf from curcumin standard and curcuminoids extract.

	Curcumin	Demethoxy curcumin	Bis-demethoxy curcumin
Standard	0.66	0.45	0.34
Turmeric extract	0.58	0.40	0.31
Ref no.13	0.75	0.55	0.27

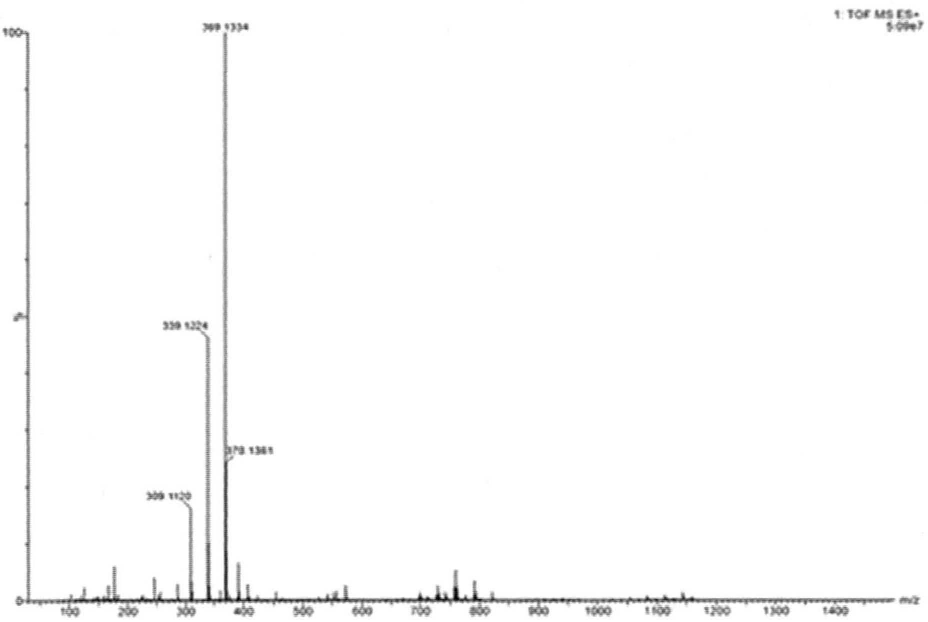

Figure 3. MS spectra of curcuminoids.

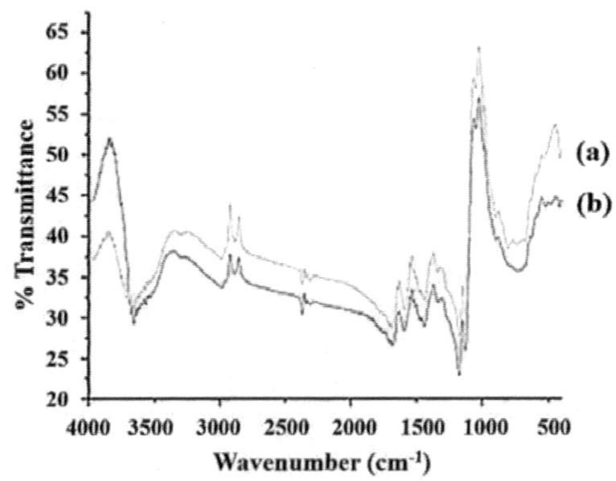

Figure 4. Comparison of FT-IR spectrum (a) Modified chitosan-curcuminoids film and (b) Control chitosan film.

in the bonding chain between the polymers so that causing the interaction between the biopolymer molecules to be reduced. With the reducing bonding between the polymers, the resulting properties of polymer will be more elastic. Modifications were also made by adding curcuminoids and ascorbic acid in various concentrations as comparison. The interaction between compounds for this modification with the polymer is by hydrogen bonding. Furthermore, the addition was analyzed by FT-IR and the spectrum is shown in figure 4.

Based on the measurement results it can be seen that chitosan-curcuminoids has broader absorbing band around 3300–3600 cm^{-1} than chitosan. It is indicating that there is a hydrogen interaction between chitosan and curcuminoids. In addition, there is a shift in C=O stretch at the wave number 1600–1750 cm^{-1} indicating that curcuminoid additive has been successfully mixed [5].

3.4. Scanning electron microscope test of film

Modified chitosan was further analyzed using SEM to see the effect of curcuminoid addition to its morphological structure. The SEM results are presented in figure 5.

Based on the observation of morphology using SEM, it can be seen that the surface of chitosan modified curcuminoids looks rougher than the control chitosan. This suggests that the presence of additional additives into chitosan may decrease the homogeneity of chitosan and increases the roughness.

3.5. In vitro antioxidant assay

Control and modified chitosan polymers were assayed with 10 ppm radical DPPH to see the magnitude of inhibition ability of the sample. DPPH measured in the range of 600–400 nm to see the maximum wavelength of DPPH and obtained maximum wave length absorption at 515 nm (figure 6). The control and sample absorption were measured with UV-Vis at 515 nm after stored an hour in dark condition (figure 7).

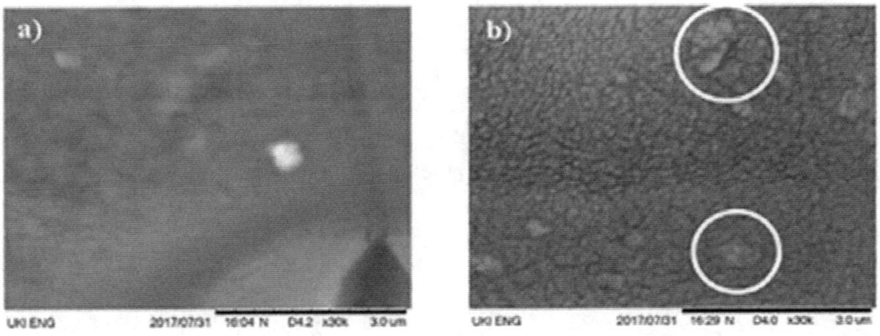

(a) (b)

Figure 5. Structure morphology of film using SEM with 30,000 times magnification, (a) control chitosan film and (b) chitosan modified with curcuminoids.

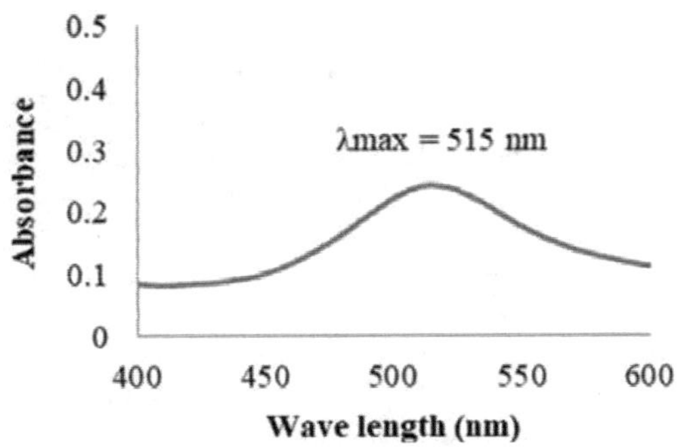

Figure 6. Graph of the absorption spectrum of DPPH.

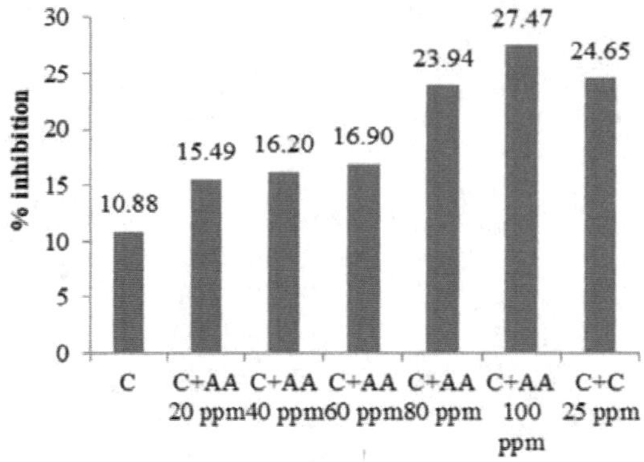

Figure 7. Comparison of inhibitory DPPH radicals on control chitosan, modified chitosan-curcuminoids 25 ppm and modified chitosan-ascorbic acid at various concentrations.

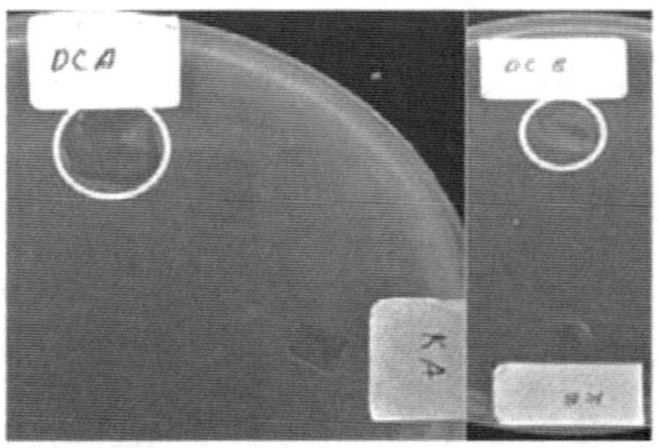

Figure 8. Antibacterial test result of chitosan modified with curcuminoids (DCA: Chitosan modified curcuminoids against E. coli bacteria. KA: Control chitosan against E. coli. DCB: Chitosan modified curcuminoids against S. aureus. KB: Control chitosan against S. aureus bacteria.

Based on the results, it can be seen that chitosan modified with curcuminoids have activity that approach ascorbic acid 80 ppm. The antioxidant activity occurs because the modifying compound is simply bonded by a fairly weak hydrogen bond. So that, when added into the radical DPPH solution, the modifying compound can release and inhibit the DPPH radical activity.

3.6. Antibacterial activity assay

Chitosan film modified with curcuminoids tested its antibacterial properties by performing antibacterial test against *E. coli* and *S. aureus* bacteria. The results of the antibacterial test can be seen in figure 8.

In figure 8, the clear zone was appeared which is indicating the presence of antibacterial activity of chitosan modified with curcuminoid. The inhibition zone of chitosan-curcuminoids against *E.coli* and *S.aureus* is 9 and 6 mm, respectively. This is due to the presence of curcuminoids that located on the surface of chitosan film so that it can make the bacteria unable to grow because of the diffusion of curcuminoids on the chitosan film modified.

4. Conclusion

Curcuminoids makes hydrogen interaction with chitosan. Basically, chitosan film has antioxidant and antibacterial activity. Modification of chitosan film with curcuminoids that isolated from turmeric can enhance the antioxidant and antibacterial activities. The curcuminoids can easily diffuse from chitosan to inhibit DPPH radicals and also the growth of S. *aureus* and E. *coli* bacteria.

Acknowledgments
This research was financially supported by Program Kreativitas Mahasiswa (PKM) 2017 from Ministry of Research, Technology and Higher Education of the Republic of Indonesia.

References

[1] Siripatrawan U and Noipha S 2012 *Food Hydrocoll.* **27** 102-8
[2] Jia B, Mei Y, Cheng L, Zhou J and Zhang L 2012 *ACS Appl. Mater. Interfaces* **4** 2897-902
[3] Fu F et al. 2015 *ACS Appl. Mater. Interfaces* **7** 2597-606
[4] Tang H et al. 2013 *Chem. Eng. J.* **234** 124-31
[5] Wang Y, Li B and Li J 2016 *J. Agric. Food Chem.* **64** 5736-41
[6] Younes I and Rinaudo M 2015 *Mar. Drugs* **13** 1133-74
[7] Rinaudo M 2014 Materials based on chitin and chitosan In *Bio-Based Plastics: Materials and Applications* edited Kabasci S (UK: John Wiley & Sons) pp. 63-80
[8] Mitelut A C, Tanase E E, Popa V I and Popa M E 2015 *AgroLife Sci. J.* **4** 52-61
[9] Sivasothy Y, Sulaiman S F, Ooi K L, Ibrahim H and Awang K 2013 *Food Control* **30** 714-20
[10] Pandharipande S L and Bhagat P H 2016 *Int. J. Sci. Eng. Technol. Res.* **5** 1378-83
[11] Hossain M S and Iqbal A 2014 *J. Bangladesh Agril. Univ.* **12** 153-60
[12] Mario F D, Rapana P, Tomati U and Galli E 2008 *Int. J. Biol. Macromol.* **4** 8-12
[13] Revathy S, Elumalai S and Antony M B 2011 *J. Exp. Sci.* **2** 21-5
[14] Verma M K et al. 2013 *Daru J. Pharm. Sci.* **21** 1-9

Application of MoS_2/reduced graphene oxide for photocatalytic degradation of rhodamine B in water environment

P T T Trang[1], T T Tam[1], V Vien[1] and N H Lien[2]

[1]Faculty of Chemistry, Quy Nhon University, Binh Dịnh, Vietnam
[2]School of Chemical Engineering, Hanoi University of Science and Technology, Hanoi, Vietnam

Corresponding author's email: lien.nguyenhong@hust.edu.vn

Abstract. In this study, MoS_2 was composited with RGO (the reduced form of graphene oxide) for enhancing the photocatalytic degradation of RhB in aqueous solution by applying visible light irradiation. A new approach to synthesize graphene was applied by reduction of graphene oxide using ascorbic acid. Factors affecting the photocatalytic degradation including pH, initial rhodamine B concentration, light source intensity and quencher existence were investigated. The results showed that MoS_2/RGO composite exhibited better photocatalytic activity in the degradation of RhB (with a conversion of 80 %) compared to pure MoS_2 (with a conversion of 30.8 %) under visible light irradiation. The higher photocatalytic performance of MoS_2/RGO was found in slight acidic environment. Among the tested free radical killers, ammonium oxalate revealed the most significant effect on the photocatalytic degradation of RhB.This research also indicated that MoS_2/RGO composite can be considered as a potential catalyst for the degradation of persistent organic dyes in wastewater.

Keywords: Rhodamine B, photocatalyst, MoS_2/RGO composite, graphene

1. Introduction

Environmental pollution is becoming serious, especially organic materials and dyes in wastewater originated from industrial plants. One of the methods currently being the interest of the scientists -thanks to its cost effectiveness, proficiency and practical perspective- is the photocatalytic treatment. The advantage of this method is that it makes use of the sun light to decompose the organic pollutants in the wastewaters.Various kinds of semiconductors such as TiO_2, ZnO, SnO_2 have been widely used as catalysts for photocatalytic water treatment processes, but they only show their high photocatalytic activity under the ultraviolet light due to large band gap [1]. Among semiconductors with low band gap, molybdenum disulphide (MoS_2) is rather new one. This semiconductor has narrow band gap of $\approx$ 1,6 eV, implying a high working efficiency in sunlight. However, the disadvantage of narrowing the band gap of semiconductors is the rapid recombination of the photo-generated electron-hole pairs, resulting in low photocatalytic activity. Therefore, finding solutions for preventing this phenomenon is of high interest [2].

Combining a semiconductor with other materials to form composites has been conducted as one of solutions for an improvement of its photocatalytic activity. Different kinds of materials have been used for making composite such as g-C_3N_4, carbon nanotube (CNT), graphene to increase the ability of electronic transmission of the original semiconductors. As the results, their photocatalytic activities have been much improved as noted by Li et al. during the degradation of methylene blue over TiO_2/graphene porous composites [3]. Among these materials, graphene and its derivatives are being emerged because of their excellent properties, such as high crystal quality, large surface area (2600 m^2/g), good conductivity, high optical transmission speed [4]. Especially, when performing the oxidation of graphene to form graphene oxide (GO), its surface holds many functional groups containing oxygen making it hydrophilic. Then, further reduction to from reduced graphene oxide (RGO) is often applied to generate a high affinity towards organic compounds [5].

Therefore, in this research, graphene will be selected for making composite MoS_2/RGO and tested for its ability in improving the photocatalytic activity of MoS_2 in the degradation of rhodamine B under visible light.

2. Materials and method

2.1. Materials

Graphite powder (Merck, Germany), ascorbic acid (India), $KMnO_4$, $NaNO_3$, H_2SO_4, HCl, H_2O_2, C_2H_5OH, $(NH_4)_6Mo_7O_{24}.4H_2O$, thiourea, rhodamine B (China) were used as origin without further purification.

2.2. Preparation of samples

GO was synthesized from graphite powder via the Hummers method [5, 6]. For the synthesis of RGO, GO powder was dispersed into 100 mL of water by ultrasonification for 1 h to achieve a homogeneous dispersion, followed by the addition of 1 g ascorbic acid and heated at 70 °C and stirred for 8 h. This solid was cleansed with water and ethanol three times and dried at 80 °C for 12 f [7]. MoS_2 was prepared via calcination of the mixture of hexaammonium heptamolybdate tetrahydrate ($[NH_4]_6Mo_7O_{24}.4H_2O$) and thiourea. For the preparation of the MoS_2/RGO composite, hydrothermal synthesis at 180 °C for 10 h was applied. After being cooled down to room temperature, the mixture in the vessel was centrifuged at 2000 rpm for 15 min to separate black solid from solution. This solid was then cleansed with water and ethanol three times before being dried at 80 °C for 10 h to obtain the MoS_2/RGO composite.

2.3. Characterization

Powder X-ray diffraction (XRD) was registered on a Siemen D-500 (Germany) diffractometer operated at 30 kV and 10 mA with Cu-Kα radiation. The surface morphology and structure of the samples were determined using a SEM-JEOL-JSM 5410 LV (Japan) and JEOL TEM–2100F microscope. Fourier transform infrared spectra (FTIR) were analyzed on a GX–Perkin Elmer spectrometer. The UV-Vis absorption spectra were registered with a GBC Instrument–2885 spectrophotometer. The XPS spectrum was obtained using ESCALab220i-XL electron.

2.4. Photocatalytic degradation

The MoS_2/RGO composite photocatalytic activity was evaluated on degradation of RhB in aqueous solution under visible light irradiation. For a specific test, 20 mg of the photocatalyst was dispersed in 80 mL RhB solution at the concentration of 20 mg/L. Before lighting, the suspension was stirred in the dark for 2–3 h to get the adsorption-desorption equilibrium between the catalyst and RhB solution. Then, the solution was illuminated by the incandescent light of 60 W and stirred constantly. Every 1 h, an aliquot of solution was taken out and centrifuged to remove the catalyst and analyzed using a UV-Vis (CE-2011) spectrophotometer for determination of the remained RhB concentration.

3. Results and discussion

3.1. Structure of MoS_2 and MoS_2/RGO

The XRD patterns of MoS_2 and MoS_2/RGO composite are shown in figure 1. In both XRD patterns, three main peaks at $2\theta = 14.1°$, $33.6°$ and $58.1°$, respectively for (002), (100) and (110) crystal facets of MoS_2 and representative for the hexagonal 2H phase of MoS_2 [8, 9] can be observed. In the XRD pattern of the composite, apart from the representative peaks for MoS_2, the peak characterized for RGO is also recorded at $2\theta = 26°$. However, the diffraction peaks of the composite are lower and larger than those of the pristine MoS_2 due to coverage RGO layers over MoS_2 crystals in the composite. This result confirms the success of synthesis of MoS_2/RGO composite by combination of calcination and hydrothermal methods.

3.2. Morphology analysis

The morphology of the as-prepared MoS_2 and MoS_2/RGO composites are presented in figure 2. MoS_2 shows scaled sheets which are stacked together (figure 2a). The morphology of MoS_2/RGO composite (figure 2b) is clearly observed with curled RGO nanosheets contacted with MoS_2 well, which is an advantage for the photo-generated carriers separation. The presence of MoS_2 in the composite has been confirmed by EDX (figure 3), and the ratio of Mo:S atoms is 1:2.3, which is almost similar to

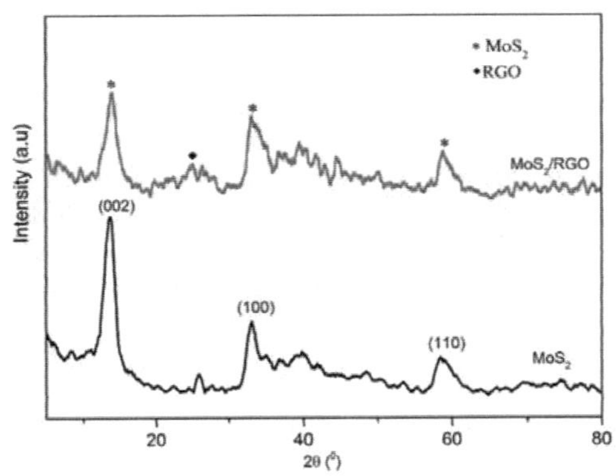

Figure 1. XRD patterns of pure MoS_2 and MoS_2/RGO composite.

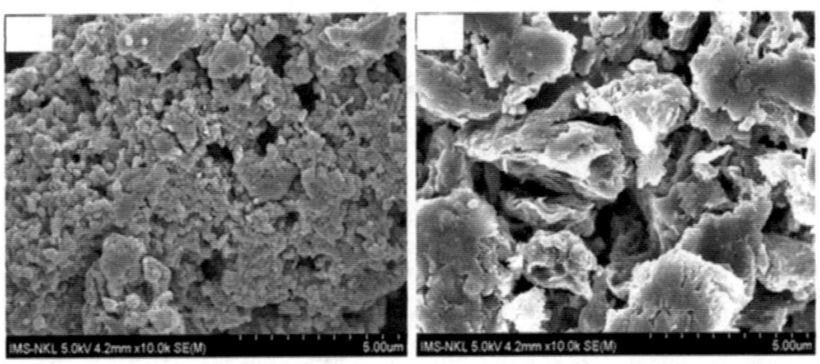

(a) (b)

Figure 2. (a) SEM images of MoS_2 and (b) MoS_2/RGO.

IOP Conf. Series: Materials Science and Engineering **902** (2020) 012041 doi:10.1088/1757-899X/902/1/012041

the theoretical proportion of MoS_2 (table 1) [1]. Moreover, the morphology of MoS_2/RGO composite was elucidated by TEM as shown in figure 4.

As being seen, the composite material was formed with layers structure. The MoS_2 was appeared in between of wrinkled RGO layers, proving the formation of composite between MoS_2 and RGO [9, 10].

3.3. X-ray photoelectron spectroscopy of MoS_2/RGO composite

XPS spectra are used to affirm the structure and composition of MoS_2/RGO composite with the results shown in figure 5.

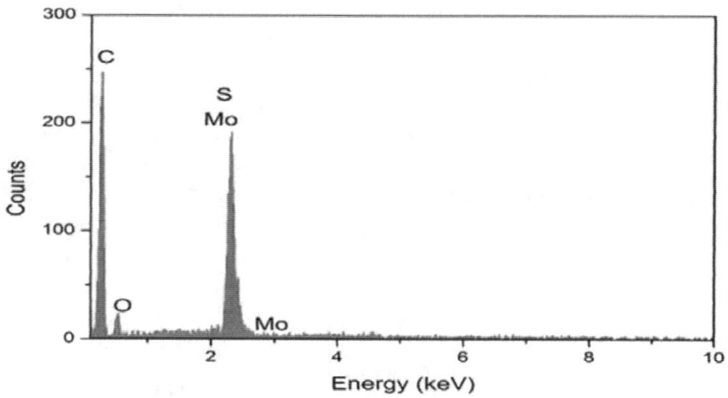

Figure 3. EDX spectra of MoS_2/RGO.

Table 1. Element composition of the MoS_2/RGO composite.

	Weight (%)				Atomic (%)			
Element	Mo	S	C	O	Mo	S	C	O
	5.94	7.51	74.43	12.12	2.57	1.08	85.85	10.50

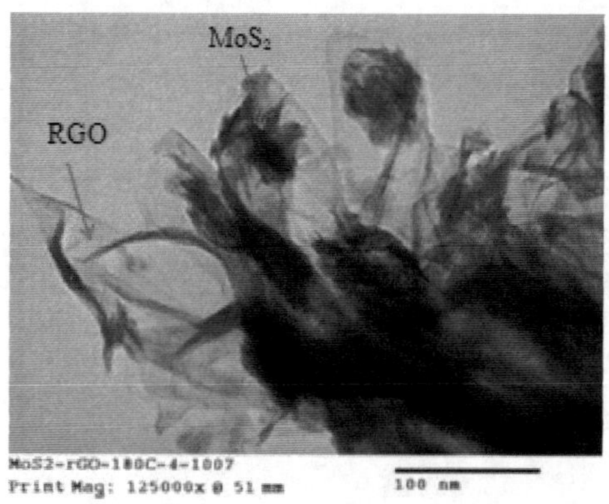

Figure 4. TEM image of MoS_2/RGO composite.

As being observed in figure 5a, peaks at 226.41 eV; 229.16 eV; 232.43 eV and 235.5 eV correspond to the S_{2s}, $Mo^{4+}{}_{3d5/2}$, $Mo^{4+}{}_{3d3/2}$ and Mo^{6+} [11]. Among them, peaks of $Mo^{4+}{}_{3d5/2}$ and $Mo^{4+}{}_{3d3/2}$ are representative for MoS_2 and S_{2s} for MoS_2. Peak at 235.5 eV of MoO_3 or MoO_4^{2-} (Mo^{+6}) is a proof of oxidizing Mo^{+4} to Mo^{+6} in the MoS_2 pattern due to the fact that oxygen has entered during the calcination. Besides, peaks appeared at 162.3 eV and 163.3 eV (figure 5b) are found to be corresponded to the $S_{2p3/2}$ and $S_{2p1/2}$ in the composite [12]. Peaks observed at 284.78 eV, 286.8 eV and 288.7 eV (figure 5c) may be attributed to the C-C or C=C, C-O and O=C-O bonds of functional groups in the RGO or the C-S bond [12, 13]. This completely confirms the presentation of S^{2-} ion in MoS_2 sample. Moreover, the O1s (figure 5d) is separated into two peaks at 530.5 eV and 532.1 eV, indicating the presence of lattice oxygen and adsorbed oxygen [13]. As a result, XPS spectra affirm that the composite material has been successfully synthesized by the combination of calcination and hydrothermal methods.

3.4. Specific surface area of composite

Figure 6 presents the N_2 adsorption-desorption isotherms of MoS_2 and MoS_2/RGO composite.

The N_2-BET specific surface areas of MoS_2 and MoS_2/RGO composite calculated from the isotherms are 14.7 m^2/g and 88.15 m^2/g, respectively. Higher specific surface area of MoS_2/RGO composite compared to pure MoS_2 demonstrates that RGO could play an effective role in improving the specific surface area of MoS_2/RGO composite, generating in high adsorption capacity of this material.

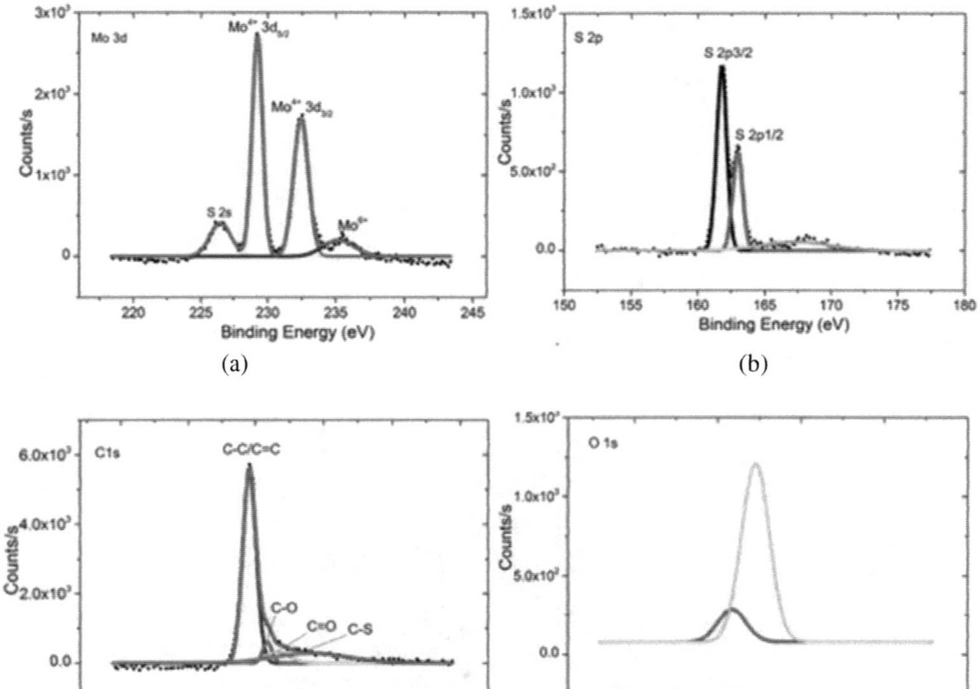

Figure 5. XPS spectra of MoS_2/RGO composite.

3.5. UV-Vis spectroscopy

For the determination of the light absorption and band gaps energy of the samples, the UV-Vis diffuse reflectance spectra of MoS_2 and MoS_2/RGO composites were taken and presented in figure 7.

The results indicate that the absorption edges of MoS_2 and MoS_2/RGO composites extend to 700 nm, expanding the light absorbability of the photocatalyst towards the visible light region, and as a result increasing its photocatalytic applications in the reality. Besides, the presence of RGO can also increase the band-gap energy of MoS_2-based catalyst, up to 2.16 eV, decreasing the photogenerated electron–hole recombination, thus improving the photocatalytic activity of the composite [12].

3.6. Photocatalytic degradation of RhB

The results of degradation of RhB with the existence of MoS_2 and MoS_2/RGO composite under visible light is described in the figure 8.

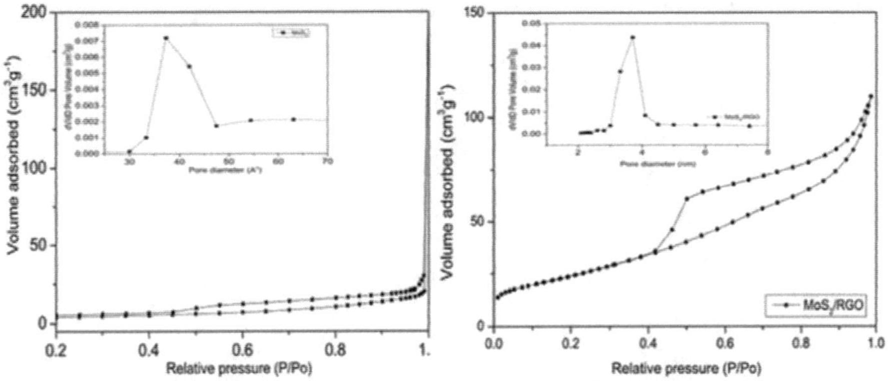

(a) (b)

Figure 6. (a) N_2 adsorption-desorption isotherms of MoS_2 and (b) MoS_2/RGO composite.

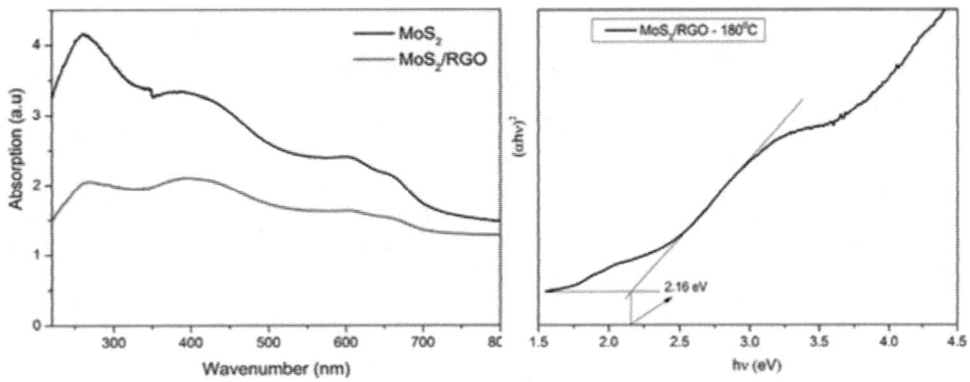

(a) (b)

Figure 7. (a) UV-Vis spectra of MoS_2 and MoS_2/RGO composite and (b) corresponding plot of the transformed Kubelka-Munk function versus the energy of the light of sample MoS_2/RGO.

The results show that the photocatalytic activity of pure MoS_2 is much lower than the one of MoS_2/RGO composite with 30.8 % and 80 % RhB degraded, respectively. These results prove that RGO plays a significant part for the enhancement of photocatalytic efficiency of MoS_2/RGO composites. By increasing the MoS_2 dispersion over RGO, the recombination of the photo-generated electron-hole pairs will be reduced. In addition, when MoS_2 is combined with RGO, the adsorption capacity of the material increases, improving the photocatalytic activity of MoS_2/RGO in the visible light.

3.7. Factors affecting the photocatalytic degradation of rhodamine B

3.7.1. Effect of pH and initial rhodamine B concentration. The results of analyzing the influence of pH and initial RhB concentration on the photocatalytic efficiency of RhB degradation are presented in figure 9.

At low concentrations, due to the strong adsorption capacity of the material, the photocatalytic role is not significant (figure 9a). As the RhB concentration increases from 20 mg/L to 100 mg/L, the efficiency of the photocatalytic degradation decreases from 80 % to 14 % after 7 h of lighting.

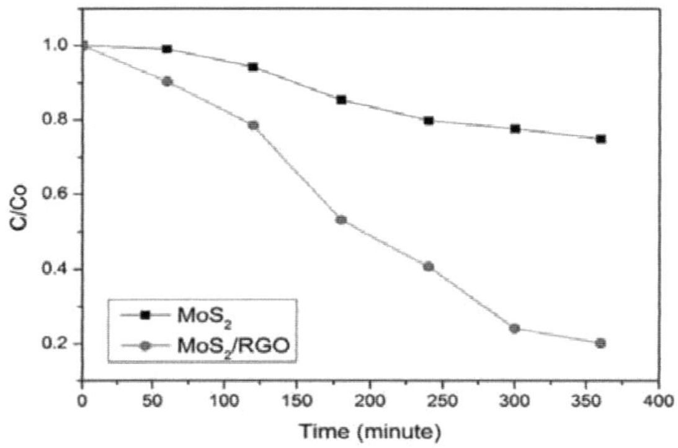

Figure 8. Photocatalytic degradation RhB of MoS_2 and MoS_2/RGO composites.

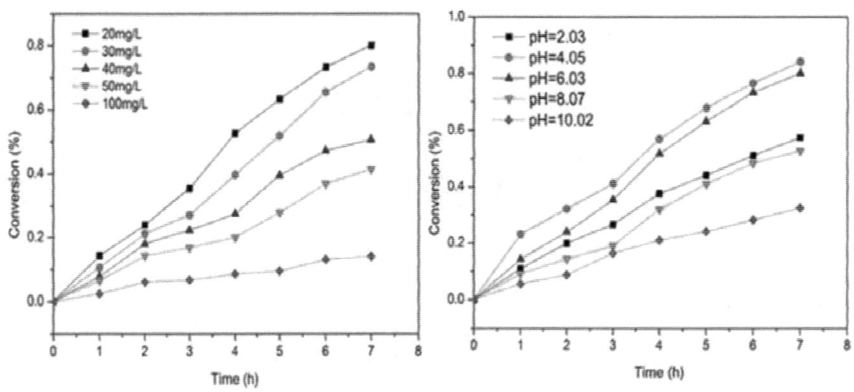

(a) (b)

Figure 9. (a) Effect of initial RhB and (b) pH concentration in photocatalytic degradation RhB on MoS_2/RGO catalyst.

Due to the increase of initial RhB concentration, the adsorption of dye molecules on the catalytic sites increases, reducing the generation of radical. In addition, an increase in the concentration of RhB leads to an increase in the density of RhB molecules, lowering the ability of MoS_2 to contact light photons [14]. In acidic region, photocatalytic efficiency of the composite elevates from 57 % to 84 % as the pH increases from 2.03 to 4.05. However, as the pH reaches 6.04, the photocatalytic efficiency slightly decreases to 80 % and lowers to 32 % as the environment is completely in the alkaline environment (pH = 10.02) (figure 9b). This effect of pH is similar to the findings reported in other research [15].

3.7.2. Effect of light source intensity and quencher. The results of investigation on the effect of light source intensity and quencher on the photocatalytic efficiency of RhB degradation are presented in figure 10.

The results in figure 10a show that as the lamp intensity increases from 40 W to 100 W, the decomposition efficiency of RhB dyes also increases from 68 % to 84 %. This can be explained that as the intensity of the light increases, more energy is provided for the electrons excitation from the valence band to the conduction band, facilitating the process of photogenerated electron-hole separation, then improving the photocatalytic degradation efficiency [16].

Besides, the presence of quencher reduces the efficiency of the RhB degradation. Among the tested quenchers, DMSO (electron) and TB ($^{\bullet}OH$) are less influential than AO (hole) and BQ ($O_2^{\bullet-}$) (figure 10b). This proves that the superoxide radical anion ($O_2^{\bullet-}$) and photogenerated holes are both active particles that have a certain contribution to the optical degradation of RhB and can be explained based on the role of photogenerated holes in direct oxidation of organic compounds [2].

The photocatalytic degradation of RhB over MoS_2/RGO can be illustrated by the graphical diagram in figure 11 [2, 14].

This mechanism is described in details as the following [14].

RhB + hν→RhB ($e_{CB}^- + h_{VB}^+$)

MoS_2/RGO + hν→ MoS_2/RGO ($e_{CB}^- + h_{VB}^+$)

Adsorbed RhB ($e_{CB}^- + h_{VB}^+$) + MoS_2/RGO → MoS_2/RGO (e_{CB}^-) + RhB (h_{VB}^+)

MoS_2/RGO (e_{CB}^-) + adsorbed O_2 (or surface O_2) →$O_2^{\cdot-}$ + MoS_2/RGO

H_2O → H^+ + OH^-

$O_2^{\cdot-}$ + H^+→ HO·

H_2O + MoS_2/RGO h_{VB}^+→ H^+ + HO·

Active species + RhB→ intermediates → CO_2 + H_2O

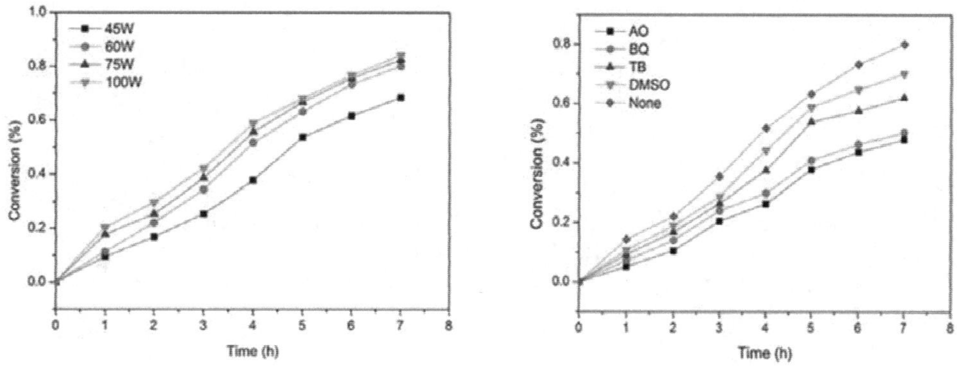

(a) (b)

Figure 10. (a) Effect of light source intensity and (b) quencher in photocatalytic degradation RhB on MoS_2/RGO catalyst.

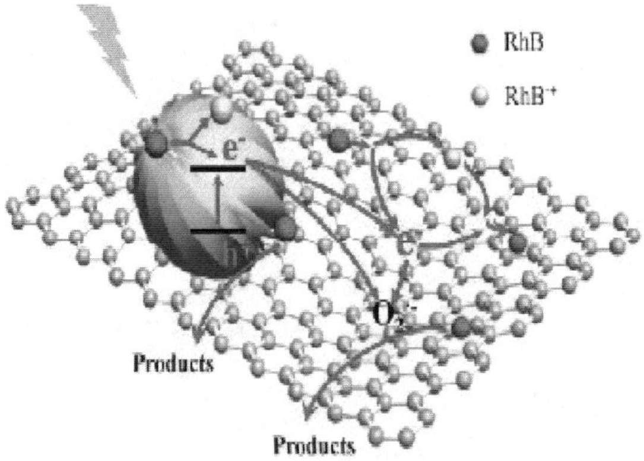

Figure 11. Suggested mechanism for the photocatalytic degradation RhB by MoS_2/RGO composites [2].

To confirm this mechanism, the intermediates and final products of RhB photocatalytic degradation are under analysis and will be published in the next article.

4. Conclusion

In this work, the MoS_2/RGO composite has been successfully synthesized via hydrothermal methods and its photocatalytic activity is much improved compared to the single MoS_2 with 84 % RhB degraded at pH = 4.05 after 7 h lighting with a 60 W–220 V incandescent lamp. Among tested quenchers, DMSO and TB have a less effect on photocatalytic degradation of RhB compared to AO and BQ. With these results, the MoS_2/RGO composite material reveals a high potential in practical application for degradation of organic compounds in the water under solar condition.

Acknowledgments

This research was funded by the National Foundation for Science and Technology Development (NAFOSTED) through the Project code 104.04-2013.66 and by the Quy Nhon University through the project code 62.52.03.01.

References

[1] Jinliang L et al. 2014 *RSC Adv.* **4** 9647-51
[2] Long Z et al. 2016 *RSC Adv.* **6** 60318-26
[3] Goloveshkin A S et al. 2013 *J. Phys. Chem. C* **117** 8509-15
[4] Yang Y, Xu L, Wang H, Wang W and Zhang L 2016 *Mater Design* **108** 632-9
[5] Hummers Jr. W S and Offerman R E 1958 *J. Am. Chem. Soc.* **80** 1339
[6] Cravanzola S, Cesano F, Magnacca G, Zecchina and Scarano D 2016 *RSC Adv.* **6** 59001-8
[7] Avouris P and Dimitrakopoulos C 2012 *Mater. Today* **15** 86-97
[8] Andrijanto E, Shoelarta S, Subriyanto G and Rifki S 2016 *AIP Conf. Proc.* **1725** 020003
[9] Xie H and Xiong X 2017 *J. Environ. Chem. Eng.* **5** 1150-8
[10] Lee J E et al. 2017 *RSC Adv.* **7** 5480-7
[11] Tronganh N et al. 2016 *RSC Adv.* **6** 74436-44
[12] Peng W, Chen Y and Li X 2016 *J. Hazard. Mater.* **309** 173-9
[13] Kumar S et al. 2017 *ChemSusChem* **10** 3588-603

[14] Isari A A, Payan A, Fattahi M, Jorfi S and Kakavandi B 2018 *Appl. Surf. Sci.* **462** 549-64
[15] Zamouche M and Hamdaoui 2012 *Energy Procedia* **18** 1228-39
[16] Muruganandham M and Swaminathan M 2006 *J. Hazard. Mater.* **135** 78-86

Methyl orange dye removal using Ni/Fe-NO_3 and Ni/Fe-[α-$SiW_{12}O_{40}$] layered double hydroxides

A Lesbani[1, 2], T Taher[2, 3], N R Palapa[3], R Mohadi[2, 3], A Yuliana[3] and Mardiyanto[4]

[1]Graduate School of Mathematics and Natural Sciences, Faculty of Mathematics and Natural Sciences (FMIPA), Universitas Sriwijaya, Ogan Ilir 30662, Indonesia
[2]Research Centre of Inorganic Materials and Coordination Complexes, Faculty of Mathematics and Natural Sciences, Universitas Sriwijaya, Ogan Ilir 30662, Indonesia
[3]Department of Chemistry, Faculty of Mathematics and Natural Sciences (FMIPA), Universitas Sriwijaya, Ogan Ilir 30662, Indonesia
[4]Department of Pharmacy, Faculty of Mathematics and Natural Sciences (FMIPA), Universitas Sriwijaya, Ogan Ilir 30662, Indonesia

Corresponding author's email: aldeslesbani@pps.unsri.c.id

Abstract. The material of layered double hydroxide (LDH) Ni/Fe-NO_3 was prepared by coprecipitation method under basic condition. Ni/Fe LDH was modified by intercalation using K_4[α-$SiW_{12}O_{40}$] to form Ni/Fe-[α-$SiW_{12}O_{40}$]. All materials were characterized by X-ray and FTIR spectroscopy. Ni/Fe LDH and intercalated Ni/Fe LDH were used as adsorbent of methyl orange dye from the aqueous solution. The parameter of adsorption such as kinetic and also thermodynamic adsorption of methyl orange on LDHs were investigated. X-Ray analysis showed that the diffraction at 11.53° [003] of Ni/Fe-NO_3 has interlayer distance 7.67 Å shift to 8.20° [003] of Ni/Fe-[α-$SiW_{12}O_{40}$] with interlayer distance 10.78 Å. FTIR spectra of Ni/Fe-NO_3 and Ni/Fe-[α-$SiW_{12}O_{40}$] show the wavenumber at 1600 cm^{-1} due to vibration of LDHs anion. Adsorption of methyl orange onto Ni/Fe-NO_3 and Ni/Fe-[α-$SiW_{12}O_{40}$] follow the pseudo second-order kinetic model, where intercalated material has a higher reactivity than Ni/Fe-NO_3 before intercalation. Isotherm model Freundlich was more fitted than isotherm Langmuir for the adsorption of methyl orange on LDHs with multilayer physical adsorption process. Adsorption capacity of methyl orange on Ni/Fe-[α-$SiW_{12}O_{40}$] as adsorbent was slightly higher (up to 7.226 $mg.g^{-1}$) than Ni/Fe-NO_3 as adsorbent (3.190 $mg.g^{-1}$).

Keywords: Layered double hydroxide, Ni/Fe, K_4[α-$SiW_{12}O_{40}$], methyl orange, adsorption

1. Introduction

The research of dyes treatment from wastewater has been conducted until this decade. Various procedures have been used to eliminate dyes from wastewater, namely coagulation [1], filtration [2], and also adsorption [3]. Adsorption is a reliable method due to the high efficiency, simple, and ease of operation. The successful adsorption process depends on properties of adsorbent. Wide range of adsorbents have been synthesized to remove dyes from wastewater such as bentonite [4], kaolinite [5], zeolite [6], and also layered double hydroxide [7]. Layered double hydroxide (LDH) is anionic layered structure material, which has empirical common formula of $[M^{2+}(1-x)M^{3+}x(OH)_2](A^{n-})x/nH_2O$. LDH has positively charged and exchangeable anion in the interlayer [8]. The structure of LDH has a unique structure due to anion exchange properties to form selective structure for application of LDH. However, the distance of basal spacing of layered materials still has small interlayer constraints and narrow layer spacing due to the small exchange ions such as nitrate, carbonate, sulfate, and chloride thus layered double hydroxides need to be modified to increase its interlayer distance [9].

Adsorption of congo red dye from aqueous solution was conducted using Mg-Al-layered double hydroxide as adsorbent. The adsorption process was controlled mainly by electrostatic interaction [10]. Anionic dye such as congo red was also removed by Mg-Fe-CO_3 as adsorbent by reusing the adsorbent of Mg-Fe-CO_3 layered double hydroxide for several times [11]. The adsorption of methyl orange, which is an anionic dye was also carried out using Mg-Al layered double hydroxide as adsorbent with 3D structure with high adsorption capacity [12]. Intercalated layered double hydroxide with heteropoly blue compound $[PW_{10}Mo_2O_{40}]^{5-}$ showed an ability for cationic dye removal such as methylene blue [13]. The intercalated layered double hydroxide showed a higher adsorption capacity of methylene blue than the pure layered double hydroxide. Thus, the research of intercalated layer double hydroxide with polyoxometalate is an interesting topic for removal of dyes from aqueous solution.

In this research, Ni/Fe-NO_3 was intercalated with Keggin ion $[\alpha\text{-}SiW_{12}O_{40}]^{4-}$ to form Ni/Fe-$[\alpha\text{-}SiW_{12}O_{40}]$ LDH. Keggin ion is used due to the stability of this polyoxometalate and easy to synthesize in acid aqueous medium. Materials were characterized using X-ray and FTIR analyses. The pHpzc of materials were also identified. Then, materials were used as adsorbent of methyl orange from aqueous solution. Factors that influence the adsorption process such as adsorption time, concentration of methyl orange, and adsorption temperature were studied in this research to obtain the kinetic and thermodynamic parameters of adsorption.

2. Materials and method

2.1. Chemicals and instrumentation

Chemicals were supplied from Merck and Sigma Aldrich as follows nickel(II) nitrate, iron(III) nitrate, sodium tungsten, sodium carbonate, sodium metasilicate, hydrochloric acid, and sodium chloride. Keggin ion of $[\alpha\text{-}SiW_{12}O_{40}]^{4-}$ was synthesized according to the literature [XX]. Water was supplied from Research Center of Inorganic Materials and Coordination Complexes, Universitas Sriwijaya. X-ray analysis was measured by Rigaku Miniflex-600 X-ray powder diffraction. Sample was scanned at scan speed 1 deg min^{-1}. FTIR spectra were obtained using Shimadzu FTIR Prestige-21 and spectra were recorded at wavenumber 400–4000 cm^{-1}. Analysis of methyl orange dye was performed using UV-Vis BioBase BK-UV 1800 PC spectrophotometer.

2.2. Synthesis of material Ni/Fe-NO_3 and Ni/Fe-$[\alpha\text{-}SiW_{12}O_{40}]$

Synthesis of Ni/Fe-NO_3 LDH was conducted using coprecipitation method according to Yang et al. [14] with slight modification. Nickel(II) nitrate was dissolved with water to form 0.3 M $Ni(NO_3)_2$. Solution of nickel nitrate was mixed with solution of iron(III) nitrate 0.1 M to form solution A. Sodium hydroxide 2 M was mixed with solution of sodium carbonate 1 M to form solution B. Solution A was mixed with solution B by slowly stirring at room temperature. The solution mixture was adjusted to pH 10–11 by the addition of sodium hydroxide solution. The solid material was formed under constant stirring. Solid material was dried for 24 h to obtain Ni/Fe-NO_3 LDH. Material Ni/Fe-NO_3 LDH was intercalated with Keggin ion to form Ni/Fe-$[\alpha\text{-}SiW_{12}O_{40}]$ using ion exchange method. Compound $K_4[\alpha\text{-}SiW_{12}O_{40}]$ was dissolved with water (solution A). Ni/Fe-NO_3 LDH was mixed with solution of sodium hydroxide (solution B). The solution was introduced with nitrogen gas then the solution A was mixed with solution B under nitrogen condition. The temperature of mixture was adjusted at 40 °C for 24 h to obtain solid material of Ni/Fe-$[\alpha\text{-}SiW_{12}O_{40}]$.

2.3. pH point zero charge (pzc) and adsorption experiments
Analysis of pHpzc was conducted using aqueous solution of sodium hydroxide or hydrochloric acid. A series of solution sodium chloride was adjusted to pH 1–10 using aqueous solution of sodium hydroxide or aqueous solution of hydrochloric acid. Then, LDH was added to the solution series. The solutions series was stirred for 24 h. The solution's pH after filtration was determined using pH meter. pHpzc graph was created by plotting the initial versus final pH.

The adsorption of methyl orange was performed under pHpzc using Ni/Fe-NO_3 and Ni/Fe-[α-$SiW_{12}O_{40}$] as adsorbents. Adsorption was studied by investigation of adsorption time, initial concentration of methyl orange and adsorption temperature. Adsorption time was varied at 5–120 min. Concentration of methyl orange was varied from 10–30 mg L^{-1} at temperature 35, 40, 45 and 50 °C.

3. Results and discussion
Materials Ni/Fe-NO_3 and Ni/Fe-[α-$SiW_{12}O_{40}$] LDHs were analyzed using X-ray analysis and the patterns are shown in figure 1. Material Ni/Fe-NO_3 has diffraction pattern at 11.53° (003), 23.06° (006), 34.83° (012), 39.70° (015) and 60.54° (110) [15]. The formation of well-ordered layer material was identified at 11.53° (003) resulting the interlayer distance of 7.67Å. Material Ni/Fe-NO_3 was intercalated with anion [α-$SiW_{12}O_{40}$]$^{4-}$ to form Ni/Fe-[α-$SiW_{12}O_{40}$]. Ni/Fe-[α-$SiW_{12}O_{40}$] has the diffraction pattern at 8.20° (003), 11°, 15–20°, 22–25°, 35–40° and 59–60°. Diffraction at (003) was shifted to a lower angle due to the insertion of Keggin ion. Interlayer distance at 8.20° (003) was 10.78 Å. The interlayer distance after intercalation was increased to 3.11 Å.

The Ni/Fe-NO_3 and Ni/Fe-[α-$SiW_{12}O_{40}$] LDH FTIR spectra are shown in figure 2a and figure 2b. Material Ni/Fe-NO_3 has vibration at 3400 cm^{-1} (ν OH), 1700 cm^{-1} (ν OH) and 1300 cm^{-1} [16]. The unique vibration at 1300 cm^{-1} was assigned to nitrate vibration. Nitrate ion was located on the interlayer distance on LDH. Intercalated material Ni/Fe-[α-$SiW_{12}O_{40}$] has similar main vibrations with Ni/Fe-NO_3. The intensity of vibration at 1300 cm^{-1} was decreased due to the insertion of Keggin ion [α-$SiW_{12}O_{40}$]$^{4-}$ on interlayer Ni/Fe LDH. Vibrations at 700–900, 1100 cm^{-1} on Ni/Fe-[α-$SiW_{12}O_{40}$] were attributed to W-O-W, W=O, and Si-O vibrations of Keggin ion.

Figure 3 shows that intersection point of Ni/Fe-NO_3 LDH was at 7 while Ni/Fe-[α-$SiW_{12}O_{40}$] at 9. There is no charges at that intersection point and charges of materials were zero. pHpzc below the intersection point is dominated by proton and material with positive charge. On the other hand, negative charge was dominated at above intersection point of pHpzc due to hydroxyl ion form solution. Adsorption of methyl orange on Ni/Fe-NO_3 and Ni/Fe-[α-$SiW_{12}O_{40}$] LDHs was conducted at pHpzc.

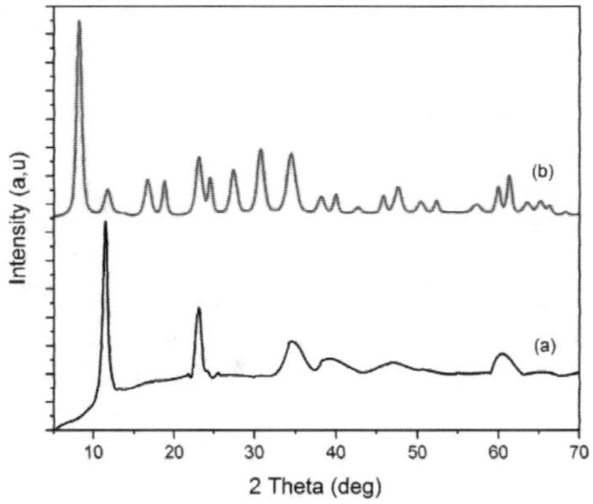

Figure 1. XRD powder pattern of Ni/Fe-NO_3 (a) and Ni/Fe-[α-$SiW_{12}O_{40}$] (b) LDHs.

Adsorption of methyl orange on Ni/Fe-NO_3 and Ni/Fe-[α-$SiW_{12}O_{40}$] LDHs was studied by investigating the adsorption time, concentration of methyl orange and adsorption temperature. Effect of adsorption time of congo red on both Ni/Fe-NO_3 and Ni/Fe-[α-$SiW_{12}O_{40}$] LDHs is shown in figure 4.

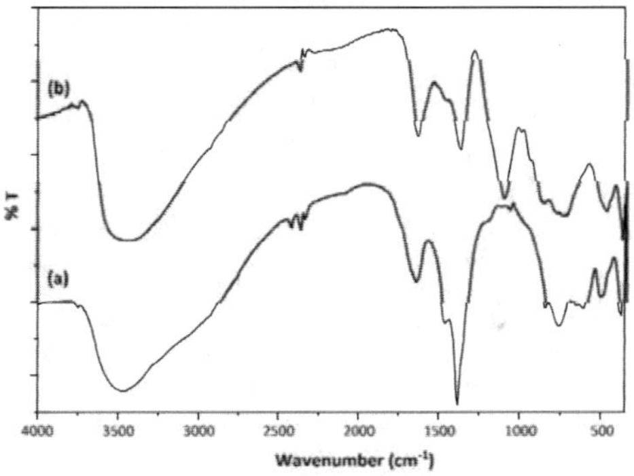

Figure 2. FTIR spectra of Ni/Fe-NO_3 (a) and Ni/Fe-[α-$SiW_{12}O_{40}$] (b) LDHs.

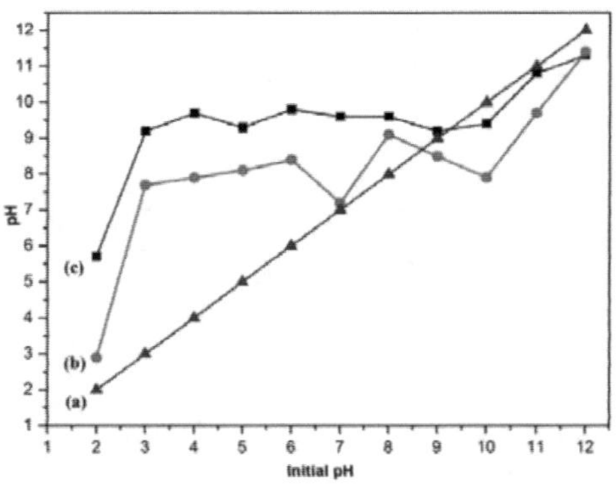

Figure 3. pHpzc graph of initial pH (a), Ni/Fe-NO_3 (b,) and Ni/Fe-[α-$SiW_{12}O_{40}$] (c) LDHs.

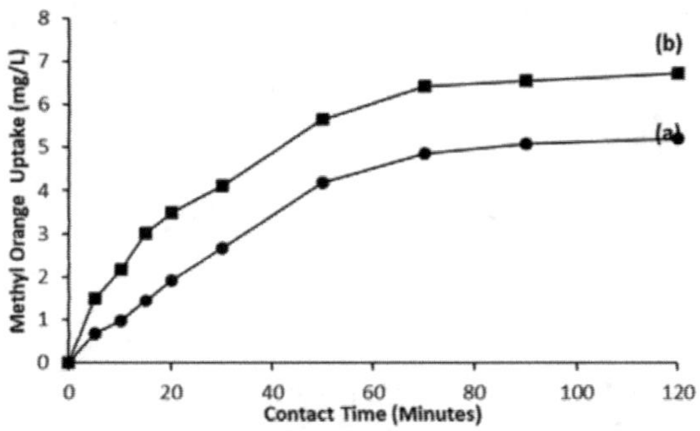

Figure 4. Effect adsorption time of Ni/Fe-NO_3 (a) Ni/Fe-[α-$SiW_{12}O_{40}$] and (b) LDHs.

Adsorption of methyl orange for both LDHs increases with the increase of adsorption time and reached the equilibrium at 70 min for both LDHs. The pattern of methyl orange adsorption on both LDHs was similar but the amount of adsorption was different. The adsorbent Ni/Fe-[α-$SiW_{12}O_{40}$] has a higher adsorption amount than Ni/Fe-NO_3. Data of adsorption time in figure 4 was processed using the pseudo first and second order kinetic models to obtain kinetic adsorption rate for methyl orange on LDH [17].

The data of kinetic model as shown in table 1 showed that methyl orange was adsorbed on both of Ni/Fe-NO_3 and Ni/Fe-[α-$SiW_{12}O_{40}$] LDHs by following the pseudo second-order kinetic model. Adsorption rate for Ni/Fe-[α-$SiW_{12}O_{40}$] was faster than Ni/Fe-NO_3 thus, the adsorbent Ni/Fe-[α-$SiW_{12}O_{40}$] has a higher reactivity than adsorbent Ni/Fe-NO_3.

The effect of methyl orange concentration and temperature adsorption on Ni/Fe-NO_3 and Ni/Fe-[α-$SiW_{12}O_{40}$] LDHs is shown in figure 5. Adsorption of methyl orange was increased with the increase of methyl orange concentration for both LDHs. The increase adsorption temperature was also increasing the adsorption amount of methyl orange. The data in figure 5 showed that adsorption of methyl orange on LDHs was a multilayer process, which was indicated at concentration 20 mg L^{-1} for both LDHs. Thermodynamic adsorption parameter of methyl orange can be obtained using the data in figure 5, which was calculated using thermodynamic equation [18]. The thermodynamic data is presented in table 2.

The data in table 2 showed the value of adsorption capacity, Gibbs energy, enthalpy, and entropy of methyl orange adsorption on LDHs. The adsorption capacity of methyl orange for both adsorbents in the range 3.190–7.226 mg g^{-1} showed an increase of adsorption capacity after intercalated Ni/Fe-NO_3

Table 1. Kinetic model of methyl orange adsorption.

Kinetic model	Value	Adsorbent	
		Ni/Fe	Ni/Fe-[α-$SiW_{12}O_{40}$]
Pseudo first-order kinetic model	Qe exp (mg g^{-1})	2.086	2.686
	Qe calc (mg g^{-1})	1.633	1.774
	R^2	0.429	0.446
Pseudo second-order kinetic model	Qe exp (mg g^{-1})	2.086	2.686
	Qe calc (mg g^{-1})	0.008	0.017
	R^2	0.831	0.964

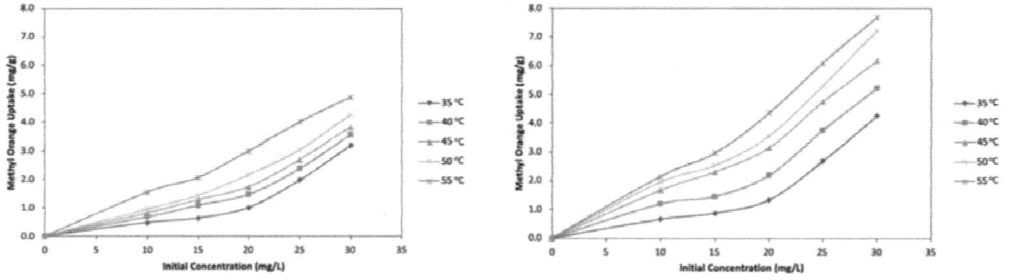

(a) (b)

Figure 5. Effect of methyl orange concentration and temperature adsorption on Ni/Fe-NO_3 (a) Ni/Fe-[α-$SiW_{12}O_{40}$] and (b) LDHs.

Table 2. Thermodynamic adsorption parameter of methyl orange on LDH.

LDHs	T (°C)	Q_e (mg g^{-1})	ΔG° (kJ.mol^{-1})	ΔH° (kJ.mol^{-1})	ΔS° (J.mol^{-1}.K^{-1})
Ni/Fe-NO_3	35	3.190	-20.693	25.707	67.269
	40	3.564	-21.055		
	45	3.826	-21.391		
	50	4.256	-21.728		
Ni/Fe-[α-$SiW_{12}O_{40}$]	35	4.256	-47.117	50.998	153.144
	40	5.221	-47.934		
	45	6.169	-48.700		
	50	7.226	-49.465		

to Ni/Fe-[α-$SiW_{12}O_{40}$]. All adsorption of methyl orange on LDHs was spontaneously occurred due to the negative value of Gibbs energy. Adsorption of methyl orange on LDH was classified as physical adsorption with energy in the range 25.707–50.998 kJ mol^{-1} and also there was a randomness increase of adsorption system on Ni/Fe-[α-$SiW_{12}O_{40}$] as adsorbent.

4. Conclusion

Adsorption of methyl orange on Ni/Fe-NO_3 and Ni/Fe-[α-$SiW_{12}O_{40}$] follows the kinetic pseudo second-order model with high reactivity of Ni/Fe-[α-$SiW_{12}O_{40}$]. Adsorption capacity of methyl orange on Ni/Fe-[α-$SiW_{12}O_{40}$] (up to 7.226 mg g^{-1}) was higher than Ni/Fe-NO_3 (up to 4.256 mg g^{-1}) and adsorption was categorized as physical adsorption with the energy range of 25.707–50.998 kJ mol^{-1}. The adsorption process was spontaneously occurred by the increase of randomness system from Ni/Fe-NO_3 to Ni/Fe-[α-$SiW_{12}O_{40}$].

Acknowledgments

Authors thank to Ministry of Research, Technology and Higher Education, Republic Indonesia for supporting this research through "Hibah Penelitian Dasar (PD)" contract number 0057.10/UN9/SB3.LP2M.PT/2019 and also to Research Center of Inorganic Materials and Coordination Complexes, Faculty of Mathematics and Natural Sciences, Universitas Sriwijaya for research facilities.

References

[1] Li H, Liu S, Zhao J and Feng N 2016 *Colloid. Surf. A Physicochem. Eng. Asp.* **494** 222-7
[2] Ong P J, Tay S W and Hong L 2018 *J. Environ. Chem. Eng.* **6** 4612-22
[3] Clark I, Smith J, Gomes R L and Lester E 2019 *J. Environ. Chem. Eng.* **7** 103175
[4] Taher T, Rohendi D, Mohadi R and Lesbani A 2018 *Chiang Mai J. Sci.* **45** 1770-81
[5] Kamel M M, Youssef B M and Kamel M M 1991 *Dyes Pigm.* **15** 175-82
[6] Alver E and Metin A Ü 2012 *Chem. Eng. J.* **200-202** 59-67
[7] Palapa N R, Mohadi R and Lesbani A 2018 *AIP Conf. Proc.* **2026** 020018
[8] Taher T et al. 2019 *Indones. J. Chem.* **19** 873-81
[9] Shafigh M, Hamidpour M and Furrer G 2019 *Appl. Clay Sci.* **170** 135-42
[10] Lafi R, Charradi K, Djebbi M A, Amara A B H and Hafiane A 2016 *Adv. Powder Technol.* **27** 232-7
[11] Ahmed I M and Gasser M S 2012 *Appl. Surf. Sci.* **259** 650-6
[12] Cao Y et al. 2019 *Mater. Lett.* **257** 126695

[13] Bi B, Xu L, Xu B and Liu X 2011 *Appl. Clay Sci.* **54** 242-7
[14] Yang Y J and Li W 2018 *J. Electroanal. Chem.* **823** 499-504
[15] Zhan T et al. 2019 *J. Colloid Interf. Sci.* **552** 671-7
[16] Khataee A et al. 2019 *Chem. Eng. J.* **375** 122102
[17] Taher T et al. 2018 *AIP Conf. Proc.* **2049** 020011
[18] Taher T, Rohendi D, Mohadi R and Lesbani A 2019 *Arab J. Basic Appl. Sci.* **26** 125-36

Lanthanum based metal organic frameworks as a new binding agent material for diffusive gradient in thin film (DGT) technique in bioavailability of phosphate removal

C Santikasari, A Saefumillah and A Zulys

Department of Chemistry, Faculty of Mathematics and Natural Sciences (FMIPA), Universitas Indonesia, Depok 16424, Indonesia

Corresponding author's email: asep.saefumillah@sci.ui.ac.id

Abstract. High concentration of bioavailable phosphorus in aquatic system has been resulting in the eutrophication phenomenon that can trigger algae bloom. The measurement of bioavailable phosphorus in water body is needed. The diffusive gradient in thin film (DGT) is a well-developed in-situ analytical tool for separation and determination if bioavailable phosphorus or dissolver reactive phosphate (DRP). Using lanthanum based metal organic frameworks
(La-MOF) with high surface area and tuneable pore might attribute the capacity of DRP adsorption in binding film. The surface area analysis of La-MOF powder was 84.957 $m^2 g^{-1}$. La-MOF then dispersed in gel solution to perform a binding film La-MOF then used in DGT tool for DRP uptake. The swelling activity of diffusive film was 2.24 and 1.93 for binding film La-MOF. The elution efficiency of DRP from binding film La-MOF was 95 % using HNO_3 0.05 $mol.L^{-1}$ as an eluent. The diffusion coefficient of DRP from La-MOF DGT was 2.216×10^{-6} $cm^2.s^{-1}$. Optimum DRP uptake capacity was up to 37.83 μg P with optimum time of deployment of 24 h. This research shows that the DRP uptake on DGT La-MOF was depended on the time deployment and DRP solution concentration.

Keywords: Phosphorus, eutrophication, lanthanum, metal-organic, frameworks

1. Introduction

Eutrophication is a major case in freshwater ecosystem, water eutrophication conveys a global challenge in environment [1]. Eutrophication is defined as the over enrichment of water ecosystem with one or more limiting growth factors needed for photosynthesis such as phosphate and nitrate, conduces in the excess growth of phytoplankton and algae [2]. Based on the regulations set by Indonesia Ministry of Environment and Forestry no 51, year 2004, the threshold value of phosphate in natural waters is in the range of 0.016 to 1.19 mgL^{-1}.

Phosphorus (P) is the substance that most often limits the biological productivity in aquatic ecosystem. Spillover consumption of P in water body can result in an eutrophication [3]. Dibasic phosphate, dihydric phosphate (HPO_4^{2-} and $H_2PO_4^-$) are the major species of P found in natural waters, which are believed as bioavailability species of P and considered to be measured as the dissolved organic phosphorus (DRP) [4]. Basic requirement is needed to reduce DRP concentration in order to counter the eutrophication and meets the government regulation. There are many measurement methods available to determine the DRP concentration in natural waters, but most of them are expensive, require an intensive labor, time consuming process and low in sensitivity [5, 6]. Hereinafter, concentration of DRP measured may not show the real concentration within the natural water because the pre-treatment procedure before the measurement. Consequently, in situ sampling is preferable applied for a more accurate measurement of DRP concentration in natural waters [7, 8].

DGT technique is an in situ sampling approach which was developed for sampling the dissolved labile species in natural water, soils, and sediments [9]. The DGT sampler is consisted of a diffusive film, binding film and a filter membrane, analytes will infiltrated the diffusion film to the binding film and bound with binding agent material with intense affinity toward the analytes [10]. The measurement of analytes relies on the affinity of binding agent applied in DGT sampler, therefore the innovation of new binding agent material for DGT has increased in recent years.

Metal organic frameworks (MOF) is organic-inorganic framework that constructed by metal ion and organic linkers which have been utilized as an efficient adsorbent due to their high surface area, tuneable pore and predictable structure [11]. Lanthanum metal organic frameworks have strong Lewis acid properties, which can demonstrate strong ligand adsorption with phosphate which have Lewis basic properties and form lanthanum phosphate complex [12].

This research was aimed to synthesize La-MOFs as a new binding agent for DRP removal. The prepared binding agent La-MOF DGT exhibited a good performance for DRP removal. The La-MOF material characteristics were determined by XRD, surface area analysis, FTIR. The effect of different factors on the DRP removal by La-MOF DGT such as elution efficiency, DRP concentration and contact time were also studied.

2. Materials and method

2.1. Materials

Lanthanum(III)nitrate hexahydrate 99.9 % (Merck), N,N-dimethylformamide (DMF) 99.8 % (Merck), terephthalic acid (BDC) 98 % (Merck), acetone (Merck), acrylamide (Sigma-Aldrich), N,N-methylenebisacrylamide (MBA) (Sigma-Aldrich), N,N,N',N'-tetramethylenediamine (TEMED) 99 % (Sigma-Aldrich, cellulose membrane nitrate 0.45 μm (Whatman), kalium dihydrogenephosphate (Merck), ascorbic acid (Merck), kalium antimonyltartrate (Merck), ammonium molybdate (Merck), sulfuric acid 98 % (Merck), natrium nitrate (Merck), ethanol 99.9 % (Merck), aqua demineralized.

2.2. Preparation of La-MOFs

Preparation of La-MOFs by solvothermal. 0.437 g $LaNO_3.12H_2O$ and 0.254 g terephthalic acid (BDC) was mixed in DMF/water solvent for 30 min then placed into a Teflon-lined autoclave and heated at 120 °C for 20 h. The filtrate then washed using ethanol and acetone three times. The solid of La-MOF was collected and dried at 60 °C for 24 h.

2.3. Preparation of diffusive and binding gel

A polyacrylamide hydrogel 40 % and 1 % MBA was used as gel solution. For diffusive film, 10 mL of gel solution was added with 70 μL of persulfate 10 % and 20 μL TEMED. For binding film, 10 mg of La-MOF powder was dispersed in gel solution. The solution was casted into mould then set to temperature 42–46 °C for 60 min. Diffusive film was stored in sodium nitrate solution and binding film stored in aqua demineralized.

2.4. La-MOF DGT assembly

A membrane cellulose nitrate with pore diameter 0.45 μm was placed at the bottom of DGT disk cap followed by diffusive film and binding film, then DGT piston spacer was fixed and the DGT disk was tightened securely.

2.5. Determination of swelling ratio

The swelling ratio of diffusive and binding film were determined by deploying the diffusive and binding film with diameter 25 mm in aqua demineralized for 2 to 60 h. Then, the weight of diffusive and binding film were measured every hour.

2.6. Determination of elution coefficient

La-MOF binding gel with diameter 25 mm was dispersed in 5 mgL^{-1} of P solution for 24 h then eluted with eluent for 24 h. Elution efficiency is described as factor elution was determined using different concentration of HNO_3 (0.025 to 1 $molL^{-1}$) as an eluent then neutralized with NaOH 0.1 mol L^{-1}.

2.7. Performance test on variation of P concentration

La-MOF DGT was placed in 250 mL P solution with the concentration ranged from 1 to 10 mgL^{-1}, the DGT was set in the middle of a beaker glass by binding with a string. P solution was stirred for 24 h. Phosphate was eluted from La-MOF binding gel using HNO_3 0.05 M.

2.8. Performance test on variation of time deployment

La-MOF DGT was placed in 250 mL P solution, the DGT was set in the middle of a beaker glass by binding with a string. P solution was stirred for 2 to 96 h. Phosphate was eluted from La-MOF binding gel using HNO_3 0.05 M.

2.9. Analytical method and DRP concentration calculation

Concentration of P captured by La-MOF DGT was analyzed by the molybdenum blue spectroscopy method. The mass (M) of DRP trapped in binding gel La-MOF was calculated according to the equation:

$$M = C_e \times (V_{bindng\ gel} + V_{HNO3})/f_e \quad (1)$$

C_e is the concentration of DRP in eluent, V binding gel is volume of binding film, V_{HNO3} is volume of eluent HNO_3 and f_e is the elution efficiency.

The accumulated DRP in the binding gel La-MOF then can be related to its solution concentration (C_{DGT}) by using this equation:

$$C_{DGT} = M \times \frac{\Delta g}{(D \times A \times t)} \quad (2)$$

where Δg is the thickness of diffusive gel and membrane filter, D is diffusion coefficient of DRP, A is area of sampling window that directly contacted with P solution and t is time of deployment.

3. Results and discussion

3.1. Structural characterization of La-MOF

The La-MOF crystal was examined by XRD. Figure 1a shows the pattern of La-MOF, the first peak at 9.50° suggesting that the pattern corresponded to the structure of MIL-53 material [12]. This peak was also observed on La-MOFs (figure 1a) at 9.5371, the position of peak signal is slightly shifted which may due to the unreacted terephthalic acid (BDC) in the reaction or residual of unreacted acetone during the washing process that trapped in La-MOF pore. Other important XRD pattern of La MOF such as 16.70°, 18.56° and 26.49° were also found in the synthesized La-MOF which corresponded with previous findings by Liu et al. [12].

Further confirmation of the La-MOF structure was defined by FTIR. In figure 1b, bands at 1563 and 1397 cm^{-1} are identified as asymmetric and symmetric stretching –COO from BDC. No bands observed at 1716 cm^{-1} for –C=O vibration from BDC indicates that the carboxyl has coordinated with lanthanum ions. Bands at 3059, 827, 675 and 753 cm^{-1} were the characteristic vibration from the aromatic frameworks of benzene ring from BDC which includes γ(=C-H) and δ(=C-H) from the face of benzene ring. Bands at 2938 and 1660 cm^{-1} is the characteristic vibration of asymmetric stretching from methyl groups and stretching from γ(C=O) of DMF which has shifted from 1685 cm^{-1} due to the amine release and ligation of DMF molecules with lanthanum ions by coordination of oxygen atom of the carbonyl group. The bands at 3482 cm^{-1} is the characteristic vibration from water molecules.

Surface area of La-MOF were analyzed by multi-point BET method was 84.957 m^2g^{-1}. The surface area found for La-MOF synthesized using solvent DMF/water was higher than the surface area of La-MOF synthesized using only DMF as solvent which is only 5.47 m^2g^{-1} [13]. By adding water to DMF, it gives a breathing effect and improve the flexibility of La-MOF structure as well for increasing the surface area of La-MOF.

3.2. Swelling activity of diffusive and binding film

The swelling activity of diffusive film and binding film La-MOF need to be determined. The film was cured in an oven at 42–46 °C for one hour, disassembled from the mould and stored in aqua demineralized for hydration treatment. Based on the properties of diffusive and binding film (figure 2), it shows that the film had to be stored for at least 3 h in aqua demineralized to swell into a dimensionally stable form. During the hydration process, excess of ammonium persulfate and TEMED were removed.

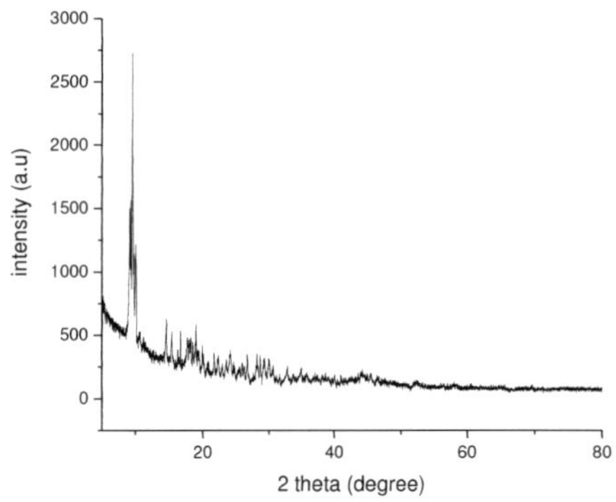

(a)

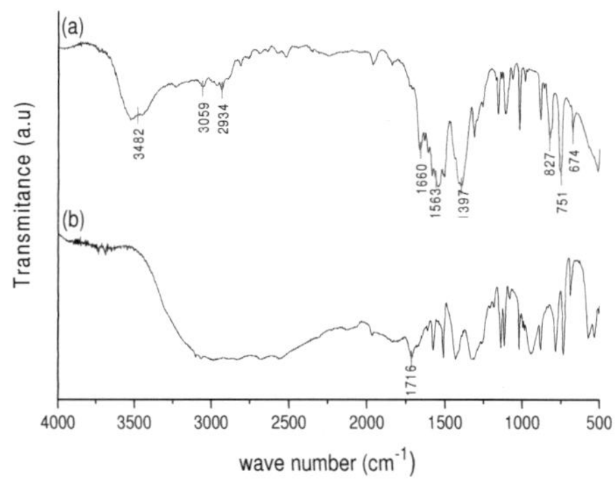

(b)

Figure 1. (a) XRD pattern of La-MOF and (b) FTIR spectra of La-MOF (top line) and BDC (bottom line).

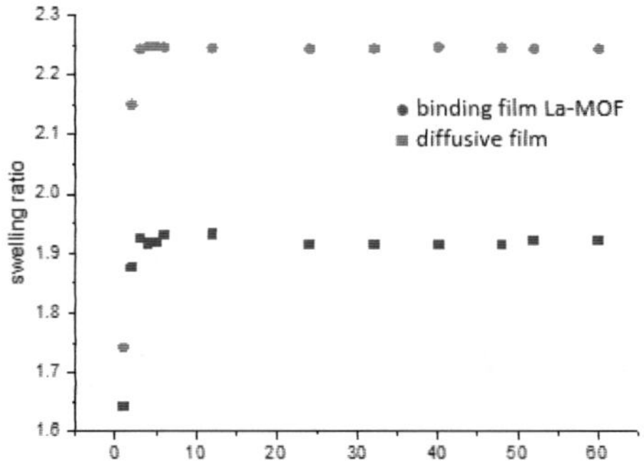

Figure 2. Swelling ratio of diffusive and binding film.

The swelling ratio of diffusive film was 2.24 and 1.93 for binding film. It showed that by adding La-MOF powder to binding film can reduce the swelling ratio because the polymeric structure in binding film became more rigid than in diffusive film. The swelling ratio in this study for diffusive film was lower than the result found by Zhang et al. This corresponded due to the different cross linker used in hydrogel formation, Zhang et al. [14] used acrylamide agarose as cross linker and this research used MBA. The different type of cross linker will affect the porosity of hydrogel which affecting the swelling ability.

3.3. Performance of elution efficiency
Elution is an extraction process of a material by washing it with an appropriate solvent. The higher the elution factor corresponded to the higher recovery value. HNO_3 was used as eluent to release phosphate ion from binding film La-MOF by ion exchange process. Ion phosphate that bound in binding film La-MOF will exchange with nitrate ion from HNO_3, hence there will be anion exchange. After eluted for 24 h, the eluent was then neutralized by 0.1 mol.L^{-1} NaOH to prevent the excess acidic condition while it analyzed by molybdenum blue method to determine the phosphate concentration. Based on the table 1, it shows that 0.05 mol.L-^{1} HNO_3 displays a higher elution factor value than 0.1 and 0.025 mol.L^{-1} HNO_3, therefore 0.05 mol.L^{-1} HNO_3 was chosen as an eluent for binding film La-MOF.

3.4. Diffusion coefficient
The diffusion coefficient (D) of DRP was measured by connecting graph of deployment time and mass of DRP adsorbed in La-MOF DGT. La-MOF DGT was deployed in 5 mg L^{-1} DRP solution for 2 to 24 h. The graph shows linear uptake of DRP in La-MOF DGT with $R^2 = 0.9982$ (figure 3).

The diffusion coefficient is calculated by intercept and slope of the linear line. The D value was 2.216×10^{-6} cm^{2}s^{-1}, which was lower than the previous diffusion coefficient found by Zhang et al. [14] which is 7.39×10^{-6} cm^{2}s^{-1}. This difference could be corresponded to the different cross linker used for hydrogel film which affect the porosity of diffusive film. This recent report was using MBA as cross linker while Zhang et al. [14] was reported to use Lancaster patent agarose as cross linker.

3.5. Capacity of La-MOF DGT
La-MOF binding gel might be saturated when it was used for quantify the high concentration of DRP or used for a long period of deployment. The uptake of DRP by La-MOF DGT was evaluated by deploying La-MOF DGT in 5 mg L^{-1} for 2 to 94 h (figure 4a). The evaluation of P uptake by

Table 1. Elution efficiency from eluent HNO_3.

Eluent	Replication	P abs. in eluent	P concentration in eluent (B)	P concentration adsorbed (A)	Elution factor (A/B)
	1	0.748	2.246	2.510	0.895
HNO_3	2	0.747	2.243	2.513	0.892
0.1 M	3	0.745	2.236	2.510	0.891
			Average		0.893
	1	0.876	2.671	2.805	0.952
HNO_3	2	0.931	2.853	2.706	0.948
0.05 M	3	0.847	2.575	2.706	0.952
			Average		0.951
	1	0.621	1.825	2.139	0.853
HNO_3	2	0.632	1.862	2.139	0.871
0.025 M	3	0.627	1.845	2.125	0.868
			Average		0.864

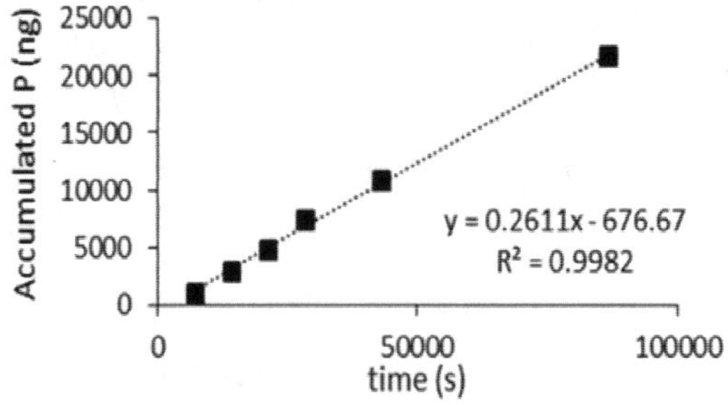

Figure 3. Diffusion coefficient of La-MOF DGT.

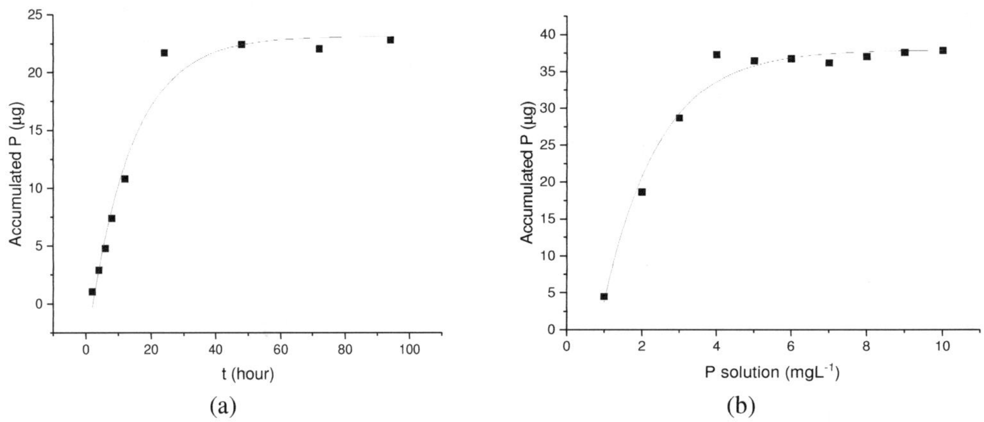

Figure 4. Accumulated DRP adsorbed by DGT La-MOF as a function of (a) time and (b) DRP concentration.

La-MOF DGT was also evaluated by deploying La-MOF DGT for 24 h with variation of DRP solution 1 to 10 mg L^{-1} (figure 4b).

Figure 5 reveals that the DRP uptake by DGT La-MOF was saturated after 24 h. Meanwhile, figure 4b shows the capacity of DRP uptake was saturated when the concentration of DRP solution was higher than 4 mg L^{-1} when the DGT La-MOF was deployed in DRP solution with range concentration of 1 to 10 mgL^{-1}. The highest DRP uptake was 37.83 µg P which was 6 times higher than the capacity of binding agent based ion oxide which is only 6.7 µg P by using 2 g of ferrihydrite slurry as a binding film [14]. Nevertheless, this finding was lower than the Metsorb binding phase with capacity 40 µg P [15], Zr-oxide binding gel [16] and ZrO-Chelex [17]. This different binding capacity might be affected by the different of binding agent amount used in DGT devices. In this research, the amount of binding agent La-MOF used was 10 mg of La-MOF dispersed in 10 ml of gel solution, while Zr-oxide gel was prepared from the mixture of 1.8 g Zirconium oxide and 3.6 mL of gel solution [3]. This results reveal that by using little amount of La-MOF as binding gel (10 mg La-MOF powder in 10 mL of solution gel) will be able to remove DRP up to 37.83 µg P.

The probability mechanism of DRP adsorption of the binding agent La-MOF inside the binding gel La-MOF in DGT system was explained by the previous research by Liu [12]. Liu stated that the proposed

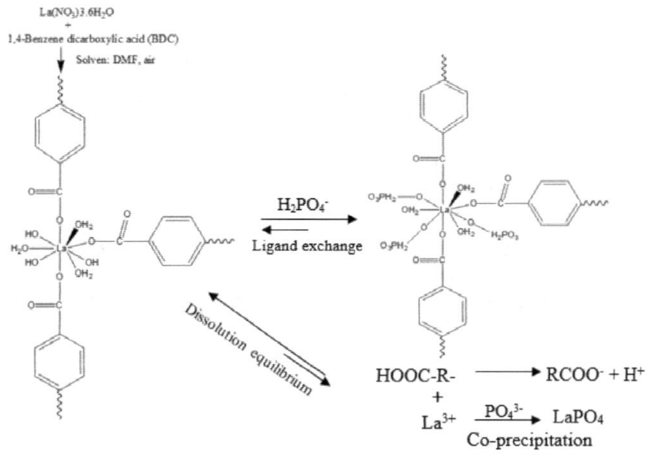

Figure 5. Probability mechanism of phosphate adsorption in binding gel La-MOF.

mechanism for phosphate removal by La-MOFs was by electrostatic attraction between La^{3+} and phosphate to perform $LaPO_4$ white precipitation and ligand exchange between the OH group from La-MOF framework and phosphate ion [12]. BDC prefers to make complexes with La by performed La-O-La and C-O-La, where the hydroxyl groups were responsible for orthophosphate adsorption by ligand exchange.

4. Conclusion

A new binding agent based on lanthanum metal organic frameworks was successfully developed by a simple solvothermal method. By combining DMF/water as a solvent for La-MOF synthesis, a higher surface area was reached. With the high surface area and high affinity to remove the dissolved reactive phosphate (DRP), La-MOF was applied as a binding agent material for binding film in DGT technique. The DGT La-MOF was able to be deployed in a long period of time without any damage in both binding and diffusive film. Furthermore, DGT La-MOF was able to remove the DRP up to 37.83 μg P.

Acknowledgments

This study was supported by Universitas Indonesia-HIBAH PITTA A period 2019 with contract no. NKB-0434/UN2.R3.1/HKP.05.00/2019.

References

[1] Conley D J et al. 2009 *Science* **323** 1014-5
[2] Hudson J J, Taylor W D and Schindler D W 2000 *Nature* **406** 54-6
[3] Ding S, Xu D, Sun Q, Yin H and Zhang C 2010 *Enviro. Sci. Technol.* **44** 8169-74
[4] Maher W and Woo L 1998 *Anal. Chim. Acta.* **375** 5-47
[5] Nollet L M L and Gelder L S P D 2007 *Handbook of Water Analysis* 2nd edition (Boca Raton, FL: CRC Press)
[6] Jarvie H P, Withers P J A and Neal C 2002 *Hydrol. Earth Syst. Sci.* **6** 113-32
[7] Worsfold P J et al. 2005 *Talanta* **66** 273-93
[8] Gardolinski P C F C et al. 2001 *Water Res.* **25** 3670-8

[9] Davison W and Zhang H 1994 *Nature* **367** 546-8
[10] Ding S et al. 2015 *Water Res.* **74** 100-9
[11] Wang C, Liu X, Chen J P and Li K 2015 *Sci. Rep.* **5** 16613-22
[12] Liu H, Guo W, Liu Z, Li X and Wang R 2016 *RSC Adv.* **6** 105282-7
[13] Min X et al. 2018 *Chem. Eng. J.* **358** 321-30
[14] Zhang H, Davison W, Gadi R and Kobayashi T 1998 *Anal. Chim. Acta* **370** 29-38
[15] Panther J G, Teasdale P R, Bennett W W, Welsh D T and Zhao H 2010 *Environ. Sci. Technol.* **44** 9419-24
[16] Ding S, Xu D, Sun Q, Yin H and Zhang C 2010 *Environ. Sci. Technol.* **44** 8169-74
[17] Xu D et al. 2013 *Environ. Sci. Technol.* **4** 10477-84

Metal organic frameworks of lanthanum and iron using BDC linker as catalysts for glucose conversion into 5-hydroxymethylfurfural (5-HMF)

N Riezzati, Y K Krisnandi and A Zulys

Department of Chemistry, Faculty of Mathematics and Natural Sciences (FMIPA), Universitas Indonesia, Depok 16424, Indonesia

Corresponding author's email: zulys@ui.ac.id

Abstract. Lanthanum and Iron-Metal Organic Frameworks (La-MOFs, Fe-MOFs) have been successfully synthesized through solvothermal method and applied as catalysts to convert glucose into 5-Hydroxymethylfurfural (5-HMF). 5-HMF is a very promising energy source which extensively used in chemical industries and polymer materials. MOF is used as source of Lewis and Brønsted acid that can improve the conformation of 5-HMF in glucose transformation. The XRD pattern, surface morphology, chemical composition, absorption spectra, and surface area were characterized by X-Ray Diffractometer (XRD), Scanning Electron Microscopy-Energy Dispersive X-Ray (SEM-EDX), Fourier Transform Infrared (FTIR) and Brunauer-Emmelt-Tellers (BET). The specific peaks of La-MOF on XRD pattern corresponded at $2\theta = 8°$ respectively, while Fe-MOF corresponded to MIL-88B structure at $2\theta = 9°$. SEM analysis of La-MOF and Fe-MOF revealed the catalysts morphology were rod-like and rhombic. The FTIR spectrum of La-MOF showed the broadening peak at 3526 cm^{-1} for O-H stretching vibrations while Fe-MOF had sharp peak intensity at 749 cm^{-1} for C-H bending vibrations. The surface area of the La-MOF and Fe-MOF were 107.107 m^2/g and 215.435 m^2/g, respectively. The activity of La-MOF and Fe-MOF as catalysts was performed in 10 mL vial glass reactor at T = 403 K with varied time from 0–8 h and catalyst loading were varied from 8–12 mg. The product analysis was carried out using High Performance Liquid Chromatography (HPLC) in which Fe-MOF showed higher activity giving yield 26.79 % rather than La-MOF with 19.2 % yield of 5-HMF in 6 h reaction time using 10 mg catalyst in DMSO as solvent.

Keywords: Fe-MOF, La-MOF, glucose conversion, BDC, 5-HMF production

1. Introduction

In the past few years, lignocellulose biomass conversion has been the center of interest to develop by the academics or industry. Lignocellulose biomass can be used as an alternative energy sources which can reduce the use of fuel cells [1]. This biomass contains such as cellulose, hemicellulose and lignin with glucose as the main component. Glucose can be converted into 5-hydroxymethylfurfural (5-HMF) as important precursor to other chemicals, polymer industry, and also biofuel as renewable energy [2]. 5-HMF can be formed by glucose dehydration using homogenous acid, metal ion, zeolite or heterogeneous catalysts [3, 4]. Metal Organic Framework (MOF) as one of the heterogeneous catalyst for glucose conversion has attracted a great attention because their properties, such as high surface area [5], interested structure, gas storage [6], and photo reactive materials. MOFs consist of metal ion as "nodes" and organic ligand as "bridges" which can be synthesized using various metal ion and organic linker. MOF, as a new class of porous materials has been reported successfully for biomass conversion which HMF yield 90 % for a full fructose conversion by using MIL-101(Cr)-SO_3H (15 %) as catalyst [7]. Based on previous report [8], in this study MOF was synthesized using Lanthanum (La) and Iron (Fe) as metal ions and benzenedicarboxylic acid (BDC) as linker. The catalysts were used for glucose conversion into 5-hydroxymethylfurfural (5-HMF). The used of MOFs as catalysts will increase the activity of glucose dehydration to 5-HMF because the amount of Lewis acid and Brønsted acid site should be significant. La-MOF and Fe-MOF were characterized to obtain information on crystal size, surface morphology, thermal stability, functional group and surface area, so it can give the best performance to convert glucose into 5-HMF.

2. Materials and method

2.1. Materials

The entire chemicals were analytical grade and used directly without further purification. The chemicals and reagents used were: lanthanum nitrate ($La(NO_3)_3.6H_2O$) 99 % (Sigma Aldrich), iron chloride ($FeCl_3.6H_2O$) 99 % (Sigma Aldrich), 1,4-benzenedicarboxylic acid ($C_8H_6O_4$) 99 % (BDC) (Sigma Aldrich), N,N-dimethylformamide (C_3H_7NO) (Merck Co), D-glucose monohydrate ($C_6H_{14}O_7$) 99 % (Merck Co), dimethyl sulfoxide 99 % (DMSO) from Sigma Aldrich and ethanol.

2.2. Synthesis of La-BDC MOF and Fe-BDC MOF

La-BDC and Fe-BDC MOF was synthesized using solvothermal method based on previous works [9]. Lanthanum nitrate (1 mmol) with 1,4-benzenedicarboxylic acid (1.5 mmol) was diluted in a solution of 5 mL DMF and a mixture of 0.25 mL H_2O [10] under vigorous stirring for 30 min and transferred into a Teflon lined stainless steel autoclave and heated over 120 °C for 20 h. Afterward, the crystals formed was separated and washed with 5-10 mL DMF and acetone, then dried at 60 °C for 24 h. Fe-BDC MOF was also synthesized using the same method and washed with ethanol. MOFs were characterized using FTIR, XRD, BET and SEM. Powder X-ray diffraction (PXRD) was registered on a Shimadzu 6000 with Cu-Kα 1/2 radiation (λ = 1.54184 Å) and 2θ = 2–30° (step width 0.02° in 2θ) and operated at 40 kV and 40 mA. FTIR analysis was also carried out at room temperature at wavelengths of 400–4000 cm^{-1} using FTIR Prestige 21 Shimadzu's. Scanning electron microscope (SEM) image taken on JEOL JSM-G510LA SEM emission field with lens semi-detector using a working distance of 3 mm and an acceleration voltage of 20 kV. The surface area and pore size were measured using the Quantachrome QuadraWin 2000–16 with the adsorption/desorption method.

2.3. Application of La-BDC and Fe-BDC MOF as catalyst to convert glucose into 5-hydroxymethylfurfural

La-BDC and Fe-BDC MOF were used as catalysts for the glucose conversion into 5-hydroxymethylfurfural (5-HMF). The catalytic tests were performed in 10 mL vial glass reactor at T = 403 K with varied time from 0, 3, 6 and 8 h and also varied amount of catalysts (0, 8, 10 and 12 mg). Then, 2 mL of 10 % wt. glucose solution in H_2O or DMSO were added to the reactor and mixed with catalyst. Blank experiment was carried out without catalyst. After heated for various reaction times, the vial was immediately placed into an ice bath at 5 °C to stop the reaction. The solution was filtered and then the product was analyzed with High Pressure Liquid Chromatography using LC200 Quantine Isochratic Pump with Inertsil ODS column packaged, UV detector with 285 cm^{-1} wavelengths (for 5-HMF) was used.

3. Results and discussion

Metal organic framework MIL-88B of La-BDC and Fe-BDC MOF were produced via solvothermal method based on Zhu et al. [9]. Figure 1 shows the XRD pattern of La-BDC MOF and Fe-BDC MOF. Based on figure 1, it can be seen that there were peaks with high intensity which indicated the crystallinity of La-BDC MOF and Fe-BDC MOF. The diffractogram of La-BDC MOF (figure 1) exhibits clear specific peaks in 2θ = 9.54° and 10.62°. The small impurity was observed at 2θ at 16.37°, 28.84°, 26.75° and 35.09°. Meanwhile, the Fe-MOF shows peaks at 9.36°, 10.01°, 16.74°, 20.30° and 24.25° which are in good agreement with the simulated MIL-88B structure [10].

SEM analysis of La-BDC and Fe-BDC materials are represented in figure 2, in which La-BDC shows morphology of rod-like with larger and agglomerated particles, while Fe-BDC was rhombic and quite uniform in small particles. It is suggested that Fe-BDC MOF with smaller particles will have a higher surface area, which confirmed by BET analysis in figure 3. La-BDC particles were agglomerated and are not uniform compared to rhombic shape Fe-BDC particles [11].

EDX analysis of La-BDC MOF and Fe-BDC MOF is confirmed by the previous research [12]. The element content of La-BDC are C 48.95 %, O 16.91 % and La 34.14 %, while the Fe-BDC consists of C 59.29 %, O 16.50 %, Cl 7.02 % and Fe 17.18 %.

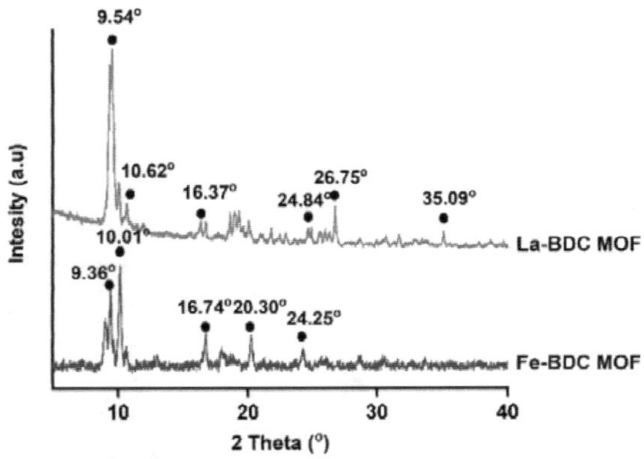

Figure 1. XRD pattern of La-BDC MOF and Fe-BDC MOF.

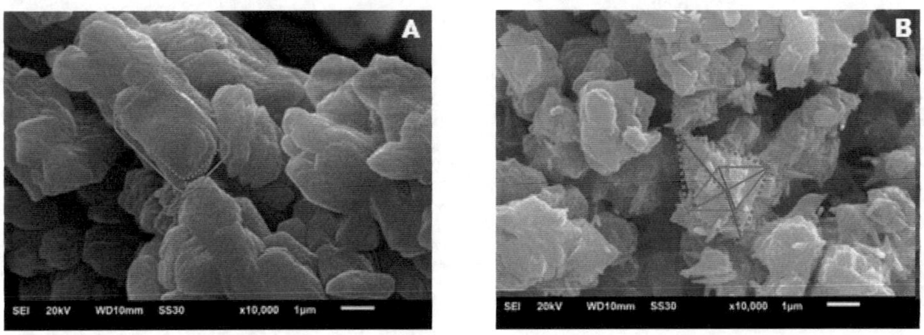

(a) (b)

Figure 2. SEM images of A. La-BDC and B. Fe-BDC for 10,000x magnitude.

Functional groups and vibrational spectra were determined using FTIR as described in figure 4. La-BDC spectrum has broad band at 3526–2666 cm^{-1} with low intensity, assigned for O–H stretching vibration of coordinated and uncoordinated water molecules, indicating that complex interaction between metal and linker was formed. There was low intensity peak at 1593 and 1395 cm^{-1} region which indicated the symmetry and asymmetry C=O stretching vibration from carboxylic in the coordinated framework with $\Delta v = 198\ cm^{-1}$. This difference indicated that the bridging mode of the carboxylic groups are found in this compound. C=C stretching vibration appears in 1400–1564 cm^{-1} area with medium intensity peak from aromatic groups [13].

Fe-BDC spectra show that at 556 cm^{-1} peak belongs to Fe-O symmetric stretching and in area 1434–1391 cm^{-1} indicated the asymmetric and symmetrical O–C–O. The asymmetric C–H bending appeared at 1505 cm^{-1} while at 1652 cm^{-1} the peak from C=O vibration from carboxylic groups appeared with weaker peak. The characteristic peaks shown at 748 cm^{-1} and 461 cm^{-1} represent the C–H bending vibrations form benzene [14].

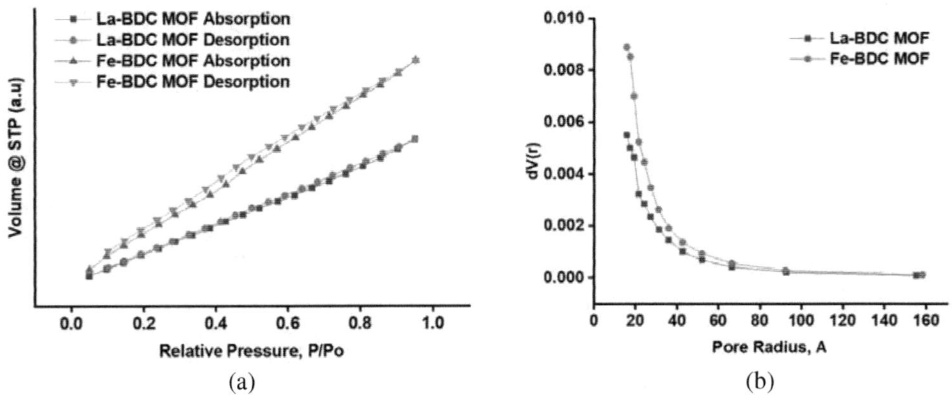

Figure 3. Graph of (a) isotherm absorption and desorption of La-BDC MOF and Fe-BDC MOF, (b) pore size distribution of La-BDC and Fe-BDC MOF.

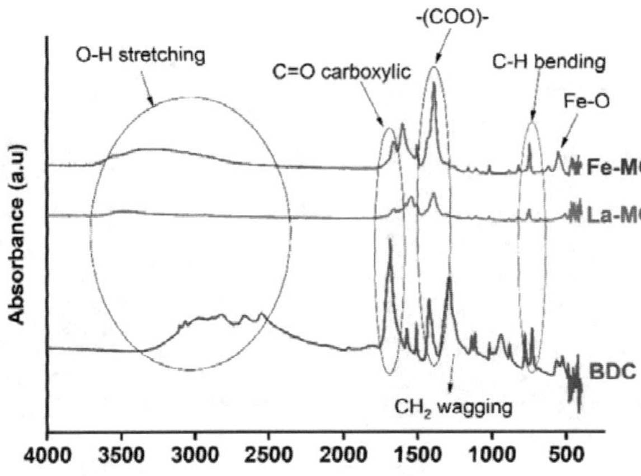

Figure 4. FTIR spectra of BDC, La-BDC MOF and Fe-BDC MOF.

BET analysis was determined to show the surface area by using Bernaeur Emmert Teller (BET) equation.

In this experiment, the surface area for La-BDC MOF was 107.107 m^2/g and for Fe-BDC MOF was 215.435 m^2/g. Lower surface area of La-BDC MOF rather than of Fe-BDC MOF indicated that the form of La-BDC MOF was a closed pore system [15]. Both La-MOF and Fe-MOF pore radius were 16-18 Å, while the pore volume of La-BDC and Fe-BDC were 0.103 cm^3/g and 0.152 cm^3/g, respectively. Based on the isotherm absorption and desorption curves, it can be suggested that the MOFs were Type I materials (microporous) which shown in figure 3. The difference of synthesis condition would give a different specific surface area, in which more irregular shape of particles will give lower surface area [16]. The higher surface area of Fe-BDC has confirmed the uniformity and smaller size of particles observed by SEM, which was expected to become a better catalyst to convert the glucose into 5-HMF rather than La-BDC MOF.

The catalytic activity of La-BDC MOF and Fe-BDC MOF was performed in 10 mL vial glass reactor by using glucose solution in water and DMSO. The first optimization was carried out by varying the reaction time, and the results are shown in figure 5a. The yield of 5-HMF shown in figure 5a with Fe-BDC MOF give a higher activity than La-MOF and blank. When the time reaction increased from 3 h to 6 h there was an increasing yield of 5-HMF from 0.04 % to 1.81 % for La-BDC MOF while Fe-BDC MOF increased from 0.97 % to 11.06 %. Thereafter, as the time reaction increased to 8 h, the yield of 5-HMF were decreased significantly to 0.57 % and 2.78 % for La-BDC MOF and Fe-BDC MOF, respectively, which is similar as the report in [16]. This may due to the decomposition of 5-HMF after 6 h and also because of the reaction pathways for 5-HMF decomposition to form levulinic and humin as by-product of this dehydration reactions as shown by the color change from clear to black [8].

The second optimization was the variation of catalyst loading. There was a significant increase in the yield of 5-HMF as the catalyst loading increased from 8 mg to 10 mg, but as the catalyst loading increased to 12 mg the yield of 5-HMF was decreased. We assumed that 10 mg is the optimum catalyst loading to produce 5-HMF. Fe-BDC also consistently has a greater yield than La-BDC MOF which only 11.22 % while compared with 1.78 % in La-BDC MOF as shown in figure 5b. It was influenced by its material characteristics such as surface area and morphology, in which the Fe-BDC MOF surface area is greater than La-BDC MOF. As the catalyst loading increased, the amount of embedded Lewis acid cites will be increased leading to more of 5-HMF production. However, the 5-HMF produced using 12 mg catalyst loading was decreased, we assumed that this is due to the compounds (including humin) were produced as the catalyst loading increased.

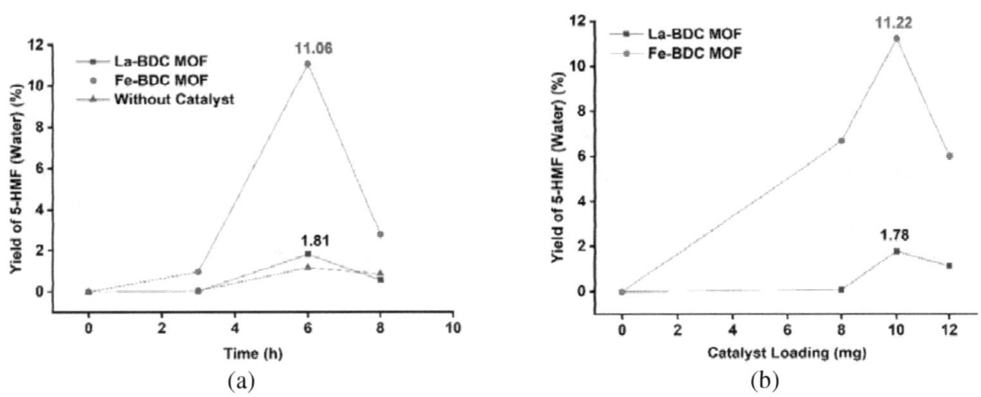

(a) (b)

Figure 5. Catalytic activity of La-BDC MOF and Fe-BDC MOF in glucose conversion to 5-HMF in water solution: (a) variation of reaction time and (b) variation of catalyst loading.

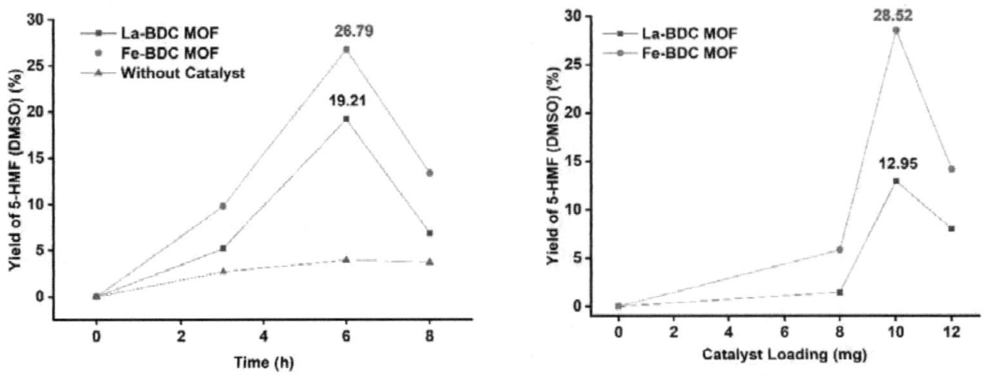

(a) (b)

Figure 6. Catalytic activity of La-BDC MOF and Fe-BDC MOF in glucose conversion to 5-HMF in DMSO solution: (a) variation of reaction time and (b) variation of catalyst loading.

The last optimization process was solvent variation by using DMSO as glucose solvent to determine the stability of MOFs as shown in figure 6. DMSO as glucose solvent was employed in the reaction with variation of reaction time and different catalyst loading. The percentage yield of 5-HMF was increased for La-BDC MOF from 5.17 % to 19.21 % while when using Fe-BDC MOF, the yield was significantly increased from 9.79 % to 26.79 % as the reaction time increased from 3 h to 6 h, but decreased after 6 h reaction time. There was no significant change of % yield over time in the blank experiment. Similarly, the experiments were carried out using water as solvent, the % yield of 5-HMF reached maximum at 10 mg catalyst loading then decreased as the catalyst loading increased.

It is suggested that water can facilitate the conversion of glucose into 5-HMF because of the increase in their solubility of the substrate. On the other hand, the use of DMSO as solvent has given a higher % yield of 5-HMF compared to water, because it can reduce the formation of by-products from the mechanism of glucose dehydration [17]. Also, DMSO will increase the formation of β-D-fructofuranose so it will increase the isomerization of glucose into fructose and also increase the yield of 5-HMF [18].

4. Conclusion

La-BDC and Fe-BDC have been successfully synthesized using DMF and H_2O as solvent via solvothermal method. Based on the results, XRD pattern of materials shows the similarity peaks with MIL-88B. SEM images show that La-BDC was a rode-like and the Fe-BDC was rhombic, while the specific surface area was 107.107 m^2/g for La-BDC and 215.28 m^2/g for Fe-BDC. Based on this work, the optimum condition for glucose conversion to 5-HMF is 10 mg of catalysts mixed with glucose in 10 mL of DMSO (10% w/v solution) and heated until 403 K for 6 h which give % yield of 19.21 % and 26.79 % for La-BDC MOF and Fe-BDC MOF, respectively. To conclude, Fe-BDC MOF is a more active catalyst for glucose conversion to form 5-HMF.

Acknowledgments

This worked is funded by International Indexed Publication for Student Research grant (PITTA B) under contract no. NKB-0607/UN2.R3.1/HKP.05.00/2019 from Directorate of Research and Community Engagement, Universitas Indonesia (DRPM UI) and assisted by Laboratory Directorate General of Customs and Excise (DGCE) in Jakarta.

References

[1] Li X et al. 2019 *Ind. Crops Prod.* **130** 184-97
[2] Herbst A and Janiak C 2016 *New J. Chem.* **40** 7958-67
[3] Qi X, Guo H, Li L and Smith R L 2012 *ChemSusChem* **5** 2215-20
[4] Su Y et al. 2016 *Am. Inst. Chem. Eng.* **62** 4403-17
[5] Pertiwi R et al. 2019 *Catalysts* **9** 1-20
[6] Li H et al. 2019 *EnergyChem* **1** 100006
[7] Chen J et al. 2014 *Green Chem.* **16** 2490-9
[8] Gong J, Katz M J and Kerton F M 2018 *RSC Adv.* **8** 31618-27
[9] Zhu T-T, Zhang Z-M, Chen W-L, Liu Z-J and Wang E-B 2016 *RSC Adv.* **6** 81622-30
[10] Shi L et al. 2015 *Adv. Sci.* **2** 1500006
[11] Pertiwi R 2017 *Investigation of Metal Organic Frameworks (MOFs): MIL-101 and MB2 as Potential Catalyst for Biomass Derived Transformation* Master Thesis (Depok: Universitas Indonesia)
[12] Wu Z et al. 2016 *Energy Fuels* **30** 10739-46
[13] Abdelbaky M S M, Amghouz Z, García-Granda S and García J R 2016 *Polymers* **8** 1-15
[14] Nehra M 2019 *Environ. Res.* **169** 229-36
[15] Pu M et al. 2018 *Appl. Catal. A Gen.* **549** 82-92
[16] Kuster B F M 1990 *Starch - Stärke* **42** 314-21
[17] Baslyman W 2015 *Microwave-assisted Dehydration of Fructose into 5-Hydroxymethylfurfural (5-HMF) over Acidic Porous Catalysts* Master Thesis (Ottawa, Ontario: University of Ottawa)
[18] Mirzaei H M and Karimi B 2016 *Green Chem.* **18** 2282-6

Hierarchical ZSM-5 synthesized from Bangka kaolin and Bayat natural zeolite for methane conversion catalysts

M Faisal, I Abdullah and Y K Krisnandi

Department of Chemistry, Faculty of Mathematics and Natural Sciences (FMIPA), Universitas Indonesia, Depok 16424, Indonesia

Corresponding author's email: yuni.krisnandi@sci.ui.ac.id

Abstract. Hierarchical ZSM-5 has been successfully synthesized from Bayat natural zeolite as alumina and silica source and Bangka kaolinite as an additional silica source using tetrapropylammonium hydroxide (TPAOH) and poly(acrylamide-co-dialyldimethylammonium chloride) (PDDA-M) as a structure and pore directing agent. XRD pattern shows an evidence of MFI (Mordenite Inverted Framework) structure with peaks at $2\theta = 8.01°$, 8.90° and 23.15°. SEM images show that the surface of as-synthesized zeolite has coffin-like structure, while EDX analysis shows it has a Si/Al molar ratio of 8.77. The N_2 isotherm adsorption exhibited a type IV H4 curve, which indicated the material has mesoporosity, although the Barrett-Joyner-Halenda analysis showed the ZSM-5 has a pore size distribution around 1.937 nm, indicated the micropore dominating its pore size. This confirms that the as-synthesized natural ZSM-5 has hierarchical properties. This as-synthesized natural ZSM-5 then was impregnated with Fe_2O_3 to improve its catalytic activity towards methane conversion reaction. XRD analysis shows an existence of Fe_2O_3 at $2\theta = 41.22°$. Both Fe-impregnated synthetic and natural ZSM-5 show a higher percentage of methane conversion (84.825 % and 86.958 %, respectively), but they give a lower percentage yield of formaldehyde (39.973 % and 32.2 %, respectively) than the parent synthetic and natural ZSM-5.

Keywords: Hierarchical ZSM-5, green synthesis, partial oxidation, methane

1. Introduction

Natural gas has a very important role in fulfilling energy needs in the future because of its enormous abundance in nature. Indonesia has very large natural gas reserves, according to PT Perusahaan Gas Negara (2015), Indonesia has proven and potential natural gas reserves of 103 Tcf (trillion standard cubic feet) or about 3 million cubic meters [1]. Most components of natural gas comprise methane (~95 %) and other gases (~5 %). On the other hand methane is one of the gases that cause the greenhouse effect if released into the atmosphere [2].

Although methane gas is a potential energy source, the production and use of methane faces several obstacles, one of which is the difficulties in terms of transportation. Because of its low density (10.75 kWh per cubic meter), methane needs to be compressed at 10–100 atm in order to be transported, this process is still relatively expensive and inefficient [3]. For this reason, it is necessary to develop catalysts that can convert methane gas into liquids in a more efficient way. One way that can be taken is to convert methane to methanol.

Methanol is a liquid that can be easily transported with the existing infrastructure. Moreover, methanol has many benefits such as being used in fuel cells, converted to gasoline or diesel fuel, and used as a precursor to various chemicals [4].

Kaolin is a very abundant mineral discovered in Indonesia. According to data from the Directorate General of Mineral and Coal in the 2015 Indonesia Mineral & Coal Information document, the abundance of kaolin in Indonesia recorded as of 2015 amounted to 1,070,015,564 tons [5]. The abundance of kaolin is spread in Banten Province, Bangka Province and West Kalimantan Province. Kaolin is a clay mineral (clay) with the molecular formula of $Al_2Si_2O_5(OH)_4$ and has a high content of Si and Al atoms which is useful in the synthesis of aluminasilicate minerals [6]. Kaolin has been widely used as a source of silica in the synthesis of zeolites, including SAPO, zeolite A, zeolite X and mordenite [7]. Kaolin has also been used as a raw material for ZSM-5 zeolite synthesis through the methacholine dealumination method resulting from calcination of kaolin at 600 °C for 6 min [8] and also with the addition of an additional silica source SiO_2 sol (40 % wt) [9]. The abundance of kaolin in Indonesia as well as the potential use of kaolin in the synthesis of aluminosilicate minerals, especially zeolites, need to be utilized.

Natural zeolites are volcanic products that freeze into volcanic rocks, sedimentary rocks and metamorphic rocks which then undergo weathering physics processes due to the external influences such as temperature. Indonesia has 500 volcanoes, of which 127 are active (Geological Agency of the Ministry of Energy and Mineral Resources). This data means that there is a potential abundance of natural zeolites from the physical processes of volcanic activity in volcanoes in Indonesia. Some types of natural zeolites formed include clinoptilolite, mordenite, phillipsite, chabazite and laumontite. Natural zeolite is found in different compositions, especially from the composition of the Si/Al ratio and the type of metal which is a minor component [9]. One of the abundant natural zeolites is natural zeolites originating from Bayat district, Klaten regency, Central Java. Bayat natural zeolite has been known to have a Si/Al ratio of 11 with a SiO_2 content of 79.1 % and Al_2O_3 of 7 % [10].

ZSM-5 synthesis has often been carried out with various alumina and silica sources such as sodium aluminate ($NaAlO_2$), tetraortosilicate (TEOS), ludox and others. However, the use of pro-analysis materials as raw material for synthesis is economically less profitable, so the use of natural materials as a source of alumina and silica is a promising alternative. This paper discusses the preparation of ZSM-5 from natural sources and its application as catalyst after modified with iron-oxides, in partial oxidation of methane. The results are compared to the similar reaction using as-synthetic ZSM-5 from pro-analysis precursors.

2. Materials and method

2.1. Materials

Indonesian kaolin from Bangka island was purchased from PT Aneka Kaoline Utama (Bangka Island, Indonesia), Bayat natural zeolite, tetrapropylammonium hydroxide (TPAOH, 40 %, Merck), sodium aluminate (Sigma Aldrich), tetraethyl orthosilicate (TEOS, 98 %, Sigma Aldrich), poly(acrylamide-co-dialyldimethylammonium chloride) (PDDA-M), CH_3COOH (glacial, Merck) and deionized water are used for ZSM-5 synthesis, while iron (III) nitrate ($Fe(NO_3)_3.9H_2O$, Sigma Aldrich) is used to prepare Fe_xO_y/ZSM-5. Methane gas (99.9 % BOC, CV Retno Gas), nitrogen gas (99.99 % UHP, CV Retno Gas) and mixed gas (nitrogen 99.5 % mixed with oxygen 0.5 %, CV Retno Gas) are used for methane partial oxidation reaction.

2.2. Natural zeolite and kaolin pre-treatment

Bayat natural zeolite and Bangka kaolin were processed through several pre-treatment steps before ready to use as starting material for ZSM-5 synthesis [11]. Each of them was sieved until 100 μm and dispersed in water (1:3 w/v). The mixture was stirred, decanted and dried at 300 °C for 2 h. Then, it was purified based on Ming and Dixon method [10], and fragmented using calcination process at 800 °C for 6 h.

The silica was collected from fragmented kaolin by using aqua-regia solution. The solution was decanted, and sediment was washed and dried in an oven at 100 °C for 2 h.

2.3. Synthesis of hierarchical ZSM-5
Synthesis of hierarchical ZSM-5 from natural sources was conducted using Bayat natural zeolite as aluminate and silicate source and Bangka kaolin as an additional silicate source. Initially, both of Bayat natural zeolite and Bangka kaolin were mixed with H_2O then stirred at room temperature for 24 h. After that, TPAOH was slowly added into Bayat natural zeolite suspension under vigorous stirring. After that, Bangka kaolin suspension was added to the mixture. Then, the pH solution was adjusted until ±11 using glacial acetic acid and then stirred at 373 K for 3 h. Afterward, the mixture was cooled down to room temperature and then a certain amount of PDDA-M solution was added dropwise into the mixture and stirred for 48 h, then hydrothermally processed in a Teflon-lined stainless steel autoclave with a volume of 200 mL at 423 K for 144 h. Then, the obtained white powder was filtered and dried at room temperature before calcined at 823 K for 3 h. On the other hand, synthetic hierarchical ZSM-5 synthesis method was adopted from the procedure by Wang et al. [12] with some modifications. $NaAlO_2$ and TEOS were used as the aluminate and silicate respectively source with a mole ratio of 1 Al_2O_3:64.3469 SiO_2:10.0768 $(TPA)_2O$:3571.66 deionized water. Both as-synthesized synthetic and natural ZSM-5 were characterized using FTIR (Bruker Transmission Alpha), XRD (Rigaku Smartlab), surface area analyzer (Quantachrome QuadraWin 2000-16) and SEM-EDX (JEOL JSM-GG510LA).

2.4. ZSM-5 surface modification
Fe_2O_3/ZSM-5 was prepared using a wet impregnation method. Initially, a desired amount of Fe^{3+} 0.2495 M (from $Fe(NO)_3.9H_2O$), was added dropwise into ZSM-5 to make 2.5 wt% of Fe content. The solution was stirred overnight at room temperature followed by calcination at 550 °C for 3 h. Both synthetic and natural Fe_2O_3/ZSM-5 were characterized using XRD (Rigaku Smartlab), surface area analyzer (Quantachrome QuadraWin 2000-16) and SEM-EDX (JEOL JSM-GG510LA).

2.5. Partial oxidation of methane reaction
Partial oxidation of methane was conducted in a batch reactor with an internal volume of 300 mL. Each reaction was carried out with 0.5 gr of catalyst, 0.75 atm of methane gas and 2 atm of nitrogen gas (containing 0.5 % oxygen) with reaction time of 120 min and reaction temperature of 150 °C. The catalyst used were as-synthetized synthetic ZSM-5, as-synthesized natural ZSM-5, and both modified Fe_2O_3/ZSM-5 derivates. Determination of methane gas before and after reaction was carried out with Gas Chromatography (GC) instrument (Shimadzu-8A), with Porapaq Q and Molecular Sieve 5A column and TCD detector. After the reaction finished, the reactor was cooled down to room temperature and 3 mL of ethanol was injected to the reactor to extract the product from the catalyst. The product was analyzed using GC instrument (Shimadzu-2014) with RTX-5 column and FID detector.

3. Results and discussion

3.1. Pre-treatment of natural zeolite

3.1.1. Pre-treatment of Bangka kaolin. Figure 1 illustrates the wide-angle powdered XRD spectra of pre-treated kaolin. The raw Bangka kaolin pattern (brown line) shown in figure 1 confirms the identity of kaolinite [13]. The material showed a strong diffraction peak at $2\theta = 10–15°$, some weaker diffraction peaks at $2\theta = 24–26°$ and weak diffraction peaks at $2\theta = 33–41°$. After thermal treatment of purified kaolin at 800 °C for 6 h yielded metakaolin, characteristic peaks for kaolinite at $2\theta = 10–15°$ and $2\theta = 33–41°$ disappear, while peaks assigned to quartz ($2\theta = 20–21°$ and $2\theta = 26–27°$) remains unchanged [14]. After acid treatment of metakaolin with aqua regia to extract the silica, characteristic peaks for quartz remains unchanged.

3.1.2. Pre-treatment of Bayat natural zeolite. Figure 2 illustrates the wide angle powdered x-ray diffractogram of pre-treated Bayat natural zeolite. The raw ZAB pattern (red line) shown in figure 2 confirms the identity of mordenite structure. The material showed a diffraction peak at 2θ = 8–11° and 2θ = 20–30°. After treatment of purified ZAB yielded fragmented ZAB, those diffraction peaks still remain unchanged, indicate there is no structural change after the treatment.

3.2. ZSM-5 synthesis from analytical grade and natural source

3.2.1. Structure and morphology. All the wide-angle powdered x-ray diffractogram of synthetic and natural ZSM-5 as well as their modified derivatives with iron oxide are shown in figure 3. Natural ZSM-5 synthesized from Bayat natural zeolite and Bangka kaolin exhibited the characteristic peaks correlated with ZSM-5, which means ZSM-5 zeolite had been successfully synthesized. After impregnation with iron oxide, there is a weak diffraction peak at 2θ = 40–42° which correlated to Fe_2O_3 species. These peaks appearance means the impregnation had been done successfully.

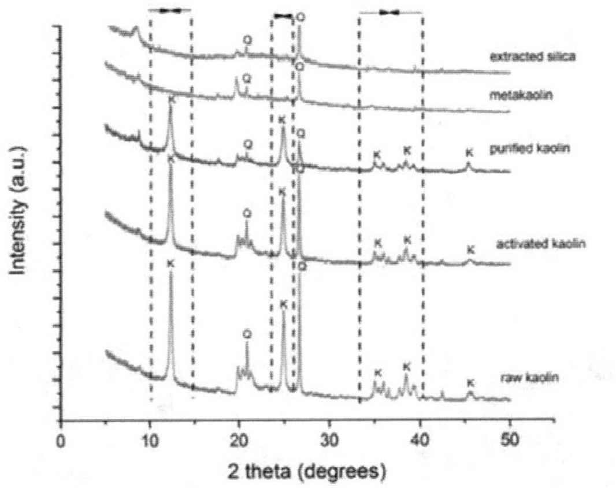

Figure 1. X-Ray Diffractogram of pre-treated kaolin. K = kaolinite, Q = quartz.

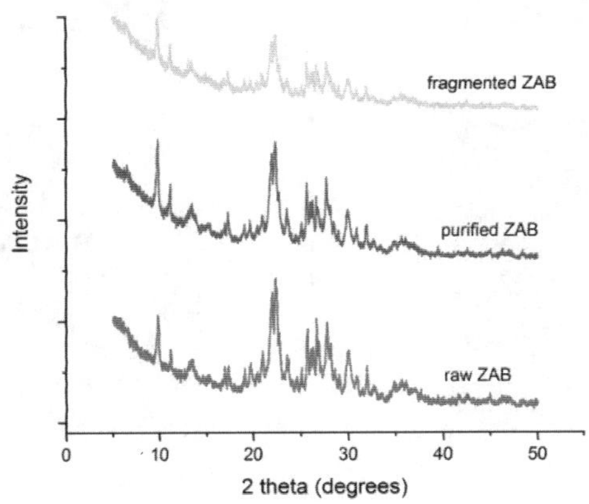

Figure 2. X-ray Diffractogram of pre-treated Bayat natural zeolite (ZAB).

IOP Conf. Series: Materials Science and Engineering **902** (2020) 012045 doi:10.1088/1757-899X/902/1/012045

SEM images of synthetic ZSM-5 and natural ZSM-5 are shown in figure 4. Both samples (a and b) show a hexagonal coffin-like morphology, which is the characteristic of ZSM-5 [15]. However, the shape shown in figure 4a is more uniform which is different from figure 4b. It can be seen in figure 4b that the shape is a little bit irregular which only happens in zeolite synthesized from natural mineral. After impregnation with iron oxide, as shown in figure 4c and figure 4d, the material still shows the same morphology which means the impregnation do not affect the material morphology.

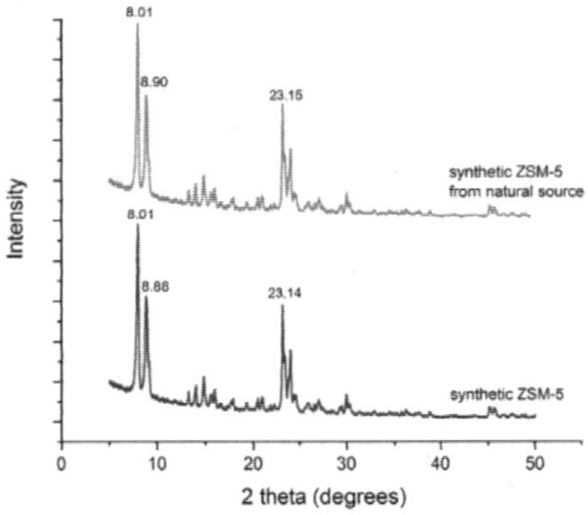

(a)

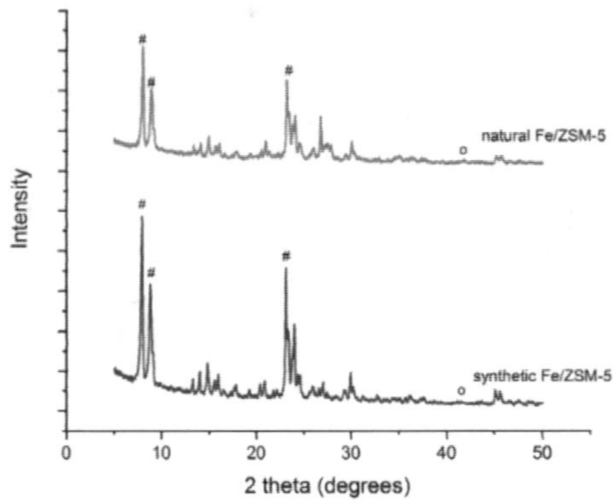

(b)

Figure 3. X-ray Diffractogram of as-synthesized (a) synthetic ZSM-5 and natural ZSM-5, (b) Fe_2O_3/ZSM-5 derivates.

(a)

(b)

(c)

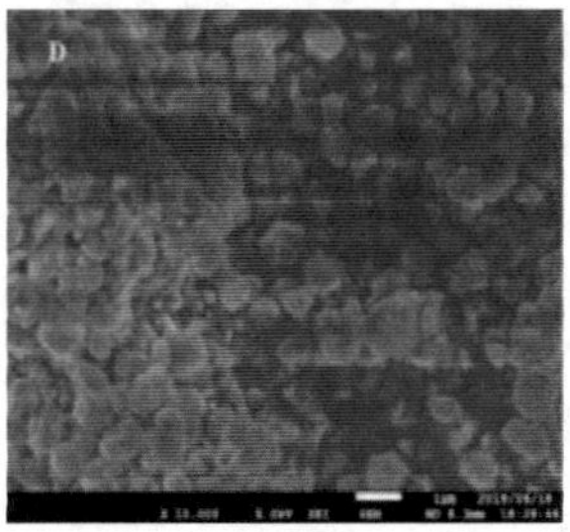

(d)

Figure 4. SEM images of (a) synthetic ZSM-5, (b) natural ZSM-5, (c) natural Fe_2O_3/ZSM-5 and (d) synthetic Fe_2O_3/ZSM-5 with a magnitude of 10,000x.

The elemental analysis of the as-synthesized synthetic ZSM-5 and natural ZSM-5 determined with EDX instrument is shown in table 1. The as-synthesized synthetic ZSM-5 and natural ZSM-5 has Si/Al ratio of 23.16 and 8.18, respectively. Table 1 also shows the atomic percent of Fe in Fe_2O_3/ZSM-5 of about 1.23 % in synthetic ZSM-5 and 2.42 % in natural ZSM-5. Higher Fe-atomic percentage in natural ZSM-5 than in synthetic ZSM-5 is correlated to its lower Si/Al ratio, because lower Si/Al ratio increases the amount of the Brønsted acid sites so it could be capable to accommodate more Fe atom [16].

BET analysis was used to determine the materials textural parameter such as surface area, pore size distribution and pore volume using Brunnaeur-Emmett-Teller equation. The surface area of synthetic ZSM-5 and natural ZSM-5 in order of 281.418 m^2/g and 168.765 m^2/g, respectively. After impregnation with iron oxide, the surface area of synthetic Fe_2O_3/ZSM-5 and natural Fe_2O_3/ZSM-5 decreased to 219.503 m^2/g and 165.875 m^2/g, respectively. This indicates that some iron oxides have grown inside the pores.

The N_2 adsorption-desorption isotherms at -196 °C for synthetic ZSM-5 and natural ZSM-5 are shown in figure 5a. As it can be seen, both of the N_2 adsorption-desorption isotherm displays a type-IV H4 profile which is the characteristic of hierarchical materials [17]. It can be also seen in figure 5a, after impregnation with iron oxide, the curve remains the same which means the impregnation do not change the pore characteristics of the materials.

The textural parameters of both zeolite materials (synthetic and natural ZSM-5) and iron oxide impregnated derivatives are summarized in table 2. As it can be seen from figure 5b, all samples show an average pore diameter of ~1.93 nm and ~2.4 nm for synthetic ZSM-5 and Fe-impregnated synthetic ZSM-5, ~2.7 nm for natural ZSM-5 and Fe-impregnated ZSM-5, which means all samples exhibited a pore of micropore and mesopore, because it is within the range for microporous (d < 2 nm) and

Table 1. Elemental composition of ZSM-5 and Fe-impregnated ZSM-5.

Material	Si	Al	Fe	Si/Al Ratio
Synthetic ZSM-5	28.73	0.62	-	23.16
Fe_2O_3/ZSM-5 Syn	33.33	1.17	1.23	14.24
Natural ZSM-5	34.20	2.09	-	8.18
Fe_2O_3/ZSM-5 Nat	29.27	3.10	2.42	4.72

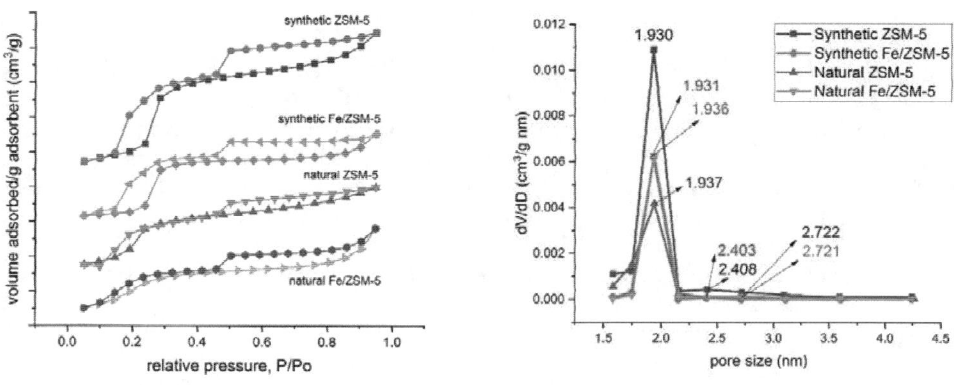

(a) (b)

Figure 5. (a) N_2 isotherm adsorption curves, (b) Pore size distribution of synthetic ZSM-5, natural ZSM-5 and Fe-impregnated synthetic and natural ZSM-5.

Table 2. Textural parameters synthetic and natural ZSM-5 before and after impregnation.

Material	BET surface area [m^2/g]	Total pore volume [cm^3/g]
Synthetic ZSM-5	281.418	0.1751
Synthetic Fe_2O_3/ZSM-5	219.503	0.1297
Natural ZSM-5	168.765	0.0963
Natural Fe_2O_3/ZSM-5	165.875	0.0947

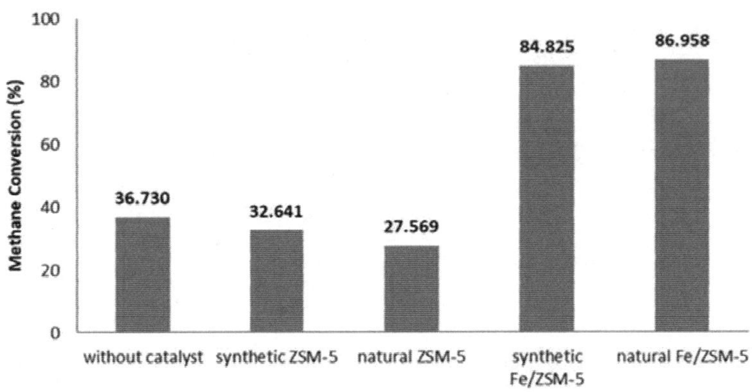

Figure 6. Percentage of methane conversion for all of the catalyst used.

mesoporous (d > 2 nm). On the other hand, natural ZSM-5 has a lower surface area than the synthetic ZSM-5, because as it can be seen from SEM image, natural ZSM-5 has a bigger morphology than synthetic ZSM-5, so its lower the surface area. Both impregnated synthetic and natural ZSM-5 showed lower surface area and pore volume than before impregnation, means the iron oxide has been successfully impregnated at the ZSM-5 surface.

3.3. Catalytic conversion of methane

The catalysts employed for the reaction were as-synthesized synthetic ZSM-5, as-synthesized natural ZSM-5, 1.23 % synthetic Fe_2O_3/ZSM-5 and 2.42 % natural Fe_2O_3/ZSM-5. Based on the experiment, it can be seen in figure 6 that both Fe-impregnated synthetic and natural ZSM-5 have higher percentage of methane conversion because the existence of iron oxide at the ZSM-5 surface makes it more reactive to break the H-CH_3 bond of the methane [18]. On the other hand, the higher percentage of methane conversion of 2.42 % natural Fe_2O_3/ZSM-5 than 1.23 % synthetic Fe/ZSM-5 could be explained by higher Fe-atomic percentage in natural Fe_2O_3/ZSM-5 (2.42 %) than synthetic Fe_2O_3/ZSM-5 (1.23 %).

The partial oxidation of methane in this experiment used O_2 as the oxidant (0.5 % O_2 in N_2 gas). The reaction was performed at 150 °C for 120 min. After the reactor was cooled down to room temperature, ethanol was injected to the reactor to extract the product from the catalyst then the products were analyzed using gas chromatography instrument with RTX-5 column and flame ionization detector.

The reaction results are shown in figure 7. It is shown that the reaction using natural ZSM-5 yielded formaldehyde whereas using synthetic ZSM-5 yielded formaldehyde and methanol. It is also shown the presence of Fe-oxide could not increase the percentage yield of formaldehyde. Both Fe-impregnated

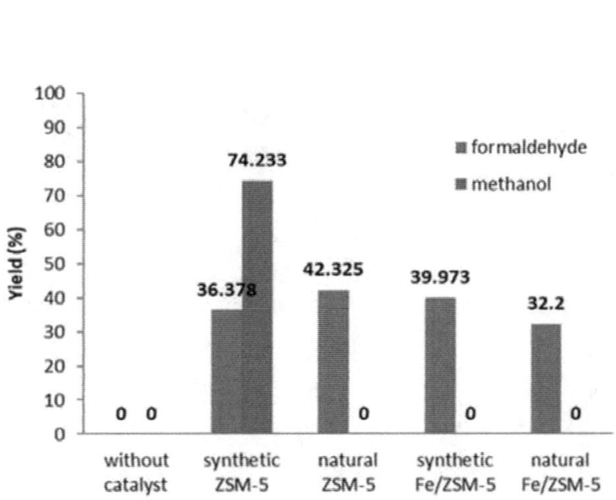

Figure 7. Percentage yield of product of partial oxidation of methane reaction.

synthetic and natural ZSM-5 show a lower percentage yield of formaldehyde (39.973 % and 32.2 %) than non-impregnated synthetic and natural ZSM-5 (36.378 % and 42.325 %).

4. Conclusion

Hierarchical ZSM-5 has been successfully synthesized from Bayat natural zeolite and kaolin Bangka, and its properties are comparable to the as-synthesized synthetic ZSM-5. Both Fe-impregnated synthetic and natural ZSM-5 shows a higher percentage of methane conversion, but they give a lower percentage yield of formaldehyde than the parent as-synthesized ZSM-5 zeolites. So it could be concluded that the synthetic and natural ZSM-5 worked better in the partial oxidation of methane reaction.

Acknowledgments

This worked is funded by International Indexed Publication for Student Research grant (PITTA-B) under contract no. NKB-0636/UN2.R3.1/HKP.05.00/2019 from Directorate of Research and Community Engagement, Universitas Indonesia (DRPM UI).

References

[1] Hakam D F and Asekomeh A O 2018 *Int. J. Energy Econ. Policy* **8** 174-81
[2] Mohajan H K 2011 *Int. J. Econ. Polit. Integr.* **1** 21-34
[3] Senthamaraikkannan G, Chakrabarti D and Prasad V 2014 Transport fuel-LNG and methane future energy *Sustainable and Clean Options for our Planet* 2nd edition improved (Amsterdam: Elsevier Ltd.) chapter 13 pp 271-88
[4] Dalena F et al. 2018 Methanol production and applications: an overview *Methanol: Science and Engineering* (Amsterdam: Elsevier Ltd.) chapter 1 pp 3-28
[5] Ministry of Energy and Mineral Resources Republic of Indonesia 2018 *Handbook of Energy & Economic Statistics of Indonesia (Final Edition)* available at https://www.esdm.go.id/assets/media/content/content-handbook-of-energy-and-economic-statistics-of-indonesia-2018-final-edition.pdf
[6] Qoniah et al. 2015 *Appl. Clay Sci.* **118** 290-4
[7] Pan et al. 2014 *Mater. Lett.* **115** 5-8
[8] Holmes S M, Khoo S H and Kovo A S 2011 *Green Chem.* **13** 1152-4

[9] Feng H, Li C and Shan H 2009 *Catal. Lett.* **129** 71-8
[10] Rohayati, Krisnandi Y K and Sihombing R 2017 *AIP Conf. Proc.* **1862** 030094
[11] Ming D W and Dixon J B 1987 *Clays Clay Miner.* **35** 463-8
[12] Wang L, Zhang Z, Yin C, Shan Z and Xiao F-S 2010 *Microporous Mesoporous Mater.* **131** 58-67
[13] Ekosse G-I E 2005 *J. Appl. Sci. Environ. Manag.* **9** 43-8
[14] Ilić B R, Mitrović A A and Miličić L R 2010 *Hem. Ind.* **64** 351-6
[15] Krisnandi Y K, Mahmuda I, Rahayu D U C and Sihombing R 2018 *IOP Conf. Ser. J. Phys. Conf. Ser.* **1095** 012044
[16] Kalamaras C et al. 2016 *Catal. Lett.* **146** 483-92
[17] Sing K S W et al. 1985 *Pure Appl. Chem.* **57** 603-19
[18] Mahyuddin M H, Staykov A, Shiota Y and Yoshizawa K 2016 *ACS Catal.* **6** 8321-31

IOP Conf. Series: Materials Science and Engineering **902** (2020) 012046 doi:10.1088/1757-899X/902/1/012046

Synthesis and characterization of lanthanide metal-organic framework with perylene 3,4,9,10-tetracarboxylate ligand

L C Christina, J Gunlazuardi and A Zulys

Department of Chemistry, Faculty of Mathematics and Natural Sciences (FMIPA), Universitas Indonesia, Depok 16424, Indonesia

Corresponding author's email: zulys@ui.ac.id

Abstract. Metal-organic framework (MOF) as a semiconductor material has unique properties and potency to be utilized as a photocatalyst. The superiority of MOF compared to other semiconductor materials is due to the more possibility of adjusting its structure and properties by varying the metal type and organic ligand as the linker to create various framework building blocks, for a number of different applications. In this study, MOFs composed of lanthanide metals (samarium and dysprosium) and PTC ligand (perylene-3,4,9,10-tetracarboxylate), were synthesized and their structural and photochemical properties were characterized. MOFs synthesis were carried out using solvothermal method at 170 ºC, with the variation of metal to ligand mole ratio of 1:1, 2:1 and 1:2 for each metal. FT-IR characterization indicates that in all samples of MOF, PTC ligands coordinate with metals in bidentate ligand mode, which is supported by the existence of peak at around 1770 cm^{-1} that resembles stretching vibration peak of conjugated C=O anhydride. XRD characterization reveals a quite good crystalline phase in at least two samples, which are Sm-PTC MOF (1:1) and Dy-PTC MOF (1:1), although their crystal types are not known yet. Characterization with UV-Vis DRS shows band gap energy for all samples of MOF ranges from 2.175 eV to 2.328 eV. The obtained band gap energy values indicate that these MOFs will have an excellent respond toward the visible or solar light and, therefore, can be considered as materials suitable for photocatalysis.

Keywords: Metal-Organic framework, lanthanide metals, perylene 3,4,9,10-tetracarboxylate, semiconductor

1. Introduction

Metal-organic framework (MOF) is a type of porous material with a crystal structure composed of metals and organic ligands as linkers between the metals so as to form its framework structure [1, 2]. MOF has more advantages compared to other porous materials due to the combination of various types of metals and organic ligands which will produce various unique MOF structures, resulting in certain physical and chemical properties. These structures can be used for several applications such as gas storage and separation, adsorption, magnetism, sensors, proton conduction, drug delivery, catalysis, and photocatalysis [3].

MOF is generally a semiconductor which has certain band gap energy, depending on the electronic properties of its constituent metals and ligands. The electronic structure of MOF produces a gap in the energy level of the valence band and the conduction band corresponding to the light energy in the range of ultraviolet and visible rays. When MOF absorbs light in accordance with its band gap energy, pairs of electrons (e^-) and positive holes (h^+) will be formed, with sufficient life time so as to initiate a redox reaction of chemicals on the surface of MOF itself. [4]. An illustration scheme of semiconductors or photocatalysts electronic system can be seen in figure 1.

In the field of photocatalysis, the value of band gap energy from photocatalyst material is of a primary concern. The smaller the value of band gap energy, the wavelength of light used to excite the electron will shift toward the visible light, which means it takes less energy to carry out the photocatalysis reaction. MOF is a potential alternative of semiconductor material for the application of photocatalysis because its tunable structure and characteristics allow the band gap energy of MOF to be controlled by modifying its constituent organic linker or metal ion [5].

Organic linker has an important role in determining the photochemical nature of MOF. By integrating organic ligand that can absorb energy in the visible light region, the band gap energy of MOF can be decreased and this MOF has the potency to be a good photocatalyst material. Organic linkers which can absorb energy in areas of visible light are those colored organic compounds whose structure have many π conjugated bonds. Several studies have shown that MOF which is incorporated with this type of organic compounds has a relatively small band gap energy compared to other semiconductor material such as TiO_2. MOF composed of dye-like ligand which has many π conjugated bonds, 3,3',5,5'-azobenzene tetracarboxylate with lanthanide metal gadolinium (Gd) has a band gap energy of 2.35 eV [6], whereas with Dy metal the band gap energy of MOF can reach up to 2.17 eV [7].

In addition to the ligand, the constituent metal of MOF also influences its photocatalytic properties. A few researches have been focused on the developing lanthanide-based MOF because it has a unique coordination geometry, magnetic properties, and good photo-response properties [8]. Lanthanide-based MOF combined with anthracene as the organic linker which has a sufficient number of π conjugated bonds shows a good band gap energy for photocatalysis, which are 2.11 eV for Ce-MOF, 2.20 eV for Tb-MOF and 2.15 eV for Dy-MOF [9] Therefore, the combination of lanthanide metal and dye-like ligand as the organic linker can produce MOF with excellent band gap energy for the photocatalysis reaction.

Perylene 3,4,9,10-tetracarboxylate (PTC) is a class of dyes consisting of aromatic fluorophore structures with many π conjugated bonds which has applications in various fields such as organic optoelectronics, photonics, and fluorescence sensors [10]. Research on MOF with PTC ligand has not been much developed. However, a paper reports that MOF composed of PTC ligand and potassium metal, $K_8(PTC)_2(H2O)_{1.5}.4H_2O$, have been successfully synthesized and proven to have semiconductor properties with band gap energy of 2.53 eV [11].

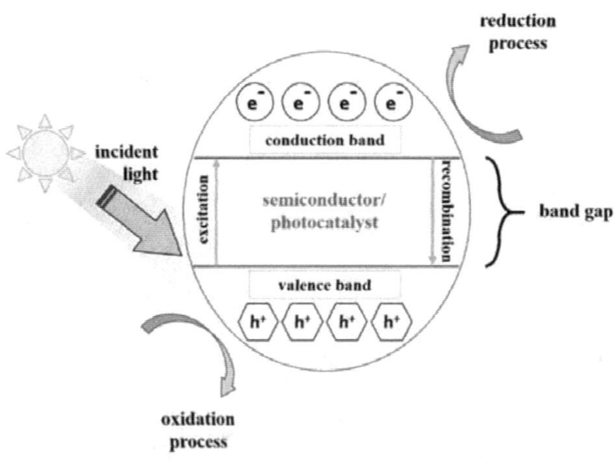

Figure 1. Illustration scheme of a photocatalytic system.

In this study, new MOFs which are constructed of dye-like ligand perylene 3,4,9,10-tetracarboxylate (PTC) combined with lanthanide metals, samarium (Sm) and dysprosium (Dy), were synthesized and their performances as photocatalyst materials were analyzed with UV-Vis Diffuse Reflectance Spectroscopy. The combination of these types of ligand and metal was chosen based on the results of previous studies which showed that each MOF composed of lanthanide metals and PTC ligand has a relatively small band gap energy and has the potential to be used as a photocatalyst material.

2. Materials and method

2.1. Materials

The materials used in this study were $Sm(NO_3)_3.6H_2O$, $Dy(NO_3)_3.6H_2O$, perylene 3,4,9,10-tetracarboxylic dianhydride obtained from Sigma-Aldrich, and NaOH, N, N-dimethylformamide, ethanol obtained from Merck, and distilled water. All materials were used directly without any further purification.

2.2. Experimental procedures

2.2.1. PTC ligand synthesis. PTC ligand preparation was carried out using a modified method of Hariharan, et al. 2017 [10]. 0.5 gr of perylene 3,4,9,10-tetracarboxylic dianhydride (1.27 mmol) was suspended in 50 mL H_2O and 0.356 gr NaOH (8.9 mmol) was added. This mixture was then stirred constantly using a magnetic stirrer for 1 h. The resulting fluorescent greenish-brown solution was then filtered and the filtrate obtained was added with excess ethanol until a yellow precipitate was formed. The yellow precipitate, namely sodium perylene 3,4,9,10-tetracarboxylate (Na_4PTC) was then filtered, washed with ethanol, and dried at room temperature in a desiccator.

2.2.2. MOF synthesis. The synthesis of Sm-MOF-PTC and Dy-MOF-PTC was carried out using the solvothermal method which refers to the research of Sun, et al. 2016 [6]. For each Sm-MOF-PTC and Dy-MOF-PTC, the metal to ligand mole ratio, volume of the solvent, and the method used were the same.

All six MOF samples were each synthesized by dissolving a particular mass of metal and ligand according to table 1 into a mixture of 11 mL H_2O and 4 mL DMF solvents. Then, the result of this dissolution was transferred into a Teflon autoclave and stirred with constant stirring using a magnetic stirrer for 45 min. After stirring was complete, the autoclave was put into an oven at 170 °C for 6 h. The results of this reaction were cooled slowly at room temperature, then filtered and washed using a mixture of H_2O and DMF solvents with the ratio of 11:4, then dried in the oven at 70 °C for 24 h.

Table 1. Summary of the amount of materials used in MOFs synthesis.

Sample	Metal mass	Ligand mass	Solvent volume
Sm-PTC (1:1)	0.4445 g	0.5163 g	15 mL
Sm-PTC (2:1)	0.8890 g	0.5163 g	15 mL
Sm-PTC (1:2)	0.4445 g	1.0326 g	15 mL
Dy-PTC (1:1)	0.4566 g	0.5163 g	15 mL
Dy-PTC (2:1)	0.9132 g	0.5163 g	15 mL
Dy-PTC (1:2)	0.4566 g	1.0326 g	15 mL

2.2.3. MOF characterization. The characterization used to analyze the two MOF samples were FTIR, UV-Vis DRS, and XRD. Measurement of band gap energy of MOFs was carried out using UV-Vis DRS (Diffuse Reflectance Spectrometry) instrument, Shimadzu UV-2450 (λ range 200 nm to 800 nm). Characterization of powder X-ray diffraction (XRD) was carried out to determine the crystallinity of MOFs structure, which used CuKα ray radiation, λ = 1.5406 Å and continuous scan type (0.0167113° per step, scan range from 4.99° to 89.99°).

3. Results and discussion

3.1. PTC ligand characterization
Perylene 3,4,9,10-tetracarboxylate (PTC) as the organic linker used in the synthesis of MOF in this study was the result of hydrolysis of the commercial compound perylene 3,4,9,10-tetracarboxylate dianhydride (PTCDA), which changed the anhydride groups in PTCDA into carboxylic anion groups in PTC ligand.

From the experimental results, a physical change where the red PTCDA turned yellow after the hydrolysis reaction was observed. Changes in the FTIR spectra of the compound before and after hydrolysis shows that PTC ligand synthesis had been successfully carried out. The FT-IR spectrum of the two compounds can be seen in figure 2. Some typical absorption peaks of anhydride groups which appear in the spectrum of PTCDA do not appear in the NaPTC spectrum. In accordance with the IR database by Sigma Aldrich, the absorption peak at around 1775 cm^{-1} shows the stretching vibration of conjugated C=O anhydride, while at around 1025 cm^{-1} the peak shows the CO-O-CO stretching vibration of anhydride group. Both of these peaks are not found in the NaPTC IR spectrum, which indicate that the product of hydrolysis reaction has no anhydride group. The absorption peaks around 1564 cm^{-1} and 1371 cm^{-1} show asymmetrical and symmetrical vibrations of the COO^- group [12], which proves that the anhydride group has been hydrolyzed into a carboxylic anion.

3.2. FT-IR characterization of MOFs
The results of FT-IR characterization for all MOF samples can be seen in figure 3. All spectra of the MOF samples show the absorption peaks at around 1770 cm^{-1} for Sm-MOF, and around 1765 cm^{-1} for Dy-MOF. There is no peak around these wavenumbers found in the IR spectrum of the PTC ligand.

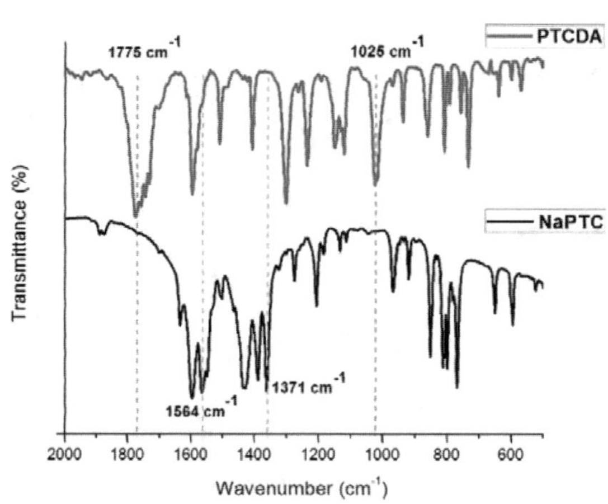

Figure 2. FT-IR spectra of PTCDA and NaPTC.

Peaks around these wavenumbers in all MOF samples are similar to the stretching vibration peak of conjugated C=O anhydride in PTCDA, with lower intensity. This might be an indication that the carboxylic groups in PTC ligand forms coordination with metal that resembles the anhydride groups. All MOF spectra also exhibit absorption peaks at around 1560 cm^{-1} and 1355 cm^{-1}, as in the PTC ligand spectrum, which show asymmetrical and symmetrical vibrations of the COO^- group [12]. However, these two absorption peaks in MOF samples have lower intensity than in PTC ligand. This might indicate that in MOF, PTC ligands coordinate with metals in bidentate ligand mode, which is also supported by the existence of peak at around 1770 cm^{-1} that resembles the stretching vibration peak of conjugated C=O anhydride.

3.3. UV-Vis DRS characterization of MOFs
Characterization with UV-Vis DRS for all synthesized MOF samples is presented in figure 4. The plot of energy hv versus $(\alpha hv)^2$ (using the Tauc equation) obtained from reflectance data are presented in figure 5, while the values of band gap energy results from this plot are shown in table 2. These band gap energy values can be categorized as relatively small when compared to other types of semiconductor materials.

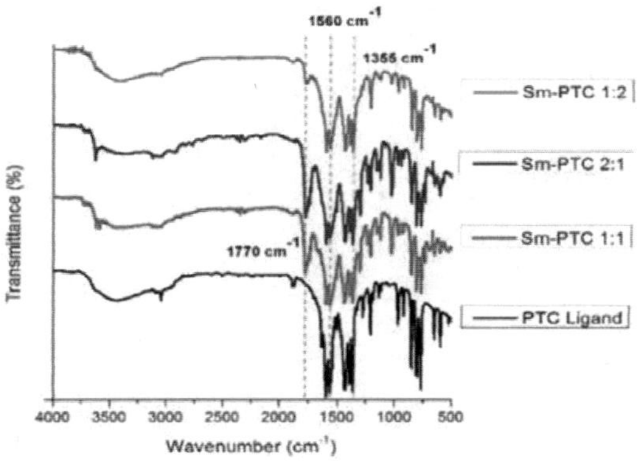

(a)

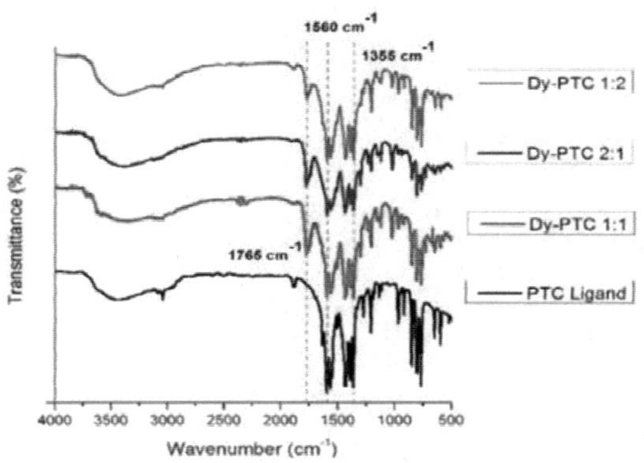

(b)

Figure 3. (a) FT-IR spectrum of Sm-MOF and PTC ligand, and (b) FT-IR spectrum of Dy-MOF and PTC ligand.

Table 2. Value of band gap energy for All MOF samples

Sample	Band gap energy (eV)
Sm-PTC (1:1)	2.217
Sm-PTC (2:1)	2.175
Sm-PTC (1:2)	2.328
Dy-PTC (1:1)	2.235
Dy-PTC (2:1)	2.210
Dy-PTC (1:2)	2.283
PTC Ligand	2.362

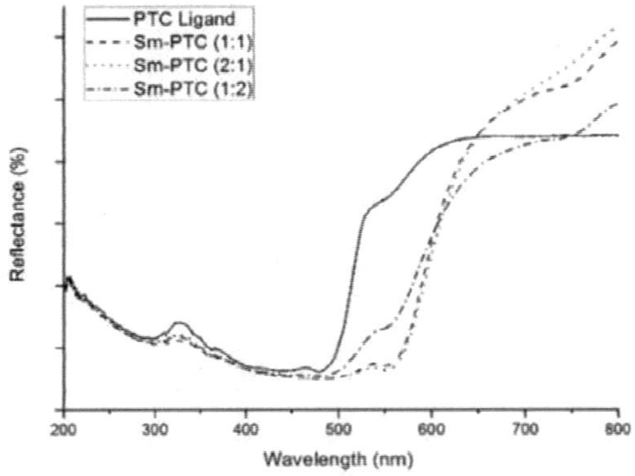

(a)

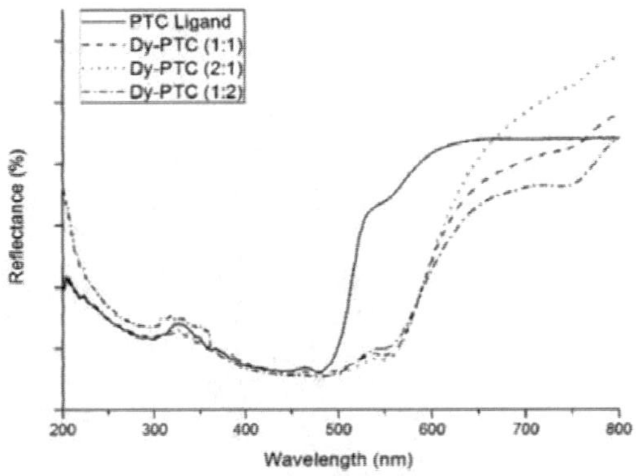

(b)

Figure 4. (a) UV-Vis DRS spectra of Sm-MOF and PTC ligand, and (b) UV-Vis DRS spectra of Dy-MOF and PTC ligand.

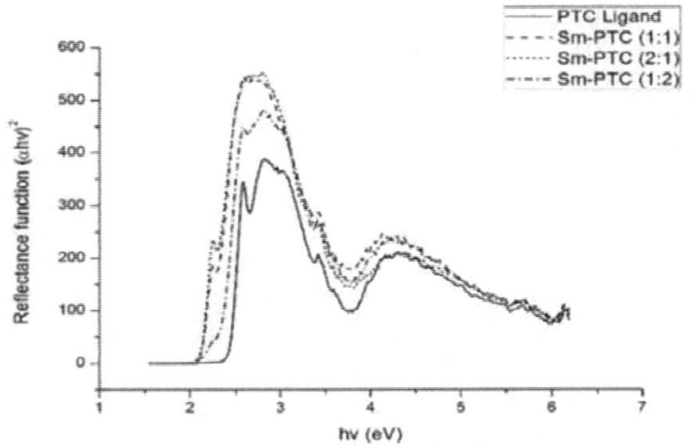

(a)

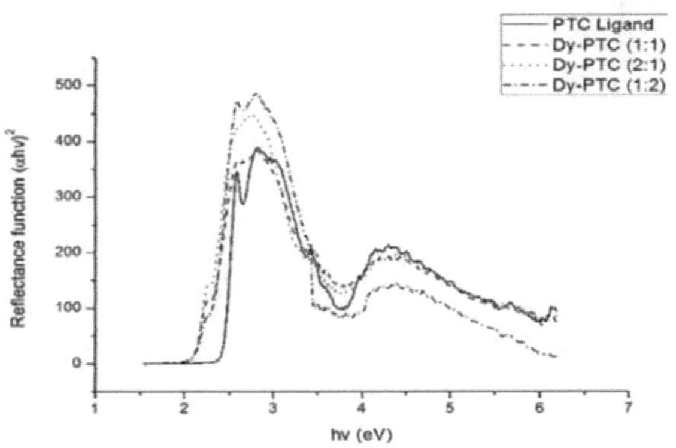

(b)

Figure 5. Tauc plot of (a) Sm-MOF and PTC ligand, (b) Dy-MOF and PTC ligand.

Compared to each other, the value of band gap energy between the samples of MOF 1:1 and 2:1 does not have a significant difference. Sample of MOF 1:2 has a band gap energy value that is slightly higher than the average of MOF 1:1 and 2:1, with a magnitude close to the band gap energy value of PTC ligand. UV-Vis DRS graphs of the MOF samples and PTC ligand are compared in figure 4. From this comparison, it can be seen that there is a shift in the curve line which was originally at a wavelength of around 500 nm for the PTC ligand, to around 590–600 nm for each MOF samples. This might be an indication that there is indeed a formation of coordination between the ligand and the metal resulting in changes in the UV-Vis DRS curve. The curve shapes for all Sm-MOF and Dy-MOF samples have similarities with a slight difference in the reflectance percentage. This difference can be seen in the area of around 700 nm, where for both Sm-MOF and Dy-MOF, the MOF ratio of 1:2 has the lowest reflectance. This might be related to the band gap energy of MOF 1:2 which has the highest value. The reason of this quite significant difference in UV-Vis DRS curve and band gap energy value might be because the formation of MOF crystals for metal to ligand mole ratio of 1:2 was not as good as MOF crystals of 1:1 or 2:1. In this sample of MOF 1:2, the possibility of many ligands remaining in the product which have not been coordinated with metals in the MOF structure, might result in a higher band gap energy value observed in the experiment.

3.4. XRD characterization of MOFs

According to the spectra of FT-IR and UV-Vis DRS for each Sm-MOF and Dy-MOF, the comparison of metal to ligand mole ratio does not provide a significant difference in the shape of the curves, but only slightly different in the intensity of the peaks. This indicates that in general, the shapes of the MOF structure for the metal to ligand mole ratio of 1:1, 2:1 and 1:2 have similarities, with the possibility of slight differences in the structure of MOF 1:2. Taking into account these similarities and also the efficiency of materials used for synthesis, therefore, MOF 1:1 can be assumed as the sample of MOF with the most optimal ratio for the structure crystallinity characterization using X-Ray Diffraction. The results of XRD characterization for Sm-MOF 1:1 and Dy-MOF 1:1 samples can be seen in figure 6, while spectrum comparison between the two samples can be seen in figure 7.

The XRD spectrum for each MOF shows a sharp peak at 2θ around 6.34 °. The presence of sharp peak indicates that the structure of the MOF is crystalline. The intensity of this peak is indeed not quite high, which means that the crystallinity of MOF is low. This is probably due to the short synthesis time (6 h) causing the incomplete crystal formation so that the resulting crystal size was still very small, with many impurities or unreacted reagents. However, all synthesized products are insoluble in various solvents such as H_2O, ethanol, DMF, and DMSO, which gives an indication that the product has the framework structure of MOF, not just a metal-ligand complex. In addition, the presence of a sharp peak in the XRD spectrum also confirms the formation of the sample's crystalline structure.

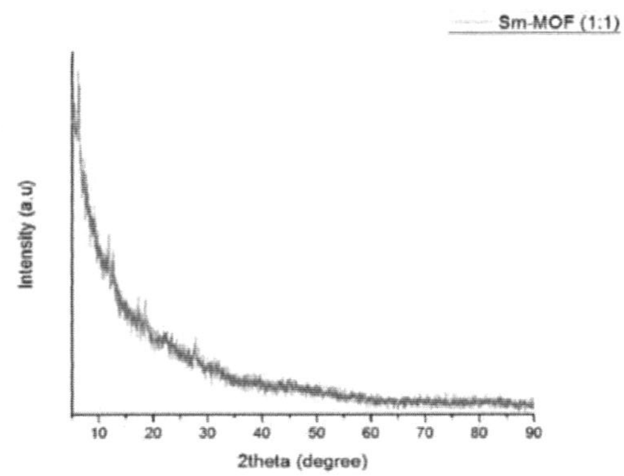

(a)

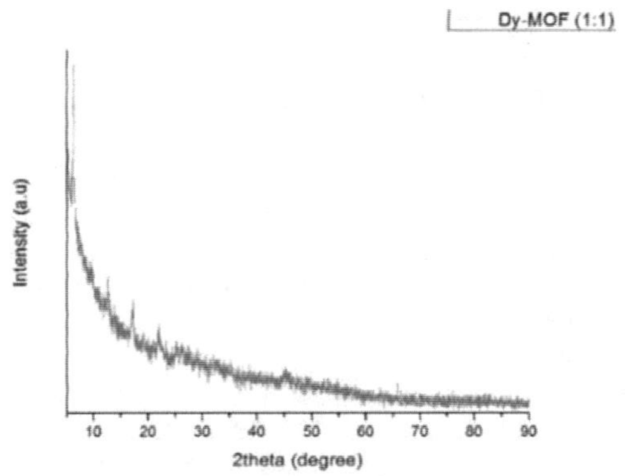

(b)

Figure 6. XRD pattern of (a) Sm-MOF 1:1 and (b) Dy-MOF 1:1.

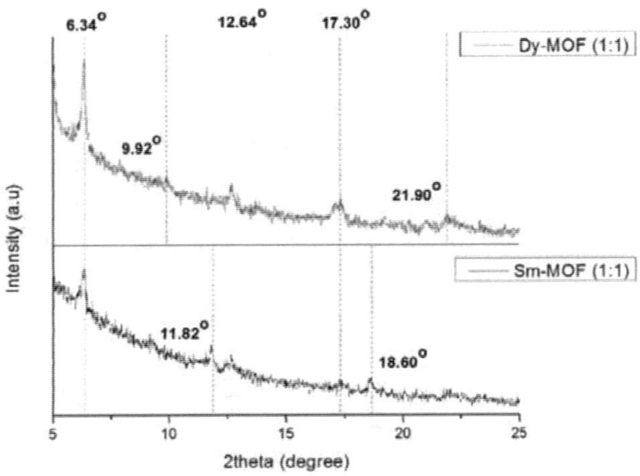

Figure 7. XRD pattern comparison of Sm-MOF 1:1 and Dy-MOF 1:1.

Some small peaks which appear in both XRD spectrum can be seen at 2θ around 12.64° and 17.30°. Peaks which only appear at Sm-MOF are at 2θ around 11.82° and 18.60°, while for Dy-MOF are at 9.92° and 21.90°. The sharpest peak in both MOF spectra is in the region of about 5° to 15°, precisely at 6.34°. The 2θ value where the sharpest peak emerged, which is below 15°, is in good agreement with XRD pattern of most MOF composed of lanthanide metals and dye-like ligands in the previous researches. [6, 7, 13]. Other small peaks may indicate specific peaks for each Sm-MOF and Dy-MOF structure. In addition, the XRD spectrum of both MOFs does not show peaks at 2θ around 25°–40°, which is the area where the typical sharp peaks of each Sm and Dy crystal will appear (according to XRD spectrum simulations for Sm and Dy metals). This indicates that under conditions of MOF synthesis, the metal did not reform the same crystal structure as its metal crystal structure, but formed another structure which was a polymer-like coordination between the metal and the ligand.

4. Conclusion

From FT-IR and UV-Vis DRS characterization results, all Sm-MOF and Dy-MOF samples with different metal to ligand ratios do not show significant differences. FT-IR characterization for MOF samples shows an absorption peak around 1770 cm^{-1}, similar to the peak of conjugated C=O anhydride, which might be an indication that PTC ligands bind to metals in bidentate mode, resembling the structure of anhydride. From the results of UV-Vis DRS characterization, band gap energy values were obtained for each MOF, which values range from 2.175 eV to 2.328 eV. MOF synthesis results in this study are known to have a crystal structure even though the crystal type is still unknown. This is proven by the presence of sharp peaks in the XRD spectrum for at least two samples, namely Sm-PTC MOF (1:1) and Dy-PTC MOF (1:1).

Acknowledgments

This work was financially supported by DRPM Universitas Indonesia under the research grant of HIBAH PITTA B 2019 (Nomor: NKB-0606/UN2.R3.1/HKP.05.00/2019).

IOP Conf. Series: Materials Science and Engineering **902** (2020) 012046 doi:10.1088/1757-899X/902/1/012046

References

[1] Tella A C and Aaron I Y 2012 *Acta Chim. Pharm. Indica* **2** 75-81

[2] Pettinari C, Marchetti F, Mosca N, Tosi G and Drozdov A 2017 *Polym. Int.* **66** 731-44

[3] Zhang Xu, Chen A, Zhong M, Zhang Z, Zhang X, Zhou Z and Bu X H 2019 *Electrochemical Energy Reviews* **2** 29-104

[4] Hagen J 2006 *Industrial Catalysis: A Practical Approach* (Germany: Weinheim, Wiley)

[5] Shen L, Liang R, Luo M, Jing F and Wu L 2015 *Phys. Chem. Chem. Phys.* **17** 117

[6] Sun X, Yu Q, Zhang F, Wei J and Yang P 2016 *Catal. Sci. Technol.* **6** 3840-44

[7] Yu Q, Dong H, Zhang X, Zhu Y X, Wang J H, Zhang F M and Sun X J 2018 *Cryst.Eng.Comm* **20** 3228

[8] Choi J R, Tachikawa T, Fujitsuka M and Majima T 2010 *Langmuir* **26** 10437-43

[9] Xia Q, Yu X, Zhao H, Wang S, Wang H, Guo Z and Xing H 2017 *Cryst. Growth* **17** 4189-95

[10] Hariharan P S, Pitchaimani J, Madhu V and Anthony S P 2016 *Optical Materials* **64** 53-7

[11] Sikdar N, Dutta D, Haldar R, Ray T, Hazra A, Bhattacharyya A J and Maji T K 2016 *J. Phys. Chem.* C **120** 13622-29

[12] Koleva B, Kolev T and Spiteller M 2010 *Spectroscopic Analysis and Structural Elucidation of Small Peptides–Experimental and Theoretical Tools; Advances in Chemistry Research* (New York: Nova Science Publishers Inc) pp 675-755

[13] Chen W, Fan R, Wang P, Dong Y and Yang Y 2018 *Chem. Asian J.* **13** 2126-34

Synthesis of hierarchical Fe_2O_3/ZSM-5 from natural minerals as catalyst in partial oxidation of bio-methane

M Rafli, M Faisal, D A Nurani, I R Saragi and Y K Krisnandi

Department of Chemistry, Faculty of Mathematics and Natural Sciences (FMIPA), Universitas Indonesia, Depok 16424, Indonesia

Corresponding author's email: yuni.krisnandi@sci.ui.ac.id

Abstract. Partial oxidation reaction of bio-methane as a source of methane using ZSM-5 heterogeneous catalyst modified with iron oxide has been carried out. In this study, the ZSM-5 catalyst was synthesized using the double template method using kaolin and natural zeolite as silicate and aluminate sources. The as-synthesized ZSM-5 nat and the Fe_2O_3/ZSM-5 nat catalysts were analyzed using XRD, XRF and BET. The characterization reveals that the synthesis of ZSM-5 was successful and the amount of iron(III) loaded was 2.98 %. The application test of Fe_2O_3/ZSM-5 nat as a catalyst was carried out in stainless steel batch reactor with a certain CH_4: N_2 feed ratio at 150 °C for 120 min. The reaction products were then analyzed with GC-FID to measure the % yield of methanol product. It is found that Fe_2O_3/ZSM-5 nat catalyst could give 54.93 % yield of methanol. Although the use of bio-methane as feed with lower CH_4 : N_2 ratio, this catalyst was also active and produced 12.26 % yield of methanol.

Keywords: Bio-methane, methanol, iron oxide, partial oxidation reaction, ZSM-5

1. Introduction

Increasing the use of renewable energy becomes the most challenging thing over the number of decades. Some major efforts have been done to investigate and develop an effective route for it. One of the most promising renewable energy is biogas due to the environment and economic factors. Especially in Indonesia, biogas with methane and carbon dioxide as the main component has a tremendous possibility and potential value for a clean-fossil-energy [1].

Biogas is a natural gas produced from an anaerobic process of organic materials such as human, household and animal wastes. The high interest to exploit this natural gas is due to the climate change issue because the flaring natural gas causes carbon dioxide [2]. Activation methane in biogas into value-added chemical such as methanol and hydrogen while reducing the greenhouse effect becomes an eminent challenge for industry through the high stability of C-H bond in CH_4 [3]. Some concepts may have been proposed but it constrained by the technology limitation, not economically feasible for local-small scale facilities and required high energy consumption [4].

Direct partial oxidation can be carried out to convert methane into methanol [5]. However, it has drawbacks to obtain the high selectivity and per cent conversion because of the extreme stability of methane. Some studies proof that catalyst is required in partial methane oxidation reaction [6]. ZSM-5 zeolites show the catalytic activity for selective oxidation of methane [7] The unique characteristic of zeolite such as pore size, activation procedure and Si/Al ratio becomes important to product selectivity.

Krisnandi et al. [8] conducted a study of two types of cobalt oxide-modified hierarchical zeolite (Co/hierarchical ZSM-5) catalysts which were modified using ion-exchange or impregnation method. Based on this study, the Co-oxide impregnated ZSM-5 catalyst produced the highest conversion compared to the Co-exchanged ZSM-5 catalyst. Furthermore, the microporous Co-oxide/ ZSM-5 counterpart also gave a lower % yield. Nurgita [9] has conducted an experiment using ZSM-5 catalysts modified by various metal oxides, and suggested that Fe and Co oxides produced the highest methanol yield, compared to other transition metals namely Mn and Ni.

Synthesis of ZSM-5 from natural resources, i.e. kaolin and natural zeolite from Indonesia has been carried out in our group [10, 11] modifying a procedure reported by Wang et al. [12]. The novelty of this study was to employ a hierarchical ZSM-5 synthesized from natural sources, modify with Fe_2O_3, in similar partial oxidation of methane to methanol. In addition, bio-methane was also introduced as feed in the reaction, in order to observe the feasibility of the partial oxidation reaction using bio-methane.

2. Materials and method

2.1. Materials

Natural Bayat zeolite, Belitung kaolin (PT. Aneka Kaoline Utama), distilled water, bio-methane (methane 83.7 % and oxygen 16.3 %, PT. Swen Inovasi Transfer), hydrogen chloride (PT. Smart Lab Indonesia), nitric acid, sodium hydroxide pellets, glacial acetic acid, 30 % hydrogen peroxide, sodium ditionite, sodium acetate, tri sodium citrate, ethanol, glacial acetic acid (Merck), tetra propyl ammonium hydroxide, poly(acrylamidedialildymethylammonium chloride), $Fe(NO_3)_3.9H_2O$ (Sigma Aldrich), methane p.a. 99.9 %, oxygen gas, nitrogen gas (CV. Retno Gas).

2.2. Pre-treatment of natural Bayat zeolite and Belitung Kaolin

Natural Bayat zeolite was treated through an activation process by dissolving zeolite in distilled water with ratio 1:3 (w/v), then the zeolite was purified by adding Na-acetate buffer solution with 1:3 (w/v). Fragmentation process was done through sub-molten salt (SMS) system by using NaOH pellets with molar ratio 2:1 (w/w) in deionized water.

2.3. Synthesis of hierarchical ZSM-5 and ZSM-5

Hierarchical ZSM-5 zeolite was synthesized using natural Bayat zeolite as the natural source of silicate and aluminate. In addition, silica extracted from kaolin was added as an additional source of silica. Hierarchical ZSM-5 synthesis was carried out by referring to the mole composition of Wang et al. [12]. Poly dimethyl diallyl ammonium chloride was added at room-temperature process as the secondary template and stirred for 24 h before transferred into Teflon-lined autoclave. The obtained zeolite was labelled as ZSM-5 nat.

2.4. Preparation and activation of natural Fe_2O_3 / ZSM-5 catalysts

The as-synthesized hierarchical ZSM-5 nat catalyst was mixed with Fe(III) solution of 0.2495 M through the impregnation method. Characterization of both as-synthesized and modified ZSM-5 was carried out using Powder X-ray diffraction (PXRD) in Shimadzu 6000, with Cu-K radiation (λ = 1.54184 Å) and 2θ = 4–70° (step width 0.02° in 2θ) and operated on 40 kV and 30 mA, and the NOVA 2200e SAA analyzer using the BET method using adsorption / desorption. The ZSM-5 which has been impregnated with iron oxide, was characterized using XRF to determine the % loading of metal impregnated into the ZSM-5.

2.5. Methane partial oxidation application test

Typical catalytic test procedure [8] was carried out in a batch reactor of 200 mL stainless steel vessel. The application of bio-methane, as a source of CH_4 gas, was also carried out with bio-methane: N_2 ratio is 0.2:2 bar. The measurement of product concentration was carried out by the standard addition method.

Analysis of reaction results was carried out using GC-FID Shimadzu 2010 which was equipped with a carbowax column to determine the methanol from the reaction product.

3. Results and discussion

3.1. XRD characterization of raw material

The natural zeolite as aluminate and silicate sources needs to be pre-treated (activation and purification) process to remove impurities. Then, the fragmentation process was performed to breakdown the zeolite structure. Figure 1a shows that the raw and purified zeolites contain a typical pattern of mordenite structure at 2θ of 9.85°; 22.21° and 25.61° [13]. All those peaks were not observed in the fragmented zeolite. It can be concluded that the fragmentation process was done successfully. Figure 1b shows the comparison between kaolin, metakaolin and extracted silica indicating that the extraction process has been done successfully because the pattern of amorphous extracted silica is different from the initial kaolin.

3.2. XRF analysis of raw material

Table 1 gives information about the composition about ratio Si and Al (Si/Al) from zeolite, kaolin and extracted silica. It can be seen that fragmented natural Bayat zeolite has 9.79, kaolin has 0.85 and extracted silica has 12.02 for Si/Al ratio.

3.3. XRD Characterization of ZSM 5 nat

Figure 2 shows that XRD result of ZSM-5 nat has typical peaks of MFI structure at 2θ of 7–10° and 22–25° [14]. The XRD result is also compared with the International Zeolite Association (IZA) Online

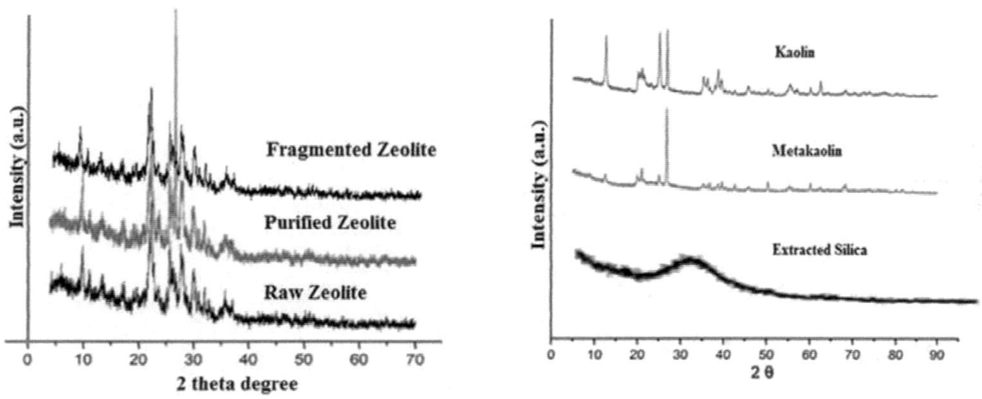

(a) (b)

Figure 1. XRD patterns of silicate and aluminate sources, before and after pre-treatment (a) Natural Bayat zeolite, and (b) Kaolin.

Table 1. XRF analysis of natural zeolite, kaolin, extracted silica and ZSM-5 nat.

Element	Fragmented natural Bayat zeolite	Kaolin	Extracted silica	ZSM-5 nat
Si	57.19	57.11	84.985	71.66
Al	5.84	66.78	7.070	4.23
Si/Al	9.79	0.85	12.02	16.94

structure database [15] and shows a similar pattern of ZSM-5. So, it can be concluded that the synthesis of ZSM-5 nat has already performed successfully. It also shows XRD pattern after impregnation iron oxides on ZSM-5 nat indicating the typical peaks for hematite iron oxide compounds on ZSM-5 material at 2θ of 30°, 42° and 64° [16].

3.4. Elemental analysis

Elemental analysis from XRF data shows that ZSM-5 nat has Si/Al ratio of 16.94. Furthermore, the percent loading of Fe(III) in Fe_2O_3/ZSM-5 nat was 2.98 %.

3.5. Surface area analysis

From table 2, it can be seen that modification of Fe_2O_3 in ZSM-5 nat did not affect the total surface area of ZSM-5 nat. However a closer look at the composition of micropore/ mesopore, there is a decrease in micropore surface area and increase in mesopores counterpart. This indicates that some of Fe_2O_3 clusters were created inside the micropore while some others aggregated on the external surface and created new mesoporosity.

3.6. Catalytic test

Figure 3 shows the chromatogram of methanol and ethanol from partial oxidation with retention time of methanol is 3.2 and ethanol is 3.4.

Table 3 shows the %yield of methanol by using methane p.a. and bio-methane as feeds in the partial oxidation reaction. The reaction using methane p.a has given a 54.93 %yield, which is comparable with the result from a similar reaction using Fe_2O_3/ZSM-5 prepared from pro analysis precursors [17].

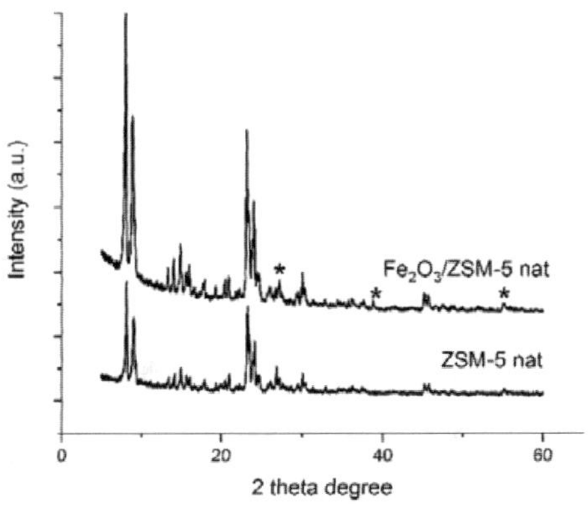

Figure 2. XRD diffractogram for ZSM-5 nat, and Fe_2O_3/ZSM-5 nat.

Table 2. Surface area analysis of ZSM-5 nat and Fe_2O_3/ZSM-5 nat.

Materials	Surface area (m^2/g)	Micropore surface area (m^2/g)	Mesopore surface area (m^2/g)	Total pore volume (cc/g)	Micropore volume (cc/g)	Mesopores volume (cc/g)	Pore radii (nm)
ZSM-5 nat	234.5	215.0	19.46	0.1611	0.1608	0.0003	1.89178
Fe_2O_3/ ZSM-5 nat	234.5	199.9	34.69	0.1404	0.09866	0.04174	1.90324

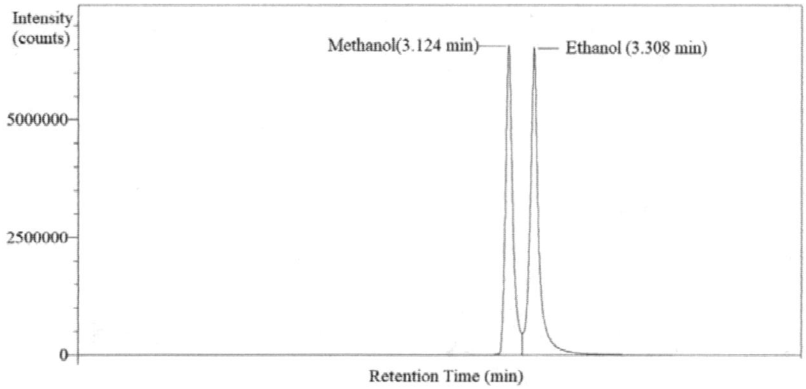

Figure 3. Chromatogram of methanol and ethanol.

Table 3. Characterization for partial oxidation of bio-methane catalytic test.

Catalyst	% yield methanol	
	Methane p.a. (0.75 atm)*	Bio-methane (0.2 atm)*
0.5 gram Fe_2O_3/ZSM-5 Nat	54.93	12.26

*N_2 feed: 2 atm

Due to the limitation of bio-gas pressure (only available at 1 atm), the bio-methane:N_2 ratio was only 0.2:2 compared to the methane p.a:N_2 ratio of 0.75:2. Nevertheless, the partial oxidation of biomethane has produced a methanol yield of 12.26 %.

4. Conclusion

Partial oxidation of bio-methane over Fe_2O_3/ ZSM-5 nat has been carried out successfully. ZSM-5 nat could be employed as support to iron(III) oxide active catalyst. Furthermore, the catalytic test shows the potential to utilize the bio-methane in methane conversion to methanol.

Acknowledgments

Authors want to acknowledge financial funding and support from Hibah Publikasi Internasional Terindeks Tugas Akhir (Hibah PITTA B) No. NKB-0686/UN2.R3.1/HKP.05.00/2019.

References

[1] Ravi M, Sushkevich V L, Knorpp A J, Newton M A, Palagin D, Pinar A B, Ranocchiari M and van Bokhoven J A 2019 *Nat. Catal.* **2** 485-94
[2] Nisbet E G, Dlugokencky E J and Bousquet P 2014 *Science* **343** 493-5
[3] Holmen A 2009 *Catal. Today* **142** 2-8
[4] Alvarez-Galvan M C, Mota N, Ojeda M, Rojas S, Navarro R M and Fierro J L G 2011 *Catal.Today* **171** 15-23
[5] Fellah M F and Onal I 2010 *J. Phys. Chem. C* **114** 3042-51

[6] Taylor S H, Hargreaves J S J, Hutchings G J, Joyner R W and Lembacher C W 1998 *Catal. Today* **42** 217-24
[7] Beznis N V, van Laak A N C, Weckhuysen B M and Bitter J H 2011 *Micropor. Mesopor. Mater.* **138** 176-83
[8] Krisnandi Y K, Samodro B A, Sihombing R and Howe R F 2015 *Indones. J. Chem.* **15** 263-8
[9] Nurgita B G, Nurani D A and Krisnandi zy K 2019 *IOP Conf. Ser. Mater. Sci. Eng.* **496** 012032
[10] Paramadini S A, Ekananda R and Krisnandi Y K 2019 *IOP Conf. Ser. Mater. Sci. Eng.* **496** 012022
[11] Rohayati, Krisnandi Y K and Sihombing R 2017 *AIP Conf. Proc.* **1862** 030094
[12] Wang L, Zhang Z, Yin C, Shan Z and Xiao F S 2010 *Micropor. Mesopor. Mater.* **131** 58-67
[13] Kovo A S, Hernandez O and Holmes S M 2009 *J. Mater. Chem.* **19** 6207-12
[14] Kokotailo G T, Lawton S L, Olson D H and Meier W M 1978 *Nature* **272** 437-8
[15] Colella C and Wise W S 2014 *Micropor. Mesopor. Mater.* **189** 4-10
[16] Kalamaras C, Palomas D, Bos R, Horton A, Crimmin M and Hellgardt K 2016 *Catal. Lett.* **146** 483-92
[17] Putrananda A T, Nurani D A and Krisnandi Y K 2020 *IOP Conf. Ser: Mater. Sci. Eng.* **763** 012029

Synthesis of macroanion $[\alpha\text{-}SiW_{12}O_{40}]^{4-}$ as pillaring agents to layered double hydroxide Zinc-M^{3+}

L Silaen, M F Azmi, N R Palapa, R Mohadi, Elfita and A Lesbani

Department of Chemistry, Faculty of Mathematics and Natural Sciences (FMIPA), Universitas Sriwijaya, Ogan Ilir 30662, Indonesia

Corresponding author's email: aldeslesbani@pps.unsri.ac.id

Abstract. Polyoxometalate intercalated layered double hydroxide was prepared by ion exchange of zinc-aluminum and zinc-chromium LDH with macro anion $[\alpha\text{-}SiW_{12}O_{40}]^{4-}$ at room temperature. The structural characterization was conducted with XRD, adsorption and desorption N_2 by BET method and FTIR analyses. An interlayer space of zinc-aluminum and zinc-chromium was increased after the intercalation process. The interlayer space at reflection (003) shows zinc-aluminum and zinc-chromium intercalated LDH is 10.36 Å and 10.49 Å from 7.57 Å and 7.68 Å, respectively. The value of surface area zinc-aluminum and zinc-chromium before intercalation are 2.10 m^2/g and 31.63 m^2/g and increase after intercalation to 14.0422 m^2/g and 128.87 m^2/g, respectively. This phenom identifies that both LDH has a great reflexibility based on anion in the interlayer.

Keywords: Polyoxometalate $[\alpha\text{-}SiW_{12}O_{40}]^{4-}$, layered double hydroxide, Zn-Cr, Zn-Al

1. Introduction

Layered double hydroxide (LDH) is a family of clay material that has positive charge with an interlayer space containing anion as counterbalance [1-3]. LDH can be written as $[M^{2+}_{(1-x)}M^{3+}x(OH)_2]^{(An-)}x/n{\bullet}nH_2O$ [4, 5]. LDH was formed by M^{2+} which is a metal cation divalent (e.g. Zinc (II), Ni (II), Cu (II), Co (II) and Mg (II)), M^{3+} is metal cation trivalent (Al (III), Cr (III) and Fe (III)) and A^{n-} is anionic (e.g. oxalate, citric, carbonate, nitrate, oxo-anion, acetic, aniline) [6-10]. LDH is well-known as an attractive mineral and has a high flexibility due to the exchangeable interlayer system [11]. Because of their ability, LDH is a considerable application in various sectors [2, 12], such as catalysis, pharmaceuticals, removal pollutant in water and also electrochemistry [13-15]. LDH is easy to synthesize [16], various methods are conducted to obtain LDH. Frequently, the methods used are co-precipitation and hydrothermal [17, 18]. Moreover, Djelali et al. reported that the nanoparticle of LDH has been synthesized by the sol-gel method. The nanoparticles LDH has extremely high interlayer anion exchangeability. So, their work represented the development of LDH potentials [19].

LDH has a low interlayer interaction and in recent decades, many of researchers have expanded the development of methods without breaking the basic structure [20, 21]. The main method adopts the reconstruction of layer and re-assembly anion in interlayers [17, 22]. Specifically, active species incorporation of materials, such as simple small-anions [23], macro-anions [24] and organic molecules [25] into the LDH interlayer space, which is an effective method of immobilization, and has been proven to be an effective source for increasing the efficiency and retesting a homogeneous system that is corresponded [13, 26]. For example, Zha et al. reported that through the intercalation process, LDH nanosheets are intercalated by acetate anion for application as supercapacitors [27]. Zhang et al reported the intercalation of phosphotungstic acid into LDH by reconstruction method [28]. The intercalation of oxo-anions has been studied by interlayer space, although the gallery height is smaller [29]. Interlayer space can be more flexible based on the anion inhabits the interlayer.

As it is known, polyoxometalate is a metal oxygen cluster compound with a high oxidation state. Polyoxometalate are widely used as a catalysator, magnetic doped into porous materials, and this work is used as pillaring agent. Since Woltermann reported polyoxometalate intercalated was used as gases storage [30], polyoxometalate have been proven to be the substantial balancing for material applications [31-33]. In 2017, Zhang et al. reported the preparation of Mg/Al LDH intercalated by $[PW_{12}O_{40}]^{4-}$ has been enlarged from 0.77 Å to 10.8 Å [28]. Then, in 2019, Taher et al. have done POM-Ca/Al LDH intercalated with enlarged interlayer up to 0.16 Å [34]. In present work, we reported the LDH intercalated macro-anion $[\alpha\text{-}SiW_{12}O_{40}]^{4-}$ Keggin type by ion-exchange method. As we expected, after intercalating LDH-$[\alpha\text{-}SiW_{12}O_{40}]^{4-}$ the interlayer is more expanded then can be widely used as a highly effective application. XRD, BET and FTIR were then characterized on the obtained LDH.

2. Materials and method

2.1. Materials and instrumentation

The chemicals such as zinc nitrate hexahydrate ($Zn(NO_3)_2.6H_2O$), aluminum nitrate nonahydrate ($Al(NO_3)_3{\cdot}9H_2O$), chromium nitrate nonahydrate ($Cr(NO_3)_3{\cdot}9H_2O$), sodium hydroxide (NaOH), sodium carbonate (Na_2CO_3), sodium metasilicate (Na_2SiO_3), sodium tungstate dehydrate (Na_2WO_4), hydrochloric acid (HCl) and potassium chloride (KCl) were purchased from Merck and Sigma Aldrich. Deionized water was obtained from the Purite instrument, and the characterization were performed using XRD Rigaku Miniflex-600, BET (ASAP Micromeritics 2020) and FTIR Shimadzu FTIR Prestige- 21.

2.2. Synthesis of polyoxometalate $K_4[\alpha\text{-}SiW_{12}O_{40}]$

The synthesis of polyoxometalate $K_4[\alpha\text{-}SiW_{12}O_{40}]$ was conducted as follows: 0.65 g of sodium metasilicate was diluted by 10 mL deionized water called A solution. B solution was made by 11 g of sodium tungstate dehydrate was added into 20 mL boiled deionized water. HCl 4 M was prepared and added dropwise into B solution with a constant stirring. The synthesis temperature was kept at 313 K, mixed A and B solution for 30 min. Then, pH was adjusted to 5 before the temperature of synthesis was setting up at 373 K for an hour. Three mL of sodium tungstate 1 M and 5 mL of HCl 4 M were added quickly into the solution. When the solution was at room temperature, 3.5 g KCl was added, kept slowly stirring and heteropoly acid salt was obtained. Polyoxometalate $K_4[\alpha\text{-}SiW_{12}O_{40}]$ was characterized using XRD and FTIR [35].

2.3. Preparation of pillaring layered double hydroxide Zn-Al and Zn-Cr

Zn-Al and Zn-Cr layered double hydroxide synthesis has been reported by Palapa et al. using the co-precipitation method [36]. Layered double hydroxide was characterized using XRD, BET and FTIR. Zn-Al and Zn-Cr layered double hydroxide were prepared to modify using pillaring agent macro anion polyoxometalate $K_4[\alpha\text{-}SiW_{12}O_{40}]$ by ion-exchange method. As much as 1 g of layered double hydroxide was diluted with 50 mL of deionized water, in other batch, 10 g of polyoxometalate $K_4[\alpha\text{-}SiW_{12}O_{40}]$ was added to 50 mL of deionized water with vigorous stirring for an hour. After an hour, dropwise of the layered double hydroxide solution was added into polyoxometalate solution. To make sure the mixed solution has constant pH at 8, dropwise of sodium hydroxide 0.1 M was added while adjusting pH. Then, kept the solution at 60 °C for 24 h. Layered double hydroxide pillared by macro anion $[\alpha\text{-}SiW_{12}O_{40}]^{4-}$ was obtained and characterized by XRD, BET and FTIR.

3. Results and discussion

The XRD pattern and FTIR spectra of the starting materials zinc-aluminum and zinc-chromium are shown in figure 1. The XRD pattern of zinc-aluminum (denotes as a) and zinc-chromium (denotes as b) show the sharp and symmetric peaks, which is the typical characteristic of both LDH. The diffraction of LDH are 10° (003), 22° (006), 33° (009) and 60° (110). Barnabas et al. (2016) has reported the main diffraction peak that appears at diffraction angles of 11.4° (003), 23.3° (006), 34° (009) and 60.8° (110) [37]. The peaks that appear are characteristic of the material which has a layered structure. In addition, the intensity of the resulting diffraction peak is relatively weak and a broader peak. The result represented (figure 1a) that zinc-aluminum and zinc-chromium were successfully synthesized. The reflection of (003) identified the interlamellar space of LDH. The difference distances of interlamellar space are the effect of ionic radii of metal trivalent and divalent. However, the ionic radii of Al 0.53 Å and Cr 0.64 Å assume that the interlamellar space of zinc-chromium is greater than zinc-aluminum. The result shows the interlamellar distance of zinc-chromium 7.68 Å and zinc-aluminum 7.57 Å. Forano et al. reported that ionic radii affects the interlamellar space, based on the study about the differences in metal cation [38].

The FTIR spectra (figure 1b) of each spectrum has a similar broad vibration at 3400 cm^{-1} showing the vibration of OH in water molecule or OH in layers metal cation. The FTIR spectra of zinc-chromium (denotes as b) shows impurities of anions in the interlayer. Broad and high vibration at 1385 cm^{-1} show the vibration of nitrate anion band and overlaps with carbonate anions band (in literature at 1370 [39]). Metal-oxide vibration is represented at under 800 cm^{-1}.

Layered double hydroxide intercalated using polyoxometalate $K_4[\alpha\text{-}SiW_{12}O_{40}]\bullet nH_2O$ was characterized by XRD and FTIR. The intercalation process of layered double hydroxide was carried out by co-precipitation method. The results of characterization are shown in figure 2. Figure 2a shows the XRD pattern of Zn-Al/[α-$SiW_{12}O_{40}$] (denotes as a) and Zn-Cr/[α-$SiW_{12}O_{40}$] (denotes as b). Zn-Al/[α-$SiW_{12}O_{40}$] and Zn-Cr/[α-$SiW_{12}O_{40}$] reflection (003) was shifted to lower 2 theta from 10° to 8°. (003) reflection was increased slightly to 10.36 Å and 10.49 Å, respectively.

The result of FTIR after intercalated is displayed in figure 2b. The strong, broad absorption vibration bands at 3400 are represented by OH vibration. In the case of polyoxometalate, has a unique vibration between 1000–800 cm^{-1} correspond to W=O, W-O-W and Si-O. This result is similar to Taher et al. layered double hydroxide intercalated by polyoxometalate is observed at Si-O 926 cm^{-1}, W=O 980 cm^{-1}, W-Oc-W 881 cm^{-1} and W-Oe-W 780 cm^{-1} [34].

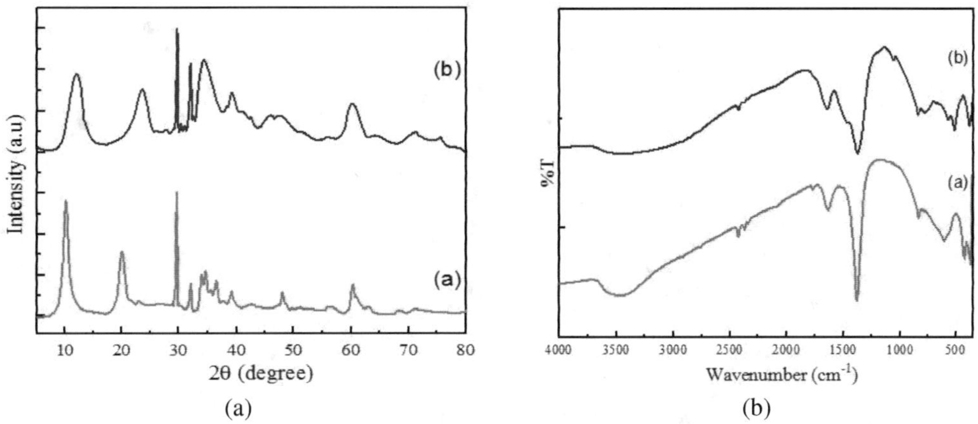

(a) (b)

Figure 1. Characterization of zinc-aluminum and zinc-chromium layered double hydroxide (a) XRD and (b) FTIR spectra.

Zinc-aluminum, zinc-chromium and also intercalated materials have been studied by surface area analyses by BET method. Nitrogen adsorption-desorption isotherm are represented in figure 3 for all materials, the isotherm type was type IV. This type indicates the mesoporous with irregular pores. The classification of hysteresis based on IUPAC, for layer materials (like montmorillonite or clay minerals) [40] usually follows H3 hysteresis loop. H3 indicates the consistency with cavitation-induced evaporation, which is when the cavitation threshold exceeds the hysteresis loop (*P/Po*) > 0.42. The H3 hysteresis explains that the binding strength of polyoxometalate anion in interlayer is more than nitrate anion. The surface and pore size are followed by Zn-Al < Zn-Al/[α-$SiW_{12}O_{40}$] and Zn-Cr < Zn-Al/[α-$SiW_{12}O_{40}$] (the data is represented in table 1).

Table 1 shows that the surface area of zinc-aluminum and zinc chromium were slightly increased after the intercalation process. Zinc-aluminum was increasing sevenfold after intercalated by polyoxometalate K_4[α-$SiW_{12}O_{40}$]•nH_2O with pore size 15.67 nm. Zinc-chromium also increased

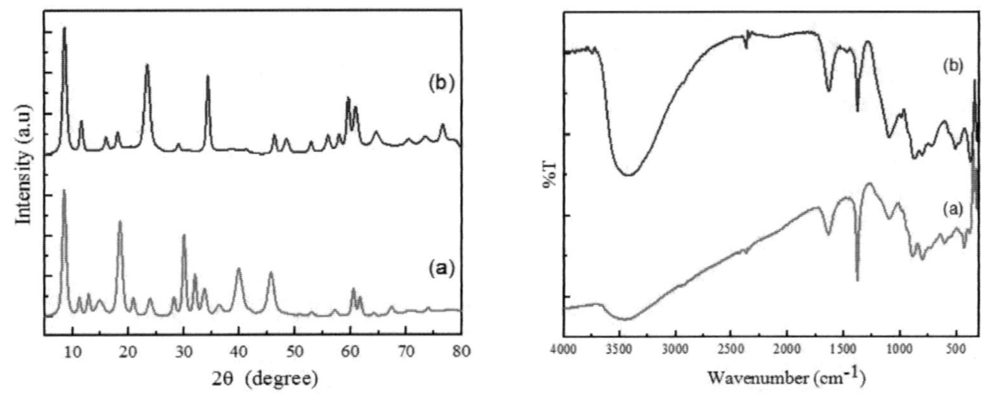

(a) (b)

Figure 2. Characterization of zinc-aluminum and zinc-chromium intercalated by polyoxometalate K_4[α-$SiW_{12}O_{40}$]•nH_2O layered double hydroxide (a) XRD and (b) FTIR spectra.

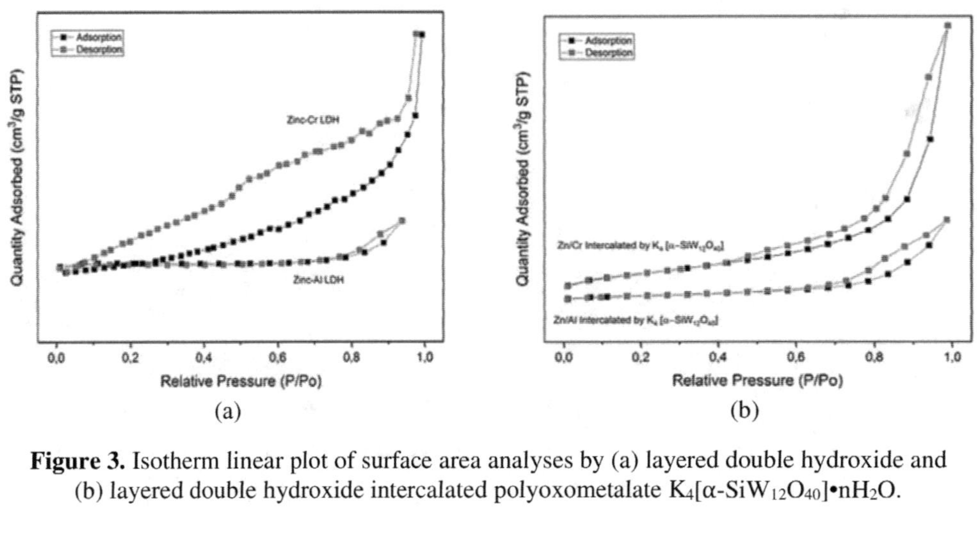

(a) (b)

Figure 3. Isotherm linear plot of surface area analyses by (a) layered double hydroxide and (b) layered double hydroxide intercalated polyoxometalate K_4[α-$SiW_{12}O_{40}$]•nH_2O.

Table 1. Surface area data of layered double hydroxide and layered double hydroxides intercalated polyoxometalate $K_4[\alpha\text{-}SiW_{12}O_{40}]\bullet nH_2O$.

LDH	S_{BET} (m^2/g)	Pore size (nm)
Zn-Al	2.10	13.51
Zn-Cr	31.63	3.69
Zn-Al/[α-$SiW_{12}O_{40}$]	14.04	15.67
Zn-Cr/[α-$SiW_{12}O_{40}$]	128.87	3.69

from 31.63 m^2/g to 128.87 m^2/g with pore size before and after intercalating has no changed. According to IUPAC, these material are in the mesopore classification because the pore size are between 2–50 nm.

4. Conclusion

Zinc-aluminum and zinc-chromium was successfully intercalated by polyoxometalate $K_4[\alpha\text{-}SiW_{12}O_{40}]\bullet nH_2O$. The result of XRD pattern shows after intercalation 2 theta was shifted to a lower degree from 10° to 8°. The shift was followed by the slightly increase of interlayer distance. The interlayer distance of zinc-aluminum and zinc-chromium before intercalation are 7.57 Å and 7.68 Å to 10.36 Å and 10.49 Å, respectively. The FTIR data shows the unique diffraction of polyoxometalate is observed after intercalation at 1000–800 cm^{-1}. The surface area analyses shows the size of these materials is mesopore.

Acknowledgments
Author thanks to Universitas Sriwijaya for supporting fin ncial this work through Hibah Profesi grant with contract number 0014/UN9/SK.LP2M.PT/2019.

References

[1] Guo Y, Zhu Z, Qiu Y and Zhao J 2013 *J. Environ. Sci. (China)* **25** 944-53
[2] Mishra G, Dash B and Pandey S 2018 *Appl. Clay Sci.* **153** 172-86
[3] Zoromba M S, Nour M A, Eltamimy H E and Abd El-Maksoud S A 2018 *Sci. Eng. Compos. Mater.* **25** 101-8
[4] Gago S, Pillinger M, Santos T M and Gonçalves I S 2004 *Ceramics-Silikaty* **48** 155-8
[5] Palapa N R, Taher T, Said M, Mohadi R and Lesbani A 2017 *Sci. Technol. Indones.* **3** 189-94
[6] Smalenskaite A, Pavasaryte L, Yang T C K and Kareiva A 2019 *Materials* **12** 1-14
[7] Zhang B, Dong Z, Sun D, Wu T and Li Y 2017 *J. Ind. Eng. Chem.* **49** 208-18
[8] Kameda T, Takeuchi H and Yoshioka T 2011 *J. Phys. Chem. Solids* **72** 846-51
[9] Delgado R R, De Pauli C P, Carrasco C B and Avena M J 2008 *Appl. Clay Sci.* **40** 27-37
[10] Yu S et al. 2017 *J. Hazard. Mater.* **321** 111-20
[11] Khan A I and O'Hare D *J. Mater. Chem.* **12** 3191-8
[12] Chen J 2011 *Host-Guest Functional Materials* (London: Elsevier) pp 405-28
[13] Li F and Duan X 2006 *Struct. Bond.* **119** 193-223
[14] Li T, Miras H and Song Y F *Catalysts* **7** 1-17
[15] Zhu J et al. 2011 *Appl. Surf. Sci.* **257** 9056-62
[16] Palapa N R, Taher T, Mohadi R, Said M and Lesbani A 2018 *AIP Conf. Proc.* **2026** 020033
[17] Szabados M, Kónya Z, Kukovecz Á, Sipos P and Pálinkó I 2019 *Appl. Clay Sci.* **174** 138-45
[18] Extremera R, Pavlovic P, Pérez M R and Barriga C 2012 *Chem. Eng. J.* **213** 392-400
[19] Djellali M, Aranda P and Ruiz-Hitzky E 2019 *Appl. Clay Sci.* **171** 65-73

[20] Cermelj K, Ruengkajorn K, Buffet J C and O'Hare D *J. Energy Chem.* **35** 88-94
[21] Ma J et al. 2017 *Appl. Clay Sci.* **150** 210-6
[22] Huang D et al. 2019 *Bioresour. Technol.* **276** 127-32
[23] Wu X et al. 2018 *Mater. Chem. Phys.* **211** 72-8
[24] Webe R S, Gallezotb P, Lefebvreb F and Suib S L 1993 *Microporous Mater.* **1** 223-7
[25] Stepanow eta al. 2004 *Nat. Mater.* **3** 229-33
[26] Bukhtiyarova M V 2019 *J. Solid State Chem.* **269** 494-506
[27] Zha D, Sun H, Fu Y, Ouyang X and Wang X 2017 *Electrochim. Acta* **236** 18-27
[28] Zhang S et al. 2018 *Polym. Degrad. Stab.* **147** 142-50
[29] Bai Z, Hu C, Liu H and Qu J 2019 *J. Colloid Interface Sci.* **539** 146-51
[30] Woltermann, G. M., 1984. US Pat. 4 454 244 to Ashland Oil
[31] Gardner E and Pinnavaia T J 1998 *Appl. Catal. A Gen.* **167** 65-74
[32] Hu C et al. 1998 *Appl. Clay Sci.* **13** 495-511
[33] Wei X et al. 2008 *J. Solid State Chem.* **181** 1292-7
[34] Taher T et al. 2019 *Bull. Chem. React. Eng. Catal.* **14** 260-7
[35] Yang R, Liu S X, Tang Q, Li S J and. Liang D D 2012 *J. Coord. Chem.* **65** 891-7
[36] Palapa N R, Saria Y, Taher T, Mohadi R and Lesbani A 2019 *Sci. Technol. Indones.* **4** 36-9
[37] Barnabas M J et al. 2016 *J. Solid State Chem.* **233** 133-42
[38] Forano C, Hibino T and Leroux F 2006 Layered Double Hydroxides in *Developments in Clay Science* (London: Elsevier) p 4352
[39] Parida K M and Mohapatra L 2012 *Chem. Eng. J.* **179** 131-9
[40] Alsamman L, Pich A and Moeller M 2017 *Development of Modified Layered Silicates with Superior Adsorption Properties for Uptake of Pollutants from Air and Water* PhD Thesis (Aachen: Aachen University)

IOP Conf. Series: Materials Science and Engineering **902** (2020) 012049 doi:10.1088/1757-899X/902/1/012049

Catalytic conversion of beef tallow with MgO derived from $MgCO_3$ for biofuels production

A Riyadhi[1, 2], Y Yulizar[1] and B H Susanto[3]

[1]Department of Chemistry, Faculty of Mathematics and Natural Sciences (FMIPA), Universitas Indonesia, Depok 16424, Indonesia
[2]Department of Chemistry, Faculty of Science and Technology, Universitas Islam Negeri Syarif Hidayatullah, Tangerang Selatan 15412, Indonesia
[3]Department of Chemical Engineering, Faculty of Engineering, Universitas Indonesia, Depok 16424, Indonesia

Corresponding author's email: yokiy@sci.ui.ac.id

Abstract. Fuels from biological sources, such as oils and fats, receive much attention. Beef tallow is a cheap and abundant feedstock that can be converted into renewable fuels. Catalytic conversion of beef tallow was accomplished with the MgO catalyst derived from $MgCO_3$. Magnesium oxide catalyst was prepared by calcination of $MgCO_3$ at temperature 700 °C. The catalyst was tested for its performance in a fixed bed reactor at temperature 300 °C for 60 min. Conversion of beef tallow to hydrocarbons biofuels was studied. The MgO catalyst was characterized by nitrogen adsorption isotherms (BET), powder X-ray diffraction (XRD), and field emission scanning electron microscopy (FESEM). The assay activity of MgO shows all fatty acids derived from beef tallow have converted into liquid and gas fraction with a black paste residue that has a boiling point of more than 300 °C. The results of the liquid fraction composition depend on the ratio of the given MgO catalyst. At the ratio of MgO catalysts and feeds 2 wt.%, the results of liquid fraction contain the alkanes (45.61%), alkenes (4.12%), alcohols (3.96%), ketones (8.55 %), esters (23.42 %), and cyclic compounds (14.34 %), while at the ratio of 4 wt.%, the liquid fraction contains the alkanes (55.56 %), alkenes (8.94 %), alcohols (5.27 %), ketones (13.73 %) and cyclic compounds (16.50 %).

Keywords: Catalytic conversion, biofuel, beef tallow, MgO and $MgCO_3$

1. Introduction

Renewable fuels such as biogasoline and biodiesel can be generated from a wide range of biomass namely animal fats, cooking oil waste, and vegetable oils [1]. Fuels derived from biological sources, among them are lipid materials such as oils and fats, have received much attention. The first-generation biodiesel or fatty acid esters are generated by the transesterification reaction of animal fat or vegetable oils with alcohols and glycerol as co-product. However, the reduced flow property at low temperature of the obtained biodiesel limits its application in engine[1, 2].

Second-generation biodiesel, also called as renewable diesel by using hydrodeoxygenation, decarboxylation, and cracking technology has become a trend. Compared to the renewable diesel in the first-generation biodiesel, the quality of the second-generation renewable diesel has a lot better performance. The second-generation renewable diesel has better fuel properties because of its higher cetane number, high energy density and lowers oxygen content [3].

The second-generation renewable diesel has high quality and environmental protection performances in the energy field [3]. The second-generation renewable diesel has chemical content equal to petrodiesel, consisting of hydrocarbon compounds such as octadecane, heptadecane, hexadecane, and pentadecane. Therefore, the renewable diesel can be used directly as fuel [3, 4].

The process of decarboxylation can convert fatty acids into linear, paraffinic hydrocarbons [5, 6]. The method of decarboxylation does not require hydrogen. It is easy to apply compared to the hydrodeoxygenation process that involves hydrogen. The process of decarboxylation does not produce water, which can cause the catalyst to be inactive [7].

Beef tallow is a byproduct of the food industry, so it is a low-cost and abundant feedstock. Beef tallow can be converted into renewable gasoline and renewable diesel at lower production costs when compared to other biofuels [8]. The main properties of the beef tallow are oleic acid (C_{18}), stearic acid (C_{18}), linoleic acid (C_{18}), palmitic acid (C_{16}), palmitoleic acid (C_{16}), and myristic acid (C_{14}) [9].

Pyrolysis of waste animal fats at the temperature of 500 °C showed a complex mixture such as alkanes, alkenes, cyclic compounds, carboxylic acids, aldehydes, ketones, and esters[10]. Catalytic decarboxylation can be done at a temperature of 300 °C, catalytic and thermal reactions can be done at the temperature of 350 °C, and thermal reaction can be done at the temperature above 400 °C [7].

The MgO catalyst is useful for the dehydrogenation of alkanes into cyclic and aromatic carbon compounds [11]. The MgO catalyst is also useful for deoxygenation of fatty acid into alkanes, alkenes, and cyclic carbon compounds [5, 6]. The MgO catalyst helps in the oxygen removal and CO_2 absorbance in the gas phase through catalytic deoxygenation [12].

Thus, the present study highlights the usage of solid base catalysts MgO prepared from low-cost $MgCO_3$ industrial grade for the conversion of beef tallow with MgO catalyst for renewable hydrocarbon biofuels production. Decomposition thermal of magnesium carbonate can produce porous magnesium oxide [13].

The research on commercial MgO as a catalyst for the catalytic beef tallow has been done by our group and have reported[14]. The resulting product still requires development because the product can freeze at 0 °C temperature, and it is not good as fuel. This research conducted the manufacture of porous MgO catalysts from the $MgCO_3$ industrial grade that costs much cheaper and environmentally friendly raw material. The MgO catalyst was characterized by XRD, BET, and SEM-EDX. Chemical compositions of products from the conversion of beef tallow into biofuels were analyzed using GC-MS.

2. Materials and method

MgO was prepared from $MgCO_3$ industrial grade with a calcination temperature of 700 °C for 90 min under atmosphere condition. Beef tallow in the study was obtained from the household industry in Kemang, Bogor, Jawa Barat, Indonesia. Tallow is a product of beef fat hydrolysis that has been separated from the residue of muscle tissue, meat and bones.

2.1. Catalyst characterization

The powder X-ray diffraction (XRD) analysis was conducted to determine the crystallography of the mixed metal oxide catalysts. The XRD analyses were carried out by the Shimadzu diffractometer model XRD-7000. The specific surface area and pore distribution of the catalysts were identified with Brunauer-Emmet-Teller (BET) method with N_2 adsorption/desorption analyzer using ASAP 2020 Micromeritics Instrument Corp. The identification of elemental compositions of Mg, O and C on the catalysts surface were determined by SEM-EDX mapping analysis using JED-2300 Analysis Station.

2.2. Activity test

Catalytic deoxygenation of beef tallow was conducted in a 150 mL stainless steel bath reactor and connected to a distillation system [14]. Around 25 g of beef tallow using 2 wt.% and 4 wt.% of MgO catalyst was inserted into the reactor. The deoxygenation reaction of beef tallow was carried out with heating at 300 °C for 60 min. The cracked/deoxygenated product was then collected in a receiver flask, and the liquid products were immediately weighed and analyzed with gas chromatography-mass spectrometry (GC-MS) Perkin Elmer Clarus 600.

3. Results and discussion

3.1. Physicochemical properties of catalysts

MgO catalyst was synthesized with calcination of industrial-grade magnesium carbonate at temperature 700 °C for 90 min under atmosphere condition. When the $MgCO_3$ heating occurs, the release of CO_2 gases causes pore formation in MgO material [13]. $MgCO_3$ is an excellent raw material for making MgO porous. The lost weight is calculated based on the material weight difference before and after material calcination. Between 500–700 °C, CO_2 escapes and emits out from the sample $MgCO_3$ [13]. From the calculation of stoichiometry, % lost weight of $MgCO_3$ to MgO is 52.20 %, and % conversion is 47.80 %. In this study, % lost weight of $MgCO_3$ calcination at 700 °C for 90 min is 53.32 %, this weight reduction occurs due to the formation of CO_2 gases and water loss.

After calcination at 700 °C, the acquired powder was characterized by XRD. as shown in figure 1. The three highest peaks at 2θ value of MgO are at 42.80; 62.19 and 78.49°, according to JCPDS 75–0447, and all peaks show the character of cubic MgO periclase. There is no other conspicuous phase in the XRD pattern, which implies the high purity of MgO particles.

Morphology and compositional analysis of the synthesized MgO derived from $MgCO_3$ were done by using electron microscopy. Figure 2 shows the scanning electron microscopy (SEM) image of MgO, which exhibits a flakes-like structure. Based on the results of SEM-EDX analysis in table 1, it shows the MgO atom composition of 43.50 % Mg and 56.50 % O, close to the atomic ratio of 1:1 from MgO.

Surface area and pore size of material were characterized using BET. In this study, MgO derived from $MgCO_3$ has a surface area of 53.42 m^2/g and an adsorption average pore width of 17.76 nm. These results show bigger than MgO being synthesized by Cai et al. [15], respectively 47.85 m^2/g and 12.15 nm is prepared using $Mg(NO_3)_2 6H_2O$.

3.2. Catalytic conversion of beef tallow

The beef tallow contents used for catalytic conversion were identified by GC-MS. The contents are shown in table 2, and the main content is octadecanoic acid, 11-octadecenoic acid and hexadecanoic acid.

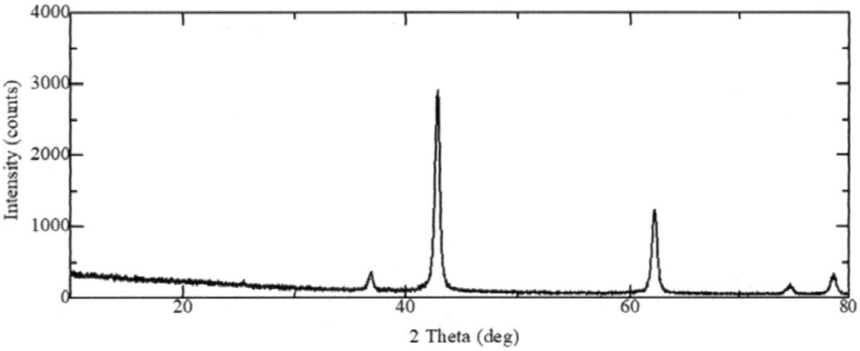

Figure 1. XRD pattern of MgO obtained from the $MgCO_3$ calcination at 700 °C.

IOP Conf. Series: Materials Science and Engineering **902** (2020) 012049 doi:10.1088/1757-899X/902/1/012049

The assay activity of 2 wt.% MgO catalyst shows all fatty acids derived from beef tallow converted into 38.24 % liquid; 27.20 % gas fraction and 34.56% black paste residue that has a boiling point of more than 300 °C, while the assay activity using 4 wt.% MgO shows the results of 41.52 % liquid, 30.96 % gas fraction and 27.52 % black paste residue, as shown in table 3. These results show more catalysts used, resulting in more liquid and gas fractions produced, on the contrary to the black paste residue.

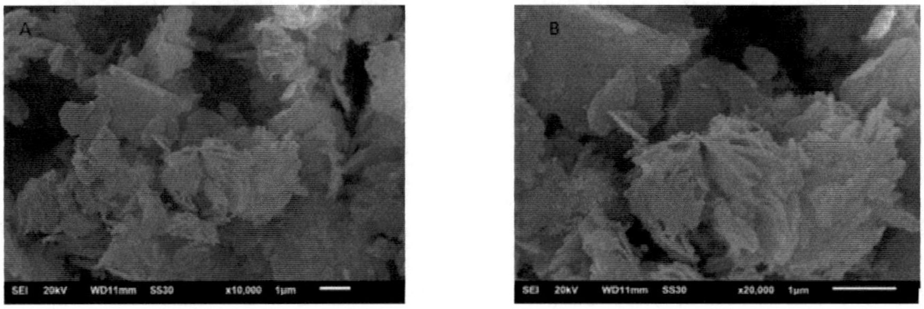

(a) (b)

Figure 2. SEM images of MgO with (a) 10,000x and (b) 20,000x magnification.

Table 1. Elemental ratio of MgO.

Element	wt.%	atom %
Mg	53.92	43.50
O	46.08	56.50

Table 2. The contents of beef tallow used for catalytic conversion.

Retention time (min)	Percentage (%)	Fatty acid	Formula
20.29	5.29	Tetradecanoic acid	$C_{14}H_{28}O_2$
23.02	1.71	9-Hexadecenoic acid	$C_{16}H_{30}O_2$
23.33	25.04	Hexadecanoic acid	$C_{16}H_{32}O_2$
25.66	1.41	9,12-Octadecadienoic acid	$C_{18}H_{32}O_2$
25.76	26.57	11-Octadecenoic acid	$C_{18}H_{34}O_2$
25.83	1.84	13-Octadecenoic acid	$C_{18}H_{34}O_2$
25.88	3.08	16-Octadecenoic acid	$C_{18}H_{34}O_2$
26.10	26.86	Octadecanoic acid	$C_{18}H_{36}O_2$

Table 3. Products distribution of beef tallow conversion using MgO catalyst.

Products (wt.%)	Catalyst	
	2 wt.% MgO	4 wt.% MgO
Liquid fraction	38.24	41.52
Gas fraction	27.20	30.96
Black paste	34.56	27.52

Table 4. Product composition based on petroleum range of liquid fraction.

Products (wt.%)	Catalyst	
	2 wt.% MgO	4 wt.% MgO
Gasoline (C_5-C_{11})	22.28	28.84
Kerosene (C_{12}-C_{15})	30.13	35.50
Diesel (C_{16}-C_{26})	47.59	35.66

Table 5. The organic compound composition of liquid fraction.

Products (wt.%)	Catalyst	
	2 wt.% MgO	4 wt.% MgO
Alkanes	45.61	55.56
Alkenes	4.12	8.94
Alcohol	3.96	5.27
Ketones	8.55	13.73
Esters	23.42	-
Cyclic	14.34	16.50

The liquid fractions products obtained from catalytic deoxygenation and cracking processes are bright yellow and do not freeze at 0 °C. The liquid fractions were analyzed using GC-MS and grouped into three lumps [16], gasoline (C_5–C_{11}), kerosene (C_{12}–C_{15}), and diesel fraction (C_{16}–C_{26}) as indicated in table 4. The resulting composition of liquid fractions depends on the percent of the given MgO catalyst. The using of 2 wt.% catalysts results in the gasoline, kerosene diesel fraction of 22.28 %; 30.13 % and 47.59 %; respectively, while by using of 4 wt.% catalyst results are 28.84 %; 35.50 % and 35.66 %, respectively. The using of 4 wt.% catalyst results in the gasoline and kerosene much more 2 wt.% catalysts, indicating many cracking reactions, of all fatty acid from beef tallow.

As a comparison with commercial MgO (5 wt.%) reported [14], catalytic cracking of tallow showed C_7–C_{11} fraction of 17.85 %, and C_{12}–C_{19} fraction of 82.15 %. The result in using MgO catalysts from $MgCO_3$ industrial-grade indicated more gasoline fraction than commercial MgO. Solid base catalysts of MgO from $MgCO_3$ industrial grade is an environmentally friendly and low-value precursor.

The product characterization based on organic compounds is shown in table 5. The using of MgO catalysts and feeds 2 wt.% results of liquid fraction contain the alkanes (45.61 %), alkenes (4.12 %), alcohols (3.96 %), ketones (8.55 %), esters (23.42 %), and cyclic compounds (14.34 %), while at the ratio of 4 wt.%, liquid fraction contain the alkanes (55.56 %), alkenes (8.94 %), alcohols (5.27 %), ketones (13.73 %) and cyclic compounds (16.50 %). The results are lower than by using the MgO catalyst and feed 4 wt.%, however, no ester product was found. The products at the ratio of MgO catalyst and feed 4 wt.% did not contain ester products. Ester products have successfully cracked either through carboxylation and carbonylation reactions.

4. Conclusion

This study has been successfully confirmed that beef tallow is valuable raw materials for bio-oil production using MgO catalyst. The assay activity of MgO synthesized from $MgCO_3$ as a catalyst at temperature 300 C for 60 min, show all fatty acids derived from beef tallow converted into gas, liquid fraction, and black paste residue. The composition of the liquid fraction depends on the percent of the given MgO catalyst. The much more catalysts used in the conversion of beef tallow will give a higher result of gasoline and kerosene but not for diesel products. The industrial-grade $MgCO_3$ can be used as a precursor for an environmentally friendly and low-value MgO production. The MgO as active catalysts for the deoxygenation and cracking hydrocarbons of tallow with different results depend on the reaction conditions.

Acknowledgments
The authors thank to The Indonesia Endowment Fund for Education Ministry of Finance Republic of Indonesia (LPDP Kementerian Keuangan RI) for the fund through the Scholarships and Research Grants.

References
[1] Bezergianni S and Dimitriadis A 2013 *Renew. Sustain. Energy Rev.* **21** 110-6
[2] Guo J-H, Xu G-Y, Shen F, Fu Y, Zhang Y and Guo Q-X 2015 *Green Chem.* **17** 2888-95
[3] Chen S, Zhou G and Miao C 2019 *Renew. Sustain. Energy Rev.* **101** 568-89
[4] Knothe G 2010 *Prog. Energy Combust. Sci.* **36** 364-73
[5] Edeh I, Overton T and Bowra S 2019 Renewable diesel production by hydrothermal decarboxylation of fatty acids over platinum on carbon catalyst *Biofuels* doi:10.1080/17597269.2018.1560554
[6] Li D, Xin H, Du X, Hao X, Liu Q and Hu C 2015 *Sci. Bull.* **60** 2096-106
[7] Roh H-S, Eum I-H, Jeong D-W, Yi B-E, Na J-G and Ko C H 2011 *Catal. Today* **164** 457-60
[8] Esteves V P P, Esteves E M M, Bungenstab D J, Feijo G L D, Araújo O de Q F and Morgado C do R V 2017 *J. Clean. Prod.* **151** 578-91
[9] Ito T, Sakurai Y, kakuta Y, Sugano M and Hirano K 2012 *Fuel Process. Technol.* **94** 47–52
[10] Ben Hassen-Trabelsi A, Kraiem T, Naoui S and Belayouni H 2014 *Waste Manag.* **34** 210-8
[11] Elkhalifa E A and Friedrich H B 2018 *Arab. J. Chem.* **11** 1154-9
[12] Asikin-Mijan N, Lee H V, Juan J C, Noorsaadah A R and Taufiq-Yap Y H 2017 *RSC Adv.* **7** 46445-60
[13] Morozov S A, Malkov A A and Malygin A A 2003 *Russ. J. Gen. Chem.* **73** 41-6
[14] Riyadhi A and Syahrullah 2016 *Integrated Lab Journal* **4** 125-38
[15] Cai Y, Li C, Wu D, Wang W, Tan F, Wang X, Wong P K and Qiao X 2017 *Chem. Eng. J.* **312** 158-66
[16] Demirbas 2015 *Petrol. Sci. Technol.* **33** 1703-11

Conversion of delignified stem sorghum bicolor into levulinic acid using micro and mesoporous ZSM-5 catalysts

Y D I Siregar[1, 2], E Saepudin[1] and Y K Krisnandi[1]

[1]Department of Chemistry, Faculty of Mathematics and Natural Sciences (FMIPA), Universitas Indonesia, Depok 16424, Indonesia
[2]Department of Chemistry, Faculty of Science and Technology (FST), UIN Syarif Hidayatullah Jakarta, Ciputat 15412, Indonesia

Corresponding author's email: yuni.krisnandi@sci.ui.ac.id

Abstract. Sorghum bicolor stem is one of the biomass wastes that can be used as raw material for synthesizing platform chemicals. Levulinic acid is one of the chemicals that is currently a substance often synthesized or isolated. Levulinic acid can be catalytically synthesized from sorghum stem waste which is rich in cellulose. Cellulose in the sorghum stem is still bound to lignin and hemicellulose, therefore pretreatment is needed to reduce lignin. The method consists of pretreatment of biomass by delignification, manufacturing of micro and mesoporous ZSM-5 catalysts (double template method) and the reaction of the conversion of stem sorghum to levulinic acid is carried out in a mini reactor. Catalyst characterizations were conducted using FTIR, XRD, SEM EDX and product analysis by HPLC. The result of delignification is the cellulose content increased from 45.10 % to 76.66 %. The result of as-synthesized ZSM-5 showed the diffractogram patterns similar to the standard ZSM-5. SEM image showed a hexagonal coffin-like shape which is the characteristic of ZSM-5. The levulinic acid produced using meso-porous ZSM-5 catalyst was observed at the reaction time of 2–6 h, while the yield of levulinic acid using micro-porous ZSM-5 catalyst was observed at the end of the reaction time, i.e. 6–10 h. It can be concluded that the pore size of the catalyst and the duration of the conversion time influenced the yield of levulinic acid.

Keywords: Sorghum, cellulose, delignification, ZSM-5 catalyst, micro and mesopores

1. Introduction

Levulinic acid is a chemical which is currently one of the most widely used raw materials in the industry [1]. Levulinic acid is structurally composed of two functional groups namely ketones and carbonyl, thus levulinic acid can be converted into other important compounds [2] in the chemical industry such as fuel, solvent, spice, cosmetic, resin, polymer, chemical intermediate and others (pharmaceutical, pesticide, fungicide) [3]. Levulinic acid was firstly synthesized by Mulder in the 1940 by heating the fructose with hydrochloric acid [4]. Fructose is one of the biomass feedstocks, the other are starch and glucose [1]. Glucose can be obtained from cellulose hydrolysis process. Cellulose is found in many biomass wastes, one of which is sorghum stems [5], in which the sorghum stem is chemically pretreated by the delignification method, because the cellulose is still bound to lignin and hemicellulose [6]. The delignification method is usually carried out using NaOH solution as the delignifying agent. The delignification method in this study refers to [7], which is used 10 % NaOH.

Previous synthesis of levulinic acid used homogeneous catalysts such as H_2SO_4, HCl, etc. [8-10]. This study used a heterogeneous catalyst so that the product is easily separated from the catalyst at the end of the reaction. Heterogeneous catalysts for levulinic acid synthesis have been reported by [11-13]. The purpose of this paper is to produce levulinic acid from delignified stem sorghum using micro and mesoporous ZSM-5 catalysts. The conditions of this study refer to [7, 11, 14].

2. Materials and method

2.1. Materials

Sorghum stems were obtained from BIOTROP SEAMEO, Tajur, Bogor, West Java. The chemicals used in this experiment were analytical grade as follows: phosphoric acid (89.0 %), hydrogen peroxide (30.0 %), sulphuric acid (96.0 %), sodium hydroxide (99.0 %) and ethanol (95.0 %) were obtained from Merck; sodium aluminate (99.0 %), tetra-orthosilicate (TEOS, 98.0 %), tetra-propylammonium hydroxide (TPAOH, 1.0 M), poly(acrylamide-co-diallyl dimethyl ammonium chloride 10 % wt., (PDD-AM), and levulinic acid ($C_5H_8O_3$) (MW1 16.12 g/mol, sg 1.134 g/mL at 25 °C were all obtained from Sigma Aldrich.

2.2. Pretreatment of sorghum stems

Sorghum stems were shelled out from the outer skin, then cut to a size of 1 cm, dried under sunrise for 3–4 days. After drying, the sorghum stem was then mashed using a grinder and sieved to give a uniform size of 100 μm. The wax from the sample was then removed through dewaxing process using n-hexane and ethanol solvents with a volume ratio of 2:1. Delignification was carried out by dispersing the dewaxed sorghum into 10 % NaOH solution with the ratio of sorghum stem to NaOH is 1:25 (w/v). The solution was then heated at 55 °C for 90 min. The residue was taken by filtering using the filter paper and neutralized with distilled water. The samples were then dried and then ready to be used as substrate in the conversion reaction into levulinic acid using micro and mesoporous ZSM-5 catalysts.

2.3. Preparation of catalysts

Mesoporous ZSM-5 was synthesized using the method as reported by [14] with some modification by [7]. The gel of mesoporous ZSM-5 was prepared from a homogeneous mixture with molar composition of 1 Al_2O_3: 64.35 SiO_2: 10.08 $(TPA)_2O$: 3571.66 H_2O. The mixture was stirred for 3 h at 100 °C, then 1.0 g of PDDAM was added and stirred for 15 h. Subsequently, the mixture was transferred to a Teflon- fined autoclave for further crystallization at 170 °C for 6 days. The product was filtered, washed, dried, calcined, and labeled as Meso ZSM-5. The microporous ZSM-5 was synthesized using a similar method as mesoporous without PDDAM as a template. It was named as Micro ZSM-5.

2.4. Reaction of delignified sorghum stem to levulinic acid using micro and mesoporous ZSM-5 catalysts

The conversion reaction was carried out in a reactor placed in an oil bath. The reaction began by transferring 0.1 g of delignified sorghum stem into the reaction vessel, then 2 mL of H_3PO_4 40 %, 0.05 mL of H_2O_2 30 % and catalyst (micro and meso porous ZSM-5 catalysts) were added. The mixture was heated for 0, 2, 4, 6, 8 and 10 h at a reaction temperature of 130 °C. The reaction results were cooled in an ice bath, and filtered to be then analyzed and identified using HPLC.

2.5. Instrumentation

XRD analysis was performed using the PanAlytical X'Pert Pro MPD. XRD instrument using Cu-Kα. Measurements made from a range of 2θ are 0–90°. SEM images were taken with SEM FEI INSPECT F50 at 20 kV and Si/Al was measured semi quantitatively with EDAX system. The specific surface area was estimated according to the BET method, while pore volume was calculated from the BJH desorption curve.

2.6. Analysis of product
HPLC analysis was carried out using the HP200 LC200 PG instrument with a UV Visible detector at λ 220 nm, 0.1 % $HClO_4$ solvent and 1 mL/min flow rate. The yield of levulinic acid (LA) was calculated using the equation:

$$\% \text{ Yield} = \frac{[weight\ of\ LA]}{[weight\ of\ delignified\ stem\ sorghum]} \times 100$$

Analysis of the chemical content of sorghum stem refers to the Klason Lignin method [15] and modified of American Society for Testing and Materials (ASTM D 1104-56), (ASTM D 1103-60) *Reapproved* 1978.

3. Results and discussion

3.1. Characterization with fourier transform infrared spectroscopy (FTIR)
Figure 1 is the result of the ZSM-5 (micro and meso) FTIR characterization before and after calcination to a temperature of 550 °C. There is a difference in peak spectra between ZSM-5 before and after calcination in the wave number region between 2960–2850 cm^{-1} which is the C-H stretching region and 1475–1350 cm^{-1} belongs to the C-H bending vibration region. The peak of the spectra contained in the wave number region as a result of the use of the TPAOH template in micro and meso ZSM-5 material and also PDD-AM in meso ZSM-5 material. ZSM-5 after calcining the peak of the spectra in the region of the wave number is lost. This shows that the TPAOH and PDD-AM organic template has decomposed due to the calcination process up to 550 °C. The calcination process also increases the vibration absorption especially in the region of wave numbers 3700–3600 cm^{-1} which is the O-H stretching vibrational region. Wave numbers between 1650–1645 cm^{-1} are the vibrational region of H-OH bending. Then, the wave number 1290–950 cm^{-1} shows the building structure of zeolites in the form of asymmetric stretching vibrations of Si-O and Al-O. Whereas, for Si-O and Al-O fingerprints, each appears at the wave numbers 1061.83 cm^{-1} and 795.65 cm^{-1} for micro ZSM-5 while 1083.99 cm^{-1} and 804.32 cm^{-1} for meso ZSM-5. At wave numbers between 1100–700 cm^{-1} are vibrations of Si-O and Al-O. And in the wave numbers between 650–500 cm^{-1} is a typical area for double ring zeolite which is an external braid on the zeolite, where the external braid is a pentacyl group of MFI type zeolite compilers [16].

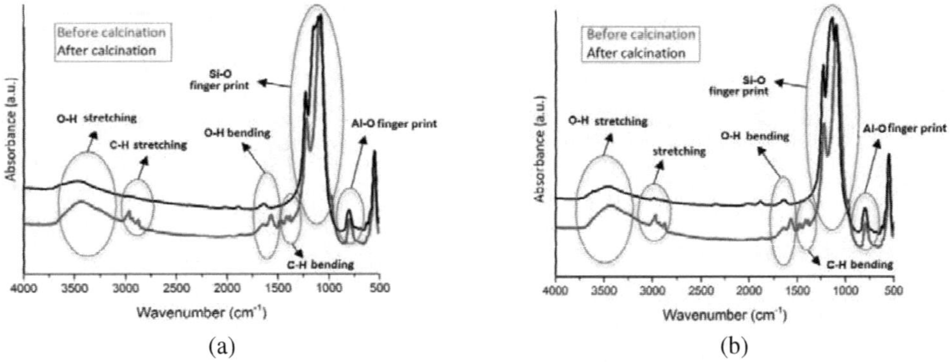

Figure 1. FTIR spectra of (a) micro ZSM-5 and (b) meso ZSM-5.

3.2. Characterization with x-ray diffraction spectroscopy (XRD)

The diffractogram of the micro and meso ZSM-5 compared to the standard ZSM-5 from IZA online is presented in figure 2. Based on the XRD pattern given by IZA online, there are characteristic peaks typical for ZSM-5 namely at 2θ = 7–9 deg and 2θ = 22–25 deg. It is clearly seen that there is a similarity between the diffractogram patterns produced by the as-synthesized ZSM-5 compared to that of the ZSM-5 standard from IZA online [17]. There was a slight shift of 2 theta to a larger number for the Micro ZSM-5 catalyst. However, the overall synthesis of Micro and Meso ZSM-5 catalysts still be said to be successful since the results of the FTIR shows the Si-O and Al-O bonds related to the framework of ZSM-5 were still visible.

3.3. Characterization using scanning electron microscopy with energy dispersed x-ray spectroscopy (SEM-EDX)

The characterization using SEM aims to see the surface morphology of the as-synthesized ZSM-5 catalysts. Figure 3 is the result of characterization using SEM showing the form of Micro ZSM-5 and Meso material is hexagonal which is a characteristic of ZSM-5, or commonly called coffin form with high crystallinity [14]. The difference is that the shape of the Micro ZSM-5 has a rather round end. The Si/Al ratios determined from EDS measurement (table 1) of the Micro and Meso ZSM-5 were 43 and 27, respectively. This result is close to the initial target of around 30 to 50. ZSM-5 micro Si/Al ratio was 43, which means the number of aluminum (Al) atoms was decreasing. This was probably due to the ZSM-5 micro synthesis process by using the sodium aluminate reagent as a source of Al, not all Al atoms were bound into the framework.

3.4. Surface area analysis of ZSM-5

Data analysis of surface area of as-synthesized Micro ZSM-5 and Meso ZSM-5 can be seen in table 1. Table 1 shows that surface area of the Micro and Meso ZSM-5 catalysts is not much different. The differences appear in the area and volume of the micropore and mesopore. The ratio of surface area and micropore volume to mesopore of Micro ZSM-5 is above 75 %. This shows that micro ZSM mostly has a micro-size area and volume. The ratio of surface area and micropore volume to mesopore of Meso ZSM-5 is up to 50 %. This shows that the Meso ZSM-5 has an almost equal number of micro-size areas and volumes, due to the use of the double template methods, the first template to direct the structure (structure directing agent) is tetrapropyl ammonium hydroxide (TPAOH) and poly (acrylamide-co-diallyldimethyl ammonium chloride (PDD-AM) is the second template [18].

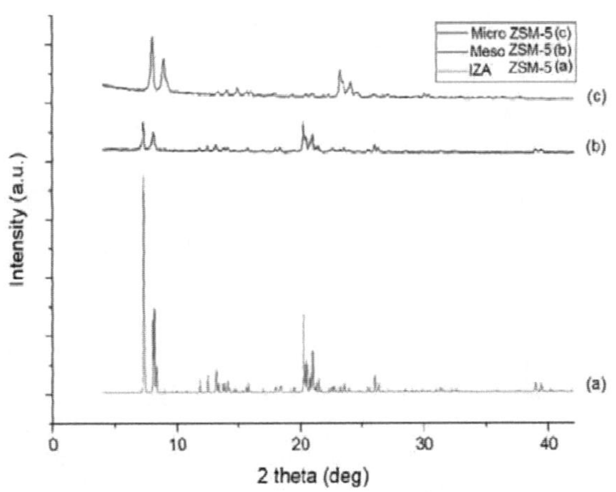

Figure 2. XRD pattern of (a) IZA ZSM-5, (b) Meso ZSM-5 and (c) Micro ZSM-5.

(a) (b)

Figure 3. SEM image of (a) Micro ZSM-5 and (b) Meso ZSM-5 (30,000 magnification).

Table 1. Physicochemical properties of synthesized micro and meso ZSM-5 catalysts.

Sample	Micro ZSM-5	Meso ZSM-5
BET surface area (m^2 g^{-1})	332.38	321.86
Micropore area (m^2 g^{-1})	304.27	187.57
Mesopore area (m^2 g^{-1})	28.10	134.29
Total pore volume (cm^3 g^{-1})	0.18	0.20
Micropore volume (cm^3 g^{-1})	0.14	0.09
Mesopore volume (cm^3 g^{-1})	0.04	0.10
Average pore radius (Å)	19.27	18.34
Surface area ratio of micropore/mesopore (%)	91.55	58.27
Volume ratio of micropore/mesopore (%)	79.05	47.70
Si/Al ratio	43.44	27

Figure 4 shows the adsorption-desorption isotherm curves of Micro ZSM-5 and Meso ZSM-5. Figure 4 shows that Meso ZSM-5 (blue) possesses the characteristic of type IV with hysterical loop [7]. On the other hand, the results of Micro ZSM-5 (red) shows that the pores produced are mostly microporous and in accordance with the type I isotherm [7]. This can be explained further by the pore size distribution graph in figure 5. This graph shows that in both the Micro and Meso ZSM-5 materials, there are two peaks of the highest pore size distribution, at 18 Å and 34 Å. The difference is that the Micro ZSM-5 has a higher distribution of pore size at 18 Å than the Meso ZSM-5, conversely the Meso ZSM-5 has more of pores sized at 34 Å. This is the consequences of using two templates in Meso ZSM-5 and only single template in Micro ZSM-5. The mesoporosity formed in Micro ZSM-5 may be resulted from the defects and combination of two or more pores to create a larger pore.

3.5. Reaction of conversion of sorghum stem to levulinic acid

The percentage yields of levulinic acid from the conversion reaction of delignified stem sorghum using Meso ZSM-5 and Micro ZSM-5 are compared in figure 6. When using Meso ZSM-5 and Micro ZSM-5, the levulinic acid as a product was observed at the beginning of the reaction until 6 h and 10 h, respectively. This results shows that the size of ZSM-5 pores affected the catalytic performance.

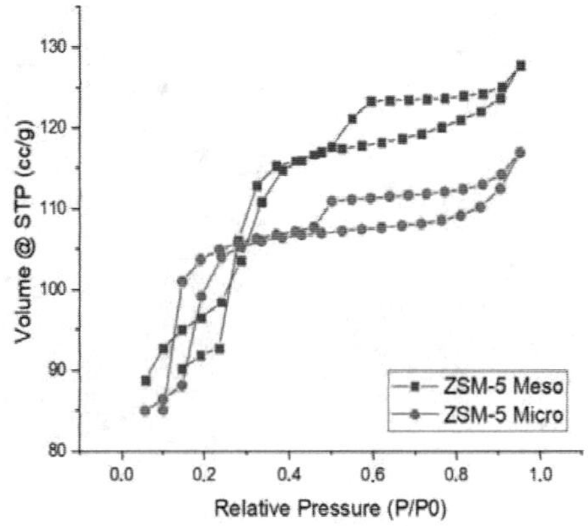

Figure 4. Isotherm adsorption-desorption of Micro ZSM-5 and Meso ZSM-5.

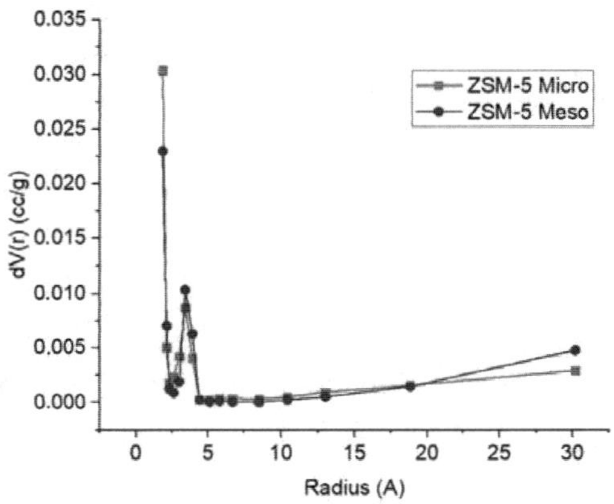

Figure 5. BJH desorption curves to measure the pore radius distribution of micro ZSM-5 and meso ZSM-5.

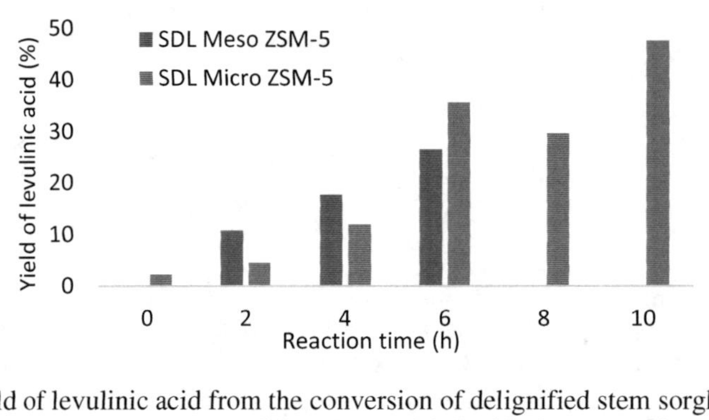

Figure 6. % Yield of levulinic acid from the conversion of delignified stem sorghum using micro and meso ZSM-5 catalysts at condition: 0.1 g of delignified stem sorghum, 0.01 g of catalyst, 40 % (v/v) H_3PO_4, 30 % (v/v) H_2O_2, at 130 °C for 10 h.

The Meso ZSM-5 give a higher % yield of levulinic acid after 2–6 h reaction compared to micro ZSM-5. The longer the reaction time, the yield of levulinic acid were increased as well, until the optimum time was achieved at 6 h. This may be due to a better access active species in the Meso ZSM-5 compared to that in Micro ZSM-5. This result is in agreement with the previous result reported in [7] when similar reaction was carried out using delignified rice husks. The % yield of levulinic acid was significantly increased when the reaction was carried out for more than 6 h using Micro ZSM-5, which may be caused by pore opening or merging to create larger pores giving more access to active sites in the Micro ZSM-5.

4. Conclusion

It is concluded that the Micro and Mesoporous ZSM-5 were successfully synthesized with the Micro ZSM-5 has a higher distribution of pore size at 18 Å than the Meso ZSM-5, conversely the Meso ZSM-5 has more of pores sized at 34 Å. The delignification result was that cellulose content increased from 45.10 % to 76.66 %. The levulinic acid produced using mesoporous ZSM-5 catalyst was observed at the reaction time of 2–6 h, while the yield of levulinic acid using Micro-porous ZSM-5 catalyst was observed at the end of the reaction time, i.e. 6–10 h. It can be concluded that the pore size of the catalyst and the duration of the conversion time influenced the yield of levulinic acid.

Acknowledgments

This research is funded by PTUTP No. NKB-1720/UN2.R3.1/HKP.05.00/2019, The Republic of Indonesia Ministry of Research, Technology and Higher Education year 2019.

References

[1] Pileidis F D and Titirici M-M 2016 *ChemSusChem* **9** 562-82
[2] Isikgora F H and Becer C R 2015 *Poly. Chem.* **6** 4497-559
[3] Kang S, Fu J and Zhang G 2018 *Renew. Sustain. Energy Rev.* **94** 340-62
[4] Mulder G J 1840 *Adv. Synth. Catal.* **21** 203-40
[5] Ndaba B 2013 *Microwave assisted pretreatment of sweet sorghum bagasse for bioethanol production* Mini-dissertation Master of Science in Chemical Engineering (South Africa: North-West University, School of Chemical and Minerals Engineering) p 145
[6] Morone A, Apte M and Pandey R A 2015 *Renew. Sustain. Energy Rev.* **51** 548-65
[7] Krisnandi Y K et al. 2019 *Indones. J. Chem.* **19** 115-23
[8] Girisuta B, Danon B, Manurung R, Janssen L P B M and Heeres H J 2008 *Bioresour. Technol.* **99** 8367-75
[9] Cha J Y and Hanna, M A 2002 *Ind. Crops Prod.* **16** 109-18
[10] Vyver S V d et al. 2011 *Energy Environ. Sci.* **4** 3601-10
[11] Chen Y, Li G, Yang F and Zhang S-M 2011 *Polym. Degrad. Stab.* **96** 863-69
[12] Zeng W, Cheng D-g, Zhang H, Chen F and Zhan X 2010 *React. Kinet. Mech. Catal.* **100** 377-84
[13] Ya'aini N, Amin N A S and Endud S 2013 *Microporous Mesoporous Mater.* **171** 14-23
[14] Wang D, Liu Z, Wang H, Xie Z and Tang Y 2010 *Microporous Mesoporous Mater.* **132** 428-34
[15] Dence C W 1992 *The Determination of Lignin* (Methodes in Lignin Chemistry) ed Lin S Y, Dence C W (Berlin: Springer-Verlag) **2.2** 33-61
[16] Gonçalves M L, Dimitrov L D, Jordão M H, Wallau M and Urquieta-González E A 2008 *Catal. Today* **133-5** 69-79
[17] IZA-SC 2017 *Database of Zeolite Structure* available at https://asia.iza-structure.org/IZA-SC/pow_pat.php?STC=MFI&ID=MFI_1
[18] Krisnandi Y K et al. 2015 *Procedia Chem.* **14** 508-15

IOP Conf. Series: Materials Science and Engineering **902** (2020) 012051 doi:10.1088/1757-899X/902/1/012051

Catalytic depolymerization of lignin from wood waste biomass over natural sourced ZSM-5 catalysts

A P Pratama, Y K Krisnandi and I Abdullah

Department of Chemistry, Faculty of Mathematics and Natural Sciences (FMIPA), Universitas Indonesia, Depok 16424, Indonesia

Corresponding author's email: yuni.krisnandi@sci.ui.ac.id

Abstract. Lignin isolated from wood waste biomass, contains 61.07 % of lignin, was converted to vanillin using ZSM-5 catalysts. Prior to use, wood waste biomass was delignified to enrich its lignin contents followed by added 10 % H_2SO_4 to isolate the lignin such as solid black material. Three types of ZSM-5 catalysts i.e. hierarchical ZSM-5, NiO/hierarchical ZSM-5 and CuO/hierarchical ZSM-5 zeolites were used as catalysts in this study. Hierarchical ZSM-5 zeolite was prepared through TPAOH template synthesis methods followed by PDDAM as secondary template. Silicate and aluminate source based on a natural source to form hierarchical ZSM-5 zeolite. Hierarchical ZSM-5 zeolite is modified with NiO and CuO by using impregnation method with Ni nitrate and Cu nitrate solution followed by calcination at 550 °C. The characterization of catalysts employed FTIR, XRD, andSEM-EDX. The reaction was carried out in 200 mL-batch reactor at 100 °C, 150 °C, 200 °C, 250 °C and 300 °C in the presence of ethanol and water (v/v) for 30 min. CuO/hierarchical ZSM-5 zeolite gave the highest % yield of vanillin as product (4.68 %) followed by NiO/hierarchical ZSM-5 zeolite (3.82 %) for the variation of temperature, compared to other ZSM-5 catalysts.

Keywords: Zeolite, ZSM-5, lignin conversion, heterogeneous catalysts, biomass

1. Introduction

The human population worldwide is continued to grow at a high rate, that implicates, the percentage of global energy used to be increased rapidly. Biomass and CO_2 are the alternatives for common sources energy form such as fossil feedstocks. Utilization of biomass could increase the energy supply in the future, reducing the CO emissions of fossil feedstock, and providing source of platform chemicals[1].

Lignocellulose biomass is mainly contained with cellulose (31–42 %), hemicellulose (10–25 %), and lignin (15–33 %) [2]. Cellulose contained in lignocellulosic biomass can be used as the raw material for producing levulinic acid [3]. The development of optimizing the use of cellulose from biomass is growing rapidly. To obtain cellulose from lignocellulosic biomass, delignification pretreatment, a process of separating lignin and cellulose was carried out. Lignin is quite dominant in lignocellulosic biomass, similar to the amount of cellulose but is still rarely used optimally.

Lignin in lignocellulose could be converted into more useful or value-added chemical products using zeolitesas catalysts. Lignin undergoes several reactions such as oxidative catalysis, pyrolysis, hydrogenation or depolimeration which produces several compounds such as vanillin, syringol, 4-ethyl gauaicol, 4-ethyl phenol, guaiacol, phenol, and phenol compounds [4]. For instance, vanillin, one of phenolic compounds which contains different functional groups, including aldehyde, hydroxyl, and ether, is the most widely used flavoring agents in the world [5].

Increasing the lignin conversion in lignocellulosic biomass requires optimization. The catalyst in lignin depolymerization must increase the conversion and suppress char formation. One of the industrial heterogeneous catalysts often used is zeolite because of its alkaline nature, having active sites, and also having porosity. Zeolites, such as ZSM-5, are ideally used as catalysts because of their uniform pore size, large surface area, and the typical structure and selectivity of their reactions [6].

Wanmolee et al. reported the degradation of lignin into monomeric phenols using homogeneous catalyst (i.e. sulfuric acid) and heterogeneous catalyst (i.e. H-Y zeolite) for the degradation of lignin produced phenol 0.68 % and vanillin 0.22 % [4]. Milovanovic et al. have identified the variations of metal oxides as catalysts supported by natural ZSM-5 in lignin conversion reactions to produce phenolic products with the best liquid phase yield of 34.8 % for NiO/ZSM-5 catalysts and the best phase yields gas by 17.6 % for the Cu_2O/ZSM-5 catalyst [7]. Furthermore, Mullen et al. reported the overall conversion of dominant content in lignocellulosic biomass using H-ZSM-5 and K-ZSM-5 as catalysts for maximum yield of around 13 % for phenols [8]. Wang et al. investigated the effect of particle on the catalytic activity of H-ZSM-5 zeolite in conversion of lignin to phenols in the supercritical ethanol-water gave 6 % yield of phenols [9]. On the other hand, in this study, the hierarchical ZSM-5 using silicate and aluminate natural source from natural Bayat zeolite with ludox as silica synthetic source was prepared [10].

This study focused on the synthesis of natural sourced ZSM-5 zeolite and prepared the other catalysts. The three types of catalysts were used, including hierarchical ZSM-5, NiO/ZSM-5, CuO/ZSM-5. It is aimed to compare the activity and selectivity of each heterogeneous catalyst in catalytic depolymerization of lignin. To the best of our knowledge, the utilization of wood waste biomass as lignin source in the production of phenolic compound such as vanilin using heterogeneous catalyst from natural sourced ZSM-5 is rarely reported.

2. Materials and method

2.1. Materials

Natural zeolite applied in this research was received Bayat Klaten, Central Java, Indonesia. Indonesian kaolin from Belitung island was supplied by PT. Aneka Kaoline Utama (Belitung Island, Indonesia). Sodium hydroxide (pellet, 99 %), hydrogen peroxide (30 %), nitric acid, acetic glacial acid, hydrochloric acid, sodium acetate trihydrate, trisodium citrate dihydrate, sodium bicarbonate, $Ni(NO_3)_2.6H_2O$, $Cu(NO_3)_2.3H_2O$, ethanol and methanol were procured from Merck (Germany). Sodium aluminate (99 %), tetrapropylammonium hydroxide (TPAOH, 1 M), and polidialydimethyl ammonium chloride acrylamide (PDD-AM, 10 %) were obtained from Sigma Aldrich, US. Phosphoric acid (97 %) was procured from PT. Smart Lab Indonesia. Distilled water was used in the preparation of each solution.

2.2. Synthesis of hierarchical ZSM-5 zeolite based on natural source

Natural zeolite and natural Belitung kaolin were sieved to the size of about 100 μm and were treated using the activation, purification, fragmentation and calcination processes [10, 11]. Synthesis of hierarchical ZSM-5 zeolite based on natural source was conducted using natural Bayat zeolite and natural Belitung kaolin as its silicate and aluminate source.Bayat natural zeolite and natural Belitung kaolin were mixed with H_2O, stirred overnight. Furthermore, TPAOH was slowly added into the reaction mixture under vigorous stirring at 100 °C. Then, the mixture pH solution was adjusted to ±11 for 3 h. Afterward, certain amount of cationic PDD-AM was slowly added into the reaction mixture at room temperature and stirred for 48 h. After stirring, the mixture was transferred into an autoclave for further crystallization at 170 °C for 144 h. The product was washed with H_2O and filtered and calcinated at 550 °C for 5 h and characterized using XRD, FTIR, SEM, EDS and SAA instruments.

2.3. Preparation of hierarchical NiO/ZSM-5 zeolite and CuO/ZSM-5 zeolite
Hierarchical NiO/ZSM-5 zeolite was prepared using wet impregnation method with Ni^{2+} solution (from nickel nitrate hexahydrate) for 24 h at 25 °C into a calcined hierarchical ZSM-5 zeolite. The mixture was dried at 110 °C for 5 h followed by calcination at 550 °C. Hierarchical CuO/ZSM-5 zeolite was prepared using wet impregnation method with Cu^{2+} solution (from copper(II) nitrate trihydrate) and similiar preparation steps with that for NiO/ZSM-5 zeolite.

2.4. Pre-treatment of wood waste biomass and Lignin isolation
Wood waste biomass was subjected to the delignification pre-treatment. Delignification was carried out by refluxing the biomass with 10 % NaOH solution (1:25, m/v) at 150 °C for 90 min while stirring. The lignin content was determined before and after pre-treatment using the procedure outlined in our previous study [3]. After that, the suspension was filtered and the black filtrate (black liquor) was obtained. Then, to the black liquor, 10 % H_2SO_4 was added until the precipitate was formed. After that, the precipitate was filtered and washed with distilled water until neutral, then the precipitate was dried.

2.5. Catalytic depolymerization lignin from wood biomass
Lignin depolymerization reaction was carried out by using a stainless steel batch reactor with a capacity of 200 mL containing pressure and temperature in an autoclave reactor. The 0.25 gram of lignin was dissolved in a solvent of water and ethanol (50:50, v/v) mixture, then it was transferred into the reactor along with a certain amount of catalyst. The reactor was flowed with N_2 gas and filled with 2 atm at room temperature. Initially, the mixture was stirred at 100 rpm until reached the desired condition at variation temperature 100–300 °C, then the stirring speed is increased at the desired time. After the reaction completed, the reactor was cooled. After that, the reaction mixture was filtered and then the obtained filtrate was analyzed using High Performance Liquid Chromatography (HPLC).

3. Results and discussion

3.1. Characterization of raw materials

3.1.1. XRD patterns. Raw natural Bayat zeolite diffraction pattern shown in figure 1a shows that there is a peak at 2θ of 9°, 22° and 25°. Fragmentation process aims to destroy the zeolite frame structure. It damages the Si-O-Si and Si-O-Al bonds into Si and Al active and tetrahedral structures XRD pattern of both raw and fragmented natural Bayat zeolite are similar to the pattern reported by Krisnandi et al. [12]. XRD patterns of raw natural Belitung kaolin in figure 1a shows that the structure of raw kaolin was still intact after pretreatment, then, it was changed in the pattern of extracted silica, leaving some high intensity peaks around 2θ 15–35°, indicating an amorphous structure. XRD patterns of kaolin before and after pretreatment are similar to the pattern reported by Krisnandi et al. [11].

3.1.2. FTIR spectra. Figure 1b presents the raw and after pretreated materials' spectra. Both natural Bayat zeolite and kaolin showed two peaks between 3800 cm^{-1} and 3600 cm^{-1}, typically for stretching hydroxyl groups structure such as -Si-OH and -Al-OH. It was typical of phyllosilicate alumina silicates [13]. The two materials also have 1650–1600 cm^{-1} band, which are attributed to the adsorbed water molecule bending vibration (H-O-H). A group of bands at1120–1020 cm^{-1} related to T-O asymmetric stretching vibration groups, and in 920-925 cm^{-1}, attributed toT-OH groups, were observed. This result shows disruptive of-Al-O-Si-in kaolin was transformed in the extracted silica, indicating the aluminum removal. The opposite occurs in the fragmented natural bayat zeolite, -Si-OH group at 3700–3490 cm^{-1} was reduced, indicating the reduced amount of silica.

3.2. Catalyst characterization

3.2.1. XRD patterns. ZSM-5 was synthesized from natural sourced fragmented zeolite of natural Bayat zeolite and extracted silica from kaolin. XRD analysis was conducted to confirm the structure of the as-synthesized ZSM-5 zeolite. Its diffraction pattern (figure 2a) has typical peaks of standard ZSM-5, there are two sharp peaks between 2θ = 7–10° and three peaks between 2θ = 22–25° [14]. For NiO characteristic of 2θ, there are two peaks between 35–40° and one peak between 47–50° (JCPDS Card No.47-1049) [15]. CuO has a typical pattern of two sharp peaks at 2θ = 35–40° and one peak at 47–50° (JCPDS Card No.01-089-2529) [16]. In general, all samples indicated that the preparation of all catalysts was successful.

3.2.2. FTIR spectra. Figure 2b shows the FTIR spectra of as-synthesized ZSM-5 and its derivatives. In the ZSM-5 FTIR spectra before calcination, peak at the wave number 2950–2850 cm^{-1} and

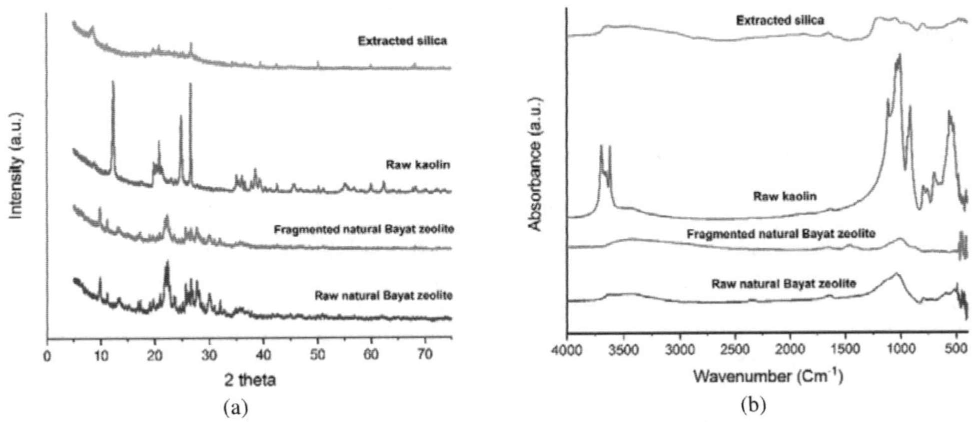

Figure 1. (a) XRD pattern and (b) FTIR spectra of raw natural Bayat zeolite, fragmented natural Bayat zeolite, raw kaolin and extracted silica from kaolin.

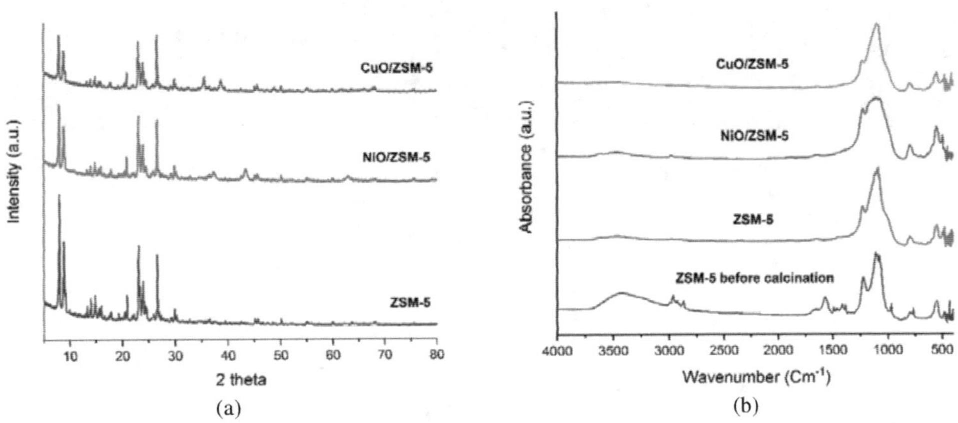

Figure 2. (a) XRD pattern and (b) FTIR spectra of catalysts.

1450–1350 cm^{-1} are the vibration and absorption band of C-H bend vibration [17]. This absorption band was from a TPAOH as primary template used in the synthesis of ZSM-5 and it was removed by the calcination process to form pores. ZSM-5 after calcination indicated the loss of organic template as the intensity of C-H vibration and C-H bending band absorption decreased. In this case, modification of both NiO and CuO to hierarchical ZSM-5 zeolite did not cause any damage to the structure of the ZSM-5 zeolite.

The FTIR spectra shows the main functional groups of the zeolite from vibration of OH stretching at around 3300 cm^{-1}, OH bending vibration at 1600 cm^{-1}, T-O-T (T = Si/Al) bending and stretching asymmetry vibration at 1300–750 cm^{-1} and double five-ring vibration at 550 cm^{-1} which are the infrared characters of ZSM-5 [14].

3.2.3. Scanning electron microscopy. SEM characterization provided morphology information of the catalysts. Figure 3 clearly shows the surface morphology and crystal size of the calcined ZSM-5. Natural sourced ZSM-5 zeolite synthesized using TPAOH template has a coffin-like morphology and has hexagonal crystal geometry [18].

3.2.4. EDX analysis. The compositionof the as-synthesized hierarchical ZSM-5, NiO/ZSM-5 and CuO/ZSM-5 were determined with EDX analysis, shown in table 1. It shows that the Si/Al ratio of the hierarchical ZSM-5 zeolite was 18.15. The elemental analysis of NiO/ZSM-5 and CuO/ZSM-5 shown in table 1 gave the mass percent of nickel and copper around 10 %, which are close to the desired amount in this work (c.a. 10 %).

3.3. Catalytic Depolymerization of lignin to phenolic compound

Lignin was isolated from raw wood waste biomass. Before isolation, the raw biomass contained 26.48 % of lignin and after isolation process, the lignin content was increased to 61.07 % (from its black liquor). In this study, vanillin was chosen as the produced of phenolic compound that was analyzed. It has some functional groups including aldehyde, hydroxyl, and ether due to the phenolic compound functional group in it [19]. Figure 4 shows that the optimal temperature to produce the highest

(a) (b) (c)

Figure 3. SEM images of (a) ZSM-5, (b) NiO/ZSM-5 and (c) CuO/ZSM-5.

Table 1. Catalysts variations of hierarchical ZSM-5 and its modification.

Material	Si	Al	Cu	Ni	Si/Al mol ratio
ZSM-5	48.01	2.82	-	-	16.36
NiO/ZSM-5	40.08	2.43	-	9.84	15.84
CuO/ZSM-5	36.55	2.49	11.03	-	14.10

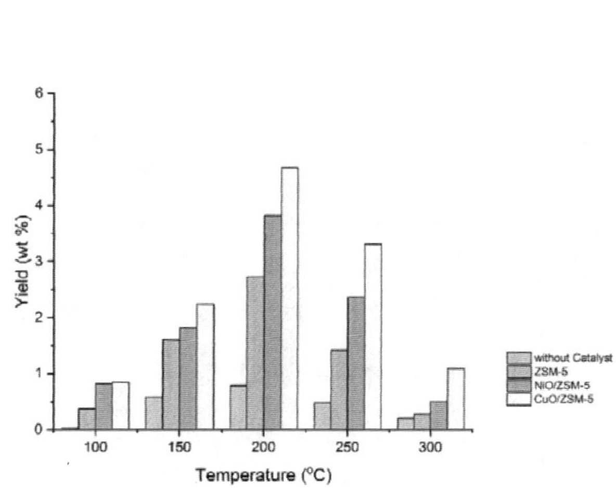

Figure 4. Percent yield of product catalytic reaction of depolimerization lignin to vanillin with temperature and catalysts variation.

vanillin yield is 200 °C. It is suggested that the catalyst used was started to actively depolymerize lignin at 200 °C [4]. However, at temperature above 200 °C the percent yield of vanillin decreased. It may be due to vanillin was degraded or oxidized further to produce other products [20].

Furthermore, figure 4 shows that the reaction without the presence of catalyst shows a lower % yield compared to the reaction with catalysts. This is because the presence of a catalyst will lower the activation energy, causing the reaction to take place faster and the optimal yield of vanillin. In this work, each type of catalyst give a different percentage yield of vanillin, depends on the elements contained in each catalyst. In addition, several studies on the vanillin conversion turned out to provide the optimal results with ZSM-5 catalysts [7] and Cu catalysts [21] followed by Ni catalysts [22], then two of the composite study here gave the highest product yield. Followed by NiO/ZSM-5 catalysts and ZSM-5 catalysts without impregnation.This supports the elemental analysis using EDS characterization that CuO/ZSM-5 has the smallest Si/Al ratio, thereby increasing its acidity, will give the greater % yield from lignin depolymerization.

4. Conclusion

Phenolic compounds such as vanillin can be obtained from lignin derived from lignocellulose biomass through the heterogeneous catalytic depolymerization reactions. The lignin depolymerization reaction can take place at a temperature of 200 °C in the presence of hierarchical ZSM-5 zeolite catalysts variation and its modification with Ni or Cu.The highest percent yield of vanillin is given by CuO/ZSM-5 followed by NiO/ZSM-5and ZSM-5

Acknowledgments

This work was financially supported by Ministry of Research and Higher Education, Republic of Indonesia under research grant Hibah Dana Riset Pendidikan Magister Menuju Doktor untuk Sarjana Unggul (Hibah PMDSU) with grant contract number NKB-3040/UN2.R3.1/HKP.05.00/2019.

References

[1] Ben H and Ragauskas A J 2013 *ACS Sustain. Chem. Eng.* **1** 316-24
[2] Galletti A M R, Antonetti C, De Luise V, Licursi D and Di Nasso N N 2012 *BioResources* **7** 1824-35

[3] Krisnandi Y K et al. 2019 *Indones. J. Chem.* **19** 115-23
[4] Wanmolee W, Daorattanachai P and Laosiripojana N 2016 *Energy Procedia* **100** 173-7
[5] Ni J, Tao F, Du H and Xu P 2015 *Sci. Rep.* **5** 13670
[6] Krisnandi Y K et al. 2015 *Procedia Chem.* **14** 508-15
[7] Milovanovic J et al. 2015 *J. Serb. Chem. Soc.* **80** 717-29
[8] Mullen C A, Tarves P C and Boateng A A 2017 *ACS Sustain. Chem. Eng.* **5** 2154-62
[9] Wang X, Du B, Pu L, Guo Y, Li H and Zhou J 2017 *J. Anal. Appl. Pyrolysis* **129** 13-20
[10] Rohayati, Krisnandi Y K and Sihombing R 2017 *AIP Conf. Proc.* **1862** 030094
[11] Krisnandi Y K et al. 2019 *Crystals* **9** 1-14
[12] Krisnandi Y K et al. 2018 *J. Phys.: Conf. Ser.* **1095** 012044
[13] Wang G et al. 2019 *Microporous Mesoporous Mater.* **284** 133-40
[14] Treacy M M and Higgins J B 2001 *Collection of Simulated XRD Powder Patterns for Zeolites* (Amsterdam: Elsevier) p 184
[15] El-kemary M, Nagy N and El-Mehasseb I 2013 *Mater. Sci. Semicond. Process.* **16** 1747-52
[16] Etefagh R, Azhir E and Shahtahmasebi N 2013 *Sci. Iran.* **20** 1055-8
[17] Zhang J, Li X, Liu J and Wang C 2019 *Catalysts* **9** 1-14
[18] Wang L, Zhang Z, Yin C, Shan Z and Xiao F-S 2010 *Microporous Mesoporous Mater.* **131** 58-67
[19] Cha T-S, Chen C-F, Yee W, Aziz A and Loh S-W 2011 *J. Microbiol. Methods* **84** 430-4
[20] Tarabanko V E and Tarabanko N 2017 *Int. J. Mol. Sci.* **18** 2421
[21] Petitjean L, Gagne R, Beach E S, Xiao D and Anastas P T 2015 *Green Chem.* **18** 150-6
[22] Mukherjee D, Singuru R, Venkataswamy P, Damma D and Reddy B M 2019 *ACS Omega* **4** 4770-8

Synthesis and characterization of zinc-aluminum and zinc-chromium LDHs intercalated citrate anion

N Hidayati, M F Azmi, M Oktriyanti, N R Palapa, R Mohadi, Elfita and A Lesbani

Department of Chemistry, Faculty of Mathematics and Natural Sciences (FMIPA), Universitas Sriwijaya, Ogan Ilir 30662, Indonesia

Corresponding author's email: aldeslesbani@pps.unsri.ac.id

Abstract. In this study, the preparation of Zinc-Al and Zinc-Cr LDH intercalated by citrate anion conducted using co-precipitation method at room temperature. The characterization of Zinc-Al and Zinc-Cr LDHs intercalated citrate anion was performed using XRD and FTIR. The result of XRD pattern and FTIR spectra confirmed that Zinc-Al and Zinc-Cr intercalated citrate anion were successfully prepared. The vibration of citrate anion observed at 1590 cm^{-1} and 1390 cm^{-1}, respectively, correspond to the characteristic of the carboxylate group (-COO-). Structure of Zinc-Al and Zinc-Cr LDHs were shown by reflection at 2 theta (003) that indicated the interlayer space was higher than before intercalation.

Keywords: Zinc-Al, Zinc-Cr, intercalation, LDH, citrate

1. Introduction

Layered double hydroxide is well-known as potential material with broad applications in various fields [1]. In recent years, these materials get attention from researchers with their potential applications such as adsorbent, catalysis, membrane, flame retardants and also anionic changes [2, 3]. Layered double hydroxide is an anionic material with lamellar structure, its ability as ion exchanger makes LDH easy for modification [4]. The layered double hydroxide has a general formula $[M^{2+}/M^{3+}(OH)\ ]_x{}^+[A^{n-}]\cdot mH_2O$ from stacking brucite-like layers metal oxide containing a positive charge due to partial substitution of divalent metal cation by trivalent metal cation [5]. The positive charge represented by anions located in the interlamellar spaces. Interlamellar space is also known as interlayer, can exchange species A^{n-} to other anions. The interlayer can be expanded to an adapted wide range of interlayer anion [6]. Therefore, A^{n-} can be exchangeable with inorganic or organic anionic compounds, such as carboxylic [7], halides [8], oxo-anion [9], and polyoxometalate [10].

Layered double hydroxide was considered as a class of materials that is easy to synthesize among several methods [11]. Various methods have been prepared depending on the intended usability and purpose for which the layered double hydroxide used. Many procedures were used to synthesize the layered double hydroxide using precipitation method by varying ratio molar and adjusting pH [12], anion exchange [13], hydrothermal synthesis [14], sol-gel process [15], structure reconstruction [16]. The most conventional method used is precipitation, which can be performed at constant pH depending on the applied conditions. The co-precipitation method was studied in this research. Co-precipitation has more advantages than other methods. The benefits of this method are easy preparation, high crystallinity, and high purity of material [15, 16]. Zhang et al. reported that Zn-Al LDH was synthesized and modified by intercalating EDTA in interlamellar space using co-precipitation method [17]. Rahmanian et al. reported Zn-Fe LDH intercalated by citrate anion and applied as sorbent [18]. The research by Meyn et al. reported that several LDHs intercalated organic acid anions using co-precipitation method has different interlayer distances [19]. In nature, citric acid is widely found. Similar with ethylenediaminetetraacetic acid (EDTA), they have advantages as an excellent chelating agent in binding toxic metal cations. Therefore, the citrate anions-intercalated LDHs are expected in various applications especially as adsorbent [18]. In this research, Zinc-Al and Zinc-Cr LDHs intercalated citrate anion using co-precipitation method with constant pH. The surface morphology and chemical structure of LDHs were analyzed by XRD and FTIR.

2. Materials and method

2.1. Materials and instrumentation
The materials used including zinc nitrate, aluminum nitrate, chromium nitrate and sodium carbonate were purchased from Merck. Sodium hydroxide and citric acid were purchased from Sigma Aldrich. Aquadest was obtained by Purite Instruments. The material was characterized by XRD Rigaku Miniflex-600, and FTIR Shimadzu FTIR Prestige- 21.

2.2. Methods

2.2.1. Synthesis of Zinc-Al LDH. The synthesis of Zinc-Al LDH was conducted by co-precipitation method [20]. A 100 mL of 0.75 M zinc nitrate (14.202 g) was added to 100 mL of 0.25 M aluminum nitrate (5.324 g) by vigorous stirring for an hour. Sodium hydroxide was prepared by diluting 0.8 g NaOH into 100 mL of water. Alkaline solution of LDHs was added dropwise to 2 M sodium hydroxide until pH reached 10 with continuous stirring at 340 rpm. The white gel was obtained and temperature synthesis was kept at 80 °C for 4 h. The material was washed and kept in the oven overnight at 60 °C. The dried Zinc-Al LDH was characterized by XRD and FTIR.

2.2.2. Synthesis of Zinc-Cr LDH. According to Sun et al., the synthesis of Zinc-Cr LDH was conducted by mixing 0.4 M zinc nitrate (7.57 g; 100 mL) and 0.2 M chromium nitrate (10.004 g; 100 mL), and added to 2.5 M sodium carbonate and 3 M sodium hydroxide under vigorous stirring for 2 h. pH of the mixture solution was adjusted to 10 using sodium hydroxide [21]. The mixture solution was kept at 60 °C overnight with moderate stirring. The gray solution was kept in the oven at 80 °C until gray gel obtained. The dried Zinc-Cr LDH was characterized by XRD and FTIR.

2.2.3. Preparation of Zinc-Al and Zinc-Cr LDH intercalated citrate anion. Under nitrogen atmosphere, 2 g of LDH gel was diluted by 50 mL deionized water. Another beaker with10 g of citric acid was diluted by 50 mL deionized water. The solution of LDH was added dropwise into citric acid with continuous stirring for 24 h, then pH was adjusted to 10. The mixture solution was kept at room temperature. The material was washed, dried and characterized by XRD and FTIR.

3. Results and discussion
Zinc-Al and Zinc-Cr LDH were successfully synthesized by co-precipitation method. The result is shown in figure 1. Figure 1 shows the powder form of LDHs. The LDHs results remained in white and grey powder, same as the color of LDH before intercalation. Zinc-Al and Zinc-Cr were intercalated by citrate anion by co-precipitation method. This method was conducted by mixing LDH gel and citric acid with a ratio of g 1:10. The mixing solution was adjusted to pH 10. The schematic of LDH intercalated by citrate anion is shown in figure 2.

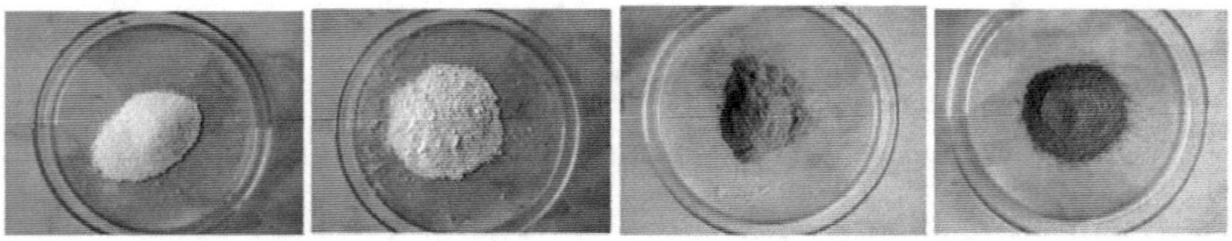

Figure 1. Powder of (a) Zinc-Al LDH, (b) Zinc-Al Intercalated, (c) Zinc-Cr LDH, and (d) Zinc-Cr Intercalated.

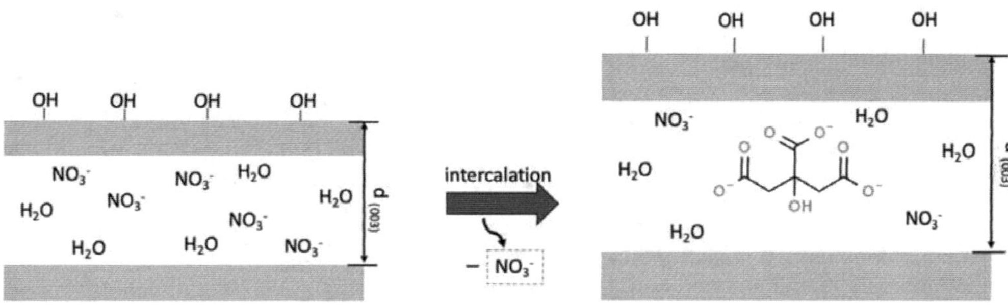

Figure 2. The scheme of LDH intercalated by citrate anion.

The FTIR spectra of the Zinc-Al, Zinc-Cr LDH, and LDH intercalated are shown in figure 3. Figure 3 shows the broad vibration at 3400 cm^{-1} which represents the O-H stretching vibration of water molecules and also hydroxide layers. The sharp peak shown in figure 3a and figure 3b at 1380 cm^{-1} are denoted as nitrate anion vibration in the interlayer space of LDH. Another lower peak at 1635 cm^{-1} corresponds to bending O-H. Figure 3 c and figure 3d show the differences prominent peaks at 1590 cm^{-1} and 1390 cm^{-1} which are the characteristics of C=O bonds and -COO- (carboxylate group). The nitrate anion stretching at 1380 cm^{-1} is not observed because of the decreased intensity and the appearance of carboxylate groups from figure 3c and figure 3d. The spectrum of Zinc-Al intercalated by citrate anion is similar to LDH, which yielded by Zhang et al. [17]. Zhang reported that after intercalation using citrate anion, Zn-Al/citrate has a vibration at 1394 and 1600, which is the characteristic of the COO- groups vibration. Based on FTIR measurements, Zinc-Al and Zinc-Cr have been successfully inserted citrate anion into the interlayer [17]. These materials were characterized by XRD powder analysis.

Figure 4, shows the XRD pattern of Zinc-Al, Zinc-Cr, and intercalated LDH. LDH materials have a similar layered structure. LDH has a unique reflection $d_{(003)}$ which represents the interlayer structure and corresponds to the distances between up and down cations layers. When the interlayer anions in a LDH changed, a corresponding change in the $d_{(003)}$ spacing should be observed. This change will be indicated if the new anions have been successfully intercalated into the LDH. Figure 4 shows Zinc-Al intercalated by citrate has a diffraction 2 theta of $d_{(003)}$, which shifting to a lower angle from 10° to 8°. Zinc-Cr intercalated citrate does not shift, but the interlayer space is higher than before intercalated. The increasing data of Zinc-Al and Zinc-Cr before and after intercalating are shown in table 1.

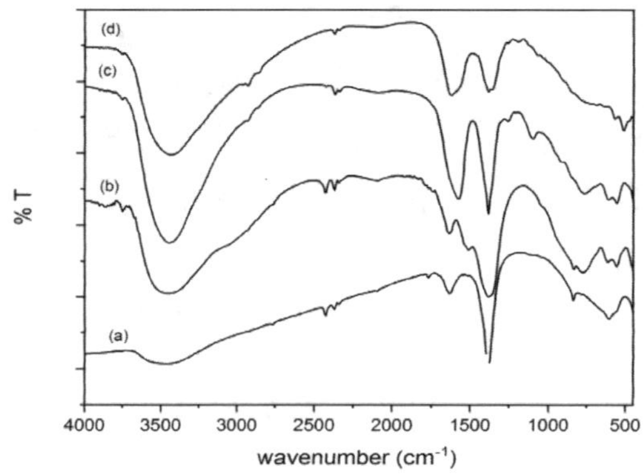

Figure 3. FTIR spectra of a) Zinc-Al LDH, b) Zinc-Cr LDH, c) Zinc-Al Intercalated, and d) Zinc-Cr Intercalated.

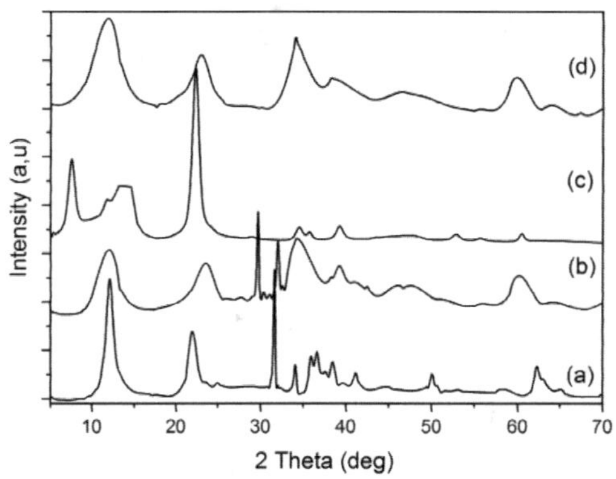

Figure 4. XRD Pattern of a) Zinc-Al LDH, b) Zinc-Cr LDH, c) Zinc-Al Intercalated, and d) Zinc-Cr Intercalated.

Table 1. Data of interlayer Zinc-Al and Zinc-Cr LDH.

	Zinc-Al LDH	Zinc-Cr LDH
Before Intercalated	7.57 Å	7.53 Å
After Intercalated	11.68 Å	10.94 Å

The increasing interlayer distances of Zinc-Al and Zinc-Cr LDH indicated the successful preparation of these materials. Figure 4 also shows other diffraction, which represents by sharp peaks at 22°, 30°, and even 60° for reflection (003), (006) and (110). This unique reflection is useful for identifying the crystallography of LDH. According to Baliarsigh et al., based on Bragg's Laws, LDH can be determined as hexagonal structure by unit cell parameters [22].

The expanded interlamellar region (003) was continued with the intercalation of citrate anions into the interlayer space of LDH. Moreover, the intensity of other reflections such (003), (006) and (110) decreased after the intercalation. As reported by Meyn et al. [19] and Perera et al. [23], the result of citrate pillaring LDH has the lowest value in this research. Meyn showed the interlayer distance after intercalated by Zn-Cr and Zn-Al 8.9 Å and 9.0 Å [19]. Perera showed the interlayer after intercalated by a citrate amount of 8.8 Å [23]. However, the materials were successfully synthesized and ready to apply in various fields.

4. Conclusion

Zinc-Al and Zinc-Cr LDH intercalation by citrate anion has been prepared by the co-precipitation method. The FTIR spectra showed the vibration of the carboxylate group at wavenumber 1390 and 1590 cm^{-1} after intercalation. The nitrate vibration was decreased after the intercalation and was not observed at 1380 cm^{-1}. XRD pattern showed that interlayer space of Zinc-Al and Zinc-Cr LDH was increased after the intercalation from 7.57 Å and 7.53 Å to 11.68 Å and 10.94 Å, respectively.

Acknowledgments

Author would like to thank Sriwijaya University for supporting financial this work through Hibah Profesi Grant with contract number 0014/UN9/SK.LP2M.PT/2019.

References

[1] Cermelj K, Ruengkajorn K, Buffet J-C and O'Hare D 2019 *J. Energy Chem.* **35** 88-94
[2] Palapa N R, Taher T, Mohadi R, Said M and Lesbani A 2018 *AIP Conf. Proc.* **2026** 020033
[3] Taher T et al. 2019 *Bull. Chem. React. Eng. Catal.* **14** 260-7
[4] Zhang B, Dong Z, Sun D, Wu T and Li Y 2017 *J. Ind. Eng. Chem.* **49** 208-18
[5] Kameda T, Saito S and Umetsu Y 2005 *Sep. Purif. Technol.* **47** 20-6
[6] Prinetto F, Ghiotti G, Graffin P and Tichit D 2000 *Micropor. Mesopor. Mat.* **39** 229-47
[7] Mohamed F, Abukhadra M R and Shaban M 2018 *Sci. Total Environ.* **640-1** 352-63
[8] Ferencz Z et al. 2016 *J. Solid State Chem.* 233 236-43
[9] Jadhav V H, Dumbre D K, Phapale V B, Borate H B and Wakharkar R D 2007 *Catal. Commun.* **8** 65-8
[10] Lesbani A et al. 2018 *AIP Conf. Proc.* **2049** 020013
[11] Li T, Miras H and Song Y-F 2017 *Catalysts* **7** 1-17
[12] Bukhtiyarova M V 2019 *J. Solid State Chem.* **269** 494-506
[13] Ahmed A A A, Talib Z A, Hussein M Z and Zakaria A 2012 *J. Solid State Chem.* **191** 271-8
[14] Shan R-r et al. 2015 *J. Ind. Eng. Chem.* **21** 561-8
[15] Yang F et al. 2016 *Appl. Clay Sci.* **123** 134-40
[16] Lei C, Pi M, Kuang P, Guo Y and Zhang F 2017 *J. Colloid Interface Sci.* **496** 158-66
[17] Zhang S, Kano N and Imaizumi H 2016 *J. Chem. Chem. Eng.* **10** 60-7
[18] Rahmanian O, Amini S and Dinari M 2018 *J. Mol. Liq.* **256** 9-15
[19] Meyn M, Beneke K and agaly G 1990 *Inorg. Chem.* **29** 5201-7
[20] Palapa N R, Mohadi R and Lesbani A 2018 *AIP Conf. Proc.* **2026** 020018
[21] Sun X et al. 2019 *Dalt. Trans.* **48** 5285-91
[22] Baliarsingh N, Parida K M and Pradhan G C 2014 *Ind. Eng. Chem. Res.* **53** 3834-41
[23] Perera J, Weerasekera M and Kottegoda N 2015 *Chem. Cent.* J. **9** 1-7

IOP Conf. Series: Materials Science and Engineering **902** (2020) 012053 doi:10.1088/1757-899X/902/1/012053

Investigation of metalloporphyrin compound in maltenes of duri crude oil

Vinny[1], Y Yulizar[1], R Ekananda[2], I Kurniawaty[2], H Haerudin[2] and D O B Apriandanu[1]

[1]Department of Chemistry, Faculty of Mathematics and Natural Sciences (FMIPA), Universitas Indonesia, Depok 16424, Indonesia
[2]Research and Development PT Pertamina (Persero), Jakarta Timur 13920, Indonesia

Corresponding author's email: yokiy@sci.ui.ac.id

Abstract. Metalloporphyrin is one of the compounds that could potentially cause deactivation of catalysts in the process of catalytic cracking and hydrotreating. Knowledge of the structure of metalloporphyrin can be used as the basic information in overcoming the problem of catalyst poisoning. This work is performed to identify metalloporphyrins in the maltenes fraction of Duri crude oil. The low content of asphaltenes in Duri crude oil with a fractional weight of 0.79 %, allows a high level of metalloporphyrin in the fraction of maltenes which has a fractional weight of 82.73 %. Metalloporphyrin in the maltenes derivate fraction, saturates, aromatics, and resin were separated using column chromatography. Saturates, aromatics and resins have fractional weight of 49.75; 34.92; and 15.33 %, respectively. Only resin contained metalloporphyrin based on the analysis of UV-Vis spectrophotometer and Fourier-transform infrared spectroscopy (FT-IR), thus metalloporphyrins in resin was further separated into fractions by column chromatography. The result of UV-Vis spectrophotometer analysis for resin separation showed an absorbance peak of metalloporphyrin at 480–700 nm, while FT-IR analysis showed the presence of pyrrole group as porphyrin base framework with vibrational region at 400–1100 cm^{-1} The results of micro plasma-atomic emission spectrometer (MP-AES) analysis showed the presence of Fe, Ni, and V metals content. The liquid chromatography-mass spectrometer (LC-MS) analysis showed the molecular structure of the obtained compound was $C_{33}H_{34}N_4VO_2$. According to the results, metalloporphyrins presented in the maltenes fraction of Duri crude oil was vanadyl porphyrin.

Keywords: Metalloporphyrin, maltenes, Duri crude oil, vanadyl porphyrin

1. Introduction

In recent years, crude oil tends to be more difficult to process due to the increasing amount of acid and the fraction of heavy crude oil [1]. In addition, the presence of metal content in crude oil has the potential to cause product contamination and a catalyst poisoning in the crude oil refinery process. The metals will attach to the active surface- of catalyst and decrease its area by clogging the pores thus the catalyst becomes inactive [2]. The low content of metals in crude oil causes a deactivation in the desulfurization process and catalytic cracking [3].

Crude oil is grouped into maltenes and asphaltenes based on their solubility in n-heptane and n-hexane solvents [4]. Maltenes is a soluble crude oil fraction in alkane solvent consisting of saturates, aromatics, and resins. Asphaltenes are a fraction in crude oil with high molecular weight and insoluble in alkane solvents. Asphaltenes tend to be soluble in aromatic solvents and form the colloidal dispersed asphalt in the maltenes fraction.

In terms of polarity distribution, metals in crude oil mostly found in resins and asphaltenes[5]. Metals in crude oil can be either found in a free state or forming a metal complex such as metalloporphyrin. There are two types of porphyrin in crude oil, the one that has a large mass and large size while settling together with asphaltenes, and other that has a lower mass and prefer to bind with resin [3].

Porphyrin compounds, particularly vanadyl porphyrins in resins, affect the amount of asphaltenes in crude oil. Asphaltenes compounds are generally separated from crude oil through precipitation with hexane and n-heptane solvents. The higher concentration of vanadyl porphyrin in the resin, the quantity of asphaltenes deposited by the n-hexane mixture will decrease and vice versa [6]. The low value of asphaltenes fractional weight of Duri crude oil indicates a high content of metalloporphyrin in the resin fraction. However, there was a research showing that metalloporphyrin contained in the fraction of crude oil, especially in asphaltenes [7, 8].

This study examined more about the content of metalloporphyrin compounds in maltenes derivate fractions, saturates, aromatics, and resins. Metalloporphyrin in maltenes was analyzed by column chromatography method, and further characterized by UV-vis spectrophotometer, FT-IR, LC-MS, and MP-AES.

2. Experimental

2.1. Materials
The materials used for this study were Duri crude oil, toluene (Merck), n-heptane (Merck), methanol (Merck), chloroform (Merck), filter paper Whatman no.42 (GE), silica gel 60 (70-230 mesh) (Merck), aluminum foil. The characterization equipment for this research were FT-IR spectrocopy Prestige 21 Shimadzu, UV-Vis spectrophotometer Shimadzu 2600, LC-MS Agilent 1200, MP-AES Agilent 4210.

2.2. Sample preparation
The sample preparation of Duri crude oil followed the procedure of ASTM D6560. Sample preparation was performed to separate maltenes and asphaltenes based on their solubility in n-heptane [9].

2.3. Separation of saturate, aromatic, and resin
The column for separation with silica gel was activated at 250 °C for 24 h using a wet method. Maltenes were separated into its derived fractions through the adsorption method on silica. The sample was orderly eluted with 150 mL of heptane for saturates, 150 mL of toluene for aromatic and 300 mL of toluene: chloroform: methanol (1:1:1) (v/v) for resins. 25 mL of each separation was collected into fractions and characterized by UV-Vis spectrophotometer and FT-IR.

2.4. Resin separation by column chromatography
The resin was mixed with the silica gel and inserted into the column as reported in our prior work [10]. The first eluent was a mixture of toluene: chloroform: methanol with a volume ratio of 1:1:0.1. The second eluent was toluene: chloroform: methanol with a volume ratio of 1:1:1 to bring the sample retained in the stationary phase. The separation results were collected into fractions every 5 mL and characterized by UV-Vis spectrophotometer, FTIR, MP-AES, and LC-MS.

3. Results and discussion
According to the sample preparation, the fractional weight of maltenes and asphaltenes in crude oil were 82.73 and 0.79 %, respectively. The composition of saturates, aromatics, and resins of each type of crude oil were varied. Based on the calculation result, the fractional weight of saturates, aromatics, and resin from maltenes of Duri crude oil were 49.75; 34.92; and 15.33 %, respectively. The porphyrins

IOP Conf. Series: Materials Science and Engineering **902** (2020) 012053 doi:10.1088/1757-899X/902/1/012053

compounds have peaks at 390–425 nm for free porphyrins and 480–700 nm for porphyrin-bound metals [11]. The UV-Vis absorption spectra of the saturates and aromatics in figure 1a and figure 1b do not show a typical uptake of porphyrin. This result indicates that saturates and aromatics contain no porphyrins. Meanwhile, figure 1c shows the presence of porphyrins in both fractions 2 and 3 by evaluating the different pattern at 50–700 nm.

The vibrational mode of pyrroles porphyrin base framework is in the wavenumber region of 450–1100 cm^{-1} [12]. Figure 2 shows the presence of pyrrole symmetric folding at 785 cm^{-1}, and pyrrole asymmetric folding with bending vibration types at 669 cm^{-1}. The column chromatography separation of 2nd and 3rd resin fractions resulted in 17 subfractions. The eluent consisting of toluene: chloroform: methanol with a volume ratio of 1:1:0.1 was used to separate the subfraction of 1–11, while the eluent consisting of toluene: chloroform: methanol with a volume ratio of 1:1:1 was used to separate the subfractions of 12–17.

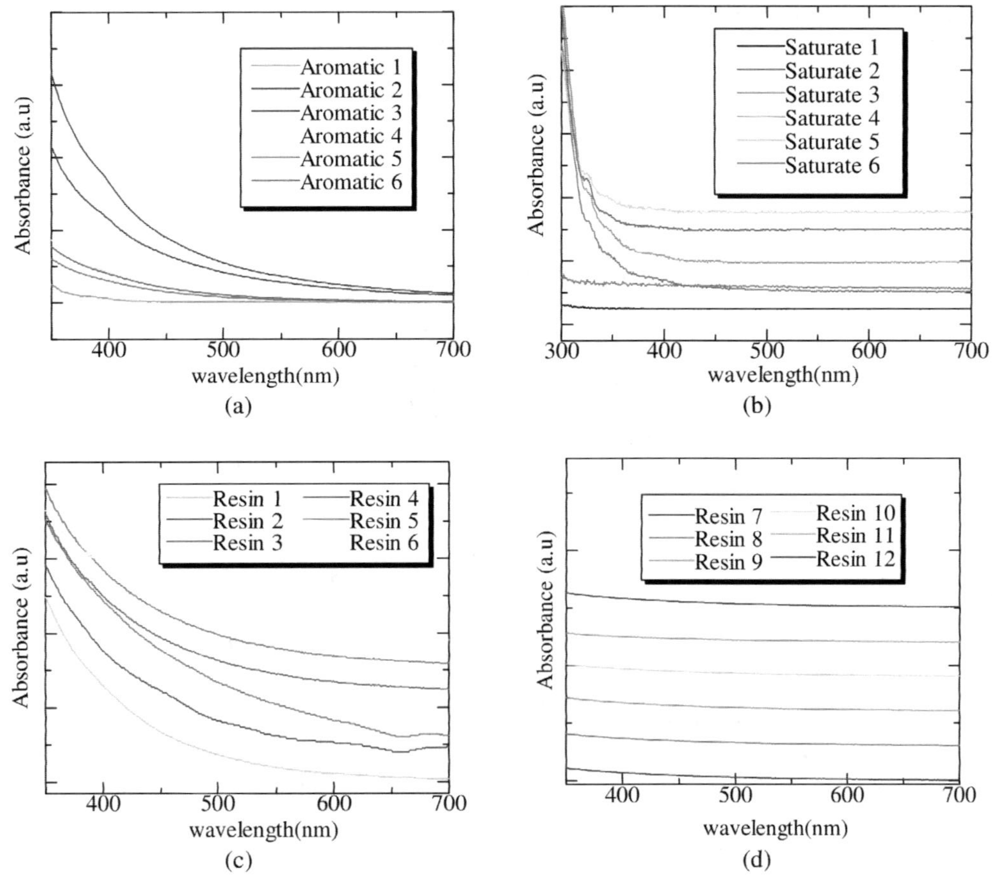

Figure 1. UV–Vis absorption spectra of (a) saturates, (b) aromatics (c) resin fractions of 1–6, and (d) resin fractions of 7–12.

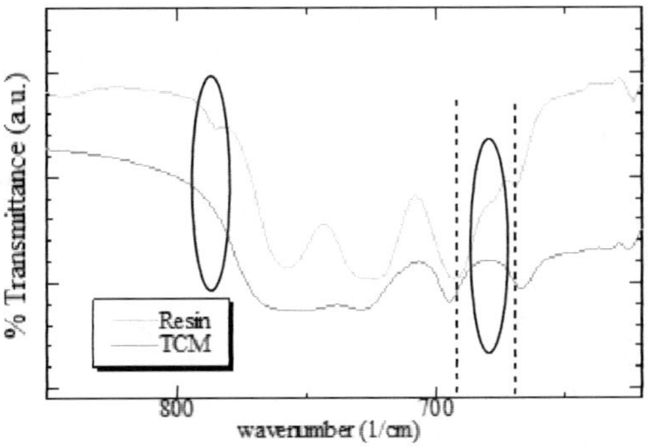

Figure 2. FTIR spectra of 2nd and 3rd resin fraction.

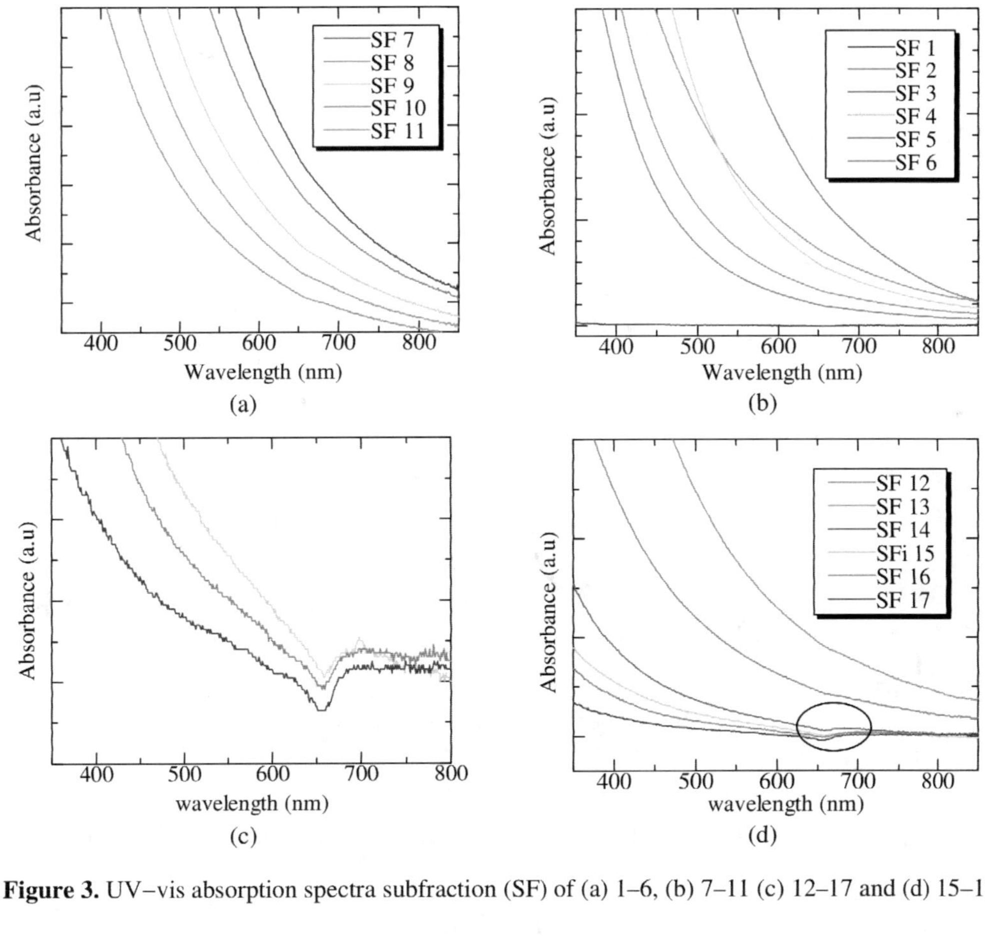

Figure 3. UV–vis absorption spectra subfraction (SF) of (a) 1–6, (b) 7–11 (c) 12–17 and (d) 15–17.

3.1. UV-vis spectrophotometer characterization of resin separation

The subfractions of resin separation obtained by column chromatography were characterized. Figure 3a and figure 3b show the absorption spectra for the subfractions of 1–11 do not show any distinctive peak of metalloporphyrin, while the analysis results for the subfractions of 12–17 show the existence of a typical absorption peak, particularly in the subfractions of 15, 16, and 17 as presented in figure 3c. These three subfractions have a different pattern from other subfraction spectra. This result indicates the presence of the same compounds in those three subfractions. Therefore, they were combined for further characterization. Absorbance peaks for fractions 15, 16, and 17 tend to go down because the blank evaporated during the measurement process as shown in figure 3d. Hence, the solvent of the sample was replaced with methanol to prove the formation of the maximum absorption peak. The result of the UV-vis spectrophotometer measurements exhibits a typical absorption peak at maximum wavelength of 515 nm and 673 nm as shown in figure 4a. This result corresponds to the wavelength range of metalloporphyrin which is 480–700 nm.

3.2. FTIR characterization of resin separation
The result of the FTIR analysis of subfractions 15, 16, and 17 indicates the presence of a pyrrole group at 677 cm^{-1} as shown in figure 4b. Thus, it can be seen that the porphyrin compound could be separated and purified with the resin separation process by column chromatography.

3.3. MPAES characterization of resin separation
Characterization with MP-AES gives information about the number of metals contained in the sample It shows that there were 3 types of metals, V, Ni, and Fe that bind to porphyrin. The subfraction of resin separation contain 3 types of metals as shown in table 1. This result demonstrates that the porphyrin contained in the resin subfractions 15–17 was in the form of metalloporphyrin.

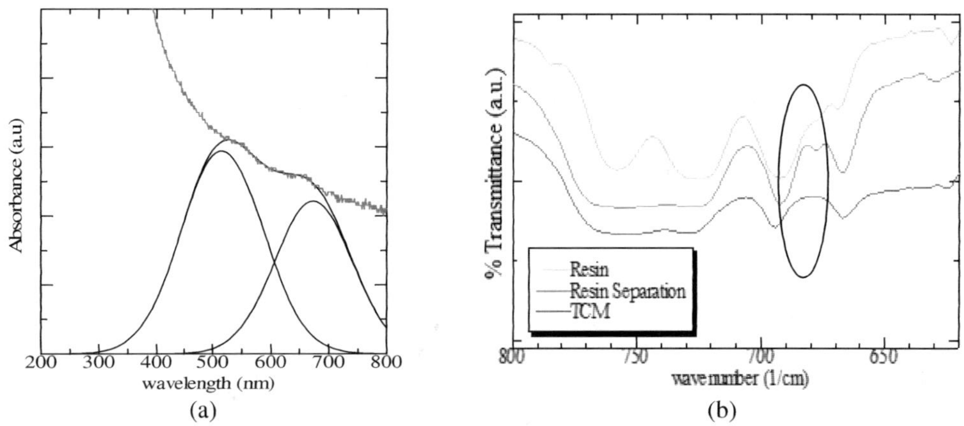

Figure 4. (a) UV-vis absorption spectrum of the mixture of resin separation fraction of 15–17 and (b) FTIR spectra of resin and resin separation.

Table 1. MP-AES analysis of resin and resin separation.

Sample	Concentration (ppm)		
	Fe	Ni	V
Resin	7.776	4.948	0.104
Resin separation	2.969	0.760	0.051

IOP Conf. Series: Materials Science and Engineering **902** (2020) 012053 doi:10.1088/1757-899X/902/1/012053

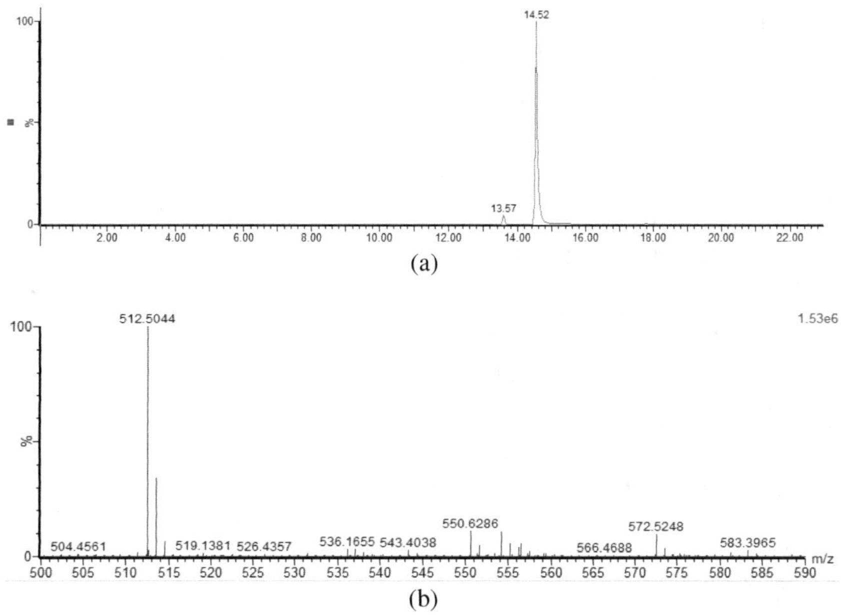

Figure 5. (a) Liquid chromatogram and (b) mass spectrogram of resin subfraction 15–17.

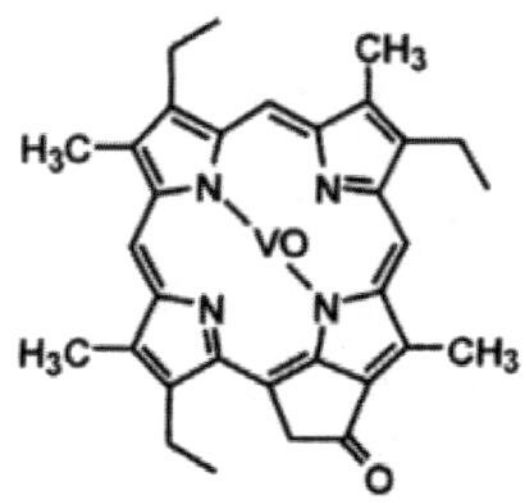

Figure 6. Structure of vanadyl porphyrin ($C_{33}H_{34}N_4VO_2$).

3.4. LC-MS characterization

Based on the LC analysis in figure 5a, there was a peak at a retention time of 14.52 min. The MS result at a retention time of 14.52 min in figure 5b has the m/z value of 572. The obtained m/z value is the positive ion of the target compound. Therefore, the molecular weight of the target compound was 571. This result was conformed to the previous result [13] that the vanadyl porphyrin compounds with the m/z value of 571 has a structure as shown in figure 6.

4. Conclusion

Metalloporphyrin have been identified in resins by column chromatography separation method. Resins are the only derived fraction of maltenes containing metalloporphyrin compounds. The results of the analysis with LC-MS showed the presence of vanadyl porphyrin in the form of $C_{33}H_{34}N_4VO_2$.

Acknowledgements
This work was funded by Pertamina Join Research Grant between Department of Chemistry Faculty of Mathematics and Natural Sciencse Universitas Indonesia and Research & Technology Center PT. Pertamina (Persero) 2018.

References

[1] Moeini F, Hemmati-Sarapardeh H, Ghazanfari M H, Masihi M and Ayatollahi S 2014 *Fluid Phase Equilib.* **375** 191-200
[2] Rodriguez M A, Elizalde I and Ancheyta J 2012 *Fuel* **100** 152-62
[3] Dechaine G P and Gray M R 2010 *Energ. Fuel.* **24** 2795-808
[4] Reisch M 1999 *Chem. Eng. News* **77** 70
[5] Zhao S, Kotlyar L S, Woods J R, Sparks B D and Chung K H 2000 *Petrol. Sci. Technol.* **18** 587-606
[6] Yakubov M R et al. 2016 *Energ. Fuel.* **30** 8997-9002
[7] Luz M S and Oliveira P V 2019 *Talanta* **199** 147-54
[8] Chainet F, Lienemann C P, Courtiade M, Ponthus J and Donard O F X 2011 *J. Anal. At. Spectrom.* **26** 30-51
[9] Affendi T, Yulizar Y, Haerudin H, Kurniawaty I and Apriandanu D O B 2020 *IOP Conf. Ser.: Mater. Sci. Eng.* **763** 012017
[10] Kurniawaty I, Yulizar Y, Gamar I, Haerudin H and Apriandanu D O B 2018 *AIP Conf. Proc.* **2023** 020087
[11] Kumolo S T, Yulizar Y, Haerudin H, Kurniawaty I and Apriandanu D O B 2019 *IOP Conf. Ser.: Mater. Sci. Eng.* **496** 012038
[12] Rio Y, Rodriguez-Morgade M S and Torres T 2008 *Org. Biomol. Chem.* **6** 1877-94
[13] Mishra S, Kaur S, Tripathi S K, Mahajan C G and Saini GS 2006 *J.Chem. Sci.* **118** 361-9.

IOP Conf. Series: Materials Science and Engineering **902** (2020) 012054 doi:10.1088/1757-899X/902/1/012054

In vivo study of 8-OHdG as a biomarker DNA damage by combining the exposure of nonyl phenol and copper using ELISA technique

Budiawan, A D Rizkita, S Handayani and I C Dani

Department of Chemistry, Faculty of Mathematics and Natural Sciences (FMIPA), Universitas Indonesia, Depok 16424, Indonesia

Corresponding author's email: dr.budiawan@sci.ui.id

Abstract. Nonylphenol (NP) is an environmental contaminants produced by alkylphenol ethoxylates which is non-ionic surfactant. NP has recently been found in foods derived from antioxidant hydrolysis of tris nonylphenyl phosphite which is used as a heat stabilizer in the manufacture of polymer food packaging materials, such as styrene, vinyl polymer, rubber and polyolefins. Copper (Cu) is a redox metal which in the normal dose is an essential mineral that needed by the body, but excess Cu will cause damage in some organs. Both NP and Cu can produce OH radicals in the body, which means that OH radicals will potentially attack DNA and thus form a biomarker of DNA damage such as 8-OHdG. This study of DNA adducts of 8-OHdG formation due to NP as a xenobiotic and Cu as metals. In this research, the eight-month-old Sprague Dawly rats weighing 200–300 g were randomly divided into three groups (Control, Cu, NP, and NP + Cu, n = 5). The tested rats were orally treated every 24 h for 28 days. Finally, the collected plasma was tested using ELISA technique to study the 8-OHdG content. Based on the results, it was shown that both NP and Cu metals could trigger the formation of 8-OHdG. The highest concentration of 8-OHdG formed was the exposure group of NP + Cu in day 14 with a concentration of 1.3412 ng/mL.

Keywords: Nonylphenol, oxidative stress, 8-OHdG, copper, ELISA

1. Introduction

In the past century, the fast growing industries and human population have led the researchers to develop various types of chemical synthesis to be applied in the foods and beverages industries, which causes pollution to be likely emerged and harming the aquatic systems, leading to the environmental damage that have impact on the aquatic biota and humans health [1]. One of the most hazardous pollutants is endocrine disrupting chemical (EDC). This substance is dangerous because it has a structural similarity to endocrine hormones. Chemicals that are included in the endocrine disruptors category are varied, namely polycyclic aromatic hydrocarbon group, pesticides, metals, solvents, plasticizers, and detergent. These substances can promote an oxidative stress which characterized by the formation of excess radicals caused by the inhibition of antioxidant enzyme activity to cell damage due to changes in metabolism of endocrine disrupting substances. In addition, high concentrations of EDC and continuous exposure will increase the risk of cancer [2].

NP is considered as a strong xenoestrogen in the aquatic environments because of its low solubility (4.5 mg/L at 25 °C) so that NP is not readily degraded in the environment and often used in the manufacture of Nonyl Phenol Ethoxylates (NPEs) [3]. NPE is a non-ionic surfactant found in plastics, solvent, detergent, paint and plasticizer. NP exposure to organisms can disrupt the endocrine system, which results in the hormonal and reproductive organ development to become imbalance [4]. Chronic exposure of NP can increase the state of oxidative stress by increasing the reactive radicals such as hydroxyl radicals •OH. This reactive species can interact with biomolecules, namely DNA which causes DNA damage which, if not repaired, can lead to carcinogenesis [5].

Oxidative stress describes a higher level of radicals than antioxidants with the ability of biological systems to detoxify the reactive intermediates or repair the damage caused. Radicals are produced naturally in the body during oxygen metabolism and has an essential role in cell signaling and homeostasis. Metal ions produce intracellular radicals directly or indirectly, where the Fenton reaction is one of the best-known mechanisms. During this reaction, the transition metal ion reacts with H_2O_2 to produce toxic •OH and oxidize the metal ion [6].

Several metals can produce free radicals, for example, Cu metal. This metal is a cofactor of many enzymes such as cytochrome c oxidase, ascorbate oxidase or SOD, which are involved in the cellular redox mechanisms [7]. Cu can cause an oxidative stress through two mechanisms: first, Cu can accelerate the formation of ROS through the Fenton or Haber-Weiss reaction [8]; second, significant increase of Cu exposure will reduce GSH levels [9].

Several guanine oxidation products at C-8 position have been used as markers of DNA damage. 8-OHdG is a biomarker of DNA damage caused by oxidative stress both in the urine and plasma due to the ease in which the guanine bases interact with hydroxyl radicals

This study was conducted to analyze the mechanism of the formation of 8-OHdG due to NP and Cu exposure to Rattus norvegicus with the autoxidation process approach through the reaction of Fenton with a simple ELISA technique.

2. Materials and method

The research was carried out using a scientific approach to analyze the phenomenon of the increase of cancer risk in the presence of biomarkers of DNA damage in-vivo using the ELISA kit method on the test animals plasma exposed to nonylphenol as xenobiotics and Cu metals as catalysts for the formation of ROS. This study was approved by the ethics committee at our institute.

2.1. Test animal preparation

The test animals were kept in cages and consisted of 5 mice in each cage representing each treatment group. Each treatment was given an identification tag so that it was easy to distinguish between treatment groups. Test animals were acclimatized for about two weeks to be able to adapt to the new cage environment. During the acclimation process, the test animals were given pellet feed and boiled water for drink. Animal cages were cleaned three times a week with disinfectant and replaced the base of the cage. The lighting of the room where was conditioned with 12 h light and 12 h dark at room temperature around 27–28 °C.

2.2. 8-OHdG formation in the control group, Cu, NP and NP + Cu

The sunflower oil, Cu solution 4 mg/kg bw, NP solution 25 mg/kg bw, and a mixture of Cu and NP solution were prepared for the oral feeding process. Rats were grouped into 4 groups (control, Cu, NP, NP + Cu) then given feed in the form of pellets, drinking ad libitum and every day each group was treated differently from one another orally as much as 10 mL/Kg bw for 28 days.

2.3. Blood sample collection

Before drawing blood from rats, the test animals were anesthetized using 0.1 mL of ketamine and xylazine for 2-4 min. The rats were weighed and then blood was drawn through the sinus orbital eye using a hematocrit pipette of 1−1.5 mL and put in a sterile tube containing EDTA. The blood was centrifuged at a speed of 3500 rpm for 20 min at a temperature of 20 °C. Plasma was separated and stored in a freezer at -20 °C until the measurement process.

2.4. Analysis of 8-OHdG in plasma

Analysis of the 8-OHdG formation in plasma samples from rat test animals was performed using the ELISA kit method from CUSABIO. First, all reagents, standards, and samples were prepared. Then, 100 μL standard and sample were added to the well plate and covered using para-film. Then, incubated for 2 h at 37 °C. Next, the liquid in each well plate was remove and added with 100 μL biotin-antibody to each well plate, then covered with para-film and incubated for 1 h at 37 °C. After incubation was complete, the liquid in the well plate was removed and washed using a wash buffer and let stood for 2 min. Washing was done for 3 times. Then, 100 μL HRP-avidin was added to each well plate and covered with para-film. Then, incubated for 1 h at 37 °C. After completion, the well plate was washed again using wash buffer 5 times. Then, 90 μL TMB substrate was added to each well plate and incubated again for 15 min at 37 °C. After incubation was complete, a stop solution was added to each well plate. Then, the optical density was determined on each well plate for 5 min using spectrophotometer at a wavelength of 450 nm.

3. Results and discussion

The in vivo study began with an acclimation process that lasted for 3 weeks. The Health Research Ethics Committee of the Faculty of Medicine, Universitas Indonesia has approved the in vivo study method to be carried out, with an ethical approval number: 1112 / UN2.F1 / ETIK / 2018.

3.1. Observation on the test animals

The exposure on rats used in this study is sub-chronic exposure, which is the exposure in low concentrations for a short period of time. Exposure of NP is 25 mg/kg bw above its NOAEL value, while Cu metal exposure is 4 mg/kg bw. Changes of rats' body weight, before and after exposure can be seen in figure 1. Based on the observations, the body weight of rats given xenobiotic exposure tends to decrease. The weight of the control group rats increased from 214.2 g to 251.8 g at the end of the exposure. However, in the Cu exposure group, the body weight of rats decreased from 229.5 g to 190.95 g. This shows that sub-chronic exposure of Cu affects the body weight of rats. The body weight of rats in the NP exposure group also decreased from 260 g to 238 g. The same thing happened in the combination of NP and Cu where the weight loss from 266 g to 236.2 g was experienced, so it was seen that the sub-chronic exposure of both NP and Cu can cause the weight lost. Previous studies of organ damage test using NP for 8 weeks showed micronuclei induction in erythrocyte cells. In that research, erythrocytes in the blood of females experienced a resistance at a dose of 25 mg/kg bw [10]. The accumulated NP will cause a decrease in sperm quality due to the nature of NP which resembles the hormone estrogen and will result in the decrease total number of sperm cells produced thereby affecting the weight of the testicles of test animals [11].

3.2. Analysis of plasma in vivo studies

The formation of 8-OHdG in the in vivo study was analyzed using a sandwich type ELISA kit derived from CUSABIO. Before analyzing using an ELISA kit, blood taken from rats was centrifuged at 20 °C at a speed of 3500 rpm for 15 min until blood plasma was obtained. Then, the plasma was stored in a freezer at -20 °C until the analysis of the 8-OHdG compound using an ELISA kit was carried out. A sandwich type ELISA was used to identify the specific antigens, 8-OHdG in the sample. A microplot surface has been prepared with bound 8-OHdG antibodies that are known to be able to capture the desired antigen. Primary antibodies were added 8-OHdG to the samples such as 'sandwich', then biotin antibodies were added to bind with the HRP-avidin enzyme (figure 2). Unbonded enzyme antibody conjugates were cleaned using a buffer. TMB substrates were added and converted enzymatically to produce quantifiable colors. One of the advantages of using specific antibodies is that they can eliminate steps to purify antigens from other antigen mixtures, making it easier to test and increase their specificity and sensitivity [12].

The 8-OHdG compound, which is a product of DNA cutting by DNA repair, will be distributed through the blood and excreted in the urine without further metabolic processes. Analysis of 8-OHdG compounds in blood plasma as a biomarker of DNA damage in the body is carried out because of the ease of sampling and handling.

3.3. 8-OHdG formation in the control group

The control group in this study was a group of rats that were only snapped using sun flower oil without the addition of any compound. Although not exposed to xenobiotics, rat can experience oxidative stress due to other factors such as the environment and the state of rat during the study. The control group also produced 8-OHdG and can be seen in figure 3.

Naturally, cells can experience oxidative stress because they can produce ROS. Each cell in the human body can process 1021 oxygen molecules each day with 1% metabolism producing ROS. Based on the excretion of 8-OHdG in urine, it can be estimated that about 20,000 nucleosides in DNA are damaged in each human cell per day [13]. Therefore, even though mice are not rounded with xenobiotics, it will still produce 8-OHdG, but tend to decrease due to metabolic processes in the body working to repair 8-OHdG products that are formed.

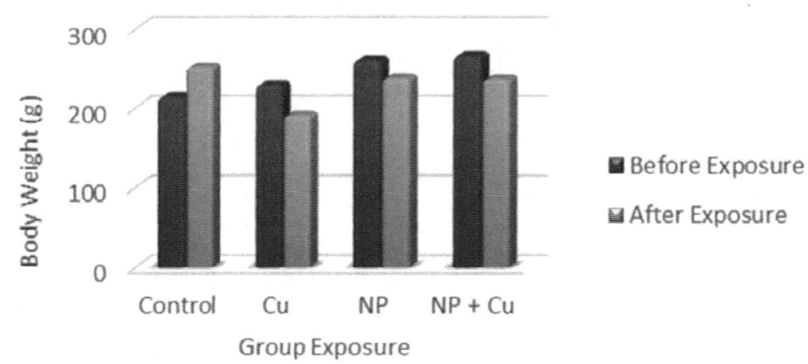

Figure 1. Body weight of rat before and after exposure.

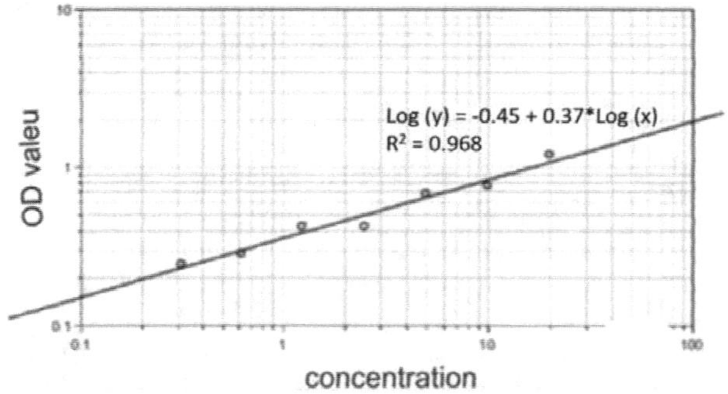

Figure 2. 8-OHdG standard calibration curve

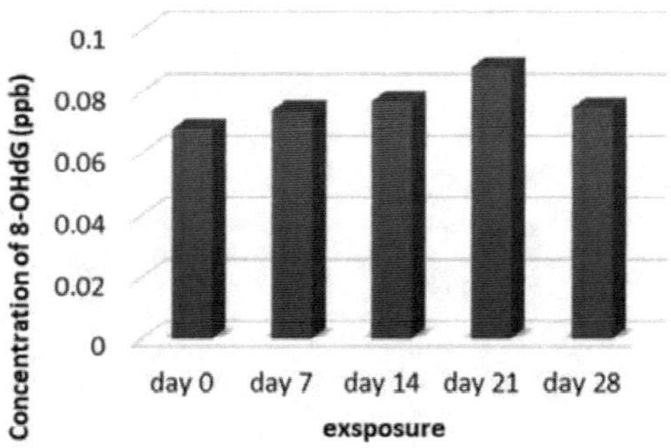

Figure 3. The control group produces 8-OHdG.

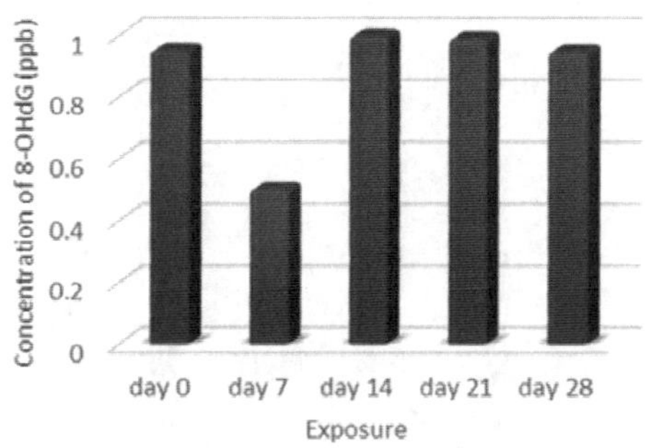

Figure 4. Effect of exposure to NP on the 8-OHdG formation.

3.4. 8-OHdG formation in the NP group
Exposure to NP in rats will produce an 8-OHdG DNA adducts as shown in figure 4. Based on the graph, NP exposure tends to decrease after the first week of exposure. This is partly due to the blood lysis that occurs during the centrifugation process, because in that week the centrifugation process is carried out at room temperature. However, it shows an increase of 8-OHdG products in week 2. In weeks 3 to 5, the 8-OHdG products produced tend to be consistent, it shows that the presence of xenobiotics that enter the body will continuously produce an excess ROS so that enzymes that act as antioxidant are no longer able to neutralize the radical products so that 8-OHdG will continue to form.

Exposure to NP has been shown to cause oxidative damage to DNA [14]. NP induces mitochondrial damage, which will lead to endocrine damage [15]. NP itself can undergo an autoxidation process by producing radicals that can trigger oxidative damage to DNA. However, as is well known, NP is rapidly hydrolyzed by the esterase enzyme.

3.5. Effects of Cu exposure on the formation of 8-OHdG
Exposure of Cu in rats will produce an 8-OHdG DNA adduct, as shown in figure 5. Based on the figure, exposure of Cu decreases after the third week during the experiment. This is partly because of the blood lysis that occurs during the centrifugation process, which can interfere with the resulting 8-OHdG concentration.

Like other studies, DNA adduct in the puppy's liver are formed due to the increase of oxygen exposure accompanied by both Fe and Cu content found in food consumed by the mother [16].

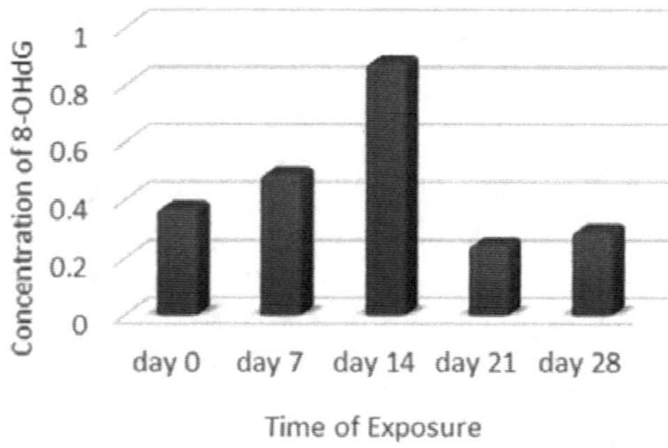

Figure 5. Effect of exposure to Cu on the 8-OHdG formation.

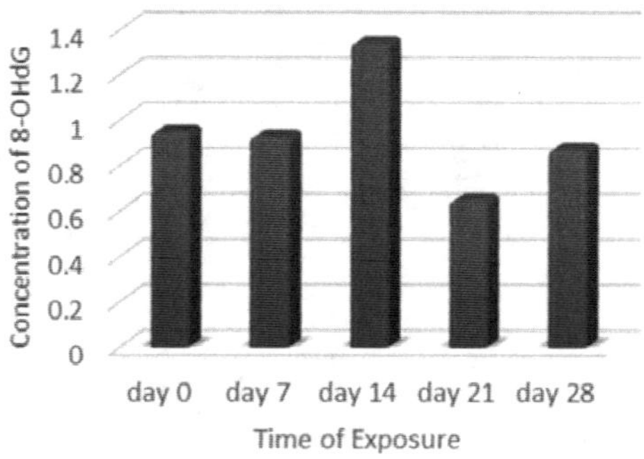

Figure 6. Effect of exposure combination of NP and Cu on the 8-OHdG formation.

3.6. The exposure effect of the combination of NP and Cu metals on 8-OHdG formation

The effect of NP and Cu metals on the formation of 8-OHdG is shown in figure 6. The exposure combination of NP and Cu results in the formation of 8-OHdG. As explained in the previous discussion, both NP and Cu can cause oxidative damage to DNA. The concentration of 8-OHdG in NP with Cu metals also decreased in the third week during the experiment because of blood lysis during the centrifugation process. This exposure group experienced the most significant increase in 8-OHdG and showed the most significant weight loss. That is because the synergy correlation between xenobiotics and metals on the production of 8-OHdG.

3.7. Comparison of the Formation of 8-OHdG in each group

Comparison of the formation of 8-OHdG in various exposure groups at 28 weeks can be seen in figure 7. Based on the pictures, the control group had the lowest 8-OHdG concentration compared to other exposure groups. This shows that both NP and Cu metal play a role in the formation of 8-OHdG.

Based on figure 7, exposure from a combination of NP and Cu causes the formation of 8-OHdG with the highest concentration compared to other groups. This is because the correlation between xenobiotics and metal is highly involved in redox reactions in the body and the excess of this metal will facilitate the formation of radicals in cells such as hydroxy radicals.

NP will inhibit the activity of SOD enzymes so that radicals formed from oxygen cannot be neutralized again by antioxidant enzymes. These radicals will interact with DNA bases and form 8-OHdG, as well as Cu, excess Cu will allow hydrogen peroxide to interact with Cu (I), thus forming

IOP Conf. Series: Materials Science and Engineering **902** (2020) 012054 doi:10.1088/1757-899X/902/1/012054

a hydroxyl radical. that is why a combination of NP and Cu exposure will increase the yield of 8-OHdG formation than other exposure groups.

Figure 8 shows the formation of radicals in the body. Oxygen will be reduced to superoxide radicals in the presence of the enzyme cytochrome P450. Superoxide radicals will be neutralized by the SOD enzyme into H_2O_2 products. H_2O_2 neutralization process with the enzyme catalase will produce H_2O and O_2 can take place. However, H_2O_2 also has the potential to react with Cu to produce OH radicals. Figure 9 shows the OH radicals that are formed to be likely interact with DNA. Guanine is one of the bases mostly attacked by OH radicals because it has the highest oxidation level compared to thymine, cytosine, and adenine. Guanine position C number 8 is the preferred position for the OH radicals because of the open planar position steric effect compared to C positions 4 and 5. So, 8-OHdG is the most effective biomarker for identifying the DNA damage.

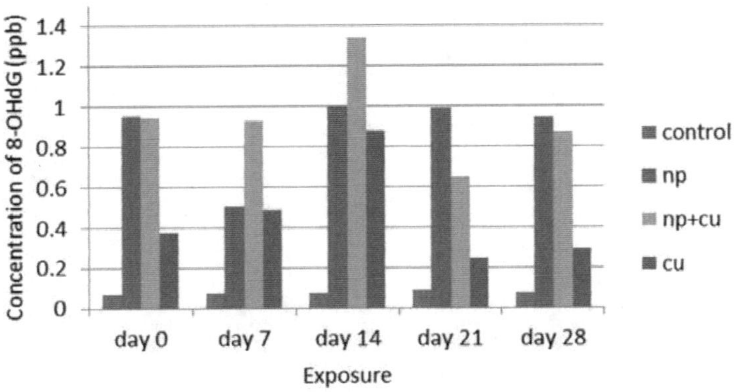

Figure 7. Comparison of the formation of 8-OHdG in various exposure groups.

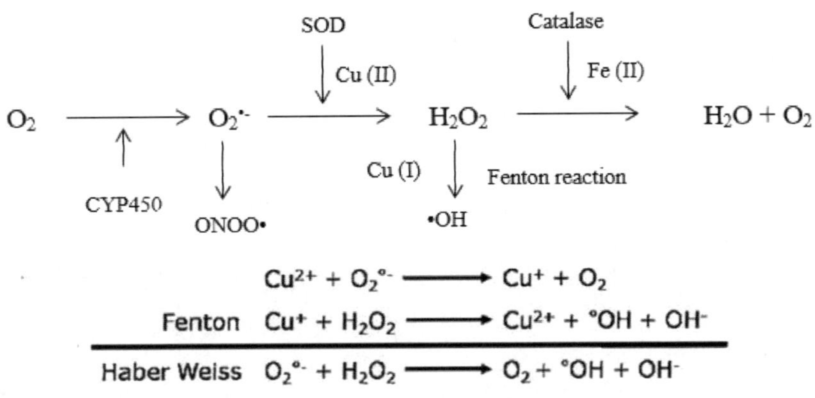

Figure 8. Radical formation scheme in the body's metabolism.

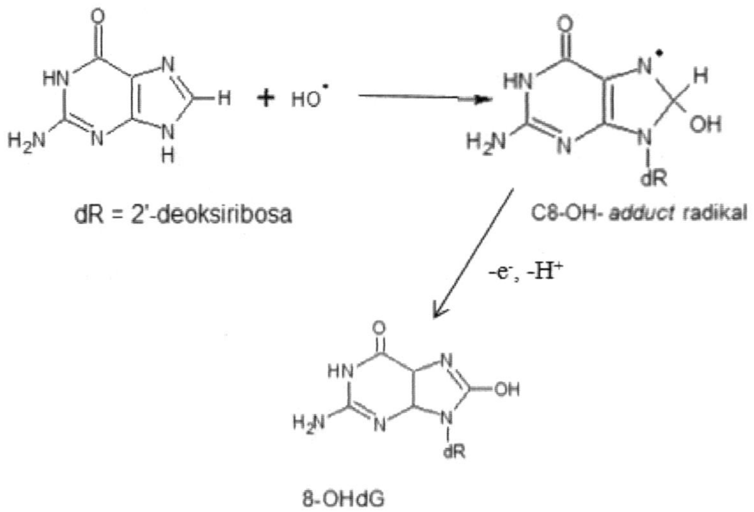

Figure 9. 8-OHdG formation scheme from radical hydroxyl exposure.

4. Conclusion

There was an increase in the levels of DNA adducts (8-OHdG) in plasma during the time of exposure in the NP exposure group and the combination of NP and Cu metal exposure group. The synergistic effect between the Fenton-like reaction with NP in the formation of DNA 8-OHdG is shown by the increase level of 8-OHdG in the plasma of the NP and metal combination groups compared to the both NP and Cu alone exposure groups.

Acknowledgments

We are thanking the Hibah PDUPT with contract number: 1/E1/KP.PTNBH/2019 and 234/PKS/R/UI/2019 for funding this research.

References

[1] Diamanti-kandarakis E et al. 2009 *Endocr. Rev.* **30** 293-342
[2] Nimrod A C and Benson W H 1996 *Crit. Rev. Toxicol.* **26** 335-64
[3] Soares A, Guieysse B, Jefferson B, Cartmell E and Lester J N 2008 *Environ. Int.* **34** 1033-49
[4] Barrera-díaz C E, Frontana-uribe B A, Rodríguez-peña M, Gomez-palma J C and Bilyeu B 2018 *Catal. Today* **305** 108-16
[5] Noorimotlagh Z, Mirzaee S A, Ahmadi M, Jaafarzadeh N and Rahim F 2018 *Ecotoxicol. Environ. Saf.* **158** 171-81
[6] Lee J-C, Son Y-O, Pratheeshkumar P and Shi X 2012 *Free Radic. Biol. Med.* **53** 742-57
[7] Kuo M T, Chen H H W, Song I-M, Savaraj N and Ishikawa T 2007 *Cancer Metastasis Rev.* **26** 71-83
[8] Leonard S S, Harris G K and Shi X 2004 *Free rad. Biol. Med.* **37** 1921-42
[9] Speisky H et al. 2008 *Bioorg. Med. Chem.* **16** 6568-74
[10] Dobrzynska M M 2014 *Mut. Res. Genet. Toxicol. Environ. Mutagen.* **772** 14-9
[11] Aly H A A, Domènech O and Banjar Z M 2012 *Toxicol. Appl. Pharmacol.* **261** 134-41
[12] Gan S D, Patel K R and Elisa S 2013 *J. Invest. Dermatol.* **133** e12

[13] Beckman K B and Ames B N 1997 *J. Biol. Chem.* **272** 19633-6
[14] Yeltekin A C and Oğuz A R 2017 *Drug Chem. Toxicol.* **41** 415-23
[15] Jambor T et al. 2017 *Environ. Pollut.* **222** 219-25
[16] Zhou G, Randerath E and Randerath K 2001 *Mut. Res-Fund. Mol. Mech. Mutagen.* **479** 71-9

IOP Conf. Series: Materials Science and Engineering **902** (2020) 012055 doi:10.1088/1757-899X/902/1/012055

Synthesis and characterization of metal organic framework using succinic acid ligand with cobalt and iron metals as methylene blue dye adsorbent

D A Nurani, B C B Butar and Y K Krisnandi

Department of Chemistry, Faculty of Mathematics and Natural Sciences (FMIPA), Universitas Indonesia, Depok 16424, Indonesia

Corresponding author's email: yuni.krisnandi@sci.ui.ac.id

Abstract. Metal-organic Framework (MOF) as a porous material has a potency for an organic dye adsorbent due to its tunable pore size. It can be a good candidate for selective separation of organic dye waste because of its organic functionality, high thermal and stability. In this study, MOF based on cobalt (Co) and iron (Fe) cations with succinic acid ligand was utilized for Methylene Blue (MB) removal from aqueous solution. Co-MOF and Fe-MOF were synthesized using solvothermal method at 120 °C for 20 h and characterized by using infrared spectroscopy, X-Ray diffraction technique, thermogravimetric analyses, N_2 adsorption-desorption isotherm, and scanning electron microscope. The adsorption of MB from aqueous solution by Co-MOF and Fe-MOF were carried out and their adsorption properties were investigated in various parameters such as pH solution, contact time, MOF loading. Moreover, their adsorption isotherm and reusable capacity were explored. As the results, the optimal adsorption was achieved at pH 9 for both MOFs, 5 mg of MOFs loading in 5 mL MB and 180 min for Co-MOF and 150 min for Fe-MOF. Both MOFs could be reused as a good adsorbent with the ratio of ethanol and H_2O (70:30) as an eluent. This adsorption activity followed the Langmuir isotherm with the maximum adsorption capacity (Qm) of 3.9494 mg/g and 4.6685 mg/g for Co-MOF and Fe-MOF respectively.

Keywords: Metal-organic framework, adsorption, methylene blue

1. Introduction

Along with the increasing necessity in life, the world's industrial has also increased. One of the very rapid industrial developments is textile industry that contains organic dyes, which have complex and stable structure and difficult to decompose under natural condition [1]. The most commonly used cationic dyes in industry is methylene blue (MB) which is easily dissolved in water, relatively easy to produce, have bright colors and are easier to use in fabrics. MB can be toxic to the environment and health. The existence of concentrated MB in the waters can prevent the penetration of sunlight into the water so that it can automatically disrupt the life of the aquatic ecosystem. Meanwhile, the impact on health, MB content can cause burning eyes which can be a permanent injury to humans or animals, difficult breathing, digestive system disorders, and can also cause nausea, vomiting, excessive sweating, and methemoglobinemia or abnormalities in the blood [2].

Many methods have been developed to eliminate dyes pollutant in industrial wastewater such as membrane filtration, photocatalysis, coagulation and adsorption [3-5]. Adsorption is considered method due to simplicity, efficiency, and low consumption to reduce dyes pollutant. The common adsorbents used in the treatment of organic dyes carbon active, zeolites, graphene, and polymeric resins [6-7] whereas some of them are lacking of selectivity and have small diameter porous.

Metal-Organic Framework (MOF) is a porous coordination polymer composed of organic compounds (can be bidentate or polydentate ligands) bound to inorganic compounds (can be metal-oxide clusters or metal cations) as secondary building units (SBUs) form an unlimited network. MOF can be used as an adsorbent because it has the properties of organic substances which are expected to be able to interact with organic dyes and have a large surface area and pore volume so that they can be used as good adsorbents [8-9]. The removing of cationic dyes (MB) using Ti-MIL has been reported by Oveisi et al which has high adsorption capacity and stability over three cycles [10]. The 1,4-benzenedicarboxylate (BDC) and 2-amino-1,4-benzenedicarboxylate (NH_2-BDC) were used as linker. The transition metal used in synthesis of MIL-101 (Fe) has high adsorption capacity 472.7 $mg.g^{-1}$ and has a good reusability [11]. In previous research, Yt-MOF has been synthesized using succinic acid as linker. The synthesized material exhibits high capacity over Rhodamine B adsorption. In this study, the transition metal (Fe and Co) and succinic acid linker based MOFs have been investigated to MB removal in aquatic system. The combination of low cost and non-toxic material (transition metal and succinic acid) could improve the surface area and increase the adsorption ability for cationic dyes. The reusability of the adsorbent is also studied for three cycle experiments.

2. Materials and method

2.1. Materials

The materials used in this study were $Fe(NO_3)_3.9H_2O$, $Co(NO_3)_2.6H_2O$ 99 %, succinic acid 99.5 %, N, N-dimethylformamide 99.5 %, NaOH 99 %, acetic acid 100 % obtained from Merck, and distilled water.

2.2. Method

2.2.1. MOF synthesis. The synthesis of Co-MOF and Fe-MOF was carried out using the solvothermal method (Wu et al. with modification) [12]. Approximately, 1.5 mmol of succinic acid ligand (0.1771 g) was dissolved into a mixture of 5 mL DMF and 0.25 mL H_2O solvents. The mixture was stirred for 10 min. Then, 1 mmol of each $Fe(NO_3)_3$ and $Co(NO_3)_2$ were added into the mixture and stirred for 30 min. The mixture was heated at 120 °C for 20 h. The resulting of brown crystal for Fe-MOF and purple crystal for Co-MOF were isolated and washed with the mixture of DMF and H_2O with ratio of 5:0.25 for several times. The obtained of products were dried in the oven at 60 °C overnight.

2.2.2. MOF characterization. The Co-MOF dan Fe-MOF were characterized by XRD, TGA, N_2 adsorption-desorption isotherm, FTIR, and SEM. X-Ray Diffraction (XRD) was used to determine the crystallinity of MOFs, which used CuKα ray radiation, λ = 1.5406 Å and continuous scan type with a Shimadzu 6000. Thermogravimetric analysis (TGA) was used to know the thermostability of MOFs and performed with a heating rate 10 °C/min and temperature ranging from 30–600 °C in a flowing atmosphere of N_2. Surface area and pore size of MOFs samples were analyzed using the BET and BJH methods. The determinations of surface area and pore size were based on the isotherm of adsorption and desorption of nitrogen at 300 °C with SAA NOVA 2200e instrumentation. The functional groups of MOFs samples were analyzed by a FTIR Prestige 21 within the wave range 400–4000 cm^{-1}. The morphological and surface characteristic of MOFs was obtained from field emission scanning electron microscope (FESEM) using FESEM JEOL JESM-6700F.333. For application, UV-Vis spectrophotometer was used to measure the concentration of MB before and after adsorption process.

2.2.3. Methylene blue adsorption. The adsorption process was carried out by dispersing 10 mg MOF into 5 mL of 5 ppm methylene blue solution, then put in a closed beaker glass and stirred using a magnetic stirrer with a pH variation of 3, 5, 7, 9 and 12 with a speed of 120 rpm for 2 h at room temperature. After that, the variation of contact time was conducted for 120, 150, 180, 240, 300 and 360 min at the optimum pH. The optimum adsorbent mass was also determined with variations of 2, 5, 10, 15, 20, 25 mg under optimum pH and contact time conditions. Langmuir and Freundlich isotherm models were explored to study the type of adsorption mechanism and to obtain the theoretical maximum adsorption capacity. The reusability of synthesized adsorbent was determined for three cycle experiments.

3. Results and discussion

3.1. Characterization of MOFs

3.1.1. FTIR characterization of MOFs. The result of the FTIR of all MOF samples can be seen in figure 1. The spectra show the absorption peak at 1700 cm^{-1} of succinic acid is shifting to the lower wavenumber and getting the smaller intensity of both Fe-MOF and Co-MOF. Peak around 1700 cm^{-1} indicates the presence of C=O. This shifting might be an indication that C=O in succinic acid ligand form coordination with metal so the bonding is getting weak. The spectra of both MOFs show a shifting of absorption peak to a higher wavenumber around 3000–3500 cm^{-1} that indicates a stronger bond of a presence O-H group from H_2O as a solvent. All MOF spectra also exhibit absorption peak at around 1490 cm^{-1} and 1220 cm^{-1}, as in the spectrum of succinic acid ligand, which shows asymmetrical and symmetrical vibration of COO^- group [13]. Therefore, all of the spectra show that the cobalt and iron metals have successfully formed a coordination with ligand.

3.1.2. XRD characterization of MOFs. The XRD patterns of Co-MOF and Fe-MOF are shown in figure 2. The diffraction peaks in 2θ area of Co-MOF matched with the previous report that has synthesized Cd-Succinate MOF [14], indicating that Co-MOF sample was well crystallized and the crystallinity of MOF is similar to the Cd-MOF reference. The diffraction peak of Fe-MOF is also nearly matched with the previous report that has synthesized Y-Succinate MOF, indicating that

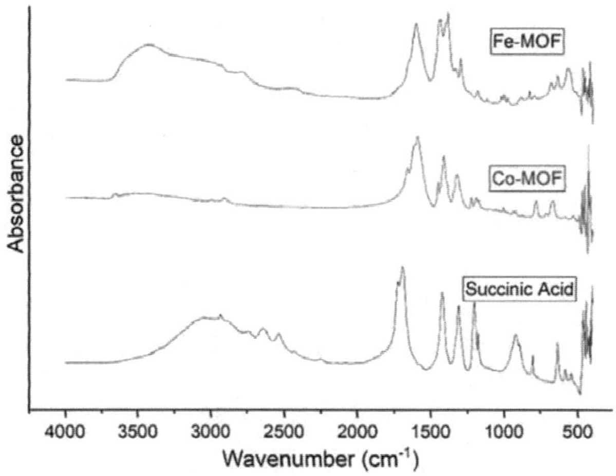

Figure 1. FT-IR spectra of Fe-succinate MOF, Co-succinate MOF and succinic acid ligand.

Fe-MOF was well crystallized and has a similar crystal structure compared with Y-MOF reference [15]. Furthermore, the synthesized MOFs are insoluble in some solvents i.e. H_2O, DMF, and ethanol, indicating the high crystallinity.

3.1.3. TGA characterization of MOFs. Figure 3 shows that the main weight loss of Co-MOF is produced in the temperature range of 322–428 °C of 47.97 % weight.
Fe-MOF also shows the main weight loss of 9.06 % in the range temperature of 249–380 °C. These weight losses indicate the organic linker decomposition, which is succinic acid. It was also known that the pyrolysis temperature of the succinate group is 320 °C [16]. The result shows the synthesized Co-MOF is having a higher thermal stability compared with Fe-MOF sample. The TGA curves also show that Co-MOF is more hydrophilic than Fe-MOF because it has more water content with comparison 2.6 to 2.2 %.

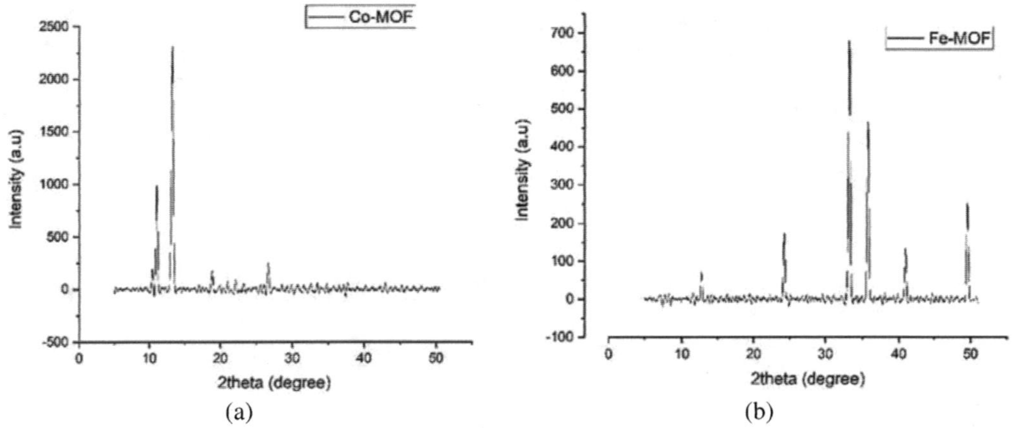

Figure 2. XRD pattern of (a) Co-MOF and (b) Fe-MOF.

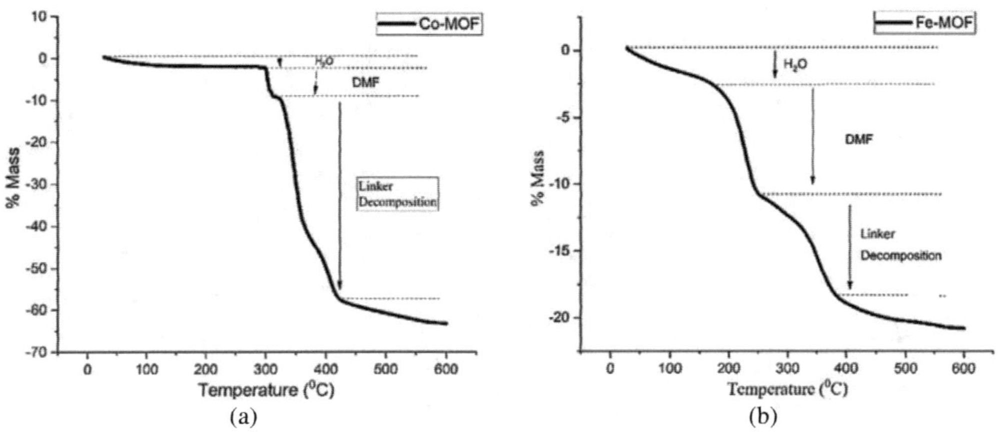

Figure 3. TGA curves of (a) Co-succinate MOF and (b) Fe-succinate MOF.

3.1.4. Surface area and pore analysis of MOFs. The surface area and pore of both MOFs were characterized by nitrogen adsorption-desorption experiments at 300 °C (figure 4). From figure 4, the synthesized Co-MOF shows type I adsorption- desorption isotherm indicating only microporous in the network. Whereas, Fe-MOF is similar to type IV isotherm of adsorption-desorption, which is the typical microporous and mesoporous network [17]. Mean pore size of Co-MOF is 1.5806 nm and Fe-MOF is 1.5816. The data of the surface area, pore-volume, and radius pore are listed in table 1.

3.1.5. FESEM characterization of MOFs. Figure 5 shows the SEM images of both Co-MOF (a) and Fe-MOF (b). Co-MOF shows spherical crystal structure as reported by Zhang et al. [18] with range crystal size 15–20 μm. Fe-MOF shows a clear cuboid shape crystal that is in good agreement with the report by Zhu et al. [11]. Fe-MOF has a crystal size in the range of 2–5 μm.

3.2. Methylene blue adsorption

3.2.1. Determination of optimum pH. pH is one of the important parameters which controls the MB adsorption [1]. Figure 6 shows that adsorption capacity increase at pH range 3 to 9 and decrease at pH 12 of both MOFs. As the increasing of pH, the negative charges in adsorbent also increase. The carboxylic group in ligand turns into anion carboxylate and can be reacted with methylene blue dyes which have a positive charge. However, a significant increase of pH exhibited the competition of adsorbent and hydroxyl ions [OH^-] [1]. Due to its optimum adsorption capacity, pH of 9 was selected for further experiments.

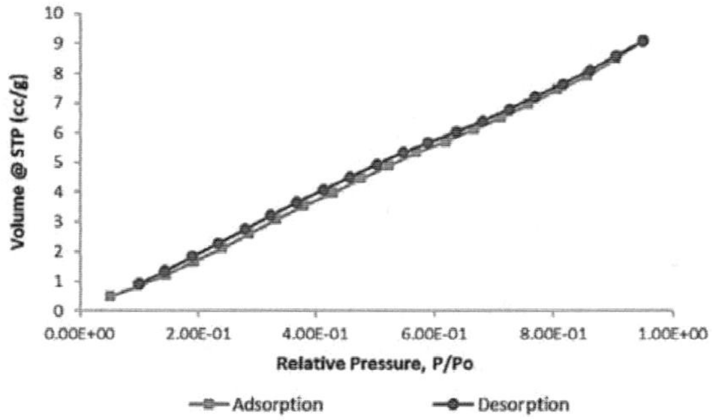

(a)

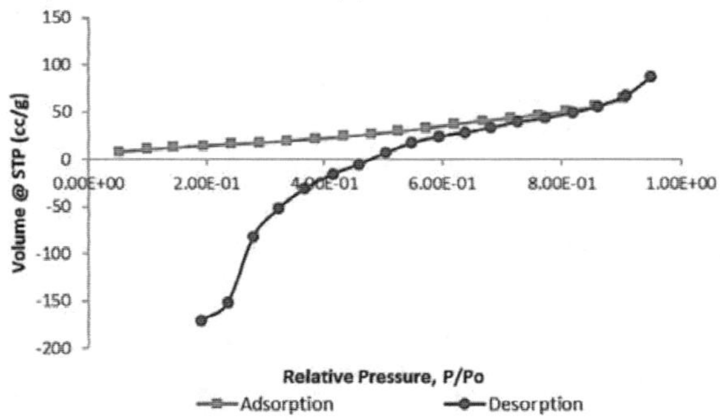

(b)

Figure 4. N_2 adsorption-desorption isotherm of (a) Co-succinate MOF and (b) Fe-succinate-MOF.

Table 1. Some properties of Co-MOF and Fe-MOF for nitrogen adsorption experiments at 300 °C.

Parameter	Co-succinate MOF	Fe-succinate MOF
Pore volume	0.013 cc/g	0.283 cc/g
Pore radius	1.5806 nm	1.5816 nm
Surface area	15.115 m²/g	61.326 m²/g

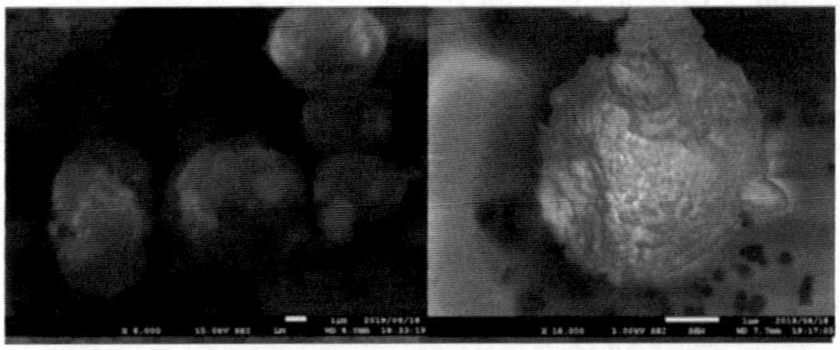

(a)

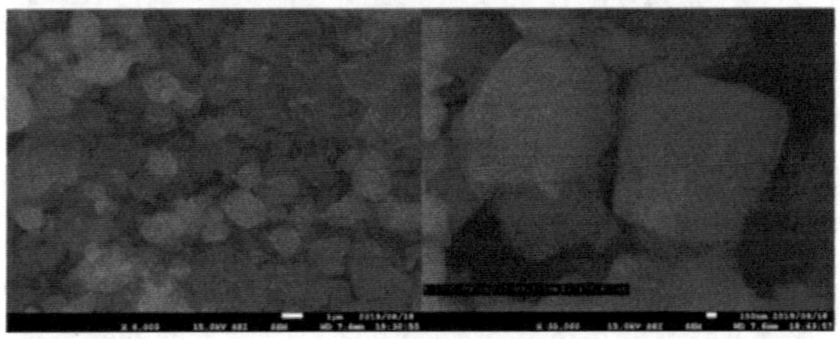

(b)

Figure 5. FESEM images of (a) Co-succinate MOF and (b) Fe-succinate MOF.

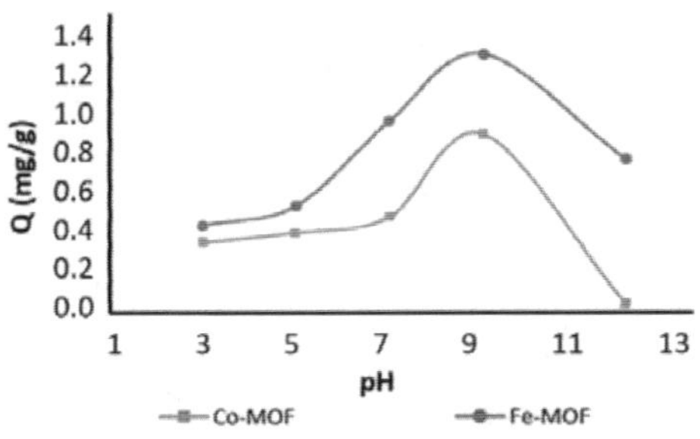

(a)

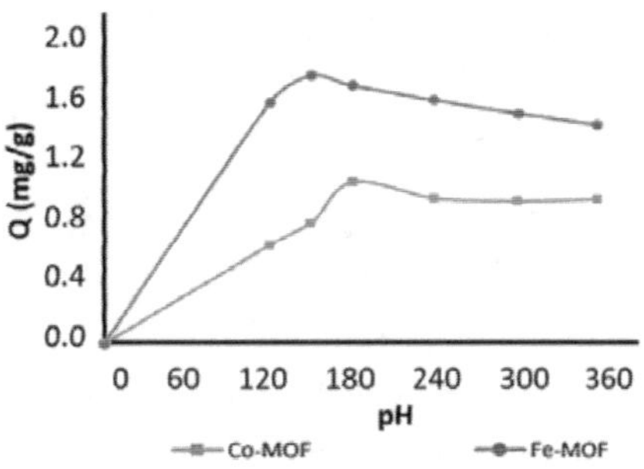

(b)

Figure 6. (a) Determination of optimum pH and (b) contact time in MB adsorption by Co- and Fe-MOFs.

As seen in figure 6b, MB adsorption increased with the increase of contact time (from 120 to 180 min) and remained unchanged at higher contact time due to the electrostatic interaction between MOFs and MB. The optimum contact time of Fe-MOF was 150 and Co-MOF adsorbed MB optimally at 180 min. As the contact time increased, the MOFs adsorbent can adsorb more MB (adsorbates) and interact with each other to reach the equilibrium point. There were two possibilities that could occur during the adsorption process: (1) the second adsorption layer occurs and forms the next layer, (2) the second layer does not form the next layer of adsorption and the adsorbate diffuses out of the pore and returns to the fluid that carries it [1].

3.2.2. Determination of adsorbent mass. It can be seen in figure 7 that the optimum adsorbent mass of both Co-MOF and Fe-MOF was 5 mg. The increasing mass of adsorbent at range of 10–25 mg caused the decreasing of adsorption capacity. It is caused by the aggregation particle due to the excessive amount of adsorbent which reduced the surface area.

3.2.3. Isotherm adsorption. Adsorption isotherm was widely used to describe the adsorption progress and investigate the mechanisms of adsorption [19]. Langmuir isotherm assumes that adsorption process forms a monolayer and homogeneous surface while Freundlich isotherm shows a multilayer adsorbent and the adsorption takes place on heterogeneous surface [1]. The Langmuir and Freundlich equations are presented in equation 1 and equation 2, respectively.

$$\frac{Ce}{Qe} = \frac{1}{Qm} \times Ce + \frac{1}{b \times Qm} \tag{1}$$

$$\log Qe = \log Kf + \frac{1}{n} \log Ce \tag{2}$$

Ce is equilibrium concentration (ppm), Qe is equilibrium concentration (mg/g), b is Langmuir Constant, Qm is optimum adsorption capacity, Kf is Freundlich Constant, and n is the intensity of adsorption [19].

As shown in figure 8 and table 2, MB adsorption on both as-synthesized MOFs were fitted to the Langmuir isotherm model based on high linear coefficient (R^2). This indicates that monolayer and homogeneous surface covers the adsorbate molecules [11]. The adsorption of MB takes place on the specific sites of MOFs with the maximum capacities Qm are 3.95 for Co-MOF and 4.67 $mg.g^{-1}$ for Fe-MOF.

3.2.4. Reusability of MOFs. Reusability is an important parameter in dyes removal application due to the ability of desorption process of adsorbent. As can be seen in figure 9, the adsorption capacity decreases slightly in two cycle experiments while it decreases significantly in three cycle experiments.

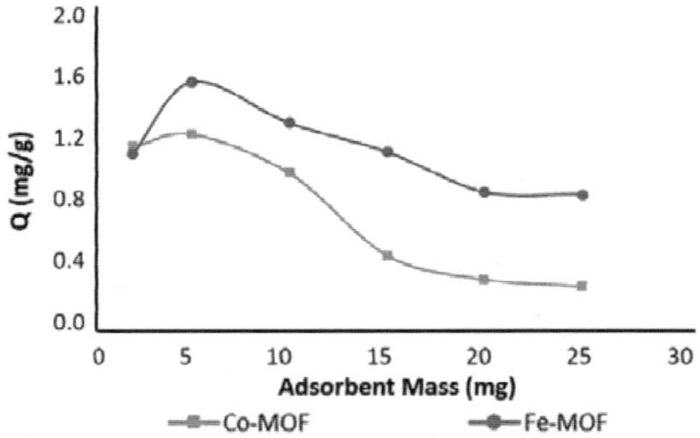

Figure 7. Determination of optimum adsorbent mass in MB adsorption by Co- and Fe- MOFs.

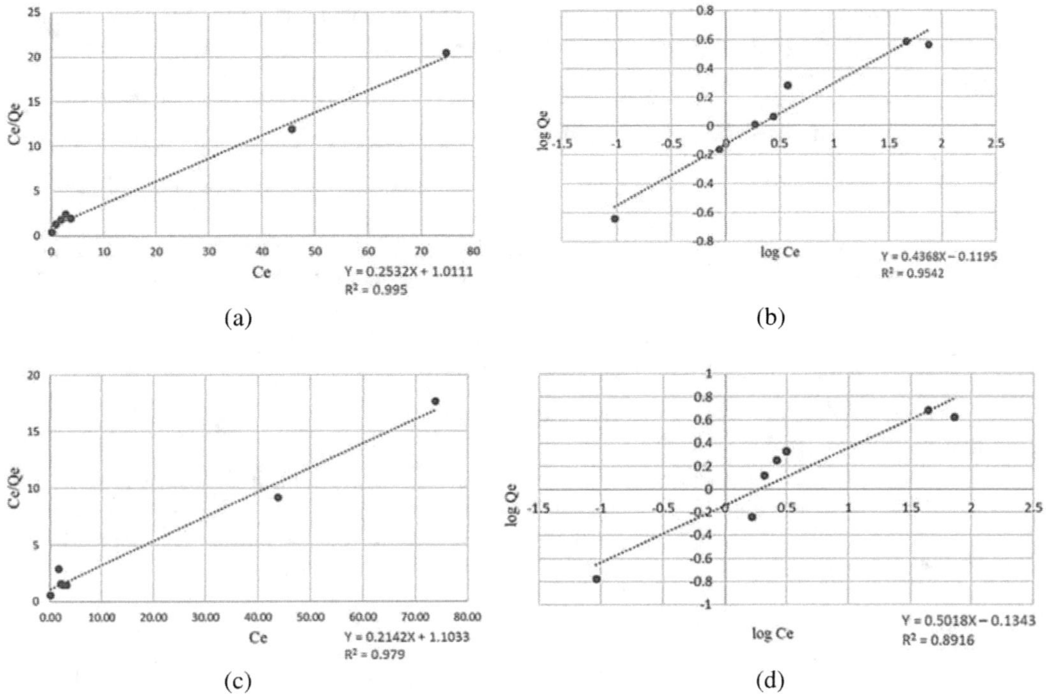

Figure 8. Comparison of Langmuir (a, c) and Freundlich (b, d) isotherm model for MB adsorption by Co-MOF (a, b) and Fe-MOF (c, d).

Table 2. Langmuir and Freundlich isotherm constants of Co-MOF and Fe-MOF.

MOF	Langmuir Isotherm			Freundlich Isotherm		
	Qm (mg/g)	B	R^2	n	Kf	R^2
Co-MOF	3.95	0.2504	0.995	2.2894	0.7595	0.9542
Fe-MOF	4.67	0.1941	0.979	1.9928	0.7340	0.8916

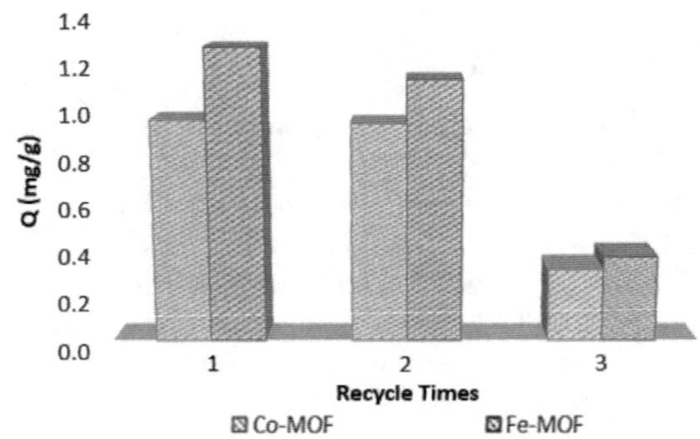

Figure 9. Reusability MOFs on the MB adsorption capacity.

This could be attributed to the desorption process where MB has high affinity to MOFs and it was hardly removed from the adsorbent [1]. The RSD of three repetition experiments was 49.9 % for Co-MOF and 52.9 % for Fe-MOF. The results show that the mixture of ethanol: water (7:3 v / v) is able to remove MB from a single dye solution whereas the desorption time and the effect of washing process to the MOFs structure should be studied further.

4. Conclusion

Synthesis of Co-MOF and Fe- MOF has been successfully carried out using the solvothermal method, based on the characterization results. MOFs also have high thermal stability with the decomposition temperature was above 200 °C. According to the best linear coefficient, the adsorption followed the Langmuir adsorption where the MOFs structure has monolayer and homogeneous surface. MOFs are also able to be reused for 2nd cycle in MB adsorption application.

Acknowledgments

This work was funded by Universitas Indonesia [Hibah Publikasi Internasional Terindeks Tugas Akhir (Hibah PITTA B) No. NKB-0686/UN2.R3.1/HKP.05.00/2019]. Dr. J. Y. Mulyana (Tokyo Metropolitan University, Division of hydrogen supply systems and carriers, Japan) was acknowledged for his generosity in hosting BCB Butar Butar during her summer research.

References

[1] Li H, Cao X, Zhang C, Yu Q, Zhao Z, Niu X, Sun X, Liu Y, Ma L and Li Z 2017 *RSC Adv.* **17** 16273-81
[2] Wainwright M and Crossley K B 2002 *J Chemother.* **14** 431-43
[3] Li X, Pi Y, Xia Q, Li Z and Xiao J 2016 *Appl. Catal B-Environ.* **191** 192-201
[4] Katheresan V, Kansedo J and Lau J 2018 *J. Environ. Chem. Eng.* **6** 4676-97
[5] Salehi E, Daraei P and A A S 2016 *Carbohydr. Polym.* **152** 419-32
[6] Kausar A, Iqbal M, Javed A, Aftab K, Nazli Z H, Bhatti H N and Nouren S 2018 *J. Mol. Liq.* **256** 395-407
[7] Chan L S, Cheung W H, Allen S J and Mckay G 2017 *HKIE Transactions* **24** 182-92
[8] Zhou Y, Lu J, Zhou Y and Liu Y 2019 *Environ. Pollut.* **252** 352-65
[9] Lee Y J, Chang Y J, Lee D J and Hsu J P 2018 *J. Taiwan Inst. Chem. E.* **000** 1-8
[10] Oveisi M, Asli M A and Mahmoodi N M 2018 *J. Hazard. Mater.* **347** 123-40
[11] Zhu T T, Zhang Z M, Chen W L and Liu Z J and Wang E B 2016 *RSC Adv.* **6** 81622-30
[12] Wu Y, Breeze M I, Clarkson G J, Millange F, Hare D O and Walton R I 2016 *Angew. Chem. Int. Ed.* **55** 4992-6
[13] Kolev T, Spiteller M and Koleva B 2010 *Amino Acids* **38** 45-50
[14] Vaidhyanathan R, Natarajan S and Rao C N R 2003 *Cryst. Growth De.* **3** 47-51
[15] Amghous Z, Roces L, Granda S G, Garcia J R, Souhail BV, Mafra L, Shi F N and Rocha J *J. Solid State Chem.* 2009 **182** 3365-73
[16] Kim Y J and Jung D Y 1999 *Bull Korean Chem. Soc.* **20** 827-30
[17] Ambroz F, Macdonald T J, Martis V and Parkin I P 2018 *Small Methods* **2** 1-17
[18] Zhang Y, Bo X, Nsabimana A, Han C, Li M and Guo L 2015 *J. Mater. Chem. A.* **3** 732-8
[19] Lin S, Song Z, Che G, Ren A, Li P, Liu C and Zhang J 2014 *Micropor. Mesopor. Mat.* **193** 27-34

Cs-137 radionuclide bioaccumulation study in gold fish (*Cyprinus carpio*) through freshwater path with variations of potassium ion concentration (K^+)

T A Baramanda[1], Budiawan[1], R Bakri[1] and H Suseno[2]

[1]Department of Chemistry, Faculty of Mathematics and Natural Sciences (FMIPA), Universitas Indonesia, Depok 16424, Indonesia
[2]Center for Technology Radiation Safety and Metrology, National Nuclear Energy Agency (BATAN), Jakarta 12710, Indonesia

Corresponding author's email: dr.budiawan@sci.ui.ac.id

Abstract. Few anthropogenic radionuclides like Cs-137 are difficult to detect. It needs a method to evaluate the environment quality based on molecular network of an organism exposed to the radionuclide arises, and the said method is called Biomonitoring. In this study, biokinetic experiment is conducted with the determination of Concentration Factor (CF) and Bioaccumulation Factor values of Cs-137 in goldfish (*Cyprinus carpio*) by means of freshwater influenced by its aquatic environment condition, which is potassium ion concentration (K^+). CF value of Cs-137 concentration in *Cyprinus carpio* after the end accumulation in media conditioned with variation of potassium concentration 2.5; 5; 7.5 and 10 ppm, respectively are 2.762; 2.645; 2.637; 2.568 $mL.g^{-1}$ and Bioconcentration Factor (BCF) for variation of K^+ ion concentration 2.5; 5; 7.5 and 10 ppm, respectively are 1.32; 2.12; 1.77; 2.07 $ml.g^{-1}$. Based on those values, goldfish (*Cyprinus carpio*) cannot be categorized as bioindicator for Cs-137.

Keywords: Cs-137, radionuclide bioaccumulation, gold fish, *Cyprinus carpio*

1. Introduction

The National Nuclear Energy Agency (BATAN) as an institution in Indonesia that deals with the application of nuclear energy in Indonesia has been operated with RSG 30Mwatt in *Pusat Penelitian Ilmu Pengetahuan dan Teknologi* (Puspitek), Serpong Area since 1987. This reactor can release elements from the fission elements. Potential contamination is feared to disturb the balance and health of the surrounding environment, especially the nearest aquatic environment such as Cisadane river. Radiocesium is a fission radionuclide, one of which comes from a nuclear reactor [1, 2].

The Serpong Puspitek Nuclear Area has an environmental monitoring system for nuclide radiation that can reach 0 to 5 km from the reactor. However, some anthropogenic radionuclides such as Cs-137 and Pu-241 are difficult to detect, so there is a need for an environmental quality evaluation method based on the analysis of the molecular tissue from the organism exposed to the radionuclides, the method is called Biomonitoring. The presence of Cs-137 radionuclides in waters can then accumulate in some freshwater biota, such as goldfish (*Cyprinus Carpio*).

Aquatic environment is very dynamic. One of the dynamics is the change in water salinity due to changes in the rainy and dry seasons. One change in salinity is caused by the change in K^+ concentration in water. This result changes the bioaccumulation ability of Cs-137. Some research related to Cs-137 bioaccumulation has been carried out including milkfish (Chanos chanos) [3] , teleost [4, 5], and zebra fish [6]. However, there is still limited study concerning the bioaccumulation of Cs-137 in fresh water biotas. In this study, biokinetic experiments were performed with the determination of CF (Concentration Factor) and biokinetic value of Cs-137 in goldfish (*Cyprinus carpio*) through freshwater channels influenced by their aquatic environment, that is, the concentration of potassium ion (K^+) by doing the bioaccumulation of the radionuclide in the laboratory.

2. Experimental

2.1. Chemical reagents

Goldfish (*Cyprinus carpio*) were collected from water around Puspitek, Serpong Area, South Tangerang. Equipment used in this research were beaker glass, gamma spectrometer, cotton, waterstone, aquarium, pump, aerator, aquarium skimmer, digital scales, surgical instrument, syringe, measuring pipette, micropipette (100 μL), stirrer, spray bottle and fishnet. The materials used in this study were N_2 liquid, KCl (2.5; 5; 7.5; 10 ppm), radionuclide Cs-137 (2 Bq/mL) and freshwater.

2.2. Preparation of aquarium and freshwater

Preparation of the aquarium was done by preparing and ensuring the aquarium with a capacity of 50 liters will be used is leak-free. When the aquarium was ready, the aquarium was filled with 40 L of fresh water.

2.3. Acclimatization

Acclimatization was done to give time to the object of research to adapt, in the form of living organisms in the research environment. Acclimatization process was done by placing the research object in the aquarium. During the acclimation process, the biota feeding was done once a day. Acclimatization process was done for 7 days. The condition of the water media is like the natural condition at room temperature, aerated for oxygen supply and lighting 12 h : 12 h.

2.4. Biota contamination through freshwater media and uptake process contaminants with the variation of potassium ion concentration (K^+)

After the acclimation process was completed, eight goldfishes were taken for the treatment of variation in potassium ion concentration. In the treatment of variations, the goldfish were divided into four glass jars, each containing 2 L of fresh water that has been irrigated, each containing 2 goldfish. Before goldfishes were placed in a glass jar, firstly, into the glass jar 4 mL of a radioactive tracer Cs-137 with activity 2 Bq/mL was added. Replacement of freshwater media was done every two days. The contamination process and contaminant uptake were performed for 9 days.

2.5. Measurement of Cs-137 activity during uptake process

The activity measurement process of Cs-137 in the biota was done every day in a period of 9 days. A HPGe detector was used for the Cs-137 analysis at energy of 666 KeV and counting time of 3 min. The biota to be measured was placed in a plastic cylindrical container containing freshwater contaminants and placed on a sample holder. Each measurement should be made under the same conditions, i.e. the distance of the container with the detector, the water level in the container, and the container geometry used. Measurements were made for 5 min for each goldfish. Measurements were made 3 times.

2.6. Depuration process

The depuration process was done by placing a goldfish in a contaminant-free aquarium completed with the filtration system and aeration for 5 days. During the process of depuration, replacement of freshwater media and goldfish feeding were done every day.

The process of measuring the activity of Cs-137 in the biota was done every day in a period of 3 days. A HPGe detector was used for the Cs-137 analysis at energy of 666 KeV and counting. The biota to be measured was placed in a plastic cylindrical container containing freshwater contaminants and placed on a sample holder. Each measurement should be carried out under the same conditions, i.e. the distance of the container with the detector, the water level in the container, and the container geometry used. Measurements were made for 5 min for each Goldfish. Measurements were made 3 times.

2.7. Preparation of biota standards

Preparation of biota standards was done by taking one goldfish that is free from contamination and weighed. Part of the goldfish (stomach) was slashed and the entrails were removed, then pieces of tissue were inserted into the stomach of goldfish and then dropped the radionuclide Cs-137 (activity 50 Bq / mL) as much as 50 μL. The traceable goldfish standard was then inserted into the plastic, sealed, and placed into a container in the same condition as the container used for the counting uptake and depuration. The prepared standards were then measured using a gamma detector.

2.8. Dissection

In the dissection stage, it was done by placing 5 goldfishes in an aquarium containing 70 L of water, then adding 140 mL of Cs-137 radionuclide contaminant with an activity of 2 Bq/mL. After that, surgery was performed to one goldfish per day. The surgery was performed for five days. In the process of surgery, the parts taken from the body of goldfish were gills, internal organs, and ivory. Then, the taken body parts of the goldfish were placed in a plastic container. Thereafter, the measurement of Cs-137 activity as performed using a gamma detector. The replacement of water media and feeding were done every other day.

3. Results and discussion

3.1. Uptake process

Experimental results of the concentration factor (CF) Cs-137 in *Cyprinus carpio* for 14 days in various concentrations of potassium ions (K^+) are shown in figure 1.

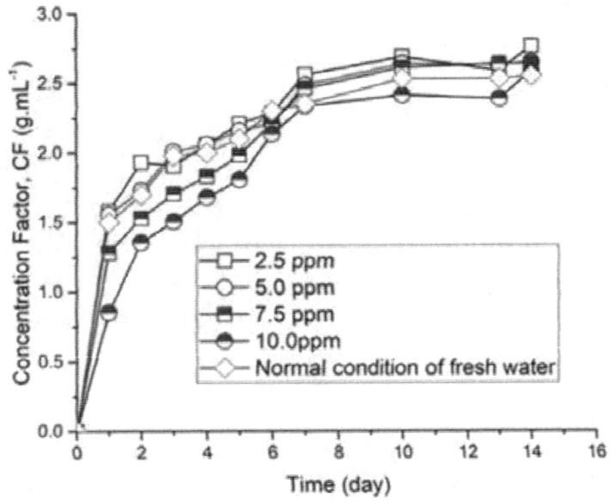

Figure 1. Concentration Factor (CF) Cs-137 in *Cyprinus carpio* in various K ions$^+$.

Bioaccumulation is a progressive increase in the concentration of a type of compound (including Cs-137) in an organism caused by the uptake rate of the compound is greater than the depuration rate) [7]. The bioaccumulation capacity of Cs-137 through freshwater lines by *Cyprinus carpio* is represented by the value of Concentration Factor (CF) which is the ratio between the concentration of contaminants in the biota body and the concentration of contaminants present in the water. CF value represents the ability of aquatic biota to increase the concentration of contaminants from water into the body.

Based on the measurement result of radionuclide Cs-137 with gamma ray detector in *Cyprinus carpio*, the data of count value of energy level in cps (count per second) were calculated. This data is further processed to determine the value of the concentration factor (CF). The concentration factor (CF) is a ratio between the pollutant concentration in the biota body to its concentration in water (mL g^{-1}).

In the environment, salinity waters tend to change with the changing seasons. In the rainy season, salinity tends to decrease and preferably in the dry season, salinity increases. Salinity is related to the concentration of K^+ ion in water. The experiment results showed an increase in K^+ ions in water will decrease the ability of Cs-137 bioaccumulation. The CF value at the addition of 2.5 ppm K shows at 2.76 g.mL^{-1}. On the other hand, the addition of K^+ ions decreases its bioaccumulation ability to 2.58 g.mL^{-1}. The main variable of the aquatic ecosystem is salinity which plays an important role in fish physiology. Fish can compensate for salt levels through an active ion excretion to maintain osmolality. Based on this the physiology, fish is strongly influenced by salinity including in the context of the contaminants bioaccumulation process [8].

The bioaccumulation capacity of a biota is also represented by biokinetic parameters such as k_u and k_e. The uptake rate of the contaminant (k_u) in a single compartment is assumed as a contaminant uptake mechanism by the entire body of the biota. However, the speed of its distribution into various types of organs is neglected. The value of k_u (mL g^{-1} day^{-1}) is an uptake rate calculated based on the slope of the CF curve to t (from t = 0 to t under steady state) [9]. The effect of K^+ ions addition to the rate of Cs-137 uptake is shown in figure 2.

Figure 2 shows that there is no relationship between the increase of K^+ ions in water and the rate at which Cs-137 is taken by *Cyprinus carpio*. In the condition of the different K^+ concentration, the uptake constants are varied greatly, which is in the range of 0. 08 to 0.13g^{-1}.mL. This lack of relationship between K concentration and uptake rate shows that bioaccumulation is more influenced by the physiological conditions of fish that live in various salinities.

3.2. Depuration process

The process of release or excretion of Cesium from the body is one of the processes for maintaining the electrolyte balance in the body of a biota. The ability to release contaminants by the biota bodies is represented by the value of % retained. Depuration process of Cs-137 is shown in figure 3.

The depuration process is done after the retrieval process stopped, and the biota is transferred to an aquarium containing contaminant-free freshwater. Depuration is the process of releasing contaminants from the body of a biota when the level of the exposed contaminant is reduced or eliminated. Naturally, this depuration process occurs when the entry of contaminants into the aquatic environment stops or decreases so that the contaminants will be removed from the biota network.

At the time of entry into the body of the biota, cesium will behave like Potassium and will be eliminated in the same way as potassium, i.e. by the mechanism of regulation of ionic composition through renal reabsorption (nephridia) and out with ammonia [10]. The mechanism of excretion or reabsorption of renal ions in general will involve an active transport mechanism so that large ions will be more difficult to excrete or reabsorb. The reduction of Cs-137 in the body of *Cyprinus carpio* is also influenced by the enabling environment, i.e., Cs-137 in low freshwater media.

3.3. Dissection

The distribution of Cs-137 in the body part of the golden fish (*Cyprinus carpio*) is represented in the activity value of Cs-137 (Bq/g) and shown in figure 4.

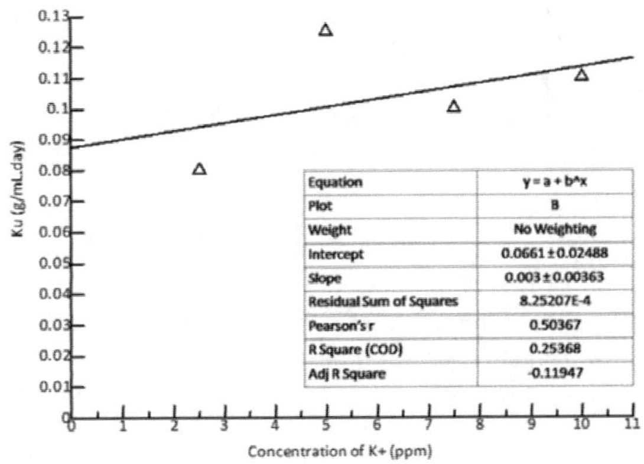

Figure 2. Influence of various concentrations of potassium ions (K^+) to uptake rate.

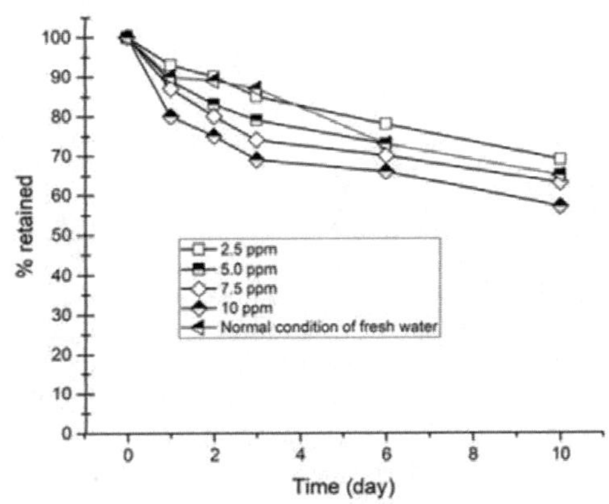

Figure 3. Loss Cs-137 (depuration) from in *Cyprinus caprio*. After 10 days of this depuration process Cs-137 is still stuck in the body of goldfish by 57 to 70 %.

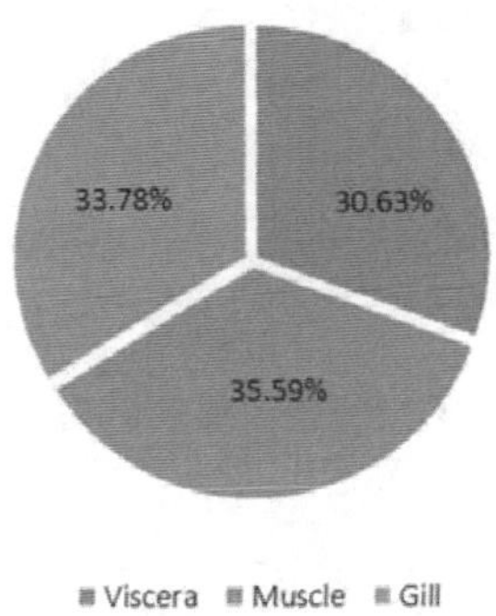

Figure 4. Percentage distribution of Cs-137 in golden fish organs (*Cyprinus carpio*) after 9 days of accumulation.

The dissection was conducted only to the important parts of the body of the golden fish that are related to the path of contaminant transfer. The distribution of Cs-137 activity in the *Cyprinus carpio* sample has the same pattern in the flesh> gill> internal organs. The activity of Cs-137 in meat is greater than the gills and internal organs. This is because, the absorption rate of metals in meat is higher than the others. The activity in the internal organs of the fish body is the smallest, since Cs-137 are distributed to other body parts.

4. Conclusion

Increasing the concentration of potassium ions (K^+) will decrease the value of steady state concentration factor (CF). The constant rate of harvest of Cs-137 in golden fish (*Cyprinus carpio*) is not significantly influenced by the variation of potassium ion concentration (K^+). Increasing the concentration of potassium ions (K^+) will increase the value of Cs-137 release rate in goldfish (*Cyprinus carpio*). The Concentration Factor (CF) Cs-137 in golden fish (*Cyprinus carpio*) is not significantly affected by the variation of potassium ion concentration. Cs-137 is distributed most often in the meat (muscle). Golden fish (*Cyprinus carpio*) cannot be used for risk assessment of Cs-137 radionuclide pollution due to the BCF <500, but with other bioconic parameters *Cyprinus carpio* can be used as a bioindicator to identify the release of Cs-137 to waters.

References

[1] Suseno H 2014 *AIP Conf Proc.* **1589** 342-5
[2] Suseno H, Wahono I B, Muslim M and Yahya M N 2017 *Reg. Stud. Mar. Sci.* **10** 81-5
[3] Prihatiningsih W R, Suseno H, Zamani N P and Soedharma D 2016 *Mar. Pollut. Bull.* **110** 647-53
[4] Sezer N, Kılıç Ö, Metian M and Belivermiş M 2018 *J. Environ. Radioact.* **192** 376-84
[5] Rinderhagen M, Ritterhoff J and Zauke G-P 2000 *Environ. Res.* **9** 161-94
[6] Reinardy H C *et al.* 2011 *Sci. Total Environ.* **409** 3771-9
[7] Al-Sid-Cheikh M, Rouleau C and Pelletier E 2013 *Mar. Environ. Res.* **86** 21-8
[8] Garnier-Laplace J, Vray F and Baudin J P 1997 *Water Air Soil Pollut.* **98** 141-66
[9] Metian M, Warneu M, Teyssie J-L and Bastamante P 2011 *J. Environ. Radioact.* **102** 543-50
[10] Wickramasooriya P L, Waduge V A, Attanayake T N and Udugala-Ganehenege M Y 2015 *Environ. Chem. Lett.* **13** 487-93

Study of bioaccumulation and heavy metal depuration of cadmium in green mussel (*Perna viridis*)

Budiawan[1], R Bakri[1], S A Aziz[1] and H Suseno[2]

[1]Department of Chemistry, Faculty of Mathematics and Natural Sciences (FMIPA), Universitas Indonesia, Depok 16424, Indonesia
[2]Center for Technology Radiation Safety and Metrology, National Nuclear Energy Agency (BATAN), Jakarta 12710, Indonesia

Corresponding author's email: dr.budiawan@gmail.com

Abstract. Green mussel (*Perna viridis*) is a favorite seafood, which has good protein content and economical price. The shellfish habitat is in a polluted area in the Jakarta Bay caused by industrial waste. Presently, the consumption of green mussels is very problematic because the habitat of green mussels has experienced pollution. Bioaccumulation of heavy metal study was conducted to determine the process of heavy metal accumulation in a biota by using a certain dosage. The heavy metal cadmium was used to be exposed to a biota as much as half LC_{50} is 0.1 ppm. The depuration process was performed to reduce the levels of cadmium metal in green mussel. A seven-day drainage depuration method was used by immersing acetic acid and citric acid with variations of concentration 0.75 %, 1.5 %, and 2.25 % in 24, 48, 72, 96, 120 min variations. Levels of cadmium in green mussel were measured using atomic absorption spectrophotometry. Concentration of cadmium decreased to 3.05 $mg.Kg^{-1}$, 1.7 $mg.Kg^{-1}$ and 0.65 $mg.Kg^{-1}$ after depuration process that used immersing by water, acetic acid and citric acid, respectively.

Keywords: Green mussel, *Perna viridis*, heavy metal accumulation, cadmium

1. Introduction

The estuarial community of Kamal has been working on the fisheries sector since the 1970's. Presently, less than 40 % of fishermen conduct marine cultivation including fish, shrimp, and mussel. The Jabotabek surrounding area that have 20.3 million of inhabitants cause the pollution in Jakarta bay [1]. In 2003, the Regional Environmental Management Board of the Province of Jakarta Capital Special Region mentioned the industrial waste that generates 2050 B3. This B3 waste is a major source of pollution in waters in Jakarta. The increase in the concentration of pollutants, both organic or inorganic, resulting in the occurrence of eutrophication and a decrease in the number of fauna as well as a decline in water quality and sediment on the waters [2]. The main polluters in a waste waters source have hazardous substances and components that are toxic, such as heavy metals. Heavy metals are classified based on the nature of the toxicity and the source of its existence which is divided into the low, medium and high categories. One example of a heavy metal which belongs to a high toxicity i.e. metal cadmium [1]. The Ministry of Environment and Forestry describes the value of a threshold based on Raw Water Quality Metal Cadmium in waters at 0.001 ppm. The metal content of the dissolved cadmium in Kamal estuary is as much as 0.02 ppm. In 2001, an advanced research was done to measure the content of the metal cadmium which was increased to 0.046 ppm [3]. In 2002, the research on back metal dissolved cadmium reached 0.785 ppm. Heavy metals contamination in waters can be accumulated in marine biota. Green mussel *(Perna viridis)* is a rich biota in the waters of the estuary of Kamal. Mussels is a favorite seafood commodity having a good protein content and economical price. Two hundred grams of mussels contained 21.9 %, protein 14.5 %, fat 4.3 %, carbohydrate 18.5 % and the remaining water 40.80 % [4]. Mussels have a filter feeder, capable of eating anything that exists in the environment and stuck in the body. Consuming shellfish became a public concern because it has been contaminated with heavy metal and gave an effect to the human metabolism.

Some bioaccumulation studies have conducted on tilapia fish [5] , milk fish [6, 7] and shrimp [8]. In this study, Cd bioaccumulation and depuration by green mussel were conducted.

2. Method

2.1. Preparation of aquarium and sea water

The aquarium of 300 L was used, filled with 250 L of clean sea water. The salinity of sea water was determined by refractometer and a temperature of 25 °C was tailored in the laboratory using a thermometer.

2.2. Retrieval of biota

Green mussels (*Perna viridis*) as bio indicator were used and taken from the waters of the Jakarta bay, Muara Kamal. Biota samples obtained were cleaned from dirt. Furthermore, the biota was collected in a plastic sack and ice gel was added to maintain the freshness while headed to BATAN PTKMR aquatic laboratory for acclimatization. Mussels used were approximately 1–3 of the month, with a length of about 3 cm.

2.3. Acclimatization biota

The Biota samples taken from Kamal were put into a clean sea water as much as 2–3 times. After that, the biota were bound with a rope and hung with a *raffia* string in the aquarium for acclimatization. The aquarium were aerated in dark and light continued to 12:12 h. During the acclimatization, biota was fed with *Artemia.*

2.4. Bioaccumulation of metal cadmium

The aquarium with 50 L capacity was filled by approximately of 40 L sea water. The aquarium was equipped with filtration, aeration and skimmer systems. Thirty of shells *Perna viridis* were immersed in this aquarium system.

2.5. Depuration

Depuration were conducted in 2 types i.e., drainage and acid immersion. Depuration drainage was carried out using the aquarium contained 40 L sea water. Aquaria systems equipped with modeling the flow of water will be drained and filtrated. Sea water was replaced every day. Fourteen biotas were used in this experiment. Depuration acid soaking was conducted on citric and acetic acid. Variation of the acid concentrations were 0.75 %, 1.5 % and 2.25 %. The time variation were 48 h, 72 h, 96 min, and 120 min. This experiment was conducted in three different containers based on concentration. Each container contains 10 shells, which will be taking as much as two in every variation of the time.

2.6. The determination of metals content

Mussel dissection was conducted to separate parts of the mussel. Each part was destructed with USGS Test Method B-9001-95. Sulfuric and nitric acid were used for destruction process. The composition of the acids was 1:5. The cadmium content in each part was determined by Atomic absorption Spectrophotometer.

3. Results and discussion

3.1. Description of location

Industrial areas around the estuary are galvanizer, metal casting, garment and Pharmaceuticals [1]. Density of industry around Muara Kamal leads to potentially disposal of waste containing heavy metals into the residential areas. The wastes were disposed directly without being processed in advance. It can interfere with the quality of the water and the ecosystem also reside in these waters. A lot of industrial waste containing heavy metals can have an impact on living things. The impact can be felt through the human by consuming the seafood that has been polluted and also the water. As the supporting data, the recording conditions (table 1) at the mouth of Kamal are presented.

3.2. Retrieval of biota

The sampling conducted in the estuary of Kamal, was taking place in the middle of the ocean for cultivation of biota. Biota is the mussels. Green shells used in this research were 1–3 months, with long range of 3 cm, 2 cm width, 1.5 cm thickness.

The chosen green mussels still have a substrate. The living mussels have the characteristics of shells, and are green to brown color. When in the water, living mussels will sink. Generally, living mussels are attached to a substrate using a byssus thread. Byssus is present on the lower leg foot clams adapted to attach to the substrate.

3.3. Acclimatization of biota

Acclimatization is the adjusting process of individuals to the changes in environmental conditions. Acclimatization aims to relieve stress in the animal experiments in conditions of the aquarium so that it can be used in the experiment of bioaccumulation as well as to reduce the risk of death in biota.

Acclimatization in the form of media is used in aquariums. Preparation of the aquarium is done a day earlier, with aeration and filtration system already ensuring a well mounted. Filtration components, namely cotton filter, rocks are to filter the impurities bioball insoluble. Impurities insoluble fibers such as fibers that attached to the body of the shellfish. Also, a skimmer is used to filter the impurities dissolved in the aquarium. Soluble impurities resulting from the process of excretion of the mussels. The working principle of the skimmer is to create air bubbles that make contact with colloids and particles–solid particles. This tool is used to create such conditions in the sea produces froth–effervescence.

The use of the mussels still affixed to the substrate is to approach the green mussel habitat to its original condition. Previously, the mussels were cleaned by dipping the clams in a clean sea water, so that the mud or dirt does not contain in the aquarium. Clean mussels are suspended inside the aquarium. The process of acclimatization is done for seven days without giving the contaminants. During the first seven days, the nutrients around it will be absorbed. Green mussel byssus will use to move and some stick to the walls of the aquarium.

3.4. Bioaccumulation of cadmium metal

Measurement of content of the metal cadmium is performed using atomic absorption spectroscopy. By calculating the radio wave electromagnetic absorbance by the free atoms an element. Contaminant uptake observations was done by analyzing the content of the metal cadmium (table 2) in a certain period. The higher levels were obtained on the seventh day. The process of bioaccumulation was reached viewing the cadmium levels that tend to reach a steady state.

Taken control mussels originating from aquarium acclimatization. Cadmium levels were measured using atomic absorption spectroscopy of 0.99 mg/Kg cadmium levels may be seen. On the seventh day (table 3) of 7.21 mg/Kg, experienced a rise of 86.27 %. The condition tends to reach a steady state starting on the sixth day and the seventh day, at cadmium level during bioaccumulation. Cadmium levels in biota of the conversion can be done on the parameters of the biokinetic CF (Concentration Factor).

Table 1. Muara Kamal data condition

Parameter	Value
Temperature (°C)	32
Salinity (°/₀₀)	33
pH	6-7

Table 2. The content of cadmium metal in the green mussels.

Time (Day)	The content of cadmium metal in The Green Mussels (mg/Kg)
1	3.15
2	3.98
3	4.78
4	6.03
5	6.66
6	6.91
7	7.21

Table 3. CF's value

Time (Day)	CF (mL.g^{-1})
1	31.5
2	39.77
3	47.78
4	60.4
5	66.63
6	69.1
7	72.1

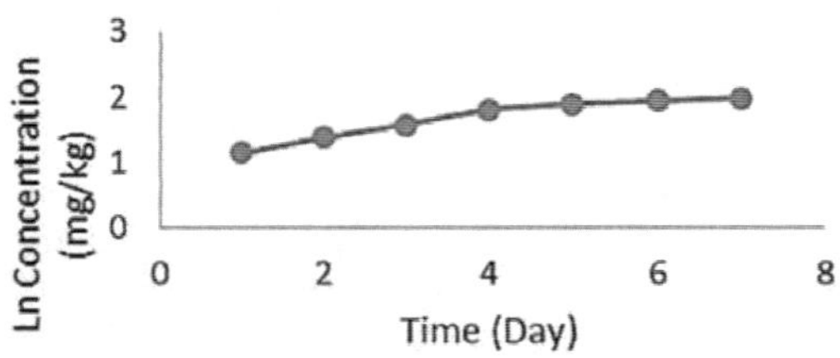

Figure 1. Bioaccumulation rate

It can be seen also from the process of concentration rate of bioaccumulation. The rate of this concentration can use a principle of first order. Depicted in a graph (figure 1) where the x is the length of exposure to contaminants and the y axis is the ln (concentration). It can be seen the influence of rate depends on the time of exposure to the contaminants. The value of the rate will be even greater at the longer time.

The ability of bioaccumulation in biota is also represented by the rate of uptake of contaminants (ku). In a single compartment, the Ku value is assumed as the mechanism of contaminant uptake by biota throughout the body. However, the speed of its distribution into the various types of organs is ignored. The Ku values (mL g^{-1} day^{-1}) is a constant rate of uptake are calculated based on the slope of the curve CFt against t (from t = 0 to t at steady state conditions).

3.5. Bioaccumulation model

The equations in this model refers to a single compartment, bioaccumulation approach to the value that is used, to the CFss and derived from the experimental results. Form change to steady state was reached on day seven, and shown with the value of the CFss. Physiological factors which include metabolic processes contaminants in the body is assumed to be constant [9]. Retrieval rate constants determined by changing the equation becomes a linear equation so that equation is obtained as follows:

$$ln\ (CF_{ss} - CF_t) = k_u . t$$

From equations, it is plotted into a graph of today vs. ln (CFss – CFt) where t is the x axis (day) while the y-axis is ln (CFss – CFt). The result of the equation will produce an equation of a line and the slope of the equation of a line is the value from the process of contaminant uptake of green mussels. Bioaccumulation modeling is shown in figure 2.

3.6. Depuration

These methods use the system of flow circulating. The depuration process was conducted for seven days, and replaced with clean water every day. During the depuration process, Cd were determined using atomic absorption spectroscopy. Cadmium levels during the process of depuration has decreased. Decrease in low cadmium metal is obtained on the seventh day (table 4), by the amount of 3.058 mg/Kg cadmium Levels. Declining the most at 57.59 % compared with levels of cadmium at present day bioaccumulation.

Retrieval process (uptake) and redemption (depuration) contaminants can be done in a single compartment (figure 3).

Bioaccumulation process, when the levels of cadmium is increasing and likely to reach steady state. The release of cadmium levels indicated at any given time. Rate of depuration in this study was conducted in the order of one (figure 4). Refers to the rate of bioaccumulation, the rate of water loss depuration was obtained. The decline in the rate based on the concentration of cadmium at any time that is increasingly low. From the rate of depuration, biological half-life can be known that illustrates the ability of mussels to eliminate Cd^{2+}contaminants accumulated in its body of 3.55.

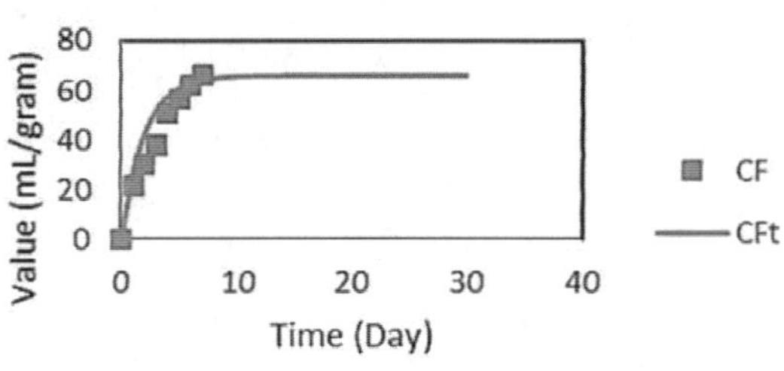

Figure 2. Bioaccumulation model

The measurement levels of cadmium on depuration was performed by acid immersion. The levels of cadmium that is stuck in the body of biota after a depuration was measured. Cadmium levels decline at any concentration. Cadmium levels decrease occurred in the body of the green shells, along with the increasing time of immersion in acetic acid solvent. The decrease levels of cadmium can be seen, too, in the retention values (table 5 and 6).

Depuration rate in acetic acid (figure 5) and citric acid (figure 6) were declined. The kinetic of reaction rate of depuration followed the first order one. The depuration reached the steady state after 120 min.

3.7. Determination of BCF value

BCF or Bioconcentration Factor is the ratio between the concentration in the organism and in the water during a state of stable equilibrium [10]. The BCF is determined by comparing the values of

Table 4. Content and retention of cadmium metal.

Time (Day)	Content of metal cadmium in Mussle (mg/Kg)	Retention (%)
1	6.031	0.97
2	5.315	0.87
3	5.07	0.84
4	4.647	0.78
5	4.42	0.75
6	3.35	0.6
7	3.058	0.56

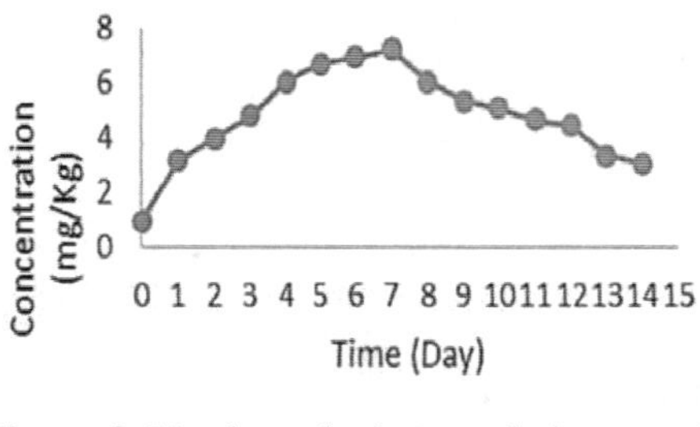

Figure 3. Uptake and release cadmium metal

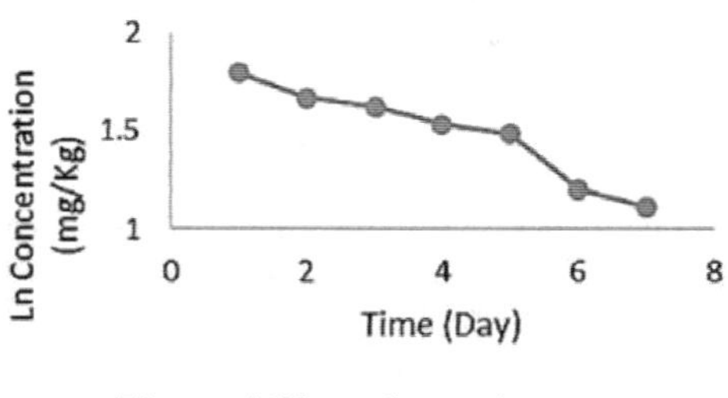

Figure 4. Water depuration rate

Table 5. Content and retention of cadmium by green mussel with acetic acid depuration.

Time (minute)	Acetic Acid 0.75 % Cd's content in mussle (mg/Kg)	Acetic Acid 0.75 % Retention (%)	Acetic Acid 1.5 % Cd's content in mussle (mg/Kg)	Acetic Acid 1.5 % Retention (%)	Acetic Acid 2.25 % Cd's content in mussle (mg/Kg)	Acetic Acid 2.25 % Retention (%)
24	6.17	0.99	5.98	0.97	6.15	0.99
48	3.16	0.58	5.73	0.93	5.59	0.91
72	2.56	0.49	5.46	0.9	5.26	0.87
96	2.18	0.44	4.77	0.8	3.8	0.66
120	2.15	0.43	1.98	0.4	1.7	0.37

the constants of uptake (ku) with constant release (to). The constant taking of value is obtained from the slope equation of a line on the rate of uptake on the process of bioaccumulation of 0.5782. The slope value of the line equation in the chart represents the constant value, the rate of contaminants (ku) by the mussels.

To analyze the concentrations of chemical substances contained in certain ecosystems/waters used a value of BCF in table 7. Can be determined with equation, i.e. BCF $= \frac{Ku}{Ke}$.

Table 6. Content and retention of cadmium in green mussels with citric acid depuration

Time (minute)	Citric Acid 0.75 % Cd's content in mussle (mg/Kg)	Citric Acid 0.75 % Retention (%)	Citric Acid 1.5 % Cd's content in mussle (mg/Kg)	Citric Acid 1.5 % Retention (%)	Citric Acid 2.25 % Cd's content in mussle (mg/Kg)	Citric Acid 2.25 % Retention (%)
24	6.15	1.11	6.48	2.6	7.27	1.04
48	5.28	0.87	4.64	3.57	5.42	0.78
72	4.78	0.8	4.18	3.72	3.22	0.72
96	3.08	0.56	2.27	4.16	0.85	0.45
120	2.25	0.45	1.75	4.25	0.65	0.38

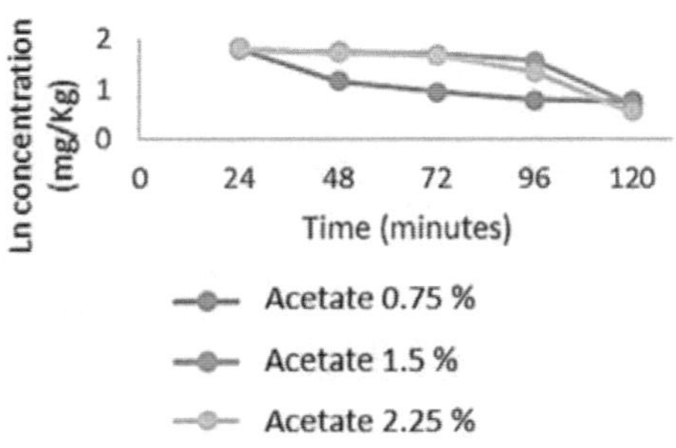

Figure 5. Acetic acid depuration rate

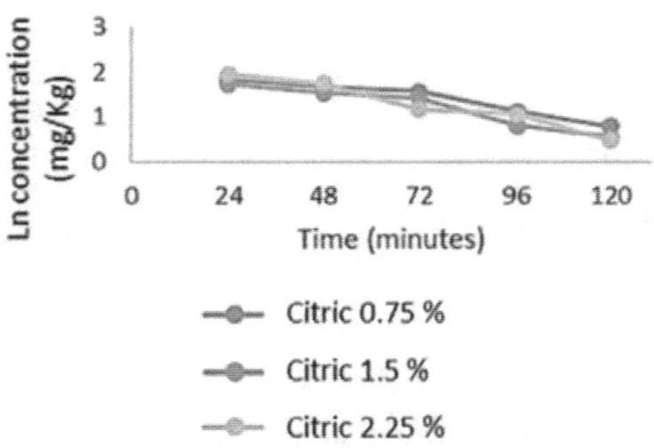

Figure 6. Citric acid depuration rate

Table 7. BCF value

	Ku	Ke	BCF
Water Depuration	0.5872	0.1954	3.01
Acetic Acid Depuration 0.75 %	0.5872	0.0141	41.64
Acetic Acid Depuration 1.5 %	0.5872	0.0141	41.64
Acetic Acid Depuration 2.25 %	0.5872	0.0169	34.75
Citric Acid Depuration 0.75 %	0.5872	0.3861	1.52
Citric Acid Depuration 1.5 %	0.5872	0.3097	1.896
Citric Acid Depuration 2.25 %	0.5872	0.5028	1.17

Table 8. Content of protein in green mussle

Sample	% Protein
Control	16.56
After Water Depuration	9.995
After Acetic Acid Depuration 0.75 %	15.169
After Acetic Acid Depuration 1.5 %	14.963
After Acetic Acid Depuration 2.25 %	12.304
After Citric Acid Depuration 0.75 %	15.818
After Citric Acid Depuration 1.5 %	12.364
After Citric Acid Depuration 2.25 %	10.516

3.8. Determination of protein with kjehdahl method

Kjehdahl method based on the calculation of the total nitrogen contained in the shellfish meat then the value converted to protein levels is obtained. This protein test is done on the shellfish control, bioaccumulation and depuration. Protein levels can be seen in table 8.

4. Conclusion

Mussels are able to accumulate cadmium metal with a value of CF to 72.09 mL g^{-1}. The ability of the cadmium bioaccumulation was 7.21 mg Kg^{-1}. The lowest concentration of cadmium metal is obtained on the mussels after the depuration on drainage method of 3.058 mg Kg^{-1}, 2.25 % acetic acid immersion during 120 min of 1.7 mg Kg^{-1}, and 2.25 % citric acid immersion during 120 min of 0.65 mg Kg^{-1}.

Acknowledgments

This research were sponsored by Dikti grant Project and performance in Aquatic Laboratory Center Technology for Radiation Safety, National Nuclear Energy Agency (BATAN).

References

[1] Wendling C C *et al.* 2013 *Mar. Pollut. Bull.* **71** 222-9
[2] Koropitan A F and Ikeda M 2016 *Procedia Environ. Sci.* **33** 532-52
[3] Riani E, Cordova M R and Arifin Z 2018 *Mar. Pollut. Bull.* **133** 664-70
[4] Al-Usmani S M P, Jagtap T G and Patil D N 2015 *Mar. Pollut. Bull.* **99** 328-31
[5] Suseno H, Hudiyono S, Budiawan B and Wisnubroto D S 2010 *Aust. J. Basic Appl. Sci.* **4** 792-9
[6] Prihatiningsih W R, Suseno H, Zamani N P and Soedharma D 2016 *Mar. Pollut. Bull.* **110** 647-53
[7] Prihatiningsih W R, Suseno H, Zamani N P and Soedharma D 2016 *Atom Indo.* **42** 129-35
[8] Suseno H, Hudiyono S and Muslim M 2016 *HAYATI J. Biosci.* **23** 117-20
[9] Landrum P F and Fisher S W 1998 *Influence of Lipids on the Bioaccumulation and Trophic Transfer of Organic Contaminants in Freshwater Ecosystems* (New York: Springer) pp. 203-34
[10] Opperhuizen A and Schrap S M 1987 *Environ. Toxicol. Chem.* **6** 335-42

The localized surface plasmon resonance on noble metal-semiconductor: Au nanosphere-ZnO nanorod

M Sujak and D Djuhana

Department of Physics, Faculty of Mathematics and Natural Science (FMIPA), Universitas Indonesia, Depok 16424, Indonesia

Corresponding author's email: dede.djuhana@sci.ui.ac.id

Abstract. In this work, we have investigated the localized surface plasmon resonance of metal-semiconductor model. The raw materials are gold nanosphere (Au-NP) and ZnO nanorod (ZnO-NR). The nanoparticles were surrounded by air with refractive index constant (1.0+0i). Moreover, the diameter and aspect ratio of Au-NP and ZnO-NR were 40 nm and 4, respectively. Then, we designed the nanoparticle in two different configurations. First, Au-NP was placed at the sides of ZnO-NR and second, Au-NP was placed at the tip of nanorod. Next, we varied the gap-distance among them from 0 nm to 40 nm with 10 nm increment. The photon wavelength used were from 300 nm to 1200 nm within two different excitations (transverse and longitudinal excitation). Furthermore, our result suggests that on the longitudinal excitation, when the gap-distance increases, the resonance peak of second configuration remains at constant value and the intensity significantly decreases compared to the first configuration. However, on the transverse excitation, the first configuration has a high intensity and the resonance peaks is slightly red-shifted. In addition, based on the near-field mapping which reflects the energy density of electromagnetic field by LSPR, the LSPR of Au-NP stands to aid the exciton production to enhance the photocatalytic activity.

Keywords: Surface plasmon resonance, ZnO, gold, photocatalysis activity

1. Introduction

The strong interaction between light and noble metals (e.g. gold, silver, and copper) on the nanoscale display extraordinary applications including bio-sensor [1], enhance-Raman scattering [2], solar cell application [3] and enhance photocatalytic activity [4-6]. The interaction light with metal has been understood in terms of Localized Surface Plasmon Resonance (LSPR), where light energy is trapped in a small length/volume/area without propagation [7]. Furthermore, the utility of LSPR effect was well-known to enhance the photocatalytic activity in semiconductor materials such as ZnO and TiO_2 [4, 5, 8]. Noble metals greatly aid to enhance the photocatalytic activity. As reported by Fauziyah et al. in order to increase the efficiency of photocatalytic activity in the case of ZnO, it could be done not only by increasing the surface area, but also the surface active site [8]. In addition, Chou and co-worker reported the sensitivity of Au modified ZnO on different refractive index medium, and showed that the resonance peak shifted to a lower energy as the refractive index increased [9]. Another noble metals such as silver (Ag) had been carried out by Ren and co-worker to modify the ZnO nanorod array for photocatalytic application [10]. Moreover, there were systematical studies about the effect of

IOP Conf. Series: Materials Science and Engineering **902** (2020) 012058 doi:10.1088/1757-899X/902/1/012058

electromagnetic coupling between exciton from ZnO and plasmon resonance from gold nanoparticle [11].

However, there was a lack of understanding on how the LSPR affects the photocatalytic activity process. In this work, we have systematically studied by the means of numerical calculation of how the LSPR effect from Au can enhance the spectral of ZnO and we also serve a near-field map of Au nanoparticle adjacent to ZnO nanorod with different gap-distances.

2. Simulation procedure

We have carried out the extinction cross section calculation by employing a metal-semiconductor model, as illustrated in figure 1. The calculation was used MNPBEM Toolbox embedded in MatLab software [12]. This toolbox was utilized a Boundary Element Method approach to build a convenience metal-semiconductor structure. Furthermore, the substraction of scattering cross section and absorption cross section leaves extinction cross section, thus it could be utilized to analyze the result and the electric field map to show the electric responses of Au-NP and ZnO-NR for different gap- distance.

Gold nanosphere with diameter *(D)* 40 nm was employed as well as ZnO nanorod profile with length *(L)* and width *(w)* are 280 nm and 70 nm, respectively or the aspect ratio is 4. These sizes were determined by TEM images as reference [6]. The dielectric function of gold was taken from Johnson and Christy [13] and for ZnO adopted by Stelling's data [14]. The refractive index of medium around metal-semiconductor is air (1+0i). Moreover, the long-axis of ZnO was situated in line with y-axis (see figure 1) while Au-NP positions were put next to the length and width of ZnO. Then, the gap position *(s)* was determined from the tip of ZnO and Au by varying *s* from 0 nm to 40 nm with 10 nm increment. The photon wavelength in range 300 nm to 1200 nm was applied within two difference polarizations, along z-axis and y-axis.

3. Results and discussion

We examined the spectral, namely, Longitudinal Plasmon (LP) mode and Transverse Plasmon (TP) mode. The LP occurred when the electrical part of photon wavelength is polarized along z-axis while TP can be exhibited when the electrical part is properly arranged to be polarized in y-axis (see the inset figure 1). Figure 2 shows the extinction spectra of metal-semiconductor model where Au-NP is situated next to the short-axis of ZnO-NR by gap distance *s*. It can be observed that, the extinction peak of ZnO is appeared around ~350 nm (~3.5 eV) due to the UV-light energy absorption that imposes the photocatalytic activity [8, 15]. According to Fauziyah et al. and Musa et al. work, the absorption in the UV region correlates with ZnO band gap energy profile, around ~3.24 eV (380 nm) and 3.378 eV (367 nm), respectively. Our simulation was noticeably different from their work in some aspects. It might be come from the dielectric function measurement treatment, meshing of metal-semiconductor model, and environment of metal-semiconductor. Since Au-NP was examined on the center within short gap-distance from nanorod, then a broad band arose around 500 nm to 550 nm. This peak is distinctly the Au-NP itself by comparing to the pure-gold spectra, indicating by dash-line. Increasing the gap distance between ZnO and gold does not affect the LP spectral significantly, both the intensity and LSPRs peak. On the other hand, the TP spectra lead to decrease their intensity as the gap distance increases and LSPRs peak position are slightly blue-shifted as the gap increases from 0 nm (λ_{LSPR} = 515 nm) to 20 nm (λ_{LSPR}= 510 nm). Then, if the gap distance is much greater than 20 nm the LSPRs peak remains stable at 509 nm.

Figure 3 presents the extinction spectra of ZnO-NR and Au-NP along z-axis as shown in the inset figure. In can be seen that both LP and TP could enhance the ZnO intensities when Au-NP is put at the tip of nanorod. The LP spectra shows that increasing the gap-distance between endcap of nanorod and Au-NP significantly decreases the intensity whilst the LSPRs peak unchanged at 514 nm. However, in TP spectra, it seems that the gap-distance increases to 40 nm, so both intensity and LSPRs peak constantly stands at around $\sim 2 \times 10^3$ nm^2 and ~507 nm, respectively. Interestingly, the LSPRs peak of

LP is red-shifted or lowering their energy compared to Au-NP's peak (~508 nm), but TP's LSPR no change was observed.

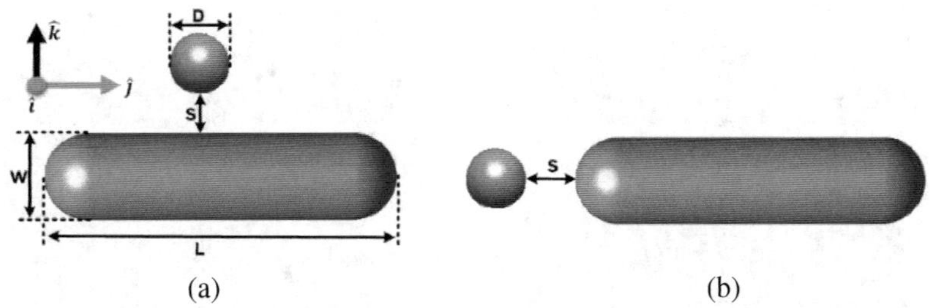

Figure 1. The schematic diagram representation of ZnO-NR with Au-NP on (a) side mode and (b) head mode. Gold's diameter *(D)* is 40 nm, gap between them *(s)* is 0 nm to 40 nm with increment 10 nm, and length *(L)* and width *(w)* of ZnO-NR are 280 nm and 70 nm, respectively.

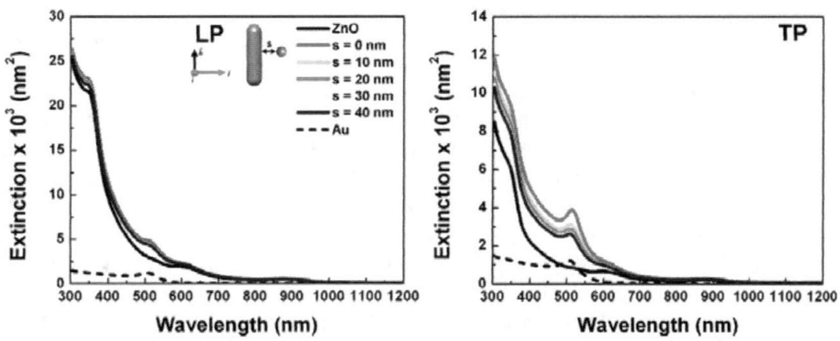

Figure 2. The extinction spectra of ZnO-NR coupled with Au-NP situated perpendicularly to long-axis of nanorod in Longitudinal Plasmon (LP) mode (left) and Transverse Plasmon (TP) mode (right).

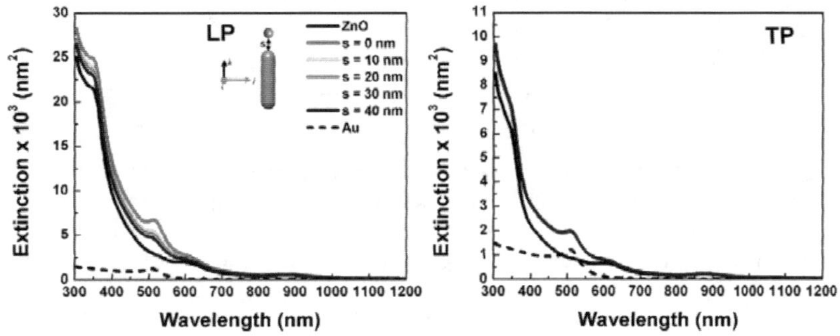

Figure 3. The extinction spectra of ZnO-NR coupled with Au-NP situated next to long-axis of nanorod in Longitudinal Plasmon (LP) mode (left) and Transverse Plasmon (TP) mode (right).

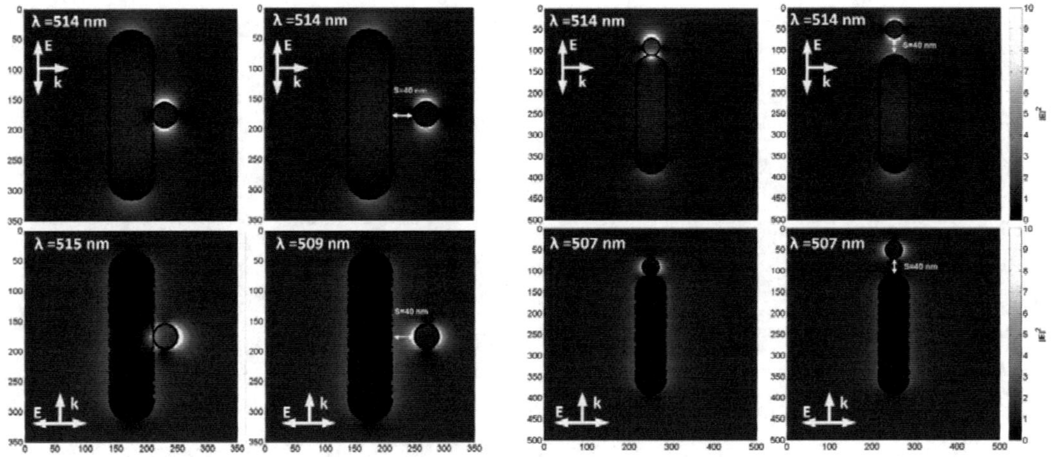

(a) (b)

Figure 4. Near-field map of metal-semiconductor modeled by Au-NP (a) at side of ZnO-NR and (b) upper side of ZnO-NR, within two excitation modes: Longitudinal Plasmon (LP) mode (upper row) and Transverse Plasmon (TP) mode (lower row).

Next, we examined the near-field map distribution of metal-semiconductor model, as shown in figure 4. The configuration of Au-NP at the tip of ZnO-NR (LP mode) was the highest energy electromagnetic (EM) density with photon energy around 2.41 eV (~514 nm). However, on TP mode where Au-NP put at the sides of ZnO exhibiting a strong EM energy density with photon energy 2.40 eV (~515 nm). High EM energy density is due to the confinement in a small region between Au-NP and ZnO-NR when their position are in line with electric part polarization. Furthermore, the confinement of local electric field surround the noble metal (Au) and semiconductor (ZnO) can effectively enhance the photocatalytic activity of ZnO to hinder the electron-hole pair recombination [16]. In this case, the LSPR effect from Au can support and enhance near electromagnetic field intensity to increase the photoreaction [17]. For example, in figure 4a (lower row) and figure 4b (upper row), it can clearly be observed that high EM energy intensity confined among Au and ZnO surface than other surface. This result is similar with Chen and co-worker, who did the gold coated and uncoated SiO_2 above TiO_2 thin film. In fact, the plasmon inducing free electron from gold to TiO_2 was prevented by SiO_2. However, their profile shows high and intense EM energy density than gold without SiO_2 shell. In other word, the LSPR from gold can aid to produce more exciton inside TiO_2 to compensate for the loss electron trap effect due to SiO_2 insulation [17].

4. Conclusion

In conclusion, we have systematically observed the LSPR phenomena of metal-semiconductor model. The results show that the spectral of ZnO-NR raises when Au-NP is placed on the near surface of ZnO. According to electric field map, the LSPR effect from Au-NP effectively enhances both spectral and electromagnetic energy intensity due to the confinement of strong electric field for the given polarizations.

Acknowledgments
We are grateful to the Research and Public Service Community, Universitas Indonesia for funding support via Hibah Publikasi Internasional Terindeks (PIT-9) grant with contract number: NKB-0023/UN2.R3.1/HKP.05.00/2019.

References

[1] Becker J, Trugler A, Jakab A, Hohenester U and Sonnichsen C 2010 *Plasmonics* **5** 161-7
[2] Gu G H, Kim M Y, Yoon H J and Suh J S 2014 *Bull. Korean Chem. Soc.* **35** 725-30
[3] Notarianni M, Vernon K, Chou A, Aljada M, Liu J and Motta N 2014 *Solar Energy* **106** 23-7
[4] Zhou N, Polavarapu L, Gao N, Pan Y, Yuan P, Wang Q and Xu Q-H 2013 *Nanoscale* **5** 4236-41
[5] Iwantono I, Saad S K M, Anggelina F, Awitdrus A, Ramli M A and Umar A A 2019 *Physica E: Low Dimens. Syst. Nanostruct.* **111** 44-50
[6] Putri A E, Fauzia V and Roza L 2019 *Mater. Res. Express* **6** 084008
[7] Liu J, He H, Xiao D, Yin S, Ji W, Jiang S, Luo D, Wang B and Liu Y 2018 *Materials* **11** 1833
[8] Roza L, Fauzia V and Rahman M Y A 2019 *Surf. Interfac.* **15** 117-24
[9] Chou H-T, Huang W-H, Wu T-M, Yu Y-K and Hsu H-C 2017 *Sens. Biosensing Res.* **14** 17-20
[10] Ren C, Yang B, Wu M, Xu J, Fu Z, lv Y, Guo T, Zhao Y and Zhu C 2010 *J. Hazard. Mater.* **182** 123-9
[11] Bertoni G, Fabbri F, Villani M, Lazzarini L, Turner S, Tendeloo G V, Calestani D, Gradečak S, Zappettini A and Salviati G 2016 *Sci. Rep.* **6** 19168
[12] Hohenester U and Trügler A 2012 *Comput. Phys. Commun.* **183** 370-81
[13] Johnson P B and Christy R W 1972 *Phys. Rev. B* **6** 4370-9
[14] Stelling C, Singh C R, Karg M, König T A F, Thelakkat M and Retsch M 2017 *Sci. Rep.* **7** 42530
[15] Musa I, Qamhieh N and Mahmoud S T 2017 *Results Phys.* **7** 3552-6
[16] Khan R, Uthirakumar P, Bae K-B, Leem S-J and Lee I-H 2016 *Mater. Lett.* **163** 8-11
[17] Chen J-J, Wu J C S, Wu P C and Tsai D P 2012 *J. Phys. Chem. C* **116** 26535-42

IOP Conf. Series: Materials Science and Engineering **902** (2020) 012059 doi:10.1088/1757-899X/902/1/012059

Calculation of anisotropic reflectance of $SrNbO_{3.4}$ within Density Functional Theory

G Parusa, A Syahroni and M A Majidi

Department of Physics, Faculty of Mathematics and Natural Sciences (FMIPA), Universitas Indonesia, Depok 16424, Indonesia

Corresponding author's email: aziz.majidi@sci.ui.ac.id

Abstract. A recent experimental study on $SrNbO_{3.4}$ material has revealed a very pronounced anisotropy of its complex dielectric function. This material shows a metallic characteristic along ***a*** crystal direction but behaves as an insulator along ***b*** and ***c*** crystal directions. To investigate this interesting phenomenon, we perform first-principles calculations within density functional theory, to theoretically produce the complex dielectric functions along those crystal directions, and predict the anisotropy of the corresponding reflectance. This result ensures its anisotropic optical property and can explain the unique feature of this material under exposure of electromagnetic radiation.

Keywords: Anisotropic reflectance, dielectric functions, density functional theory

1. Introduction

Among the research interests in plasmonic technology is the search of material which can be used as a good optical switch. The capability to play as a good optical switch may require a low switching power density with yet a high switching contrast. Investigations on isotropic materials such as gold (Au) show that the optical switching functionality of that material demands a high switching power density but performs a low switching contrast [1–6]. Those findings suggest that one needs to search for materials with anisotropic optical characterstics.

Rather than using isotropic materials, one can consider a material that has different optical response characteristic along each crystal direction. Materials which exhibit this feature are called the anisotropic materials, and one of them is the $SrNbO_{3+\delta}$ family. The layered perovskite-related niobates ($n = 5$) $SrNbO_{3.4}$ has been studied and found that it has an interesting optical properties. This feature is firstly recognized by the experimental study using Angle-Resolved Photoemission Spectroscopy (ARPES) which informs that perovskite-related $Sr_{1-y}La_yNbO_{3.5-x}$ has one-dimensional Fermi surface along Γ-X direction [7]. Furthermore, experimental study of its electronic and vibrational properties also show that $SrNbO_{3.41}$ and $SrNbO_{3.45}$ exhibit a metallic behavior at room temperature with strongly dispersing band along ***a*** crystal direction, and insulating with non-dispersive character along ***b*** direction [8]. Recently, another experimental study also reveals that under a very low power density of pump-probe microscopy, this material displays the reflectance reduction only if the pumping electric field is polarized along the ***b*** crystal direction while the probing field is polarized along the ***a*** crystal axis [9]. By these findings, one can conclude that the $SrNbO_{3.4}$ has a highly anisotropic property, where along ***a*** crystal axis it behaves as a metal, while along ***b*** and ***c*** it is an insulator.

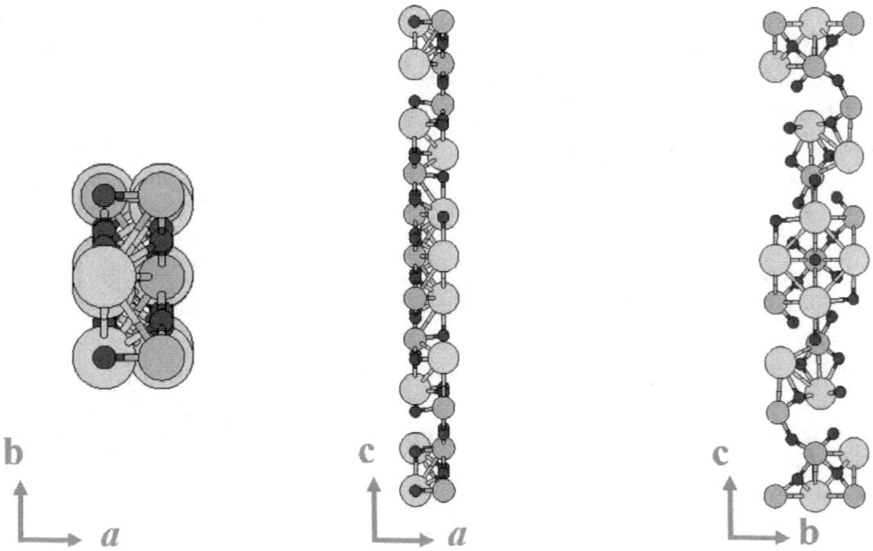

Figure 1. Crystal structure of $SrNbO_{3.4}$. The cyan balls represent the strontium atom, the gray balls represent the niobium atom and the red balls represent the oxygen atom.

Here, our study is aimed to explore the optical properties of such an anisotropic material. For this purpose, we propose to perform first-principles calculation to generate the reflectance of $SrNbO_{3.4}$, as this quantity is easily understood to interpret the characteristic of any materials. By considering its reflectance curve along each crystal direction, one would expect to know how this material responds to the external electromagnetic radiation and understand the meaning of anisotropic property of the crystal itself.

2. Computational method

Since reflectance depends on the real and imaginary parts of complex dielectric function, we need to calculate those spectra before going beyond. We use Density Functional Theory (DFT) [10, 11] implemented in Quantum ESPRESSO package [12] to get the complex dielectric function within independent particle level. An important thing to note in this calculation is that the complex dielectric function, especially the imaginary part one, is sensitive to the number of k-points sampling [13–15]. Therefore, we do the convergence study first before generating the reflectance spectra. We use the $N \times N \times 3$ Γ-centered Monkhorst-Pack k-points meshes and take $N = 12, 18, 24, 30$ in each calculation.

For DFT calculation, we implement the Optimized Norm-Conserving Vanderbilt Pseudopotential (ONCV) [16] from Schlipf-Gygi norm-conserving pseudopotential library [17]. Those pseudopotentials treat Sr-4*s*4*p*5*s*, Nb-4*s*4*p*5*s*4*d*, and O-2*s*2*p* as valence states. The cut-off kinetic energy of the wavefunctions is set to 50 Ry while cut-off kinetic energy of charge density and potential is set to the default value. The atomic position of perovskite ($n = 5$) $SrNbO_{3.4}$ is taken from ref. [18], as in figure 1, and the Fermi level is adjusted by using variable `tot_charge` in order to get the same result as the experiment [7, 8, 19]. We use `tot_charge` $= 0.72$ as this value leads to a minimum criterion to capture the anisotropic characteristic of this material.

Because of the anisotropic property of the material, we cannot consider the optical response function only along a certain direction. We must also investigate the complex dielectric function along other crystal axes and then conclude which k-points grid yields the converged spectra. We then generate the reflectance spectra after having the optimum k-points grid in hand.

3. Results and discussion

The results of convergence study are summarized in figure 2a to figure 2c. One important thing is that these spectrum are plotted from $\omega = 0.025$ eV instead since the resulting data are not well-defined at $\omega = 0$ eV. By considering these results, one can notice that along ***a*** crystal axis it has the Drude peak which indicates the metallic behavior, while along the other directions this characteristic does not appear. This remarkable result yields the conclusion that this material has a highly anisotropic property. Focusing attention on the convergence study of ε_2, the converged spectra is obtained by using $N = 18$ k-points grid for ε_2 along ***b*** and ***c*** crystal axes. However, this value is surprisingly different if one considers the ε_2 along ***a*** crystal axis.

Along this direction, there are several discrepancies as the effect of varying k-points sampling. First to note is the peak around 0.4 eV. Under $N = 12$, this peak is still split into two peaks. By increasing the number of k-points sampling, one can achieve the converged spectra by using $N = 18$. The second to note is the peak around 1.0 eV. Increasing the number of k-points sampling leads to the red-shifted peak even after using $N = 30$. This trend suggests us to not using the smaller number of k-points, and by carefully inspecting the peak position under k-points grid variations one can regard whether $N = 24$ or $N = 30$ as the converged spectra. The third thing to consider is the Drude peak. Because the limited range of data, one cannot actually investigate the Drude peak position. However, the position can be

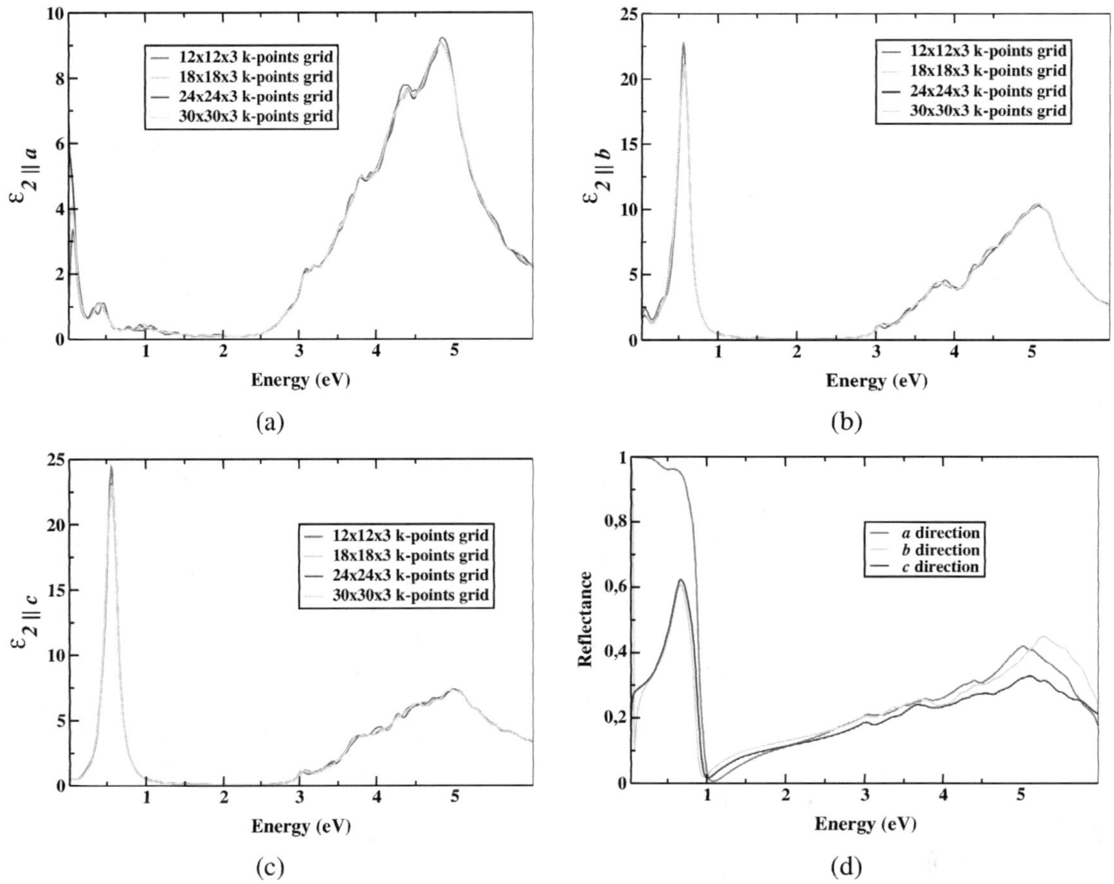

Figure 2. Optical response functions of $SrNbO_{3.4}$. Panels (a) to (c) show the results of convergence study of complex dielectric function along each crystal axis with respect to the number of k-points sampling. Panel (d) shows the resulting reflectance spectra of the material under optimum k-points grid.

regarded as the extrapolation of the well-defined data in the spectra. From this point of view, it can be noticed that there is an anomaly as the number of k-points increases. At $N = 24$, the peak position reaches the highest value while at $N = 30$ the peak position decreases. It means that we must consider higher number of k-points in the convergence study, but it is not possible since we are limited by our computing resources. Hence, we must choose between $N = 24$ or $N = 30$ which yields the better spectra. From physical point of view, calculation within DFT is done at independent particle level and therefore one should encounter the independent particle characteristic in its dielectric function. The $N = 24$ spectra shows a good agreement with that because it displays the similar crossing behavior with y axis of the graph, unlike the $N = 30$ spectra which exhibits the very rare behavior to be called as the Drude peak. By this argument, and concluding the overall discussion of convergence study, one can state that in order to get the converged spectra, the $N = 24$ is the trustworthy value physically and numerically.

By using the results of the convergence study described above, the reflectance spectra is generated as in figure 2d. This optical quantity ensures the anisotropic property of this material, that it perfectly reflects the photon which has the frequency below the plasma frequency ($\omega_p \approx 1.0$ eV) along the $\boldsymbol{a}$ crystal direction and poorly reflects the photon on the other directions. In other words, we have the material which is metal and insulator at the same time which depends on what direction we consider. Nevertheless, this calculation still lacks corrections. As stated earlier, the DFT calculates the complex dielectric function independent of particle level and therefore we do not include the effect of electron-electron and electron-hole interactions yet. However, as a first study, this reflectance spectra is a good reference to interpret the physical property of highly anisotropic material.

4. Conclusion

In conclusion, we have presented a DFT study to calculate the dielectric functions and the reflectance of $SrNbO_{3.4}$ as a representative of a highly anisotropic material. The converged spectra of complex dielectric function which is sensitive to the number of k-points sampling is achieved after using the $24 \times 24 \times 3$ Monkhorst-Pack k-points meshes. This choice is obtained not only by considering the numerical cost, but also the physical meaning behind it. This number of k-points sampling yields the reflectance spectra which becomes the strong evident that the $SrNbO_{3.4}$ material has a highly anisotropic property. Despite the meaningful results obtained in this study, we acknowledge that our results are subject to some corrections as we have not yet included either electron-electron or electron-hole interactions in this study.

Acknowledgments

We thank Universitas Indonesia for providing us a full funding for this project through PITTA A Research Grant NKB-0442/UN2.R3.1/HKP.05.00/2018. We also acknowledge the Indonesian Institute of Sciences (LIPI) for HPC facilities which are used to perform the computational calculation in addition to our own computing cluster at the Theoretical and Computational Condensed-Matter Physics (TCMP) Laboratory, Department of Physics, Universitas Indonesia.

References

[1] MacDonald K F, Sámson Z L, Stockman M I and Zheludev N I 2009 *Nat. Photonics* **3** 55–8
[2] Sámson Z L, MacDonald K F and Zheludev N I 2009 *J. Opt. A: Pure Appl. Opt.* **11** 114031
[3] Wurtz G A, Pollard R, Hendren W, Wiederrecht G P, Gosztola D J, Podolskiy V A and Zayats A V 2011 *Nat. Nanotechnol.* **6** 107–11
[4] Abb M, Albella P, Aizpurua J and Muskens O L 2011 *Nano Lett.* **11** 2457–63
[5] Vasa P, Wang W, Pomraenke R, Lammers M, Maiuri M, Manzoni C, Cerullo G and Lienau C 2013 *Nat. Photonics* **7** 128–32
[6] Wu H Y *et al.* 2016 *Sci. Rep.* **6** 24293
[7] Kuntscher C A, Gerhold S, Nücker N, Cummins T R, Lu D H, Schuppler S, Gopinath C S, Lichtenberg F, Mannhart J and Bohnen K P 2000 *Phys. Rev. B* **61** 1876–83

[8] Kuntscher C A, Schuppler S, Haas P, Gorshunov B, Dressel M, Grioni M and Lichtenberg F 2004 *Phys. Rev. B* **70** 245123
[9] Asmara T C et al Photoinduced metastable exciton-driven metal-insulator transition in quasi-one-dimensional transition metal oxides, *under reviewed* (2020)
[10] Hohenberg P and Kohn W 1964 *Phys. Rev.* **136** B864–71
[11] Kohn W and Sham L J 1965 *Phys. Rev.* **140** A1133–38
[12] Giannozzi P et al 2009 *J. Phys.: Condens. Matter* **21** 395502
[13] Rocca D, Ping Y, Gebauer R and Galli G 2012 *Phys. Rev. B* **85** 045116
[14] Reining L, Olevano V, Rubio A and Onida G 2002 *Phys. Rev. Lett.* **88** 066404
[15] Aguilera I, Vidal J, Wahnón P, Reining L and Botti S 2011 *Phys. Rev. B* **84** 085145
[16] Hamann D R 2013 *Phys. Rev. B* **88** 085117
[17] Schlipf M and Gygi F 2015 *Comput. Phys. Commun.* **196** 36–44
[18] Abrahams S C, Schmalle H W, Williams T, Reller A, Lichtenberg F, Widmer D, Bednorz J G, Spreiter R, Bosshard C and Günter P 1998 *Acta Cryst. B* **54** 399–416
[19] Kuntscher C A, Schuppler S, Haas P, Gorshunov B, Dressel M, Grioni M, Lichtenberg F, Herrnberger A, Mayr F and Mannhart J 2002 *Phys. Rev. Lett.* **89** 236403

The diameter effect on the magnetization switching time of sphere-shaped ferromagnets using micromagnetic approach

C Kurniawan[1, 2], A T Widodo[1] and D Djuhana[1]

[1]Department of Physics, Faculty of Mathematics and Natural Sciences (FMIPA), Universitas Indonesia, Depok 16424, Indonesia
[2]Research Center for Physics, Indonesian Institute of Sciences (LIPI), Puspiptek Office Area Tangerang Selatan 15314, Indonesia

Corresponding author's email: dede.djuhana@sci.ui.ac.id

Abstract. In this work, the magnetization switching time has been studied as the diameter variation of sphere-shaped ferromagnets by means of micromagnetic simulation. Some ferromagnetic elements such as cobalt, iron, nickel, and permalloy were numerically simulated with diameter size variation from 50 nm to 100 nm. The micromagnetic simulation was performed by public software Object Oriented Micromagnetic Framework (OOMMF) based on the Landau-Lifshitz-Gilbert (LLG) equation. The ferromagnetic nano-sphere was induced by the quasi-static magnetic field to observe its magnetization response. Generally, it is observed that the switching time increases as the diameter size increases on the ferromagnetic elements. However, the switching time are relatively insensitive as the diameter increases in the Cobalt element for the diameter range from 60 nm to 90 nm. This behavior is contributed to the distribution of magnetization easy axis on the ferromagnetic elements. The understanding of domain structures during magnetization switching process is important in the development of nano-patterned magnetic memory storage.

Keywords: Magnetization switching, sphere, ferromagnetics, micromagnetics

1. Introduction

The advancement of the magnetic recording technology in the last 20 years have made the dynamic of ferromagnetic nanomaterials are of the important research topics on the developing of Magnetoresistive Random-Access Memory (MRAM), microwave oscillators, and magnetic nanosensors as the potential applications [1-4]. One of the interesting issues in the development of granular magnetic recording media is the understanding of magnetization reversal in individual grain or element [5, 6]. It is known that the advantages of using individual magnetic grain for storing data is the reduction of bit transition jitter and increase the signal to noise ratio compared to the regular perpendicular magnetization media [7]. Another important attempt to improve the data density is by using the high magnetic anisotropy materials in the exchange-coupled system [8]. Therefore, the magnetic domain formation and evolution of individual grain with magnetic anisotropy variation are of the interesting approaches to improve our understanding in granular storage media application.

In magnetic materials, the magnetic properties are also contributed by the distribution of magnetic element size, shape, and the interelement interaction. There are many studies using different magnetic element shape, such as circle [9], square [10], nanoring [11] and Reuleaux's triangle [12]. However, the

IOP Conf. Series: Materials Science and Engineering **902** (2020) 012060 doi:10.1088/1757-899X/902/1/012060

magnetization switching of sphere-shaped ferromagnetic element was rarely studied, although many of the nanoelement ferromagnetic material was a spherical shape.

In this work, the magnetization switching time has been studied as the diameter variation of sphere-shaped ferromagnets by means of micromagnetic simulation. Some kinds of basic ferromagnetic materials, such as cobalt, iron, nickel, and permalloy were numerically simulated under a quasi-static magnetic field induction to obtain the magnetic hysteresis curve. The switching phenomena are investigated by observing the domain structure evolution and time consumed during magnetization switching process, which called as the switching time.

2. Micromagnetic model

The micromagnetic simulation based on the Landau-Lifshitz-Gilbert equation was performed by public software OOMMF with damping constant (α) of 0.1 and the cell size of 2.5 × 2.5 × 2.5 nm^3 by considering the exchange length (l_{ex}) of the materials [13, 14]. Some ferromagnetic elements used in the simulation, such as cobalt (Co), iron (Fe), nickel (Ni) and permalloy (Py) with the material parameters are depicted in table 1. The materials Co, Fe, Ni, and Py were chosen as they are the basic elements (and alloy for Py) of ferromagnetic materials. The sphere-shaped model was numerically simulated with diameter size varied from 50 nm to 100 nm as shown in figure 1. Such the size range was modeled as the representation of single-domain particle (smaller) and multi-domain particle (larger), as predicted by Kittel [15]. The ferromagnetic nano-sphere was induced by the quasi-static magnetic field to observe its magnetization response and produce a hysteresis curve. The domain structure was investigated using the time-resolve imaging technique during reversal process. The switching magnetization field and time were observed based on the produced hysteresis curve and compared to the time-resolved images.

Table 1. The material parameters used in the micromagnetic simulation.

Materials	M_s [A/m]	A [J/m]	K [J/m^3]	l_{ex} [nm]	Crystal structure	Ref.
Cobalt (Co)	1,400 × 10^3	14 × 10^{-12}	53 × 10^4	4.7	hcp	[16]
Iron (Fe)	1,700 × 10^3	21 × 10^{-12}	4.8 × 10^4	3.3	bcc	[12]
Nickel (Ni)	490 × 10^3	9 × 10^{-12}	-0.57 × 10^4	7.6	fcc	[12]
Permalloy (Py)	860 × 10^3	13 × 10^{-12}	0	5.1	fcc	[17]

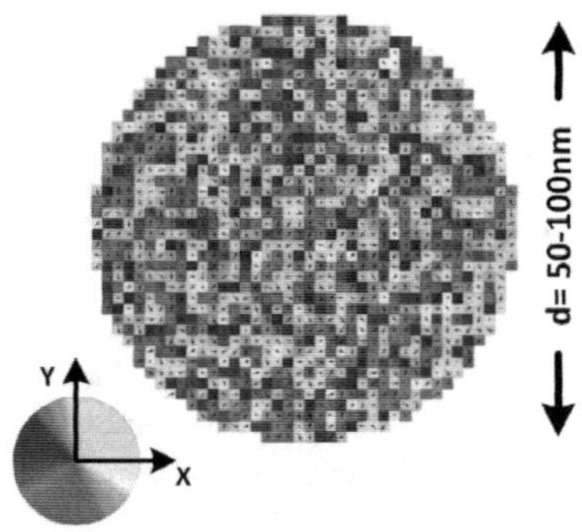

Figure 1. Sphere-shaped ferromagnets with random initial magnetization modeled in the micromagnetic simulation.

3. Results and discussion

The hysteresis curves of the sphere-shaped ferromagnets produced from the simulation are displayed in figure 2. The quasi-static magnetic field was induced in the *x*-direction, which simulated the hard-axis orientation (HAO) of the sphere-shaped ferromagnets. There are some point marked with number 1–4 in the hysteresis curves, that are the saturation point, remanence point, coercivity point, and opposite saturation point. It can be seen that the hysteresis curves are relatively slanted for all ferromagnets as the diameter increases. However, the materials with smaller anisotropy, such as Nickel and Permalloy showed the square shaped hysteresis loop in the smaller size as depicted in figure 2a. It means that the slanted hysteresis curve was depended on the magnetocrystalline anisotropy of the materials. In the larger diameter, the slanted level of hysteresis curves is relatively higher. It means that the magnetic field needs to switch to uniform the saturation magnetization is also increased. At this point, the switching phenomena was expected to have a linear behavior with the slanted level of hysteresis curve.

To improve the understanding of switching phenomena, the evolution of domain structures during switching magnetization process was also investigated. As shown in figure 3, the domain structure of all sphere-shaped ferromagnets at saturation (and the opposites), remanence and coercivity condition are observed. Generally, the switching processes are followed by the vortex domain structure formation, which can be seen at remanence and coercivity condition (point 2 and 3). However, it is observed that the different domain structure can be formed at remanence condition. At smaller diameter size of Co and Ni, the vortex core is not observed at the remanence condition. For Co sphere, this effect is due to the higher anisotropy value, that might be not allowing the formation of vortex core at the smaller size close to single domain particle size. On the other hand, the absence of vortex core for Ni sphere might be contributed by the squareness effect of hysteresis curve. As compared with the figure 2a, the Ni sphere has a higher squareness of hysteresis curve. This behavior caused the domain wall nucleation did not happen in the remanence condition and made the Ni sphere has the higher magnetization remanence value at smaller size. We have expected that the domain nucleation is the important factor during the switching process.

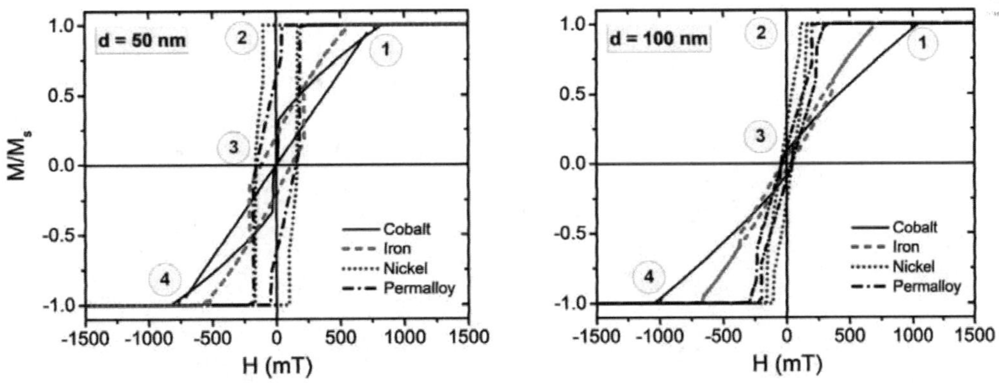

(a) (b)

Figure 2. Hysteresis curve of sphere-shaped ferromagnets at (a) d = 50 nm and (b) d = 100 nm. The bullet number pointing the domain structure observation during switching process.

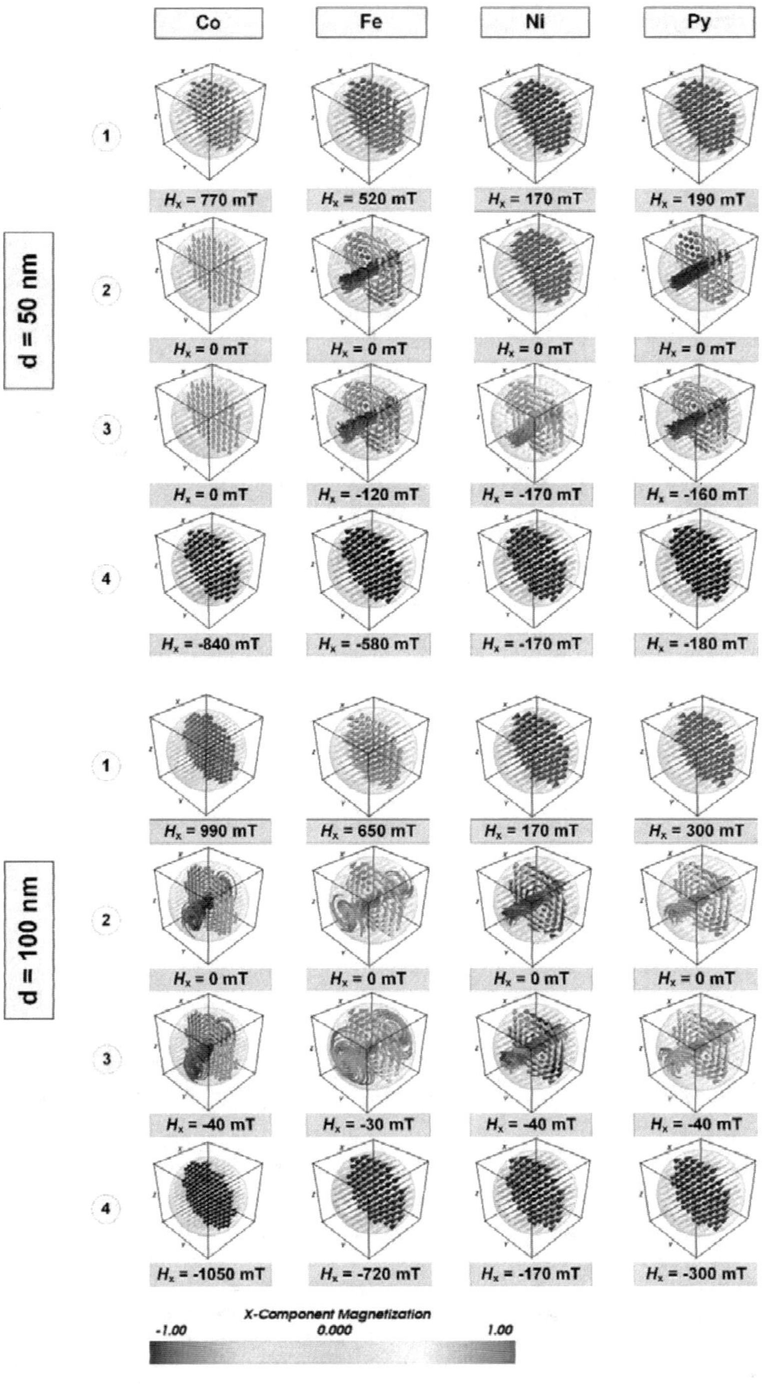

Figure 3. Domain evolution of sphere-shaped ferromagnets during switching process for d = 50 nm and d = 100 nm. Point 1 and 4 show the opposite direction of saturation magnetization. Point 2 and point 3 show the remanence and coercive condition, respectively.

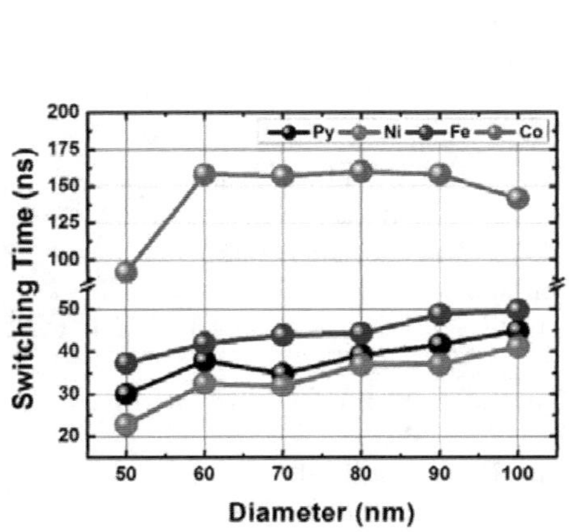

Figure 4. Switching time for sphere-shaped ferromagnets with the diameter variation.

As displayed in figure 4, the switching time of sphere-shaped ferromagnets are plotted as the diameter size increases. Generally, it is observed that the switching time increases as the diameter size increases on the ferromagnetic elements. This result agrees with the previous result of magnetization reversal in hexagonal-shaped $BaFe_{12}O_{19}$ materials [18]. However, the switching time are relatively insensitive as the diameter increases in the Cobalt element for diameter ranges from 60 nm to 90 nm. This behavior is contributed to the distribution of magnetization easy axis on the ferromagnetic elements. It also observed that the switching time for Co has one order value larger compared to the other sphere-shaped ferromagnets. This result agrees with the hysteresis curves produced in the simulation. It can be compared that the slanted hysteresis is linearly correlated with the switching time. Therefore, the larger magnetocyrstalline anisotropy material produced a larger switching time value at hard-axis orientation. This study showed that the understanding of domain structures evolution during the magnetization switching process could be an important step in the development of nano-patterned magnetic memory storage.

4. Conclusion

In this paper, the magnetization switching time has been studied as the diameter variation of sphere-shaped ferromagnets, such as cobalt, iron, nickel, and permalloy by means of micromagnetic simulation using public software OOMMF based on the LLG equation. It was observed that the switching processes were followed by the vortex domain structure formation, which can be seen at remanence and coercivity condition. It was observed that the switching time increases as the diameter size increases on the ferromagnetic elements. However, the switching time are relatively insensitive as the diameter increases in the cobalt element for diameter ranges from 60 nm to 90 nm contributed to the distribution of magnetization easy axis on the ferromagnetic elements. It can be compared that the slanted hysteresis is linearly correlated with the switching time. Therefore, the larger magnetocyrstalline anisotropy material produced a larger switching time value at hard-axis orientation. The understanding of domain structures evolution during the magnetization switching process is important in the development of nano-patterned magnetic memory storage.

Acknowledgments

This work is supported by Indexed International Publication Grant (PIT 9) year 2019 through Directorate of Research and Community Engagement (DRPM) Universitas Indonesia under contract number: NKB-0023/UN2.R3.1/HKP.05.00/2019.

References

[1] Sun L, Zhai Y, Niu D, Xu Y and Zhai H 2015 *IEEE Int. Magn. Conf.* **51** 1-4
[2] Bhatti S, Sbiaa R, Hirohata A, Ohno H, Fukami S and Piramanayagam S N 2017 *Mater. Today* **20** 530-48
[3] Joshi V K 2016 *Eng. Sci. Technol. Int. J.* **19** 1503-13
[4] Sbiaa R and Piramanayagam S N 2017 *Phys. Status Solidi RRL-Rapid Res. Lett.* **11** 1700163
[5] Mu C, Jing J, Dong J, Wang W, Xu J, Nie A, Xiang J, Wen F and Liu Z 2019 *J. Magn. Magn. Mater.* **474** 301-4
[6] Ali S R, Naz F, Akber H, Naeem M, Ali S I, Basit S A, Sarim M and Qaseem S 2018 *Chin. Phys. B* **27** 097503
[7] Oezelt H, Kovacs A, Fischbacher J, Matthes P, Kirk E, Wohlhüter P, Heyderman L J, Albrecht M and Schrefl T 2016 *J. Appl. Phys.* **120** 093904
[8] Oezelt H, Kovacs A, Wohlhüter P, Kirk E, Nissen D, Matthes P, Heyderman L J, Albrecht M and Schrefl T 2015 *J. Appl. Phys.* **117** 17E501
[9] Ding J, Kakazei G N, Liu X, Guslienko K Y and Adeyeye A O 2015 *Sci. Rep.* **4** 4796
[10] Cowburn R P and Welland M E 1998 *Phys. Rev. B* **58** 9217-26
[11] Torres-Heredia J J, López-Urías F and Muñoz-Sandoval E 2006 *J. Magn. Magn. Mater.* **305** 133-40
[12] López-Urías F, Torres-Heredia J J and Muñoz-Sandoval E 2005 *J. Magn. Magn. Mater.* **294** e7-12
[13] Donahue M J and Porter D G 1999 *OOMMF User's Guide, Version 1.0* (Gaithersburg, MD: National Institute of Standards and Technology)
[14] Miltat J E and Donahue M J 2007 Numerical micromagnetics: finite difference methods *Handbook of Magnetism and Advanced Magnetic Materials Volume 2: Micromagnetism* Ed. Kronmuller H et al. (Hoboken: John Wiley & Sons, Ltd.) pp 742-64
[15] Kittel C 1949 *Rev. Mod. Phys.* **21** 541-83
[16] Dennis C L et al. 2002 *J. Phys. Condens. Matter* **14** R1175-262
[17] Piao H-G, Djuhana D, Shim J-H, Lee S-H, Jun S-H, Oh S K, Yu S-C and Kim D-H 2010 *J. Nanosci. Nanotechnol.* **10** 7212-6
[18] Djuhana D, Oktri D and Kurniawan C 2018 *Makara J. Sci.* **22** 198-204

IOP Conf. Series: Materials Science and Engineering **902** (2020) 012061 doi:10.1088/1757-899X/902/1/012061

Electronic structure of 9 quintuple layers Bi_2Se_3 within Density Functional Theory

M Hasan and M A Majidi

Department of Physics, Faculty of Mathematics and Natural Sciences (FMIPA), Universitas Indonesia, Depok 16424, Indonesia

Corresponding author's email: aziz.majidi@sci.ui.ac.id

Abstract. Bismuth Selenide (Bi_2Se_3) is a well-known 3D-type topological insulator. The interior part of this material is insulating, while its surface state is metallic which is protected by time-reversal symmetry. The experimental study of this material by varying its thickness reveals that the metallic surface state starts to appear at a minimum thickness of six Quintuple Layer (QL). This study was performed within the Density Functional Theory (DFT) in which a vacuum gap is added to the multilayer structure in order to capture the surface states of the material. While a previous study has explored the electronic structure of this material upto a thickness of 8 QL, a more detailed information of how the system evolves from this thickness to a fully bulk limit has not much been explored. With this motivation, we propose to calculate the electronic structure of this material with thickness of 9 QL. We perform the DFT calculation using Quantum Espresso (QE) package. Our results show that the band structure of this system reveals the surface metallic states with many additional bands as compared to those of thinner layer systems.

Keywords: Topological insulator, Bi_2Se_3, DFT, surface states, electronic structure

1. Introduction

A topological insulator is a material that has opposite electrical properties in the inside and the surface of the material. The inside part of the material is insulating, but there is a metallic state that occurs on the surface because of the topological invariance of the wave function of the system. This phenomenon occurs due to spin-orbit coupling that drives the electron inside the material. The first study of the topological insulator starts from a simple system called Quantum Hall Droplet. Later, physicists found that topological insulator material should necessarily be built from atoms of nearly-heavy elements and small-band gap semiconductor. The requirement of a nearly heavy element is based on the fact that spin-orbit coupling would become strong enough to modify the band structure of the material, while the small-band gap is needed as the large-band gap material is hard to be modified by spin-orbit coupling [1].

The search of topological insulator was peaked on the material bismuth selenide (Bi_2Se_3) and bismuth telluride (Bi_2Te_3), which have fulfilled the requirements above. Bismuth Selenide is determined to be three dimensional (3D) kind of topological insulator. The boundary of this material is a metallic state, and the bulk is an insulator. There are many experimental and theoretical studies of this material [1, 2].

One of the experimental studies of this material was performed by varying the thickness of the material [3]. The experiment took samples of 1-6 Quintuple Layer (QL) thikness, except 4 QL. The sample with thicknesses of 1-3 QL exhibits small gaps between the conduction band and the valence band. Here, the surface states still appear as a pair of lower and upper band with a small gap separating them. These surface states have to do with rashba-type spin-orbit coupling. The gap at low thicknesses appear due to hybridization between two surface states of opposite surfaces of the material [3, 4]. As the system thickness increases from 2 QL to 3 QL, these opposite surfaces get farther apart resulting in the splitting of the surface states. This splitting continues and evolves to form a band crossing at 5 QL, similar to the Dirac cone in graphene bandstructure. At 6 QL the formation of a nearly linear band crossing at the Γ point becomes more apparent. Thus, the system can be considered to become a well-defined topological insulator starting at 6 QL.

The experimental study of Ref. [3] has motivated a number of further theoretical studies, especially to calculate the electronic structure using DFT. DFT is a common method to study electronic structure of solids or molecules [5]. Of our particular interest is the DFT study of the thickness dependence of bismuth selenide (Bi_2Se_3) up to 8 QL and its influence on the electronic structure and optical properties of the material [4].

This study has simulated the material with thickness variation from 1 up to 8 QL and reveals the metallic surface states in the band structure starting at the thickness of 3 QL. In this present study we would like to extend their study by investigating the electronic structure of the material with the thickness of 9 QL.

2. Computational method

We implement the DFT calculation using Quantum Espresso (QE) package [6]. WIthin the QE package there are two important calculations: SCF (self-consistent field) and NSCF (non self-consistent field). The SCF calculation essentially produces the self-consistent electron density, while NSCF yields the eigenvectors and eigenvalues of the system Hamiltonian. When we run the SCF calculation, we only need small number of k-points (here we use $8 \times 8 \times 1$). While for the NSCF calculation we need larger number of k-points (which in this case we use $32 \times 32 \times 1$).

We use primitive hexagonal cells to model the thin film of Bi_2Se_3 with lattice parameter $a = b = 4.13$ Å and $c = 26.32$ Å [7]. As the algorithm has been designed essentially to handle three-dimensional system, for calculation of quasi two-dimensional system one has to add vacuum spaces on top and bottom of the corresponding supercell. The length of the vacuum space on each side of the supercell needs to be made large enough such that when such a unit is repeated along the direction where the vacuum spaces are inserted it would generate a system with many quasi two-dimensional system with a vacumm gap separating every two adjacent slabs. If each vacuum space is large enough the entire system forms as a collection of independent quasi two-dimensional systems. This way, we can safely consider each individual slab as a good representative of independent quasi two-dimensional system. In our model, we construct a vacuum space with a thickness of 10 times 3 QL on each side of our super cell such that every two adjacent slabs is separated by 20 times 3 QL long vacuum gap. In figure 1, we illustrate how we construct our supercell. Note, however, that the figure only shows a small part of the length of the vacuum space on each side just for the illustration.

The unavailability of the detailed information of atomic positions of the system leads us to determine the atomic positions solely based on the general information of the crystal structure [7]. After doing so we obtain the crystal structure, where we illustrate only for 3 QL thickness as shown in figure 2a. For 3 QL thickness, there are effectively 15 atoms contained in the supercell. The bismuth selenide structure is arranged in one quintuple layer composed of 5 atoms consisting of two Bi atoms and three Se atoms arranged alternately. As the standard procedure in DFT to ensure that we obtain the correct, i.e. stable, atomic positions we need to do structural optimization by *relax* calculation. After the relax calculation, we obtain the optimal structure as shown in figure 2b.

The resulting atomic positions after structural optimization are presented in table 1. In this study we aim to calculate the Bi_2Se_3 system with a thickness of 9 QL. In our calculations we use the cutoff energy of wave function about 40 Ry. We set this value after the convergence test of the cutoff energy (*ecutwfc*) on the system with 3 QL thickness, as shown in figure 3. We accommodate possible Van der Waals interaction between adjacent QL layers in a unit cell by setting the DFT-D3 correction on the input of

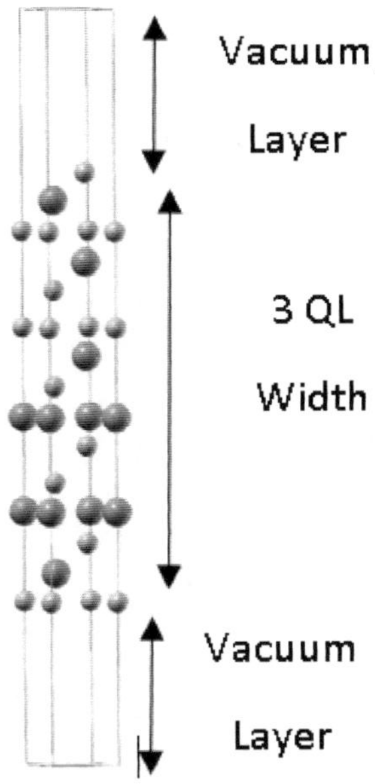

Figure 1. Unit supercell with thickness of 3 QL and vacuum gaps on top and bottom.

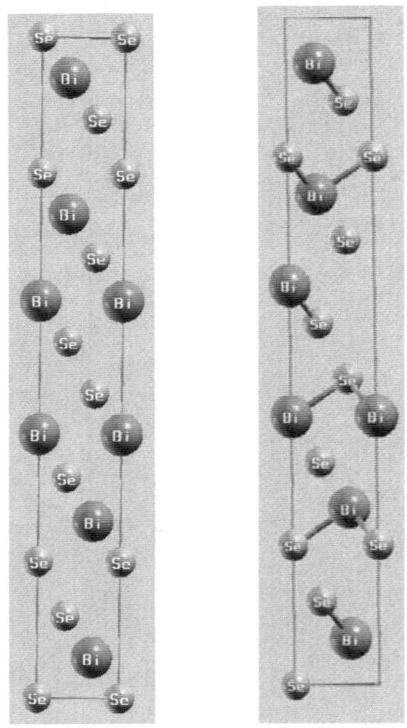

Figure 2. Crystal structure (a) before and (b) after relax calculation.

Table 1. Atomic positions after relax calculation.

Atom	x(Å)	y(Å)	z(Å)
Bi	0.000000000	2.389013277	26.381185952
Se	2.069000000	1.194459769	24.786420167
Se	0.000000000	-0.000006300	22.413978240
Bi	0.000000000	2.389084152	20.819254311
Se	2.069000000	1.194527324	18.880458959
Bi	0.000000000	-0.000011495	16.941658632
Se	0.000000000	2.389061005	15.346917468
Se	2.069000000	1.194534286	12.973383870
Bi	0.000000000	0.000003074	11.378522723
Se	0.000000000	2.389070870	9.439411840
Bi	2.069000000	1.194546022	7.500293317
Se	0.000000000	-0.000006908	5.905492677
Se	0.000000000	2.388992782	3.533618949
Bi	2.069000000	1.194469511	1.938900504
Se	0.000000000	-0.000237370	0.000102391

QE. We use the norm-conserving pesudopotential [8]. The spin-orbit coupling is included by setting-up the non-collinear calculation and using the relativistic pseudopotential. Band structure calculation is started by SCF calculation using automatic mode k-points generation, where in our case we use $8 \times 8 \times 1$ k-poins. We then continue with NSCF calculation using k-points in `crystal_b` mode for Γ , K, and M high symmetry points that are listed in table 2. This high symmetry points of the hexagonal Brillouine

IOP Conf. Series: Materials Science and Engineering **902** (2020) 012061 doi:10.1088/1757-899X/902/1/012061

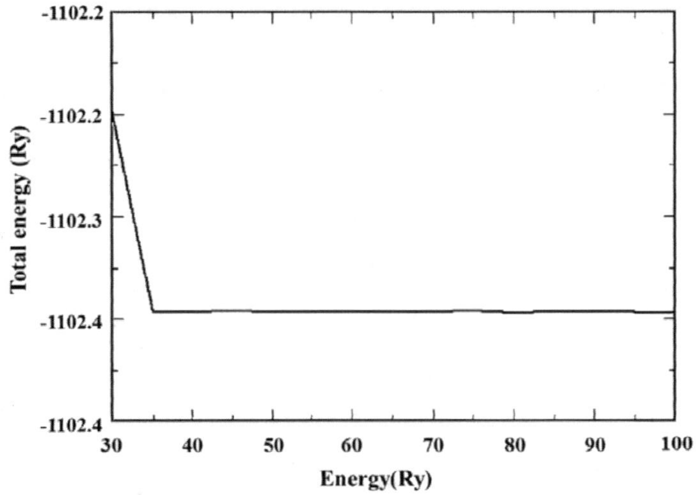

Figure 3. Convergence test for cutoff energy of plane wave.

Table 2. Three high symmetry points of the Brillouin zone of 9 QL Bi_2Se_3.

Points	x	y	z
Γ	0.000	0.000	0.000
K	0.000	0.500	0.500
M	0.500	0.000	0.500

zone are illustrated in figure 4. As seen in this figure, the Γ point is the center of the Brillouin zone, while the K point resides in the middle of the line connecting between the corners of the Brillouin zone, and the M point is located at the corner of the Brillouin zone.

3. Results and discussion

After running SCF, NSCF, and bands calculation, we obtain the band structure of 9 QL bismuth selenide as shown in figure 5. There are 518 bands calculated through quantum espresso package, and we include all bands to be plotted. The Fermi energy of this system is found to be 5.7934 eV. This Fermi energy line crosses a band, namely it does not lie inside a gap, which indicates that the states represented by this band of the system is metallic. However, it is important to note that the band in which the Fermi energy is located only comes from the surface states, whereas all the bands corresponding to the bulk states form groups of lower and upper bands that are very well separated by a gap, indicating that the bulk part of the system is insulating. The surface band exhibits a crossing at the Γ point similar to that of the Dirac cone in graphene band structure. This Dirac cone-like strcuture confirms the topological insulator behavior of the system, as explained before. Other features of this band structure are inverted bands near the Γ point. These inverted bands occur as a consequence of the presence of spin-orbit coupling included in the calculation consistent with that explained explained in Ref. [2]. The result of this study confirms that the thin film with 9 QL thickness exhibits metallic surface states with the bulk states having well defined energy gap, characterizing the topological insulator. Of course we obtain more bands as compared to the systems with less number of QL layers.

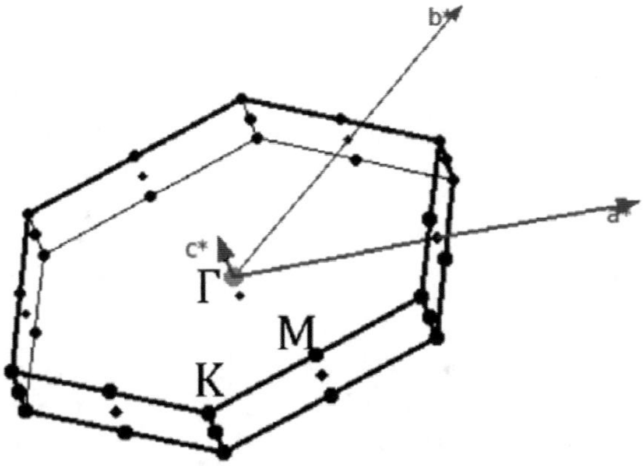

Figure 4. Brillouin zone of thin film with thickness of 9 QL.

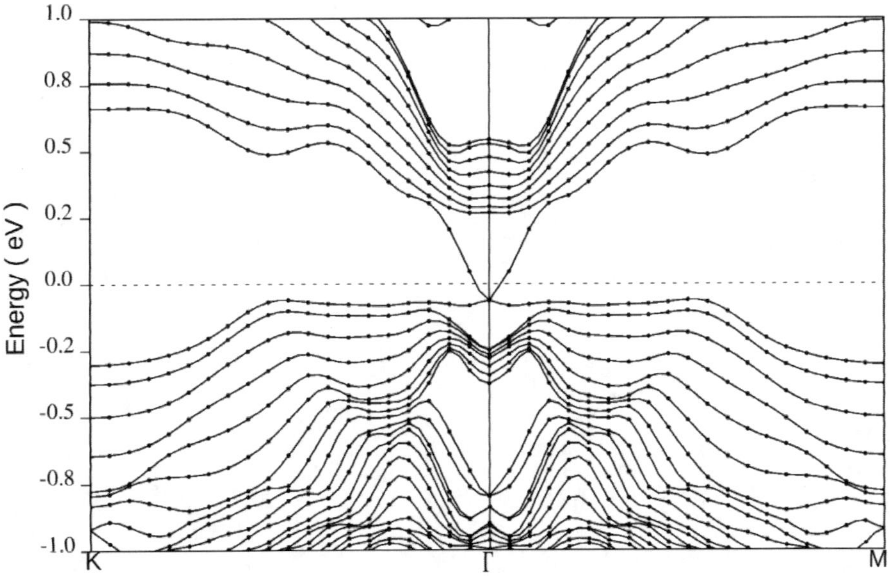

Figure 5. Band structure of Bi_2Se_3 thin film with thickness of 9 QL.

4. Conclusion

We have presented our DFT study for a 9 QL thickness Bi_2Se_3. From our calculation results, it is clear that the system we study exhibits metallic states on the surface of the system while maintaining the bulk part of the band structure with a well-defined energy gap. In addition, we obtain all the expected features of the topological insulating characteristics of Bi_2Se_3, such as the Dirac cone-like structure and the inverted bands. Our overall results confirm the properness of the computational method we implement with the Quantum Espresso package.

Acknowledgments
We are very grateful to Universitas Indonesia for providing us full funding for this project through PITTA Research Grant No. NKB-0644/UN2.R3.1/HKP.05.00/2019. The computational work is partially supported by HPC-LIPI service and the Theoretical and Computational Condensed-Matter Physics (TCMP) Lab at the Department of Physics of FMIPA Universitas Indonesia.

References
[1] Moore J E 2010 *Nat. Phys.* **464** 194-8
[2] Zhang H, Liu C X, Qi X L, Dai X, Fang Z and Zhang S C 2009 *Nat. Phys.* **5** 438-42
[3] Zhang Y et al 2010 *Nature Physics* **6** 584-8
[4] Li Z, Chen S, Sun J, Li X, Qiu H and Yang J 2018 *J. Phys.: Condens. Matter* **30** 065503
[5] Kohn W and Sham L J 1965 *Phys. Rev.* **140** A1133-8
[6] Giannozzi P et al. 2009 *J. Phys.: Condens. Matter* **21** 395502
[7] Jensen M R 2014 *Understanding the low energy physics of bismuth selenide: A three-dimensional topological insulator* Masters thesis (København: Niels Bohr Institute, Faculty of Science, University of Copenhagen)
[8] Hamann D R, Schlüter M and Chiang C 1979 *Phys. Rev. Lett.* **43** 1494-7

Simulation of critical temperature of FeTe, FeSe and $FeTe_{0.5}Se_{0.5}$ using equation McMillan

D P U Asmarakusuma and D Suyanto

Department of Physics, Faculty of Mathematics and Natural Sciences (FMIPA), Universitas Indonesia, Depok 16424, Indonesia

Corresponding author's email: dsuyanto@fisika.ui.ac.id

Abstract. This study uses simulation to produce the critical temperature of FeTe, FeSe, and $FeTe_{0.5}Se_{0.5}$ superconductors using McMillan equation which show shifts in λ (electron-phonon coupling strength), ω_{log} (logarithmic average of phonon frequency), and Density of States (DOS) at the Fermi level compared to its composing elements. Comparison of critical temperatures shows that Fe mixed with Se and/or Te experience some changes that could possibly increase of the critical temperature.

Keywords: Critical temperature, McMillan equation, simulation

1. Introduction

Since the discovery of superconductivity on FeSe [1] , many studies have been conducted in order to understand the superconductivity of Fe-chalcogenides. Fe-chalcogenides have several types of crystal structures such as the anti-PbO-type and the NiAs-type. The common compounds of Fe-chalcogenides that have been mostly observed is the one with the anti-PbO-type crystal structures. Several compounds of Fe-chalcogenides that have this type of crystal structure are FeS, FeSe and FeTe [2]. In this phase, FeSe exhibits superconductivity while FeTe does not show any superconductivity [3]. However, if Se is doped into the material, $FeTe_{1-x}Se_x$ shows superconductivity. Thus, in this study we aim to calculate the critical temperature of FeSe, FeTe, $FeTe_{0.5}Se_{0.5}$ and its constituent element (Fe, Te, and Se). Furthermore, we also want to take into account: changes of electron-phonon coupling strength, logaritmic average of phonon frequency, and the density of states to the magnitude of the critical temperature. Here, we use simulation then the apply the McMillan equation to obtain the critical temperature of each material.

2. Method

McMillan equation is based on the Eliashberg theory [4] which is an extension from BCS theory [5]. The main point of Eliashberg theory is the Coulomb repulsion which is defined as,

$$\mu = N(0)|V_c| \quad (1)$$

where μ is the instantaneous Coulomb repulsion, $N(0)$ is the density of single spin from the electronic state at the Fermi level and $|V_c|$ is the potential. Besides the Coulomb repulsion, another main point of Eliashberg theory is the spectral function of electron phonon which defined as,

$$\alpha^2F(\omega) = N(0)\frac{\sum_{\boldsymbol{kk}'}|M_{\boldsymbol{kk}'}|^2\delta(\omega-\omega_{\boldsymbol{Q}!})\delta(\epsilon_{\boldsymbol{k}})\delta(\epsilon_{\boldsymbol{k}'})}{\sum_{\boldsymbol{kk}'}\delta(\epsilon_{\boldsymbol{k}})\delta(\epsilon_{\boldsymbol{k}'})} \quad (2)$$

where $\alpha^2F(\omega)$ is the spectral function of electron-phonon, $\boldsymbol{Q} = \boldsymbol{k} - \boldsymbol{k}'$ with $\boldsymbol{Q}$ and $\boldsymbol{k}$ being the wave number and the band quantum number for phonon and electron, $\omega_{\boldsymbol{Q}}$ and $\epsilon_{\boldsymbol{k}}$ is the energy of phonon and electron respectively and $M_{\boldsymbol{kk}'}$ is the matrix element of electron-sphonon coupling.

McMillan then found a solution of Eliashberg theory for finite temperature to determine the critical temperature, T_c, for any cases. McMillan defined the critical temperature, T_c, as,

$$T_c = \frac{\Theta_D}{1.45}exp\left(-\frac{1.04(1+\lambda)}{\lambda-\mu^*(1+0.62\lambda)}\right). \quad (3)$$

Here, Θ_D is the Debye temperature, λ is the strength of electron-phonon spectral function $\alpha^2F(\omega)$ and μ^* is the effective Coulomb repulsion. However, Dynes [6] then introduced further changes to equation 3 where $\frac{\Theta_D}{1.45}$ is corrected to $\frac{\langle\omega\rangle}{1.20}$ with $\langle\omega\rangle$ is the average phonon energy. Thus, equation 3 is then transformed into,

$$T_c = \frac{<\omega>}{1.20}exp\left(-\frac{1.04(1+\lambda)}{\lambda-\mu^*(1+0.62\lambda)}\right). \quad (4)$$

Then, using McMillan equation, the critical temperature can be calculated as,

$$T_c = \frac{\omega_{log}}{1.2}e^{\left(\frac{-1.04(1+\lambda)}{\lambda(1-0.62\mu^*)-\mu^*}\right)}. \quad (5)$$

where

$$\mu^* = \frac{V_C N(0)}{1 + V_C N(0) ln\left(\frac{E_F}{\omega_D}\right)} \quad (6)$$

and

$$\omega_{log} = e^{\left(\frac{2}{\lambda}\int\frac{d\omega}{\omega}\alpha^2F(\omega)log\omega\right)}. \quad (7)$$

We then perform the simulation of Fe, Te, Se, FeTe, FeSe and $FeTe_{0.5}Se_{0.5}$. Using Quantum Espresso 6.0 in the simulation, we obtain several variables such as density of states, Fermi energy, ω_{log}, λ and the ratio between λ and T_c. Then, the critical temperature for each material is calculated using equation 5 to equation 7.

3. Results and discussion

Using Quantum Espresso, we have simulated the Fe, Te, Se, FeSe, FeTe, and $FeTe_{0.5}Se_{0.5}$ and have obtained the critical temperature (T_c), λ, ω_{log}, density of states at the Fermi level (DOS (E_F)) and the energy Fermi as shown in table 1. Our simulation of Fe shows that Fe has T_c of 0.01 K and a BCC structure. Our results are different with the previous study [7]. The study have shown that Fe at ambient pressure which has a BCC structure do not exhibit superconductivity phenomenon. We also obtain that at the Fermi level the density of states of Fe is 2.43 $eV^{-1}m^{-3}$ while the λ is 1.20. So, the difference between these values for Fe is relatively small. The value of ω_{log} which equals to 25.93 THz is small, thus, it does not give a significant contribution in enhancing the value of the critical temperature of Fe.

The simulations of Te and Se shows that the critical values are 0.64 K and 2.23 K, respectively, and both have the hexagonal crystal structure. Similar to Fe, the results are different with previous study [7] which shows that Te and Se are not superconductor at normal pressure. The values of ω_{log} for both elements is quite large, that is, 73.61 THz and 62.18 THz, respectively. However, the large value of ω_{log} of those elements does not correspond to a large value of critical temperature, T_c. This is due to the small value of λ of both elements in which 0.19 for Te and 0.27 for Se.

For FeTe, we found that the critical temperature, T_c, of FeTe is 0.006 K and the crystal structure of FeTe is tetragonal. Our simulation which shows that FeTe has a critical temperature is different with previous study [8] which shows that FeTe is not a superconductor. FeTe has DOS at the Fermi level at 20.17 $eV^{-1}m^{-3}$ with $\lambda = 0.11$ and $\omega_{log} = 83.23$ THz. Compare to its constituent element, Fe and Te, λ is smaller. λ of Fe is greater than 1 which means that Fe has a strong coupling while λ value of Te which smaller than 1 correspond to the weak coupling for Te. When Fe and Te are combined, the λ values drop dramatically to the area of weak coupling ($\lambda < 1$). At Fermi Level, the density of states of FeTe is greater than Te but smaller than Fe. The Fermi energy of FeTe decrease so its value is smaller than the Fermi energy of Fe, although it is still greater than the Fermi energy of Te. We could see that there is a slight enhancement on the ω_{log} value of FeTe compared to its constituent elements however the small value of λ induce the small value of T_c.

Previous study of FeSe has observed that the T_c of this material is about 8–13 K [9, 10] which in agreement with our results. Our simulations shows that the T_c of FeSe is 12.4 K with tetragonal structure which also in agreement with the literature [9]. The critical temperature of FeSe excess its constituent elements. The DOS of FeSe at the Fermi level and its value of ω_{log} is greater than Fe and Se, differ from FeTe where both values are smaller than Fe. The value of λ of FeSe do not decrease dramatically although the value is still smaller than Fe. However, the λ value of FeSe is greater than Se as one of its constituent element. For FeSe, $\lambda = 1.19$ which greater than 1 correspond to a strong coupling that occur in this material. The values of λ, ω_{log}, and DOS at Fermi level of FeSe contribute to a critical temperature that far exceeds the critical temperature of its constituent elements.

For $FeTe_{0.5}Se_{0.5}$, the simulation obtain a hexagonal crystal structure and $T_c = 12.2$ K which is smaller than the experimental result [11]. The magnitude of the critical temperature of this material is greater than its constituent element but almost the same as critical temperature of FeSe. The value of Fermi energy of $FeTe_{0.5}Se_{0.5}$ is smaller than FeSe but greater than FeTe. Meanwhile its λ is greater than the value of λ of its constituent elements and FeTe also FeSe. This value pass over the limit of

Table 1. Simulation results

	T_c (K)	λ	ωlog	DOS(E_F) ($eV^{-1}m^{-3}$)	E_F (eV)
Fe	0.01	1.2	25.93	2.43	16.81
Te	1.77	0.38	73.62	10.5	1.92
Se	2.23	0.27	62.18	7.02	5.17
FeTe	0.06	0.11	83.23	20.17	9.34
FeSe	12.4	1.19	105.15	16.65	16.34
$FeTe_{0.5}Se_{0.5}$	12.2	1.49	89.13	16.49	15.97

the weak coupling, so it has a strong interaction coupling. The density of states at the Fermi level for $FeTe_{0.5}Se_{0.5}$ is greater than those of its constituent elements and close to the value of FeSe while the value of ωlog is greater than its constituent elements and FeTe but is smaller than FeSe. Our results shows that a partial substitution of Te with Se contributes to the λ, ωlog and DOS at the Fermi level so that the critical temperature of $FeTe_{0.5}Se_{0.5}$ exceed the critical temperature of FeTe and slightly smaller than the critical temperature FeSe.

4. Conclusion

We have simulated the Fe, Te, Se, FeTe, FeSe, and $FeTe_{0.5}Se_{0.5}$ using quantum espresso. From the discussion, it can be concluded that the critical temperature of FeSe and $FeTe_{0.5}Se_{0.5}$ has far exceeded its constituent elements. FeTe has lower T_c than its constituent elements. The simulation results show the value of T_c of FeSe, and $FeTe_{0.5}Se_{0.5}$ correspond to the experimental results.

References

[1] Hsu F C 2008 *PNAS* **105** 14262-4
[2] Feng Z et al. 2018 *Sci. Rep.* **8** 4039
[3] de Siqueira F N, Monteiro J F H L, Jurelo A R, Dias F T and Junior J L P 2019 *Brazillian Journal of Physics* **49** 646-53
[4] Eliashberg G M 1960 *Zh. Eksp. Teor. Fiz.* **38** 966
[5] Bardeen J, Cooper L N and Schrieffer J R 1957 *Phys. Rev.* **208** 1175
[6] Dynes R C 1972 *Solid State Commun.* **10** 615-8
[7] Buzea C and Robbie K 2004 *Supercond. Sci. Tech.* **18** R1-R8
[8] Fang M H, Pham H M, Qian B, Liu T J, Vehstedt E K, Liu Y, Spinu L and Mao Z Q 2008 *Phys. Rev. B* **78** 224503
[9] Pomjakushina E, Conder K, Pomjakushin V, Bendele M and Khasanov R 2009 *Phys. Rev. B* **80** 024517
[10] Jiao L, Huang C L, Robler S, Koz C, Robler U K, Schwarz U and Wirth S 2017 *Sci. Rep.* **7** 44024
[11] Wang W et al. 2014 *Appl. Phys. Lett.* **105** 232602

Simple isolation method of cannabinol from *Cannabis sativa* to produce secondary reference standard analysis material

N Hidayati[1, 2], A Saefumillah[1] and A H Cahyana[1]

[1]Department of Chemistry, Faculty of Mathematics and Natural Sciences (FMIPA), Universitas Indonesia, Depok 16424, Indonesia
[2]National Agency of Drug and Food Control-Republic of Indonesia, Jakarta 10560, Indonesia

Corresponding author's email: asep.saefumillah@sci.ui.ac.id

Abstract. The simple isolation method of Cannabinol compounds from cannabis plants by maceration combined with ultrasonication assisted extraction, fractionation, separation and purification was carried out to obtain Cannabinol isolates which can be used as a reference standard analysis material. Ultrasonication was proven to shorten the extraction time, where extraction for 5 h with ultrasonication assisted for 15 min produced 5.536 % yield greater than the 24 h extraction. The results of extraction were fractionated using n-hexane and then chromatographically isolated with a column containing Silica Gel 60, with 2.5 cm diameter and a 15.3 cm height, eluted by n-hexane-ethyl acetate (90:10) solution. The fraction containing Cannabinol was purified using HPTLC preparative with eluent n-hexane-ethyl acetate (80:20). The purification of the Cannabinol isolate was further characterized by Spectrophotometer UV, FTIR, DCS, GCMS and LCMSMS and compared its profile to the reference standard of Cannabinol from Lipomed. The characterization results showed that the purified isolates had UV spectra with λ max at 219.0 nm and 284.5 nm, FTIR spectra at wave numbers 1620.21 cm^{-1}, 1051.20 cm^{-1}, 1581.63 cm^{-1}, 1026.13 cm^{-1}, 1128.36 cm^{-1} and 1232.51 cm^{-1}. The DSC thermogram shows the melting point of compound is 74.36 °C with 99.35 % purity, GCMS fragmentation at m/z 295, 296, 238 and 310, LCMSMS with $[M + H]^+$ at 311.1 and MS^2 at 222.95, 292.95 and 240.95 confirmed the chemical structure of the compound. The results of the characterization of pure isolates indicate that the compound produced was Cannabinol in accordance with the standard characterization profile of the reference standard of Cannabinol used.

Keywords: Cannabinol, isolation, ultrasonication assisted extraction, purification, reference standard

1. Introduction

Reference standard analysis material is used to guarantee of the validity of the testing method. Narcotics reference standard material is difficult to obtain in Indonesia, because of the high price and import bureaucracy. So far, the evidence from abuse and illegal trafficking cases of cannabis have been destroyed after the case is finished, even though it can be used as a raw material to produce some reference standard material for cannabis testing. The aging and improperly storing condition of cannabis evidence causes the Δ^9-Tetrahydrocannabinol acid as the major compound of cannabis undergoes a decarboxylation then oxidized to Cannabinol [1]. Cannabinol is one of marker testing of the cannabis compound that recommended by United Nations Office on drugs and Crime (UNODC) [2]. So, the aging cannabis evidence is potential to be applied as raw material to produce Cannabinol secondary reference standard analysis material.

Nowadays, isolation of cannabinoid compounds by Supercritical Fluid Extraction (SFE) extraction techniques continued by HPLC sample preparative are widely used because they have advantages including selective, no residual solvent, and automation. However, these techniques have drawbacks including very expensive equipment, and complex operating systems [3, 4]. Conventional techniques for extraction, such as maceration are widely used and easy to do, but require large amounts of solvents and time consuming, while Ultrasonication Assisted Extraction (UAE) technique gives significantly different results [5, 6]. For dry biological material, alcoholic solvents are used because it can break down the cell membranes and extract more endocellular materials [7]. Accordingly, it requires simple and safe extraction method to get good yields in a shorter time of extraction. The extract of cannabinoid must be purified to get almost 100 % purity of isolate. Purification of natural product using column chromatography and High Performed Thin Layer Chromatography (HPTLC) Preparative have been reported to have a satisfactory result [8, 9].

In this study, isolation method of Cannabinol compounds from cannabis plants by maceration combined with ultrasonication assisted extraction, fractionation, separation and purification was carried out to obtain Cannabinol isolates which can be used as a reference standard analysis material. The purified Cannabinol isolate was further characterized by UV Densitometer, FTIR, DCS, GCMS and LCMSMS and compared its profile to the reference standard of Cannabinol from Lipomed.

2. Materials and method

2.1. Materials

Methanol LC grade, acetonitrile LCMS grade, formic acid, n-hexane, toluene, and ethyl acetate were purchased from Merck, Germany. TLC aluminum Plate 20 × 20 cm and HPTLC glass plate 5 × 10 cm containing Silica gel 60 F_{254} and Silica gel 60 were obtained from Merck. Cannabinol RS from Lipomed and raw material dry cannabis (retained sample from Police Department of Aceh, Mei 2018) were obtained from National Agency of Drug and Food Control-Republic of Indonesia. Glass Column 2.5 cm id, Ultrasonic Branson 5510, Mechanical shaker Combi shaker NB-101 MT. LCMS column × Select HSS T3 Waters 1.8 µm, 1 mm × 100 mm, Elite-5 MS Perkin Elmer 30 m × 0.25 mm id × 0.25 µm df. The instruments used in this study were UV Densitometer (Camag TLC Scanner 3), FTIR (Shimadzu IR Prestige-21), DSC (Shimadzu-60A), GCMS (Agilent 5977B MSD) and LCMSMS (Waters Acquity).

2.2. Maceration combined with ultrasonication assisted extraction of cannabinol (CBN)

Dry cannabis was grounded. Each 1 g cannabis powder was added with 30 mL methanol, sonicated for 5, 10, 15, 20, 25, 30 min, respectively, then macerated under mechanical shaker at 150 rpm for 5 h. For control solution, each 1 g cannabis powder was added with 30 mL methanol then macerated under mechanical shaker with 150 rpm for 5 h and 24 h. The extracts were spotted on TLC plate, then developed using n-hexane-ethyl acetate (90:10) to a height of 8 cm. UV Densitometer was used for plate analysis [10].

2.3. Isolation and purification of cannabinol

CBN extracts were fractionated by water-n-hexane (50:50). N-hexane fraction was then allowed to dry. Each 1g dry n-hexane fraction was separated using column chromatography containing the silica gel 60, with 2.5 cm diameter and 15.3 cm height and n-hexane-ethyl acetate (90:10) as eluent. Fraction was collected at each 7 mL, until 30 fraction. The purification of CBN fraction was performed using

IOP Conf. Series: Materials Science and Engineering **902** (2020) 012063 doi:10.1088/1757-899X/902/1/012063

HPTLC plate with n-hexane-ethyl acetate (80:20). Then, the spots of CBN were collected and dissolved in methanol, filtered then the solvent was evaporated.

2.4. Characterization of CBN isolate

CBN isolate was characterized using UV Densitometer which was scanned from 200 to 400 nm. FTIR were recorded in the range of 4000–400 cm^{-1} using dried KBr powder into a mortar with 1 % (w/w) of the sample. The melting point and purity of the sample were determined by DCS. The molecular weight was confirmed by GCMS and LCMSMS. GCMS scan mode using HP-5 MS column 30 m × 250 mm × 0.25 mm, with injection temperature of 290 °C, MSD transfer line 290 °C, gas flow 1.5 mL/min, column temperature 100 °C with 1 min hold, rate 10 °C/min until 290 °C and held for 5 min. The analysis was conducted for 25 min with mass scan of 40–550. LCMSMS, using HSS T3 column, mobile phase A (Water containing 0.1 % formic acid) and B (acetonitrile containing 0.1 % formic acid), with the composition of solution A start on 80 % to 15 % during 6 min of the running analysis. The mass spectrometer was operated in MRM and Daughter Scan Mode, with capillary voltage at 3.5 kV ESI positive ion mode.

3. Results and discussion

3.1. Maceration combined with ultrasonication assisted extraction of cannabinol (CBN)

TLC and densitometer UV profile of CBN extract by maceration combined with ultrasonication assisted extraction is shown in figure 1. Based on figure 1 and table 1, it shows that ultrasonication can shorter the extraction time, because of the solvent cavitation production, which can accelerate dissolution and diffusion of solutes and heat transfer, which increases the extraction efficiency [11].

Result of maceration combined with ultrasonication assisted extraction of CBN is shown in figure 1. Based on figure 1 and table 1, it shows the yield is increased by adding sonication in maceration technique.

Figure 2 shows that yield of extraction is increased by ultrasonication assisted, which 15 min ultrasonication before 5 h maceration can increase 5.536 % yield, compared by 24 h maceration without sonication. Ultrasonication can shorter the extraction time, that was because of producing solvent cavitation which can accelerate dissolution and diffusion of solutes and heat transfer, which increases extraction efficiency [11]. (repeating sentences from above). The result after 15 min sonication is not significantly different which indicates an equilibrium was reached. The increasing time of sonication will not affect the extraction after the equilibrium of the solute is reached inside and outside the solid material [11].

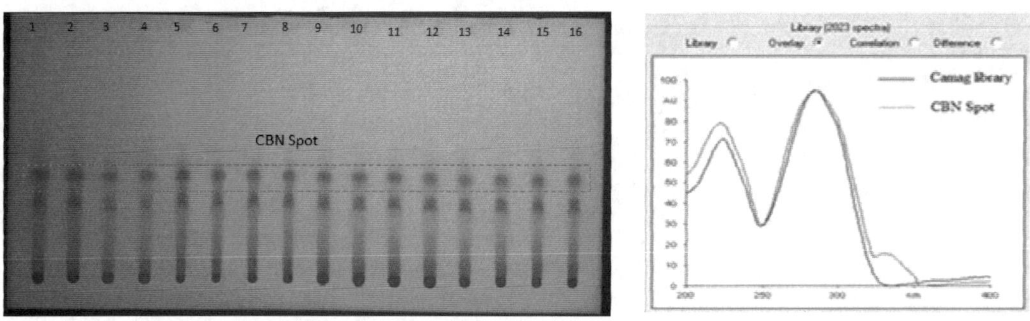

(a) (b)

Figure 1. (a) TLC and (b) densitometer UV profiles of extract CBN.

IOP Conf. Series: Materials Science and Engineering **902** (2020) 012063 doi:10.1088/1757-899X/902/1/012063

Table 1. Effect of sonication in CBN extraction.

No. spot	Time of maceration (M) and sonication (S)	Concentration (g/mL)	Spotted volume (μL)	Rf	Height (AU)	Average height (AU)	Increasing yield (%)
1	24 h M	0.17	20	0.49	552.0	544.6	
2	25 h M	0.17	20	0.49	537.2		
3	5 h M 0 min S	0.17	20	0.49	537.2	538.6	-1.111
4	5 h M 0 min S	0.17	20	0.49	539.9		
5	5 h M 5 min S	0.17	20	0.50	565.0	565.2	3.783
6	5 h M 5 min S	0.17	20	0.50	565.4		
7	5 h M 10 min S	0.17	20	0.50	565.4	562.8	3.342
8	5 h M 10 min S	0.17	20	0.50	560.2		
9	5 h M 15 min S	0.17	20	0.50	574.0	574.8	5.536
10	5 h M 15 min S	0.17	20	0.50	575.5		
11	5 h M 20 min S	0.17	20	0.50	574.5	574.5	5.481
12	5 h M 20 min S	0.17	20	0.50	574.4		
13	5 h M 25 min S	0.17	20	0.49	575.7	574.0	5.389
14	5 h M 25 min S	0.17	20	0.49	572.2		
15	5 h M 30 min S	0.17	20	0.49	570.3	570.7	4.783
16	5 h M 30 min S	0.17	20	0.50	571.0		

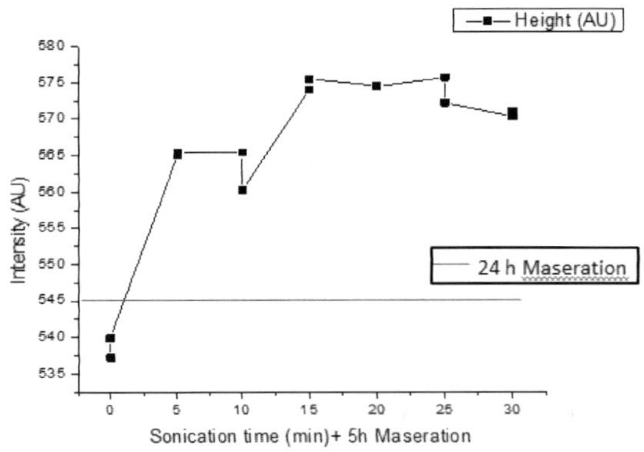

Figure 2. Yield increasing by ultrasonication.

3.2. Isolation and purification of cannabinol

CBN extract was fractionated with n-hexane-water (50:50). Because CBN is insoluble in water, so CBN will be dissolved in n-hexane fraction [2]. CBN in n-hexane fraction was then isolated by column chromatography. Silica gel column is a polar stationary phase with eluent n-hexane-ethyl acetate (90:10) which is less polar. The analytes are separated on the basis of their distribution coefficients as the eluent flows through the column [12]. CBN is eluted in the beginning and more polar compound stays longer in the column. Fraction was collected at each 7 mL until 30 fraction. Total eluent volume was about three times of the volume column, to ensure the target compound is eluted properly [13]. Figure 3 shows the TLC profile of fractions. CBN compounds were in the fraction no. 10–14.

Based on the TLC profile, fraction no. 10–14 still contain another compound, therefore, purification was carried out to get the required purity compound as a reference standard material. HPLTC Preparative was selected as the purification technique, because it is easy to do, low solvent consumption and quiet efficient. Figure 4 shows a different profile of CBN isolate before and after purification. After purification with HPTLC Preparative CBN isolate only has one spot indicates an increasing purity of the CBN compound.

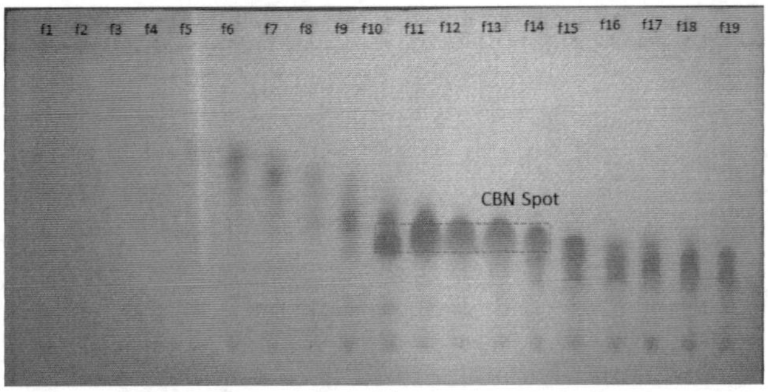

(a)

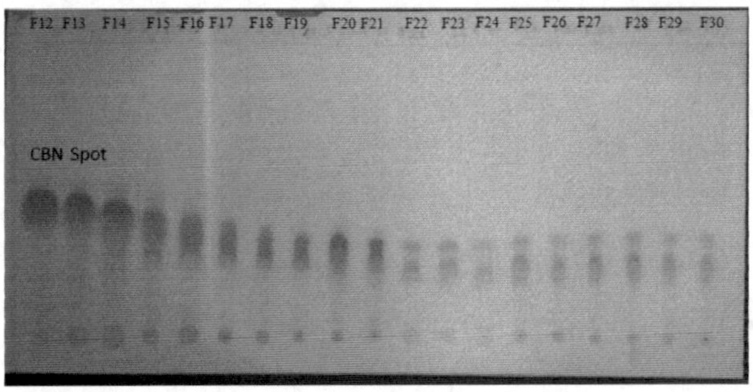

(b)

Figure 3. TLC Profile of fractions (a) fraction no. 1–19 and (b) fraction no. 12–30.

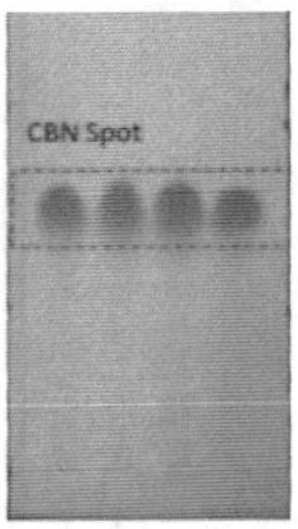

(a)

(b)

(c)

Figure 4. TLC profile (a) isolate CBN from separation in column chromatography, (b) purification by HPTLC preparative and (c) pure CBN isolate.

3.3. Characterization of CBN isolate

The purified CBN isolate was then characterized using UV Densitometer, FTIR, DCS, GCMS and LCMSMS and compared its profile with the reference standard of Cannabinol from Lipomed, except for DSC analysis. Figure 5 shows the UV spectrum of purified CBN isolate and CBN Lipomed as the reference standard. Table 2 shows λ max of the compound.

Purified CBN isolate and CBN Lipomed have a similar λ max, with less than 2 nm difference indicating that both are similar compounds [14, 15].

FTIR spectrum of purified CBN isolate and CBN lipomed is shown in figure 6. FTIR profile indicates that both compounds have the same constituent functional group.

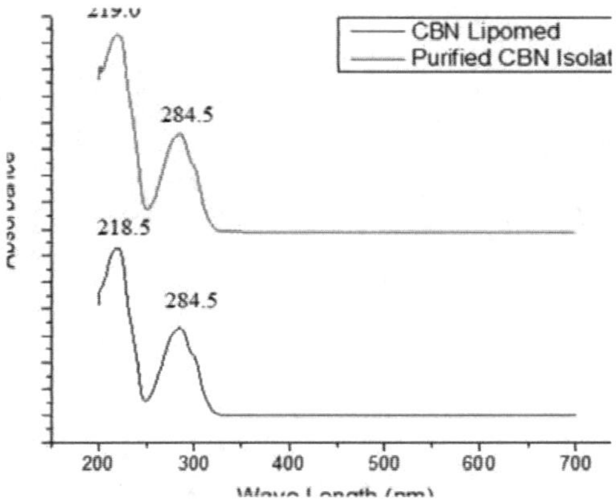

Figure 5. UV Spectrum of CBN purified isolate and CBN lipomed.

Table 2. λ max of CBN purified isolate and CBN lipomed.

λ max CBN lipomed	Absorbance	λ max purified CBN isolate	Absorbance
284.5	0.660	284.5	0.737
218.5	1.255	219.0	1.484

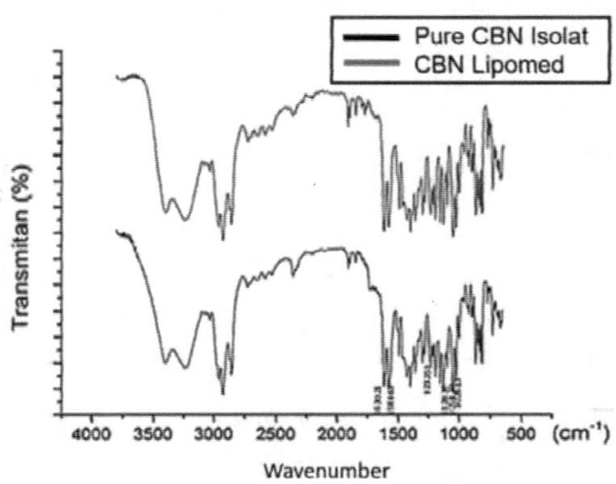

Figure 6. FTIR spectrum of CBN purified isolate and CBN lipomed.

Based on figure 6 and table 3, FTIR profile of purified CBN isolate, functional group of -C=C- of the aromatic compound that appears at the wave number area 1620.21 cm^{-1}, wave number 1581.63 cm^{-1} of -C-O- bond, wave number 1026.13 cm^{-1} of -C-O-C- symmetric and 1232.51 cm^{-1} of -C-O-C- asymmetric, that confirmed with CBN Lipomed and CBN literature profile, indicated the similar structure of purified CBN isolate and CBN Lipomed. Figure 7 shows the melting point of purified CBN isolate, which the melting point of compound is 74.36 °C and 99.35 % purity.

Confirmation of the structure and molecular weight were determined by GCMS and LCMS/MS. Based on figure 8, the purified CBN isolate has only one peak appears in retention time 18.307 min, same as the retention time of CBN peak from CBN lipomed. Fragmentation of the compound, m/z 310.1, 296.2, 295.2, 251.1, 239.1, 238.1, 223.1 and base peak at 295.2 are similar between purified CBN isolate and CBN Lipomed, indicating a structure similarity of the compounds, that is CBN with 310 molecular weight.

Relative abundance of the m/z fragmentation is also used as qualification of identification of the compound [17]. Based on table 4, the relative abundance of fragmentation of purified CBN isolate is similar to the CBN Lipomed, which is slightly different at m/z 296.2, with the difference value of 2.17 point, indeed it is still in range of tolerance value.

LCMSMS data in figure 9 show that molecular weight of purified CBN isolate is 310, confirmed with daughters scan result, that shows [M+1] m/z 311 fragmented to 293, 241, 223, 222, 195, 169, 119 and 71, similar profile with CBN Lipomed. MRM Data in table 5 show that relative abundance of MSMS fragmentation are in accordance to the qualification of identification point [17].

Table 3. λ max of CBN purified isolate and CBN lipomed.

Wave number observation (cm^{-1})		Wave number literature	Functional grup [15]
CBN lipomed	Purified CBN isolate	(cm^{-1}) [16]	
1620.21	1620.21	1620	-C=C-, Aromatic
1051.20	1051.20	1050	
1581.63	1581.63	1580	-C-O-
1026.13	1026.13	1030	-C-O-C- symmetric
1128.36	1128.36	1120	
1232.51	1232.51	1228	-C-O-C- asymmetric

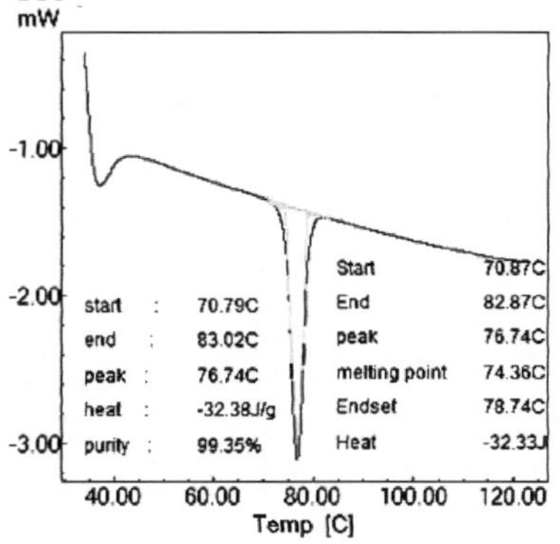

Figure 7. Thermogram of CBN purified isolate.

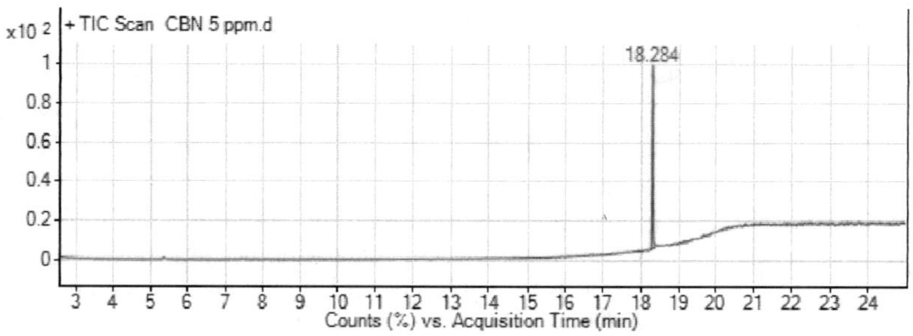

(a)

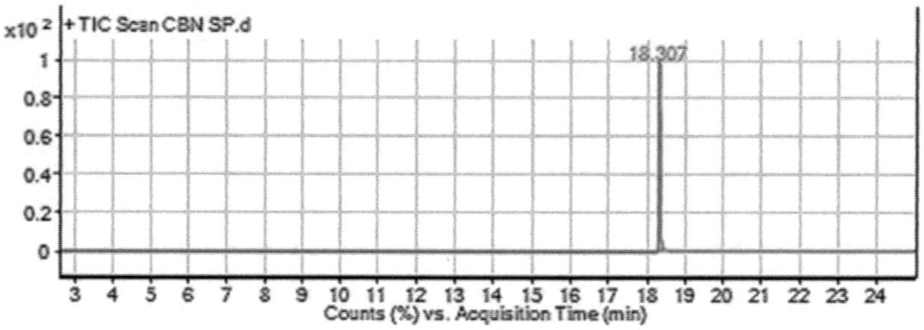

(b)

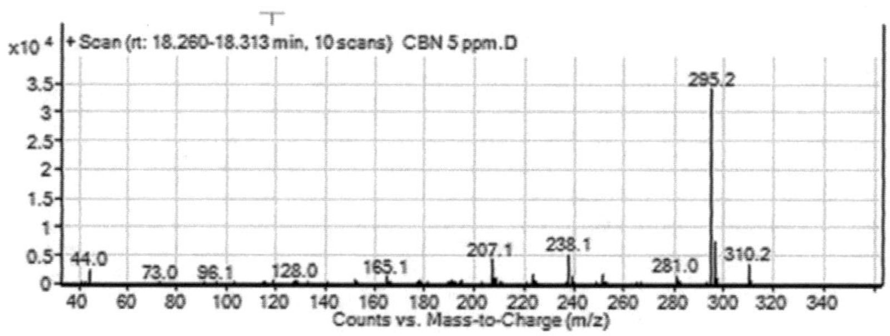

(c)

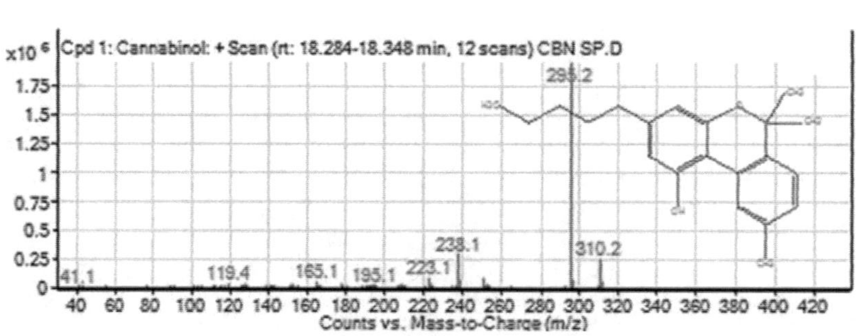

(d)

Figure 8. GCMS profile (a) chromatogram of CBN lipomed, (b) chromatogram of purified CBN isolate, (c) fragmentation of CBN lipomed, (d) fragmentation of purified CBN isolate.

Table 4. GCMS Fragmentation and m/z abundance of CBN purified isolate and CBN lipomed.

CBN Lipomed			Purified CBN Isolate			Tolerance of relative abundance different [17]
m/z	Abundance	Relative abundance (%)	m/z	Abundance	Relative abundance (%)	
223.1	1653.54	4.84	223.1	88809.41	4.54	± 50 %
238.1	5220.6	15.28	238.1	296073.97	15.14	± 20 %
239.1	1283.69	3.76	239.1	72898.01	3.73	± 50 %
251.1	1554.73	4.55	251.1	85172.15	4.35	± 50 %
295.2	34160.38	100	295.2	1956120.4	100	
296.2	7681.78	22.49	296.2	482341.88	24.66	± 15 %
310.2	3409.84	9.98	310.2	252629.33	12.91	± 20 %

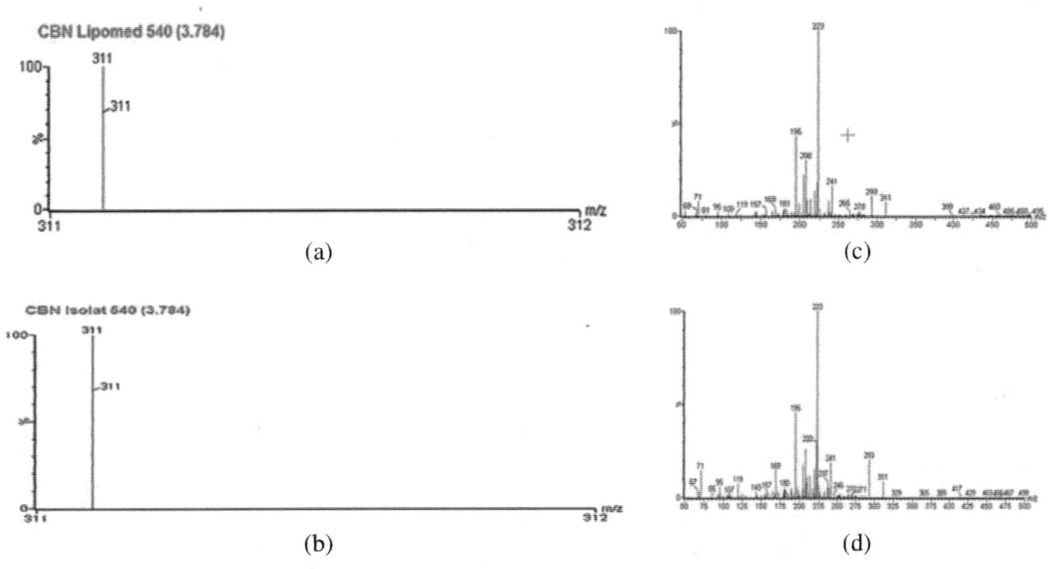

Figure 9. LCMSMS profile (a) parent scan of CBN lipomed, (b) parent scan of purified CBN isolate, (c) daughter scan of CBN lipomed, (d) daughter scan of purified CBN isolate.

Table 5. LCMSMS MRM mode: MSMS fragmentation profile and m/z abundance of purified CBN isolate and CBN lipomed.

CBN Lipomed			Purified CBN Isolate			Tolerance of relative abundance different [17]
m/z	Abundance	Relative abundance (%)	m/z	Abundance	Relative abundance (%)	
222.95	216756	100	222.95	58375	100	
292.95	122415	56.48	292.95	33688	57.71	± 20 %
240.95	77707	38.35	240.95	22389	38.35	± 25 %
194.95	80558	37.17	194.95	22601	38.72	± 25 %

4. Conclusion

CBN Compound has been extracted from cannabis plants by maceration combined with ultrasonication assisted extraction. Extraction time reduced from 24 h to 5 h by adding ultrasonication before maceration, while 15 min ultrasonication before 5 h maceration could increase 5.536 % yield relative compared to 24 h maceration. The isolation of CBN compound from CBN extract by column chromatography with n-hexane-ethyl acetate (90:10) eluent and purification by HPTLC preparative with n-hexane-ethyl acetate (80:20) eluent produce a CBN isolate with high purity, indicating by single spot of CBN in HPTLC system. The purified CBN isolate was characterized using Spectrophotometer UV, FTIR, DCS, GCMS and LCMSMS and compared the profile with the teference standard of Cannabinol from Lipomed. The characterization results showed that the resulting isolate was Cannabinol with molecular weight of 310, melting point 74.36 °C and purity 99.35 %, therefore, it could be a candidate for Cannabinol secondary reference standard material.

Acknowledgments

The authors would like to thank Universitas Indonesia and National Agency of Drug and Food Control-Republic of Indonesia.

References

[1] Hartsel J A, Eades J, Hickory B and Makriyannis A 2016 Cannabis sativa and hemp *Nutraceuticals* (Netherlands: Elsevier Inc.) chapter 53
[2] UNODC 2009 *Recommended Methods for the Identification and the Analysis of Cannabis and Cannabis Products* (United Nations, New York: Manual for use by national drug analysis laboratories) pp 25-47
[3] Aizpurua-Olaizola O et al. 2014 *Anal. Bioanal. Chem.* **406** 7549-60
[4] Leite J d A et al. 2018 *Sci. Justice* **58** 355-65
[5] Agarwal C, Máthé K, Hofmann T and Csóka L 2018 *J. Food Sci.* **83** 700-10
[6] Brighenti V, Pellati F, Steinbach M, Maran D and Benvenuti S 2017 *J. Pharmaceut. Biomed. Anal.* **143** 228-36
[7] Colegate S M and Molyneux R J 2008 *Bioactive Natural Products : Detection, Isolation and Structural Determination* 2nd edition (UK: CRC Press) pp 12-55
[8] Do T K T, Hadji-Minaglou F, Antoniotti S and Fernandez X 2014 *J. Chromatogr. A* **1325** 256-60
[9] Wohlfarth A, Mahler H and Auwärter V 2011 *J. Chromatogr. B* **879** 3059-64
[10] Fischedick J T, Glas R, Hazekamp A and Verpoorte R 2009 *Phytochem. Anal.* **20** 421-6
[11] Zhang Q-W, Lin L-G and Ye W-C 2018 *Chin. Med.* **13** 1-26
[12] Priyadarshini R, Raj G M and Shewade D G 2016 *Eur. J. Biomed. Pharm. Sci.* **3** 366-77
[13] Teledyne 2010 *Effective Organic Compound Purification: Guidelines & Tactics for Flash Chromatography* 4th edition (United Kingdom: Teledyne) pp 8-16
[14] Skoog D A, Holler F J and Nieman T A 1998 *Principles of Instrumental Analysis* 5th edition (Philadelphia: Harcourt Brace College Publishers)
[15] Kosela S 2010 *Cara Mudah dan Sederhana Penentuan Struktur Molekul berdasarkan Spectra Data (NMR, Mass, IR, UV)* (Depok: Fakultas Ekonomi, Universitas Indonesia)
[16] Kapp R W 2004 *Clarke's Analysys of Drugs and Poisons* 3rd edition Ed. Moffat A C et al. (London: The Pharmaceuticals Press) pp 45-6
[17] Stolker A A M, Dijkman E, Niesing W and Hogendoorn E A 2003 *Am. Chem. Soc.* **2003** 32-49

The growth rate and chlorophyll content of water hyacinth under different type of water sources

A A Hasibuan, R Yuniati and W Wardhana

Department of Biology, Faculty of Mathematics and Natural Sciences (FMIPA), Universitas Indonesia, Depok 16424, Indonesia

Corresponding author's email: ratnayuniati@sci.ui.ac.id

Abstract. Agathis Lake is one of the six lakes around Universitas Indonesia, Depok. This lake is a water body whose water quality has been damaged due to pollution from domestic and industrial waste around Beji area, Depok. Water hyacinth (*Eichhornia crassipes* (Mart.) Solms.) is a plant that has the ability to absorb contaminants in aquatic environment. This study was designed to determine the growth rate and chlorophyll content of water hyacinth under different type of water sources as growing media. These two physiological parameters are used to determine whether the contaminants contained in the Lake Agathis interfere the growth of water hyacinth. The research was conducted at the greenhouse of Departement of Biology, FMIPA, Universitas Indonesia for six days, after 1-week acclimatization by culturing water hyacinth in 30 L capacity plastic tanks. Sixteen plants of water hyacinth, each of approximately 28–31 g wet weight, were placed into four plastic tanks. Two tanks filled with 10 L tap water, and another two filled with 10 L lake water (Agathis Lake). The results showed that the Relative Growth Rate (RGR) of water hyacinth in the tap water was higher than the lake water. The chlorophyll content of water hyacinth cultured in the lake water was higher than the plant cultured in tap water.

Keywords: Chlorophyll, relative growth rate, water hyacinth, lake agathis

1. Introduction

Water hyacinth (*Eichhornia crassipes*) is one of the aquatic plants that has rapid rate of growth and so are regarded as weed. The rapid growth of water hyacinth resulted in low dissolved oxygen in water and it can cover large areas of water bodies [1]. Besides the negative effect, water hyacinth has several benefits such as absorbing heavy metals, providing livestock feed, and its fiber can be used to make handicrafts [2].

Due to the ability to absorb heavy metals, water hyacinth can get poisoned because of the binding of element on heavy metals to cell enzyme, thus inhibiting the enzymes that regulate the process of respiration, photosynthesis, and growth. Several studies show that heavy metals such as Cd and Cu cause the reductions in growth [3].

Water hyacinth also has the ability to absorb contaminants in the aquatic environment. This can have an impact on morphological and physiological characteristics. Different environment can lead to the different morphological and physiological characters of water hyacinth. Also, it is not yet known whether this ability affects the physiology of water hyacinth. Agathis Lake is one of the six lakes around Universitas Indonesia, Depok. This lake is a water body whose the water quality has been damaged due to pollution from domestic and industrial waste around Beji area, Depok. This study was designed to determine the growth rate and chlorophyll content of water hyacinth under different type of water sources as growing media. These two physiological parameters are used to determine whether the contaminants contained in Lake Agathis interfere the growth of water hyacinth.

2. Methodology

2.1. Plant and experimental design

Water hyacinth of similar size was collected from Jannati Firdaus Farm, Sawangan, Depok. The research was conducted at the Greenhouse of Department of Biology, FMIPA, Universitas Indonesia for six days. The collected plant which amounted sixteen plants of water hyacinth were placed into four plastic tanks filled with tap water for acclimatization. After acclimatization for 7 days, water hyacinth was put into two plastic tanks filled with 10 L tap water and two plastic tanks filled with 10 L lake water. Environmental parameters such as temperature and pH were observed every day. After the experiment was over, the water hyacinth sample was taken to the physiology laboratory to measure its chlorophyll content.

2.2. Relative Growth Rate

Relative Growth Rate of plants was measured on day 2, day 4, and day 6 during the experiment. Relative Growth Rate (RGR) was determined using the following equation 1,

$$RGR = (\ln W2 - \ln W1)/t \; x \; 100\% \quad (1)$$

where, *W1* is the mean initial fresh weight, *W2* is the fresh weight in the final status and *t* is the time interval (the experiment period) [4].

2.3. Chlorophyll content

Chlorophyll a, chlorophyll b, and total chlorophyll were measured in the leaves of water hyacinth. Leaves of water hyacinth from tap water and lake water were grinded using mortar by adding 85 % acetone as a solvent. Then, the solution was filtered using Whatman Filter No. 1. The filtrate was then added with the same solvent until it becomes 25 mL. After that, the obtained solution was measured for its absorbance using a spectrophotometer. Chlorophyll a was read at wavelength 645 nm and wavelength 663 nm for chlorophyll b [5]. Total chlorophyll was determined using the following equation 2 [6],

$$\begin{aligned} \text{Chlorophyll a} &= 12.7\ D663 - 2.69\ D645\ (mg/L) \\ \text{Chlorophyll b} &= 22.9\ D645 - 4.68\ D663\ (mg/L) \quad (2) \\ \text{Total chlorophyll} &= 20.2\ D645 + 8.02\ D663\ (mg/L) \end{aligned}$$

2.4. Nutrient content in Lake Agathis

Nutrient testing was carried out at *Laboratorium Penguji Balai Penelitian Tanaman Rempah dan Obat (Balittro)*, Cimanggu, Bogor, 16111, Indonesia. Water from Lake Agathis and tap water were both tested. The tested water sample was 600 mL.

3. Results and discussion

In the table 1, it can be seen that during acclimatization the environmental parameters did not change significantly. Range of water temperature during the acclimatization is 28–30 °C, environmental temperature is 33–34 °C, humidity from 44–55 %, stable pH at 5, and range of light intensity is 507–645 cd.

From table 2, it can be seen that there is a different pH between lake water and tap water. pH of lake water is 7 while pH of tap water is 6. Water temperature of the lake water ranges from 28–29 °C while tap water is 27–29 °C. Based on the literatures, water hyacinth has an ideal growing condition i.e. water temperature below 34 °C, environmental temperature above 0 °C [7].

In the table 3 and figure 1, it can be seen that water hyacinth grown in tap water has a higher RGR value compared to the one in lake water. In day 2, both water hyacinth which is grown in tap water and lake water have the same value of RGR. But, when entering the day 3, water hyacinth grown in tap water has a higher RGR compared to the lake water. Likewise, for the day 6.

In the table 4, the total chlorophyll is higher in water hyacinth grown in the lake water. Total chlorophyll content and chlorophyll a/b ratio are often used as indicators of stress in plants and also parameters for photosynthetic activity [8].

As shown in the table 5, lake water contains higher values of P, Na, Ca, Mg, C-organic than tap water. But, in tap water, Fe, Zn, and Co are detected, while in lake water they are not detected. When nutrient availability increases, some species have the ability to increase their growth [9].

Table 1. Environmental parameters during acclimatization.

Date	Water temperature	Environment temperature	Humidity	pH	Light intensity
27 June 2019	30 °C	34 °C	45 %	5	645 cd
28 June 2019	28 °C	34 °C	44 %	5	545 cd
29 June 2019	28 °C	34 °C	49 %	5	539 cd
1 July 2019	29 °C	33 °C	51 %	5	534 cd
2 July 2019	28 °C	33 °C	55 %	5	507 cd
3 July 2019	29 °C	38 °C	54 %	5	540 cd

Table 2. Environmental parameters during experiment.

Day	Lake water		Tap water	
	Water temperature	pH	Water temperature	pH
1	28 °C	7	27 °C	6
2	28 °C	7	28 °C	6
3	28 °C	7	28 °C	6
4	29 °C	7	29 °C	6
5	29 °C	7	29 °C	6
6	28 °C	7	28 °C	6

Table 3. Relative growth rate of water hyacinth.

Day	Relative growth rate ($g.day^{-1}$)	
	Lake water	Tap water
2	0.011	0.011
4	0.011	0.037
6	0.037	0.059

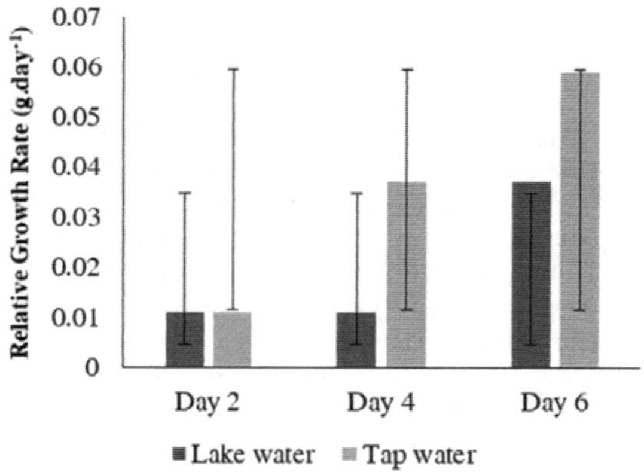

Figure 1. Relative growth rate of water hyacinth.

Table 4. Chlorophyll content

	Chlorophyll a	Chlorophyll b	Total chlorophyll
Lake water	1.143	0.608	1.751
Tap water	0.914	0.385	1.300

Table 5. Nutrient content

	Lake water	Tap water
N (%)	0.011	0.014
P (ppm)	20.43	11.00
K (ppm)	2.70	3.79
Na (ppm)	43.82	21.50
Ca (ppm)	45.72	27.38
Mg (ppm)	10.44	8.78
C-org (%)	0.10	0.01
Fe (ppm)	5.25	2.44
Cu (ppm)	Not detected	Not detected
Mn (ppm)	Not detected	Not detected
Zn (ppm)	Not detected	0.20
Pb (ppm)	Not detected	Not detected
Cd (ppm)	Not detected	Not detected
Co (ppm)	Not detected	0.60

The result showed that different environmental condition will produce a different physiological character. Relative Growth Rate of water hyacinth has a higher value when it grows in tap water. It can happen because in tap water there are macronutrients i.e. Fe, Zn and Co, on the contrary to lake water

IOP Conf. Series: Materials Science and Engineering **902** (2020) 012064 doi:10.1088/1757-899X/902/1/012064

Figure 2. Yellowing on leaves of water hyacinth.

where Fe, Zn and Co are not detected. Iron (Fe) is important for plant because Fe plays a role in cellular processes and also is necessary for plant growth and productivity. In plant, Fe are in the center of on chloroplast, mitochondria and vacuoles. Zinc (Zn) has several functions such as it is important for carbohydrate metabolism, protein metabolism, auxin and pollen formation in plants. Zn is also able to control biological membranes and performs a defence mechanism against the harmful pathogens. Cobalt (Co) also plays an important role for the plant growth and biochemical processes in plant. But, if the amount is too high it will cause toxicity [10]. Plant with high growth rates are indicated to have a good ability to mobilize and store nutrients in the tissues and assimilate the large amounts of carbon dioxide (CO_2) [11]. During the study, the water temperature also supports the water hyacinth growth. Previous study showed that water hyacinth RGR is increased when nutrient and temperature is increased. Those results showed a positive correlation between nutrient and temperature in water [9].

Chlorophyll refers to the green pigment found in plants, algae, and photosynthetic bacteria. Chlorophyll in plant takes an important part in the process of photosynthesis in which chlorophyll will absorb the light and convert it into chemical energy. Chlorophyll contain chloroplasts. Chloroplasts is an organelle in a plant cell which has a double membrane. Higher plants have two kinds of chlorophyll, namely chlorophyll a and chlorophyll b [12].

Based on the result, chlorophyll content in leaves of water hyacinth which grown in lake water has a higher value compared with tap water. Nutrient test resulted that Co is detected in tap water which makes the total chlorophyll decreases. Heavy metals such as Co can inhibit some enzymes that play a role in metabolic processes, and this is the main cause of chlorophyll degradation. Decrease in chlorophyll content is also a sign that plant is under stress. These results are confirmed by visual characteristics of the plant i.e. yellowing on leaves (figure 2). One study also proves that total chlorophyll content decreases when plant is exposed to Co [13].

High chlorophyll content in water hyacinth grown in lake water is a plant response to water stress. A few studies have been conducted in this regards. Water stress makes chlorophyll content decrease and this is reported in mung bean. But, some studies show the opposite where the chlorophyll content actually increases during water stress [14]. A difference in results indicates that the effect of drought and water stress on plant chlorophyll content depends on the plant genotype and several other factors [15].

4. Conclusion

The result of this study indicates that different type of water sources affect the physiological characters of water hyacinth. The results show that the Relative Growth Rate (RGR) of water hyacinth in the tap water was higher than the lake water. The chlorophyll content of water hyacinth cultured in the lake water was higher than plant cultured in tap water. The difference of physiological characters is due to the nutrient content in lake water and tap water. These results also show that contaminants in Lake Agathis interfere the growth of water hyacinth. The ability of water hyacinth to absorb heavy metals and pollutants has been proven in several studies. But, apparently it has an impact on the physiology of water hyacinth. When physiology is disturbed, it is feared that it will have a negative effect to water hyacinth, such as reducing the ability to absorb heavy metals.

Acknowledgments

This work was financially supported by Universitas Indonesia under research grant PITTA 2019.

References

[1] Greco M K B and de Freitas J R 2002 *Braz. J. Biol* **62** 463-71
[2] Rakotosarisoa T F, Richter T, Rakotondramanana H and Mantila-Contrereras J 2016 *Econ. Bot* **70** 365-79
[3] Kay S H, Haller W T and Garrand L A 1984 *Aquat. Toxicol.* **5** 117-28
[4] Gong Y, Zhou X, Ma X and Chen J 2018 *J. Clean. Prod* **181** 1-7
[5] Duman F, Otzurk F and Aydin Z 2010 *Ecotoxicology* **19** 983-93
[6] Porra R J 2002 *Photosynth. Res.* **73** 149-56
[7] David G, Henry N, Julia B, Sam D 2015 *SPAC* **3** 59-69
[8] Putra R S, Novarita D and Cahyana F 2016 *AIP Conf. Proc.* **1744** 020052
[9] Wenhua Y, Dan Y, Dong X, Lingfei Y, Wen X and Cuimin H 2014 *Aquat Bot* **119** 51-6
[10] Tripathi D K, Singh V P, Jumar D S and Chauhan D K 2012 *Acta Physiol. Plant.* **34** 279-89
[11] Junior E S O et al. 2018 *Aquat. Bot.* **145** 1-9
[12] Masato N 2018 *Plant Cell Physiol* **59** 1111-9
[13] Ibrahim M H, Kong Y C and Zain N A M 2017 *Molecules* **22** 1-16
[14] Rambabu B, Padma V, Thatikunta R and Sunil N *Nature Environment and Pollution Technology* **15** 1205-8
[15] Syamsia et al. 2018 *IOP Conf. Ser.: Earth Environ. Sci.* **156** 012040

Acetylation of curcuminoids extract from Turmeric Rhizomes *(Curcuma longa)* as antibacterial compounds against *S. aureus* and *E. coli*

A Fadhlurrahma, E Saepudin and D U C Rahayu

Department of Chemistry, Faculty of Mathematics and Natural Sciences (FMIPA), Universitas Indonesia, Depok 16424, Indonesia

Corresponding author's email: endang.saefudin@sci.ui.ac.id

Abstract. Curcuminoid are active compounds of turmeric rhizome that widely known to have antibacterial activities. The low activity of curcuminoid as antimicrobial may be due to the presence of phenol moiety and hence it has low lipophilicity and low bioavailability. Antibacterial activities of curcuminoid derivatives may be improved by increasing their lipophilicity, one of which is replacing the hydrogen atom on the phenol groups with acetyl group by acetylation. The curcuminoids were extracted from turmeric rhizomes by Soxhlet method yielded of 10.24 %. This curcuminoids were structurally modified by acetylation using acetic anhydride with Ni/SiO_2 catalyst. The products were separated through column chromatography then characterized using thin layer chromatography (TLC), UV-Vis, FTIR and LC-MS spectrometers. The results showed that the best condition of this reaction found when 15 % (w/w) catalyst was used with product conversion of 90.44 %. Diameter of inhibitory zone of acetyl curcuminoid derivatives showed the highest antibacterial activity at a concentration of 500 ppm against *S. aureus* of 18 mm and against *E. coli* of 13 mm. On the other hand, at the same concentration, the curcuminoid had inhibitory zone of 7.5 mm and 8 mm against *S. aureus* and *E. coli*, respectively. Therefore, the acetylation of curcuminoids has increased their antibacterial activities against Gram positive bacteria *S. aureus* up to 2.4 fold, while against Gram negative bacteria *E. coli* by only 1.6 fold.

Keywords: Curcuminoid, acetylation, Ni/SiO_2 catalyst, antibacterial, *Staphylococcus aureus*, *Escherichia coli*

1. Introduction

Curcumin with the chemical formula of ((1*E*,6*E*)-1,7-bis (4-hydroxy-3-methoxyphenyl)-1,6-heptadiene-3,5-dione) is the main yellow compound of phytophenol isolated from *Curcuma* sp. [1, 2]. Curcumin in nature is found together with two other compounds, demethoxycurcumin and bis-demetoxycurcumin, known as curcuminoids. Curcumin has pharmacological activities as antibacterial, anti-inflammatory, antioxidant, antiprotozoal, antitumor, the immune system enhancer (immunomodulators), neuroprotective and antiaging [3, 4].

Infectious diseases such as respiratory tract and diarrhea or disorders are one of the biggest causes of death in developing countries such as Indonesia, caused by pathogenic microorganisms [5]. Common treatments for infectious diseases are the use of antibiotics. However, irrational use of antibiotics can make pathogenic microbes become resistant and develop a major cause of infection [6]. The increasing number of bacterial resistances to antibiotics today, has triggered the need to find antibacterial that are more effective. Curcumin has been shown to have antibacterial activities against *S. aureus* with MIC value of 187.5 μg/mL [7], against *E. coli* with MIC values of 250 μg/mL [8]. These low antibacterial activities may be due to the presence of the phenol moiety which cause it difficult to be transported across the membrane due to its low lipophilicity and therefore has a low bioavailability. Converting the phenol group to its less polar derivatives, mono-*O*-acetylcurcumin and di-*O*-acetylcurcumin, using pyridine catalyst proved to increase curcuminoid antimicrobial activities against *M. tuberculosis* with MIC values of 50 and 100 μg/mL, respectively [9]. Here, we report the modification of curcuminoids by acetylation using a heterogeneous catalyst, Ni/SiO_2. As heterogeneous catalyst, Ni/SiO_2 has many advantages over homogeneous catalysts commonly used for esterification of hydroxy groups. It is environmentally friendly, economical, easy to be separated, reusable and avoid the use of high temperatures to minimize thermal risk [10].

2. Materials and method

2.1. Materials and apparatus

Turmeric *(Curcuma longa)* was obtained from a traditional local Market in Depok, West Java, Indonesia. n-hexane, ethanol, dichloromethane, acetone, diethyl ether, ethyl acetate, $Ni(NO_3)_2.6H_2O$, silica gel and DMSO were purchased from E. Merck Indonesia, amoxicillin was a gift from Faculty of Pharmacy, UI, *Staphylococcus aureus, Escherichia coli* (from Biochemistry Laboratory, Department of Chemistry UI). Nutrient broth and agar were purchased from Sigma. The instruments used were FT-IR spectrometer Shimadzu IR Prestige21, UV-VIS spectrophotometer Shimadzu UV-2450 and LC/MS/MS ABSCIEX 3200 Q TRAP.

2.2. Preparation of Ni/SiO2 catalysts

Nickel nitrate hexahydrate (2.51 g) was dissolved in distilled water (20.0 mL) and silica gel (5.0 g) was added and stirred for 2 h using a magnetic stirrer at room temperature. The mixture was left to stand at room temperature overnight. The excess water was removed by heating the mixture on an oil bath at 100 °C and using a rotary evaporator under vacuum to evaporate the water. The catalyst material was dried in an oven at 100–120 °C for 12 h.

2.3. Curcuminoids isolation

Dried powdered of turmeric (40 g) was wrapped in a filter paper and placed into the Soxhlet apparatus with ethanol as a solvent. The Soxhlet was heated on an oil bath for 9 h. The extract was then concentrated in the rotatory evaporator to obtain a brownish red curcuminoid extract. The curcuminoid extract was washed with n-hexane followed by dichloromethane. The curcuminoid derivatives then were characterized using TLC on silica plate and eluted with chloroform: methanol (19:1). The curcuminoid derivatives were then analyzed by using FTIR dan UV-Vis.

2.4. Acetylation of curcuminoids

The curcuminoid (0.5 g) was placed in a two-neck round bottom flask. Acetic anhydride (6.5 mmol) was added to the flask, then Ni/SiO_2 catalyst was added with varying amounts of catalyst. A condenser was attached to the flask and the reaction mixture was stirred with a magnetic stirrer at 70 °C for 5 h. The reaction was checked every hour using TLC to see if all the starting materials have been converted to their acetyl derivatives. Upon completion, the mixture was transferred into a separatory funnel and extracted with ethyl acetate twice and the water phase was discarded. The organic phase was washed with distilled water three times. The solution was placed in a round bottom flask and the remaining water was removed with the addition of anhydrous Na_2SO_4. The solution was filtered and the solvent was removed using a rotary evaporator.

2.5. Separation with column chromatography

Acetyl curcuminoids were separated from the unreacted curcuminoids using column chromatography on silica gel and eluted with n-hexane: ethyl acetate (8:2). The fractions were collected in 5 mL vials. Each fraction was monitored by TLC analysis and visualized under UV light at 366 nm.

2.6. Antibacterial activity test

Antibacterial activity of curcuminoid extract and their acetyl derivatives were evaluated using disc diffusion methods. The suspension of test bacteria was transferred aseptically into a petri dish. Warm liquid medium (± 40 °C) was poured into the petri dishes. Petri dishes containing bacteria and media were stirred slowly, so the bacteria can diffuse evenly. The media was then allowed to solidify. Sterile disc papers (6 mm diameter) was placed on the surface of the media. Each compound to be tested (31.25, 62.5, 125, 250 and 500 ppm) was dropped into the disc paper. The petri dish and its contents were incubated for 24 h in an incubator at 37 °C.

3. Results and discussion

3.1. Isolation of curcuminoids from Turmeric Rhizomes

Turmeric rhizome (8 kg), after chopping, grinding, and drying, resulted only 364 g of turmeric powder. The dried turmeric powder was then extracted using the Soxhlet extraction method. The initial extract result was obtained in dark red. After 9 h, the extraction was stopped, and the extract was concentrated in rotary evaporator followed by washing with a mixture of hexane and dichloromethane to remove unwanted materials. Curcuminoids extract was obtained as much as 20.487 g with a yield % of 10.24 % out of 200 g of turmeric powder.

3.2. Characterization of curcuminoids

The curcuminoid extract was characterized by TLC analysis. Eluent used was chloroform: methanol (19: 1) visualized under UV light at λ = 254 nm. The TLC result of the curcuminoid extract showed 3 major compounds, which is consistence with the one that has been reported elsewhere [3, 4, 9] (consisting of curcumin, demethoxycurcumin, and bisdemethoxycurcumin.

FTIR spectrum shows a broad peak around 3300 cm^{-1} which corresponds to the stretch vibration of hydroxyl groups (figure 1). The carbonyl, ethylene, and conjugated aromatic double bond showed peaks at 1626, 1587 and 1513 cm^{-1}, respectively. Other absorption bands show the vibration of the C-O-C ether bond at 1032 cm^{-1} and an enol C-O bond absorption occurs at a wave number of 1283 cm^{-1}, which

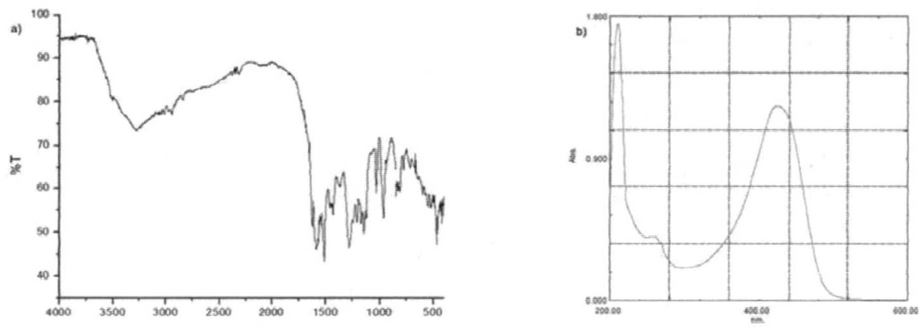

(a) (b)

Figure 1. (a) FTIR and (b) UV spectra of curcuminoids.

are in accordance with the structure of curcuminoids. In the UV-Vis spectrophotometer the absorption peak is shown at a maximum wavelength of 425 nm.

3.3. Acetylation of Curcuminoids with Ni/SiO_2 catalyst

Ni/SiO_2 was chosen as a heterogenous catalyst in the acetylation reaction of curcuminoids for its advantages in catalyst separation, which is easier, environmentally friendly, economical, reusable and to avoid the use of high temperatures to minimize thermal risk. In addition, the operational procedures performed is simple, safe and energy efficient (figure 2). The Ni/SiO_2 catalyst, apparently, plays a major role to maximize the activation of acetic anhydride.

According to the mechanism, the formation of Ni-acylium ion was facilitated by the hydroxyl group presents on silica surface [10]. Presumably, the reaction proceeds via an acyl-oxygen cleavage followed by nucleophilic attack of the hydroxyl group of phenol moiety leading to formation of the product (figure 3).

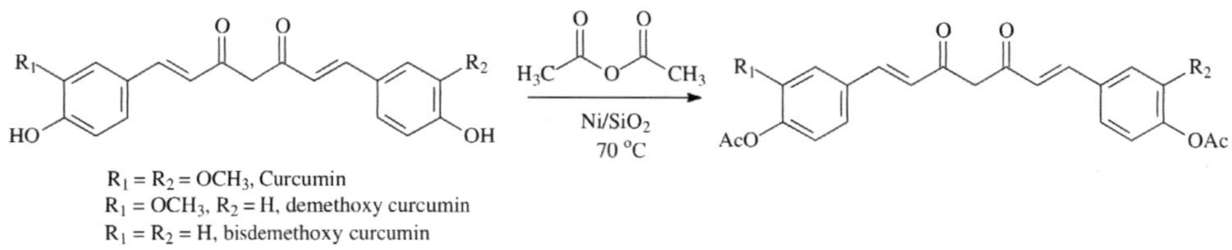

Figure 2. Reaction scheme of curcuminoids acetylation.

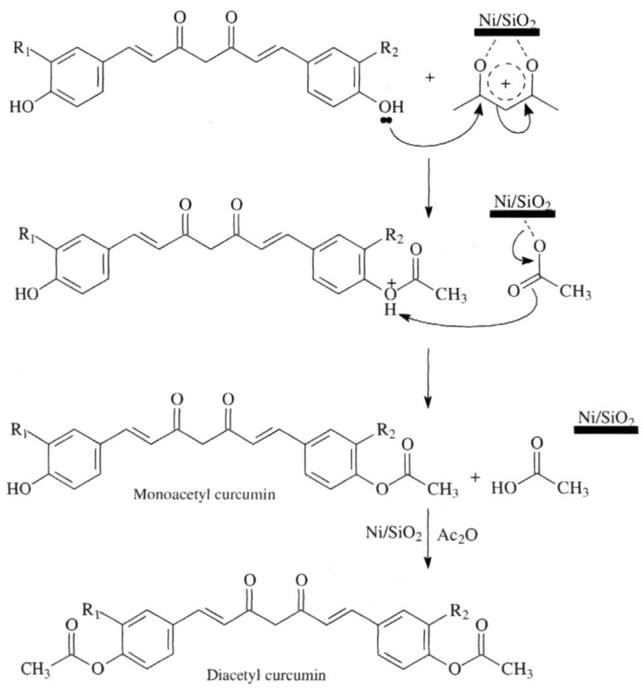

Figure 3. The proposed mechanism of acetylation.

TLC analysis on silica plate with eluent of dichloromethane : methanol (98:2), viewed under UV light at λ = 366 nm, showed that after 5 h of reaction, almost all the starting materials have been converted to their acetyl derivatives indicating by their greater rf values (figure 4). After the reaction has completed, the solid catalyst was separated by filtration. Water was added to the solution and the product was extracted with ethyl acetate.

When the amount of catalyst was varied at 5, 10, 15 and 20 %, the 15 % of catalyst produced the highest product conversion as much as 90.44 %.

The FTIR spectrum of acetylation reaction results showed that the hydroxyl group of curcuminoid at wave number regions of 3200–3600 cm^{-1} have disappeared compared to the curcuminoid spectrum, indicating that the acetylation of curcuminoid in the present of Ni/SiO_2 catalyst has successfully replaced the hydroxyl groups of curcuminoid side chain (figure 4).

The UV-Vis spectrum of the products with variations in the amount of catalyst of 5 %, 10 % and 15 % showed a shift to the lower maximum wavelength when compared to the starting material (curcuminoid extract with a maximum wavelength of 425 nm) (figure 5). The spectrum of the product of 15 % catalyst showed the highest intensity and the least noticeable shoulder peak at 425 nm, attributable to the highest conversion of the reaction.

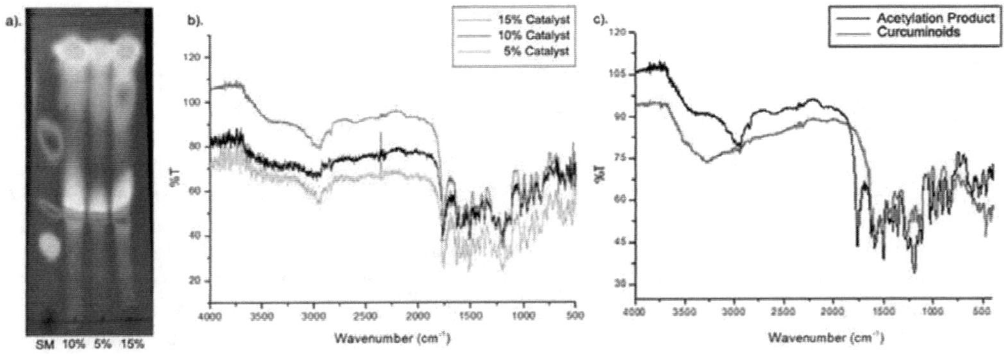

(a) (b) (c)

Figure 4. The analysis results of (a) TLC and (b) FTIR spectra on acetylated reaction products with varying amounts of catalyst (c) the two IR spectra showing the disappearance of the hydroxyl group after acetylation.

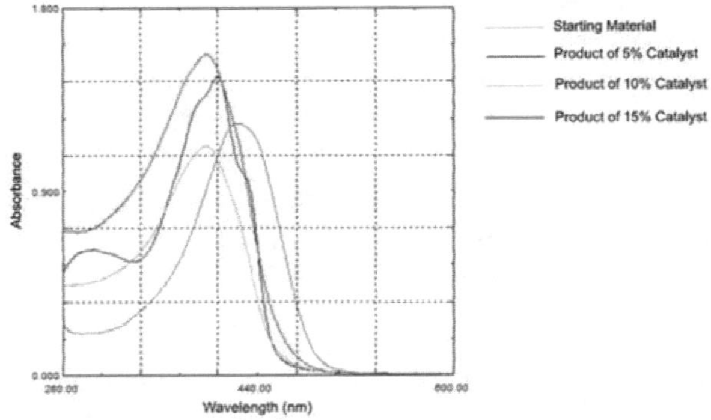

Figure 5. The UV-Vis spectra of curcuminoid acetylation with varying amount of catalysts.

Table 1. Inhibitory zone of acetyl curcumin and curcuminoids against *S. aureus* and *E. coli.*

Concentration (ppm)	Acetyl curcumin		Curcuminoids	
	Average inhibitory Zone (mm)			
	S. aureus	*E. coli*	*S. aureus*	*E. coli*
31.25	7.5	6	6	6
62.5	9.5	6	6	6
125	12.5	6	6,5	6
250	15.5	8	7	7.5
500	18	13	7.5	8

3.4. Purification using column chromatography

The product of acetylation reaction was purified by column chromatography on silica gel to remove the unreacted starting material. The column was eluted isocratically with the mixture of solvent [n-hexane: ethyl acetate (8:2).

3.5. Antibacterial activity test

Disk diffusion method was used to test the antibacterial activity of curcuminoid and acetyl curcuminoid. Various concentrations of 31.25; 62.5; 125; 250 and 500 ppm were used to see the inhibition of the compounds to the growth of *S. aureus* and *E. coli* by measuring the diameter of the clear zone. The results of the test showed that diacetyl curcumin exhibited a higher inhibitory zone than that of curcuminoids both on *E. coli* and *S. aureus*, where *S. aureus* experienced higher inhibition compared to the growth of *E. coli*.

Although it is premature to generalize this statement, our finding (table 1) showed that the diacetyl curcuminoid seems to have a greater effect on Gram positive bacteria (*S. aureus*) than that of Gram negative bacteria (*E. coli*). In all concentrations, diacetyl curcuminoid exhibited linear inhibitory zones on *S. aureus*, the higher the concentration, the greater the inhibitory zones, while on *E. coli*, significant increase of inhibitory zones was only noticeable at the concentration of 250 ppm or higher. The data on table 1 also tell us that the acetylation of curcuminoids has increased the antibacterial activity of the compounds. At 500 ppm, the increase was 2.4 fold against *S. aureus* and 1.6 fold against *E. coli*.

4. Conclusion

Acetylation of curcuminoids using Ni/SiO_2 catalyst was successfully carried out to produce acetyl curcumin with optimum conditions at 15 % (wt.) catalyst with conversion of 90.44 %. Acetylation of curcuminoids has improved the antibacterial activity of curcuminoid by 2.4-fold against Gram-positive *S. aureus* and 1.6-fold against Gram-negative bacterium *E. coli*.

Acknowledgments

This work was funded by Hibah Publikasi Internasional Terindeks Untuk Tugas Akhir Mahasiswa (PITTA 2265/UN2.R3.1/HKP.05.00/2018), Universitas Indonesia 2018.

References

[1] Changtam C, Hongmanee P and Suksamrarn A 2010 *Eur. J. Med. Chem.* **45** 4446-57
[2] Jawetz E, Melnick J L and Adelberg E 2005 *Mikrobiologi Kedokteran* (Jakarta: Salemba Medika)
[3] Araujo C A C and Leon L L 2001 *Mem. Inst. Oswaldo Cruz* **96** 723-8

[4] Majeed M, Badmaev V, Shirakumar U, Rajerdran R and Passwater R 1995 *Curcuminoids Antioxidant Phytonutrients* (New Jersey: NutriScience Publisher Inc.)
[5] Ibrahim T A, Opawale B O, Oyinloye J M A 2011 *Life Sci. Leafl.* **15** 490-8
[6] Hayakawa H, Minaniya Y, Ito K, Yamamoto Y and Fukuda T 2011 *Am. J. Plant Sci.* **2** 111-9
[7] Rahman A and Jonnalagadda S B 2008 *Catal. Lett.* **123** 264-8
[8] Tajbakhsh S et al. 2008 *Afr. J. Biotechnol.* 7 3832-5
[9] Verghese J and Joy M T 1989 *Flavour Fragr. J.* **4** 31-2
[10] Alam M, Rahman A, Alandis N M and Shaik M R 2014 *Arab. J. Chem.* **7** 53-6

Phytochemical and antiacne investigation of Indonesian White Turmeric *(Curcuma zedoaria)* Rhizomes

D A Setyani[1], D U C Rahayu[1], S Handayani[1] and P Sugita[2]

[1]Department of Chemistry, Faculty of Mathematics and Natural Sciences (FMIPA), Universitas Indonesia, Depok 16424, Indonesia
[2]Department of Chemistry, Faculty of Mathematics and Natural Sciences (FMIPA), Institut Pertanian Bogor, Bogor 16680, Indonesia

Corresponding author's email: dyahutamicr@sci.ui.ac.id

Abstract. White turmeric (*Curcuma zedoaria*) is one of plant species belongs to Zingiberaceae from Himalaya, India. Previous research showed that the rhizomes of this plant contained several secondary metabolites such as alkaloid, phenolic, and terpenoid which commonly known to have antibacterial activities. This study aims to investigate the phytochemical of white turmeric rhizomes growth in Indonesia and evaluate its antibacterial activity against *Propionibacterium acnes* and *Staphylococcus epidermidis* as acne-causing bacteria. In this study, secondary metabolites from Indonesian white turmeric rhizomes were successfully extracted in methanol and partitioned in polarity increasement of several organic solvents. Furthermore, crude and all partitioned extracts were subjected to phytochemical investigation. The ability of crude and ethyl acetate extracts to inhibit the growth of *P. acnes* and *S. epidermidis* were evaluated using disk diffusion method. According to the antibacterial activity results, all of samples did not exhibit antibacterial activity against both acne-causing bacteria. The results of this study will provide a useful information regarding the phytochemical and antiacne investigation of Indonesian white turmeric rhizomes.

Keywords: White turmeric rhizomes, *Curcuma zedoaria*, Zingiberaceae, antiacne, antibacterial

1. Introduction

Acne is a chronic inflammatory disease of pilosebaceous which commonly occurs in adolescence. Acne in teenagers occupies a prevalence rate of 85 % with the age range of 12–25 years [1]. In Indonesia, acne ranks at the third most diseases according to the number of visitors in Department of Skin Health and Venereology in several hospitals and skin clinics. One of factors causing acne is bacterial infection of *Propionibacterium acnes* and *Staphylococcus epidermidis* which belong to a normal flora bacteria on skin and Gram-positive bacteria. These two bacteria are pathogenic and often resistant to various types of antibiotics. *P. acnes* was resistant to doxycycline and minocycline in 100 % usage, while 10 % usage was found to be resistant to erythromycin, clindamycin, and tetracycline. Moreover, the side effect from using antibiotics such as tetracycline can cause liver damage for pregnancy women [2]. Therefore, research in discovering new compounds having potential as antibacterial activity, especially derived from natural products, are urgently needed.

White turmeric (*Curcuma zedoaria*) is one of the plant species belongs to Zingiberaceae from Himalaya, India. In Indonesia, white turmeric is usually used as herbal medicine, especially its rhizomes

IOP Conf. Series: Materials Science and Engineering **902** (2020) 012066 doi:10.1088/1757-899X/902/1/012066

which is commonly well-known to have several bioactivities such as antioxidant, anti-inflammatory, anticancer, and antibacterial activities [3]. These bioactivities in white turmeric rhizomes are due to its secondary metabolites. Previous study had been reported secondary metabolites from white turmeric rhizomes grown in Faisalabad, Pakistan consisted of terpenoid [4]. Flavonoid and curcuminoid were also reported in Indonesian white turmeric rhizomes from Jimbaran, Bali and Batu, East Java [5-6]. Furthermore, antibacterial activity of white turmeric rhizomes extract was previously reported against various bacteria i.e. *S. aureus, Escherichia coli, Streptococcus pyogenes, Symphoricarpos albus, Pseudomonas aeruginosa,* and *S. mutans* [7-8]. However, research on phytochemical investigation of Indonesian white turmeric rhizomes from Bogor, West Java together with its bioactivity towards acne-causing bacteria are not yet reported. To the best of our knowledge, differences in geographical location will influence the differences in secondary metabolites [9]. Therefore, the aims in this present study are to investigate the phytochemical and evaluate antiacne activity of Indonesian white turmeric rhizomes. This study was aligned with our interest on Indonesian medicinal plants during the recent years [10-16].

2. Materials and method

2.1. General

White turmeric (*C. zedoaria*) rhizomes were obtained as dried powder from Pusat Studi Biofarmaka Tropika (Trop. BRC), Institut Pertanian Bogor. Organic solvents used for extraction and partition were in technical grade i.e. methanol, *n*-hexane, and ethyl acetate (EtOAc). Extracts were monitored by using thin layer chromatography (TLC) and performed on pre-coated silica gel 60 GF_{254} plates (Merck). Chemicals used for phytochemical screening were purchased from Merck, i.e. NaOH, Mg, HCl, NH_4OH, chloroform, $(CH_3CO)_2O$ anhydrous, H_2SO_4, $FeCl_3$, Wagner, and Dragendorff reagents. Moreover, reagents and equipment used for antibacterial activity assay using disk diffusion method were also purchased form Merck, i.e. paper disk (diameter $\pm$ 6 mm), beef extract, peptone, nutrient agar, and dimethyl sulfoxide (DMSO), while clindamycin was purchased from drug store in Depok, West Java. Acne-causing bacteria such as *Propionibacterium acnes* and *Staphylococcus epidermidis* were obtained from Biochemistry Laboratory, Department of Chemistry, Universitas Indonesia.

2.2. Extraction and partition of white turmeric rhizomes (C. zedoaria)

Dried and powdered white turmeric rhizomes (2 kg) were extracted at 3 x 24 h in methanol at room temperature. After filtering the residue, filtrate was evaporated to remove the solvent under reduced pressure at 40°C using rotatory evaporator yielded the crude extract. The crude extract then was partitioned by using organic solvents with increasing polarity from *n*-hexane, EtOAc, and methanol. Each partition extract was also subjected under similar work up to remove the solvent. Both crude and partition extracts then were calculated for its yields following previous report [17] and subjected to analyze for its compounds using TLC with *n*-hexane:EtOAc = 7:3 (v/v) as mobile phase.

2.3. Phytochemical investigation of crude and all partitioned extracts

Phytochemical screening such as flavonoid, tannin, saponin, alkaloid, terpenoid, and steroid tests was conducted for crude and all partitioned (*n*-hexane, EtOAc, and methanol) extracts of white turmeric rhizomes following standardized procedures [18]. Flavonoid test was carried out using NaOH and Mg-HCl while tannin test was conducted using $FeCl_3$. Moreover, the presence of alkaloid was studied using Wagner and Dragendroff reagents after adding chloroform, NH_4OH, and HCl to the extracts. Terpenoid and steroid tests were investigated using $(CH_3CO)_2O$ and H_2SO_4 after dissolving the extracts with chloroform. The color changing on flavonoid, tannin, alkaloid, terpenoid, and steroid test were observed after each reagent was added. Furthermore, saponin was also identified by shaking the extracts after dissolving in aquadest and observed when the stable foam was formed.

2.4. Antibacterial activity assay of crude and EtOAc extracts
The ability to inhibit acne-causing bacteria growth of crude and EtOAc extracts were carried out with various concentrations of 1000, 500, 250, 125 and 62.5 ppm using disk diffusion method following established procedure [19]. Clindamycin (250 ppm) was selected as positive control while DMSO was used for negative control.

3. Results and discussion

3.1. Extraction and partition of white turmeric rhizomes (C. zedoaria)
In this present study, 2 kg of white turmeric rhizomes were extracted by maceration in methanol for 3 x 24 h. Maceration involved soaking plant material in container with solvent for a minimum period of 3 days with frequent agitation to soften and break the plant's cell wall to release soluble phytochemicals [20]. Methanol filtrate was separated from the residue and subjected to evaporate to obtain crude extract. Crude extract was established as blackish brown gummy with the weighted mass of 73.68 g with 3.684 % in yield.

A total 65 g of crude extract then was partitioned by using *n*-hexane, EtOAc, and methanol, successively. Table 1 summarizes the partition data together with the weighted mass and yield. According to table 1, the highest mass was obtained in EtOAc extract with the weighted mass of 30.59 g (47.06 %). This result indicated that compounds contained in white turmeric rhizomes were classified as semi polar compounds. It was also proven by TLC analysis showed in figure 1. Based on figure 1, EtOAc extract exhibited the most spot (compounds) among others, with semi polar compounds.

Table 1. The partition data of crude extract from white turmeric rhizomes.

Solvent	Extract (g)	Yield (%)
n-Hexane	7.17	11.03
EtOAc	30.59	47.06
Methanol	4.31	6.63

Figure 1. TLC analysis of crude and all partitioned extracts using *n*-hexane:EtOAc = 7:3 (v/v) visualized under UV 254 nm. C: crude extract, H: *n*-hexane extract, E: EtOAc extract, and M: methanol extract.

3.2. Phytochemical investigation of crude and all partitioned extracts

Preliminary phytochemicals screening of crude and all partitioned extracts are summarized in table 2. According to table 2, crude extract of white turmeric rhizomes was found to be composed by alkaloid, saponin, terpenoid, flavonoid, and tannin. Partition of crude extract distributed the compounds in three organic solvent having difference in polarity. For example, *n*-hexane extract contained terpenoid and flavonoid, EtOAc extract contained alkaloid, terpenoid, and flavonoid, while methanol extract contained saponin, flavonoid, and tannin. In comparison with phytochemical analysis from other regions of white turmeric rhizomes, summarized results of several literature will be further described.

Previous reported showed that phytochemical analysis of dichloromethane extract of Brazilian white turmeric rhizomes revealed the presence of terpenoid as dominance compound and confirmed by FTIR and NMR spectral data as curcumenol and mixture of phytosterols [21]. Essential oil obtained from hydro-distillation of Northeast Indian white turmeric rhizomes consisted of curzerenone, 1,8-cineole, and germacrone after analyzing with GC/MS [22]. Moreover, another white turmeric rhizomes essential oil from India also reported to have predominate terpenoid such as 1,8-cineole, cymene, α-phellandrene, and β-eudesmol [23]. Meanwhile, essential oil from Indian white turmeric leaves exhibited mainly mono- and sesquiterpenoid [24].

3.3. Antibacterial activity assay of crude and EtOAc extracts

Antibacterial activity assay was carried out on crude and EtOAc extracts against acne-causing bacteria, i.e. *P. acnes* and *S. epidermidis*. Clindamycin was chosen as positive control since it classified as lincomycin derived antibiotics which can inhibit protein synthesis at 50 S ribosome affecting the initial phase of peptide formation in bacteria [25]. The results of antibacterial activity assay are exhibited in table 3. According to table 3, both crude and EtOAc extracts did not showed antibacterial activity

Table 2. Preliminary phytochemical screening of crude and all partitioned extracts from white turmeric rhizomes.

Phytochemicals	Crude extract	*n*-Hexane extract	EtOAc extract	Methanol extract
Alkaloids	Present	Absent	Present	Absent
Saponins	Present	Absent	Absent	Present
Terpenoids	Present	Present	Present	Absent
Flavonoids	Present	Present	Present	Present
Tannins	Present	Absent	Absent	Present
Steroids	Absent	Absent	Absent	Absent

Table 3. Antibacterial activity assay results of crude and EtOAc extracts.

Concentrations (ppm)	Average diameter of *P.acnes* bacteria inhibition zone (mm)		Average diameter of *S. epidermidis* bacteria inhibition zone (mm)	
	Crude	EtOAc extract	Crude	EtOAc extract
62.5	0	0	0	0
125	0	0	0	0
250	0	0	0	0
500	0	0	0	0
1000	0	0	0	0

*The average diameter of clindamycin inhibitory zone (+) for *P. acnes* and *S. epidermidis* were 9.5 and 9.0 mm, respectively. Meanwhile, the average diameter of DMSO inhibitory zone (-) for both two bacteria were 0 mm towards both acne-causing bacteria, as proven by the absence of clear zone produced around the paper disc. These results indicated that all of samples cannot inhibit the growth of acne-causing bacteria. However, previous study had been reported that methanol extract of white turmeric rhizomes showed a medium antibacterial activity against *S. aureus* with inhibitory zone of 13 mm [26].

4. Conclusion

Secondary metabolites from Indonesian white turmeric rhizomes (*C. zedoaria*) can be extracted through maceration method in methanol with 3.684 % in yield. Further partition processes of crude extract gave three partitioned extracts with EtOAc extract which is having the most compounds and weighed mass. However, crude and EtOAc extracts exhibited inactive in inhibiting the growth of acne-causing bacteria, i.e. *P. acnes* and *S. epidermidis*. Further research should be carried out to fractionate the extract together with examine its antibacterial activity in other bacteria.

Acknowledgments

This work was financially supported by Universitas Indonesia under PUTI Prosiding 2020 research grant No. NKB-963/UN2.RST/HKP.05.00/2020.

References

[1] Dreno B 2017 *J. Eur. Acad. Dermatol. Venereol.* **31** 8-12
[2] Sitohang I B S, Fatgan H, Effendi E and Wahid M 2019 *Med. J. Indones.* **28** 21-7
[3] Marliani L, Budiana W and Anandari Y 2017 *Indones. J. Pharm. Sci. Technol.* **4** 57-63
[4] Tariq S, Imran M, Mushtaq Z and Asghar N 2016 *Lipids Health Dis.* **15:39** 1-10
[5] Ashfahani E D, Wiratmini N I and Sukmaningsih A A S A 2010 *Jurnal Biologi Udayana* **14** 20-3
[6] Ongko N X, Chiuman L and Ginting C N 2019 *Am. Sci. Res. J. Eng. Technol. Sci.* **55** 69-74
[7] Chachad D P, Talpade M B and Jagdale S P 2015 *Int. J. Sci. Res.* **5** 938-40
[8] Wirahmi S D, Busman and Edrizal 2019 *Menara Ilmu* **13** 19-28
[9] Sugita P, Anggraini D, Syahbirin G, Rahayu, D U C and Ilmiawati A 2019 *J. Indones. Chem. Soc.* **2** 114-20
[10] Rahayu D U C, Adilah S N and Sugita P 2018 *Eur. J. Pharm. Med. Res.* **5** 582-8
[11] Sugita P, Octaviana N, Wukirsari T and Rahayu D U C 2018 *J. Pharm. Res.* **12** 293-7
[12] Rahayu D U C, Hartono and Sugita P 2018 *Rasayan J. Chem.* **11** 762-5
[13] Sugita P, Firdaus S O, Ilmiawati A and Rahayu D U C 2018 *J. Chem. Pharm. Res.* **10** 68-75
[14] Purwantiningsih, Hartono, Firdaus S O, Kurniawanti, Rahayu D U C, Nurhayati L, Ilmiawati A and Wukirsari T 2019 *Drug Invent. Today* **11** 386-90
[15] Rahayu D U C, Nurfadhilah D and Sugita P 2019 *Int. J. Phram. Sci. Res.* **10** 3354-8
[16] Purwantiningsih, Jannah N and Rahayu D U C 2020 *Rasayan J. Chem.* **13** 322-6
[17] Cespedes C L, Avila J G, Martinez A, Serrato B, Calderon-Mugica J C and Salgado-Garciglia 2006 *J. Agric. Food. Chem.* **54** 3521-7
[18] Tiwari P, Kumar B, Kaur M, Kaur G and Kaur H 2011 *Internationale Pharmaceutica Sciencia* **1** 98-106
[19] Dogan B, Bektore B, Karabacak E and Ozyurt M 2017 *Turkderm-Turk. Arch. Dermatol. Venereol.* **51** 32-6
[20] Azwanida N N 2015 *Med. Aromat. Plants* **4** 1-6
[21] Navarro D de F, de Souza M M, Neto R A, Golin V, Niero R, Yunes R A, Monache F D and Filho V C 2002 *Phytomedicine* **9** 427-32
[22] Purkayastha J, Nath S C and Klinkby N 2011 *J. Essent. Oil Res.* **18** 154-5
[23] Singh G, Singh O P and Maurya S 2002 *Prog. Cryst. Growth Charact. Mater.* **45** 75-81

IOP Conf. Series: Materials Science and Engineering **902** (2020) 012066 doi:10.1088/1757-899X/902/1/012066

[24] Garg S N, Naquvi A A, Bansal R P, Bahl J R and Kumar S 2005 *J. Essent. Oil Res.* **17** 29-31
[25] Smilack J D, Wilson W R and Cockerill F R 1991 *Mayo Clin. Proc.* **66** 1270-80
[26] Das K and Rahman M A 2012 *Int. J. Pharm. Pharm. Sci.* **4** 322-8

IOP Conf. Series: Materials Science and Engineering **902** (2020) 012067 doi:10.1088/1757-899X/902/1/012067

The capacity of water hyacinth as biofilter and bioaccumulator based on its size

J Elizabeth, R Yuniati and W Wardhana

Department of Biology, Faculty of Mathematics and Natural Sciences (FMIPA), Universitas Indonesia, Depok 16424, Indonesia

Corresponding author's email: ratnayuniati@sci.ui.ac.id

Abstract. Nowadays, aquatic environmental contamination caused by human activities is increasing. One of several types of contaminants is heavy metal. Water hyacinth, *Eichhornia crassipes* (Mart.) Solms. is an aquatic plant that has been known to have a capacity to absorb heavy metals through its roots. Previous study reported that water hyacinth can reduce the amount of the suspended solid in the water. The objective of this study is to evaluate the performance of three sizes of water hyacinth in filtering the suspended solid from turbid water. The experiment was conducted in a greenhouse over a period of seven days on August 2019. The plant material was obtained from water body in Sawangan, Depok area. We used fifteen plants which categorized into three sizes of water hyacinth, i.e. small, medium and large. Those plants were first acclimatized for a period of seven days in tap water without addition of nutrients. For the next seven days, the water hyacinths were transferred to a fiberglass rectangular fish tank filled with 50 L water from Situ (Lake) Agathis, Depok. From the results, it was found that the accumulation capacity of lead (Pb), cadmium (Cd) and copper (Cu) by water hyacinth decreased by the increase of plant size. Another result showed that compared to leaves and stems, roots accumulate more heavy metals. Under seven days of growing period, all the fifteen water hyacinth plants can reduce the total dissolved solids (TDS) from 261 ppm to 204 ppm, and total suspended solids (TSS) from 0.0449 ppm to 0.0151 ppm.

Keywords: Biofilter, bioaccumulator, heavy metal, total dissolved solids, total suspended solids, water hyacinth

1. Introduction

Aquatic environment is already contaminated by heavy metals. Some of heavy metals that can be a damage for environment are lead (Pb), cadmium (Cd), copper (Cu) and zinc (Zn). Generally, heavy metals come from industrial waste that discarded without any treatment. This act causes heavy metals go with the flow of water to any river or other water source [1].

Eichhornia crassipes (Mart.) Solms. known as water hyacinth is a plant that lives on water. Water hyacinth can grow fast by its vegetative pieces that brought by water flow can grow into adult water hyacinth. Water hyacinth may harm the water environment because it can grow up to 1.9 % each day [2]. Water that covered by water hyacinth is going to be lack of oxygen, and it also takes effect on carbon dioxide that contained in water. Too much water hyacinth on the surface of water also can prevent the sunlight to penetrate the water [3].

Beside all the negative effects that can be caused by water hyacinth, this plant can absorb heavy metals that contained in the water. Usually, water hyacinth is used as bioaccumulator of waste water [4]. The role of water hyacinth can purify water and retard a water spot to silting up. It depends on the composition and amount of polluter substance in water [5]. If water hyacinth died, the waste will become an underwater deposit. By a good superintendence, water hyacinth can be a good bioaccumulator in the water [6].

Situ Agathis might be already contaminated by heavy metals. It might come from water that sourced from the residents of citizens around Universitas Indonesia. The existence of water hyacinth in Universitas Indonesia is so many. Usually, it can be found at inlet and outlet of the water near Universitas Indonesia, especially Situ Agathis [7]. But, lately we found that water hyacinth population was destroyed by the university guards because its existence can ruin water environment. However, people do not know that water hyacinth can accumulate sedimentation of water so that the water can be clearer than before. This research aims to know the capacity of water hyacinth as a bioaccumulator and biofilter in the water based on its size.

2. Materials and method

2.1. Collecting water hyacinth

The plant material was obtained from a water body in Sawangan, Depok area on June 25, 2019. Before that, we measured the environment parameter, such as pH (using pH meter), environment temperature and water temperature (using thermometer), light intensity (using lux meter), also wind acceleration of the field (using anemometer). We used fifteen plants which categorized into three sizes of water hyacinth, i.e. small, medium and large.

2.2. Acclimatization and treatment

The plants were first acclimatized for a period of seven days in tap water without addition of nutrients using water pump with capacity of 40W. Gross weight of water hyacinths was measured using a scale. For the next seven days, the water hyacinths were transferred into a fiberglass rectangular fish tank filled with 50 L water from Situ (Lake) Agathis, Depok (figure 1). During the acclimatization and treatment, the environment parameter was also measured, such as pH, environment, and water temperature, light intensity, also wind acceleration. After seven days of treatment, the gross weight of water hyacinth using scale was measured to see the difference between before and after treatment.

2.3. Heavy metals amount, TDS and TSS measurement

July 16th, 2019 we started to measure the total suspended solids (TSS) of 1 L of Situ Agathis water before and after treatment by filtering the water using Whatman filter paper number 1. Also, we measured the weight of filter paper before and after filtering the water. After that, we calculated the result into the TSS formula.

Figure 1. The system that we made to treat Situ Agathis' water with water hyacinth.

$$TSS = (\text{mass of dry filter paper after used} - \text{mass of dried filter paper before used}) \times \frac{1000}{sample\ volume}$$

The Total dissolved solids (TDS) was measured by heat up 1 L of Situ Agathis water before and after treatment. After its getting warm, we measured the TDS using TDS meter. Then, we brought all water hyacinths to PT. Mutuagung Lestari Laboratory to measure the heavy metals (Pb, Cd and Cu) contained in the water hyacinths. We chose lead (Pb), cadmium (Cd) and copper (Cu) because they are easier to be found in water [8].

3. Results and discussion

The classification of plant size was measured by the length of water hyacinth from the edge of roots to the edge of the tallest leaf of the plant. Small water hyacinth has the length of 15–20 cm, medium water hyacinth has the length of 25–30 cm, and the large water hyacinth has the length of 35–40 cm. In table 1 we can see that mostly, the bigger size is the water hyacinth, the heavier the gross weight. But, some of small water hyacinth has the same weight as medium sizes because of small plant has more roots than the medium sizes. Water hyacinths' gross weight average before treatment is mostly heavier than the gross weight before treatment. It caused by the growth of water hyacinths day by day.

Table 2 and table 3 show that water temperature, environment temperature, and humidity are mostly stable, but the light intensity increases drastically on day 2 of the treatment. Also, pH of water on the treatment day is higher than acclimatization day.

Water hyacinth's organ has different capability to absorb heavy metals. Water hyacinth also can localize the amount of heavy metals at certain organ [9]. Based on the result, roots of water hyacinth can absorb heavy metals better than the stem and leaf (table 4, figure 2). Heavy metals can be absorbed by water hyacinth because there is rhizospherous microbes on the surface of water hyacinth roots. This microbe is able to change a non-organic Cu to be an organic Cu so that can be used by the water hyacinth as metalloenzim from plastocyanin enzyme, which can be used on photosynthesis to stimulate cell division of water hyacinth. That is why water hyacinth can still grow richly in the contaminated water and the most absorbed heavy metal is copper (Cu) [10]. The ability of root to absorb the substances is called Rhizofiltration. Rhizofiltration can be used to rejuvenate water that contaminated by harmful substances [11]. A larger water hyacinth has a shorter root (figure 2). While, a smaller water hyacinth has a longer root (figure 3). Based on table 4, heavy metals mostly can be absorbed by the small water hyacinth better than the large water hyacinth. The longer root of small water hyacinth may be because of small plant can absorb more nutrition and heavy metals than the large water hyacinth.

Table 1. The comparable data of gross weight of water hyacinth.

Plant size	t0			t7		
	Small (gr)	Medium (gr)	Large (gr)	Small (gr)	Medium (gr)	Large (gr)
1	40	35	220	40	40	245
2	30	40	190	40	55	200
3	30	30	205	25	35	240
4	25	25	245	25	25	230
5	25	35	125	25	40	145
Average	30	33	197	31	39	212
SD	6.12	5.70	45.08	8.22	10.84	41.32

Table 2. The comparable data of environment parameter on the acclimatization day.

Acclimatization day	1	2	3	4	5	7
Water temperature (°C)	30	28	28	29	28	29
Environment temperature (°C)	34	34	34	33	33	38
Humidity (%)	45	44	49	51	55	44
pH	5	5	5	5	5	5
Light intensity (cd)	645	545	539	534	507	540

Table 3. The comparable data of environment parameter on the treatment day.

Treatment day	1	2	3	4	5	7
Water temperature (°C)	29	28	28	28	27	27
Environment temperature (°C)	36	33	36	35	34	34
Humidity (%)	47	51	45	46	51	45
pH	6	6	6	7	7	7
Light intensity (cd)	426	780	948	886	878	838

Table 4. The comparable data of heavy metals and water amount that contained in water hyacinth.

Parameter	Small			Medium			Large		
	Root	Stem	Leaf	Root	Stem	Leaf	Root	Stem	Leaf
Pb (ppm)	1.34	0.15	0.06	2.17	0.08	0.01	0.00938	< 0.00448	< 0.00448
Cd (ppm)	0.047	0.05	0.05	0.89	0.02	0.03	0.805	0.0181	0.00071
Cu (ppm)	53.5	16.7	17	83.8	15.1	19.5	84.4	11.2	17.6
Water content (%)	7.69	7.66	6.56	7.69	6.58	7.67	7.69	6.58	5.02

Figure 2. Large water hyacinth has short roots.

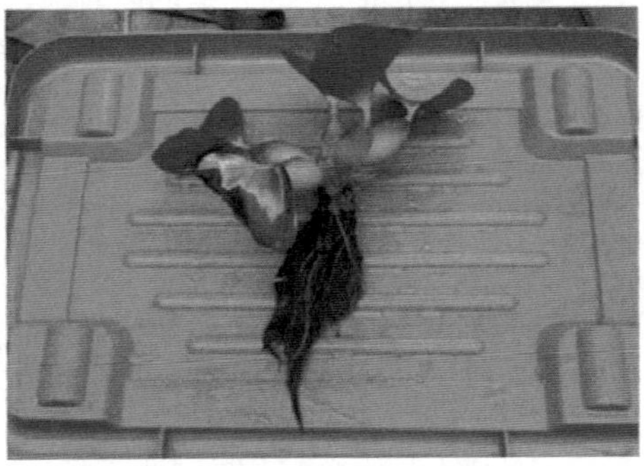

Figure 3. Small water hyacinth has long roots.

From table 4 and figure 2 we can see that mostly heavy metals that can be absorbed well by water hyacinth is copper (Cu). Heavy metals that contained in the root are mostly in higher amount than stem and leaf (figure 4).

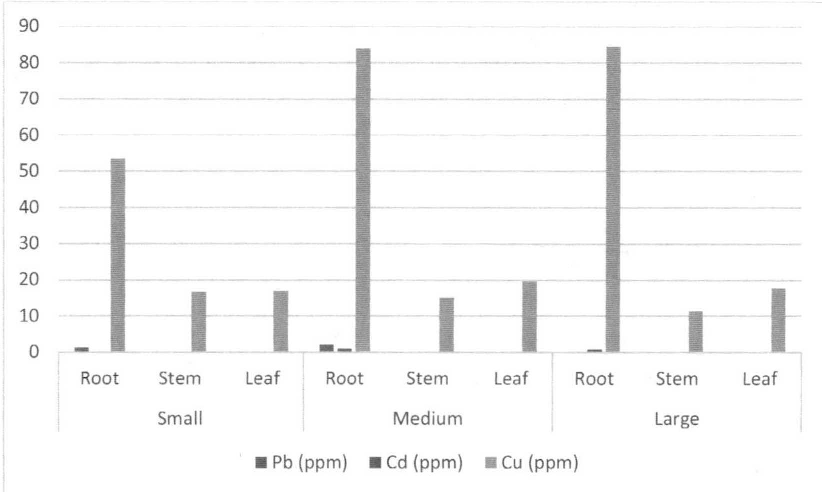

Figure 4. The comparable data of heavy metals amount that contained in water hyacinth.

Table 5. The comparable data of Total Dissolve Solid (TDS) and Total Suspended Solid (TSS) before and after treatment.

	Total Dissolve Solid (TDS)	Total Suspended Solid (TSS)
Before treatment	261 ppm	0.0449 mg/l
After treatment	204 ppm	0.0151 mg/l

Table 5 shows that both Total Dissolve Solid (TDS) and Total Suspended Solid (TSS) before treatment are higher than the after treatment.

Water that contaminated by heavy metals usually has a higher acidity than the uncontaminated one. From table 2 and table 3, we can see that pH on the acclimatization day is lower that pH on the treatment day. It means that the acidity of water is decreasing during the water treatment by water hyacinth. The higher amount of heavy metals is in the water, the lower pH of the water. If water has a lower pH (higher acidity), the solubility of heavy metal is bigger. Also, if water has a higher pH (lower acidity), the solubility of heavy metal is getting smaller. Usually, the process is followed by the deposition on the base of water [12]. The other environment parameters are mostly stable at both situations, before and during the treatment.

The Total suspended solid (TSS) shows the amount of residue that contained in water. The higher is the number of TSS, the higher amount of residue that contained in water. Total dissolve solid (TDS) shows the turbidity of water. The higher is the number of TDS, the bigger particle contained in the water and becomes more turbid [13]. Table 5 shows that the presence of water hyacinth can reduce TDS from 261 ppm to 204 ppm and TSS from 0.0449 ppm to 0.0151 ppm. It is all because the capability of water hyacinth that can absorb substance in the water and makes the water becomes clearer.

4. Conclusion

The results of this study show that heavy metals which can be well-absorbed by water hyacinth is copper (Cu). Water hyacinth's root can accumulate heavy metal in larger amount than its stem and leaf.

The size of water hyacinth does not determine the capability to absorb heavy metals. Water hyacinth is effective to be applied as a bioaccumulator and biofilter, as shown by the number of TSS and TDS that reduced after the placement of water hyacinth on the water. Water hyacinth can make the water to be clearer and reduced the substance or heavy metals that can damage the water environment.

Acknowledgments
We would like to acknowledge Directorate of Research and Community Service (DRPM) Universitas Indonesia for financial support of this research through Hibah PITTA 2019.

References

[1] Rochyatun E, Kaisupy M T and Rozak A 2006 *Makara* **10** 35-40
[2] Yonathan A, Prasetya A R and Pramudono B 2013 *J. Tek. Kim. Industri* **1** 412-6
[3] Ratnani R D 2012 *Momentum* **8** 1-5
[4] Sittadewi E H 2007 *J. Tek. Ling.* **3** 229-34
[5] Rukmi D P, Ellyke and Pujiati R S 2013 *Artikel Ilmiah Hasil Penelitian Mahasiswa* **2013** 1-7
[6] Stefhany C A, Sutisna M and Pharmawati K 2013 *J. Institut Tek. Nas.* **1** 1-11
[7] Kartika S 2017 *Al-Kauniyah; Journal of Biology* **10** 17-25
[8] Siswoyo E, Kasam and Widyanti D 2009 *J. Sains Tek. Ling.* **1** 68-76
[9] Indrasti N S, Suprihatin, Burhanudin, Noviati A 2006 *J. Tek. Ind. Pert.* **16** 44-50
[10] Setyowati S, Suprapti N H and Wiryani E 2005 *Bioma* **7** 1-8
[11] Rawat K, Fulekar M H and Pathak B 2012 *Int. J. Innov. Bio-sci.* **2** 139-99
[12] Tangio J S 2013 *J. Entropi* **8** 500-6
[13] Peavy H S, Rowe D R and Tchobanoglous G 1985 *Environmental Engineering* (New York: McGraw-Hill)

IOP Conf. Series: Materials Science and Engineering **902** (2020) 012068 doi:10.1088/1757-899X/902/1/012068

Activity of hydrogenated curcuminoid on Pd/C catalyst and its antibacterial activity against *Staphylococcus aureus* and *Streptococcus mutans*

N H Adriani, D U C Rahayu and E Saepudin

Department of Chemistry, Faculty of Mathematics and Natural Sciences (FMIPA), Universitas Indonesia, Depok 16424, Indonesia

Corresponding author's email: endang.saefudin@sci.ui.ac.id

Abstract. Turmeric (*Curcuma longa*) has been known for its benefit as one of medicinal herbs. Curcuminoids, the active compounds in turmeric, consist of three structurally related compounds, curcumin, demethoxycurcumin, and bisdemethoxycurcumin. Curcuminoid has been reported to be effective for the treatment of chronic gingivitis and hence it can be a potential candidate to be used as an active ingredient for antimicrobial in a mouthwash. However, its yellow color has hampered its application as a mouthwash as it may cause the aesthetical problem to the teeth. In this study, curcuminoids, obtained from Soxhlet extraction of turmeric powder, were reduced by hydrogenation on Pd/C catalyst to remove the color. TLC analysis revealed that the products consist of three compounds as expected. The product was purified from unreacted curcuminoids by column chromatography on silica gel, to obtain a colorless product with percent conversion of 23.45 %. These three compounds were characterized by Uv-Vis spectrophotometer resulted in three peaks in three different wavelengths: 282, 283 and 280 nm. Characterization with FT-IR spectrometer showed an increase intensity of CH-sp3 peak (2932 cm^{-1}) when compared to the IR spectrum of standard curcuminoids, and the shift in C-O absorption (1231 cm^{-1}), indicated the existence of methoxy groups that can distinguish one to the other compounds. These three compounds then were evaluated for their antibacterial activity against *Staphylococcus aureus* and *Streptococcus mutans*. The result showed that the tetrahydrocurcumin had moderate antibacterial activity against *S. aureus* with inhibition zone of 5.5 mm and tetrahydro desmethoxycurcumin also showed moderate antibacterial activity against *S. mutans* with inhibition zone of 5.5 mm.

Keywords: Antibacterial, curcuminoid, hydrogenation, *Staphylococcus aureus*, *Streptococcus mutans*

1. Introduction

Turmeric plant (*curcuma longa*) is a tropical plant that has been long used as spice and dye, both as skin dye and clothes dye. Turmeric is believed to be originated from India and India is still the largest producer of turmeric in the world. In the year of 800 AD, turmeric plants have spread to almost all regions in Asia, including China and Africa [1].

The nature of the turmeric plant that has healing properties is known because of the curcuminoid content contained in turmeric. Various studies have been conducted over the past two decades which are related to the benefits of curcuminoid [2, 3]. It is known that curcuminoids have anti-inflammatory properties [4, 5], antioxidants [6, 7], antiprotozoal [8, 3] and antibacterial [9].

Maintaining the health of the oral cavity is very important, because the oral cavity is the "entrance" of various kinds of bacteria that can stay in the oral cavity. One of the diseases of the oral cavity caused by bacteria is dental caries. Caries is a disease of the hard part of the tooth where this part is demineralized by the activity of microorganisms in a carbohydrate that can be distributed. If left unchecked, it will cause total damage to the affected tooth [10, 11].

Curcuminoid contained in turmeric plant with antibacterial properties has been shown can be used as a mouthwash. Curcumin is comparable to chlorhexidine as an anti-inflammatory mouthwash [12]. However, the yellow color of curcuminoid compounds in the turmeric plant has hampered their use as they may damage the aesthetic of the teeth if used as a mouthwash. The yellow color of curcuminoid is known due to the presence of double bonds conjugated with the carbonyl moiety (figure 1). Thus, removing the double bonds would also remove the color. This research aimed to remove the color of curcuminoid by hydrogenation on Pd/C as a catalyst. The antibacterial activity of hydrogenated curcuminoids (tetrahydrocurcuminoids) was evaluated using disc diffusion method. The inhibition diameter of bacterial growth in media containing tetrahydrocurcuminoids was compared with that of curcumin extract.

2. Materials and method

2.1. Materials and apparatus

Turmeric rhizomes (*Curcuma longa*) were collected from local market, ethanol, n-hexane, dichloromethane, nitrogen gas, hydrogen gas, Pd / C catalyst 10 % (Sigma Aldrich), acetone, ethyl acetate, chloroform, methanol, silica gel 60 (E. Merck), TLC plate, KBr, Listerine, DMSO, *Streptococcus mutans* strains (Faculty of Dentistry of Universitas Indonesia), *Staphylococcus aureus* strains (Department of Chemistry, Universitas Indonesia), beef extract, nutrient broth and nutrient agar. The instruments used were UV-VIS and FT-IR spectrophotometers.

2.2. Extraction of curcuminoid

Dried and powdered turmeric rhizomes was extracted by Soxhlet extraction method using ethanol as solvent for 8 h. The solvent was evaporated using a rotatory evaporator resulting in red brown extract of curcuminoids. The extract was washed with n-hexane and dichloromethane. The remaining solid was analyzed using TLC with chloroform:methanol (19:1) as a mobile phase, and characterized with Uv-Vis and FT-IR spectrophotometers.

2.3. Hydrogenation of curcuminoid

Curcuminoid extract (0.5 g) was placed in a two-neck flask fitted with a balloon on one neck and a three-way stopper on the other neck. Pd/C 10 % (25 mg) and dry acetone (25 mL) were added into the flask. Prior to the addition of hydrogen gas, a slow stream of nitrogen gas was passed into the flask to remove any gas and moisture in the flask, then hydrogen gas was added into the flask until the balloon inflated to about half size of its maximum size. The mixture was stirred at 30 °C for 210 min until the orange color turned to very pale yellow. The formation of the product was monitored by TLC, and the product was characterized using Uv-Vis and FT-IR.

2.4. Separation of tetrahydrocurcuminoids from curcuminoids

Tetrahydrocurcuminoids were separated from curcuminoids by column chromatography on silica gel eluted with *n*-hexane: ethyl acetate (6:4) isocratically. The fractions were collected in 10 mL vial bottles and monitored by TLC. Fractions that have the same spot were combined and characterized using Uv-Vis and FT-IR.

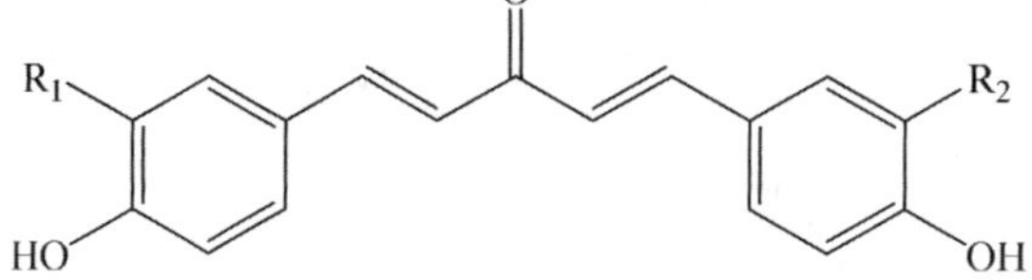

Figure 1. The structure of curcuminoids.

Table 1. Rf values of each curcuminoid components.

Curcuminoids	*Rf* (A)
Curcumin	0.950
Demethoxycurcumin	0.750
Bis-demethoxycurcumin	0.525

2.5. Purification of tetrahydrocurcuminoid derivatives
Tetrahydrocurcuminoids were purified into each of its derivatives using column chromatography on silica gel. The column was eluted isocratically with the mixture of chloroform: methanol: *n*-hexane (19:1:7.5). The output was collected in 10 mL vial bottles and analyzed with TLC method. Fractions that have the same spot were combined and characterized using Uv-Vis and FT-IR.

2.6. Antibacterial activity test using disc diffusion method
A suspension of diluted bacteria (200 μL) was dropped into a petri dish using a micropipette with a sterilized tip. Into the petri dish, warm and liquid nutrient agar (20 mL) was added. The NA medium was allowed to solidify. A sterile disc paper (6 mm diameter) was dipped into the test solution. The test solutions were curcuminoid extract and tetrahydrocurcuminoids. The concentrations were varied from 62.5; 125; 250 and 500 ppm in the DMSO. The disc papers placed on 1 petri dish were 6 pieces consist of positive and negative controls as well as solutions of curcuminoid extract and tetrahydrocurcuminods. Listerine was used as a positive control. DMSO solution was used as a negative control. The petri dish was then incubated for 24 h in the incubator at 37 °C. The clear zones of each disc paper were measured.

3. Results and discussion

3.1. Isolation of curcuminoids
TLC analysis on the curcuminoid extract using chloroform: methanol (19:1) eluent showed that curcuminoid compounds were separated into three different compounds, they are curcumin, demethoxycurcumin, and bis-demethoxycurcumin. The *Rf* value of each component is shown in table 1.

The data in table 1 shows that the three compounds have a noticeable difference in polarity. Having two methyl groups attached to the phenolic groups, curcumin shows the highest Rf value, hence it is the most non-polar. On the other hand, both demethoxycurcumin (DMC) and bis-demethoxycurcumin (BDMC) show lower Rf values, which can be correlated to the absence of methoxy groups. The less the methoxy group, the higher the polarity. Therefore, the nonpolar nature of curcumin is influenced by its structure which has two methoxy groups on its side chain. preventing the phenol groups to lose their proton when ionized. Thus, the methoxy groups can hold the proton through hydrogen bonding, keeping the proton stays closer to its phenoxy group and can easily rejoin (figure 2).

Uv-Vis spectrum of curcuminoid extract in methanol (figure 3) showed that curcuminoid extract has two strong absorption bands, one in the UV region with maximum wavelength at 265 nm, and the other in the visible region at 418 nm. The absorption peak, presumably, is caused by the presence of electron dipoles which can form the excitation type $\pi \rightarrow \pi^*$ of the conjugate π system.

In the IR spectrum results (figure 4), a strong and broad peak at the wave number 3310 cm^{-1} indicates the presence of stretch vibration from the O-H bond. The absorption peak at the wave number 1627 cm^{-1} shows the presence of C=O bond and the peak at 1513 cm^{-1} originates from the stretch vibration of conjugated double bond of benzene, while the peak at 1281 cm^{-1} indicates the presence of CO enol bond.

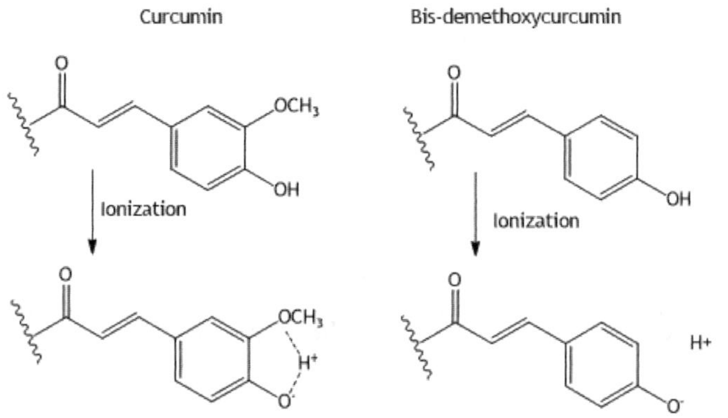

Figure 2. A scheme representing the polarity of curcumin and bis-demethoxycurcumin.

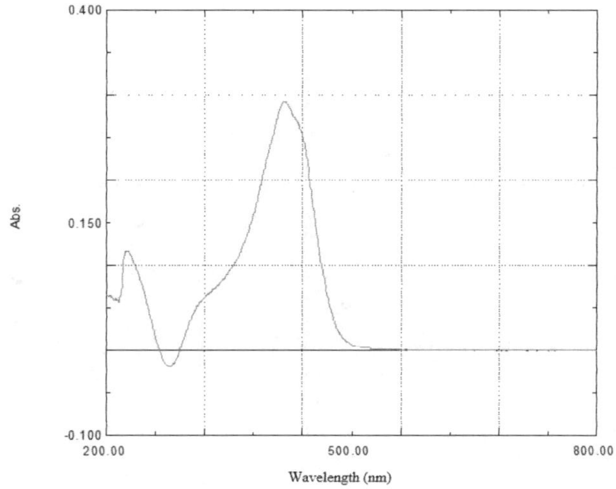

Figure 3. Uv-Vis spectrum of curcuminoid extract.

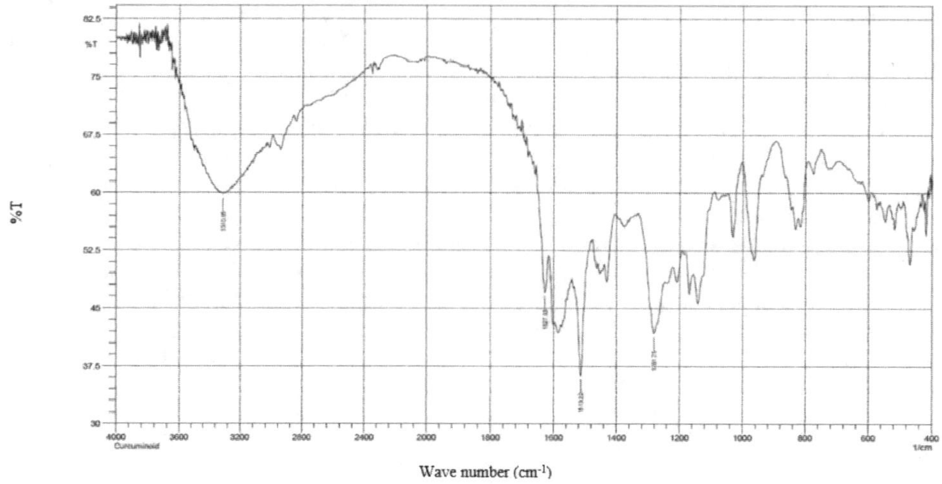

Figure 4. IR spectrum of curcuminoid extract.

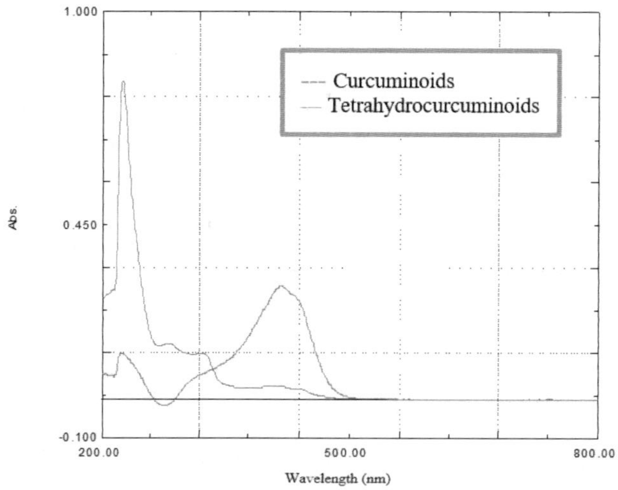

Figure 5. Uv-Vis spectra of curcuminoid extract and tetrahydrocurcuminoids.

3.2. Hydrogenation of curcuminoids

The reduction of curcuminoid by hydrogenation was achieved with % conversion of 71.84 %. The yellow orangish color of the curcuminoids solution has mostly gone indicating the loss of aliphatic double bonds. TLC analysis shows that three new spots appear above the weak spots of curcuminoids.

Figure 5 shows that most of the curcuminoids with the absorption peak at 418 nm has disappeared and shifted to 281 nm, indicating the loss of the aliphatic double bond in tetrahydrocurcuminoid derivatives.

The FT-IR spectrum confirms the formation of tetrahydrocurcuminoids with the appearance of a new absorption peak at 2932 cm^{-1} indicating the presence of C-H sp3 (figure 6).

3.3. Separation of tetrahydrocurcuminoids from curcuminoids
The separation was done by column chromatography on silica gel with a mixture of 6:4 ratio of solvent of n-hexane: ethyl acetate and the polarity were increased to 5:5. The separation was monitored by TLC. Using this method, all of the unreacted curcuminoid can be removed resulting the colorless solution of tetrahydro curcuminoids. In the Uv-Vis spectrum (figure 7), after separation, there is an absorption peak at the same wavelength, which is at 281 nm but has a greater intensity.

3.4. Purification of tetrahydrocurcuminoid derivatives
The colorless solution of THC was concentrated in a vacuum evaporator and loaded back to column chromatography on silica gel and eluted with 8:2 solvent ration of hexane: ethyl acetate followed by gradient increased polarity. TLC analysis clearly showed a single spot of each compound with different Rf values indicating that the three curcuminoids have been separated. Tetrahydro curcumin (THC) was first appeared at fraction F5, followed by tetrahydro demethoxycurcumin (THDC) at F7 and then tetrahydro bis-demethoxy curcumin (THBDC) at F10.

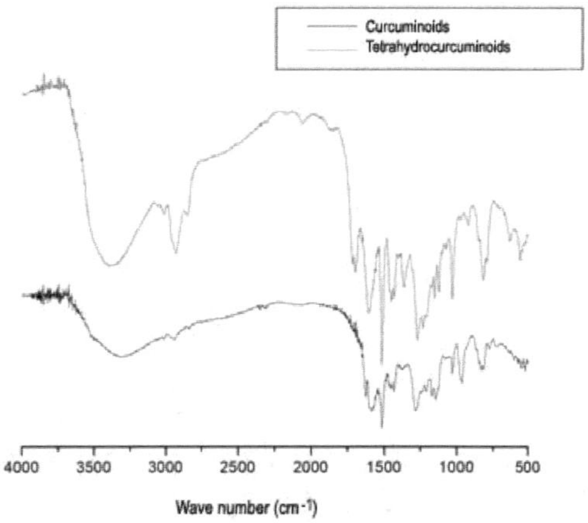

Figure 6. FT-IR Spectrum of curcuminoid extract and tetrahydrocurcuminoids.

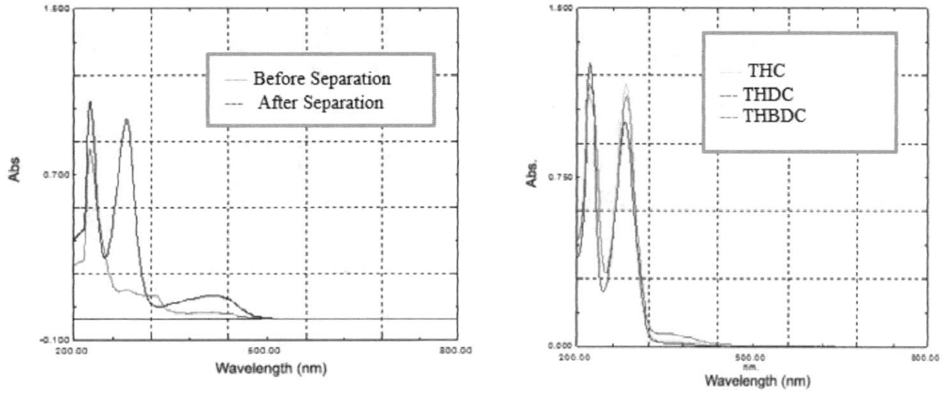

Figure 7. Uv-Vis (spectra before and after separation left) and spectra of THC, THDC and THBDC.

IOP Conf. Series: Materials Science and Engineering **902** (2020) 012068 doi:10.1088/1757-899X/902/1/012068

Figure 7 shows that the three tetrahydrocurcuminoids have almost identical absorption peaks but slightly different maximum wavelength. THC shows a maximum wavelength at 282 nm, the THDC at 283 nm, and the THBDC at a wavelength of 280 nm. This small difference in absorption peaks clearly shows that the methoxy groups attached to the benzene rings have almost no contribution to the absorption peak.

The absorption pattern of IR spectra of the three compounds looked almost identical (figure 8). The only difference was found at wavenumber 1231 cm^{-1} (asymmetric C-O-C stretch). Both THC and THDC appear to have this peak, where this peak is missing in THBDC spectrum confirming the absence of methoxy groups in THBDC.

3.5. Antibacterial activity test

Figure 9 shows comparative studies among the activities of the test concentrations to the growth of bacteria tested. The data show that the higher the concentration of the test solution, the higher the diameter of the clear zone which means the greater inhibition to the test bacteria. These data revealed that curcuminoids showed moderate activity against *Streptococcus mutans* and low activity

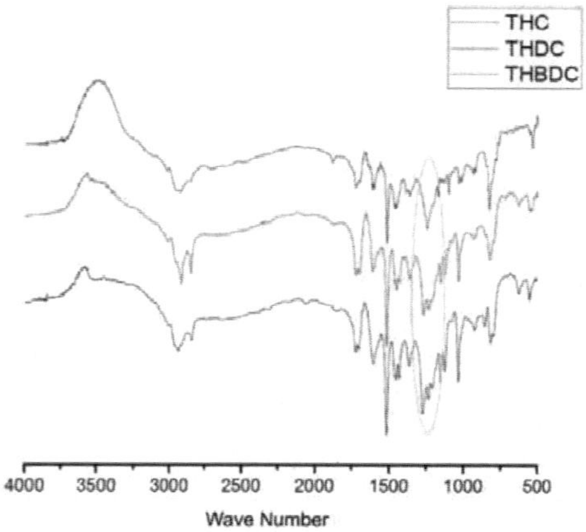

Figure 8. IR spectra of THC, THDC and THBDC.

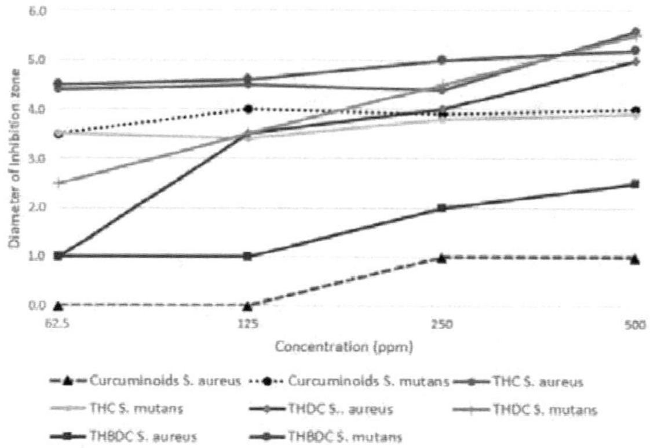

Figure 9. The effect of curcuminoid and tetrahydrocurcuminoids concentration against inhibition zones of the growth of *S. aureus* and *S. mutans* bacteria.

against *Staphylococcus aureus*. Their tetrahydro derivatives showed interesting results. All tetrahydro curcuminoids, tetrahydro curcumin (THC), tetrahydro demethoxycurcumin (THDC) and tetrahydro bis-demethoxycurcumin (THBDC) showed higher activities against *S. aureus* compared to curcuminoid extract, where THC showed the highest activity. On the other hand, the activity against *S. mutans* showed mix results. At concentration 125 ppm, their activities are comparable to curcuminoid extract. However, at the concentration of 250 ppm or higher, the activity of THDC and THBDC appeared higher than that of curcuminoids. This finding implies that the loss of double bonds in curcuminoids increases its antibacterial activity, particularly against *S. aureus*.

Among these test compounds, THDC shows the highest activity against *S. mutans*, while against *S. aureus*, THC has the greatest activity.

4. Conclusion

The reduction of curcuminoids was successfully carried out by hydrogenation reaction on 10 % Pd/C catalyst for 210 min with percent conversion of 71.84 %. All three tetrahydrocurcuminoid derivatives showed the better antibacterial activities than the curcuminoid extract.

Tetrahydrourcumin compounds (500 ppm) have moderate levels of antibacterial activity in *Staphylococcus aureus* with inhibition zone of 5.5 mm, while tetrahydrodemethoxycurcumin (500 ppm) has moderate levels of antibacterial activity on *Streptococcus mutans* with inhibition zone of 5.5 mm. Loss of double bonds in curcuminoids increases the antibacterial activity of curcuminoids, therefore tetrahydrocurcuminoids can be a potential candidate for a saver antibacterial agent in mouthwash solution.

Acknowledgments

We thank Faculty of Dentistry, Universitas Indonesia for providing us with *Streptococcus mutants.* This work was financially supported by Hibah PITTA 2018 No. 2265 / UN2.R3.1 / HKP.05.00 / 2018 from DRPM Universitas Indonesia.

References

[1] Benzie I F F and Wachtel-Galor S *Herbal Medicine: Biomolecular and Clinical Aspects* 2nd edition (Boca Raton: CRC Press/Taylor & Francis) chapter 13
[2] Gunes H, Gulen D, Mutlu R, Gumus A, Tas, T and Topkaya A E 2016 *Toxicol. Ind. Health* **32**, 246-50
[3] Rasmussen H B, Christensen S B, Kvist L P and Karazmi A 2000 *Planta Med.* **66** 396-8
[4] Chandra D and Gupta S S 1972 *Indian J. Med. Res.* **60** 138-42
[5] Huang H C, Jan T R and Yeh S F 1992 *Eur. J. Pharmacol.* **221** 381-4
[6] Erenoglu C, Kanter M, Aksu B, Sagiroglu T, Ayvaz S, Aktas C and Erboga M 2011 *Balkan Med. J.* **28** 352-7
[7] Reddy A C and Lokesh B R 1992 *Mol. Cell. Biochem.***111** 117-24
[8] Araújo A C and Leon L L 2001 *Mem Inst. Oswaldo Cruz.* **96** 723-8
[9] Chopra RN, Gupta JC and Chopra GS 1941 *Ind. Jour. Med. Res.* **29** 769-72
[10] Setijanto R D, Hadi A C, S L M, Palupi R, Bramantoro T, Berniyanti T, Sosiawan A and Wening G R S 2017 *TIMNAS7-JSMID4* **1** 126-129
[11] Reddy A C and Lokesh B R 1992 *Mol. Cell. Biochem.***111** 117-24
[12] Muglikar S, Patil K C, Shivswami S and Hegde R 2013 *Oral Health Prev Dent.* **11** 81-6

IOP Conf. Series: Materials Science and Engineering **902** (2020) 012069 doi:10.1088/1757-899X/902/1/012069

Microwave-assisted synthesis of 7-hydroxy-4-methyl coumarin and its bioactivity against acne-causing bacteria

A W Winata[1], D U C Rahayu[1], S Handayani[1] and H Dianhar[2]

[1]Department of Chemistry, Faculty of Mathematics and Natural Sciences (FMIPA), Universitas Indonesia, Depok 16424, Indonesia
[2]Chemistry Study Program, Faculty of Mathematics and Natural Sciences (FMIPA), Universitas Negeri Jakarta, Jakarta 13220, Indonesia

Corresponding author's email: dyahutamicr@sci.ui.ac.id

Abstract. Acne is a problem that often occurs in adolescents, especially at puberty phase. One of the acne-causing factors is proliferation of *Propionibacterium acnes* and *Staphylococcus epidermidis* bacteria. Acne is usually treated with antibiotics, but the resistance of antibiotics is also found. Therefore, research on new antibacterial compounds continues to be developed. One of the compounds reported to have antibacterial activity is coumarin derivatives. In this research, 7-hydroxy-4-methyl coumarin had been successfully synthesized from resorcinol and ethyl acetoacetate with the presence of *p*-toluene sulfonic acid (PTSA) as catalyst and solvent free reaction using 800W microwave-assisted method. This compound obtained in 48.36 % in yield with the reaction condition as follow: resorcinol/ethyl acetoacetate molar ratio of 1:1, 10 mol% PTSA, and 150s reaction time. Synthesized compound was further elucidated using FTIR, UV-Vis, and ^{1}H-NMR spectral data. Unfortunately, study on the ability to inhibit the growth of *P. acnes* and *S. epidermidis* showed that the compound was inactive towards both acne-causing bacteria. Therefore, it can be concluded that 7-hydroxy-4-methyl coumarin did not have potential antibacterial activity against acne-causing bacteria.

Keywords: Microwave-assisted synthesis, 7-hydroxy-4-methyl coumarin, antibacterial, acne-causing bacteria

1. Introduction

Acne is a problem that often occurs in adolescents, especially during puberty phase. Acne infects at least 85 % prevalence in the 12–24 years age range and can also cause long-term scarring and have a large psychological impact on individuals [1]. Several factors causing acne are inflammation and proliferation of *Propionibacterium acnes*. Not only the *P. acnes*, but also there is another acne-causing bacterium, i.e. *Staphylococcus epidermidis* which is classified as opportunistic pathogenic bacteria [2]. Tetracycline is a bacteriostatic agent with a broad-spectrum antimicrobial activity against gram-positive and negative bacteria both aerobic and anaerobic [3]. Based on its structure, tetracycline has a conjugated multicyclic framework with a lot of hydroxyl groups, at least five hydroxyl groups on simple tetracycline. In order to figure out the analogue compound of tetracycline, in terms of its structure, coumarins have a similar structural framework compared with tetracycline.

Previous study on nine standard natural products, coumarin modified with triterpenoid showed a strong resistance to the growth of *P. acnes* [4]. Coumarins are usually synthesized in a laboratory scale having substitution at C4 and C7 by Pechmann condensation using an acid catalyst with several methods [5-8]. Among all methods, microwave irradiation is proven as a simple and powerful method in preparing coumarins in a good yield with a shorter reaction time [8]. Following our interest on microwave-assisted synthesis and searching for potential antibacterial compound [9-14], herein we describe the microwave-assisted synthesis using *p*-toluene sulfonic acid (PTSA) catalyst and structure elucidation of 7-hydroxy-4-methyl coumarin. We also evaluate the ability of its compound using disk diffusion method against acne-causing bacteria in human skin tissue, i.e. *P. acnes* and *S. epidermidis*.

2. Materials and method

2.1. General

Domestic microwave oven Samsung ME-731K/XSE was used for microwave irradiation. Coumarin was measured for its melting point by using Electrothermal Melting Temp Apparatus and elucidated for its structure by using UV-Vis Shimadzu UV-2450 spectrophotometer, FTIR Shimadzu IR Prestige 21 spectrometer, and ^{1}H-NMR Agilent 500 MHz spectrometer. The chemicals used for synthesis were purchased from Merck, i.e. resorcinol, ethyl acetoacetate, PTSA, ethanol, *n*-hexane, ethyl acetate, and TLC plates (Si 60 GF_{254}, 0.25 mm thickness), while for antibacterial activity, nutrient agar, DMSO, clindamycin, and tetracycline were used. Acne-causing bacteria (*P. acnes* and *S. epidermidis*) were obtained from Biochemistry Laboratory, Department of Chemistry, FMIPA Universitas Indonesia.

2.2. Microwave-assisted synthesis of 7-hydroxy-4-methyl coumarin

7-Hydroxy-4-methyl coumarin was synthesized by reacting resorcinol (1 mmol) with ethyl acetoacetate (1 mmol) and PTSA (10 mol% of resorcinol) in solvent-free. The mixture then was irradiated in 800 W microwave for 150 s. The reaction temperature was controlled using an infrared thermometer and maintained at ±80 °C. The mixture was purified by recrystallization using warm ethanol and identified by TLC analysis. The product then was measured for its melting point and characterized for its structure using FT-IR, UV-Vis, and ^{1}H-NMR.

2.3. Antibacterial activity assay against acne-causing bacteria using disk diffusion method

Into a sterile petri dish, 200 μL of suspension for each bacterium test (1×10^8 cells/mL, 0.5 McFarland standard) were added with 20 mL nutrient agar at ±45–50 °C. The petri dish then was shaken until homogeneous mixture was produced then left to freeze. Sterile disc papers were placed on the media and 4 μL coumarin in various concentrations was dropped onto disc paper. The petri dish then was incubated at 37 °C for 24 hours. Inhibition zones indicated by clear areas around paper discs were measured in diameter. As a positive control, clindamycin 1 % and tetracycline 1 % were used, while DMSO was used for negative control. Antibacterial activity was repeated twice.

3. Results and discussion

3.1. Microwave-assisted synthesis of 7-hydroxy-4-methyl coumarin

In this present work, 7-hydroxy-4-methyl coumarin has been successfully synthesized from resorcinol and ethyl acetoacetate following the reaction scheme shown in figure 1. This compound is classified as aromatic heterocyclic with the molecular formula of $C_{10}H_8O_3$ and the IUPAC name of 7-hydroxy-4-methyl-2*H*-chromene-2-one. This compound belongs to the group of cyclic phenylpropanoid with the presence of methyl groups at C4 and hydroxy groups at C7.

The product was obtained in a pale-yellow color and odorless solid resembling flake with 48.36 % in yield. This yield was not as greater as previous report using $InCl_3$ as Lewis acid catalyst [15]. This result showed that Brønsted acid catalyst is less reactive compared to Lewis acid catalyst in synthesizing coumarin according to the product yield. TLC identification showed a new spot different from its precursors in *n*-hexane: ethyl acetate (3:2, v/v) with R_f of resorcinol, ethyl acetoacetate, and product were 0.52, 0.80, and 0.46, respectively. Melting point measurement showed the value of 188–189 °C, in line with the previous study [16].

3.2. FTIR characterization of 7-hydroxy-4-methyl coumarin

The results of FTIR characterization of synthesized compound are presented in figure 2. According to figure 2, FTIR spectrum showed the presence of several functional groups, i.e. hydroxyl (-OH) group at 2400–3500 cm^{-1} with typically broad-peak, carbonyl (C=O) group at 1705–1725 cm^{-1}, and alkene double bond (C=C) at 1620–1680 cm^{-1}. The presence of alkene double bond peak is an evidence that the synthesized product has been formed, since both precursors and catalyst do not have this functional group. Therefore, the alkene double bond is a unique functional group in the product.

3.3. UV-Vis characterization of 7-hydroxy-4-methyl coumarin

Figure 3 exhibits the UV-Vis spectrum in methanol of synthesized compound. Based on figure 3, the product has a maximum wavelength at 322 nm corresponds to cinnamoyl chromophore. Compared with the maximum wavelength of both precursors (data not shown), the maximum wavelength of product is shifted to bathochromic due to the extended conjugated system. Therefore, it leads to the formation of more stable products [17]. This data is also strengthened by previous study on 7-hydroxy-4-methyl coumarin synthesis having the maximum absorbance at λ_{max} 337 nm [16].

3.4. ^{1}H-NMR characterization of 7-hydroxy-4-methyl coumarin

Characterization using ^{1}H-NMR (500 MHz, acetone-d_6) was conducted to ensure the formation in the synthesized product. Table 1 summarizes the ^{1}H-NMR data in comparison to the previous study [16].

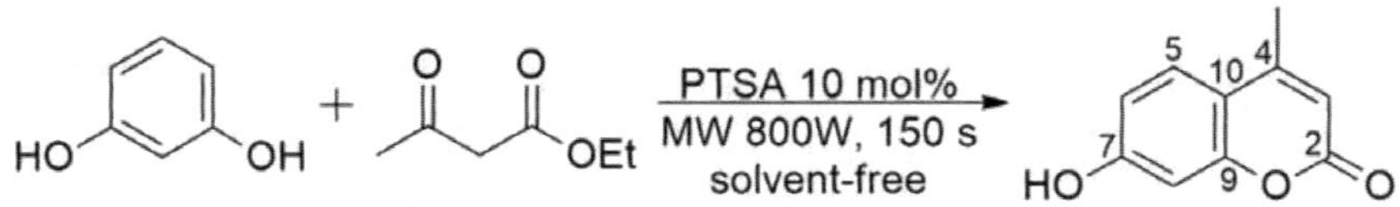

Figure 1. Reaction scheme for the formation of 7-hydroxy-4-methylcoumarin compound.

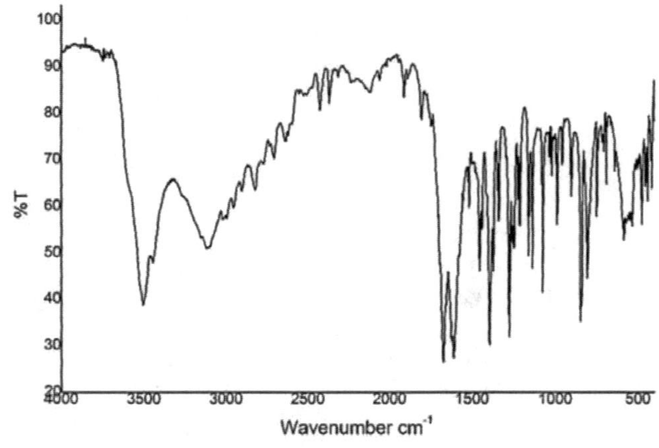

Figure 2. FTIR spectrum of 7-hydroxy-4-methyl coumarin.

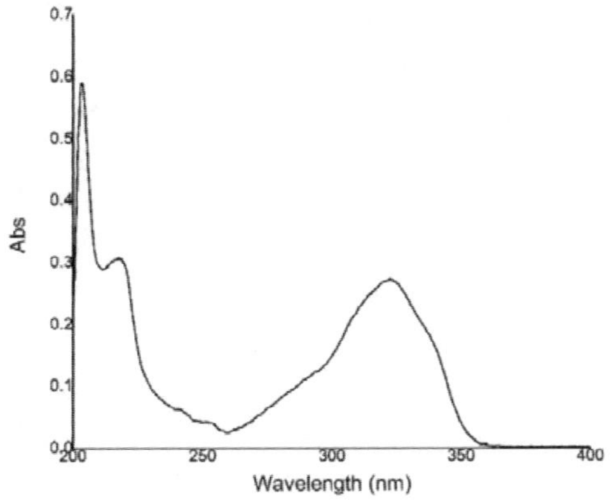

Figure 3. UV-Vis spectrum of 7-hydroxy-4-methylcoumarin.

Table 1. ^{1}H-NMR data of 7-hydroxy-4-methyl coumarin compared with previous literature.

No. C	Synthesized product (acetone-d_6) δ_H, ppm (int, *m*, $J_{coupling}$)	7-Hydroxy-4-methyl coumarin (DMSO-d_6) δ_H, ppm (int, *m*, $J_{coupling}$)
3	6.09 (1H, *s*)	6.10 (1H, *s*)
5	7.62 (1H, *d*, 8.7 Hz)	7.56 (1H, *d*, 8.7 Hz)
6	6.75 ppm (1H, *d*, 2.4 Hz)	6.69 (1H, *d*, 2.4 Hz)
8	6.87 ppm (1H, *dd*, 8.7; 2.4 Hz)	6.78 (1H, *dd*, 2.4; 8.7 Hz)
11	2.43 ppm (3H, *s*)	2.34 ppm (3H, *s*)
7-OH	9.41 ppm (1H, *s*)	10.51 (1H, *s*)

The product formation is indicated by the presence of three proton signals in aromatic ring chemical shift around 6–8 ppm. The peaks at chemical shift of 6.75 ppm (1H, *d*, J=2.4 Hz); 6.87 ppm (1H, *dd*, J=8.7; 2.4 Hz); and 7.62 ppm (1H, *d*, J=8.7 Hz) belongs to proton signals from trisubstituted benzene ring following ABX system for C8, C6, and C5, respectively. The resulted coupling constant also confirmed that the synthesized compound has ABX system with values of 2.4 and 8.7 Hz for positions 1,3 (meta) and 1,2 (ortho), respectively. A peak at 6.09 ppm (1H, *s*) is a proton at C3 which specifically known for proton bonded to alkenes. Singlet multiplicity of proton at C3 means that it has no neighbour protons and trisubstituted alkene. Moreover, a peak at chemical shift of 2.34 ppm (3H, *s*) revealed the deshielding proton signal typically for protons across the alkene double bond (H-C-C=C). The presence of one substituted hydroxyl group in C7 is indicated by the presence of a signal in 9.41 ppm (1H, *s*). According to ^{1}H-NMR spectrum, these six proton signals deduce a benzopyrone skeleton with double bond alkene at C3 with the presence of a methyl group at C4 and a hydroxyl group at C7. Based on table 1, it can be confirmed that the synthesized product matched with 7-hydroxy-4-methyl coumarin. Slightly differences in some of chemical shifts were found due to different solvent used in the sample measurement.

3.5. Antibacterial activity assay against acne-causing bacteria using disk diffusion method

The ability of the synthesized compound in inhibiting the growth of acne-causing bacteria, i.e. *P. acnes* and *S. epidermidis*, was conducted in this study using disk diffusion method. The various of coumarin

Table 2. The average diameter of 7-hydroxy-4-methyl coumarin inhibition zone against acne-causing bacteria.

Concentration (mg/mL)	The average diameter of the zone of inhibition of bacterial growth (mm)*	
	P. Acnes	*S. Epidermidis*
5	0	0
7.5	0	0
10	0	0
12.5	0	0
15	0	0
20	0	0

* Inhibition zone of both clindamycin and tetracycline as positive control were 10 and 40 mm, respectively, for both acne-causing bacteria. Negative control (DMSO) did not show any activity against all bacteria.

concentrations (5, 7.5, 10, 12.5, 15 and 20 mg/mL) with DMSO as negative control, clindamycin 10 mg/mL and tetracycline 10 mg/mL as positive control were used in this assay. Clear area produced around the disc paper indicates the antibacterial activity of the compound [11].

The average diameter of synthesized compound inhibition zone against both acne-causing bacteria is summarized in table 2. According to table 2, 7-hydroxy-4-methyl coumarin did not show any antibacterial activity against *P. acnes* or *S. epidermidis* at all concentrations. This result indicated that the coumarin-derived synthesized compound cannot inhibit the growth of *P. acnes* and *S. epidermidis* bacteria. However, recent study reported that coumarin derivatives, i.e. 6-methyl and 6,7-dihydroxy coumarin exhibited an antibacterial activity towards gram-negative bacteria, i.e. *Escherichia coli and Pseudomonas aeruginosa*, and gram-positive bacteria, i.e. *S. aureus* and *Bacillus cereus* [18].

4. Conclusion

In conclusion, 7-hydroxy-4-methyl coumarin was successfully synthesized from resorcinol and ethyl acetoacetate with PTSA 10 mol % under 800 W microwave irradiation at 150 s reaction time with 48.36 % yield. However, this compound did not exhibit an activity towards the acne-causing bacteria. Further research is needed to modify its structure together with evaluation on its antibacterial activity for other bacteria.

Acknowledgments

This work was financially supported by Universitas Indonesia under PUTI Prosiding 2020 research grant No. NKB-963/UN2.RST/HKP.05.00/2020.

References

[1] Bhate K and Williams H C 2013 *Brit. J. Dermatol.* **168** 474-85
[2] Winston M H and Shalita A R 1991 *Ped. Clin. North Am.* **38** 889-903
[3] Klein N C and Cunha B A 1995 *Med. Clin. N. Am.* **79** 789-801
[4] Kim S *et al.* 2018 *Molecules* **23** 3001
[5] De S K and Gibbs R A 2005 *Synthesis* **8** 1231-3
[6] Li S, Qi X and Huang B 2016 *Catal. Today* **276** 139-44
[7] Merza J 2018 *Chem. Mater. Res.* **10** 16-23
[8] Prajapati D and Gohain M 2007 *Catal. Lett.* **119** 59-63
[9] Lestari I I, Rahayu D U C, Nurani D A, Krisnandi Y K and Budianto E 2019 *AIP Conf. Proc.* **2168** 020066

[10] Yuniastuti A, Rahayu D U C, Nurani D A, Krisnandi Y K and Budianto E 2019 *AIP Conf. Proc.* **2168** 020067
[11] Rahayu D U C, Hartono and Sugita P 2018 *Rasayan J. Chem.* **11** 762-5
[12] Sugita P, Firdaus S O, Ilmiawati A and Rahayu D U C 2018 *J. Chem. Pharm. Res.* **10** 68-75
[13] Purwantiningsih *et al.* 2019 *Drug Invent. Today* **11** 386-90
[14] Purwantiningsih, Jannah N and Rahayu D U C 2020 *Rasayan J. Chem.* **13** 322-6
[15] Bose D S, Rudradas A P and Babu M H 2002 *Tetrahedron Lett.* **43** 9195-7
[16] Wahyuningrum D, Zulqarnaen M and Suendo V 2014 *AIP Conf. Proc.* **1589** 169-73
[17] Meier H, Gerold J, Kolshorn H and Muhling B 2004 *Chem.: Eur. J.* **10** 360-70
[18] Souza S M de, Monache F D and Junior A S 2005 *Z. Naturforsch. C.* **60** 693-700

The Effect of leaf surface character on the ability of water hyacinth, *Eichhornia crassipes* (Mart.) Solms. to transpire water

E D Maylani, R Yuniati and W Wardhana

Department of Biology, Faculty of Mathematics and Natural Sciences (FMIPA), Universitas Indonesia, Depok 16424, Indonesia.

Corresponding author's email: ratnayuniati@sci.ui.ac.id

Abstract. The objectives of this study were to evaluate the effect of leaf surface character on the ability of water hyacinth (*Eichhornia crassipes* (Mart.) Solms.) to transpire water. This experiment was performed using a healthy and acclimatized water hyacinth collected from Sawangan, Depok area. Five individual plants with the same size (6 leaves per plant), leaf length (10.4 ± 0.8 cm), leaf width (7.8 ± 0.7 cm) and root length (24.8 ± 7.4 cm) were put in each 20 L plastic tank containing tap water and allowed to grow in a greenhouse for six days. The transpiration rate of each plant was measured using a weighing method every day. The number of stomata, stomatal density and stomatal index were measured; amount of water loss from the individual plant through transpiration was analyzed. Based on the results of the study, it was found that the lowest leaf surface area is correlated with the number of stomatal and transpiration rates in the individual measured plants.

Keywords: *Eichhornia crassipes*, transpiration, number of stomata, stomata index

1. Introduction

Plants absorb water to fulfill their metabolic needs. Water absorption is influenced by several factors, one of which is the transpiration rate where the higher the transpiration rate, the higher amount of water will be absorbed by the roots [1]. Transpiration is the loss of water from the body of plants in the form of steam through stomata, cuticles or lenticels [2]. Approximately 90 % of the water contained in plants is transmitted out through the stomata. This is due to the breadth of the leaf surface and also the leaf surface which is more exposed to sunlight than other parts of a plant. In most plant species, transpiration via the cuticle is only 10 % [3].

The process of transpiration is influenced by various factors, namely internal and external factors. Internal factors include leaf size, the thickness of leaves, the presence or absence of a wax layer on the leaf surface, the number of trichome on the leaf surface, the number of stomata, and the shape and location of the stomata. External factors include the light intensity, temperature, air humidity, wind speed and groundwater content, potential groundwater gradients-tissue-atmosphere, and the presence of toxic substances in the environment [4]. Transpiration plays a role in the process of transporting water to the leaves and water diffusion between cells, absorption, and transport of water and nutrients, transport of assimilates, removing excess water, regulating stomata openings, and maintaining leaf temperature. Transpiration is closely related to stomata which are an internal factor that has the most influence on the process of transpiration rate. The greater the number of stomata, the higher the rate of transportation [5]. The number of water hyacinth stomata on the upper surface of the leaf is more than the bottom surface of the leaf. This is one form of adaptation of aquatic plants to accelerate the evaporation process so that physiological temperatures in plants are better maintained [4].

Water hyacinth has the ability to breed very quickly around 3%/day so that the water hyacinth is able to cover the entire surface of the water [6]. Water hyacinth can survive in waters for six years and has a high ability to grow in highly polluted water to absorb nutrients, metal ions, and organic pollutants. Besides the ability to absorb heavy metals, water hyacinth has some adverse effects including increase sedimentation and evapotranspiration. The loss of water by water hyacinth can reach three times greater than the rate of natural evaporation of surface water that does not have water hyacinth [7]. Studies have been conducted, evaporation of open water averaging 4.3 mm per day and evapotranspiration of aquatic plants has an average of 7.8 mm per day [8]. The increase of water loss through evapotranspiration of water hyacinth is considered as one of the most important weed problems in water bodies. Therefore, it is necessary to conduct a research about the leaf surface character on the ability of water hyacinth (*Eichhornia crassipes* (Mart.) Solms.) to transpire water. So, it can be estimated how long the water hyacinth can be used effectively without causing sedimentation.

2. Materials and method

2.1. Plant materials

The study was conducted in the greenhouse of FMIPA UI in June–July 2019. Water hyacinth were collected from Sawangan area, Depok. Five individual plants with the same number of leaves, leaf width, and stem height were selected for the experiments. After the selection, they were rinsed with tap water and placed in a 20 L plastic tank filled with tap water for 8 days to let them adapt to the new environment.

2.2. Measurement of transpiration

The rate of transpiration can be investigated by measuring the decrease in mass due to water loss. The experiment was done by weighing method conducted in the greenhouse of FMIPA UI. Each of five plants was placed individually in a plastic tank filled with 6 L of tap water. A thin film of oil was added to the surface of the water to prevent loss of water by evaporation. After that, the initial weight was calculated using a digital weighing scale (50 kg capacity). The transpiration was carried out for 6 days. During the experiments, temperature, wind speed, humidity, and light intensity were recorded at the exact time i.e. 11.00–12.00 a.m. On the 6th day, the final weight of the water hyacinth was measured to determine the amount of water loss from every tank. Any difference in weight will indicate the loss of water by the plant (due to transpiration). The transpiration rate is measured as the amount of water lost expressed in units of $g/cm^2/h$. After completing the transpiration process, each individual leaf was laid on a 1-mm grid paper and trace the outlines. All the leaf patterns were cut and the weight of the leaf patterns was measured using analytical scales.

2.3. Measurement of the number of stomata, stomatal density, and stomatal index

Each water hyacinth leaf was cut into three pieces and checked for the number of stomata both on the top and bottom of the leaf. Checking the amount of stomata was done by slashing the bottom and top layers of the water hyacinth leave that have been cleaned with alcohol before. After that the incision was transferred to the glass object that has been dripped with distilled water, then, one drop of safranin was added and covered with a glass cover [4]. The preparations were then examined using a Leica bl 500 microscopes with a magnification of 40 x 10. After that, a scale was given and numbered stomata and epidermal cells were used using ImageJ. The field of view of the microscope is 0.0798 mm^2 at a magnification of 40 x 10 with a scale of 50 μm. Stomata density can be calculated with the following formula [9]:

$$\text{Stomatal density} = \frac{\textit{Number of stomata}}{\textit{Field of view}}$$

Stomatal Index formula:

$$\text{Stomata Indeks} = \frac{\text{Number of stomatal}}{\text{Number of stomatal + number of epidermis cell}} \text{ x } 100\ \%$$

3. Results and discussion

Based on table 1, it is known that the average number of stomata in the adaxial is higher than in the abaxial section. The number of stomata is correlated to the density of stomata, higher number of stomata, higher density of stomata. The stomata index is a comparison of the number of epidermal cells in a certain unit of area. The highest stomata index was shown in the 5th test with the lowest number of epidermal cells, while the lowest stomatal index was shown in the 2nd test with the highest number of epidermal cells. The highest transpiration rate was shown on the 3rd test and the lowest was on the 4th test. At the lowest transpiration rate, it is correlated to the number of stomata, stomata density, weight difference and lowest leaf area.

Based on the research that has been done, the obtained results of external factors, internal factors, number of stomata/field of view, number of epidermal cells/field of view, stomata index/field of view and density of stomata/field of view can influence the transpiration of *Eichhornia crassipes* plants. The process of transpiration is influenced by various factors, namely internal factors, and external factors. Internal factors include leaf area, the thickness of leaves, the presence or absence of a waxy layer on the leaf surface, the number of trichome on the leaf surface, the number of stomata, and the shape and location of the stomata.

Based on the study of the number of stomata on adaxial (9.6 ± 1.1) is more dominant than abaxial (9.4 ± 0.5) of the leaves, leaf area 327.91–468.93 cm2, root length (24.8 ± 7.4 cm), leaf thickness of 0.028–0.034 μm, shape of stomata is paracytic, size of the upper surface stomata is 42.5 x 39.6 μm^2 and size of the lower surface stomata is 39.70 x 37.5 μm^2, have waxy layer and do not have trichome. The results of this study were consistent with other studies which stated that aquatic plants have a way to adapt to their environment such as large leaf area 532.96 cm^2, roots that are not too long 22.46 cm,

Table 1. The average number of stomata, epidermis, stomatal index, stomatal density, water weight differences, transpiration rate and leaf area.

Replication	Leaf surface	Number of stomata/ field of view	Number of epidermal cell/ field of view	Stomatal index (%)/ field of view	Stomatal density/ field of view	W0-W1 (kg)	Transpiration rate (g/cm^2/hour)	Leaf area (cm^2)
1	Adaxial	10	81	10.98	125.31	0.575	0.0204	468.93
	Abaxial	9	80	10.11	112.78			
2	Adaxial	9	84	9.67	112.78	0.485	0.0201	378.91
	Abaxial	10	85	10.52	125.31			
3	Adaxial	11	73	13.09	137.84	0.655	0.0275	399.87
	Abaxial	10	79	11.23	125.31			
4	Adaxial	8	60	11.76	100.25	0.39	0.0173	327.21
	Abaxial	9	71	11.25	112.78			
5	Adaxial	10	59	14.49	125.31	0.705	0.035	394.95
	Abaxial	9	66	12	112.78			

have a waxy layer, and stomata are generally large 47.5 μm x 35 μm and the location of the stomata is higher above the surface of the leaf [4, 10].

External factors include light intensity, temperature, air humidity, and wind speed [4]. Based on study, the average temperature is 29.41 °C, air humidity 49 %, wind speed 0–1.5 m/s and light intensity 974 Lux. The growth of water hyacinth in this study has met the optimal requirements. The fastest growth of water hyacinth is at 240,000 lux, and its minimum requirement is 24,000 lux, but it can grow under a broad range of light intensities, relative humidity of 15–75 %, a pH between 4.5–7.5, temperatures between 20–30 °C [11, 12].

3.1. Internal factors that affect transpiration rates

The leaf area influences the rate of transpiration. The wider the leaf area, the greater the rate of transpiration because wide leaves tend to have more stomata [5]. Figure 1 shows that the highest transpiration rate is shown in the 5th test with a value of 0.035, leaf surface area 394.5 cm^2 and the number of stomata 19. The lowest transpiration rate is shown on the 4th test with a value of 0.0173, leaf area 327.21 cm^2 and the number of stomata 17. Correlation analysis (figure 1) shows a positive relationship between the leaf area to the number of stomata with the correlation coefficient value of 0.505303. This shows that the correlation coefficient is moderate. The R square value obtained is 0.2553. These results indicate that the rate of transpiration which influenced by the number of stomata of 25.53 %. While the remaining 74.47 % is influenced by other factors.

The number and size of stomata are influenced by genotype and environment factors. The level of stomata density is influenced by the environmental factors such as: water availability, light intensity, temperature, and CO_2 concentration. The higher the intensity of light, the frequency of stomata on both leaf surfaces will be also increased, although the increase in frequency is not significant. Plants that grow in dry environments with high light intensity tend to have a lot of stomata, but their size is small compared to plants that grow in wet and protected environments. If the stomata are too close, it will inhibit the evaporation. Maximum water discharge occurs if the distance between stomata is 20 times its diameter. If the holes are too close together, then evaporation from one hole will inhibit the evaporation of the nearby holes. This is because the path taken by the water molecules passing through the hole is not straight but turn aside due to the influence of the corners of the closing cells [13].

Transpiration is closely related to stomata. Nearly 97 % of water from plants is lost through stomatal transpiration, so the more stomata, the higher the transpiration rate [3]. Correlation analysis (figure 2) regarding the relationship of the number of stomata to the rate of transpiration shows a positive relationship with the correlation coefficient value of 0.502162. This shows that the correlation coefficient is moderate. The R square value obtained is 0.2522. These results indicate that the rate of transpiration is influenced by the number of stomata of 25.22 %. While the remaining 74.78 % is influenced by other external and internal factors.

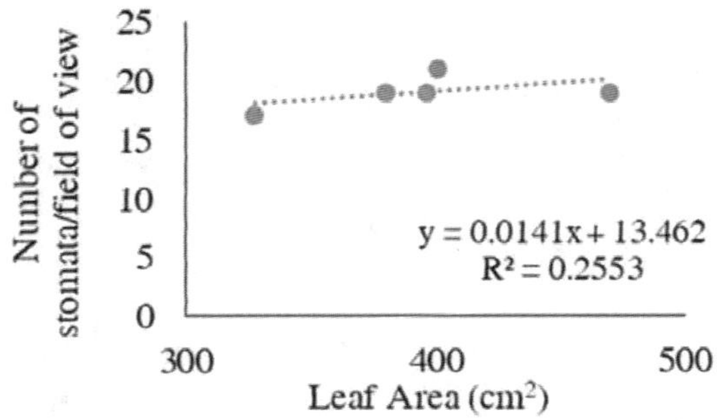

Figure 1. Linear regression graph between leaf surface area and number of stomata.

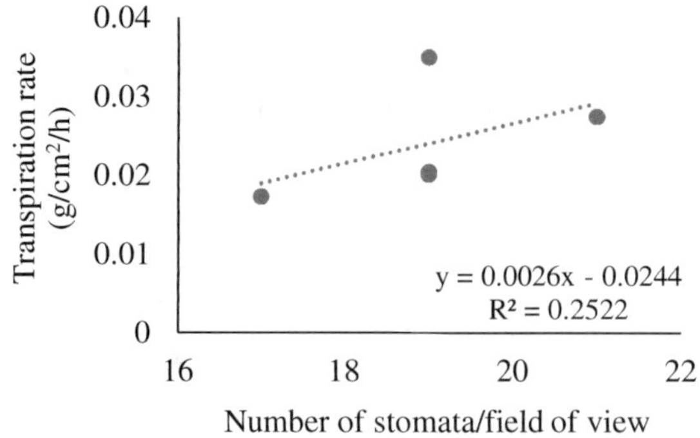

Figure 2. Linear regression graph between the number of stomata/field of view with the rate of transpiration.

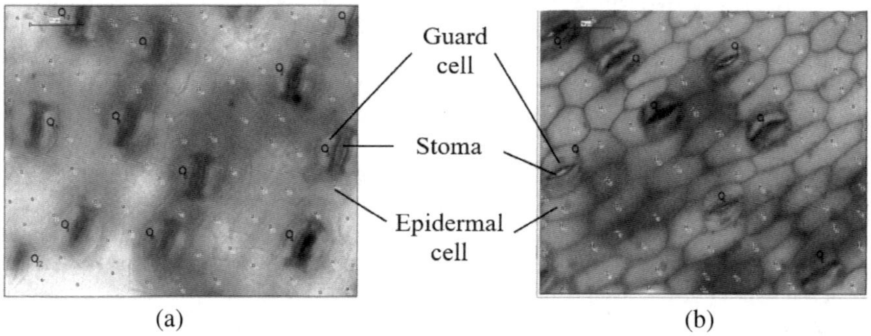

Figure 3. Paradermal section of the water hyacinth leaf showing the stomata. (a) Upper surface, and (b) Lower surface. (Bar = 50 μm).

Figure 3 shows stomata on the upper and lower surface of water hyacinth leaf. Based on table 1, it shows that on the 4th test, the smaller the leaf surface area, the less the number of stomata, which then influences the lower rate of transpiration. The same event was also shown in the 2nd test. If seen based on the number of stomata, the possibility of the rate of transpiration is low because the number of stomata on the upper surface of the leaf is less than on the lower surface of the leaf. This is less consistent with the literature which says the number of water hyacinth stomata on the upper surface of the leaf is more than the subsurface of the leaf [5]. This study has also been proven (figures 1 and 2 that show a positive correlation. Based on the 3rd and 5th tests, it showed a high stomata index due to the number of stomata is more than epidermal cells so that it can accelerate the evaporation of water from plants. Table 1 shows that the number of stomata at 1st, 3rd and 5th replications are in accordance with the literature which shows that the number of stomata is more on the leaf surface. This is one form of adaptation of aquatic plants to accelerate the evaporation process so that physiological temperatures in plants are better maintained [4]. In addition to leaf surface area, length roots affect the transpiration which is then related to the amount of water absorption by the root [14]. Trichomes on the leaves can prevent an excessive evaporation. In addition, evaporation can be reduced by the formation of a cuticle layer on the leaf surface which is quite thick. However, the water hyacinth plant does not have trichomes on the leaf surface. The waxy layer develops preventing the loss of water because of an intense heat [15].

Besides the number, the size of the stomata also determines the rate of transpiration. Stomata with a large size have a relatively higher transpiration rate [16]. Based on the research [4], the water hyacinth stomata has a length of 0.0475 mm and a width of 0.035 mm with the type of paracytic. The size of the water hyacinth stomata is the largest compared to other aquatic plants such as water spinach, lotus, yellow velvetleaf, and water jasmine. Stomatal opening is influenced by carbon dioxide, light, humidity, temperature, wind, leaf water potential and the rate of photosynthesis. In most plants, light and moisture in the low leaves make the guards cells lose their turgor, resulting in closure of the stomata [17]. Stomata are very helpful in terms of CO_2 absorption for photosynthesis. The distribution of stomata is closely related to the speed and intensity of transpiration in the leaves. For example, the location of one another with a certain distance within a certain limit, the more portions the faster the evaporation. If the holes are too close together, then evaporation from one hole will inhibit the evaporation of the nearby holes. This is because the path taken by the water molecules passing through the hole is not straight but turn aside due to the influence of the corners of the closing cells [13].

3.2. External factors that affect transpiration rates

The process of opening and closing the stomata is greatly influenced by light. Sunlight can affect the rate of transpiration in two ways, first, the light will affect the temperature of the leaf, so that it can affect transpiration activity and second, it can affect transpiration through its effect on opening and closing the stomata [14]. Rays contain heat, especially infra-red rays which can raise the temperature. The higher the intensity of sunlight, the higher the air temperature, so that the stomata will open wide which makes the transpiration rate higher.

In sunny conditions, the air does not contain much water. In these conditions the vapor pressure inside the leaf is much higher than the vapor pressure outside the leaf, so that water molecules diffuse from a high concentration (inside the leaf) to a low concentration (outside the leaf) so as to launch a transpiration [18]. Therefore, humidity affects the transpiration rate greatly. The lower RH air will further accelerate the rate of transpiration because water vapor will move from having high pressure (leaves) to low pressure (air) [16]. If the leaves have sufficient water content and open stomata, the rate of transpiration depends on the difference between the concentrations of water vapor molecules in the intercellular cavity in the leaves with the concentration of water vapor molecules in the air [19].

The wind has a dual influence that tends to conflict with the rate of transpiration. The wind carries the transpiration water vapor, so the wind decreases the humidity of the air above the stomata. However, if the wind sweeps the leaf surface, it will affect the temperature of the leaf. The temperature of the leaves will decrease and this can reduce the rate of transpiration. The rate of transpiration can be influenced by the availability of groundwater and the rate of absorption of water in the roots. During the day, water is usually transmitted faster than water absorption from the ground. This causes a deficit of water in the leaves, resulting in a large absorption at night. If the availability of groundwater decreases due to absorption by the roots, then the movement of water through the soil into the roots becomes slow. This tends to increase the water deficit in the leaves and decrease the rate of transpiration. Deeper rooting increases water availability and root proliferation increases the uptake of water from a unit of soil volume before plant withering [19]. In addition, the pH of water also affects the opening and closing of stomata. An increase in environmental pH stimulates phosphorylase to convert starch into glucose. An increase in osmose so that water enters from neighboring cells to cover cells, increasing volume causes turgor, so porous stomata are opened [13].

4. Conclusion

Based on the results of the study, the effect of leaf surface character on the ability of water hyacinth to transpire water can be summarized as: (1) the lowest leaf surface area is correlated with number of stomatal and transpiration rates in the individual measured plants. (2) External factors: the intensity of sunlight, temperature, humidity and wind speed affect the internal factors such as the number of stomata, the size of the stomata and the process of opening and closing the stomata which can affect the speed of the transpiration rate.

Acknowledgments
This work was financially supported by Universitas Indonesia under research grant PITTA 2019.

References

[1] Prijono S and Laksamana M T S 2016 *J-PAL* **7** 15-24
[2] Izza F and Laily A N 2015 *Seminar Nasional Konservasi Dan Pemanfaatan Sumberdaya Alam (Surakarta, Indonesia)* pp 17780
[3] Lovelles A R 1991 *Prinsip-Prinsip Biologi Tumbuhan Untuk Daerah Tropik* (Jakarta: Gramedia Pustaka Utama Indonesia)
[4] Firdaus M and Maasawet E T 2015 *Prosiding Seminar Nasional I Biologi, Sains, Lingkungan, dan Pembelajaran (Samarinda, Indonesia)* pp 81-92
[5] Papuangan N, Nurhasanah, Djurumudi M 2014 *Jurnal ßIOêduKASI* **3** 287-92
[6] Jaikumar M 2012 *Int. J. Recent Sci.* **3** 95-102
[7] Osmond R and Petroeschhevsky A 2013 *Water hyacinth Control Modules Control options for water hyacinth (Eichhornia crassipes) in Australia* available at https://www.dpi.nsw.gov.au/__data/assets/pdf_file/0005/505706/waterhyacinth-control-modules-full-accessible.pdf.
[8] Stan F-I, Neculau G, Zaharia L, IoanaToroimac G and Mihalache S 2016 *Procedia Environmental Sciences* **32** 281-9
[9] Lestari E G 2006 *Biodiversitas* **7** 44-8
[10] Dewi Y S 2012 *J. Tek. Ling.* **13** 151-8
[11] Tham H T 2012 *Water hyacinth (Eichhornia crassipes)-Biomass production, sensilability and feeding value to growing cattle* PhD Thesis (Uppsala: Department of Animal Nutrition and Management, Faculty of Veterinary Medicine and Animal Science, Swedish University of Agricultural Sciences)
[12] Téllez T R, López E M de R, Granado G L, Pérez E A, López R M and Guzmán J M S 2008 *Aquat. Invasions* **3** 42-53
[13] Haryanti S 2010 *Buletin Anatomi dan Fisiologis* **18** 21-8
[14] Salisbury F B and Ross C W 1995 *Plant Physiology* (Bandung: ITB Press, Indonesia)
[15] Akendo I C O, Gumbe L O and Gitau A N 2008 *Agricultural Engineering International: CIGR Journal* **10** 1-11
[16] Hamim 2012 *Fisiologi Tumbuhan* (Jakarta: Univeristas Terbuka, Indonesia)
[17] Gardner F P, Pearce R B and Mitchell R L 2008 *Terjemahan: Physiology of Crop Plants* (Jakarta: Penerbit Universitas Indonesia, Indonesia (UI-Press))
[18] Dwidjoseputro D l980 *Pengantar Fisiologi Tumbuhan* (Jakarta: Gramedia, Indonesia)
[19] Taiz L and Zeiger E 2003 *Ann. Bot.* **91** 750-1

Utilization of monosodium glutamate (MSG) for the synthesis of 1,8-dioxoacridine and polyhydro-quinoline derivates with SiO2-AA-glutamate catalyst and antioxidant ability test

A H Cahyana, T Andika and A R Liandi

Department of Chemistry, Faculty of Mathematics and Natural Sciences (FMIPA), Universitas Indonesia, Depok 16424, Indonesia

Corresponding author's email: herrykim@ui.ac.id

Abstract. Derivatives of 1,8-dioxoacridine and polyhydro-quinoline compounds are active compounds that have several beneficial bioactivities. The purpose of this research is to synthesize 1,8-dioxoacridine and polyhydro-quinoline derivatives supported by SiO_2-AA-Glu as the catalyst and test their antioxidant activity. The reaction was conducted at the optimum condition with ethanol solvent at boiling point temperature (78 °C) for 60 min with a catalyst concentration of 3 % mol with a yield of 88.16 %. Analysis of formation products was confirmed using FTIR, UV-Vis Spectrophotometer, and GC-MS. The results of the analysis produced three 1,8-dioxoacridine and polyhydro-quinoline derivatives, namely 3,3,6,6-tetramethyl-9-phenyl-3,4,6,7,9,10-hexahydro-2H,5H-acridine-1,8-dione (compound 1), 9-(4-hydroxy-phenyl)-3,3,6,6-tetramethyl-3,4,6,7,9,10-hexahydro-2H,5H-acridine-1,8-dione (compound 2), and 7,7-dimethyl-5-oxo-4-phenyl-1,4,5,6,7,8-hexahydro-quinoline-3-carboxylic acid ethyl ester (compound 3). Compound 2 shows the best activity as an antioxidant compared to the other two compounds with an IC_{50} value of 83.72 ppm.

Keywords: 1,8-dioxoacridine, polyhydro-quinoline, SiO_2-AA-Glu catalyst, antioxidant

1. Introduction

The acridine compound was first discovered in the 19th century [1]. The structure of this compound has similarities with 1,4-dihydropyridines (DHP) which is an analog of nicotinamide adenine dinucleotide (NADH) [2-6]. Research on the development of 1,8-dioxoacridine derivatives is still under investigation. Based on previous research reports, the compound 1,8-dioxoacridine has beneficial bioactivity in the health field. Among them, antimicrobial, carbonic anhydrase inhibitors for glaucoma treatment, anti-tumors, fungicides, as anti-multidrug-resistant, cytotoxic, and cardiovascular disease [7-9].

The use of catalysts in the synthesis of 1,8-dioxoacridine compounds was carried out to streamline the reaction. In this study, synthesized SiO_2-AA-glutamate made from monosodium glutamate (msg) was used to support the reaction. The use of msg as a catalyst is possible because of the dual properties as acids and bases of the structure [10]. In this study, the modification of msg was conducted to improve its capabilities. Heterogeneous catalyst SiO_2-AA-Glu has advantages compared to homogeneous catalysts which are more effective, non-corrosive, easily separated from the product produced, and can be reused.

IOP Conf. Series: Materials Science and Engineering **902** (2020) 012071 doi:10.1088/1757-899X/902/1/012071

2. Materials and method

2.1. Chemical reagents and instrumentations

In this research, the tools needed for synthesis and analysis structure were glassware, magnetic stirrer, thermometer, filter paper, TLC, stirring hot plate IKA C-MAG HS 7 shaker Julabo SW22. The analytical instruments used were the Fourier Transform Infrared (FTIR) IRPrestige-21 Shimadzu Spectrophotometer, the Shimadzu UV-Vis UV-2450 Spectrophotometer, and the GC-MS 6890 spectroscopy. The materials used in this study were ethyl acetate, n-hexane, SiO_2, Monosodium Glutamate (MSG) Viet Shen brand, sodium alginate, ammonium acetate, variations of aldehydes (benzaldehyde, hydroxy benzaldehyde), dimedone, ethyl acetoacetate, aquadest, ethanol 96 %, HCl 1 M, DPPH.

2.2. Synthesis of glutamic acid from monosodium glutamic

Monosodium Glutamate (MSG) was weighed as much as 2.5 g and placed in a beaker then added with 15 mL of HCl or CH_3COOH 1 M. The mixture was stirred for 10 min at 70 °C. The precipitated sludge was filtered and recrystallized with ethanol 96 %. The synthesized product was characterized by TLC, FTIR and UV-Vis spectrophotometer.

2.3. Synthesis of SiO_2-AA-Glu catalyst

Two beakers were prepared and the first beaker was filled with 1 g of sodium alginate dissolved in 20 mL of distilled water. The second beaker was filled with 1 g of glutamic acid that has been synthesized, then each beaker was added with 3 g of silica and stirred. Then, 3 mL of 1M HCl was added to form sediments. The sediment was then heated to vaporize H_2O and washed with distilled water to neutral pH. After that, the product was dried and formed 4.8 g of white powder SiO2-AA-Glu and characterized by FTIR.

2.4. Synthesis of 1,8-dioxoacridine derivatives

Aldehyde compounds (1 mmol), dimedone (2 mmol), ammonium acetate (1.5 mmol) and SiO_2-AA-Glu catalyst were mixed (figure 1). The mixture was heated in an oil bath and under reflux conditions. The reaction process was monitored using TLC (Ethyl acetate:n-hexane = 3:7). Then, the solvent was evaporated to get the product. The product was purified by recrystallization using ethanol 96 %. Product characterization analysis was performed with (FTIR), UV-Vis spectrophotometer and GCMS.

2.5. Antioxidant activity test

The sample solution was prepared in a series concentrations of 500, 300, 200 100, 50, 10 and 5 ppm in a methanol solvent. Then, 2 mL each solution was reacted with 1mL DPPH 0.1 mM. The reaction mixture was then incubated at room temperature for 30 min. After that, the absorbance of the solution was measured at a wavelength of 518 nm using a UV/Vis spectrophotometer. Radical scavenging activity was expressed as a percent of inhibition which can be calculated with the following formula:

$$\%\text{inhibition} = \frac{(\text{absorbance of control} - \text{absorbance of sample})}{\text{absorbance of control}} \; x \; 100\%$$

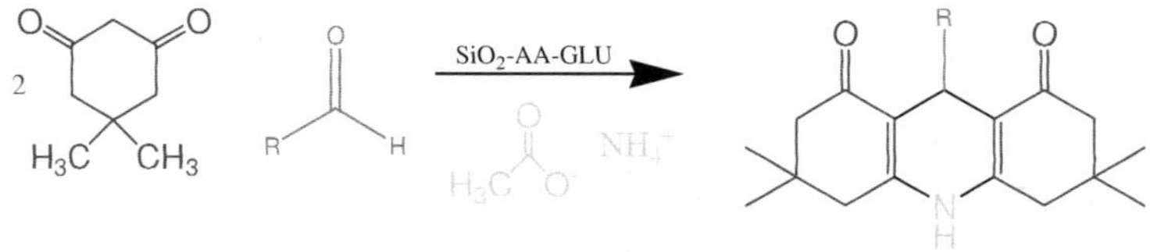

Figure 1. Reaction scheme for the synthesis of 1,8-dioxoacridine derivatives.

The percentage of inhibition was then plotted with a logarithmic line to get IC_{50} with the x-axis as concentration (ppm) and the y-axis as a percent of inhibition [11-12].

3. Results and discussion

Synthesis of glutamic acid using the acidification method is considered the most effective because it can properly substitute Na^+ ions to H^+ ions. HCl acid is an acid that can be well hydrolyzed in solution because it is a strong acid. To see its characteristics, an analysis with FTIR was carried out to find out the functional groups of compounds. Figure 2a shows a comparison between the FTIR spectrum of glutamic acid and monosodium glutamic. Both FTIR spectra show peaks in wave numbers broadened around 2750–3250 cm^{-1} which is O-H stretching vibrations, but the O-H peak of MSG is different from glutamic acid due to the influence of ionic substituents found in MSG. In the wavenumber between 3500 and 3300 cm^{-1} indicate the existence of N-H primary vibrations. The N-H peak of glutamic acid is indicated by the overlapping with the O-H peak so that it covers the N-H peak. In addition, there is also a stretching vibration C=O at wavenumber 1750 cm^{-1} with strong intensity and C-H at 3000 cm^{-1}. Characterization using UV-Vis spectrophotometry (figure 2b) showed the maximum wavelength of glutamic acid was 262 nm. However, in MSG there is an increase in wavelength uptake to 267 (rightward shift), which is called the bathochromic shift due to the auxochrome substitution of O-H with O-Na. MSG yield obtained from the synthesis is 92.88 %.

SiO_2-AA-Glu catalyst is a heterogeneous catalyst based on green chemistry. In this catalyst structure, sodium alginate which is reacted with glutamic acid forms hydrogen bonds that are bound to oxygen groups with oxygen and nitrogen in the catalyst structure. SiO_2 acts as a binder between sodium alginate and glutamic acid. The addition of HCl aims to cut the alginate polymer bonds to form sediments consisting of catalyst layers which are then dried to form powders. The catalyst yield obtained by 96 %. The characterization of SiO_2-AA-Glu with FTIR (figure 3) produced several absorptions, including O-H stretching vibrations of carboxylic acid at wave number 3445 cm^{-1}, vibrations of stretching carbonyl (C =O) at wave number 1627 cm^{-1} and stretching vibration of Si-O-Si at wave number 1097 cm^{-1}.

The evaluation of the ability of SiO_2-AA-Glu as a catalyst was carried out by synthesizing 1,8-dioxoacridine and polyhydro-quinoline derivatives. Two 1,8-dioxoacridine compounds were synthesized by reacting benzaldehyde (compound 1)/4-hydroxybenzaldehyde (compound 2), dimedone, and ammonium acetate. While the synthesis of polyhydro-quinoline derivative was carried out by reacting dimedone, benzaldehyde, ammonium acetate and ethyl acetoacetate (compound 3). The three synthesized compounds were analyzed by functional group vibration, maximum wavelength,

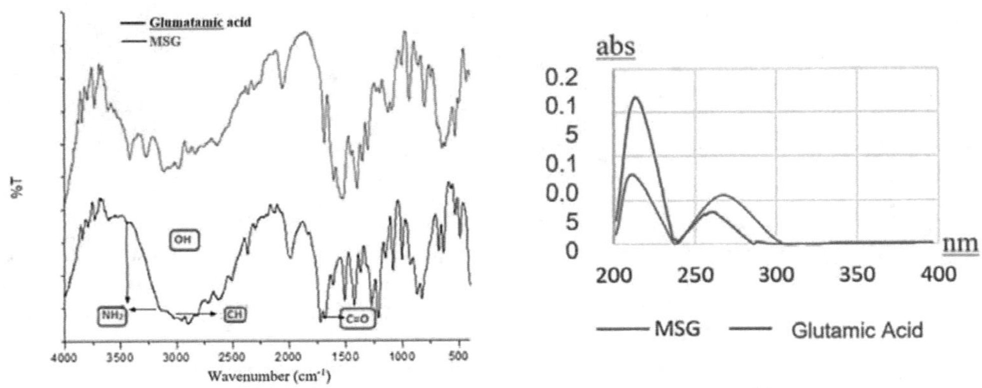

(a) (b)

Figure 2. (a) FTIR and (b) Uv-Vis spectra of MSG and glutamic acid.

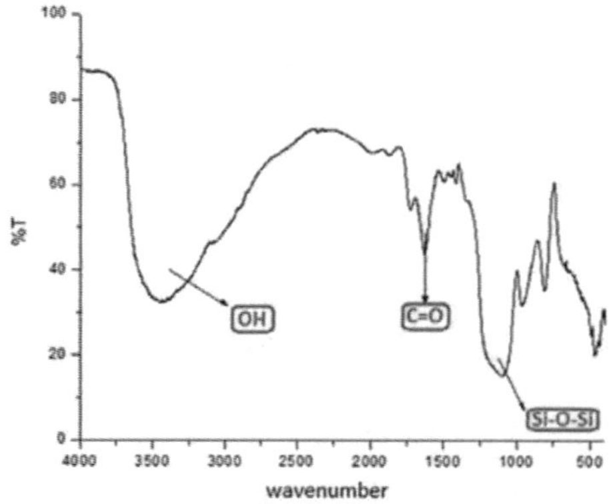

Figure 3. FTIR spectrum of SiO_2-AA-Glu catalyst.

Table 1. Characterization data and yield of 1,8-dioxoacridine and polyhydro-quinoline derivatives.

No	Product	Characterization data	Yield (%)	Photo
1	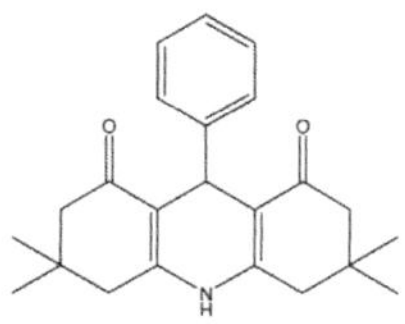	Yellowish white powder. IR (cm^{-1}): 3201 (N-H stretching), 1495 (N-H bending), 2996 (C-H sp^2), 2902 (C-H sp^3), 1404 (C-N), 15140 (C=C), and 1623 (C=O). UV-Vis (nm): 266. MS (m/z): 349	88.16	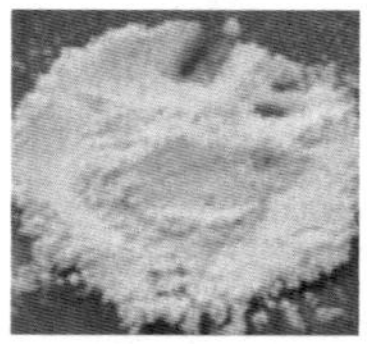
2	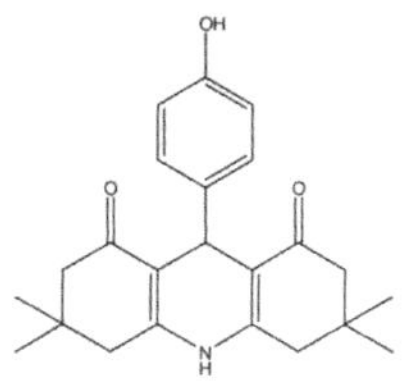	Yellow powder. IR (cm^{-1}): 3234 (N-H stretching), 1485 (N-H bending), 2987 (C-H sp^2), 2900 (C-H sp^3), 1413 (C-N), 1506 (C=C), and 1664 (C=O). UV-Vis (nm): 265. MS (m/z): 364.9	85.11	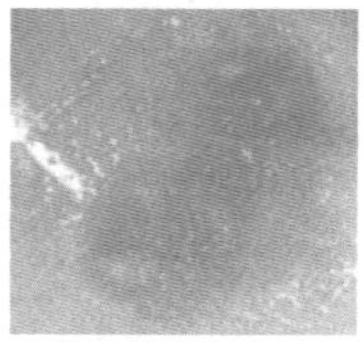
3	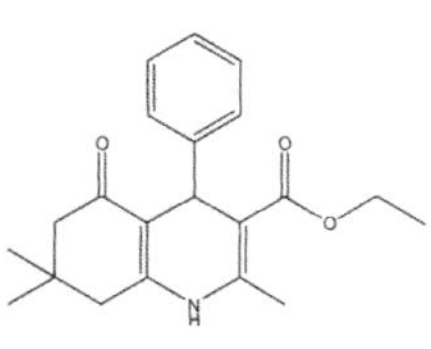	Yellowish white powder. IR (cm^{-1}): 3252 (N-H stretching), 1514 (N-H bending), 3001 (C-H sp^2), 2930 (C-H sp^3), 1423 (C-N), 1572 (C=C), and 1744 (C=O). UV-Vis (nm): 364. MS (m/z): 339.2.	71.07	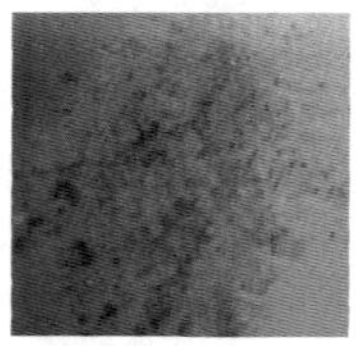

and molecular weight by FTIR, UV-Vis spectrophotometer, and GCMS. The results of analysis with these instruments, it was found that 1,8-dioxoacridine and polyhydro-quinoline derivatives have been successfully synthesized as 3,3,6,6-Tetramethyl-9-phenyl-3,4,6,7,9,10-hexahydro-2H,5H-acridine-1,8-dione (compound 1), 9-(4-Hydroxy-phenyl)-3,3,6,6-tetramethyl-3,4,6,7,9,10-hexahydro-2H,5H-acridine-1,8-dione (compound 2), and 7,7-Dimethyl-5-oxo-4-phenyl-1,4,5,6,7,8-hexahydro-quinoline-3-carboxylic acid ethyl ester (compound 3). Product characterization and yield data are illustrated in table 1.

The optimum conditions for the reaction to obtain the optimum yield as shown in table 1 was carried by varying the reaction conditions. Determination of the optimum reaction conditions was carried out by synthesizing compound 1 by varying the reaction temperature, reaction time, catalyst concentration, and type of solvent (table 2). The optimum conditions obtained for the synthesis of compound 1 in this study were with ethanol solvent at boiling point temperature (78 °C) for 60 min with a catalyst concentration of 3 % mol with a yield 88.16 %. In green chemistry, it would be more beneficial to use water as a solvent (entry 11) in this reaction. This is caused by the high yield produced, more environmentally friendly, and requires lower costs.

Experimental evidence in support of the formation of 1,8-dioxoacridine and polyhydro-quinoline derivatives through FTIR obtained several specific peaks. Figure 4a shows the absorption at 3201 cm^{-1} which is N-H stretching vibration, 1495 cm^{-1} is N-H bending vibration, 2996 cm^{-1} is the absorption of C-H sp^2 stretching vibration (benzene ring), while 2902 cm^{-1} is of C-H sp^3 stretching vibration. In addition, there is a peak at 1404 cm^{-1} is as stretching vibration of C-N, 15140 cm^{-1} is stretching vibration of C = C, and 1623 cm^{-1} is the absorption of carbonyl groups (C = O). UV-Vis Spectrophotometer analysis (figure 4b) showed the maximum wavelength of compound 1 was 266 nm. Peak compound 1 shifts to the right (bathochromic shift) compared with its precursors.

Table 2. Optimization of the reaction condition for the synthesis of compound 1.

Entry	Catalyst	Time (min)	Temp. (°C)	Solvent	% yield
1	0	60	78	Ethanol	40.45 %
2	1	60	78	Ethanol	61.63 %
3	2	60	78	Ethanol	74.12 %
4	3	60	78	Ethanol	88.16 %
5	5	60	78	Ethanol	85.55 %
6	10	60	78	Ethanol	84.11 %
7	3	45	78	Ethanol	85.00 %
8	3	30	78	Ethanol	72.12 %
9	3	60	50	Ethanol	65.20 %
10	3	60	Room temperature	Ethanol	62.20 %
11	3	60	70	water	75.15 %
12	3	60	78	Ethanol	85.11 %
13	3	60	78	Ethanol	86.75 %

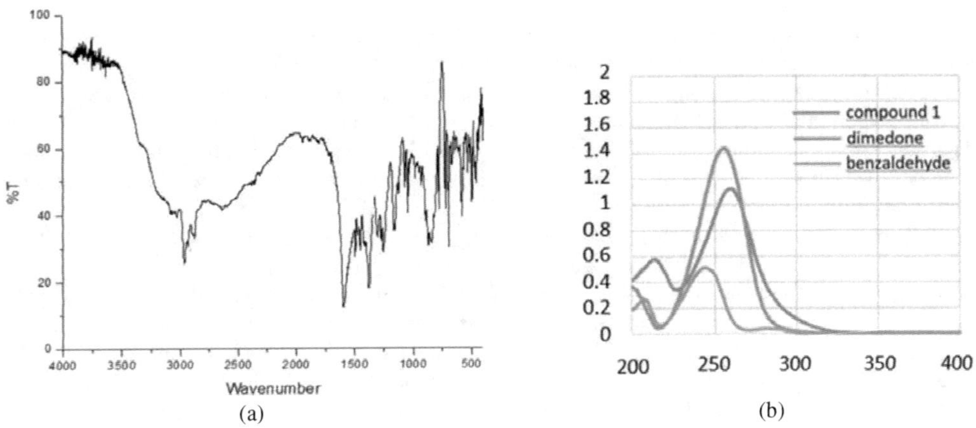

Figure 4. FTIR (a) dan UV-Vis, and (b) spectra of compound 1.

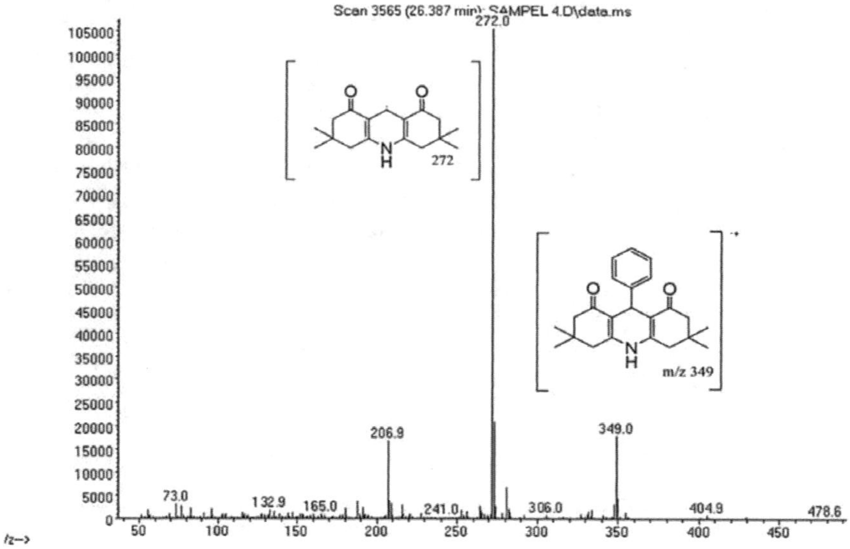

Figure 5. Mass spectrum of compound 1.

Table 3. Antioxidant activity test of 1,8-dioxoacridine and polyhydro-quinoline derivatives.

Sample	IC_{50} (ppm)
Compound 1	90.37
Compound 2	83.72
Compound 3	95.77
Benzaldehyde	715.68
4-Hidroxybenzaldehyde	135.31

The determination of molecular structure from the aspect of molecular weight has been carried out using GCMS. Based on the results of the peak on chromatogram obtained from the chromatographic process, there is a peak with m/z 349. This peak was then analyzed with a mass spectrum as shown in figure 5. Compound 1 with the molecular formula $C_{23}H_{27}NO_2$ has a molecular weight of 349 g/mol which is appropriate with the results of the analysis of this instrument. Some important fragments found in the mass spectrum of compound 1 are m/z 272. In this condition, it is estimated that compound 1 loses the benzene ring.

DPPH compounds are free radicals that are stable due to electron delocalization. The delocalization causes the purple color in the DPPH solution. The mechanism of the antioxidant activity of a compound is by donating hydrogen atoms to DPPH radicals. The DPPH radical compound will change to its yellow form after receiving the hydrogen atom from the antioxidant compound. The antioxidant activity value is determined as the IC_{50} value, which states the minimum concentration needed to inhibit 50 % of free radical activity. IC_{50} values of each product are shown in table 3. The addition of the O-H group which is a differentiator of compound 2 with compound 1, makes compound 2 has a better antioxidant bioactivity than compound 1.

4. Conclusion

In summary, the utilization of monosodium glutamate (MSG) for the synthesis of 1,8-dioxoacridine and polyhydro-quinoline derivates were successfully synthesized using SiO_2-AA-Glu as a catalyst. The determination and characterization of products were analyzed by FTIR, Uv-Vis, and GC-MS and showed the properties of the target compound. The analysis identified that the compound formed was 1,8-dioxoacridine and polyhydro-quinoline derivates namely 3,3,6,6-Tetramethyl-9-phenyl-3,4,6,7,9,10-hexahydro-2H,5H-acridine-1,8-dione (compound 1), 9-(4-Hydroxy-phenyl)-3,3,6,6-tetramethyl-3,4,6,7,9,10-hexahydro-2H,5H-acridine-1,8-dione (compound 2), and 7,7-Dimethyl-5-oxo-4-phenyl-1,4,5,6,7,8-hexahydro-quinoline-3-carboxylic acid ethyl ester (compound 3). The best condition of the reaction was carried out with ethanol solvent at boiling point temperature (78 °C) for 60 min with a catalyst concentration of 3 % mol with a yield of 88.16 %. Compound 2 shows the best activity as an antioxidant compared to the other two compounds with an IC_{50} value of 83.72 ppm.

Acknowledgments

This research was financially supported by Directorate General of Higher Education (DIKTI), Republic of Indonesia through Penelitian Dasar Unggulan Perguruan Tinggi (PDUPT) 2019 with contract No. NKB-1586/UN2R3.1/HKP.05.00/2019.

References

[1] Ulus R, Yesildag I, Elmastas M and Kaya M 2015 *Med. Chem. Res.* **24** 3752-9
[2] Hempel A, Hull SE, Bogucka-Ledochowska M and Dauter Z 1979 *Acta Cryst.* **B35** 474-6
[3] Srividya N, Ramamurthy P and Ramakrishnan VT 1997 *Spectrochim. Acta* **53** 1743-53
[4] Singh S, Chhina S, Sharma V K and Sachev S S 1982 *J. Chem. Soc.* 453-4
[5] Venkatachalapathy B, Ramamurthy P, Ramakrishnan VT 1997 *J. Photochem.Photobiol. A* **111** 163-9
[6] Saygılı R et al. 2015 *J. Mol. Struct.* **1083** 252-9
[7] Aday B, Pamuk H, Kaya M and Sen F 2016 *J. Nanosci. Nanotechnol.* **16** 6498-504
[8] Kaya M, Basar E, Çakir E, Tunca E and Bulbul M 2012 *J. Enzyme Inhib. Med. Chem.* **27** 509-14
[9] Ulus R, Yesildag I, Tanc M, Bülbül M, Kaya M and Supuran CT 2013 *Bioorg. Med. Chem.* **21** 5799-805
[10] Kolsepatil S R, Pahade N and Lingampalle D 2017 *Der Pharma Chemica* **9** 52-5
[11] Subbareddy C V and Sumathy S 2017 *New J. Chem.* **41** 9388-96
[12] Febriantini D, Cahyana A H, Yunarti R T, Liandi A R and Ardiansah B 2019 *AIP Conf. Proc.* **2168** 020068

IOP Conf. Series: Materials Science and Engineering **902** (2020) 012072 doi:10.1088/1757-899X/902/1/012072

Reduction of curcuminoid extract from turmeric (*Curcuma longa*) rhizomes and its antibacterial activities

S Q Dewi, E Saepudin and D U C Rahayu

Department of Chemistry, Faculty of Mathematics and Natural Sciences (FMIPA), Universitas Indonesia, Depok 16424, Indonesia

Corresponding author's email: endang.saefudin@sci.ui.ac.id

Abstract. Curcuminoid is an active compound of turmeric rhizomes commonly used as yellow pigment. In some cases, the presence of pigment in beauty product is not desirable due to its aesthetics effect. In this study, yellow color removal of curcuminoid was carried out by two steps hydrogenation, i.e. using $LiAlH_4$ followed by Pd-C catalyst to form dihydrocurcuminol and hexahydrocurcuminol derivatives, respectively. Dihydrocurcuminol derivatives had successfully synthesized by observing a wavelength shift from 419 to 428 nm in UV-Vis spectra and FTIR spectra also showed a reduction intensity of the carbonyl groups at 1600 cm^{-1}. Moreover, hexahydrocurcuminol derivatives were identified at 283 nm in UV-Vis spectra while in FTIR spectra showed an absorption band of C-H sp^3 at 2942 cm^{-1}. Curcuminoid and hexahydrocurcuminol derivatives were evaluated for their antibacterial activities against *Staphyloccocus aureus* with 2 and 2.5 mm and *Fusobacterium nucleatum* with 4 and 2.75 mm inhibitory zone diameter, respectively. It can be concluded that curcuminoid and hexahydrocurcuminol derivatives had a weak antibacterial activity against both bacteria tests.

Keywords: hydrogenation, curcuminoid, turmeric, antibacterial activities

1. Introduction

Turmeric (*Curcuma longa*) is a herb plants belongs to Zingiberaceae family that commonly used as herbal and traditional medicine for various diseases such as rheumatoid arthritis, chronic anterior uveitis, conjunctivitis, skin cancer, small pox, chicken pox, wound healing, urinary tract infections, and liver ailments [1]. Curcuminoid, an active compound contained in turmeric, is known to have antioxidants, antitumor, anticancer, antimicrobial, anti-Alzheimer's, and antitoxic activities [2]. According to Muhammad et al., spices and herbs contain antibacterial compounds [3]. Turmeric contains some active compounds that can be used as an antibacterial agent such as essential oil, alkaloid, flavonoid, tannin, curcuminoid, and terpenoid derivatives [4].

Due to its color, curcuminoid is rarely used as antibacterial in some beauty product, such as toothpaste, mouthwash, and soap. To decolorize its color, structure transformations by reduction of carbonyl ketone group and double bond alkene group should be conducted to remove the conjugated double bond. Furthermore, tetrahydrocurcuminoid, hexahydrocurcuminoid, and octahydrocurcuminoid (hexahydro-curcuminol) as the reduced form of curcuminoid are more hydrophilic than curcuminoid. Therefore, they have better stability in aqueous media [5] and this property is good for antibacterial and antioxidant activities.

In this research, curcuminoid extract from turmeric rhizome was successfully isolated using Soxhlet. Reduction of curcuminoid extract was performed with $LiAlH_4$ followed by Pd-C catalyst. Curcuminoid and hexahydrocurcuminol derivatives were then tested for their antibacterial activities towards mouth bacteria, i.e. *Staphylococcus aureus* and *Fusobacterium nucleatum*. Hexahydrocurcuminol derivatives were expected to have a better antibacterial activity than curcuminoid derivatives due to their better stability [6]. In addition, hexahydrocurcuminol derivatives were also expected to have a wider application due to the loss of the conjugated double bond which caused curcuminoid derivatives losing their colors. Therefore, it can be used as an active substance in cosmetic products, toothpaste, and mouthwash for further application.

2. Material and methods

2.1. Apparatus and materials
Turmeric (*Curcuma longa*) was obtained from Kramatjati Market, East Jakarta in August 2017. The chemicals used were purchased from Sigma Aldrich i.e. $LiAlH_4$ and 10 % (w/w) Pd-C while the rest were obtained from Merck i.e. *n*-hexane, ethanol, dichloromethane, acetone, diethylether, ethyl acetate, thin layer chromatography (TLC) plate, chloroform, methanol, dimethylsulfoxide (DMSO), sodium bromide, H_2 gas, and N_2 gas. For antibacterial activities, the chemicals used were Listerine®, *Staphyloccocus aureus* (Biochemistry Laboratory, Department of Chemistry, Universitas Indonesia), *Fusobacterium nucleatum* (Microbiology Laboratory, Faculty of Dentistry, Universitas Indonesia), and nutrient broth and agar from Merck. The instruments used were general laboratory glass equipment, rotatory evaporator, Soxhlet, vacuum pump, hydrogenation equipment, hotplate, analytical scales, magnetic stirrer, Shimadzu IR Prestige 21 spectrometer, Shimadzu UV-Vis spectrophotometer, petri dish, test tube, filter paper, vortex, water bath, and TLC equipment.

2.2. Curcuminoid derivatives isolation
Dried and powdered turmeric (40 g) was wrapped in filter paper and inserted into Soxhlet apparatus with ethanol as the solvent. The Soxhlet apparatus was heated at 60 °C in an oilbath for 9 hours. The extract then was concentrated using rotatory evaporator to obtain a brownish-red curcuminoid extract. The curcuminoid extract was washed with *n*-hexane followed by dichloromethane. The curcuminoid extract then was analyzed using TLC with silica as a stationary phase and chloroform: methanol (19:1, v/v) as a mobile phase. The curcuminoid extract was characterized by using FTIR dan UV-Vis.

2.3. Reduction of curcuminoid extract using $LiAlH_4$
Curcuminoid extract (0.5 g) and $LiAlH_4$ (0.05 g) were dissolved in diethylether (31.25 mL). 3 h stirring at room temperature was applied to the mixture. The reaction was stopped by adding aquadest into the solution. The solution then was extracted two times to separate the organic and inorganic phase. The organic phase was collected and diluted with 0.5 M HCl to obtain neutral condition. The product, dihydrocurcuminol derivatives, then was analyzed using TLC with silica as a stationary phase and *n*-hexane:ethyl acetate (9:7, v/v) as a mobile phase. The product then was further characterized using FTIR and UV-Vis.

2.4. Hydrogenation of dihydrocurcuminol derivatives using Pd-C 10 % (w/w) catalyst
Dihydrocurcuminol derivatives (0.5 g) were dissolved in 25 mL acetone and Pd-C 10 % (w/w, 25 mg) was added into 100 mL two-necked round bottom flask in H_2 atmosphere. The mixture then was stirred and maintained in a water bath at a constant temperature of 30 °C for 2.5 hours. After the reaction was completed, Pd-C catalyst was filtered from the mixture. The product, hexahydrocurcuminol derivatives, then was analyzed using TLC with silica as a stationary phase and *n*-hexane:ethyl acetate (1:1, v/v) as a mobile phase. The product then was further characterized using FTIR and UV-Vis.

2.5. Antibacterial activity test using disc diffusion method
Antibacterial activity of curcuminoid extract and hexahydrocurcuminol derivatives were evaluated using disc diffusion method [7]. Suspension of bacteria test was poured aseptically into a petri dish followed by adding the media at temperature ± 40 °C. Petri dishes containing bacteria and test media were stirred slowly and allowed to solid. Sterile disc paper with diameter of 6 mm was put in the petri dish and each tested compound (62.5, 125, 250, 500 and 1000 ppm) was dropped into disc paper and incubated at 37 °C for 24 hours. Listerine® was used as positive control while DMSO was used as negative control.

3. Results and discussion

3.1. Curcuminoid derivatives isolation
Curcuminoid extract (3.306 g, 8.27 % yield) was obtained as brownish-red gummy after Soxhlet extraction of turmeric rhizomes followed by washing with *n*-hexane and dichloromethane. TLC analysis of crude extract, *n*-hexane-soluble, dichloromethane-soluble fractions, and curcuminoid extract (figure 1) indicated that after washed with *n*-hexane and dichloromethane, the spot of curcuminoid derivatives in the extract can be clearly seen.

UV-Vis and FTIR spectra of curcuminoid extract are shown in figure 2. According to figure 2a, curcuminoid extract showed an absorption peak at a maximum wavelength of 419 nm. The resulting curcuminoid extract has a theoretical conformity with the maximum wavelength of standard curcuminoid at 428 nm [8]. FTIR spectra of curcuminoid extract (figure 2b) exhibited the absorption band for hydroxyl, carbonyl, double bond aromatic, and ether groups at the wavenumber of 3296, 1639, 1512, and 1282 cm^{-1}, respectively. All functional groups in FTIR spectra are related to curcuminoid structure.

3.2. Reduction using $LiAlH_4$
Reaction scheme of curcuminoid extract reduction using $LiAlH_4$ is illustrated in figure 3. $LiAlH_4$ is a reductive agent that converts carbonyl group into hydroxyl group. An orange solid was obtained (0.210 g, 48 % yield) and subjected to TLC analysis with *n*-hexane:ethyl acetate (9:7, v/v) as a solvent and further characterized using FTIR and UV-Vis (figure 4). TLC analysis exhibited that *Rf* value of curcuminoid and dihydrocurcuminol derivatives were 0.625 and 0.375, respectively. The lower value of dihydrocurcuminol *Rf* is due to the increase of polarity.

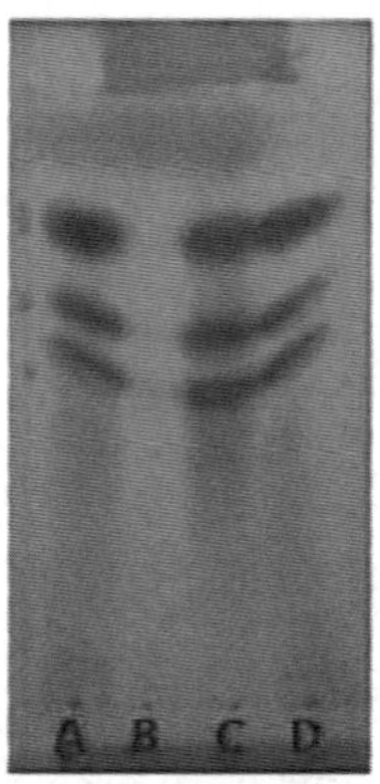

Figure 1. TLC analysis in chloroform:methanol (19:1, v/v) of crude curcuminoid (A), *n*-hexane-soluble (B), dichloromethane-soluble (C), curcuminoid extract (D) visualized under UV 254 nm.

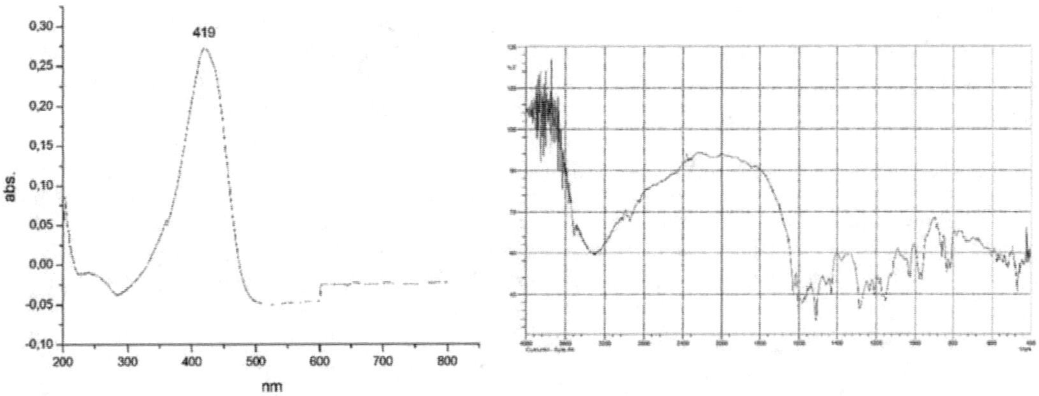

(a) (b)

Figure 2. (a) UV-Vis and (b) FTIR spectra of curcuminoid.

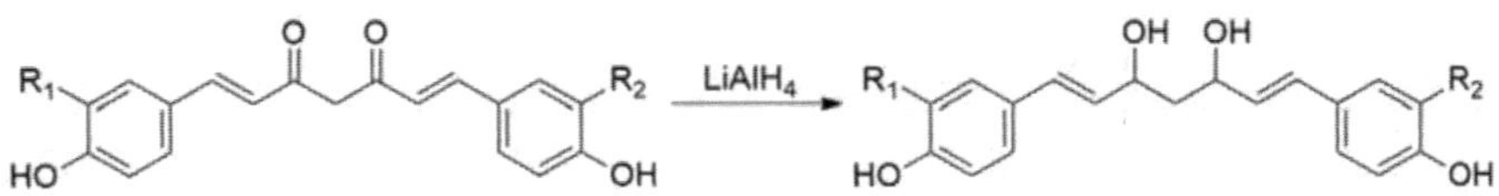

Figure 3. Reaction scheme of reduction using $LiAlH_4$.

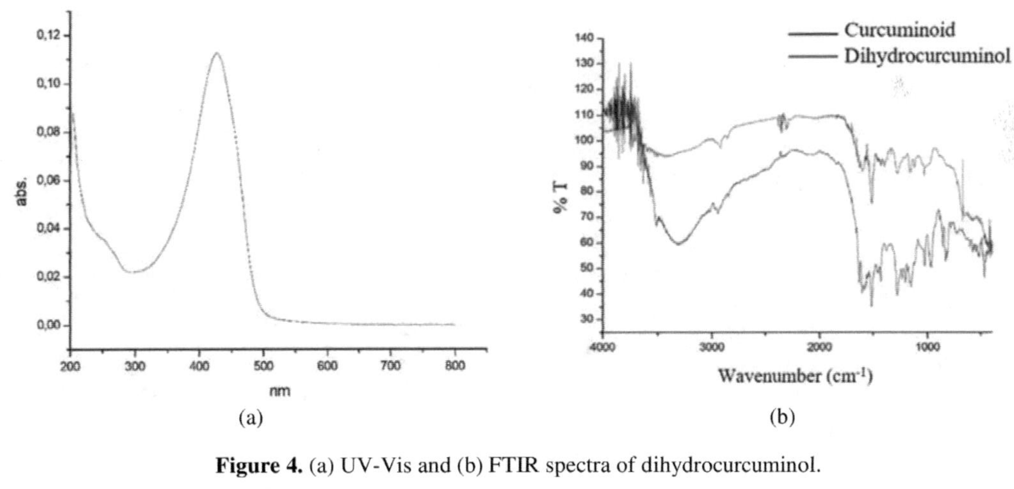

(a) (b)

Figure 4. (a) UV-Vis and (b) FTIR spectra of dihydrocurcuminol.

According to UV-Vis spectra (figure 4a), λ_{max} of dihydrocurcuminol was appeared at 428 nm. Comparing with curcuminoid extract (λ_{max} 419 nm), it indicates that there is a bathochromic shift due to the addition of hydroxyl groups as auxochrome. Therefore, it can be concluded that dihydrocurcuminol derivatives had been formed.

FTIR spectra of the compared curcuminoid and dihydrocurcuminol (figure 4b) showed the vibration peaks at wavenumber of 3446 cm^{-1} from the intramolecular bond of hydroxyl group. In addition, the increasing intensity of hydroxyl group peak compared with curcuminoid, it is estimated that more hydroxyl group will be produced. Moreover, the reduced intensity for carbonyl group at the wavenumber of 1690–1760 indicated that carbonyl group is also transformed to hydroxyl group. Based on FTIR spectra, it can be concluded that dihydroxycurcuminol had been formed.

3.3. Hydrogenation using Pd-C 10 % (w/w) catalyst

The hydrogenation process is the addition reaction of double bond C=C to yield single bond C-C which illustrated in figure 5. After the hydrogenation, hexahydrocurcuminol was obtained a colorless to yellow solid (0.187 g, 91.21 % yield). Hexahydrocurcuminol derivatives then were analyzed by TLC using *n*-hexane:ethyl acetate (1:1, v/v) as a solvent. TLC analysis was also performed by comparing with dihydrocuruminol. The *Rf* value of hexahydrocurcuminol was found at 0.2 while dihydrochurcuminol showed *Rf* value of 0.25. Based on *Rf* value, it can be concluded that hexahydrocurcuminol is more polar compound compared to dihydro-curcuminol.

Subsequent characterization was performed using UV-Vis (figure 6a) to measure wavelength absorbed by hexahydrocurcuminol derivatives. According to Figure 6a, the maximum wavelength of hexahydrocurcuminol was 285.4 nm. Compared with dihydrocurcuminol and curcuminoid, there was a hypsochromic shift which signifies the lost of conjugate double bond. Moreover, FTIR spectra of hexahydrocurcuminol (figure 6b) was performed to identify the changes of functional groups in this reaction. Based on figure 6b, hexahydrocurcuminol derivatives were successfully formed by the marked increase of C-H sp^3 intensity at the wave number of 2942 cm^{-1} and the reduced increase of the C-O bond of alcohol due to the loss of the conjugated double bond.

3.4. Antibacterial activity of curcuminoid extract and hexahydrocurcuminol

The antibacterial activity of curcuminoid and hexahydrocurcuminol were examined using disc diffusion method in various concentrations of 62.5, 125, 250, 500 and 1000 ppm (figure 7).

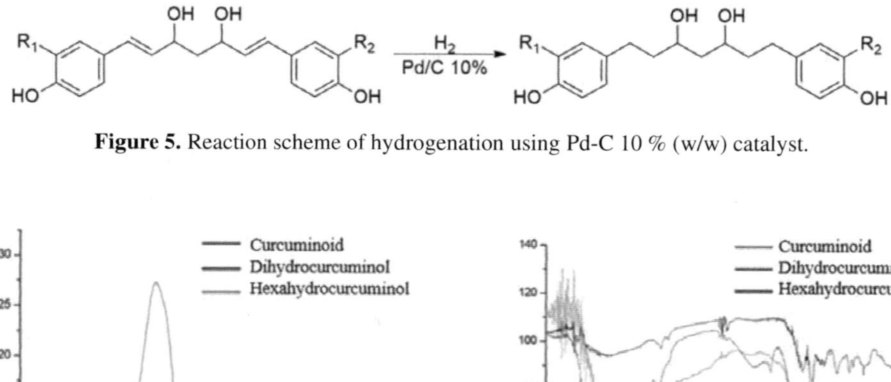

Figure 5. Reaction scheme of hydrogenation using Pd-C 10 % (w/w) catalyst.

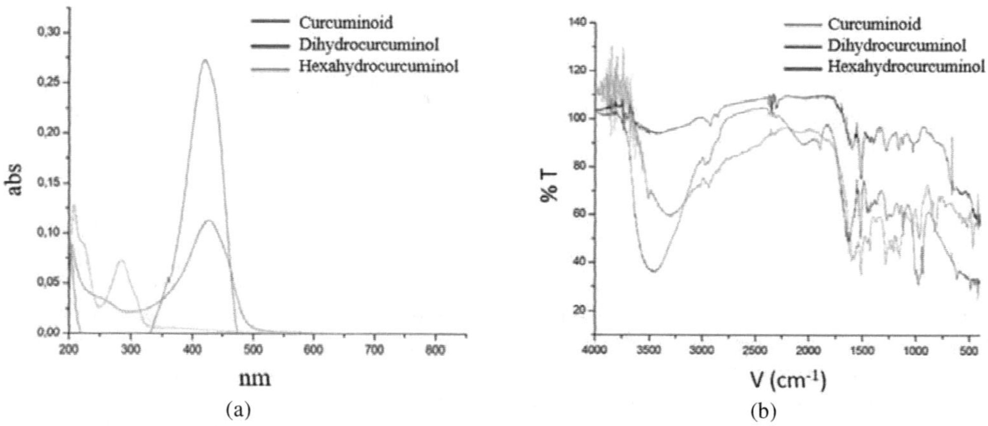

Figure 6. (a) UV-Vis and (b) FTIR spectra of hexahydrocurcuminol.

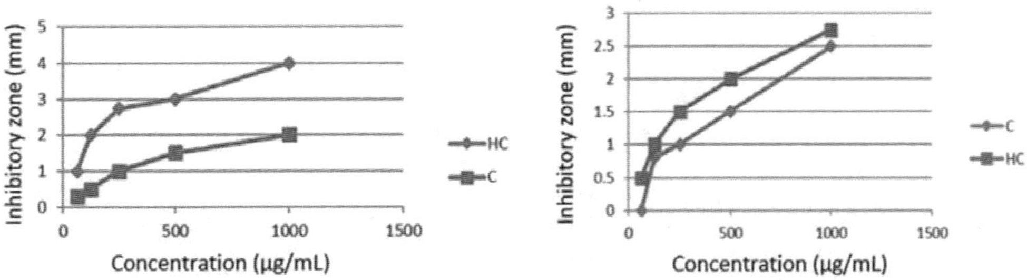

(a) (b)

Figure 7. Antibacterial activities of hexahydrocurcuminol (HC) and curcuminoid (C) towards (a) *S. aureus* and (b) *F. nucleatum.*

Disc diffusion method was intended to evaluate the inhibitory zone diameter of *S. aureus* and *F. nucleatum* growth compared to the negative control (DMSO) and positive control (Listerine®). According to figure 7, the greater concentration of the compound, the greater inhibitory zone will be produced. Hexahydrocurcuminol was found to be more effective than curcuminoid extract in terms of its antibacterial properties. It can be concluded that hexahydrocurcuminol can inhibit the development of gram-positive bacteria (*S. aureus*) and gram-negative bacteria (*F. nucleatum*).

4. Conclusion

Curcuminoid extract was successfully isolated from turmeric using Soxhlet extraction with 8.26 % yield (3.306 g). Carbonyl groups on curcuminoid derivatives were successfully reduced to hydroxyl groups using $LiAlH_4$ (3 h, 30 °C) to yield dihydrocurcuminol derivatives (0.210 g, 42 % yield). Further reduction of dihydrocurcuminol derivatives using Pd-C 10 % (w/w) catalyst (2.5 h, 30 °C) was carried out to obtain hexahydrocurcuminol derivatives (0.187 g, 91.21 % yield). Further separation and characterization need to be carried out to purify the compounds and establish the certain structures. Antibacterial activities of both curcuminoid and hexahydrocurcuminol derivatives were evaluated against *S. aureus* with 2 and 2.5 mm, and *F. nucleatum* with 4 and 2.75 mm inhibitory zone diameter, respectively. It can be concluded that the curcuminoid and hexahydrocurcuminol derivatives had a weak antibacterial activity against both bacteria tests.

References

[1] Dong S, Luo X, Liu Y, Zhang M, Li B and Dai W 2018 *Phytochem. Lett.* **27** 148-53
[2] Li R, Xiang C, Ye M, Li H F, Zhang X and Guo D A 2011 *Food Chem.* **126** 1890-95
[3] Muhammad S, Iqra Y, Muhammad I K, Muhammad N, Muhammad A S and Muhammad A 2016 *Pak. J. Food Sci.* **26** 153-60
[4] Qin N Y, Yang F Q, Wang Y T and Li S P 2007 *J. Pharm. Biomed. Anal.* **43** 486-92
[5] Sneharani A H, Singh S A, Srinivas P and Rao A G A 2011 *Eur. Food Res. Technol.* **233** 561-68
[6] Manjunatha J R, Bettadaiah B K, Negi P S and Srinivas P 2013 *Food Chem.* **139** 332-8
[7] Berberis C M, Sandoval E, Rodriguez C H, Ramirez M S, Famiglietti A, Almuzara M and Vay C 2018 *J. Glob. Antimicrob. Resist.* **14** 246-52
[8] Waranyoupalin R, Wongnawa S, Wongnawa M, Pakawatchai C, Panichayupakaranant P and Sherdshoopongse P 2008 *Cent. Eur. J. Chem.* **7** 388-94

IOP Conf. Series: Materials Science and Engineering **902** (2020) 012073 doi:10.1088/1757-899X/902/1/012073

Mercury effects on the early seedling of *Paraserianthes falcataria* (L.) Nielsen grew in hydroponic culture

R Yuniarti and R Yuniati

Department of Biology, Faculty of Mathematics and Natural Sciences (FMIPA), Universitas Indonesia, Depok 16424, Indonesia

Corresponding author's email: ratnayuniati@sci.ui.ac.id

Abstract. The purpose of this research was to investigate the impact of various level of mercury on the early seedling growth of sengon, *Paraserianthes falcataria* (L.) Nielsen using hydroponic culture. The study was carried out in twenty replications under greenhouse condition. The mercury used in this study were provided as chloride salts. Eight concentrations (30, 40, 50, 60, 70, 80, 90 and 100 μM) were used in this study. For each treatment, the pH was adjusted to 4.5. Fourteen-day-old seedlings were exposed to $HgCl_2$. Shoot and roots system of control and mercury-stress plants were harvested after 7 days of treatment. Seedling response to mercury stress was observed in the plant parts on the 3rd day after the treatment, and become severe on the 7th day. Results showed that 30 and 40 μM, the concentrations had minimum inhibitory effects on the seedling growth, respectively. The results also revealed both shoot and root dry weight of the treated seedling were sharply reduced as the mercury concentration increased to 60 μM. In a concentration of 70–100 μM the biomass dry weight was not much different. Research results show this plant can be considered as a strong candidate for a phytoremediation agent.

Keywords: Mercury effects, *Paraserianthes falcataria* (L.), hydroponic culture

1. Introduction

Plants in nature have to deal with various types of abiotic and biotic stresses. Abiotic stress is one of the factors that cause a decrease in crop productivity. At present, a very concerning environmental problem is the heavy metal pollution. Soil, water, and even air are polluted by heavy metals due to the rapid growth of industrialization, the use of different fertilizers, pesticides, and herbicides in the agricultural fields [1]. Heavy metals such as Fe, Mn, Cu, Ni, Co, Cd, Zn, Hg and arsenic accumulated in the soils through human activities [2].

Mercury (Hg) is an element that is less abundant on earth [3]. Typical normal soils usually contain 20–150 ppb mercury [4]. Increasing levels of mercury in the soil are often caused by the use of fertilizers, lime, and livestock manures. The amount of mercury presents in the soil and its absorption by plants depends on several variables such as soil pH, soil aeration, cation-exchange capacity, and plant species. The absorption can be reduced when the soil's pH is high [5, 6].

Plants have the ability to extract, concentrate, and accumulate various heavy metals from soil and water whether or not they are needed for growth [7]. Factors that influence the absorption and accumulation of heavy metals in plant tissues include temperature, moisture, organic matter, pH, nutrient availability and plant species [8, 9]. The stage of plant development and the field condition will determine the sensitivity of plants to heavy metals [10].

Paraserianthes falcataria (L.) Nielsen or sengon is a fast-growing, nitrogen-fixing, easily cultivated plant. This plant can grow on a wide range of soils. It can grow well on dry or damp soils and saline to acid soils as long as the drainage is sufficient [11]. As a legume plant that can fix nitrogen, this plant usually planted for reforestation [12]. Based on those characteristics, we assume this plant can be used as a phytoremediator in Hg-polluted soils. This study aims to examine the mercury effects on the early seedling of *P. falcataria* L. Nielsen that could become a supporting data for the potential use of this plant as a phytoremediator.

2. Materials and method

The *P. falcataria* pre-germinated seeds were imbibed in warm water at room temperature overnight. The seeds were sown on plastic pots containing zeolite media for fourteen days. After fourteen days in the zeolite, the seedlings were transferred into a rectangular tank filled with 20 L of full-strength Hoagland media solution containing $Ca(NO_3)_2$, KNO_3, KH_2PO_4, $MgSO_4$, Fe-EDDHA, $MnSO_4$, $ZnSO_4$. $7H_2O$, $CuSO_4.5H_2O$, H_3BO_3 and $(NH_4)_6 Mo_7 O_{24}.4H_2O$. The pH was adjusted to 5.6 and replaced with fresh solution every two days. Each plastic tank contained 20 seedlings. After seven days of adaptation, the seedlings were treated with different $HgCl_2$ concentrations: 30, 40, 50, 60, 70, 80, 90 and 100 µM for 7 days. Each concentration of $HgCl_2$ was added to half-strength Hoagland nutrients solution in an individual plastic tank.

This study was performed in the greenhouse at the Faculty of Mathematics and Natural Sciences, Universitas Indonesia from April to May 2019. The average daytime temperature during April to May 2019 is 28.1 °C. Observations were conducted daily to record changes in leaf color and condition, such as curling and wilting. Fallen leaves were gathered for further analysis. After the completion of the study, *P. falcataria* seedlings were harvested by separating the shoots and roots for weighing the wet weight (WW). And then, the samples were dried in an oven for 48 h at 65 °C to obtain dry weight (DW). Relative water content was calculated as, RWC (%) = [(WW - DW)/WW] x 100 [13].

3. Results and discussion

Hg-induced toxicity in plants causes a decrease in growth. *P. falcataria* seedlings treated to various levels of $HgCl_2$ demonstrated a slowing off of root and shoot growth after 7 days of treatment. Results showed that the 30 and 40 µM Hg concentrations had minimum inhibitory effects on the seedling growth, respectively. Plant biomass is a good indicator for assessing the growth conditions of plants exposed to heavy metal. The resulting study showed that mean root and shoot weight were different in the Hg treatment plants as compared with the control. The decreased root and shoot and of the treated plants (figure 1) demonstrated the severe impact of Hg concentrations on growth (both in shoots and roots) of *P. falcataria* seedlings, comparable to the wheat plant biomass decline patterns reported in Sahu et al. [14].

Plants exposed to heavy metal toxicity show symptoms of stunted growth, leaf chlorosis, and root browning [15]. Symptoms of damage will depend on the affected area . Mercury ions can enter together with water and bind aquaporin to prevent water entry to plants. Plants experience oxidative stress due to membrane damage that is triggered by disruption of the electron transport chain in the mitochondria and chloroplast [16]. The addition of 30 µM Hg did not cause changes in leaf color or seedling conditions. Plants exposed to Hg do not indicate symptoms of toxicity up to 30 µM Hg. At higher Hg concentrations (60–100 µM) toxicity symptoms appear such as chlorosis, drying of leaf margins, curly leaves, browning of roots, and falling leaves (figure 2). The seedlings developed leaf chlorosis on day 4 after the administration of $HgCl_2$. On day 5, all leaves begin to wilt and experience an earlier aging. After day 7 the administration of $HgCl_2$ was stopped because the plant died. The ability to accumulate

IOP Conf. Series: Materials Science and Engineering **902** (2020) 012073 doi:10.1088/1757-899X/902/1/012073

and tolerance of plants to heavy metals varied from one species to another, some shows symptoms of toxicity earlier than others.

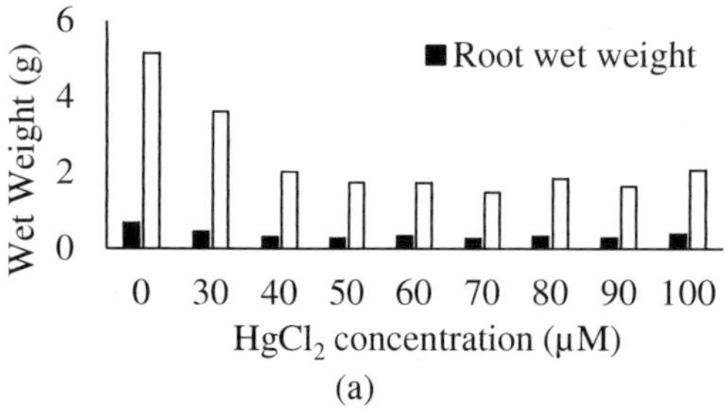

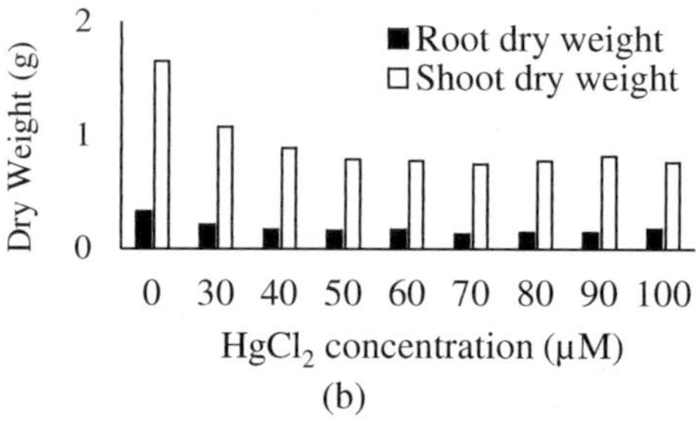

Figure 1. The wet and dry weight of shoot and root of *P. falcataria* seedlings which grown for 7 days in presence of $HgCl_2$ in hydroponic cultures.

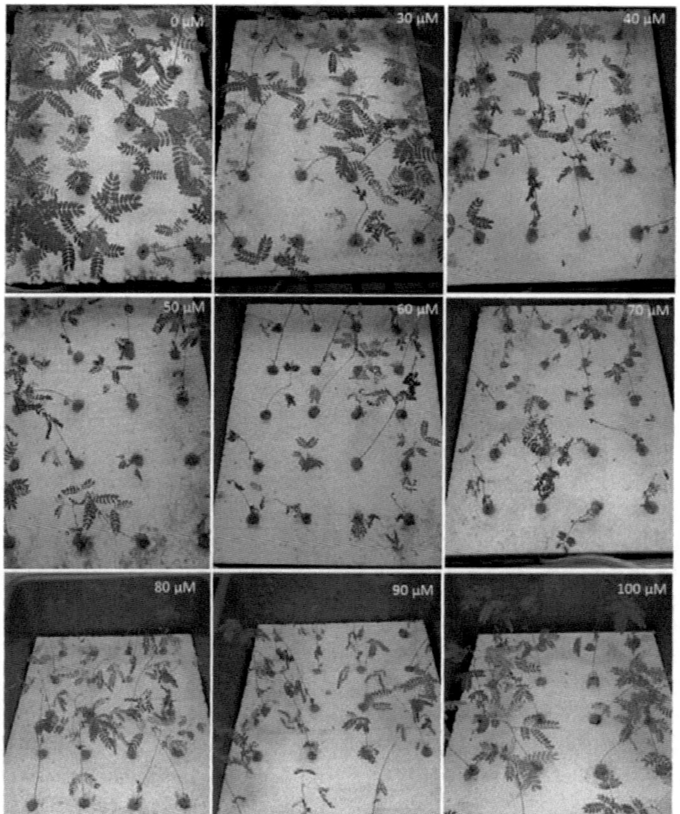

Figure 2. Mercury toxicity symptom in *P. falcataria* seedling grown in half-strength Hoagland solution with the addition of 30–100 μM $HgCl_2$.

Abiotic stress that strikes plants induces metabolic and physiological disruption [17], thereby reducing plant fitness productivity [18]. Abiotic stress is one of the stressors that have an impact on growth and is responsible for decreasing yield. In most plant species the percentages of growth reduction can reach more than 50 [19]. In this study, the decline in biomass indicates that photosynthesis is disrupted. Hg exposure can reduce chlorophyll synthesis, photosynthesis rate, transpiration, and water absorption because it is related to Hg which causes loss of potassium, magnesium, manganese, and iron. Magnesium and iron are needed to synthesize chlorophyll [20].

Figure 3 shows that the addition of higher levels of Hg caused the lower biomass production of *P. falcataria* plants. Hg^{2+} at concentrations from 40 to 100 μM caused a 66 % reduction in biomass.

The growth of the roots of the plant being in direct contact with the media is more affected compared to the shoot. In the present study, the plant biomass fresh and dry weight started to decline at a concentration of 30 μM, but there was only a slight different between 40 until 90 μM.

Plant growth and development are reduced because plants are lack of water, triggered by the accumulation of Hg ions in cells [21]. As an indicator of phytotoxicity, we measured the Relative Water Content (RWC). The RWC of our samples appears to be fluctuating but shows a downward trend. Starting at the treatment of 30 μM Hg there was a decrease in the range of 2–4 % (table 1). Relative water content is slightly higher in 30 μM Hg-treated plants than in the control. Similar patterns also occur in grass pea *Lathyrus sativus* which treated with lead. The addition of Pb might cause the stomata to close, and the atmospheric carbon fixing activities is disrupted [22].

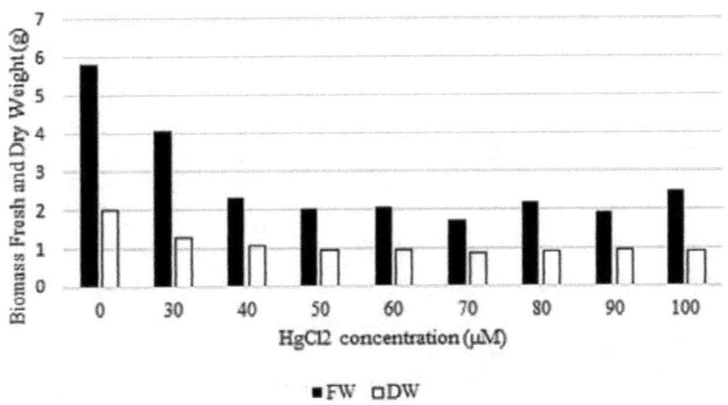

Figure 3. Mercury ($HgCl_2$) effect on fresh and dry weight biomass of *Paraserianthes falcataria* L. seedlings.

Table 1. Biomass fresh and dry weight and Relative Water Content of *P. falcataria* seedlings grown for 7 days in presence of $HgCl_2$ in hydroponic cultures.

Hg concentration (μM)	Biomass Fresh Weight (g)	Biomass Dry Weight (g)	Relative Water Content (%)
0	5.82	1.98	65.98
30	4.05	1.3	67.90
40	2.33	1.07	54.08
50	2.02	0.95	52.97
60	2.06	0.95	53.88
70	1.74	0.87	50.00
80	2.17	0.93	57.14
90	1.94	0.97	50.00
100	2.47	0.93	62.35

At the cellular level, Hg metal ions will bind with ligands containing sulfur and nitrogen (amino acids) to enter the cell. The damage due to mercury includes blocking the functional groups of enzymes, polynucleotides, or transportation systems for nutrient ions, denaturing and inactivating enzymes, damaging cells and disrupting the integrity of the organelle's membrane. One of the causes of mercury toxicity is the change in cell membrane permeability and the replacement of essential ions with Hg ions [23].

With enzyme denaturation, membrane damage, water deficit, and disruption of carbon fixation activities, photosynthesis are greatly reduced, which results in a decrease in plant biomass. But from the observations, not all plants show a stress response. Some individual plants are still healthy, namely the treatment of 30 μM and 100 μM (figure 2). This shows that some sengon plants can adapt and tolerate Hg

4. Conclusion

The present research indicated that increasing Hg^{2+} concentrations in the nutrient solution decreases both root and shoot biomass. The results also revealed both shoot and root dry weight of the treated seedling were reduced as the mercury concentration increases to 60 μM. At a concentration of 70 –100 μM the biomass dry weight was not much different. Based on the results of the study, more research is needed to ensure the possibility of *P. falcataria* (L.) Nielsen to become a phytoremediation agent because this study uses plants that are still in the early seedling phase.

Acknowledgments

The authors gratefully acknowledge Directorate Research and Community Service (DRPM) Universitas Indonesia to provide the grant from Hibah PITTA 2019.

References

[1] Zheng N, Liu J, Wang Q and Liang Z 2010 *Sci. Total Environ.* **40** 726-33
[2] Aydinalp C, Marinova S 2009 *Bulg. J. Agric. Sci.* **15** 347-50
[3] Goldwater L J 1971 *Sci. Amer.* **224** 15-21
[4] WHO [World Health Organization] 1976 *Environmental Health Criteria World Health Organization, Geneva* available at http://www.inchem.org/ documents/ehc/ehc/ehc001.htm
[5] Patra M, Bhowmik N, Bandopadhyay B and Sharma A 2004 *Environ. Exp. Bot.* **52** 199-223
[6] Patra M and Sharma A 2000 *Bot. Rev.* **66** 379-422
[7] Ozturk M, Altay V and Karahan F 2017 *Int. J. Plant Env.* **3** 1-7
[8] Mittler R and Blumwald E 2010 *Ann. Rev. Plant Biol.* **61** 443-62
[9] Khan S, Cao Q, Zheng Y M, Huang Y Z and Zhu Y G 2008 *Env. Pollution.* **152** 686-92
[10] Tangahu B V, Abdullah S R S, Idris H B M, Anuar N and Mukhlisin M 2011 *Int. J. Chem. Eng.* **2011** 939161
[11] Soerianegara I and Lemmens R H M J 1994 *Timber Trees: Major Commercia Timbers* (Maharashtra: Universal Book Serv)
[12] Heyne T 1988 *Tumbuhan Berguna Indonesia* (Jakarta: Forestry Research and Development Agency)
[13] Chen J, Shiyab S, Han F X, Monts D L, Waggoner A W and Su Z Y 2009 *Ecotoxicology* **18** 110-21
[14] Sahu G K, Upadhyay S and Sahoo B B 2012 *Physiol. Mol. Biol. Plants* **18** 21-31
[15] Ozturk M, Yucel E, Gucel S, Sakcali S and Aksoy A 2008 *Plants as Biomonitors of Trace Elements Pollution in Soil* (New York: Wiley & Sons) pp. 723-44
[16] Nagajyoti P C, Lee K D and Sreekanth T V M 2010 *Environ. Chem. Lett.* **8** 199-216

[17] Massad T J, Dyer L A and Vega C G 2012 *PLoS One* **7** 7554
[18] Shao H B, Chu L Y, Jaleel C A and Zhao C X 2008 *C. R. Biol.* **331** 215-25
[19] Wang W, Vinocur B and Altman A 2003 *Planta* **218** 1-14
[20] Boening D W 2000 *Chemosphere* **40** 1335-51
[21] Gao S, Ou-yang C, Tang L, Zhu J, Xu Y, Wang S and Chen F 2010 *J. Hazard Mater.* **182** 591-7
[22] Brunet J, Repellin A, Varrault G, Terrync N and Zuily-fodil Y 2008 *C. R. Biologies* **331** 859-64
[23] Sarma H 2011 *J. Env. Sci. Tech.* **4** 118-38

Antibacterial activity of curcuminoid derivatives resulted from hydrogenation reaction with Pd-C catalyst from turmeric (*Curcuma longa*) rhizomes extract

L D Sudewo, E Saepudin and D U C Rahayu

Department of Chemistry, Faculty of Mathematics and Natural Sciences (FMIPA), Universitas Indonesia, Depok 16424, Indonesia

Corresponding author's email: endang.saefudin@sci.ui.ac.id

Abstract. Turmeric (*Curcuma longa*) is one of the most commonly used plants for traditional medicine especially for antibacterial activity. However, because of its color, it is necessary to conduct a transformation reaction that can change the color of curcuminoid into colorless but still have antibacterial activity. The aim of this research was to analyze the antibacterial activity of tetrahydrocurcuminoid, derived from hydrogenation of curcuminoid isolated from turmeric extract towards *Staphylococcus aureus* and *Fusobacterium nucleatum*. Hydrogenation reaction was carried out using H_2 gas with Pd-C catalyst to remove the conjugated double bond in curcuminoid to form tetrahydrocurcuminoid. Tetrahydrocurcuminoid then was purified by using silica gel column chromatography and characterized using UV-Vis and FTIR spectral data. Antibacterial activity of curcuminoid and tetrahydrocurcuminoid were tested against *S. aureus* and *F. nucleatum* using disc diffusion method. The antibacterial activity showed that tetrahydrocurcuminoid had a better antibacterial activity against *S. aureus* and *F. nucleatum* with average diameter of clear zone were 5.5 and 2.75 mm, respectively, than curcuminoid with 2 and 2.5 mm, respectively.

Keywords: turmeric (*Curcuma longa*), hydrogenation, antibacterial, *Staphylococcus aureus,* and *Fusobacterium nucleatum*

1. Introduction

The floral resources in Indonesia are estimated around 30–40 thousand species which are categorized as medicinal plants [1]. On some areas in Indonesia, turmeric is widely used as herbs or traditional medicine. Over the past two-decade, the research on curcuminoid as an active ingredient for several diseases has been conducted. Among these studies, various studies reported that curcumin contained in turmeric proved a series of health benefits such as anti-inflammatory, antitumor, anti-cancer, and anti-bacterial [2-5].

Curcuminoid is diarylheptanoid derivatives which commonly exists in turmeric as three main components, namely curcumin, demethoxycurcumin, and bis-demethoxycurcumin. Curcumin belongs to polyphenol compound class that was well-known to have antibacterial activity based on altering membrane permeability leading to a nutritional leakage in bacterial cells [6]. However, the only weakness of turmeric is the yellow dye from curcuminoid that inhibits this plant in the utilization as an alternative for herbal mouthwash. Therefore, some modification should be carried out to treat the color of turmeric.

In this research, hydrogenation reaction was carried out using Pd-C catalyst to remove double bonds on curcuminoid structures and then further purified using silica gel column chromatography. The result of purified hydrogenation reaction was tetrahydrocurcuminoid which was characterized using FTIR and UV-Vis. Evaluation of its ability to inhibit the growth of *Staphylococcus aureus* and *Fusobacterium nucleatum* was also conducted using disc diffusion method. From this research, it is expected to use tetrahydrocurcuminoid as an alternative herbal mouthwash for further application.

2. Experimental method

2.1. Materials

The materials used in isolation were turmeric (*Curcuma longa*) rhizomes, ethanol, silicon oil, *n*-hexane, and dichloromethane. For hydrogenation reaction, the required materials were anhydrous sodium sulfate (Na_2SO_4), nitrogen gas (N_2), hydrogen gas (H_2), Pd-C 10 % (Sigma Aldrich), and acetone. For separation and purification, the materials used were ethyl acetate, $CHCl_3$, methanol, and silica gel 60 (0.063–0.200 mm, 230 mesh ASTM Merck), while the chemicals used in characterization were methanol pro analysis and potassium bromide (KBr). The materials used for antibacterial activity test were aquadest, nutrient agar medium (NA) (Microbiology Nutrient Agar Nutritive, Merck), beef extract, peptone (peptone from casein pancrecially digested for microbiology, Merck), 70% alcohol, mouthwash (Listerine), dimethylsulfoxide (DMSO), denaturated alcohol, *Staphylococcus aureus* (Department of Chemistry, Universitas Indonesia), and *Fusobacterium nucleatum* (Dental Laboratory, Universitas Indonesia).

2.2. Extract preparation

Turmeric rhizomes were obtained from Fresh Market Kota Wisata (Cibubur, West Java). Fresh rhizomes were cleaned, cut into small pieces, air-dried for 7 days, and ground into powder.

2.3. Isolation of curcuminoid

Turmeric powder (40 g) was put in a porous container which is made from filter paper and stuffed with cotton, then in a 250 mL round bottom flask ethanol (260 mL) was added. The mixture was extracted using Soxhlet system at 80 °C for 8–10 h. The filtrates then were concentrated using reduced pressure using a rotatory evaporator. The extract was next separated using *n*-hexane and dichloromethane. The total curcuminoid portion extracted from turmeric rhizomes was identified by using TLC and analyzed by FTIR and UV-Vis.

2.4. Synthesis of tetrahhydrocurcuminoids with hydrogenation reaction using Pd-C catalyst

The isolated curcuminoid (0.5 g) was dissolved in acetone (25 mL) with subsequent addition of 10 % Pd/C (25 mg) [6-7]. The mixture then was poured in the reactor and H_2 gas (1 atm) was passed in different time intervals (30, 60, 90, 120 and 150 min). Filtration was applied to the mixture, and the result was concentrated using N_2 gas to produce a crude product. The crude product was purified by using silica gel column chromatography to acquire the tetrahydrocurcuminoid. The total tetrahydrocurcuminoid was also identified by TLC and analyzed by FTIR and UV-Vis.

2.5. Anti-bacterial assay using disc diffusion method

2.5.1. Cultivation of microorganism. Bacteria cultures were grown on nutrient broth at concentration 10^6 CFU/ml which was monitored by using UV-Vis, kept at 37 °C for 24 h, then preserved and maintained on nutrient agar slants at 37 °C before usage [8].

2.5.2. Antibacterial activity test. Samples (curcuminoid and tetrahydrocurcuminoid) were dissolved in DMSO in various concentrations of 62.5, 125, 250, 500 and 1000 ppm [9]. Sterile disc having 6 mm diameter was impregnated with 10 mL of each serial dilution of extract solution. The impregnated discs which contained various concentrations were added onto nutrient agar surface spread with 0.1 mL of bacterial culture (standardized to 0.5 McFarland standards (10^6 CFU/ml). The plates were incubated at 37 °C for 24 h. The results were registered by determining the growth inhibition zone around the discs [10]. Negative control was DMSO while positive control was standard antibacterial mouthwash (Listerine). The experiments were conducted in duplicate.

3. Results and discussion

The isolated curcuminoid was obtained as reddish-yellow gummy. TLC was used to determine the content of each component of curcuminoid before and after addition of *n*-hexane and dichloromethane. The results of TLC are summarized in table 1. The *Rf* value intense spot obtained was corresponded to *Rf* value of the reported curcumin [11]. Moreover, the other spots were observed as demethoxycurcumin and bis-demethoxycurcumin analyzed by a functional group of each structure. The chemical structure of curcuminoid is illustrated in figure 1 which gives an idea about various functional groups presents the compound.

Table 1. *Rf* of curcuminoid TLC in $CHCl_3$:MeOH = 19:1.

	Before	After
Bis-demethoxycurcumin	0.57	0.57
Demethoxycurcumin	0.67	0.67
Curcumin (intense spot)	0.80	0.82

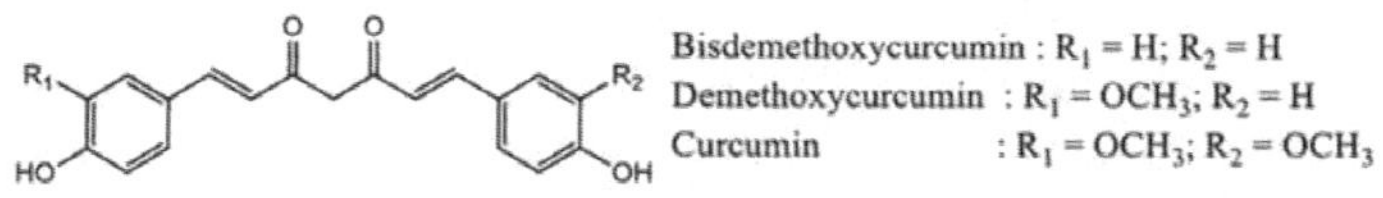

Figure 1. Chemical structure of curcuminoid.

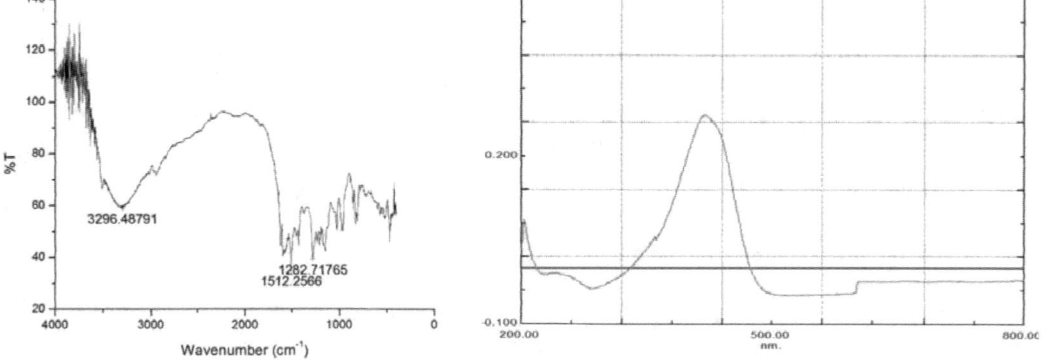

Figure 2. (a) FTIR spectrum and (b) the UV-Vis spectrum of isolated curcuminoid.

IOP Conf. Series: Materials Science and Engineering **902** (2020) 012074 doi:10.1088/1757-899X/902/1/012074

Identification of isolated curcuminoid was further confirmed by FTIR spectral data (figure 2a). FTIR spectrum of curcuminoid showed a unique stretching band of O-H at 3296 cm^{-1}. Furthermore, the peak at 1600 cm^{-1} was assigned to C=C symmetric aromatic ring stretching while the peak at 1512 cm^{-1} and 1282 cm^{-1} represent C=O and enol C-O, respectively. The FTIR spectrum of the curcuminoid showed a similarity with FTIR spectrum reported in previous research [11]. The isolated curcuminoid was also analyzed by using UV-Vis (figure 2b) and showed similarity with UV-Vis spectrum reported in the previous study [12]. The high absorbance caused by the existence of the conjugated bonds in the form of double bonds intermittent with a single bond this also affects the yellow color of the compound of curcuminoid, where later electron phi will be delocalized advanced so that the energy level of phi* decreases and reduces anti-bond [13].

Utilizing the curcuminoid as a substrate for hydrogenation reaction was carried out to perform a reaction with an accompanying change in color (figure 3). The result of the hydrogenation reaction in different interval time reaction is represented in table 2. According to table 2, it can be seen that there was another spot corresponded to tetrahydrocurcuminoid which was slightly higher than curcuminoid spot. However, hydrogenation reaction results did not eliminate the yellow color completely. The result of the hydrogenation reaction after 150 min was still remained yellow. This indicates that the results of the hydrogenation reaction of curcuminoid were not all converted to tetrahydrocurcuminoid. Data of the TLC analysis indicated that tetrahydrocurcuminoid has been formed after 90 min of hydrogenation, evidenced by the presence of the new spot above the curcuminoid spots. It can be concluded that the minimum time required for producing tetrahydrocurcuminoid was within 90 min. The results from hydrogenation were also analyzed by FTIR and UV-Vis spectral data (figure 4).

FTIR spectrum (figure 4a) supported UV-Vis spectrum (figure 4b) which both spectrum showed differences in 90 min of reaction. According to figure 4(a), in 30 and 60 min no peak at 2900 cm^{-1} which correspond to C-H sp^3 resulted from the loss of double bonds on curcuminoid (figure 3)

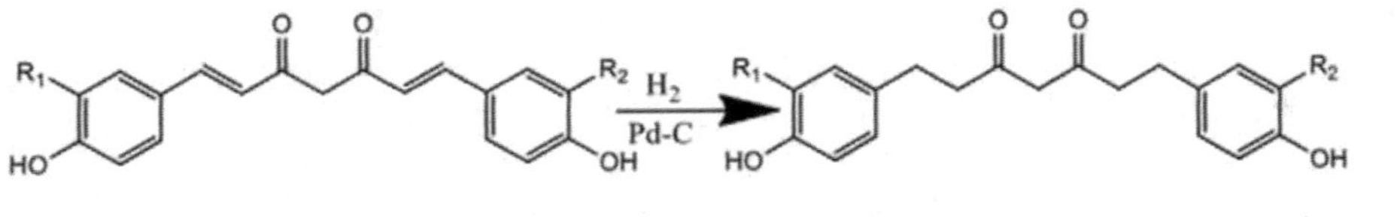

Bisdemethoxycurcumin : $R_1 = H$; $R_2 = H$
Demethoxycurcumin : $R_1 = OCH_3$; $R_2 = H$
Curcumin : $R_1 = OCH_3$; $R_2 = OCH_3$

Tetrahydrobisdemethoxycurcumin : $R_1 = H$; $R_2 = H$
Tetrahydrodemethoxycurcumin : $R_1 = OCH_3$; $R_2 = H$
Tetrahydrocurcumin : $R_1 = OCH_3$; $R_2 = OCH_3$

Figure 3. Hydrogenation reduction reaction of curcuminoid into tetrahydrocurcuminoid.

Table 2. Comparison of time and TLC analysis results.

Time Interval	30 min	60 min	90 min	120 min	150 min
Change of color	Intense reddish brown	Intense dark red	Clear reddish orange	Clear orange	Clear orange slightly yellow
TLC Analysis (*n*-hexane: EtOAc = 1:1) under UV 254 nm	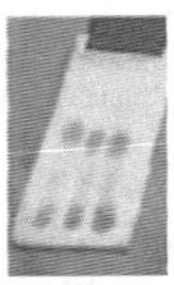	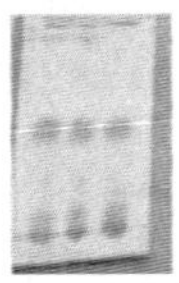		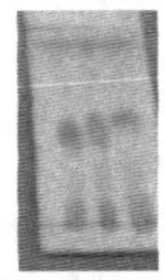	

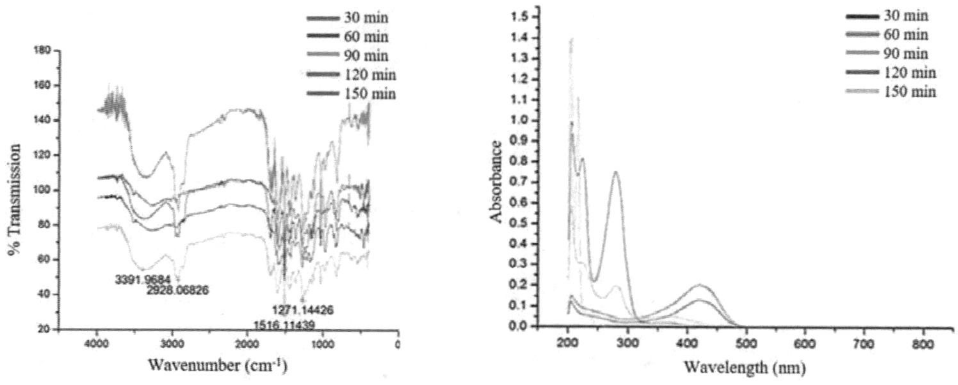

(a) (b)

Figure 4. (a) FTIR spectrum and (b) UV-Vis spectrum of tetrahydrocurcuminoid.

had been formed. Furthermore, based on figure 4b, the 30 and 60 min of reaction resulted UV-Vis spectrum there was still a similar peak to curcuminoid 418 nm (figure 2b) while the 90, 120 and 150 min reaction it can be seen that there was a shift to a lower wavelength of 240–280 nm. This UV-Vis spectral data was correlated with TLC analysis (table 2) that the tetrahydrocurcuminoid begins to form at 90 min reaction time and the changes continue to increase within the time. Subsequently, separation using silica gel column chromatography with n-hexane:EtOAc (8:2) was conducted to separate tetrahydrocurcuminoid with the unreacted curcuminoid.

FTIR spectrum (figure 4a) supported UV-Vis spectrum (figure 4b) which both spectrum showed differences in 90 min of reaction. According to figure 4a, in 30 and 60 min no peak at 2900 cm^{-1} which correspond to C-H sp^3 resulted from the loss of double bonds on curcuminoid (figure 3) had been formed. Furthermore, based on figure 4b, the 30 and 60 min of reaction resulted UV-Vis spectrum there was still a similar peak to curcuminoid 418 nm (figure 2b) while the 90, 120 and 150 min reaction it can be seen that there was a shift to a lower wavelength of 240–280 nm. This UV-Vis spectral data was correlated with TLC analysis (table 2) that the tetrahydrocurcuminoid begins to form at 90 min reaction time and the changes continue to increase within the time. Subsequently, separation using silica gel column chromatography with n-hexane:EtOAc (8:2) was conducted to separate tetrahydrocurcuminoid with the unreacted curcuminoid.

The results of column chromatography yielded 50 fractions. Among them, fraction 26–41 showed tetrahydrocurcuminoid spot when identified using TLC. Therefore, further characterization was conducted by using FTIR and UV-Vis (figure 5). According to figure 5a, FTIR spectrum showed that tetrahydrocurcuminoid after separation had more visibility of CH-sp^3 peak at 2900 cm^{-1} due to the absence of curcuminoid. This result was also strengthened with UV-Vis spectrum (figure 5b). It can be concluded that the separation using the column chromatography can purify tetrahydro-curcuminoid from the mixture of unreacted curcuminoid.

Antibacterial activities of both curcuminoid and tetrahydrocurcuminoid were evaluated against *S. aureus* and *F. nucleatum* by using disc diffusion method (table 3). According to table 3, tetrahydrocurcuminoid had a larger mean inhibitory zone diameter than curcuminoid towards *S. aureus* and *F. nucleatum.* The best antibacterial activity effect was shown by tetrahydrocurcumin at 1000 ppm against *S. Aureus* (5.5 mm). Tetrahydrocurcuminiod was one of the curcuminoid derivatives which exhibited more nonpolar than curcuminoid. Because it was influenced by the loss of double bonds contained in the compound, so it can be more easily used in the use of drugs for human,

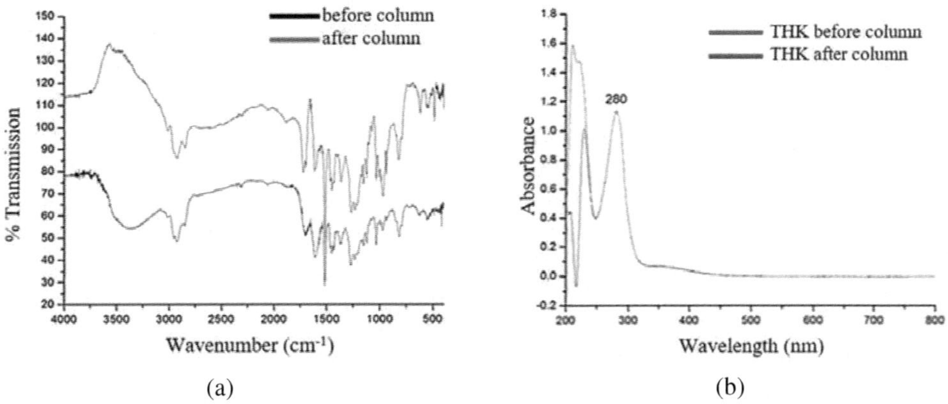

Figure 5. FTIR spectrum of tetrahydrocurcuminoid (a) before and (b) after separation.

Table 3. Average measurement of inhibitory zone of curcuminoid and tetrahydrocurcuminoid towards *S. aureus* and *F. nucleatum* together with its classification[a].

	Average diameter of clear zone (mm)			
	S. aureus		*F. nucleatum*	
	Tetrahydro-curcuminoid	Curcuminoid	Tetrahydro-curcuminoid	Curcuminoid
1000 ppm	5.5 (M)	2 (W)	2.75(W)	2.5(W)
500 ppm	3.25(W)	1.5(W)	2.25(W)	1.5(W)
250 ppm	2.25(W)	1.0(W)	1.75(W)	1.0(W)
125 ppm	1.75(W)	0.5(W)	1.25(W)	0.8(W)
62.5 ppm	0.5(W)	0.3(W)	0.75(W)	0
Listerine (+)	6	0.5	4.5	0.5
DMSO (-)	0	0	0	0

[a]Classification of clear zone [13-14]: > 20 mm (SS: super strong), 10–20 mm (S: strong), 5–10 mm (M: medium), and < 5 mm (W: weak)

the solubility of organic compounds in fat relates to the easy or not to penetrate the cell membrane. Non-polar compounds are fat-soluble, having a large fat/water partition coefficient value that easily penetrates cell membranes in a passive diffusion [14-15]. Supported by the loss of yellow dye from the compound and leaving no marks or stains that are detrimental to humans, tetrahydrocurcuminoid can be explored as alternative herbal mouthwash for further application.

4. Conclusion

Tetrahydrocurcuminoid can be produced through the hydrogenation reaction step using Pd-C catalyst in which the compound was separated by using column chromatography. Tetrahydrocurcuminoid showed medium and weak antibacterial activity towards *S. aureus* and *F. nucleatum*, respectively, at 1000 ppm with the average diameter of clear zone were 5.5 and 2.75 mm, respectively. This study found that tetrahydrocurcuminoid synthesized from curcuminoid showed a better antibacterial activity than curcuminoid which can be used as an alternative herbal mouthwash for further application.

Acknowledgments
This work was financially supported by Universitas Indonesia under research grant PITTA 2018 with grant contract number 2265/UN2.R3.1/HKP.05.00/2018.

References

[1] Wijayakusuma M H 2007 *Penyembuhan dengan Temulawak* (Jakarta: Sarana Pustaka Prima)
[2] Fadus M C, Lau C, Bikhchandani J and Lynch H T 2016 *J. Tradit. Complement. Med.* **7** 1-8
[3] Sirohi V K, Popli P, Sankhwar P, Kaushal J B, Gupta K, Manohar M and Dwivedi A 2017 *J. Nutr. Biochem.* **44** 60-70
[4] Aggarwal B B, Kumar A and Bharti A C 2003 *Anticancer Res.* **23** 363-98
[5] Pangemanan A, Fatimawali and Budiarso F 2016 *EBIOMEDIK* **4** 81-5
[6] Oomah B D and Mazza G 1997 *LWT – Food Sci. Technol.* **30** 135-40
[7] Singh R P and Jain D A 2012 *J. Pharm. Res.* 3650-3
[8] Hudzicki J 2016 *MBio* 1-23
[9] Chairandy C M, Seaforth C, Phelps R H, Pollard B P S and Khambey 1999 *J. Ethonopharmocol.* **64** 265-70
[10] Teow S Y, Liew K, Ali S A, Khoo A S B and Peh S C 2016 *J. Trop. Med.* 1-10
[11] Pawar H, Karde M, Mundle N, Jadhav P and Mehra K 2014 *J. Med. Chem.* **4** 588-91
[12] Waranyoupalin R, Wongnawa S, Wongnawa M, Pakawatchai C, Panichayupakaranant P and Sherdshoopongse P 2009 *Cent. Eur. J. Chem.* **7** 388-94
[13] Paiva P M G, Pontual E V, Coelho L C B B and Napoleao T H 2013 *FORMATEX* 641-9
[14] Bauer A W, Kirby W M M, Sherris J C and Turk M 1966 *Am. J. Clin. Pathol.* **45** 493-6
[15] Davis W W and Stout T R 1971 *Appl. Microbiol.* **22** 659-65

AUTHOR INDEX